Lecture Notes in Computer Science 9786

Commenced Publication in 1973
Founding and Former Series Editors:
Gerhard Goos, Juris Hartmanis, and Jan van Leeuwen

More information about this series at http://www.springer.com/series/7407

Editors

Osvaldo Gervasi
University of Perugia
Perugia
Italy

Beniamino Murgante
University of Basilicata
Potenza
Italy

Sanjay Misra
Covenant University
Ota
Nigeria

Ana Maria A.C. Rocha
University of Minho
Braga
Portugal

Carmelo M. Torre
Polytechnic University
Bari
Italy

David Taniar
Monash University
Clayton, VIC
Australia

Bernady O. Apduhan
Kyushu Sangyo University
Fukuoka
Japan

Elena Stankova
Saint Petersburg State University
Saint Petersburg
Russia

Shangguang Wang
Beijing University of Posts
    and Telecommunications
Beijing
China

ISSN 0302-9743   ISSN 1611-3349 (electronic)
Lecture Notes in Computer Science
ISBN 978-3-319-42084-4   ISBN 978-3-319-42085-1 (eBook)
DOI 10.1007/978-3-319-42085-1

Library of Congress Control Number: 2016944355

LNCS Sublibrary: SL1 – Theoretical Computer Science and General Issues

Printed on acid-free paper

This Springer imprint is published by Springer Nature
The registered company is Springer International Publishing AG Switzerland

Osvaldo Gervasi · Beniamino Murgante
Sanjay Misra · Ana Maria A.C. Rocha
Carmelo M. Torre · David Taniar
Bernady O. Apduhan · Elena Stankova
Shangguang Wang (Eds.)

Computational Science and Its Applications – ICCSA 2016

16th International Conference
Beijing, China, July 4–7, 2016
Proceedings, Part I

 Springer

Preface

These multi-volume proceedings (LNCS volumes 9786, 9787, 9788, 9789, and 9790) consist of the peer-reviewed papers from the 2016 International Conference on Computational Science and Its Applications (ICCSA 2016) held in Beijing, China, during July 4–7, 2016.

ICCSA 2016 was a successful event in the series of conferences, previously held in Banff, Canada (2015), Guimares, Portugal (2014), Ho Chi Minh City, Vietnam (2013), Salvador, Brazil (2012), Santander, Spain (2011), Fukuoka, Japan (2010), Suwon, South Korea (2009), Perugia, Italy (2008), Kuala Lumpur, Malaysia (2007), Glasgow, UK (2006), Singapore (2005), Assisi, Italy (2004), Montreal, Canada (2003), (as ICCS) Amsterdam, The Netherlands (2002), and San Francisco, USA (2001).

Computational science is a main pillar of most present research as well as industrial and commercial activities and it plays a unique role in exploiting ICT innovative technologies. The ICCSA conference series has been providing a venue to researchers and industry practitioners to discuss new ideas, to share complex problems and their solutions, and to shape new trends in computational science.

Apart from the general tracks, ICCSA 2016 also included 33 international workshops, in various areas of computational sciences, ranging from computational science technologies to specific areas of computational sciences, such as computer graphics and virtual reality. The program also featured three keynote speeches and two tutorials.

The success of the ICCSA conference series, in general, and ICCSA 2016, in particular, is due to the support of many people: authors, presenters, participants, keynote speakers, session chairs, Organizing Committee members, student volunteers, Program Committee members, Steering Committee members, and many people in other various roles. We would like to thank them all.

We would also like to thank our sponsors, in particular NVidia and Springer for their very important support and for making the Best Paper Award ceremony so impressive.

We would also like to thank Springer for their continuous support in publishing the ICCSA conference proceedings.

July 2016
Shangguang Wang
Osvaldo Gervasi
Bernady O. Apduhan

Organization

ICCSA 2016 was organized by Beijing University of Post and Telecommunication (China), University of Perugia (Italy), Monash University (Australia), Kyushu Sangyo University (Japan), University of Basilicata (Italy), University of Minho, (Portugal), and the State Key Laboratory of Networking and Switching Technology (China).

Honorary General Chairs

Junliang Chen	Beijing University of Posts and Telecommunications, China
Antonio Laganà	University of Perugia, Italy
Norio Shiratori	Tohoku University, Japan
Kenneth C.J. Tan	Sardina Systems, Estonia

General Chairs

Shangguang Wang	Beijing University of Posts and Telecommunications, China
Osvaldo Gervasi	University of Perugia, Italy
Bernady O. Apduhan	Kyushu Sangyo University, Japan

Program Committee Chairs

Sen Su	Beijing University of Posts and Telecommunications, China
Beniamino Murgante	University of Basilicata, Italy
Ana Maria A.C. Rocha	University of Minho, Portugal
David Taniar	Monash University, Australia

International Advisory Committee

Jemal Abawajy	Deakin University, Australia
Dharma P. Agarwal	University of Cincinnati, USA
Marina L. Gavrilova	University of Calgary, Canada
Claudia Bauzer Medeiros	University of Campinas, Brazil
Manfred M. Fisher	Vienna University of Economics and Business, Austria
Yee Leung	Chinese University of Hong Kong, SAR China

International Liaison Chairs

Ana Carla P. Bitencourt	Universidade Federal do Reconcavo da Bahia, Brazil
Alfredo Cuzzocrea	ICAR-CNR and University of Calabria, Italy
Maria Irene Falcão	University of Minho, Portugal

Robert C.H. Hsu   Chung Hua University, Taiwan
Tai-Hoon Kim   Hannam University, Korea
Sanjay Misra   University of Minna, Nigeria
Takashi Naka   Kyushu Sangyo University, Japan
Rafael D.C. Santos   National Institute for Space Research, Brazil
Maribel Yasmina Santos   University of Minho, Portugal

Workshop and Session Organizing Chairs

Beniamino Murgante   University of Basilicata, Italy
Sanjay Misra   Covenant University, Nigeria
Jorge Gustavo Rocha   University of Minho, Portugal

Award Chair

Wenny Rahayu   La Trobe University, Australia

Publicity Committee Chair

Zibing Zheng   Sun Yat-Sen University, China
Mingdong Tang   Hunan University of Science and Technology, China
Yutao Ma   Wuhan University, China
Ao Zhou   Beijing University of Posts and Telecommunications,
                                China
Ruisheng Shi   Beijing University of Posts and Telecommunications,
                                China

Workshop Organizers

Agricultural and Environment Information and Decision Support Systems (AEIDSS 2016)

Sandro Bimonte   IRSTEA, France
André Miralles   IRSTEA, France
Thérèse Libourel   LIRMM, France
François Pinet   IRSTEA, France

Advances in Information Systems and Technologies for Emergency Preparedness and Risk Assessment (ASTER 2016)

Maurizio Pollino   ENEA, Italy
Marco Vona   University of Basilicata, Italy
Beniamino Murgante   University of Basilicata, Italy

Advances in Web-Based Learning (AWBL 2016)

Mustafa Murat Inceoglu   Ege University, Turkey

Bio- and Neuro-Inspired Computing and Applications (BIOCA 2016)

Nadia Nedjah   State University of Rio de Janeiro, Brazil
Luiza de Macedo Mourell   State University of Rio de Janeiro, Brazil

Computer-Aided Modeling, Simulation, and Analysis (CAMSA 2016)

Jie Shen   University of Michigan, USA and Jilin University,
                                  China
Hao Chenina   Shanghai University of Engineering Science, China
Xiaoqiang Liun   Donghua University, China
Weichun Shi   Shanghai Maritime University, China
Yujie Liu   Southeast Jiaotong University, China

Computational and Applied Statistics (CAS 2016)

Ana Cristina Braga   University of Minho, Portugal
Ana Paula Costa Conceicao   University of Minho, Portugal
  Amorim

Computational Geometry and Security Applications (CGSA 2016)

Marina L. Gavrilova   University of Calgary, Canada

Computational Algorithms and Sustainable Assessment (CLASS 2016)

Antonino Marvuglia   Public Research Centre Henri Tudor, Luxembourg
Mikhail Kanevski   Université de Lausanne, Switzerland
Beniamino Murgante   University of Basilicata, Italy

Chemistry and Materials Sciences and Technologies (CMST 2016)

Antonio Laganà   University of Perugia, Italy
Noelia Faginas Lago   University of Perugia, Italy
Leonardo Pacifici   University of Perugia, Italy

Computational Optimization and Applications (COA 2016)

Ana Maria Rocha   University of Minho, Portugal
Humberto Rocha   University of Coimbra, Portugal

Cities, Technologies, and Planning (CTP 2016)

Giuseppe Borruso   University of Trieste, Italy
Beniamino Murgante   University of Basilicata, Italy

Databases and Computerized Information Retrieval Systems (DCIRS 2016)

Sultan Alamri   College of Computing and Informatics, SEU,
                                  Saudi Arabia
Adil Fahad   Albaha University, Saudi Arabia
Abdullah Alamri   Jeddah University, Saudi Arabia

Data Science for Intelligent Decision Support (DS4IDS 2016)

Filipe Portela	University of Minho, Portugal
Manuel Filipe Santos	University of Minho, Portugal

Econometrics and Multidimensional Evaluation in the Urban Environment (EMEUE 2016)

Carmelo M. Torre	Polytechnic of Bari, Italy
Maria Cerreta	University of Naples Federico II, Italy
Paola Perchinunno	University of Bari, Italy
Simona Panaro	University of Naples Federico II, Italy
Raffaele Attardi	University of Naples Federico II, Italy

Future Computing Systems, Technologies, and Applications (FISTA 2016)

Bernady O. Apduhan	Kyushu Sangyo University, Japan
Rafael Santos	National Institute for Space Research, Brazil
Jianhua Ma	Hosei University, Japan
Qun Jin	Waseda University, Japan

Geographical Analysis, Urban Modeling, Spatial Statistics (GEO-AND-MOD 2016)

Giuseppe Borruso	University of Trieste, Italy
Beniamino Murgante	University of Basilicata, Italy
Hartmut Asche	University of Potsdam, Germany

GPU Technologies (GPUTech 2016)

Gervasi Osvaldo	University of Perugia, Italy
Sergio Tasso	University of Perugia, Italy
Flavio Vella	University of Rome La Sapienza, Italy

ICT and Remote Sensing for Environmental and Risk Monitoring (RS-Env 2016)

Rosa Lasaponara	Institute of Methodologies for Environmental Analysis, National Research Council, Italy
Weigu Song	University of Science and Technology of China, China
Eufemia Tarantino	Polytechnic of Bari, Italy
Bernd Fichtelmann	DLR, Germany

7th International Symposium on Software Quality (ISSQ 2016)

Sanjay Misra	Covenant University, Nigeria

International Workshop on Biomathematics, Bioinformatics, and Biostatisticss (IBBB 2016)

Unal Ufuktepe	American University of the Middle East, Kuwait

Land Use Monitoring for Soil Consumption Reduction (LUMS 2016)

Carmelo M. Torre	Polytechnic of Bari, Italy
Alessandro Bonifazi	Polytechnic of Bari, Italy
Valentina Sannicandro	University of Naples Federico II, Italy
Massimiliano Bencardino	University of Salerno, Italy
Gianluca di Cugno	Polytechnic of Bari, Italy
Beniamino Murgante	University of Basilicata, Italy

Mobile Communications (MC 2016)

Hyunseung Choo	Sungkyunkwan University, Korea

Mobile Computing, Sensing, and Actuation for Cyber Physical Systems (MSA4IoT 2016)

Saad Qaisar	NUST School of Electrical Engineering and Computer Science, Pakistan
Moonseong Kim	Korean Intellectual Property Office, Korea

Quantum Mechanics: Computational Strategies and Applications (QM-CSA 2016)

Mirco Ragni	Universidad Federal de Bahia, Brazil
Ana Carla Peixoto Bitencourt	Universidade Estadual de Feira de Santana, Brazil
Vincenzo Aquilanti	University of Perugia, Italy
Andrea Lombardi	University of Perugia, Italy
Federico Palazzetti	University of Perugia, Italy

Remote Sensing for Cultural Heritage: Documentation, Management, and Monitoring (RSCH 2016)

Rosa Lasaponara	IRMMA, CNR, Italy
Nicola Masini	IBAM, CNR, Italy Zhengzhou Base, International Center on Space Technologies for Natural and Cultural Heritage, China
Chen Fulong	Institute of Remote Sensing and Digital Earth, Chinese Academy of Sciences, China

Scientific Computing Infrastructure (SCI 2016)

Elena Stankova	Saint Petersburg State University, Russia
Vladimir Korkhov	Saint Petersburg State University, Russia
Alexander Bogdanov	Saint Petersburg State University, Russia

Software Engineering Processes and Applications (SEPA 2016)

Sanjay Misra	Covenant University, Nigeria

Social Networks Research and Applications (SNRA 2016)

Eric Pardede	La Trobe University, Australia
Wenny Rahayu	La Trobe University, Australia
David Taniar	Monash University, Australia

Sustainability Performance Assessment: Models, Approaches, and Applications Toward Interdisciplinarity and Integrated Solutions (SPA 2016)

Francesco Scorza	University of Basilicata, Italy
Valentin Grecu	Lucia Blaga University on Sibiu, Romania

Tools and Techniques in Software Development Processes (TTSDP 2016)

Sanjay Misra	Covenant University, Nigeria

Volunteered Geographic Information: From Open Street Map to Participation (VGI 2016)

Claudia Ceppi	University of Basilicata, Italy
Beniamino Murgante	University of Basilicata, Italy
Francesco Mancini	University of Modena and Reggio Emilia, Italy
Giuseppe Borruso	University of Trieste, Italy

Virtual Reality and Its Applications (VRA 2016)

Osvaldo Gervasi	University of Perugia, Italy
Lucio Depaolis	University of Salento, Italy

Web-Based Collective Evolutionary Systems: Models, Measures, Applications (WCES 2016)

Alfredo Milani	University of Perugia, Italy
Valentina Franzoni	University of Rome La Sapienza, Italy
Yuanxi Li	Hong Kong Baptist University, Hong Kong, SAR China
Clement Leung	United International College, Zhuhai, China
Rajdeep Niyogi	Indian Institute of Technology, Roorkee, India

Program Committee

Jemal Abawajy	Deakin University, Australia
Kenny Adamson	University of Ulster, UK
Hartmut Asche	University of Potsdam, Germany
Michela Bertolotto	University College Dublin, Ireland
Sandro Bimonte	CEMAGREF, TSCF, France
Rod Blais	University of Calgary, Canada
Ivan Blečić	University of Sassari, Italy
Giuseppe Borruso	University of Trieste, Italy
Yves Caniou	Lyon University, France

José A. Cardoso e Cunha	Universidade Nova de Lisboa, Portugal
Carlo Cattani	University of Salerno, Italy
Mete Celik	Erciyes University, Turkey
Alexander Chemeris	National Technical University of Ukraine KPI, Ukraine
Min Young Chung	Sungkyunkwan University, Korea
Elisete Correia	University of Trás os Montes e Alto Douro, Portugal
Gilberto Corso Pereira	Federal University of Bahia, Brazil
M. Fernanda Costa	University of Minho, Portugal
Alfredo Cuzzocrea	ICAR-CNR and University of Calabria, Italy
Florbela Maria da Cruz Domingues Correia	Intituto Politécnico de Viana do Castelo, Portugal
Vanda Marisa da Rosa Milheiro Lourenço	FCT from University Nova de Lisboa, Portugal
Carla Dal Sasso Freitas	Universidade Federal do Rio Grande do Sul, Brazil
Pradesh Debba	The Council for Scientific and Industrial Research (CSIR), South Africa
Hendrik Decker	Instituto Tecnológico de Informática, Spain
Adelaide de Fátima Baptista Valente Freitas	University of Aveiro, Portugal
Carina Soares da Silva Fortes	Escola Superior de Tecnologias da Saúde de Lisboa, Portugal
Frank Devai	London South Bank University, UK
Rodolphe Devillers	Memorial University of Newfoundland, Canada
Joana Dias	University of Coimbra, Portugal
Prabu Dorairaj	NetApp, India/USA
M. Irene Falcao	University of Minho, Portugal
Cherry Liu Fang	U.S. DOE Ames Laboratory, USA
Florbela Fernandes	Polytechnic Institute of Bragança, Portugal
Jose-Jesús Fernandez	National Centre for Biotechnology, CSIS, Spain
Mara Celia Furtado Rocha	PRODEB-Pós Cultura/UFBA, Brazil
Akemi Galvez	University of Cantabria, Spain
Paulino Jose Garcia Nieto	University of Oviedo, Spain
Marina Gavrilova	University of Calgary, Canada
Jerome Gensel	LSR-IMAG, France
Mara Giaoutzi	National Technical University, Athens, Greece
Andrzej M. Goscinski	Deakin University, Australia
Alex Hagen-Zanker	University of Cambridge, UK
Malgorzata Hanzl	Technical University of Lodz, Poland
Shanmugasundaram Hariharan	B.S. Abdur Rahman University, India
Tutut Herawan	Universitas Teknologi Yogyakarta, Indonesia
Hisamoto Hiyoshi	Gunma University, Japan
Fermin Huarte	University of Barcelona, Spain
Andrés Iglesias	University of Cantabria, Spain
Mustafa Inceoglu	Ege University, Turkey
Peter Jimack	University of Leeds, UK

Qun Jin	Waseda University, Japan
Farid Karimipour	Vienna University of Technology, Austria
Baris Kazar	Oracle Corp., USA
Maulana Adhinugraha Kiki	Telkom University, Indonesia
DongSeong Kim	University of Canterbury, New Zealand
Taihoon Kim	Hannam University, Korea
Ivana Kolingerova	University of West Bohemia, Czech Republic
Dieter Kranzlmueller	LMU and LRZ Munich, Germany
Antonio Laganà	University of Perugia, Italy
Rosa Lasaponara	National Research Council, Italy
Maurizio Lazzari	National Research Council, Italy
Cheng Siong Lee	Monash University, Australia
Sangyoun Lee	Yonsei University, Korea
Jongchan Lee	Kunsan National University, Korea
Clement Leung	United International College, Zhuhai, China
Chendong Li	University of Connecticut, USA
Gang Li	Deakin University, Australia
Ming Li	East China Normal University, China
Fang Liu	AMES Laboratories, USA
Xin Liu	University of Calgary, Canada
Savino Longo	University of Bari, Italy
Tinghuai Ma	NanJing University of Information Science and Technology, China
Isabel Cristina Maciel Natário	FCT from University Nova de Lisboa, Portugal
Sergio Maffioletti	University of Zurich, Switzerland
Ernesto Marcheggiani	Katholieke Universiteit Leuven, Belgium
Antonino Marvuglia	Research Centre Henri Tudor, Luxembourg
Nicola Masini	National Research Council, Italy
Nirvana Meratnia	University of Twente, The Netherlands
Alfredo Milani	University of Perugia, Italy
Sanjay Misra	Federal University of Technology Minna, Nigeria
Giuseppe Modica	University of Reggio Calabria, Italy
José Luis Montaña	University of Cantabria, Spain
Beniamino Murgante	University of Basilicata, Italy
Jiri Nedoma	Academy of Sciences of the Czech Republic, Czech Republic
Laszlo Neumann	University of Girona, Spain
Irene Oliveira	University of Trás os Montes e Alto Douro, Portugal
Kok-Leong Ong	Deakin University, Australia
Belen Palop	Universidad de Valladolid, Spain
Marcin Paprzycki	Polish Academy of Sciences, Poland
Eric Pardede	La Trobe University, Australia
Kwangjin Park	Wonkwang University, Korea
Telmo Pinto	University of Minho, Portugal

Maurizio Pollino	Italian National Agency for New Technologies, Energy and Sustainable Economic Development, Italy
Alenka Poplin	University of Hamburg, Germany
Vidyasagar Potdar	Curtin University of Technology, Australia
David C. Prosperi	Florida Atlantic University, USA
Maria Emilia F. Queiroz Athayde	University of Minho, Portugal
Wenny Rahayu	La Trobe University, Australia
Jerzy Respondek	Silesian University of Technology, Poland
Ana Maria A.C. Rocha	University of Minho, Portugal
Maria Clara Rocha	ESTES Coimbra, Portugal
Humberto Rocha	INESC-Coimbra, Portugal
Alexey Rodionov	Institute of Computational Mathematics and Mathematical Geophysics, Russia
Jon Rokne	University of Calgary, Canada
Octavio Roncero	CSIC, Spain
Maytham Safar	Kuwait University, Kuwait
Chiara Saracino	A.O. Ospedale Niguarda Ca' Granda - Milano, Italy
Haiduke Sarafian	The Pennsylvania State University, USA
Jie Shen	University of Michigan, USA
Qi Shi	Liverpool John Moores University, UK
Dale Shires	U.S. Army Research Laboratory, USA
Takuo Suganuma	Tohoku University, Japan
Sergio Tasso	University of Perugia, Italy
Parimala Thulasiraman	University of Manitoba, Canada
Carmelo M. Torre	Polytechnic of Bari, Italy
Giuseppe A. Trunfio	University of Sassari, Italy
Unal Ufuktepe	American University of the Middle East, Kuwait
Toshihiro Uchibayashi	Kyushu Sangyo University, Japan
Mario Valle	Swiss National Supercomputing Centre, Switzerland
Pablo Vanegas	University of Cuenca, Equador
Piero Giorgio Verdini	INFN Pisa and CERN, Italy
Marco Vizzari	University of Perugia, Italy
Koichi Wada	University of Tsukuba, Japan
Krzysztof Walkowiak	Wroclaw University of Technology, Poland
Robert Weibel	University of Zurich, Switzerland
Roland Wismüller	Universität Siegen, Germany
Mudasser Wyne	SOET National University, USA
Chung-Huang Yang	National Kaohsiung Normal University, Taiwan
Xin-She Yang	National Physical Laboratory, UK
Salim Zabir	France Telecom Japan Co., Japan
Haifeng Zhao	University of California, Davis, USA
Kewen Zhao	University of Qiongzhou, China
Albert Y. Zomaya	University of Sydney, Australia

Reviewers

Abawajy, Jemal	Deakin University, Australia
Abuhelaleh, Mohammed	Univeristy of Bridgeport, USA
Acharjee, Shukla	Dibrugarh University, India
Andrianov, Sergei Nikolaevich	Universitetskii prospekt, Russia
Aguilar, José Alfonso	Universidad Autónoma de Sinaloa, Mexico
Ahmed, Faisal	University of Calgary, Canada
Álberti, Margarita	University of Barcelona, Spain
Amato, Alba	Seconda Universit degli Studi di Napoli, Italy
Amorim, Ana Paula	University of Minho, Portugal
Apduhan, Bernady	Kyushu Sangyo University, Japan
Aquilanti, Vincenzo	University of Perugia, Italy
Asche, Hartmut	Posdam University, Germany
Athayde Maria, Emlia Feijão Queiroz	University of Minho, Portugal
Attardi, Raffaele	University of Napoli Federico II, Italy
Azam, Samiul	United International University, Bangladesh
Azevedo, Ana	Athabasca University, USA
Badard, Thierry	Laval University, Canada
Baioletti, Marco	University of Perugia, Italy
Bartoli, Daniele	University of Perugia, Italy
Bentayeb, Fadila	Université Lyon, France
Bilan, Zhu	Tokyo University of Agriculture and Technology, Japan
Bimonte, Sandro	IRSTEA, France
Blecic, Ivan	Università di Cagliari, Italy
Bogdanov, Alexander	Saint Petersburg State University, Russia
Borruso, Giuseppe	University of Trieste, Italy
Bostenaru, Maria	"Ion Mincu" University of Architecture and Urbanism, Romania
Braga Ana, Cristina	University of Minho, Portugal
Canora, Filomena	University of Basilicata, Italy
Cardoso, Rui	Institute of Telecommunications, Portugal
Ceppi, Claudia	Polytechnic of Bari, Italy
Cerreta, Maria	University Federico II of Naples, Italy
Choo, Hyunseung	Sungkyunkwan University, South Korea
Coletti, Cecilia	University of Chieti, Italy
Correia, Elisete	University of Trás-Os-Montes e Alto Douro, Portugal
Correia Florbela Maria, da Cruz Domingues	Instituto Politécnico de Viana do Castelo, Portugal
Costa, Fernanda	University of Minho, Portugal
Crasso, Marco	National Scientific and Technical Research Council, Argentina
Crawford, Broderick	Universidad Catolica de Valparaiso, Chile

Cuzzocrea, Alfredo	University of Trieste, Italy
Cutini, Valerio	University of Pisa, Italy
Danese, Maria	IBAM, CNR, Italy
Decker, Hendrik	Instituto Tecnológico de Informática, Spain
Degtyarev, Alexander	Saint Petersburg State University, Russia
Demartini, Gianluca	University of Sheffield, UK
Di Leo, Margherita	JRC, European Commission, Belgium
Dias, Joana	University of Coimbra, Portugal
Dilo, Arta	University of Twente, The Netherlands
Dorazio, Laurent	ISIMA, France
Duarte, Júlio	University of Minho, Portugal
El-Zawawy, Mohamed A.	Cairo University, Egypt
Escalona, Maria-Jose	University of Seville, Spain
Falcinelli, Stefano	University of Perugia, Italy
Fernandes, Florbela	Escola Superior de Tecnologia e Gest ão de Bragança, Portugal
Florence, Le Ber	ENGEES, France
Freitas Adelaide, de Fátima Baptista Valente	University of Aveiro, Portugal
Frunzete, Madalin	Polytechnic University of Bucharest, Romania
Gankevich, Ivan	Saint Petersburg State University, Russia
Garau, Chiara	University of Cagliari, Italy
Garcia, Ernesto	University of the Basque Country, Spain
Gavrilova, Marina	University of Calgary, Canada
Gensel, Jerome	IMAG, France
Gervasi, Osvaldo	University of Perugia, Italy
Gizzi, Fabrizio	National Research Council, Italy
Gorbachev, Yuriy	Geolink Technologies, Russia
Grilli, Luca	University of Perugia, Italy
Guerra, Eduardo	National Institute for Space Research, Brazil
Hanzl, Malgorzata	University of Lodz, Poland
Hegedus, Peter	University of Szeged, Hungary
Herawan, Tutut	University of Malaya, Malaysia
Hu, Ya-Han	National Chung Cheng University, Taiwan
Ibrahim, Michael	Cairo University, Egipt
Ifrim, Georgiana	Insight, Ireland
Irrazábal, Emanuel	Universidad Nacional del Nordeste, Argentina
Janana, Loureio	University of Mato Grosso do Sul, Brazil
Jaiswal, Shruti	Delhi Technological University, India
Johnson, Franklin	Universidad de Playa Ancha, Chile
Karimipour, Farid	Vienna University of Technology, Austria
Kapcak, Sinan	American University of the Middle East in Kuwait, Kuwait
Kiki Maulana, Adhinugraha	Telkom University, Indonesia
Kim, Moonseong	KIPO, South Korea
Kobusińska, Anna	Poznan University of Technology, Poland

Oliveira, Irene   University of Trás-Os-Montes e Alto Douro, Portugal
Panetta, J.B.   Tecnologia Geofísica Petróleo Brasileiro SA,
                          PETROBRAS, Brazil
Papa, Enrica   University of Amsterdam, The Netherlands
Papathanasiou, Jason   University of Macedonia, Greece
Pardede, Eric   La Trobe University, Australia
Pascale, Stefania   University of Basilicata, Italy
Paul, Padma Polash   University of Calgary, Canada
Perchinunno, Paola   University of Bari, Italy
Pereira, Oscar   Universidade de Aveiro, Portugal
Pham, Quoc Trung   HCMC University of Technology, Vietnam
Pinet, Francois   IRSTEA, France
Pirani, Fernando   University of Perugia, Italy
Pollino, Maurizio   ENEA, Italy
Pusatli, Tolga   Cankaya University, Turkey
Qaisar, Saad   NURST, Pakistan
Qian, Junyan   Guilin University of Electronic Technology, China
Raffaeta, Alessandra   University of Venice, Italy
Ragni, Mirco   Universidade Estadual de Feira de Santana, Brazil
Rahman, Wasiur   Technical University Darmstadt, Germany
Rampino, Sergio   Scuola Normale di Pisa, Italy
Rahayu, Wenny   La Trobe University, Australia
Ravat, Franck   IRIT, France
Raza, Syed Muhammad   Sungkyunkwan University, South Korea
Roccatello, Eduard   3DGIS, Italy
Rocha, Ana Maria   University of Minho, Portugal
Rocha, Humberto   University of Coimbra, Portugal
Rocha, Jorge   University of Minho, Portugal
Rocha, Maria Clara   ESTES Coimbra, Portugal
Romano, Bernardino   University of l'Aquila, Italy
Sannicandro, Valentina   Polytechnic of Bari, Italy
Santiago Júnior, Valdivino Instituto Nacional de Pesquisas Espaciais, Brazil
Sarafian, Haiduke   Pennsylvania State University, USA
Schneider, Michel   ISIMA, France
Selmaoui, Nazha   University of New Caledonia, New Caledonia
Scerri, Simon   University of Bonn, Germany
Shakhov, Vladimir   Institute of Computational Mathematics
                          and Mathematical Geophysics, Russia
Shen, Jie   University of Michigan, USA
Silva-Fortes, Carina   ESTeSL-IPL, Portugal
Singh, Upasana   University of Kwa Zulu-Natal, South Africa
Skarga-Bandurova, Inna   Technological Institute of East Ukrainian National
                          University, Ukraine
Soares, Michel   Federal University of Sergipe, Brazil
Souza, Eric   Universidade Nova de Lisboa, Portugal
Stankova, Elena   Saint Petersburg State University, Russia

Stalidis, George	TEI of Thessaloniki, Greece
Taniar, David	Monash University, Australia
Tasso, Sergio	University of Perugia, Italy
Telmo, Pinto	University of Minho, Portugal
Tengku, Adil	La Trobe University, Australia
Thorat, Pankaj	Sungkyunkwan University, South Korea
Tiago Garcia, de Senna Carneiro	Federal University of Ouro Preto, Brazil
Tilio, Lucia	University of Basilicata, Italy
Torre, Carmelo Maria	Polytechnic of Bari, Italy
Tripathi, Ashish	MNNIT Allahabad, India
Tripp, Barba	Carolina, Universidad Autnoma de Sinaloa, Mexico
Trunfio, Giuseppe A.	University of Sassari, Italy
Upadhyay, Ashish	Indian Institute of Public Health-Gandhinagar, India
Valuev, Ilya	Russian Academy of Sciences, Russia
Varella, Evangelia	Aristotle University of Thessaloniki, Greece
Vasyunin, Dmitry	University of Amsterdam, The Netherlans
Vijaykumar, Nandamudi	INPE, Brazil
Villalba, Maite	Universidad Europea de Madrid, Spain
Walkowiak, Krzysztof	Wroclav University of Technology, Poland
Wanderley, Fernando	FCT/UNL, Portugal
Wei Hoo, Chong	Motorola, USA
Xia, Feng	Dalian University of Technology (DUT), China
Yamauchi, Toshihiro	Okayama University, Japan
Yeoum, Sanggil	Sungkyunkwan University, South Korea
Yirsaw, Ayalew	University of Botswana, Bostwana
Yujie, Liu	Southeast Jiaotong University, China
Zafer, Agacik	American University of the Middle East in Kuwait, Kuwait
Zalyubovskiy, Vyacheslav	Russian Academy of Sciences, Russia
Zeile, Peter	Technische Universitat Kaiserslautern, Germany
Žemlička, Michal	Charles University, Czech Republic
Zivkovic, Ljiljana	Republic Agency for Spatial Planning, Belgrade
Zunino, Alejandro	Universidad Nacional del Centro, Argentina

Sponsoring Organizations

ICCSA 2016 would not have been possible without the tremendous support of many organizations and institutions, for which all organizers and participants of ICCSA 2016 express their sincere gratitude:

Springer International Publishing AG, Switzerland
(http://www.springer.com)

NVidia Co., USA
(http://www.nvidia.com)

Beijing University of Post and Telecommunication, China
(http://english.bupt.edu.cn/)

State Key Laboratory of Networking and Switching Technology, China

University of Perugia, Italy
(http://www.unipg.it)

University of Basilicata, Italy
(http://www.unibas.it)

Monash University, Australia
(http://monash.edu)

Kyushu Sangyo University, Japan
(www.kyusan-u.ac.jp)

Universidade do Minho, Portugal
(http://www.uminho.pt)

Contents – Part I

Short Papers

Computational Methods, Algorithms and Scientific Applications

A Nonlinear Multiscale Viscosity Method to Solve Compressible Flow Problems

Sérgio Souza Bento[(⊠)], Leonardo Muniz de Lima, Ramoni Zancanela Sedano,
Lucia Catabriga, and Isaac P. Santos

High Performance Computing Lab, Federal University of Espírito Santo,
Av. Fernando Ferrari, 514, Goiabeiras, Vitória, ES 29075-910, Brazil
{sergio.bento,isaac.santos}@ufes.br, lmuniz@ifes.edu.br,
{rsedano,luciac}@inf.ufes.br

Abstract. In this work we present a nonlinear multiscale viscosity method to solve inviscid compressible flow problems in conservative variables. The basic idea of the method consists of adding artificial viscosity adaptively in all scales of the discretization. The amount of viscosity added to the numerical model is based on the YZβ shock-capturing parameter, which has the property of being mesh and numerical solution dependent. The subgrid scale space is defined using bubble functions whose degrees of freedom are locally eliminated in favor of the degrees of freedom that live on the resolved scales. This new numerical formulation can be considered a free parameter and self adaptive method. Performance and accuracy comparisons with the well known method SUPG combined with shock capturing operators are conducted based on benchmark 2D problems.

Keywords: Finite element method · Multiscale stabilized formulation · Compressible flow problems

1   Introduction

The numerical solution of the compressible flows may exhibit global spurious oscillations, especially near shock regions. More accurate and stable results can be obtained using stabilized formulations, either linear or nonlinear approach [4,5,12,20]. In the 1990s it was shown that the stabilized finite element methods could be derived from the variational multiscale framework, which consists of a consistent decomposition of the approximation space into resolved (coarse) and unresolved (subgrid) scales subspaces via a variational projection. The numerical oscillations originated by the standard Galerkin method can be related to scales that are not represented by the discretization, that is, the unresolved scales. In this case, those unresolved scales or their effect may be inserted into the problem formulation to be solved on the resolved scales, represented by the chosen discretization. Examples of multiscale methods can be found in [6,9,11, 13,14]. It is important to highlight that those stabilization/multiscale techniques prevent numerical oscillations and other instabilities in solving problems with high Reynolds and/or Mach numbers and shocks or strong boundary layers [17].

© Springer International Publishing Switzerland 2016
O. Gervasi et al. (Eds.): ICCSA 2016, Part I, LNCS 9786, pp. 3–17, 2016.
DOI: 10.1007/978-3-319-42085-1_1

Santos and Almeida [15] presented a nonliner multiscale method to solve advection dominated transport problem where a nonlinear artificial diffusion is added on the subgrid scale. The amount of artificial diffusion added to the numerical model was stablished considering a two-level decomposition of the function space and the velocity field into the resolved (coarse) and unresolved (subgrid) scales. The subgrid velocity field was determined by requiring the minimum of the associated kinetic energy for which the residue of the resolved scale solution vanishes on each element of the discretization. The idea of adding a nonlinear diffusion in both scales (subgrid and coarse) of the discretization was considered in [2] through the Dynamic Diffusion (DD) method, where the subgrid space is constructed by bubbles functions defined into elements and the amount of nonlinear diffusion is similar to the method presented in [15]. This methodology was extended to the compressible Euler equations in [16], where comparisons with the well known SUPG method coupled with the two shock capturing operators: the Consistent Approximate Upwind Petrov-Galerkin (CAU) [7] and the YZβ [19], were made. Although the DD method offered good results, it did not live up to expectations compared to the SUPG formulations with shock capturing operators as CAU and YZβ.

The SUPG method coupled with the YZβ shock capturing operator has offered numerical solution with good accuracy for compressible problems. Moreover, Tezduyar [17] has been proposing adaptive ways for calculations of the local length scale (also known as "element length") present in the stabilization parameters. The calculus of the local length scale parameter is made taking into account the directions of high gradients and the spatial discretization domain. The stabilization parameter resulting acts adaptively and is useful to avoid excessive viscosity helping to maintain smaller numerical dissipations.

In this paper we propose a new numerical formulation, named, Nonlinear Multiscale Viscosity (NMV) method, to solve inviscid compressible flow problems in conservative variables. As the DD method, the basic idea is to add a nonlinear artificial viscosity in all scales of the discretization, but the amount of artificial viscosity is defined by the stabilization parameter of the YZβ method, as proposed in [19, 20]. The nonlinear artificial viscosity added to the numerical formulation is made adaptively, leading the NMV to a self adaptive methodology.

The remainder of this work is organized as follows. Section 2 briefly addresses the governing equations and the variational multiscale formulation. Numerical experiments are conducted in Sect. 3 to show the behavior of the new multiscale finite element method for a variety of benchmark Euler equations problems. Section 4 concludes this paper.

2 Governing Equations and Variational Multiscale Formulation

The two-dimensional Euler equations in conservative variables, $\mathbf{U} = (\rho, \rho u, \rho v, \rho e)$, without source terms are an inviscid system of conservation laws represented by

$$\frac{\partial \mathbf{U}}{\partial t} + \frac{\partial \mathbf{F}_x}{\partial x} + \frac{\partial \mathbf{F}_y}{\partial y} = \mathbf{0}, \quad \text{on } \Omega \times [0, T_f], \tag{1}$$

where ρ is the fluid density, $\mathbf{u} = (u, v)$ is the velocity vector, e is the total energy per unit mass, $\mathbf{F}_x$ and $\mathbf{F}_y$ are the Euler fluxes, Ω is a domain in $\mathbb{R}^2$, and T_f is a positive real number, representing the final time. Alternatively, Eq. (1) can be written as

$$\frac{\partial \mathbf{U}}{\partial t} + \mathbf{A_x}\frac{\partial \mathbf{U}}{\partial x} + \mathbf{A_y}\frac{\partial \mathbf{U}}{\partial y} = \mathbf{0}, \quad \text{on } \Omega \times [0, T_f], \tag{2}$$

where $\mathbf{A}_x = \frac{\partial \mathbf{F}_x}{\partial \mathbf{U}}$ and $\mathbf{A}_y = \frac{\partial \mathbf{F}_y}{\partial \mathbf{U}}$. Associated to Eq. (2) we have a proper set of boundary and initial conditions.

To define the finite element discretization, we consider a triangular partition $\mathcal{T}_H$ of the domain Ω into n_{el} elements, where: $\Omega = \bigcup_{e=1}^{n_{el}} \Omega_e$ and $\Omega_i \cap \Omega_j = \emptyset$, $i, j = 1, 2, \cdots, n_{el}$, $i \neq j$. We introduce the space $\mathcal{V}_E$, that is written as the direct sum

$$\mathcal{V}_E = \mathcal{V}_h \oplus \mathcal{V}_B, \tag{3}$$

where the subspaces $\mathcal{V}_h$ and $\mathcal{V}_B$ are given by

$$\mathcal{V}_h = \{\mathbf{U}_h \in [H^1(\Omega)]^4 \mid \mathbf{U}_h|_{\Omega_e} \in [\mathbb{P}_1(\Omega_e)]^4, \mathbf{U}_h \cdot \mathbf{e}_k = g_k(t) \text{ in } \Gamma_{g_k}\};$$

$$\mathcal{V}_B = \{\mathbf{U}_B \in [H_0^1(\mathcal{T}_H)]^4 \mid \mathbf{U}_B|_{\Omega_e} \in [span(\psi_B)]^4, \quad \forall \Omega_e \in \mathcal{T}_H\},$$

where $\mathbb{P}_1(\Omega_e)$ represents the set of first order polynomials in Ω_e, ψ_B is a bubble function ($0 \leq \psi_B \leq 1$ and $\psi_B \in H_0^1(\mathcal{T}_H)$) and H^1, H_0^1 are Hilbert spaces [3]. The space $\mathcal{V}_h$ represents the resolved (coarse) scale space whereas $\mathcal{V}_B$ stands for the subgrid (fine) scale space (Fig. 1).

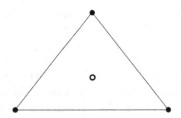

Fig. 1. $\mathcal{V}_E$ Representation: $\bullet$ stands for $\mathcal{V}_h$ nodes and $\circ$ stands for $\mathcal{V}_B$ nodes.

The NMV method for the Euler equation consists of find $\mathbf{U}_E = \mathbf{U}_h + \mathbf{U}_B \in \mathcal{V}_E$ with $\mathbf{U}_h \in \mathcal{V}_h$, $\mathbf{U}_B \in \mathcal{V}_B$ such that

$$\int_\Omega \mathbf{W}_E \cdot \left(\frac{\partial \mathbf{U}_E}{\partial t} + \mathbf{A}_x^h \frac{\partial \mathbf{U}_E}{\partial x} + \mathbf{A}_y^h \frac{\partial \mathbf{U}_E}{\partial y}\right) d\Omega +$$

$$\sum_{e=1}^{nel} \int_{\Omega_e} \delta_h \left(\frac{\partial \mathbf{W}_E}{\partial x} \cdot \frac{\partial \mathbf{U}_E}{\partial x} + \frac{\partial \mathbf{W}_E}{\partial y} \cdot \frac{\partial \mathbf{U}_E}{\partial y}\right) d\Omega = \mathbf{0} \quad \forall \mathbf{W}_E \in \mathcal{V}_E, \tag{4}$$

where $\mathbf{W}_E = \mathbf{W}_h + \mathbf{W}_B \in \mathcal{V}_E$ with $\mathbf{W}_h \in \mathcal{V}_h$, $\mathbf{W}_B \in \mathcal{V}_B$ and the amount of artificial viscosity is calculated on the element-level by using the YZβ shock-capturing viscosity parameter [19]

$$\delta_h = \|\mathbf{Y}^{-1}R(\mathbf{U}_h)\| \left(\sum_{i=1}^{2} \left\| \mathbf{Y}^{-1}\frac{\partial \mathbf{U}_h}{\partial x_i} \right\|^2 \right)^{\frac{\beta}{2}-1} \|\mathbf{Y}^{-1}\mathbf{U}_h\|^{1-\beta}h^\beta, \qquad (5)$$

where

$$R(\mathbf{U}_h) = \frac{\partial \mathbf{U}_h}{\partial t} + \mathbf{A}_x^h \frac{\partial \mathbf{U}_h}{\partial x} + \mathbf{A}_y^h \frac{\partial \mathbf{U}_h}{\partial y}$$

is the residue of the problem on Ω_e, $\mathbf{Y}$ is a diagonal matrix constructed from the reference values of the components of $\mathbf{U}$, given by

$$\mathbf{Y} = diag\left((U_1)_{\text{ref}}, (U_2)_{\text{ref}}, (U_3)_{\text{ref}}, (U_4)_{\text{ref}}\right), \qquad (6)$$

h is the local length scale defined as in [17] by

$$h = \left(\sum_a |\mathbf{j} \cdot \nabla N_a| \right)^{-1}, \qquad (7)$$

$\mathbf{j}$ is a unit vector defined as

$$\mathbf{j} = \frac{\nabla \rho}{\|\nabla \rho\|}$$

and N_a is the interpolation function associated with node a. It is important to note that, the local length h is defined automatically taking into account the directions of high gradients and spatial discretization domain.

Generally, the parameter β is set as $\beta = 1$ for smoother shocks and $\beta = 2$ for sharper shocks. The compromise between the $\beta = 1$ and $\beta = 2$ selections was defined in [17–19] as the following average expression for δ_h:

$$\delta_h = \frac{1}{2}\left(\delta_h|_{\beta=1} + \delta_h|_{\beta=2}\right).$$

The numerical solution is obtained using iterative procedures for space and time. The iterative procedure for space is defined of the following way: given $\mathbf{U}_E^i$ at iteration i, we find $\mathbf{U}_E^{i+1}$ satisfying the formulation (4) with $\delta_h = \delta_h(\mathbf{U}_E^i) = \delta_h^i$, for $i = 0, 1, \cdots, i_{MAX}$. The formulation (4) can be partitioned in two subproblems, one related to the resolved scale, given by

$$\int_\Omega \mathbf{W}_h \cdot \left(\frac{\partial \mathbf{U}_h^{i+1}}{\partial t} + \mathbf{A}_x^h \frac{\partial \mathbf{U}_h^{i+1}}{\partial x} + \mathbf{A}_y^h \frac{\partial \mathbf{U}_h^{i+1}}{\partial y} \right) d\Omega +$$

$$\int_\Omega \mathbf{W}_h \cdot \left(\frac{\partial \mathbf{U}_B^{i+1}}{\partial t} + \mathbf{A}_x^h \frac{\partial \mathbf{U}_B^{i+1}}{\partial x} + \mathbf{A}_y^h \frac{\partial \mathbf{U}_B^{i+1}}{\partial y} \right) d\Omega +$$

$$\sum_{e=1}^{nel} \int_{\Omega_e} \delta_h^i \left(\frac{\partial \mathbf{W}_h}{\partial x} \cdot \frac{\partial \mathbf{U}_h^{i+1}}{\partial x} + \frac{\partial \mathbf{W}_h}{\partial y} \cdot \frac{\partial \mathbf{U}_h^{i+1}}{\partial y} \right) d\Omega = \mathbf{0}, \quad \forall \mathbf{W}_h \in \mathcal{V}_h, \qquad (8)$$

and another, representing the subgrid scale is written as

$$\int_{\Omega} \mathbf{W}_B \cdot \frac{\partial \mathbf{U}_B^{i+1}}{\partial t} d\Omega + \int_{\Omega} \mathbf{W}_B \cdot \Big(\frac{\partial \mathbf{U}_h^{i+1}}{\partial t} + \mathbf{A}_x^h \frac{\partial \mathbf{U}_h^{i+1}}{\partial x} + \mathbf{A}_y^h \frac{\partial \mathbf{U}_h^{i+1}}{\partial y} \Big) d\Omega +$$

$$\sum_{e=1}^{nel} \int_{\Omega_e} \delta_h^i \Big(\frac{\partial \mathbf{W}_B}{\partial x} \cdot \frac{\partial \mathbf{U}_B^{i+1}}{\partial x} + \frac{\partial \mathbf{W}_B}{\partial y} \cdot \frac{\partial \mathbf{U}_B^{i+1}}{\partial y} \Big) d\Omega = \mathbf{0}, \quad \forall \mathbf{W}_B \in \mathcal{V}_B, \quad (9)$$

where some terms were omitted, once they are zero.

Applying the standard finite element approximation on Eqs. (8) and (9), we arrive at a local system of ordinary differential equations:

$$\begin{bmatrix} M_{hh} & M_{hB} \\ M_{Bh} & M_{BB} \end{bmatrix} \begin{bmatrix} \dot{U}_h \\ \dot{U}_B \end{bmatrix} + \begin{bmatrix} K_{hh} & K_{hB} \\ K_{Bh} & K_{BB} \end{bmatrix} \begin{bmatrix} U_h \\ U_B \end{bmatrix} = \begin{bmatrix} 0_h \\ 0_B \end{bmatrix}, \tag{10}$$

where U_h and U_B are, respectively, the nodal values of the unknowns $\mathbf{U}_h$ and $\mathbf{U}_B$ on each element Ω_e, whereas $\dot{U}_h$ and $\dot{U}_B$ are its time derivative.

The numerical solution is advanced in time by the implicit predictor-multicorrector algorithm given in [10] and adapted for the DD method in [16] for the Euler equations. The degrees of freedom related to the subgrid space are locally eliminated in favor of the ones of the macro space using a static condensation approach. Algorithm 1 shows the implicit predictor-multicorrector steps, considering second order approximations in time for the micro and macro scales subproblems, where Δt is the time-step; subscripts $n+1$ and n mean, respectively, the solution on the time-step $n+1$ and n; $\alpha = 0.5$ is the time advancing parameter; i is the iteration counter and N_2 is a nonsingular diagonal matrix. The resulting linear systems of equations are solved by the GMRES method considering all matrices stored by the well know strategy element-by-element [10].

3   Numerical Experiments

In this section we present numerical experiments considering three well known 2D benchmark problems: 'oblique shock', 'reflected shock' and 'explosion', discretized by unstructured triangular meshes using Delaunay triangulation through the software Gmsh [8]. The first and second problems used GMRES with 5 vectors to restart, tolerance equal to 10^{-1}, the number of multicorrections fixed to 3, the time-step size is 10^{-3} and the simulation is run until 3000 steps. The third problem used GMRES with 30 vectors to restart, tolerance equal to 10^{-5}, the number of multicorrections fixed to 3, the time-step size is 10^{-3} and the simulation is run until 250 steps. We compare the NMV method with SUPG formulation and two shock capturing operators, the CAU and the YZβ, named here, respectively, as SUPG + CAU and SUPG + YZβ. The tests were performed on a machine with an Intel Core i7-4770 3.4 GHz processor with 16 GB of RAM and Ubuntu 12.04 operating system.

Algorithm 1. NMV Predictor Multicorrector Algorithm

Step 1. $i = 0$

Step 2. Predictor phase:

$$U_h^{n+1,0} = U_h^n + (1 - \alpha)\Delta t \dot{U}_h^n,$$
$$\dot{U}_h^{n+1,0} = 0$$
$$U_B^{n+1,0} = U_B^n + (1 - \alpha)\Delta t \dot{U}_B^n,$$
$$\dot{U}_B^{n+1,0} = 0$$

Step 3. Multicorrector phase:

Residual Force:

$$R_1^{n+1,i} = F_h^{n+1} - \left(M_{hh}\dot{U}_h^{n+1,i} + M_{hB}\dot{U}_B^{n+1,i} \right)$$
$$- \left(K_{hh}U_h^{n+1,i} + K_{hB}U_B^{n+1,i} \right)$$

$$R_2^{n+1,i} = F_B^{n+1} - \left(M_{Bh}\dot{U}_h^{n+1,i} + M_{BB}\dot{U}_B^{n+1,i} \right)$$
$$- \left(K_{Bh}U_h^{n+1,i} + K_{BB}U_B^{n+1,i} \right)$$

Solve:

$$M^* \Delta \dot{U}_h^{n+1,i+1} = F^*,$$

with $M^* = M_1 - N_1 N_2^{-1} M_2$ and $F^* = R_1 - N_1 N_2^{-1} R_2$

where $M_1 = M_{hh} + \alpha \Delta t K_{hh}$, $N_1 = M_{hB} + \alpha \Delta t K_{hB}$

$M_2 = M_{Bh} + \alpha \Delta t K_{Bh}$ and $N_2 = M_{BB} + \alpha \Delta t K_{BB}$

Corrector:

$$U_h^{n+1,i+1} = U_h^{n,i} + \alpha \Delta t \Delta \dot{U}_h^{n+1,i+1},$$
$$\dot{U}_h^{n+1,i+1} = \dot{U}_h^{n+1,i} + \Delta \dot{U}_h^{n+1,i+1}$$
$$U_B^{n+1,i+1} = U_B^{n,i} + \alpha \Delta t \Delta \dot{U}_B^{n+1,i+1},$$
$$\dot{U}_B^{n+1,i+1} = \dot{U}_B^{n+1,i} + \Delta \dot{U}_B^{n+1,i+1}$$

with $\Delta \dot{U}_B^{n+1,i+1} = N_2^{-1} \left(R_2^{n+1,i} - M_2 \Delta \dot{U}_h^{n+1,i+1} \right)$

3.1   2D Oblique Shock Problem

The first problem is a Mach 2 uniform flow over a wedge, at an angle of $-10°$ with respect to a horizontal wall. The solution involves an oblique shock at an angle of $29.3°$ emanating from the leading edge of the wedge, as shown in Fig. 2. The computational domain is a square with $0 \le x \le 1$ and $0 \le y \le 1$. Prescribing the following inflow data on the left and top boundaries results in a solution with the following outflow data:

$$\text{Inflow} \begin{cases} M &= 2.0 \\ \rho &= 1.0 \\ u &- \cos 10^0 \\ v &= -\sin 10^0 \\ p &= 0.17857 \end{cases} \qquad \text{Outflow} \begin{cases} M &= 1.64052 \\ \rho &= 1.45843 \\ u &= 0.88731 \\ v &= 0.0 \\ p &= 0.30475 \end{cases} \qquad (11)$$

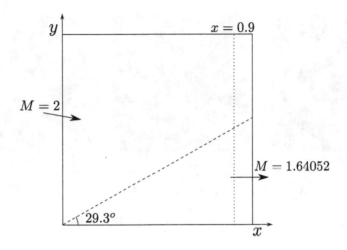

Fig. 2. Oblique shock problem description

Here M is the Mach number and p is the pressure. Four Dirichlet boundary conditions are imposed at the left and the top boundaries, the condition $v = 0$ is set at the bottom boundary, and no boundary condition is imposed at the outflow (right) boundary.

For all simulations we consider an unstructured mesh consisting of 462 nodes and 846 elements. For the reference values used in Eq. (6), we consider the initial condition values for the left domain. Figure 3 shows the 2D density distribution obtained with all methods. Figure 4 shows the density profile along $x = 0.9$, obtained with SUPG + CAU, SUPG + YZβ and NMV methods. The solution obtained with the SUPG + YZβ is slightly better than the NMV on the left of the shock, whereas the solution with NMV is better on the right of the shock. The SUPG + CAU method clearly exhibit more dissipation.

On the other hand, the NMV method needs less GMRES iterations and CPU time than the others, as we can see in Table 1. The NMV method need less than half the number of GMRES iterations required by SUPG + CAU and SUPG + YZβ methods. Furthermore, the NMV method requires approximately 60 % and 55 %, respectively, of the CPU time required by the SUPG + CAU and the SUPG + YZβ methods.

Table 1. Oblique shock problem: computational performance

Methods	GMRES iterations	CPU time (s)
SUPG + CAU	58,514	54.225
SUPG + YZβ	69,863	60.302
NMV	18,020	33.361

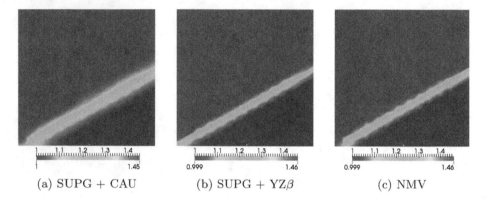

(a) SUPG + CAU   (b) SUPG + YZβ (c) NMV

Fig. 3. Oblique shock problem: density distribution 2D solution at time $t = 3$. (Color figure online)

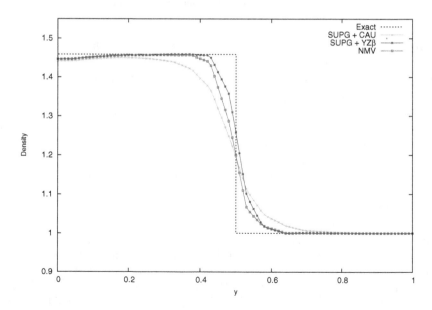

Fig. 4. Oblique shock problem: density profile along $x = 0.9$. (Color figure online)

3.2   2D Reflected Shock Problem

This problem consists of three regions (R1, R2 and R3) separated by an oblique shock and its reflection from a wall, as shown in Fig. 5. Prescribing the following Mach 2.9 inflow data in the first region on the left (R1), and requiring the incident shock to be at an angle of 29°, leads to the following exact solution at the other two regions (R2 and R3):

$$
\text{R1}
\begin{cases}
M = 2.9 \\
\rho = 1.0 \\
u = 2.9 \\
v = 0.0 \\
p = 0.714286
\end{cases}
\quad
\text{R2}
\begin{cases}
M = 2.3781 \\
\rho = 1.7 \\
u = 2.61934 \\
v = -0.50632 \\
p = 1.52819
\end{cases}
\quad
\text{R3}
\begin{cases}
M = 1.94235 \\
\rho = 2.68728 \\
u = 2.40140 \\
v = 0.0 \\
p = 2.93407
\end{cases}
\quad (12)
$$

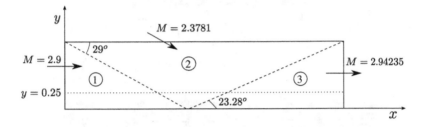

Fig. 5. Reflected shock problem description.

The computational domain is a rectangle with $0 \leq x \leq 4.1$ and $0 \leq y \leq 1$. We prescribe the density, velocities and pressure at the left and top boundaries, the slip condition with $v = 0$ is imposed at the bottom boundary, and no boundary condition is imposed at the outflow (right) boundary.

For all simulations we consider an unstructured mesh consisting of 1,315 nodes and 2,464 elements. For the reference values used in Eq. (6), we consider the initial condition values for the left domain. Figure 6 shows the 2D density distribution obtained with all methods. Figure 7 shows the density profile along $y = 0.25$, obtained with SUPG + CAU, SUPG + YZβ and NMV methods. We may observe a good agreement between the SUPG + YZβ and NMV solutions, clearly exhibit less dissipation than the SUPG + CAU solution.

One more time, the NMV method needs less GMRES iterations and CPU time than the others, as we can see in Table 2. The NMV method need less than half the number of GMRES iterations required by SUPG + CAU and SUPG + YZβ methods. Additionally, the NMV method requires approximately 48 % and 77 %, respectively, of the CPU time required by the SUPG + CAU and the SUPG + YZβ methods.

Table 2. Reflected shock problem: computational performance

Methods	GMRES Iterations	CPU Time (s)
SUPG + CAU	80,948	198.365
SUPG + YZβ	35,023	124.468
NMV	17,482	96.139

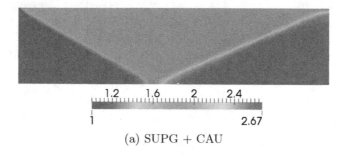

(a) SUPG + CAU

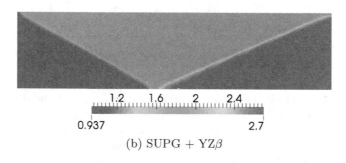

(b) SUPG + YZβ

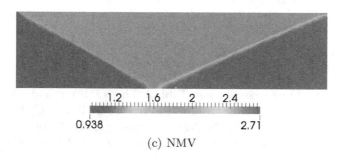

(c) NMV

Fig. 6. Reflected shock problem: density distribution 2D solution at time $t = 3$. (Color figure online)

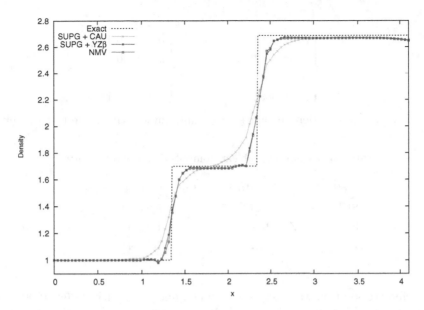

Fig. 7. Reflected shock problem: density profile along $y = 0.25$. (Color figure online)

3.3 2D Explosion Problem

We consider the explosion problem for an ideal gas with $\gamma = 1.4$ as described by [1]. The 2D Euler equations are solved on a 2.0×2.0 square domain in the $xy-$plane. The initial condition consists of the region inside of a circle with radius $R = 0.4$ centered at $(1,1)$ and the region outside the circle, see Fig. 8. The flow variables are constant in each of these regions and are separated by a circular discontinuity at time $t = 0$. The two constant states are chosen as

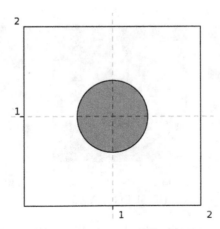

Fig. 8. Explosion problem description.

$$\text{ins} \begin{cases} \rho &=& 1.0 \\ u &=& 0.0 \\ v &=& 0.0 \\ p &=& 1.0 \end{cases} \quad \text{out} \begin{cases} \rho &=& 0.125 \\ u &=& 0.0 \\ v &=& 0.0 \\ p &=& 0.1 \end{cases} \tag{13}$$

Subscripts ins and out denote values inside and outside the circle respectively.

Table 3. Explosion problem: computational performance

Methods	GMRES Iterations	CPU Time (s)
SUPG + CAU	1,273,137	9,965.661
SUPG + YZβ	15,012	178.953
NMV	11,228	70.029

A reference solution was used considering a fine mesh with 1000×1000 computing cells by WAF method and it is in good agreement with the analytical solution as described in [21]. In our simulation, we consider an unstructured mesh with 13,438 nodes and 26,474 elements. For the reference values used in Eq. (6), we consider the initial condition values for inside the circle. Figure 9 shows the 2D density distribution and Fig. 10 shows the 3D density distribution obtained with all methods. Figure 11 compares the radial variations of the density obtained using SUPG + CAU, SUPG + YZβ and NMV methods. The solution obtained with NMV is slightly more accurate than the SUPG + YZβ solution, whereas the SUPG + CAU solution clearly exhibit more dissipation.

Again, the NMV method needs less GMRES iterations and CPU time than the others, as we can see in Table 3. The NMV method needed less GMRES iterations than required by SUPG + CAU and SUPG + YZβ methods. In addition,

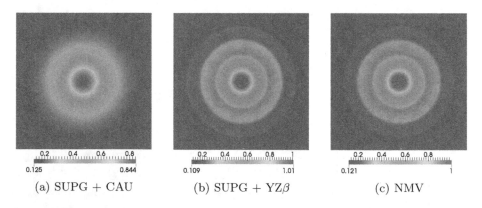

(a) SUPG + CAU   (b) SUPG + YZβ (c) NMV

Fig. 9. Explosion problem: density distribution 2D solution at time $t = 0.25$. (Color figure online)

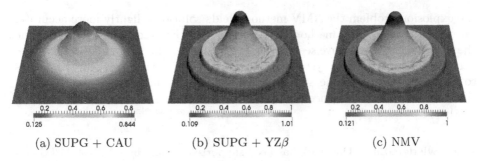

(a) SUPG + CAU   (b) SUPG + YZβ (c) NMV

Fig. 10. Explosion problem: density distribution 3D solution at time $t = 0.25$. (Color figure online)

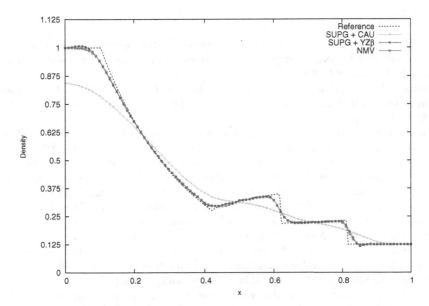

Fig. 11. Explosion problem: comparisons of radial variations of density obtained using SUPG + CAU, SUPG + YZβ and NMV, with the reference solution. (Color figure online)

the NMV method requires approximately 0.7% and 39%, respectively, of the CPU time required by the SUPG + CAU and the SUPG + YZβ methods.

4   Conclusions

We presented a new nonlinear multiscale finite element formulation self adaptive for the inviscid compressible flows in conservative variables, where the amount of artificial viscosity is determined by the YZβ shock-capturing parameter. Solutions obtained with the NMV method is comparable with those obtained with the SUPG + YZβ method in oblique shock and reflected shock problems, whereas in the explosion problem the NMV method yields solutions slightly more accurate. Furthermore, the NMV method requires less GMRES iterations and CPU time than the others, as we have seen in the experiments. The NMV method improve conditioning of the linear system, coupled nonlinear equation system that needs to be solved at every time step of a flow computation, which makes substantial difference in convergence of the iterative solution and computationally less costly.

Acknowledgments. This work has been supported in part by CNPq, CAPES and FAPES.

References

1. Abbassi, H., Mashayek, F., Jacobs, G.B.: Shock capturing with entropy-based artificial viscosity for staggered grid discontinuous spectral element method. Comput. Fluids **98**, 152–163 (2014)
2. Arruda, N.C.B., Almeida, R.C., do Carmo, E.G.D.: Dynamic diffusion formulations for advection dominated transport problems. Mecânica Computacional **29**, 2011–2025 (2010)
3. Brenner, S.C., Scott, L.R.: The Mathematical Theory of Finite Element Methods. Texts in Applied Mathematics. Springer, New York, Berlin, Paris (2002)
4. Brooks, A.N., Hughes, T.J.R.: Streamline upwind petrov-galerkin formulations for convection dominated flows with particular emphasis on the incompressible navier-stokes equations. Comput. Methods Appl. Mech. Eng. **32**, 199–259 (1982)
5. Catabriga, L., de Souza, D.F., Coutinho, A.L., Tezduyar, T.E.: Three-dimensional edge-based SUPG computation of inviscid compressible flows with YZβ shock-capturing. J. Appl. Mech. **76**(2), 021208–021208-7 (2009). ASME
6. Franca, L.P., Neslicurk, A., Stynes, M.: On the stability of residual-free bubbles for convection-diffusion problems and their approximation by a two-level finite element method. Comput. Methods Appl. Mech. Eng. **166**, 35–49 (1998)
7. Galeão, A., Carmo, E.: A consistent approximate upwind petrov-galerkin method for convection-dominated problems. Comput. Methods Appl. Mech. Eng. **68**, 83–95 (1988)
8. Geuzaine, C., Remacle, J.F.: Gmsh: a 3-D finite element mesh generator with built-in pre- and post-processing facilities. Int. J. Numer. Methods Eng. **79**, 1309–1331 (2009)
9. Gravemeier, V., Gee, M.W., Kronbichler, M., Wall, W.A.: An algebraic variational multiscale-multigrid method for large eddy simulation of turbulent flow. Comput. Methods Appl. Mech. Eng. **199**(13), 853–864 (2010)
10. Hughes, T.J.R., Tezduyar, T.E.: Finite element methods for first-order hyperbolic systems with particular emphasis on the compressible euler equations. Comput. Methods Appl. Mech. Eng. **45**(1), 217–284 (1984)
11. Hughes, T.J.: Multiscale phenomena: Green's functions, the Dirichlet-to-Neumann formulation, subgrid scale models, bubbles and the origins of stabilized methods. Comput. Methods Appl. Mech. Eng. **127**(1–4), 387–401 (1995)
12. John, V., Knobloch, P.: On spurious oscillations at layers diminishing (sold) methods for convection-diffusion equations: part i-a review. Comput. Methods Appl. Mech. Eng. **196**(17), 2197–2215 (2007)

13. Nassehi, V., Parvazinia, M.: A multiscale finite element space-time discretization method for transient transport phenomena using bubble functions. Finite Elem. Anal. Des. **45**(5), 315–323 (2009)
14. Rispoli, F., Saavedra, R., Corsini, A., Tezduyar, T.E.: Computation of inviscid compressible flows with the V-SGS stabilization and YZβ shock-capturing. Int. J. Numer. Methods Fluids **54**(6–8), 695–706 (2007)
15. Santos, I.P., Almeida, R.C.: A nonlinear subgrid method for advection-diffusion problems. Comput. Methods Appl. Mech. Eng. **196**, 4771–4778 (2007)
16. Sedano, R.Z., Bento, S.S., Lima, L.M., Catabriga, L.: Predictor-multicorrector schemes for the multiscale dynamic diffusion method to solve compressible flow problems. In: CILAMCE2015 - XXXVI Ibero-Latin American Congress on Computational Methods in Engineering, November 2015
17. Tezduyar, T.E.: Determination of the stabilization and shock-capturing parameters in supg formulation of compressible flows. In: Proceedings of the European Congress on Computational Methods in Applied Sciences and Engineering, ECCOMAS (2004)
18. Tezduyar, T.E.: Finite elements in fluids: stabilized formulations and moving boundaries and interfaces. Comput. Fluids **36**(2), 191–206 (2007)
19. Tezduyar, T.E., Senga, M.: Stabilization and shock-capturing parameters in SUPG formulation of compressible flows. Comput. Methods in Appl. Mech. Eng. **195**(13–16), 1621–1632 (2006)
20. Tezduyar, T.E., Senga, M.: SUPG finite element computation of inviscid supersonic flows with YZβ shock-Capturing. Comput. Fluids **36**(1), 147–159 (2007)
21. Toro, E.F.: Riemann Solvers and Numerical Methods for Fluid Dynamics: A Practical Introduction. Springer Science & Business Media, Heidelberg (2009)

"Extended Cross-Product" and Solution of a Linear System of Equations

Vaclav Skala[✉]

Faculty of Applied Sciences, University of West Bohemia,
Univerzitni 8, 30614 Plzen, Czech Republic
skala@kiv.zcu.cz
http://www.VaclavSkala.euss

Abstract. Many problems, not only in computer vision and visualization, lead to a system of linear equations $Ax = 0$ or $Ax = b$ and fast and robust solution is required. A vast majority of computational problems in computer vision, visualization and computer graphics are three dimensional in principle. This paper presents equivalence of the cross–product operation and solution of a system of linear equations $Ax = 0$ or $Ax = b$ using projective space representation and homogeneous coordinates. This leads to a conclusion that division operation for a solution of a system of linear equations is not required, if projective representation and homogeneous coordinates are used. An efficient solution on CPU and GPU based architectures is presented with an application to barycentric coordinates computation as well.

Keywords: Linear system of equations · Extended cross-product · Projective space computation · Geometric algebra · Scientific computation

1 Introduction

Many applications, not only in computer vision, require a solution of a homogeneous system of linear equations $Ax = 0$ or a non-homogeneous system of linear equations $Ax = b$. There are several numerical methods used implemented in standard numerical libraries. However, the numerical solution actually does not allow further symbolic manipulation. Even more, solutions of equations $Ax = 0$ and $Ax = b$ are considered as different problems and especially $Ax = 0$ is not usually solved quite correctly as users tend to use some additional condition for x unknown (usually setting $x_k = 1$ or so).

In the following, we show the equivalence of the extended cross-product (outer product or progressive product) with a solution of both types of linear systems of equations, i.e. $Ax = 0$ and $Ax = b$.

Many problems in computer vision, computer graphics and visualization are 3-dimensional. Therefore specific numerical approaches can be applied to speed up the solution. In the following extended cross-product, also called outer product or progressive product, is introduced in the "classical" notation using "×" symbol.

© Springer International Publishing Switzerland 2016
O. Gervasi et al. (Eds.): ICCSA 2016, Part I, LNCS 9786, pp. 18–35, 2016.
DOI: 10.1007/978-3-319-42085-1_2

2 Extended Cross Product

Let us consider the standard cross-product of two vectors $a = [a_1, a_2, a_3]^T$ and $b = [b_1, b_2, b_3]^T$. Then the cross-product is defined as:

$$a \times b = \det \begin{bmatrix} i & j & k \\ a_1 & a_2 & a_3 \\ b_1 & b_2 & b_3 \end{bmatrix} \tag{1}$$

where: $i = [1, 0, 0]^T, j = [0, 1, 0]^T, k = [0, 0, 1]^T$.

If a matrix form is needed, then we can write:

$$a \times b = \begin{bmatrix} 0 & -a_3 & a_2 \\ a_3 & 0 & -a_1 \\ -a_2 & a_1 & 0 \end{bmatrix} \begin{bmatrix} b_1 \\ b_2 \\ b_3 \end{bmatrix} \tag{2}$$

In some applications the matrix form is more convenient.

Let us introduce the extended cross-product of three vectors $a = [a_1, \ldots, a_n]^T$, $b = [b_1, \ldots, b_n]^T$ and $c = [c_1, \ldots, c_n]^T$, $n = 4$ as:

$$a \times b \times c = \det \begin{bmatrix} i & j & k & l \\ a_1 & a_2 & a_3 & a_4 \\ b_1 & b_2 & b_3 & b_4 \\ c_1 & c_2 & c_3 & c_4 \end{bmatrix} \tag{3}$$

where: $i = [1, 0, 0, 0]^T, j = [0, 1, 0, 0]^T, k = [0, 0, 1, 0]^T, l = [0, 0, 0, 1]^T$.

It can be shown that there exists a matrix form for the extended cross-product representation:

$$a \times b \times c = (-1)^{n+1} \begin{bmatrix} 0 & -\delta_{34} & \delta_{24} & -\delta_{23} \\ \delta_{34} & 0 & -\delta_{14} & \delta_{13} \\ -\delta_{24} & \delta_{14} & 0 & -\delta_{12} \\ \delta_{23} & -\delta_{13} & \delta_{12} & 0 \end{bmatrix} \begin{bmatrix} c_1 \\ c_2 \\ c_3 \\ c_4 \end{bmatrix} \tag{4}$$

where: $n = 4$. In this case and δ_{ij} are sub-determinants with columns i, j of the matrix T defined as:

$$T = \begin{bmatrix} a_1 & a_2 & a_3 & a_4 \\ b_1 & b_2 & b_3 & b_4 \end{bmatrix} \tag{5}$$

e.g. sub-determinant $\delta_{24} = \det \begin{bmatrix} a_2 & a_4 \\ b_2 & b_4 \end{bmatrix}$ etc.

The extended cross-product for 5-dimensions is defined as:

$$\boldsymbol{a} \times \boldsymbol{b} \times \boldsymbol{c} \times \boldsymbol{d} = \det \begin{bmatrix} \boldsymbol{i} & \boldsymbol{j} & \boldsymbol{k} & \boldsymbol{l} & \boldsymbol{n} \\ a_1 & a_2 & a_3 & a_4 & a_5 \\ b_1 & b_2 & b_3 & b_4 & b_5 \\ c_1 & c_2 & c_3 & c_4 & c_5 \\ d_1 & d_2 & d_3 & d_4 & d_5 \end{bmatrix} \tag{6}$$

where: $\boldsymbol{i} = [1,0,0,0,0]^T$, $\boldsymbol{j} = [0,1,0,0,0]^T$, $\boldsymbol{k} = [0,0,1,0,0]^T$, $\boldsymbol{l} = [0,0,0,1,0]^T$, $\boldsymbol{n} = [0,0,0,0,0,1]^T$.

It can be shown that there exists a matrix form as well:

$$\boldsymbol{a} \times \boldsymbol{b} \times \boldsymbol{c} \times \boldsymbol{d} = (-1)^{n+1} \begin{bmatrix} 0 & -\delta_{345} & \delta_{245} & -\delta_{235} & \delta_{234} \\ \delta_{345} & 0 & -\delta_{145} & \delta_{135} & -\delta_{134} \\ -\delta_{245} & \delta_{145} & 0 & -\delta_{125} & \delta_{124} \\ \delta_{235} & -\delta_{135} & \delta_{125} & 0 & -\delta_{123} \\ -\delta_{234} & \delta_{134} & -\delta_{124} & \delta_{123} & 0 \end{bmatrix} \begin{bmatrix} d_1 \\ d_2 \\ d_3 \\ d_4 \\ d_5 \end{bmatrix} \tag{7}$$

where $n = 5$. In this case and δ_{ijk} are sub-determinants with columns i, j, k of the matrix T defined as:

$$\boldsymbol{T} = \begin{bmatrix} a_1 & a_2 & a_3 & a_4 & a_5 \\ b_1 & b_2 & b_3 & b_4 & b_5 \\ c_1 & c_2 & c_3 & c_4 & c_5 \end{bmatrix} \tag{8}$$

e.g. sub-determinant δ_{245} is defined as:

$$\delta_{245} = \det \begin{bmatrix} a_2 & a_4 & a_5 \\ b_2 & b_4 & b_5 \\ c_2 & c_4 & c_5 \end{bmatrix} = a_2 \det \begin{bmatrix} b_4 & b_5 \\ c_4 & c_5 \end{bmatrix} - a_4 \det \begin{bmatrix} b_2 & b_5 \\ c_2 & c_5 \end{bmatrix} + a_5 \det \begin{bmatrix} b_2 & b_4 \\ c_2 & c_4 \end{bmatrix}$$

$$\tag{9}$$

In spite of the "complicated" description above, this approach leads to a faster computation in the case of lower dimensions, see Sect. 7.

3   Projective Representation and Duality Principle

Projective representation and its application for computation are considered to be mysterious or too complex. Nevertheless we are using it naturally very frequently in the form of fractions, e.g. a/b. We also know that fractions help us to express values, which cannot be expressed precisely due to limited length of a mantissa, e.g. $1/3 = 0{,}33\ldots\ldots.333\ldots = 0.\bar{3}$.

In the following we will explore projective representation, actually rational fractions, and its applicability.

3.1 Projective Representation

Projective extension of the Euclidean space is used commonly in computer graphics and computer vision mostly for geometric transformations. However, in computational sciences, the projective representation is not used, in general. This chapter shortly introduces basic properties and mutual conversions. More detailed description of projective representation and applications can be found in [12, 15, 20].

The given point $X = (X,Y)$ in the Euclidean space E^2 is represented in homogeneous coordinates as $x = [x, y : w]^T$, $w \neq 0$. It can be seen that x is actually a line in the projective space P^3 with the origin excluded. Mutual conversions are defined as:

$$X = \frac{x}{w} \qquad Y = \frac{y}{w} \tag{10}$$

where: $w \neq 0$ is the homogeneous coordinate. Note that the homogeneous coordinate w is actually a scaling factor with no physical meaning, while x, y are values with physical units in general.

The projective representation enables us nearly double precision as the mantissa of x, resp. y and w are used for a value representation. However we have to distinguish two different data types, i.e.

- Projective representation of a n-dimensional value $X = (X_1, \ldots, X_n)$, represented by one dimensional array $x = [x_1, \ldots, x_n : x_w]^T$, e.g. coordinates of a point, that is fixed to the origin of the coordinate system.
- Projective representation of a n-dimensional vector (in the mathematical meaning) $A = (A_1, \ldots, A_n)$, represented by one dimensional array $a = [a_1, \ldots, a_n : a_w]^T$. In this case the homogeneous coordinate a_w is actually just a scaling factor. Any vector is not fixed to the origin of the coordinate system and it is "movable".

Therefore a user should take an attention to the correctness of operations. Another interesting application of the projective representation is the rational trigonometry [19].

3.2 Principle of Duality

The projective representation offers also one very important property – principle of duality. The principle of duality in E^2 states that any theorem remains true when we interchange the words "point" and "line", "lie on" and "pass through", "join" and "intersection", "collinear" and "concurrent" and so on. Once the theorem has been established, the dual theorem is obtained as described above [1, 5, 9, 14]. In other words, the principle of duality says that in all theorems it is possible to substitute the term "point" by the term "line" and the term "line" by the term "point" etc. in E^2 and the given theorem stays valid. Similar duality is valid for E^3 as well, i.e. the terms "point" and "plane" are dual etc. it can be shown that operations "join" a "meet" are dual as well.

This helps a lot to solve some geometrical problems. In the following we will demonstrate that on very simple geometrical problems like intersection of two lines, resp. three planes and computation of a line given by two points, resp. of a plane given by three points.

4 Solution of $Ax = B$

Solution of non-homogeneous system of equation $AX = b$ is used in many computational tasks.

For simplicity of explanation, let us consider a simple example of intersection computation of two lines p_1 a p_2 in E^2 given as:

$$p_1 : A_1X + B_1Y + C_1 = 0 \qquad p_2 : A_2X + B_2Y + C_2 = 0 \qquad (11)$$

An intersection point of two those lines is given as a solution of a linear system of equations: $Ax = b$:

$$\begin{bmatrix} a_1 & b_1 \\ a_2 & b_2 \end{bmatrix} \begin{bmatrix} X \\ Y \end{bmatrix} = \begin{bmatrix} -c_1 \\ -c_2 \end{bmatrix} \qquad (12)$$

Generally, for the given system of n liner equations with n unknowns in the form $AX = b$ the solution is given:

$$X_i = \frac{\det(A_i)}{\det(A)} \quad i = 1, \dots, n \qquad (13)$$

where: A is a regular matrix $n \times n$ having non-zero determinant, the matrix A_i is the matrix A with replaced i^{th} column by the vector b and $X = [X_1, \dots, X_n]^T$ is a vector of unknown values.

In a low dimensional case using general methods for solution of linear equations, e.g. Gauss-Seidel elimination etc., is computational expensive. Also division operation is computationally expensive and decreasing precision of a solution.

Usually, a condition **if** $\det(A) < eps$ **then** EXIT is taken for solving "close to singular cases". Of course, nobody knows, what a value of *eps* is appropriate.

5 Solution of $Ax = 0$

There is another very simple geometrical problem; determination of a line p given by two points $X_1 = (X_1, Y_1)$ and $X_2 = (X_2, Y_2)$ in E^2. This seems to be a quite simple problem as we can write:

$$aX_1 + bY_1 + c = 0 \qquad aX_2 + bY_2 + c = 0 \qquad (14)$$

i.e. it leads to a solution of homogeneous systems of equations $AX = 0$, i.e.:

$$\begin{bmatrix} X_1 & Y_1 & 1 \\ X_2 & Y_2 & 1 \end{bmatrix} \begin{bmatrix} a \\ b \\ c \end{bmatrix} = 0 \tag{15}$$

In this case, we obtain one parametric set of solutions as the Eq. (15) can be multiplied by any value $q \neq 0$ and the line is the same.

There is a problem – we know that lines and points are dual in the E^2 case, so the question is why the solutions are not dual. However if the projective representation is used the duality principle will be valid, as follows.

6   Solution $Ax = b$ and $Ax = 0$

Let us consider again intersection of two lines $p_1 = [a_1, b_1 : c_1]^T$ a $p_2 = [a_2, b_2 : c_2]^T$ leading to a solution of non-homogeneous linear system $AX = b$, which is given as:

$$p_1 : a_1 X + b_1 Y + c_1 = 0 \qquad p_2 : a_2 X + b_2 Y + c_2 = 0 \tag{16}$$

If the equations are multiplied by $w \neq 0$ we obtain:

$$p_1 : a_1 X + b_1 Y + c_1 \triangleq \qquad p_2 : a_2 X + b_2 Y + c_2 \triangleq$$
$$a_1 x + b_1 y + c_1 w = 0 \qquad a_2 x + b_2 y + c_2 w = 0 \tag{17}$$

where: $\triangleq$ means "projectively equivalent to" as $x = wX$ and $y = wY$.

Now we can rewrite the equations to the matrix form as $Ax = 0$:

$$\begin{bmatrix} a_1 & b_1 & -c_1 \\ a_2 & b_2 & -b_2 \end{bmatrix} \begin{bmatrix} x \\ y \\ w \end{bmatrix} = \begin{bmatrix} 0 \\ 0 \end{bmatrix} \tag{18}$$

where $x = [x, y : w]^T$ is the intersection point in the homogeneous coordinates.

In the case of computation of a line given by two points given in homogeneous coordinates, i.e. $x_1 = [x_1, y_1 : w_1]^T$ and $x_2 = [x_2, y_2 : w_2]^T$, the Eq. (14) is multiplied by $w_i \neq 0$. Then, we get a solution in the matrix form as $Ax = 0$, i.e.

$$\begin{bmatrix} x_1 & y_1 & w_1 \\ x_2 & y_2 & w_2 \end{bmatrix} \begin{bmatrix} a \\ b \\ c \end{bmatrix} = 0 \tag{19}$$

Now, we can see that the formulation is leading in the both cases to the same numerical problem: to a solution of a homogeneous linear system of equations.

However, a solution of homogeneous linear system of equations is not quite straightforward as there is a one parametric set of solutions and all of them are projectively equivalent. It can be seen that the solution of Eq. (18), i.e. intersection of two lines in E^2, is equivalent to:

$$x = p_1 \times p_2 \tag{20}$$

and due to the principle of duality we can write for a line given by two points:

$$p = x_1 \times x_2 \tag{21}$$

In the three dimensional case we can use extended cross-product [12, 15, 16].

A plane $\rho : aX + bY + cY + d = 0$ given by three points $x_1 = [x_1, y_1, z_1 : w_1]^T$, $x_2 = [x_2, y_2, z_2 : w_2]^T$ and $x_2 = [x_3, y_3, z_3 : w_3]^T$ is determined in the projective representation as:

$$\rho = [a, b, c : d]^T = x_1 \times x_2 \times x_2 \tag{22}$$

and the intersection point x of three planes points $\rho_1 = [a_1, b_1, c_1 : d_1]^T$, $\rho_2 = [a_2, b_2, c_2 : d_2]^T$ and $\rho_3 = [a_3, b_3, c_3 : d_3]^T$ is determined in the projective representation as:

$$x = [x, y, z : w]^T = \rho_1 \times \rho_2 \times \rho_2 \tag{23}$$

due to the duality principle.

It can be seen that there is no division operation needed, if the result can be left in the projective representation. The approach presented above has another one great advantage as it allows symbolic manipulation as we have avoided numerical solution and also precision is nearly doubled.

7 Barycentric Coordinates Computation

Barycentric coordinates are often used in many engineering applications, not only in geometry. The barycentric coordinates computation leads to a solution of a system of linear equations. However it was shown, that a solution of a linear system equations is equivalent to the extended cross product [12–14]. Therefore it is possible to compute barycentric coordinates using cross product which is convenient for application of SSE instructions or for GPU oriented computations. Let us demonstrate the proposed approach on a simple example again.

Given a triangle in E^2 defined by points $x_i = [x_i, y_i : 1]^T$, $i = 1, \ldots, 3$, the barycentric coordinates of the point $x_0 = [x_0, y_0 : 1]^T$ can be computed as follows:

$$\begin{aligned} \lambda_1 x_1 + \lambda_2 x_2 + \lambda_3 x_3 &= x_0 \\ \lambda_1 y_1 + \lambda_2 y_2 + \lambda_3 y_3 &= y_0 \\ \lambda_1 + \lambda_2 + \lambda_3 &= 1 \end{aligned} \tag{24}$$

For simplicity, we set $w_i = 1$, $i = 1, \ldots, 3$. It means that we have to solve a system of linear equations $Ax = b$:

$$
\begin{bmatrix} x_1 & x_2 & x_3 \\ y_1 & y_2 & y_3 \\ 1 & 1 & 1 \end{bmatrix} \begin{bmatrix} \lambda_1 \\ \lambda_2 \\ \lambda_3 \end{bmatrix} = \begin{bmatrix} x_0 \\ y_0 \\ 1 \end{bmatrix}
\tag{25}
$$

if the points are given in the projective space with homogeneous coordinates $x_i = [x_i, y_i : w_i]^T$, $i = 1, \ldots, 3$ and $x_0 = [x_0, y_0 : w_0]^T$. It can be easily proved, due to the multilinearity, we need to solve a linear system $Ax = b$:

$$
\begin{bmatrix} x_1 & x_2 & x_3 \\ y_1 & y_2 & y_3 \\ w_1 & w_2 & w_3 \end{bmatrix} \begin{bmatrix} \lambda_1 \\ \lambda_2 \\ \lambda_3 \end{bmatrix} = \begin{bmatrix} x_0 \\ y_0 \\ w_0 \end{bmatrix}
\tag{26}
$$

Let us define new vectors containing a row of the matrix A and vector b as:

$$
x = [x_1, x_2, x_3, x_0]^T \quad y = [y_1, y_2, y_3, y_0]^T \quad w = [w_1, w_2, w_3, w_0]^T
\tag{27}
$$

The projective barycentric coordinates $\xi = [\xi_1, \xi_2, \xi_3 : \xi_w]^T$ are given as:

$$
\lambda_1 = -\frac{\xi_1}{\xi_w} \quad \lambda_2 = -\frac{\xi_2}{\xi_w} \quad \lambda_3 = -\frac{\xi_3}{\xi_w}
\tag{28}
$$

i.e.

$$
\lambda_i = -\frac{\xi_i}{\xi_w} \quad i = 1, \ldots, 3
\tag{29}
$$

Using the extended cross product, the projective barycentric coordinates are given as:

$$
\xi = x \times y \times w = \det \begin{bmatrix} i & j & k & l \\ x_1 & x_2 & x_3 & x_0 \\ y_1 & y_2 & y_3 & y_0 \\ w_1 & w_2 & w_3 & w_4 \end{bmatrix} = [\xi_1, \xi_2, \xi_3 : \xi_w]^T
\tag{30}
$$

where $i = [1, 0, 0, 0]^T$, $j = [0, 1, 0, 0]^T$, $k = [0, 0, 1, 0]^T$, $l = [0, 0, 0, 1]^T$

Similarly in the E^3 case, given a tetrahedron in E^3 defined by points $x_i = [x_i, y_i, z_i : w_i]^T$, $i = 1, \ldots, 3$ and the point $x_0 = [x_0, y_0, z_0 : w_0]^T$:

$$
\begin{aligned}
x &= [x_1, x_2, x_3, x_4 : x_0]^T \quad y = [y_1, y_2, y_3, y_4 : y_0]^T \\
z &= [z_1, z_2, z_3, z_4 : z_0]^T \quad w = [w_1, w_2, w_3, w_4 : w_0]^T
\end{aligned}
\tag{31}
$$

Then projective barycentric coordinates are given as:

$$\boldsymbol{\xi} = \boldsymbol{x} \times \boldsymbol{y} \times \boldsymbol{z} \times \boldsymbol{w} = [\xi_1, \xi_2, \xi_3, \xi_4 : \xi_w]^T \tag{32}$$

The Euclidean barycentric coordinates are given as:

$$\lambda_1 = -\frac{\xi_1}{\xi_w} \quad \lambda_2 = -\frac{\xi_2}{\xi_w} \quad \lambda_3 = -\frac{\xi_3}{\xi_w} \quad \lambda_4 = -\frac{\xi_4}{\xi_w} \tag{33}$$

i.e.

$$\lambda_i = -\frac{\xi_i}{\xi_w} \quad i = 1, \ldots, 4 \tag{34}$$

How Simple and Elegant Solution! The presented computation of barycentric coordinates is simple and convenient for GPU use or SSE instructions. Even more, as we have assumed from the very beginning, there is no need to convert projective values to the Euclidean notation. As a direct consequence of that is, that we are saving a lot of computational time also increasing robustness of the computation, especially due to division operation elimination. As a result is represented as a rational fraction, the precision is nearly equivalent to double mantissa precision and exponent range.

Let us again present advantages of the projective representation on simple examples.

8   Intersection of Two Planes

Intersection of two planes ρ_1 and ρ_1 in E^3 is seemingly a simple problem, but surprisingly computationally expensive, Fig. 1. Let us consider the "standard" solution in the Euclidean space and a solution using the projective approach.

Given two planes ρ_1 and ρ_2 in E^3:

$$\boldsymbol{\rho}_1 = [a_1, b_1, c_1 : d_1]^T = [\boldsymbol{n}_1^T : d_1]^T \quad \boldsymbol{\rho}_2 = [a_2, b_2, c_2 : d_2]^T = [\boldsymbol{n}_2^T : d_2]^T \tag{35}$$

where: $\boldsymbol{n}_1$ and $\boldsymbol{n}_2$ are normal vectors of those planes.

Then the directional vector s of a parametric line $\boldsymbol{X}(t) = \boldsymbol{X}_0 + st$ is given by a cross product:

$$s = \boldsymbol{n}_1 \times \boldsymbol{n}_2 \equiv [a_3, b_3, c_3]^T \tag{36}$$

and point $\boldsymbol{X}_0 \in E^3$ of the line is given as:

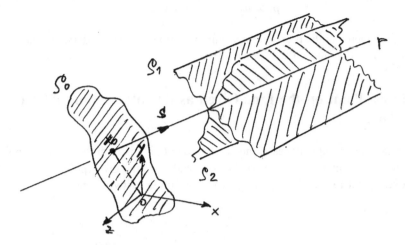

Fig. 1. A line as the intersection of two planes

$$X_0 = \frac{d_2 \begin{vmatrix} b_1 & c_1 \\ b_3 & c_3 \end{vmatrix} - d_1 \begin{vmatrix} b_2 & c_2 \\ b_3 & c_3 \end{vmatrix}}{DET} \qquad Y_0 = \frac{d_2 \begin{vmatrix} a_3 & c_3 \\ a_1 & c_1 \end{vmatrix} - d_1 \begin{vmatrix} a_3 & c_3 \\ a_2 & c_2 \end{vmatrix}}{DET}$$

$$Z_0 = \frac{d_2 \begin{vmatrix} a_1 & b_1 \\ a_3 & b_3 \end{vmatrix} - d_1 \begin{vmatrix} a_2 & b_2 \\ a_3 & b_3 \end{vmatrix}}{DET} \qquad DET = \begin{vmatrix} a_1 & b_1 & c_1 \\ a_2 & b_2 & c_2 \\ a_3 & b_3 & c_3 \end{vmatrix} \tag{37}$$

It can be seen that the formula above is quite difficult to remember and its derivation is not simple. It should be noted that there is again a severe problem with stability and robustness if a condition like $|DET| < eps$ is used. Also the formula is not convenient for GPU or SSE applications. There is another equivalent solution based on Plücker coordinates and duality application, see [12, 16].

Let us explore a solution based on the projective representation explained above.

Given two planes ρ_1 and ρ_2. Then the directional vector s of their intersection is given as:

$$s = n_1 \times n_2 \tag{38}$$

We want to determine the point x_0 of the line given as an intersection of those two planes. Let us consider a plane ρ_0 passing the origin of the coordinate system with the normal vector n_0 equivalent to s, Fig. 1. This plane ρ_0 is represented as:

$$\rho_0 = [a_0, b_0, c_0 : 0]^T = [s^T : 0]^T \tag{39}$$

Then the point x_0 is simply determined as an intersection of three planes ρ_1, ρ_2, ρ_0 as:

$$\boldsymbol{x}_0 = \boldsymbol{\rho}_1 \times \boldsymbol{\rho}_2 \times \boldsymbol{\rho}_0 = [x_0, y_0, z_0 : w_0]^T \tag{40}$$

It can be seen that the proposed algorithm is simple, easy to understand, elegant and convenient for SEE and GPU applications as it uses vector-vector operations.

9 Closest Point on the Line Given as an Intersection of Two Planes

Another example of advantages of the projective notation is finding the closest point on a line given as an intersection of two planes ρ_1 and ρ_2 to the given point $\xi \in E^3$, Fig. 2. The closest point to the given point on an intersection of two planes

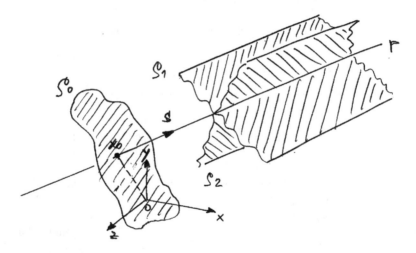

Fig. 2. The closest point to the given point on an intersection of two planes

A solution in the Euclidean space, proposed in [8], is based on a solution of a system of linear equations using Lagrange multipliers, leading to a matrix of (5×5):

$$\begin{bmatrix} 2 & 0 & 0 & n_{1x} & n_{2x} \\ 0 & 2 & 0 & n_{1y} & n_{2y} \\ 0 & 0 & 2 & n_{1z} & n_{2z} \\ n_{1x} & n_{1y} & n_{1z} & 0 & 0 \\ n_{2x} & n_{2y} & n_{2z} & 0 & 0 \end{bmatrix} \begin{bmatrix} x \\ y \\ z \\ \lambda \\ \mu \end{bmatrix} = \begin{bmatrix} 2\xi_x \\ 2\xi_y \\ 2\xi_z \\ \boldsymbol{p}_1 \boldsymbol{n}_1 \\ \boldsymbol{p}_2 \boldsymbol{n}_2 \end{bmatrix} \tag{41}$$

where: $\boldsymbol{p}_1$, resp. $\boldsymbol{p}_2$ are points on planes ρ_1, resp. ρ_2, with a normal vector $\boldsymbol{n}_1$, resp. $\boldsymbol{n}_2$. Coordinates of the closest point $\boldsymbol{x} = [x, y, z]^T$ on the intersection of two planes to the point $\xi = (\xi_x, \xi_y, \xi_z)$ are given as a solution of this system of linear equations. Note that the point ξ is given in the Euclidean space.

Let us consider a solution based on the projective representation. The proposed approach is based on basic geometric transformations with the following steps:

1. Translation of planes ρ_1, ρ_2 and point $\xi = [\xi_x, \xi_y, \xi_z : 1]^T$ so that the point ξ is in the origin of the coordinate system, i.e. using transformation matrix T for the point translation and matrix $(T^T)^{-1} = T^{-T}$ for translation of planes [11, 14, 16].

2. Intersection computation of those two translated planes; the result is a line with the directional vector s and point x_0

3. Translation of the point x_0 by inverse translation using the matrix T^{-1}

The translation matrices are defined as:

$$T = \begin{bmatrix} 1 & 0 & 0 & -\xi_x \\ 0 & 1 & 0 & -\xi_y \\ 0 & 0 & 1 & -\xi_z \\ 0 & 0 & 0 & 1 \end{bmatrix} \qquad T^{-T} = \begin{bmatrix} 1 & 0 & 0 & 0 \\ 0 & 1 & 0 & 0 \\ 0 & 0 & 1 & 0 \\ \xi_x & \xi_y & \xi_z & 1 \end{bmatrix}$$

$$T' = \begin{bmatrix} \xi_w & 0 & 0 & -\xi_x \\ 0 & \xi_w & 0 & -\xi_y \\ 0 & 0 & \xi_w & -\xi_z \\ 0 & 0 & 0 & \xi_w \end{bmatrix} \qquad (42)$$

If the point ξ is given in the projective space, i.e. $\xi = [\xi_x, \xi_y, \xi_z : \xi_w]^T$, $w \neq 1 \,\&\, w \neq 0$, then the matrix T is given as T'.

It can be seen that the computation is more simple, robust and convenient for SSE or GPU oriented applications. It should be noted that the formula is more general as the point ξ can be given in the projective space and no division operations are needed.

10 Symbolic Manipulations

Symbolic manipulations are very important and help to find or simplify computational formulas, avoid singularities etc. As the extended cross-product is an associative and anti-commutative as the cross-product in E^3 similar rules are valid, i.e. in E^3:

$$a \times (b + c) = a \times b + a \times c$$
$$a \times b = -b \times a \qquad (43)$$

In the case of the extended cross-product, i.e. in the projective notation P^3 we actually formally have operations in E^4:

$$a \times (b+c) \times d = a \times b \times d + a \times c \times d$$
$$a \times b \times c = -b \times a \times c$$
(44)

This can be easily proved by applications of rules for operations with determinants.

However, for general understanding more general theory is to be used – Geometric Algebra [2–4, 6, 7, 10, 18], in which the extended cross-product is called outer product and the above identities are rewritten as:

$$a \wedge (b+c) \wedge d = a \wedge b \wedge d + a \wedge c \wedge d$$
$$a \wedge b \wedge c = -b \wedge a \wedge c$$
(45)

where: "$\wedge$" is an operator of the *outer product*, which is equivalent to the cross-product in E^3. There is also an operator "$\vee$" for the *inner product* which is equivalent to the dot product in E^3.

In geometric algebra *geometric product* is defined as:

$$ab = a \vee b + a \wedge b$$
(46)

i.e. in the case of E^3 we can write:

$$ab = a \cdot b + a \times b$$
(47)

and getting some "strange", as a scalar and a vector (actually a bivector) are summed together. But it is a valid result and ab is called *geometric product* [18].

However, if the projective representation is used, we need to be a little bit careful with equivalent operations to the standard operations in the Euclidean space.

11   Example of Application

Let us consider a simple example in 3-dimensional space. Assume, that $Ax = b$ is a system of linear equations, i.e.:

$$\begin{bmatrix} a_{11} & a_{12} & a_{13} \\ a_{21} & a_{22} & a_{23} \\ a_{31} & a_{32} & a_{33} \end{bmatrix} \begin{bmatrix} x_1 \\ x_2 \\ x_3 \end{bmatrix} = \begin{bmatrix} b_1 \\ b_2 \\ b_3 \end{bmatrix}$$
(48)

and we want to explore $\xi = c \cdot x$, where $c = [c_1, c_2, c_3]^T$.

In the "standard" approach a system of linear equations has to be solved numerically or symbolic manipulation has to be used. We can rewrite the Eq. (48) using the projective representation as:

$$\begin{bmatrix} a_{11} & a_{12} & a_{13} & -b_1 \\ a_{21} & a_{22} & a_{23} & -b_2 \\ a_{31} & a_{32} & a_{33} & -b_3 \end{bmatrix} \begin{bmatrix} \bar{x}_1 \\ \bar{x}_2 \\ \bar{x}_3 \\ \bar{x}_w \end{bmatrix} = \begin{bmatrix} 0 \\ 0 \\ 0 \end{bmatrix} \ \& \ x_i = \frac{\bar{x}_i}{\bar{x}_w} \tag{49}$$

The conversion to the Euclidean space is given as:

$$x_i = \frac{\bar{x}_i}{\bar{x}_w} \quad i = 1,\ldots,3 \tag{50}$$

Then using equivalence of the extended cross-product and solution of a linear system of equations we can write:

$$\bar{x} = \bar{a}_1 \times \bar{a}_2 \times \bar{a}_3 \tag{51}$$

where: $\bar{x} = [\bar{x}_1, \bar{x}_2, \bar{x}_3 : \bar{x}_w]^T$, $\bar{a}_i = [a_{i1}, a_{i2}, a_{i3} : -b_i]^T$, $i = 1,\ldots,3$. It should be noted that the result is actually in the 3-dimensional projective space.

In many cases, the result of computation is not necessarily to be converted to the Euclidean space. If left in the projective representation, we save division operations, increase precision of computation as the mantissa is actually nearly doubled (mantissa of $\bar{x}_i$ and $\bar{x}_w$). Also robustness is increased as well as we haven't made any specific assumptions about collinearity of planes. Let a scalar value $\xi \in E^1$ is given as:

$$\xi = c \cdot x \tag{52}$$

The scalar value ξ can be expressed as a homogeneous vector $\bar{\xi}$ in the projective notation as:

$$\bar{\xi}^T = [\xi : \bar{\xi}_w] \ \& \ \bar{\xi}_w = 1 \tag{53}$$

Generally, the value in the Euclidean space is given as $\xi = \frac{\bar{\xi}}{\bar{\xi}_w}$. Extension to the 3-dimensional case is straightforward.

As an example let us consider a test if the given point $\bar{\xi} = [\bar{\xi}_1, \bar{\xi}_2, \bar{\xi}_3 : \bar{\xi}_w]^T$ lies on a plane given by three points $x_i, i = 1, \ldots, 3$ using projective notation. A plane p is given:

$$\rho = x_1 \times x_2 \times x_3 = [a, b, c : d]^T \tag{54}$$

and the given point has to fulfill condition $\bar{\xi} \cdot \rho = a\bar{\xi}_1 + b\bar{\xi}_2 + c\bar{\xi}_3 + d\bar{\xi}_w = 0$.

We know that:

$$
\boldsymbol{a} \times \boldsymbol{b} \times \boldsymbol{c} = \det
\begin{bmatrix}
\boldsymbol{i} & \boldsymbol{j} & \boldsymbol{k} & \boldsymbol{l} \\
a_1 & a_2 & a_3 & a_4 \\
b_1 & b_2 & b_3 & b_4 \\
c_1 & c_2 & c_3 & c_4
\end{bmatrix}
= -
\begin{bmatrix}
0 & -\delta_{34} & \delta_{24} & -\delta_{23} \\
\delta_{34} & 0 & -\delta_{14} & \delta_{13} \\
-\delta_{24} & \delta_{14} & 0 & -\delta_{12} \\
\delta_{23} & -\delta_{13} & \delta_{12} & 0
\end{bmatrix}
\begin{bmatrix}
c_1 \\ c_2 \\ c_3 \\ c_4
\end{bmatrix}
\tag{55}
$$

where: $\boldsymbol{i} = [1,0,0,0]^T$, $\boldsymbol{j} = [0,1,0,0]^T$, $\boldsymbol{k} = [0,0,1,0]^T$, $\boldsymbol{l} = [0,0,0,1]^T$. Then, the test $\boldsymbol{\xi} \cdot \boldsymbol{\rho} = 0$ is actually:

$$
[\xi_1, \xi_2, \xi_3 : \xi_w]
\begin{bmatrix}
0 & -\delta_{34} & \delta_{24} & -\delta_{23} \\
\delta_{34} & 0 & -\delta_{14} & \delta_{13} \\
-\delta_{24} & \delta_{14} & 0 & -\delta_{12} \\
\delta_{23} & -\delta_{13} & \delta_{12} & 0
\end{bmatrix}
\begin{bmatrix}
x_3 \\ y_3 \\ z_3 \\ w_3
\end{bmatrix}
= 0
\tag{56}
$$

It means that we are getting a bilinear form:

$$
\bar{\boldsymbol{\xi}}^T \boldsymbol{B} \boldsymbol{x}_3 = 0
\tag{57}
$$

where: $\boldsymbol{B}$ is an antisymmetric matrix with a null diagonal. So we can analyze such conditions more deeply in an analytical form. It means that we can explore the formula on a symbolic level. It is also possible to derive some additional information for the ξ value, resp. $\bar{\xi}$ value, if the projective notation is used. This approach can be directly extended do the d-dimensional space using geometry algebra [18].

12 Efficiency of Computation and GPU Code

Let us consider reliability and the cost of computation of the "standard" approach using Cramer's rule using determinants. For the given system of n liner equations with n unknowns in the form $\boldsymbol{Ax} = \boldsymbol{b}$ the solution is given as:

$$
X_i = \frac{\det(\boldsymbol{A}_i)}{\det(\boldsymbol{A})} \quad i = 1, \ldots, n
\tag{58}
$$

In the projective notation using homogeneous coordinates we can actually write $\boldsymbol{x} = [x_1, \ldots, x_n : w]^T$, where: $w = \det(\boldsymbol{A})$ and $x_i = \det(\boldsymbol{A}_i)$, $i = 1, \ldots, n$

The projective representation not only enables to postpone division operations, but also offers some additional advantages as follows. Computing of determinants is quite computationally expensive task. However for 2–4 dimensional cases there are some advantages using the extended cross-product as explained below (Table 1).

Table 1. Cost of determinant computation

Operation	$\text{Det}_{2\times2}$	$\text{Det}_{3\times3}$	$\text{Det}_{4\times4}$	$\text{Det}_{5\times5}$
±	1	6	24	120
×	2	12	48	240

Table 2. Cost of cross-product computation

Operation	$a \times b$	$a \times b \times c$	$a \times b \times c \times d$
"±"	3	27	159
"×"	6	52	173

Table 3. Cost of cross-product computation with subdeterminants

	$a \times b$	$a \times b \times c$	$a \times b \times c \times d$
±	3	14	60
×	6	24	77

Generally the computational expenses are given as:

$$\text{Det}_{(k+1)\times(k+1)} = k\,\text{Det}_{k\times k} + k('' \pm '') \tag{59}$$

Total cost of computation if Cramer's rule for generalized is used (Table 2):

Computational expenses for the generalized cross-product matrix based formulation, if partial intermediate computations are used (Table 3).

It means, that for the 2-dimensional and 4-dimensional cases, the expected speed up v is:

$$v \cong \frac{Cramer's\,rule}{partial\,summation} \doteq 2 \tag{60}$$

In real implementations on CPU the SSE instructions can be used which are more convenient for vector-vector operations and some steps can be made in parallel. Additional speed up can be achieved by GPU use for computation.

In the case of higher dimension modified standard algorithms can be used including iterative methods [17]. Also as the projective representation nearly doubles precision of computation, if a single precision on GPU is used (only few processors compute in a double precision), the result after conversion to the Euclidean representation is equivalent to the double precision.

13   GPU Code

Many today's computational systems can use GPU support, which allows fast and parallel processing. The above presented approach offers significant speed up as the "standard" cross-product is implemented in hardware as an instruction and the extended cross-product for 4D can be implemented as:

```
float4 cross_4D(float4 x1, float4 x2, float4 x3)
{float4 a;
   a.x = dot(x1.yzw, cross(x2.yzw, x3.yzw));
   a.y = -dot(x1.xzw, cross(x2.xzw, x3.xzw));
   a.z = dot(x1.xyw, cross(x2.xyw, x3.xyw));
   a.w = -dot(x1.xyz, cross(x2.xyz, x3.xyz));
   return a}
```

In general, it can be seen that a solution of linear systems of equations on GPU for a small dimension n is simple, fast and can be performed in parallel.

14   Conclusion

Projective representation is not widely used for general computation as it is mostly considered for as applicable to computer graphics and computer vision field only. In this paper the equivalence of cross-product and solution of linear system of equations has been presented. The presented approach is especially convenient for 3-dimensional and 4 dimensional cases applicable in many engineering and statistical computations, in which significant speed up can be obtained using SSE instructions or GPU use. Also, the presented approach enables symbolic manipulation as the solution of a system of linear equations is transformed to extended cross-product using a matrix form which enables symbolic manipulations.

Direct application of the presented approach has also been demonstrated on the barycentric coordinates computation and simple geometric problems.

The presented approach enables avoiding division operations as a denominator is actually stored in the homogeneous coordinate w. It which leads to significant computational savings, increase of precision and robustness as the division operation is the longest one and the most decreasing precision of computation.

The presented approach also enables derivation of new and more computationally efficient formula in other computational fields.

Acknowledgment. The author would like to thank to colleagues at the University of West Bohemia in Plzen for fruitful discussions and to anonymous reviewers for their comments and hints which helped to improve the manuscript significantly. Special thanks belong also to SIGGRAPH and Eurographics tutorials attendee for their constructive questions, which stimulated this work.

This research was supported by the MSMT CZ - project No. LH12181.

References

1. Coxeter, H.S.M.: Introduction to Geometry. Wiley, New York (1961)
2. Doran, Ch., Lasenby, A.: Geometric Algebra for Physicists. Cambridge University Press, Cambridge (2003)

3. Dorst, L., Fontine, D., Mann, S.: Geometric Algebra for Computer Science. Morgan Kaufmann, San Francisco (2007)
4. Calvet, R.G.: Treatise of Plane Geometry through Geometric Algebra (2007)
5. Hartley, R., Zisserman, A.: MultiView Geometry in Computer Vision. Cambridge University Press, Cambridge (2000)
6. Hildenbrand, D.: Foundations of Geometric Algebra Computing. Geometry and Computing. Springer, Heidelberg (2012)
7. Kanatani, K.: Understanding Geometric Algebra. CRC Press, Boca Raton (2015)
8. Krumm, J.: Intersection of Two Planes, Microsoft Research, May 2000. http://research. microsoft.com/apps/pubs/default.aspx?id=68640
9. Johnson, M.: Proof by duality: or the discovery of "new" theorems. Math. Today **32**(11), 171–174 (1996)
10. MacDonald, A.: Linear and Geometric Algebra. CreateSpace, Charleston (2011)
11. Skala, V.: A new approach to line and line segment clipping in homogeneous coordinates. Vis. Comput. **21**(11), 905–914 (2005)
12. Skala, V.: Length, area and volume computation in homogeneous coordinates. Int. J. Image Graph. **6**(4), 625–639 (2006)
13. Skala, V.: Barycentric Coordinates Computation in Homogeneous Coordinates. Comput. Graph. **32**(1), 120–127 (2008). ISSN 0097-8493
14. Skala, V.: Projective geometry, duality and precision of computation in computer graphics, visualization and games. In: Tutorial Eurographics 2013, Girona (2013)
15. Skala, V.: Projective Geometry and Duality for Graphics, Games and Visualization - Course SIGGRAPH Asia 2012, Singapore (2012). ISBN:978-1-4503-1757-3
16. Skala, V.: Intersection computation in projective space using homogeneous coordinates. Int. J. Image Graph. **8**(4), 615–628 (2008)
17. Skala, V.: Modified gaussian elimination without division operations. In: ICNAAM 2013, AIP Conference Proceedings Rhodos, Greece, no. 1558, pp. 1936–1939. AIP Publishing (2013)
18. Vince, J.: Geometric Algebra for Computer Science. Springer, London (2008)
19. Wildberger, N.J.: Divine Proportions: Rational Trigonometry to Universal Geometry. Wild Egg Pty, Sydney (2005)
20. Yamaguchi, F.: Computer Aided Geometric Design: A totally Four Dimensional Approach. Springer, Tokyo (2002)

Dynamical Behavior of a Cooperation Model with Allee Effect

Unal Ufuktepe[(✉)], Burcin Kulahcioglu, and Gizem Yuce

Department of Mathematics, Izmir University of Economics, Izmir, Turkey
{unal.ufuktepe,burcin.kulahcioglu}@ieu.edu.tr, gizem.yuce@std.ieu.edu.tr

Abstract. Mutualism is an interaction between two or more species, where species derive a mutual benefit. We study the model of M.R.S. Kulenovic and M. Nurkanovic [7] by adding one of the most best-understood mechanism of Allee effect to that system which is called mate-finding. In this paper, we interpret mate-limitation of Allee effect between mutualistic species from mathematical and ecological points of view and stability analysis of the new model.

Keywords: Mutualism · Allee effect · Mate limitation · Lyapunov · Global asymptotic

1 Introduction

Mutualism is a positive relationship between two or more species in a community that benefits all individuals of those species. It drives evolution and most organisms are mutualistic in some way. These interactions are essential for life. One well known example of a mutualistic relationship is oxpecker and zebra. Oxpeckers land on zebras and eat ticks and other parasites that live on their skin. The oxpeckers get food and the beasts get pest control. Also, when there is danger, the oxpeckers fly upward and scream a warning, which helps the symbiont. Another one is bacteria and the human. A certain kind of bacteria lives in the intestines of humans and many other animals. The human cannot digest all of the food that it eats. The bacteria eats the food that the human cannot digest and partially digest it, allowing the human to finish the job. The bacteria benefit by getting food, and the human benefits by being able to digest the food it eats.

The members of many species cooperate; they get help for hunting or deceiving predators. They come together to survive negative conditions or in other way to find sexual reproduction. When there are a few individuals, it looks like they will take advantage of more welding, however, they will also suffer from a lack of conspecific. The balance changes if this negativity has more power, that is, population may extinct at low reproduction. Their fitness will be less when the population size is getting smaller. This is, in essence, Allee effect.

There are several mechanisms for Allee effects. Well-known mechanisms include fertilization efficiency in sessile organisms, mate finding in mobile organisms and cooperative breeding. In our model we study the mate limitation factor

© Springer International Publishing Switzerland 2016
O. Gervasi et al. (Eds.): ICCSA 2016, Part I, LNCS 9786, pp. 36–44, 2016.
DOI: 10.1007/978-3-319-42085-1_3

which is the situation that harder to find a (compatible and receptive) mate at low population size or density. For example; cod, gypsy moth, Glanville fritillary butterfly, alpine marmot.

The most popular one for mutualism is the interaction between flowering plants and their pollinators or between fruit-producing plants and seed dispersers. Amarasekare [1] studied both obligatory and facultative mutualisms in which one of the mutualists was non-mobile (such as a plant) and the other mobile (such as a pollinator or seed disperser), using a metacommunity framework (a set of local communities connected by dispersal). If obligate mutualisms and only one local community are considered, both species go extinct from any initial conditions if the colonization rate of the mobile mutualist is low, and a strong Allee effect arises if it is high. For facultative mutualisms, a strong Allee effect arises if the fitness reduction in absence of the mutualist drops below a critical value, otherwise both species coexist from any initial conditions. In a metacommunity composed of two or more local communities, dispersal of the mobile mutualist from source communities can rescue sink communities from extinction and thus maintain regional persistence of both species [8].

Among obligate mutualists, system bistability arises as a direct consequence of the mutualistic interaction; that is, we have an emergent Allee effect, simply because neither species can live without the other. Among facultative mutualists, however, the Allee effect is not emergent, because the decline in fitness when the second species is absent is incorporated explicitly into the model. Sexual reproduction can also be considered a sort of (within-species) obligate mutualism.

A coupled discrete logistic model is used in [2] by R. Lpez-Ruz and D. Fourner-Prunaret:

$$x_{n+1} = \mu(y_n)x_n(1 - x_n)$$
$$y_{n+1} = \mu(x_n)y_n(1 - y_n)$$

which is symbiotic interaction between both species provokes that the growth rate $\mu(z)$ is varying with time and must be positive.

In [3] W. Krawcewicz, T.D. Rogers and in [4] H.L. Smith studied a dynamical model for cooperation between two species, each of which benefits in a symmetric manner from the other. Such an idealized relationship is expressed through the family of two-dimensional recursions

$$x_{n+1} = x_n exp[r(1 - x_n) + sy_n]$$
$$y_{n+1} = y_n exp[r(1 - y_n) + sx_n]$$

where the parameters r and s are nonnegative.

A simple, autonomous cooperative system is discussed in [5] by K. Yang, X. Xie and F. Chen,

$$x_{n+1} = x_n exp[r_1 \left(\frac{K_1 + a_1 y_n}{1 + y_n}\right) - x_n]$$
$$y_{n+1} = y_n exp[r_2 \left(\frac{K_2 + a_2 x_n}{1 + x_n}\right) - y_n]$$

where r_i, K_i, i = 1, 2 are all positive constants.

In [6] by Cruz Vargas-De-Len, the global stability in continuous time cooperative models is studied to describe facultative mutualism. He subjected to the Lotka-Volterra mutualism model with proportional harvesting

$$\frac{dx}{dt} = r_1 x [1 - \frac{x}{K_1} + b_{12}\frac{y}{K_1}] - e_1 x$$
$$\frac{dy}{dt} = r_2 y [1 - \frac{y}{K_2} + b_{21}\frac{x}{K_2}] - e_2 y$$

where constants e_1 and e_2 are harvesting efforts on respective populations, r_i, K_i, b_{12} and b_{21} ($i = 1, 2$) are all positive constants, r_i are the linear birth rates, K_i are the carrying capacities, b_{12} and b_{21} measure the cooperative effect of x_1 and x_2.

M.R.S. Kulenovic and M. Nurkanovic studied the global asymptotic behaviour of the following system [7]:

$$x_{n+1} = A x_n \frac{y_n}{1+y_n}$$
$$y_{n+1} = B y_n \frac{x_n}{1+x_n} \tag{1}$$

with parameters A, B > 0 and initial values $x_0, y_0 > 0$.

In a modeling setting, the system (1) of nonlinear difference equations represents the rule by which two discrete, cooperating populations reproduce from one generation to the next. The phase variables x_n and y_n denote population sizes during the n-th generation and the sequence or orbit (x_n, y_n), n = 0,1,2,... depicts how the populations evolve over time. Cooperation between two populations is reflected by the fact that the transition function for each population is an increasing function of the other population size.

In this study we add Allee effect to the first component in model (1) and get a new one:

$$x_{n+1} = A x_n \frac{y_n}{1+y_n} \frac{x_n}{u+x_n}$$
$$y_{n+1} = B y_n \frac{x_n}{1+x_n} \tag{2}$$

where $u > 0$ denotes the Allee effect constant that determines the strength of Allee effect.

2   Analysis of Model (1)

2.1   Fixed Points and Their Stability

The system

$$x_{n+1} = A x_n \frac{y_n}{1+y_n}$$
$$y_{n+1} = B y_n \frac{x_n}{1+x_n}$$

has two fixed points (0, 0) and $(\frac{1}{B-1}, \frac{1}{A-1})$ that is positive when $A > 1$ and $B > 1$.

The positive fixed point $(\frac{1}{B-1}, \frac{1}{A-1})$ is always saddle (Fig. 3). The other fixed point $(0, 0)$ is always locally asymptotically stable because

$$J(0,0) = \begin{pmatrix} 0 & 0 \\ 0 & 0 \end{pmatrix}$$

Since both eigenvalues are zero (Fig. 1, Fig. 2 and Fig. 3).

2.2   Stability via Lyapunov Function

In order to check the global stability, we use a Lyapunov function of the map

$$F\begin{pmatrix} x \\ y \end{pmatrix} = \begin{pmatrix} Ax\frac{y}{1+y} \\ By\frac{x}{1+x} \end{pmatrix}$$

Let $V(x,y) = xy$ is the corresponding positive definite Lyapunov function (Fig. 6). Then

$$\Delta V\begin{pmatrix} x \\ y \end{pmatrix} = V(F\begin{pmatrix} x \\ y \end{pmatrix}) - V(x,y)$$

$$\Delta V(x,y) = (Ax\frac{y}{1+y})(By\frac{x}{1+x}) - xy < 0$$

Since $x, y > 0$ and if $AB \leq 1$ then

$$\Delta V(x,y) = AB\frac{(xy)^2}{(1+x)(1+y)} - xy = \frac{xy(ABxy - 1 - x - y - xy)}{(1+x)(1+y)} < 0$$

Since $|X| \to \infty$, $V(X) \to \infty$, then $(0, 0)$ is globally asymptotically stable.

3   Numerical Simulations of Model (1)

3.1   Phase Plane Diagrams

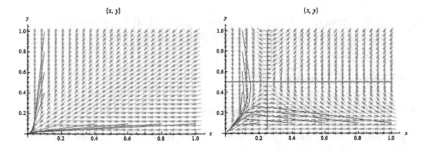

Fig. 1. Phase Plane Diagrams: the first one is the phase diagram with initial point (0.1,0.1) where $A = 0.7$, $B = 0.4$ for which there is no positive fixed point and (0,0) is globally stable. For the second one, the initial point is (0.1,0.1), $A = 3$, $B = 5$, there are two fixed points: (0,0) and (0.25,0.5), which are stable and saddle respectively

3.2 Time Series Diagrams

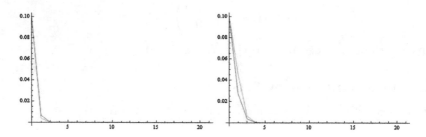

Fig. 2. Time Series Diagrams: the first one is the time series diagram with initial point (0.1,0.1) where A = 0.7, B=0.4. For the second one, the initial point is (0.1,0.1) again with A = 3, B=5. In both cases the population go to extinction in time.

3.3 Basin of Attraction

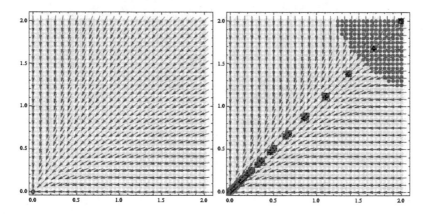

Fig. 3. Basin of Attraction: for the first one A=0.2, B=0.3 and there is no positive fixed point. For the second one, A=1.5, B=1.5, the positive fixed point is (2,2).

4 Analysis of Model (2)

4.1 Fixed Points and Their Stability

We get the following system by adding the mate limitation Allee effect:

$$x_{n+1} = Ax_n \frac{y_n}{1+y_n} \frac{x_n}{u+x_n}$$
$$y_{n+1} = By_n \frac{x_n}{1+x_n}$$

Fixed points of this system are (0,0) and $\left(\frac{1}{B-1}, \frac{1-u+Bu}{-1+A+u-Bu}\right)$.

The second fixed point is positive when $B > 1$ and $A > 1 + u(B - 1)$ or equally $B > 1$, $A > 1$ and $0 < u < \frac{A-1}{B-1}$.

$(0, 0)$ is again always locally asymptotically stable because

$$J(0,0) = \begin{pmatrix} 0 & 0 \\ 0 & 0 \end{pmatrix}$$

So both eigenvalues are zero again.

For positive fixed point $(\frac{1}{B-1}, \frac{1-u+Bu}{-1+A+u-Bu})$ Jacobian matrix is,

$$J = \begin{pmatrix} 2 + \frac{1}{-1+u-Bu} & \frac{(-1+A+u-Bu)^2}{A(-1+B)(1+(-1+B)u)} \\ -\frac{(-1+B)^2(1+(-1+B)u)}{B(1-A+(-1+B)u)} & 1 \end{pmatrix}$$

and corresponding eigenvalues are,

$$\lambda_1 = \frac{\sqrt{A}\sqrt{B}(2+3(-1+B)u)+\sqrt{-1+B}\sqrt{-4(1+(-1+B)u)^3+A(4+(-1+B)u(8-4u+5Bu))}}{2\sqrt{A}\sqrt{B}(1+(-1+B)u)}$$

$$\lambda_2 = \frac{\sqrt{A}\sqrt{B}(2+3(-1+B)u)-\sqrt{-1+B}\sqrt{-4(1+(-1+B)u)^3+A(4+(-1+B)u(8-4u+5Bu))}}{2\sqrt{A}\sqrt{B}(1+(-1+B)u)}.$$

Solving inequalities together with $B > 1$, $A > 1$ and $0 < u < \frac{A-1}{B-1}$, $|\lambda_1| \leq 1$ has infeasible solution while $|\lambda_2| \leq 1$ holds true everywhere in the domain. That means the positive fixed point is saddle point again (Fig. 4, Fig. 5 and Fig. 6).

4.2 Stability via Lyapunov Function

In order to check the global stability of (0,0), we use a Lyapunov function of the map. Let $V(x,y) = xy$ be the corresponding positive definite Lyapunov function. If $AB < 1$, $A > 1$ and $1 < u$ then

$$\Delta V \begin{pmatrix} x \\ y \end{pmatrix} = V \left(F \begin{pmatrix} x \\ y \end{pmatrix} \right) - V(x, y)$$
$$\Delta V(x, y) = \left(\frac{Ax^2 y}{(1+y)(u+x)} \right) \left(\frac{Bxy}{1+x} \right) - xy < 0$$

So (0,0) is globally asymptotically stable.

5   Numerical Simulations of Model (2)

In this section we provide some numerical evidences for the qualitative dynamic of the Model 2, the phase portraits, time series diagrams, and basin of attractions by using the codes of [9];

5.1   Phase Plane Diagrams

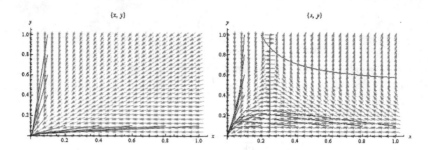

Fig. 4. Phase Plane Diagrams: the first one is the phase diagram with initial point (0.1,0.1) where A = 0.7, B = 0.4 and u = 0.25 for which there is no positive fixed point and (0,0) is globally stable. For the second one, the initial point is (0.1,0.1), A = 3, B = 5 and u = 0.1 there are two fixed points: (0,0) and (0.25,0.875), which are stable and saddle respectively.

5.2   Time Series Diagrams

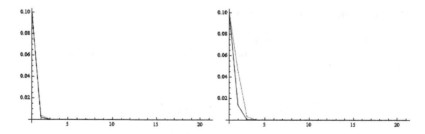

Fig. 5. Time Series Diagrams: the first one is the time series diagram with initial point (0.1,0.1) where A = 0.7, B = 0.4, u = 0.25. For the second one, the initial point is (0.1,0.1) again with A = 3, B = 5, u = 0.1. In both cases the population go to extinction in time.

5.3   Basin of Attraction

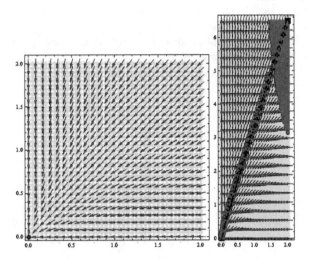

Fig. 6. Basin of Attraction: for the first one $A = 0.2$, $B = 0.3$, $u = 0.5$. For the second one, $A = 1.5$, $B = 1.5$, $u = 0.6$ with positive fixed point $(2, 6.5)$.

6   Conclusion

In this paper, chaotic dynamic and the stability of the fixed points of a nonlinear discrete-time cooperation model with Allee Effect have been investigated. Global stability of the fixed point is investigated by a Lyapunov function. Nevertheless, identifying complicated, possibly chaotic dynamics in population data, stability region of the positive fiexed point, and bifurcation of the system have remained.

References

1. Amarasekare, P.: Spatial dynamics of mutualistic interactions. J. Anim. Ecol. **73**, 128–142 (2004). Wiley
2. Lpez-Ruz, R., Fourner-Prunaret, D.: Complex behaviour in a discrete coupled logistic model for the symbiotic interacton of two species. Math. Biosci. Eng. **1**, 307–324 (2004)
3. Krawcewicz, W., Rogers, T.D.: Perfect harmony: discrete dynamics of cooperation. J. Math. Biol. **28**, 383–410 (1990). Springer
4. Smith, H.L.: Planar competitive and cooperative difference equations. J. Differ. Eqn. Appl. **3**, 335–357 (1998). Taylor & Francis
5. Yang, K., Xie, X., Chen, F.: Global stability of a discrete mutualism model. Abstarct Appl. Anal. **2014**, 1–7 (2014). Hindawi Publishing Corporation
6. Vargas-De-Len, C.: Lyapunov functions for two-species cooperative systems. Appl. Math. Comput. **219**, 2493–2497 (2012). Taylor & Francis

7. Kulenovic, M.R.S., Nurkanovic, M.: Global asymptotic behavior of a two dimensional system of difference equations modeling cooperation. J. Differ. Eqn. Appl. **9**, 149–159 (2003). Elsevier
8. Courchamp, F., Beree, L., Gascoigne, J.: Allee Effects in Ecology and Conservation. Oxford University Press, Oxford (2008)
9. Ufuktepe, U., Kapcak, S.: Applications of discrete dynamical systems with mathematica. RIMS Kyoto Proceeding (2013)

Stability of a Certain 2-Dimensional Map with Cobweb Diagram

Sinan Kapçak$^{(\boxtimes)}$

American University of the Middle East, Egaila, Kuwait
sinan.kapcak@aum.edu.kw

Abstract. In this paper, we investigate the following discrete-time map:

$$x_{n+1} = \phi(y_n),$$
$$y_{n+1} = \psi(x_n).$$

We introduce a novel method to determine the stability of the given two-dimensional map by using a one-dimensional map. A cobweb-like diagram is also introduced in order to analyze the stability of the system. We show that the stability of a fixed point in cobweb diagram implies the stability in phase diagram for the given system.

In addition, an application of the system to a non-hyperbolic fixed point is also given.

Keywords: Discrete dynamical systems · Isoclines · Global stability · Cobweb diagram · Root finding algorithm

1 Introduction

In this paper, we investigate the dynamics of a nonlinear planar map. One of the main goals of this work is to analyze the stability of the following planar map by using a cobweb-like diagram.

$$x_{n+1} = \phi(y_n),$$
$$y_{n+1} = \psi(x_n). \tag{1}$$

A *cobweb*, or *Verhulst diagram* is a visual method used in the dynamical systems to investigate the qualitative behavior of one-dimensional maps. Using a cobweb diagram, it is possible to analyze the long term status of an initial condition under repeated application of a map. Use of cobweb diagram can be found in Devaney (1989); Elaydi (2000).

Cobweb diagram is usually used for one-dimensional discrete dynamical systems; however, in this paper, we propose a novel method with cobweb diagram to investigate the dynamics of the two-dimensional discrete system (1), assuming ϕ^{-1} exists.

© Springer International Publishing Switzerland 2016
O. Gervasi et al. (Eds.): ICCSA 2016, Part I, LNCS 9786, pp. 45–53, 2016.
DOI: 10.1007/978-3-319-42085-1_4

For convenience, we will take $\phi = f^{-1}$ and $\psi = g$. Hence, we have the following system:

$$\begin{aligned} x_{n+1} &= f^{-1}(y_n), \\ y_{n+1} &= g(x_n). \end{aligned} \qquad (\star)$$

In this section, we give a lemma for a root finding algorithm which allows us to determine the stability of the given system by using cobweb-like diagram. In Sect. 2, we give a theorem for the stability condition of system $(\star)$ and apply the cobweb-like diagram in order to determine the stability.

The general two dimensional autonomous discrete system is given by

$$\begin{aligned} x_{n+1} &= \alpha(x_n, y_n), \\ y_{n+1} &= \beta(x_n, y_n), \end{aligned} \qquad (2)$$

whose isocline equations are

$$\begin{aligned} x &= \alpha(x, y), \\ y &= \beta(x, y). \end{aligned} \qquad (3)$$

It is easy to see that the isocline equations do not uniquely determine the dynamics of systems. A simple example for that is the following system which has the same isoclines as system (2) does but they have different dynamics simply because they have different eigenvalues of the Jacobian matrices:

$$\begin{aligned} x_{n+1} &= \frac{1}{2}(x_n + \alpha(x_n, y_n)), \\ y_{n+1} &= \beta(x_n, y_n). \end{aligned} \qquad (4)$$

In contrary, for system $(\star)$, since x_{n+1} and y_{n+1} depend only on y_n and x_n, respectively, there is a unique representation of the isoclines and they determine the dynamics uniquely. We investigate the dynamics of the system, just by focusing on the isoclines which are $y = f(x)$ and $y = g(x)$

System $(\star)$ might find many applications in engineering, game theory, and particularly competition models in economics and biology.

1.1   A Root Finding Algorithm

In this section, we give a lemma for finding the intersection points of two curves. However, we will use the lemma not for finding the intersection points but to construct the cobweb diagram and investigate the stability of the system in a rectangular region.

Lemma 1. *Let $f : A \to B = f(A)$ and $g : C \to D = g(C)$ be continuous functions, where $A \subset C$ and $D \subset B$. Assume that $f(\bar{x}) = g(\bar{x}) = \bar{y}$ for some $\bar{x}$ and one of the following four conditions is satisfied:*

$$(1) \begin{cases} f(x) < g(x) < \bar{y}, & \text{if} \quad x < \bar{x}, \\ \bar{y} < g(x) < f(x), & \text{if} \quad x > \bar{x}. \end{cases}$$

$$(2) \begin{cases} \bar{y} < g(x) < f(x), & \text{if } x < \bar{x}, \\ f(x) < g(x) < \bar{y}, & \text{if } x > \bar{x}. \end{cases}$$

$$(3) \begin{cases} f(x) < g(2\bar{x} - x) < \bar{y}, & \text{if } x < \bar{x}, \\ \bar{y} < g(2\bar{x} - x) < f(x), & \text{if } x > \bar{x}. \end{cases}$$

$$(4) \begin{cases} \bar{y} < g(2\bar{x} - x) < f(x), & \text{if } x < \bar{x}, \\ f(x) < g(2\bar{x} - x) < \bar{y}, & \text{if } x > \bar{x}. \end{cases}$$

Then, for any $x_0 \in C$, $(f^{-1} \circ g)^n(x_0) \to \bar{x}$ as $n \to \infty$, provided f^{-1} exists.

Proof. For each of the conditions, we can consider the theorem separately. We will prove the theorem only for Condition (1). By using similar approach, one can show the statement for the other conditions.

Assume that Condition (1) is satisfied. We will show that, for any $x_0 \in C$,

$$(f^{-1} \circ g)^n(x_0) \to \bar{x} \quad \text{as} \quad n \to \infty.$$

Since $f : A \to B = f(A)$ is invertible, it must be strictly monotone on A (either strictly increasing or strictly decreasing). However, it cannot be decreasing, because $f(x) < f(\bar{x}) = \bar{y}$ when $x < \bar{x}$. Hence, f is strictly increasing on A. Therefore, $f^{-1} : B \to A$ is also strictly increasing.

Now take any $x_0 > \bar{x}$. By assumption, we have $\bar{y} < g(x_0) < f(x_0)$. Since f^{-1} is strictly increasing, we obtain $f^{-1}(\bar{y}) < f^{-1}(g(x_0)) < f^{-1}(f(x_0))$ or $\bar{x} < (f^{-1} \circ g)(x_0) < x_0$.

Let us call $x_1 = (f^{-1} \circ g)(x_0)$ and apply the same procedure to x_1 to obtain $x_2 = (f^{-1} \circ g)^2(x_0)$. Applying the same procedure over and over, we obtain

$$\bar{x} < \ldots < x_3 < x_2 < x_1 < x_0.$$

By Monotone Convergence Theorem, the limit of the sequence $(f^{-1} \circ g)^n(x_0)$ exists. Let $F = f^{-1} \circ g$, $\lim F^n(x_0) = L$, and consider the following difference equation:

$$x_{n+1} = F(x_n) \tag{5}$$

The only fixed point of the equation is $\bar{x}$, since the only solution of the equation $F(x^*) = x^*$ is $x^* = \bar{x}$.

Then, by continuity of F, we have the following:

$$L = \lim F^{n+1}(x_0) = F(\lim F^n(x_0)) = F(L).$$

Hence the limit must be $\bar{x}$ which is the only fixed point of Eq. (5).

The case when $x_0 < \bar{x}$ can be done similarly. Therefore, $\lim F^n(x_0) \to \bar{x}$ for any $x_0 \in C$ if Condition (1) holds.

Remark 1. Note that, in Lemma 1, the functions f and g are not necessarily differentiable.

Remark 2. For the special case when $A = B = C = D = \mathbb{R}$, we give a simplified version of the lemma in Appendix whose proof is very similar.

Figure 1 shows the case for Condition 1 of Lemma 1 and how the algorithm works to find the intersection point of the two curves $y = f(x)$ and $y = g(x)$.

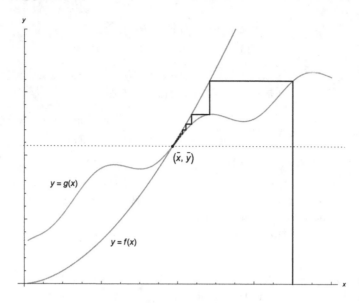

Fig. 1. Root Finding Algorithm

2 Stability of System ($\star$)

In this section, we analyze the stability of the system ($\star$). We first investigate the dynamics of the system in a rectangular region and give the stability condition.

2.1 Dynamics of System ($\star$) in a Rectangular Region

Theorem 1. *Consider the discrete dynamical system*

$$
\begin{aligned}
x_{n+1} &= f^{-1}(y_n), \\
y_{n+1} &= g(x_n),
\end{aligned} \qquad (\star)
$$

where $f : A \to B = f(A)$ and $g : C \to D = g(C)$ are continuous functions, with $A \subset C$ and $D \subset B$. Let $(\bar{x}, \bar{y})$ be a fixed point of system ($\star$) and one of the following conditions be satisfied:

$$(1) \begin{cases} f(x) < g(x) < \bar{y}, & \text{if } \ x < \bar{x}, \\ \bar{y} < g(x) < f(x), & \text{if } \ x > \bar{x}. \end{cases}$$

$$(2) \begin{cases} \bar{y} < g(x) < f(x), & \text{if } \ x < \bar{x}, \\ f(x) < g(x) < \bar{y}, & \text{if } \ x > \bar{x}. \end{cases}$$

$$(3) \begin{cases} f(x) < g(2\bar{x} - x) < \bar{y}, & \text{if } \ x < \bar{x}, \\ \bar{y} < g(2\bar{x} - x) < f(x), & \text{if } \ x > \bar{x}. \end{cases}$$

$$(4) \begin{cases} \bar{y} < g(2\bar{x} - x) < f(x), & \text{if } \ x < \bar{x}, \\ f(x) < g(2\bar{x} - x) < \bar{y}, & \text{if } \ x > \bar{x}. \end{cases}$$

Then $(\bar{x}, \bar{y})$ *is asymptotically stable fixed point on the rectangular region* $C \times B$.

Proof. Let one of the four conditions given in the theorem be satisfied. Then, by Theorem 1, $(f^{-1} \circ g)^n(x_0) \to x$ as $n \to \infty$, where $x_0 \in C$.

Now, let $F = f^{-1} \circ g$ and start with the point $(x_0, y_0) \in C \times B$. Then, for the x-components of the orbit, we have

$$x_0 \to x_1 \to F(x_0) \to F(x_1) \to F^2(x_0) \to F^2(x_1) \to F^3(x_0) \to F^3(x_1) \to \cdots$$

Therefore, $x_2 = F(x_0)$, $x_4 = F^2(x_0)$, and in general $x_{2n} = F^n(x_0)$; whereas $x_3 = F(x_1)$, $x_5 = F^2(x_1)$, and in general $x_{2n+1} = F^n(x_1)$. Hence, by Theorem 1, $x_{2n} = F^n(x_0) \to \bar{x}$ as $n \to \infty$. Since $x_1 \in A \subset C$, by using the same theorem, we obtain $x_{2n+1} = F^n(x_1) \to \bar{x}$ as $n \to \infty$. Therefore, $x_n \to \bar{x}$ as $n \to \infty$.

Since g is continuous, we have $\lim y_{n+1} = \lim g(x_n) = g(\lim x_n) = g(\bar{x}) = \bar{y}$, which proves that $(\bar{x}, \bar{y})$ is an attracting fixed point.

We have

$$x_0 \to x_1 \to x_2 \to \cdots \to \bar{x} \quad \text{and} \quad y_0 \to y_1 \to y_2 \to \cdots \to \bar{y}.$$

By Theorem 3, since $\bar{x} \in C$ and $\bar{y} \in B$ are attracting points, they are stable.

Remark 3. For system $(\star)$, Theorem 1 works also for the non-hyperbolic case. Since the trace of the Jacobian matrix is always zero, the case when determinant of the Jacobian matrix at the fixed point equals 1 is the borderline where the Neimark-Sacker bifurcation might occur. For this critical case, for which $\lambda_{1,2} = \pm i$, we can analyze the stability by applying Theorem 1. Geometrically, this is the case when the slopes of the tangent lines to the isoclines at the fixed point, say m_1 and m_2, have the property $m_1 = -m_2$. Note that, since $\lambda^4 = 1$, this is not necessarily a Neimark-Sacker bifurcation (Kuznetsov 1995).

Remark 4. Note that, in Theorem 1, the functions f and g are not necessarily differentiable.

Remark 5. For the special case when $A = B = C = D = \mathbb{R}$, the fixed point is globally asymptotically stable. We have a simplified version of the above theorem in Appendix.

2.2   Stability of System $(\star)$ with Cobweb Diagram

In the theory of Discrete Dynamical Systems, we usually use cobweb diagram in order to understand the dynamics of one-dimensional maps.

To apply the cobweb diagram for the two dimensional system $(\star)$, we take function $y = f(x)$ instead of the diagonal line $y = x$ and apply the same procedure as we do in the usual cobweb diagram. Starting with x_0, we have

$$(x_0, 0) \to (x_0, g(x_0)) \to ((f^{-1} \circ g)(x_0), g(x_0)) \to ((f^{-1} \circ g)(x_0), g((f^{-1} \circ g)(x_0))) \to \cdots$$

By Theorem 1 and the visual representation of the above sequence we can conclude that, for system ($\star$), the stability on cobweb diagram implies the stability on phase diagram. In fact, the above sequence itself is one of the orbits of the system if we start at $(x_0, g(x_0))$.

Remark 6. Theorem 1 gives the stability condition for system ($\star$) by using Theorem 1 and Theorem 1 allows us to use the cobweb diagram. However, cobweb diagram can also be used for the system (1). One can confirm that by following the orbit starting with $(x_0, 0)$. For this case, the iteration might lead to significant different future behavior including chaos.

Example 1. Consider the discrete system ($\star$) with $f : [-\frac{1}{2}, 2) \to [-2, 8)$, $f(x) = 4x$ and $g : [-1, 2) \to [-1, 8)$, $g(x) = x^3$. Hence, we have

$$x_{n+1} = \frac{1}{4}y_n,$$
$$y_{n+1} = x_n^3. \tag{6}$$

Both f and g are continuous on their domains. The only fixed point in the given region is $(0, 0)$. It is clear that the first condition of Theorem 1 holds. Therefore, $(0, 0)$ is asymptotically stable on the region $[-1, 2) \times [-2, 8)$. Figure 2 represents the cobweb diagram and the phase diagram for the system.

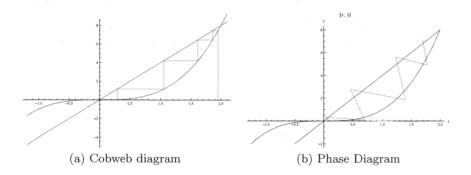

(a) Cobweb diagram   (b) Phase Diagram

Fig. 2. Stability with cobweb diagram for the 2-dimensional map in Example 1

Example 2. Consider the following discrete-time system:

$$x_{n+1} = \arctan y_n,$$
$$y_{n+1} = -\frac{1}{3}(x_n + \sin x_n). \tag{7}$$

$(0, 0)$ is a fixed point. We have $f : (-\frac{\pi}{2}, \frac{\pi}{2}) \to \mathbb{R}$, $f(x) = \tan x$ and $g : \mathbb{R} \to \mathbb{R}$, $g(x) = -\frac{1}{3}(x + \sin x)$ and the functions are continuous. Since

$$0 < \frac{1}{3}(x + \sin x) < \frac{2x}{3} < x < \tan x$$

for $x > 0$, the third condition of Theorem 1 is satisfied. Therefore, the fixed point $(0,0)$ is globally asymptotically stable on $\mathbb{R}^2$.

Main Theorem 2. *Consider the two-dimensional map*

$$\begin{aligned} x_{n+1} &= f^{-1}(y_n), \\ y_{n+1} &= g(x_n). \end{aligned} \qquad (\star)$$

and the following one-dimensional map

$$x_{n+1} = F(x_n), \qquad (8)$$

where $f : A \to B = f(A)$ *and* $g : C \to D = g(C)$ *are continuous functions, with* $A \subset C$, $D \subset B$, *and* $F = f^{-1} \circ g$. *Let* $(\bar{x}, \bar{y})$ *be a fixed point of system* $(\star)$.

Then, $\bar{x}$ *is a fixed point of system (8) and* $\bar{x}$ *of system (8) is asymptotically stable if and only if* $(\bar{x}, \bar{y})$ *of system* $(\star)$ *is asymptotically stable.*

Remark 7. Theorem 2 is the direct conclusion of Lemma 1 and Theorem 1. In order to analyze the stability of system $(\star)$, we simply take difference Eq. (8) and investigate the stability which is much easier.

Example 3. Consider the 2-dimensional map with one parameter

$$\begin{aligned} x_{n+1} &= -k \arctan y_n, \\ y_{n+1} &= x_n e^{-x_n}, \end{aligned} \qquad (9)$$

where $k > 0$. For $k = 1$, fixed point $(0,0)$ is non-hyperbolic with $\lambda_{1,2} = \pm i$. Applying Theorem 2, we have one-dimensional map $x_{n+1} = -k \arctan(x_n e^{-x_n})$ and the fixed point $x^* = 0$ of this map is globally asymptotically stable when $k \leq 1$. Therefore the fixed point $(x^*, y^*) = (0,0)$ of two-dimensional system (9) is also globally asymptotically stable when $k \leq 1$. Figure 3 displays the phase diagram of system (9) before $(k < 1)$ and after $(k > 1)$ the bifurcation.

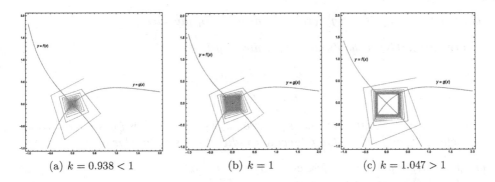

(a) $k = 0.938 < 1$ (b) $k = 1$ (c) $k = 1.047 > 1$

Fig. 3. System in Example 3

Details about types of bifurcation can be found in Elaydi (2008); Kuznetsov (1995).

3   Conclusions

We gave an analytical method to analyze the stability of 2-dimensional discrete time systems of the form $(\star)$. Stability condition was defined in a rectangular region. By using geometric approach, we can easily determine the stability of the system. Even for the non-hyperbolic case, the condition for stability is valid. Note that the stability conditions in the rectangular region does not tell anything about the case when the initial point is outside the region. As a further study, we will investigate more general systems and basin of attractions of fixed points. Also, we will study the case where f^{-1} does not necessarily exist. Although this case is geometrically clear, it needs further work.

We will also study the local and global stable/unstable manifolds of the system which requires solving some functional equations.

Another issue to be investigated is converting discrete systems to system $(\star)$. The main question is which systems are appropriate for that.

A   Related Lemmas/Theorems

Theorem 3. *Let z be an attracting fixed point of a continuous map $f : I \to \mathbb{R}$, where I is an interval. Then z is stable.*

Proof of the theorem can be found in Elaydi (2008).

Lemma 2. *Let $f : \mathbb{R} \to \mathbb{R}$ and $g : \mathbb{R} \to \mathbb{R}$ are continuous functions and $f(\bar{x}) = g(\bar{x}) = \bar{y}$ for some $\bar{x} \in \mathbb{R}$. Assume that one of the following conditions is satisfied for all real $\alpha > 0$:*

(1) $f(\bar{x} - \alpha) < g(\bar{x} - \alpha) < \bar{y} < g(\bar{x} + \alpha) < f(\bar{x} + \alpha)$
(2) $f(\bar{x} + \alpha) < g(\bar{x} + \alpha) < \bar{y} < g(\bar{x} - \alpha) < f(\bar{x} - \alpha)$
(3) $f(\bar{x} - \alpha) < g(\bar{x} + \alpha) < \bar{y} < g(\bar{x} - \alpha) < f(\bar{x} + \alpha)$
(4) $f(\bar{x} + \alpha) < g(\bar{x} - \alpha) < \bar{y} < g(\bar{x} + \alpha) < f(\bar{x} - \alpha)$

Then, for any x_0, $(f^{-1} \circ g)^n(x_0) \to \bar{x}$ as $n \to \infty$, provided f^{-1} exists.

Theorem 4. *Given the discrete dynamical system*

$$x_{n+1} = f^{-1}(y_n),$$
$$y_{n+1} = g(x_n), \qquad (\star)$$

where $f, g : \mathbb{R} \to \mathbb{R}$ are continuous functions. Assume that $(\bar{x}, \bar{y})$ is a fixed point of system $(\star)$ and one of the following conditions is satisfied for all real $\alpha > 0$:

(1) $f(\bar{x} - \alpha) < g(\bar{x} - \alpha) < \bar{y} < g(\bar{x} + \alpha) < f(\bar{x} + \alpha)$
(2) $f(\bar{x} + \alpha) < g(\bar{x} + \alpha) < \bar{y} < g(\bar{x} - \alpha) < f(\bar{x} - \alpha)$
(3) $f(\bar{x} - \alpha) < g(\bar{x} + \alpha) < \bar{y} < g(\bar{x} - \alpha) < f(\bar{x} + \alpha)$
(4) $f(\bar{x} + \alpha) < g(\bar{x} - \alpha) < \bar{y} < g(\bar{x} + \alpha) < f(\bar{x} - \alpha)$

Then $(\bar{x}, \bar{y})$ is globally asymptotically stable on $\mathbb{R}^2$.

References

Elaydi, S.: An Introduction to Difference Equations. Springer, New York (2000)

Elaydi, S.: Discrete Chaos: With Applications in Science and Engineering, 2nd edn. Chapman & Hall/CRC, Boca Raton (2008)

Devaney, R.L.: An Introduction to Chaotic Dynamical Systems, vol. 13046. Addison-Wesley, Reading (1989)

Kuznetsov, Y.A.: Elements of Applied Bifurcation Theory. Springer, New York (1995). ISBN 0-387-94418-4

A New Heuristic for Bandwidth
and Profile Reductions of Matrices
Using a Self-organizing Map

Sanderson L. Gonzaga de Oliveira[1]([⊠]), Alexandre A.A.M. de Abreu[1],
Diogo Robaina[2], and Mauricio Kischinhevsky[2]

[1] Universidade Federal de Lavras, Lavras, Minas Gerais, Brazil
sanderson@dcc.ufla.br, alexandregrandeabreu@gmail.com
[2] Universidade Federal Fluminense, Niterói, Rio de Janeiro, Brazil
drobaiana@gmail.com, kisch@ic.uff.br

Abstract. In this work, a heuristic for bandwidth and profile reductions of symmetric and asymmetric matrices using a one-dimensional self-organizing map is proposed. Experiments and comparisons of results obtained were performed in relation to results of the Variable neighborhood search for bandwidth reduction. Simulations with these two heuristics were performed with 113 instances of the Harwell-Boeing sparse matrix collection and with 2 sets of instances with linear systems composed of sparse symmetric positive-definite matrices. The linear systems were solved using the Jacobi-preconditioned Conjugate Gradient Method. According to the results presented here, the best heuristic in the simulations performed was the Variable neighborhood search for bandwidth reduction. On the other hand, when the vertices of the corresponding graph were originally ordered in a sequence given by a space-filling curve, no gain was obtained when applying a heuristic for reordering the graph vertices.

Keywords: Bandwidth reduction · Profile reduction · Self-organizing maps · Conjugate gradient method · Graph labeling

1 Introduction

Various problems in science and engineering, such as heat, fluid flow, or elasticity, are described by partial differential equations. A popular technique for solving partial differential equations is the finite element method. It consists of transforming a continuous differential equation into a system of algebraic equations. In addition, sparse large-scale linear systems of this kind can also be obtained by applying the finite volume method. Such systems of linear equations have the form $Ax = b$, where A is the matrix of coefficients, x is the vector of unknowns and b is the vector of independent terms. The resolution of this kind of linear systems occurs at the step where the most computation cost is demanded to solve the problem. This type of linear systems also occurs in applications not

© Springer International Publishing Switzerland 2016
O. Gervasi et al. (Eds.): ICCSA 2016, Part I, LNCS 9786, pp. 54–70, 2016.
DOI: 10.1007/978-3-319-42085-1_5

governed by partial differential equations, such as design and analysis of digital circuits, economic models, and industrial production systems [1].

The resolution of a linear system can be accomplished by either direct or iterative methods. As direct methods have low scalability regarding the processing and use of memory, especially in three-dimensional problems arising from partial differential equations, linear systems composed of large sparse matrices are often solved by iterative methods. The boundaries between direct and iterative methods have become less clear with the use of various ideas and techniques of direct methods in the field of iterative methods. As a result, iterative methods have become increasingly reliable [1].

A linear system composed of a large-scale symmetric and positive-definite matrix can be solved efficiently by the Conjugate Gradient Method (CGM). This method is amply used in practical applications. Additionally, the CGM can also be used to solve any linear systems by transforming $Ax = b$ to $A^{T}Ax = A^{T}b$. Specifically, the original systems of linear equations tested in this work are comprised of symmetric and positive-definite matrices.

With a local ordering of graph vertices associated with a matrix A, one intends to reduce the computational cost demanded to resolve a linear system by an iterative method. In addition, one aims to reduce the storage cost of the linear system, depending on the storage scheme employed. Local orderings of graph vertices associated with the matrix A may reduce the computational cost of the CGM [8] by improving the number of cache hits [4,6].

Reordering algorithms are used for bandwidth and profile reductions. The bandwidth of the matrix A can be defined as $\beta = \beta(A) = max\{i - min\{j \mid a_{ij} \neq 0\} \mid 1 \leq i \leq n\}$ and its profile can be defined as $profile(A) = \sum_{i=1}^{n}\left(i - \min_{1 \leq j < i}(j \mid a_{ij} \neq 0)\right)$. The bandwidth and profile minimization problems are hard [14,16].

Although heuristics for bandwidth and profile reductions have been proposed since the 1960s, to the best of our knowledge no one has ever published any research on the design of a heuristic based on artificial neural networks (ANNs) for these problems. More specifically, systematic reviews related to these problems were performed, 130 heuristics were found, and no publication related to ANNs applied to these problems was found [2,5,9].

In this paper, a new heuristic based on ANNs for bandwidth and profile reductions of symmetric and asymmetric matrices is proposed. With the use of a self-organizing map, our intention was to perform low computational-cost bandwidth and profile reductions.

Experiments and comparisons of the results were performed between the new self-organizing map for bandwidth reduction (SOM-band) heuristic, and with the heuristic described as the possible state of the art for the bandwidth reduction problem: the Variable neighborhood search for bandwidth reduction (VNS-band) [15] heuristic.

Section 2 presents definitions of self-organizing maps and of the winner-takes-most (WTM) learning approach. Section 3 describes the SOM-band heuristic.

Section 4 presents the application of a self-organizing map for bandwidth and profile reductions. Section 5 shows the results and analyses of the experiments. Finally, Sect. 6 addresses the conclusions and proposals for future studies.

2   Self-organizing Map with Winner-takes-most Approach

An ANN can be understood as a computational technique that presents a mathematical model based on the neural structure of intelligent organisms. An ANN can be defined as a computer system comprising of processing units, known as neurons.

The two main learning processes in an ANN are *supervised learning* and *unsupervised learning*. Unsupervised learning is used when one cannot establish a set of input samples and the expected result. In the unsupervised learning, the ANN training cannot be accomplished by an algorithm that uses the error to adjust the network weights, as is carried out in supervised learning.

In this work, a type of ANN that uses unsupervised learning was implemented, known as a self-organizing map (SOM), or Kohonen map or network [13], which consists of an interconnected ANN. Due to the property of self-organization, SOMs can be applied to problems of spatial clustering and data ordering. In this case, the SOM neuron arrangement is mapped from the input of the algorithm. Moreover, a Kohonen network is an arrangement of neurons generally restricted to spaces of size 1 or 2, in which it seeks to establish and maintain concepts of neighborhood. In the case of a one-dimensional SOM, one has an ordered sequence of neurons and the number of weights is equal to the number of entries. The learning process for the unsupervised SOM is described as

$$w(i, j) = w(i, j) + \eta(x_j - w(i, j)). \tag{1}$$

The learning process consists of updating weights $w(i, j)$ of the winner neuron i, with $1 \leq i \leq n$ and $1 \leq j \leq m$, where n is the number of neurons, m is the number of entries, x_j is an input value and the learning rate is η, defined in the interval $(0, 1)$. Furthermore, the learning rate η is a parameter of the algorithm.

The WTM approach is a competitive scheme to unsupervised learning. Usually, it converges rapidly and allows activation of neurons that would be rarely activated [3]. Activation is the process of modifying the synaptic weights of a neuron. In the WTM approach, in more than one neuron, their weights are modified in a single iteration. In this case, the winner neuron and its neighbors will be activated. A simplified way to implement the WTM approach is to add, in Eq. 1, the activation equations

$$w(i - 1, j) = w(i - 1, j) + \eta(x_j - w(i - 1, j)) \tag{2}$$

and

$$w(i + 1, j) = w(i + 1, j) + \eta(x_j - w(i + 1, j)). \tag{3}$$

Algorithm 1 shows the general steps for training a SOM. The learning rate η is used in line 7 to modify the synaptic weights of the winner neuron and its neighbors by the Eqs. 1, 2, and 3.

Input: a vector with entries set with values used in training the neural network, maximum number of iterations *iter_max*, and learning rate η.
Output: Kohonen neural network trained.

1   *iter* $\leftarrow$ 1;
2   assign random weights to the neurons;
3   **while** (*iter* $\leq$ *iter_max*) **do**
4   select a random value in the training set;
5   find the winner neuron that has a higher correlation to the selected value;
6   find neighboring neurons of the winner neuron;
7   the synaptic weights of neurons are modified, considering the learning rate η;
8   reduce η;
9   *iter* $\leftarrow$ *iter* + 1;

10   **return** *Kohonen neural-network trained*;

Algorithm 1. Training a self-organizing map.

3   Self-organizing Map Applied to Bandwidth and Profile Reductions

The self-organizing map heuristic for bandwidth and profile reductions (SOM-band) generates a one-dimensional SOM with a simple neighborhood. Each neuron receives 2 values representing the nonzero indices i and j of the input matrix. More specifically, each neuron represents an edge of the graph. The neurons of the SOM-band heuristic have no activation function. The WTM learning approach was used because it has better convergence than the winner-takes-all approach.

Algorithm 2 shows the steps performed by the SOM-band heuristic. Its input parameters are: a set E of edges, the learning rate η, the maximum training epochs, the number of graph vertices, and an optional parameter of execution timeout. The set of graph vertices renumbered, the new bandwidth β, and the final profile found are the output of the algorithm.

A quantity $e = (n \times 0.025) + 5$ of epochs is set in line 4 of Algorithm 2, where n is the number of graph vertices. It was defined by exploratory experiments and, clearly, is proportional to the number of vertices of an instance.

Adjacent vertices with a large difference in the final numbering can only be checked at the end of the SOM-band heuristic training. Because of this, the training is carried out in a small number of epochs so that the solution found can be evaluated. However, the maximum time may be exceeded in the first training set for large instances.

Input: set E of edges, η, *epochs_max*, number n of vertices, *time_max*.
Output: vertices renumbered, β, *profile*.

```
1  β ← E.EstimateBandwidth();
2  profile ← E.EstimateProfile();
   // a one-dimensional SOM with simple neighborhood is created
3  SOM ← CreateSOMwithSimpleNeighborhood(1, E);
4  e ← n * 0.025 + 5; // quantity of epochs
5  epoch ← 1;
6  terminate_execution ← false;
7  while ((epoch ≤ epochs_max) and (terminate_execution ≠ true)) do
       // the WTM learning algorithm is performed e times
8      SOM.WTM(η, e);
       // η is reduced in 2% in order to refine the learning process
9      η ← η * 0.98;
10     Ec ← SOM.ReturnEdges();
11     β_current ← Ec.EstimateBandwidth();
12     profile_current ← Ec.EstimateProfile();
       // if β or profile are better in the current solution,
       // both are updated, even if one of them is worse
13     if ((β_current < β) or (profile_current < profile)) then
14         β ← β_current;
15         profile ← profile_current;
16         epoch ← epoch + e;
           // execution is terminated
           // if the maximum execution time is exceeded
17         if (time_max > 0 and ExecTime() > time_max) then
18             terminate_execution ← true;
19     else
           // execution is terminated if no bandwidth or profile
20         terminate_execution ← true; // reduction is obtained
21 vertices_renumbered ← SOM.VerticesRenumbered();
22 return vertices_renumbered, β, profile;
```

Algorithm 2. SOM-band.

4   Description of Tests

The SOM-band heuristic was implemented in the C++ programming language. Specifically, the g++ 4.8.2 compiler was used.

An executable file of the VNS-band heuristic implemented in the C programming language and compatible with the Linux operating system was used. This executable program was kindly provided by one of the VNS-band heuristic's authors. It only runs with instances up to 500,000 vertices. It was implemented originally only for bandwidth reduction and the profile estimation was made possible by using the ordering found.

The Slackware 14.1 64-bit operating system was used. The GNU MPFR library with 256-bit precision was used to make it possible to obtain high precision in the calculations. The workstation used in the execution of the simulations contains an Intel® Core™ i5-3570 CPU @ 3.40 GHz with 6144 KB of cache memory and 16 GB of main memory (Intel; Santa Clara, CA, United States). It should be noticed that all tests were sequential.

5   Results and Analysis

This section shows the results and analyses of the simulations performed. Section 5.1 presents results obtained in the simulations with the VNS-band and SOM-band heuristics performed in 113 instances of the Harwell-Boeing sparse matrix collection (http://math.nist.gov/MatrixMarket/data/Harwell-Boeing) [7]. This set contains small symmetric or asymmetric matrices. These simulations were preliminary because the main objective of performing the reordering of vertices was to obtain computational-cost reduction in solving linear systems using the Jacobi-preconditioned Conjugate Gradient Method (JPCGM).

Sections 5.2 and 5.3 show simulations with linear systems arising from discretizations of the heat conduction and Laplace equations by finite volumes, respectively. A precision of 10^{-16} to the JPCGM was used in all tests and the computational costs are shown in seconds. In tables shown in Sects. 5.2 and 5.3, $\bar{\beta}$ and $\overline{profile}$ represent, respectively, the average bandwidth and profile found in the runs of each heuristic.

One second was the preliminary execution timeout set to the SOM-band and VNS-band heuristics in the simulations shown in Sects. 5.2 and 5.3. In particular, this parameter needs to be an integer due to a restriction of the VNS-band executable program. Clearly, the SOM-band and VNS-band heuristics need to compute much longer than 1 s to complete one iteration with large instances. For example, more than 1 min of computation time was used in the instance of 492,853 vertices, such as is shown in the tables below. Times shorter than 1 min did not change the execution time with the final result found in this example. It was found that even by increasing the time of the VNS-band heuristic, no reduction in the total cost of the whole simulation was obtained, noting that the total cost of the simulation is the JPCGM computational cost added to the computational time of the heuristic for bandwidth and profile reductions. In another example, with instances of about 1,000 to 2,000 vertices, the JPCGM execution times were approximately 0.5 and 1.5 s, respectively. One should not set the VNS-band timeout for 2 s or more because in such cases this would only increase the total computational cost of the entire simulation.

5.1   Tests with the Harwell-Boeing Sparse Matrix Collection

Each of the heuristics were executed 10 times with each of the 113 Harwell-Boeing sparse matrix collection instances. These 113 instances were divided into 2 subsets, such as performed by Mladenovic et al. [15]: (*i*) 33 instances, ranging from 30 to 237 vertices, accordingly as stated in the Harwell-Boeing sparse matrix collection, but each of these 33 instances has less than 200 vertices when

Table 1. Simulations with the VNS-band and SOM-band heuristics for the set composed of 33 instances of the Harwell-Boeing sparse matrix collection. The smaller bandwidths and profiles found are highlighted.

Inst.	Dim.	VNS-band							SOM-band						
		β_{min}	β_{max}	β	P_{min}	P_{max}	P	Time(s)	β_{min}	β_{max}	β	P_{min}	P_{max}	P	Time(s)
arc130	130	63	63	**63**	2607	2607	**2607**	2.02167	107	115	107	6806	7714	6934	0.5930
ash85	85	14	14	**14**	542	542	**542**	2.14721	53	69	59	1465	1911	1642	0.0734
bcspwr01	39	7	7	**7**	94	94	**94**	2.15331	24	30	25	283	395	305	0.0081
bcspwr02	49	9	9	**9**	170	170	**170**	2.23293	27	38	29	419	461	441	0.0152
bcspwr03	118	10	10	**10**	304	304	**304**	2.02476	87	101	89	3179	3474	3267	0.1050
bcsstk01	48	24	24	**24**	592	592	**592**	2.34059	37	40	38	845	876	859	0.0304
bcsstk04	132	23	23	**23**	1024	1024	**1024**	2.02436	111	115	112	7764	8037	7860	0.7770
bcsstk05	153	34	34	**34**	1787	1787	**1787**	2.02573	133	135	133	9673	10268	9841	0.5860
bcsstk22	138	52	52	**52**	1375	1375	**1375**	2.01365	108	112	108	5304	5781	5590	0.1610
can_144	144	13	13	**13**	1043	1043	**1043**	2.01614	117	124	118	7983	8531	8299	0.3280
can_161	161	18	18	**18**	1791	1791	**1791**	2.01694	131	134	131	9482	10261	9546	0.3840
curtis54	54	16	16	**16**	514	514	**514**	2.21831	37	45	37	805	1033	814	0.0443
dwt_234	234	28	28	**28**	712	712	**712**	2.02469	186	207	190	13095	13920	13111	0.4350
fs_183_1	183	81	81	**81**	6534	6534	**6534**	2.01911	152	163	153	9582	12714	10106	0.6880
gent113	113	27	27	**27**	1406	1406	**1406**	2.02029	89	97	92	4181	4332	4221	0.2220
gre_115	115	23	23	**23**	1502	1502	**1502**	2.02182	89	96	89	4339	4753	4371	0.1470
gre_185	185	21	21	**21**	2822	2822	**2822**	2.02461	152	159	154	12483	12830	12525	0.6330
ibm32	32	17	17	**17**	240	240	**240**	2.15781	22	24	22	341	370	356	0.0107
impcol_b	59	25	25	**25**	694	694	**694**	2.44516	44	49	44	1184	1299	1204	0.0540
impcol_c	137	30	30	**30**	1684	1684	**1684**	2.02195	107	117	107	5546	5958	5767	0.1820
lns_131	131	20	20	**20**	1149	1149	**1149**	2.02078	106	109	107	4474	4884	4546	0.2290
lund_a	147	16	16	**16**	921	921	**921**	2.02276	125	126	125	9243	9507	9354	0.5420
lund_b	147	23	23	**23**	2532	2532	**2532**	2.03341	120	127	124	9068	9508	9144	0.6010
mcca	180	85	85	**85**	6559	6559	**6559**	2.02281	143	159	146	11732	12710	12026	1.5700
nos1	237	5	5	**5**	11	11	**11**	2.02582	196	198	196	15791	15950	15825	0.5050
nos4	100	10	10	**10**	639	639	**639**	2.02726	75	85	76	2931	3466	2966	0.0963
pores_1	30	7	7	**7**	149	149	**149**	2.37606	21	23	21	333	349	345	0.0145
steam3	80	11	11	**11**	600	600	**600**	2.37211	63	71	64	2452	2600	2474	0.2260
west0132	132	34	34	**34**	2214	2214	**2214**	2.02233	105	110	106	6120	6529	6204	0.1660
west0156	156	57	57	**57**	4122	4122	**4122**	2.02147	124	136	125	7652	8790	7917	0.1800
west0167	167	48	48	**48**	4006	4006	**4006**	2.02050	139	144	139	9660	10467	9772	0.2780
will199	199	65	65	**65**	7972	7972	**7972**	2.02561	167	174	167	15088	15517	15184	0.4820
will57	57	9	9	**9**	265	265	**265**	2.18289	41	48	42	721	868	756	0.0481

disregarding vertices without adjacencies; and (*ii*) 80 instances, ranging from 207 to 1104 vertices. Moreover, the executions of each heuristic in each instance finished in similar amounts of time, i.e. standard deviations for all the execution times of all four heuristics were less than 0.0953.

For the VNS-band heuristic, the average runtimes of the SOM-band heuristic were used: 2 s for the set composed of 33 instances and 20 s for the set composed of 80 instances. For the SOM-band heuristic, 500 epochs were used as the maximum limit; however, the execution terminates earlier if no improvement in the result is found. The learning rate used was $\eta = 0.9$ so that the learning process could be faster at the beginning of the iterations, and in the course of learning, this value was decremented in order to refine learning, as described in Algorithm 2. In general, better results were obtained in exploratory experiments using this value.

Tables 1 and 2 show the dimension (Dim.) of the input matrix, the smallest (β_{min}) and the largest (β_{max}) bandwidths encountered, the average bandwidth ($\overline{\beta}$), the smallest (P_{min}) and the largest (P_{max}) profiles encountered, and the

Table 2. Simulations with the VNS-band and SOM-band heuristics for the set composed of 80 instances of the Harwell-Boeing sparse matrix collection. The smaller bandwidths and profiles found are highlighted.

Instance	Dim.	VNS band β_{min}	β_{max}	β	P_{min}	P_{max}	P	Time (s)	SOM band β_{min}	β_{max}	β	P_{min}	P_{max}	P	Time (s)
494_bus	494	30	35	**34**	4654	5613	**5421**	20.057410	406	434	415	56460	61624	57899	4.230
662_bus	662	51	51	**51**	12526	12526	**12526**	20.061070	553	579	568	114491	115393	115302	10.100
685_bus	685	39	39	**39**	10335	10335	**10335**	20.040370	585	607	587	136583	143048	137229	12.500
ash292	292	19	19	**19**	2159	2159	**2159**	20.057580	245	245	245	30269	30269	30269	2.350
bcspwr04	274	24	25	**24**	2405	2710	**2527**	20.064260	234	242	234	23304	23449	23405	1.860
bcspwr05	443	29	29	**29**	4580	4580	**4580**	20.060720	361	369	364	50212	51561	50603	4.070
bcsstk06	420	45	45	**45**	14078	14078	**14078**	20.065700	372	372	372	73219	73552	73252	13.000
bcsstk19	817	14	14	**14**	5768	5768	**5768**	20.108310	718	724	718	244985	251012	246118	28.000
bcsstk20	485	13	13	**13**	1314	1314	**1314**	20.070430	412	428	415	75927	81131	77889	6.910
bcsstm07	420	45	45	**45**	13842	13842	**13842**	20.055390	368	372	370	72943	75216	73620	12.500
bp__0	822	254	254	**254**	103522	103522	**103522**	20.070080	708	734	715	243865	253150	244793	25.400
bp__200	822	273	273	**273**	117117	117117	**117117**	20.288900	725	734	727	254985	263961	257677	30.900
bp__400	822	292	292	**292**	127701	127701	**127701**	20.067340	722	735	725	258853	261434	259111	31.300
bp__600	822	342	342	**342**	146178	146178	**146178**	20.054420	721	722	721	258426	275762	260982	32.800
bp__800	822	290	290	**290**	132325	132325	**132325**	20.065760	711	732	715	266339	283989	272423	35.600
bp_1000	822	308	308	**308**	131267	131267	**131267**	20.074190	712	727	716	265417	277897	266665	36.400
bp_1200	822	296	296	**296**	122734	122734	**122734**	20.082480	720	729	723	268313	272304	270308	36.900
bp_1400	822	320	320	**320**	140033	140033	**140033**	20.209990	720	730	723	273355	281906	276805	34.100
bp_1600	822	362	362	**362**	151668	151668	**151668**	20.070040	721	729	722	271571	278094	274349	37.800
can_292	292	39	39	**39**	7911	7911	**7911**	20.059000	253	257	253	31522	31522	31522	2.960
can_445	445	58	58	**58**	16523	16523	**16523**	20.065860	385	396	386	74766	76722	74961	7.910
can_715	715	108	108	**108**	29481	29481	**29481**	20.071670	626	627	626	192899	202552	196219	24.100
can_838	838	91	91	**91**	41679	41679	**41679**	20.087320	738	748	739	287338	293679	288783	37.800
dwt_209	209	23	23	**23**	2126	2126	**2126**	20.057630	175	180	176	16368	16591	16390	1.410
dwt_221	221	13	13	**13**	1833	1833	**1833**	20.048960	189	196	193	17053	18000	17235	1.270
dwt_245	245	21	21	**21**	1527	1527	**1527**	20.078250	207	208	207	18878	19737	18963	1.560
dwt_310	310	10	10	**10**	203	203	**203**	20.071830	262	262	262	34322	36542	34800	3.070
dwt_361	361	16	16	**16**	4670	4670	**4670**	20.057440	303	317	308	48218	48600	48396	4.150
dwt_419	419	28	28	**28**	5743	5743	**5743**	20.058480	354	364	358	65390	68086	66402	6.460
dwt_503	503	48	48	**48**	14445	14445	**14445**	20.042800	438	441	438	100174	101403	101238	13.500
dwt_592	592	29	29	**29**	8995	8995	**8995**	20.040820	513	517	516	131105	132578	131671	14.400
dwt_878	878	27	27	**27**	2471	2471	**2471**	20.072790	765	781	770	290660	292215	291292	35.200
dwt_918	918	37	37	**37**	11316	11316	**11316**	20.068500	798	811	799	312033	318123	312642	38.100
dwt_992	992	73	73	**73**	60682	60682	**60682**	20.079390	875	882	878	427840	433546	430145	89.300
fs_541_1	541	270	270	**270**	82803	82803	**82803**	20.083190	471	483	474	110566	114507	111553	20.000
fs_680_1	680	18	18	**18**	10126	10126	**10126**	20.068520	592	606	594	141849	145013	143747	16.700
fs_760_1	760	38	38	**38**	23996	23996	**23996**	20.061880	668	675	669	231571	234737	233239	44.300
gr_30_30	900	59	59	**59**	36319	36319	**36319**	20.065120	791	804	792	303349	308550	304703	40.000
gre_343	343	28	28	**28**	6850	6850	**6850**	20.055890	288	300	292	41735	43185	42167	3.530
gre_512	512	38	38	**38**	14058	14058	**14058**	20.068090	446	454	448	94134	96623	94382	8.590
gre_216a	216	21	21	**21**	3223	3223	**3223**	20.054740	176	176	176	16061	16061	16061	1.210
hor_131	434	55	55	**55**	18909	18909	**18909**	20.056720	374	375	374	73660	75377	74228	16.300
impcol_a	207	32	32	**32**	3264	3264	**3264**	20.037900	175	177	175	14388	14821	14547	0.860
impcol_d	425	40	40	**40**	10491	10491	**10491**	20.045320	363	366	364	61230	63353	61601	4.530
impcol_e	225	42	42	**42**	4133	4133	**4133**	20.080110	188	198	194	19632	19698	19665	2.100
jagmesh1	936	31	31	**31**	11451	11451	**11451**	20.057160	803	828	810	304899	315938	310816	29.900
jpwh_991	991	226	226	**226**	122646	122646	**122646**	20.081590	869	881	872	332098	346453	334206	56.600
lnsp_511	511	45	45	**45**	12808	12808	**12808**	20.060990	442	442	442	80405	81970	80689	12.000
mbeacxc	496	437	437	**437**	85845	85845	**85845**	20.125360	441	444	441	117204	117204	117204	206.000
mbeaflw	496	437	437	**437**	85845	85845	**85845**	20.105340	443	444	443	117462	118981	117932	194.000
mbeause	496	417	417	**417**	80434	80434	**80434**	20.099140	441	445	441	118382	121801	118723	160.000
mcfe	765	385	385	**385**	150623	150623	**150623**	20.058160	673	686	674	264899	264899	264899	167.000
nnc261	261	24	24	**24**	4233	4233	**4233**	20.052560	223	225	224	22691	22746	22729	3.240
nnc666	666	56	56	**56**	26099	26099	**26099**	20.067380	582	584	583	149606	157308	150376	22.300
nos2	957	3	3	**3**	1908	1908	**1908**	146.301700	823	823	823	268394	268865	268817	22.000
nos3	960	79	79	**79**	46502	46502	**46502**	20.065500	837	856	843	391071	394046	392558	82.200
nos5	468	64	64	**64**	23567	23567	**23567**	20.679200	411	416	412	86300	88769	86862	9.100
nos6	675	20	20	**20**	8956	8956	**8956**	20.053500	568	598	574	138814	142724	139748	12.800
nos7	729	78	78	**78**	14789	14789	**14789**	20.048410	631	635	631	184603	190029	187567	16.000
orsirr_2	886	95	95	**95**	67152	67152	**67152**	20.087230	756	787	765	273696	277702	274096	48.500
plat362	362	36	36	**36**	5510	5510	**5510**	20.077280	315	321	316	55681	58288	56541	8.470
plskz362	362	239	239	**239**	27491	27491	**27491**	20.070230	309	309	309	43361	45864	43904	2.420
pores_3	532	13	13	**13**	5319	5319	**5319**	20.050960	452	452	452	100168	102349	100618	14.800
saylr1	238	14	14	**14**	2977	2977	**2977**	20.056870	202	207	202	17074	17562	17178	1.810
saylr3	1000	56	56	**56**	28970	28970	**28970**	20.075930	869	870	869	213129	222639	215214	38.400
sherman1	1000	56	56	**56**	28970	28970	**28970**	20.059670	877	880	877	211879	216197	213523	38.600
sherman4	1104	36	36	**36**	16885	16885	**16885**	25.025720	936	949	945	198926	200978	200696	48.100
shl__0	663	231	231	**231**	56453	56453	**56453**	20.064780	568	592	572	135349	189368	146468	9.710
shl_200	663	240	240	**240**	56999	56999	**56999**	20.031620	569	590	578	139972	143038	140585	9.950
shl_400	663	237	237	**237**	57962	57962	**57962**	20.035680	568	593	576	143320	187801	149969	9.910
steam1	240	47	47	**47**	8904	8904	**8904**	20.079400	206	206	206	24663	25131	24731	6.780
steam2	600	63	63	**63**	31884	31884	**31884**	20.062410	529	538	530	161680	163992	162323	67.900
str__0	363	120	120	**120**	20957	20957	**20957**	20.057030	315	319	316	48960	48960	48960	6.290
str_200	363	127	127	**127**	25582	25582	**25582**	20.057920	317	320	317	52166	53964	52525	8.730
str_600	363	133	133	**133**	25466	25466	**25466**	20.061900	312	319	315	54450	54554	54543	8.800
west0381	381	153	153	**153**	35645	35645	**35645**	20.035010	333	339	334	56986	57845	57626	6.360
west0479	479	124	124	**124**	30177	30177	**30177**	20.069340	416	428	418	84049	84559	84202	7.040
west0497	497	86	86	**86**	22312	22312	**22312**	20.055870	418	441	429	84586	91906	87586	7.100
west0655	655	253	253	**253**	82617	82617	**82617**	20.043900	567	567	567	159756	163496	160130	14.500
west0989	989	335	335	**335**	172238	172238	**172238**	20.039000	855	882	861	347936	347936	347936	34.600

Table 3. Resolution of linear systems by the JPCGM without reordering rows and with reordering rows according to the VNS-band and SOM-band heuristics. Discretizations of the heat conduction equation by finite volumes originated these linear systems. The best JPCGM computational-cost reduction for each instance is highlighted in the "Reduction" column.

Vertices	Heuristic	$\overline{\beta}$	$\overline{profile}$	Heuristic (s)	JPCGM (s)	Iterations	Reduction
942	without reordering	918	337600	–	0.447	150	–
	VNS-band (1s)	65	39395	1.05103	0.450	150	-235.80 %
	SOM-band	929	325811	0.01136	0.442	150	-1.42 %
2165	without reordering	2120	1818156	–	1.580	221	–
	VNS-band (1s)	94	143168	1.01666	1.621	221	-66.94 %
	SOM-band	2141	1737751	0.04043	1.749	221	-13.26 %
4846	without reordering	4769	9116750	–	5.897	319	–
	VNS-band (1s)	153	496101	1.03874	6.380	319	-25.81 %
	SOM-band	4797	8727741	0.19058	6.442	319	-12.47 %
10728	without reordering	10626	45314579	–	22.571	463	–
	VNS-band (1s)	246	1490854	1.37914	23.889	463	-11.95 %
	SOM-band	10664	42851180	1.14216	30.266	463	-39.15 %
23367	without reordering	23167	216212086	–	84.405	666	–
	VNS-band (1s)	3425	16625796	1.85177	84.683	666	-2.52 %
	SOM-band	23276	204059064	6.01002	103.434	666	-29.67 %
50592	without reordering	50461	1020411959	–	280.458	973	–
	VNS-band (1s)	578	18984571	3.04496	263.354	973	**5.01%**
	SOM-band	50491	957689576	26.00135	335.666	973	-28.96 %
108683	without reordering	108216	4725435534	–	883.092	1411	–
	VNS-band (1s)	8354	75297775	8.31570	792.340	1412	**9.33%**
	SOM-band	108536	4425630101	117.77850	1019.436	1408	-28.78 %
232052	without reordering	231672	21652820640	–	2913.121	2030	–
	VNS-band (20s)	28057	445824204	31.06000	2848.496	2043	**1.15%**
	SOM-band	231829	20200695713	533.73375	3354.255	2034	-33.46 %
492853	without reordering	492100	97893937993	–	8325.150	2942	–
	VNS-band (1s)	89719	1226201509	90.93720	6829.813	2946	**16.87%**
	SOM-band	492663	91172829064	2315.11500	9432.782	2948	-41.11 %
965545	without reordering	964827	377848438952	–	24394.600	4358	–
	SOM-band	965113	350154727398	8125.33670	26258.800	4361	-40.95 %

average profile ($\overline{P}$). The best heuristic with respect to bandwidth and profile reductions were the VNS-band heuristic in this set of 113 matrices. Despite not running as fast as the SOM-band heuristic, the VNS-band heuristic yielded better bandwidth and profile reductions for these instances.

5.2 Simulations with Linear Systems Originated from Discretization of the Heat Conduction Equation by Finite Volumes

This section shows simulations applied to linear systems resulting from a two-dimensional discretization of the heat conduction equation by finite volumes with meshes generated by Delaunay refinement, and hence Voronoi diagrams were used as control volumes. The heat conduction equation can be defined as $\frac{\partial \phi}{\partial t} = \nabla \cdot (\alpha \nabla \phi)$, where ϕ is the temperature in the physical problem of

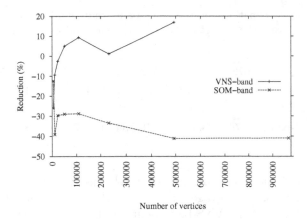

Fig. 1. Reduction of JPCGM computational cost in tests shown in Table 3 with instances ranging from 4,846 to 965,545 unknowns.

temperature variation, α is the thermal diffusivity, and t is time. The boundary conditions used were Dirichlet boundary conditions: the discrete domain was a unit square, with the values of the boundary conditions on north, west, and east equal to ten and south set as zero [12].

Ten experiments were performed for each instance without reordering of graph vertices and with reordering of graph vertices by the VNS-band, and SOM-band heuristics. Table 3 shows these results. Executions for each heuristic in each instance finished in similar amounts of time: the standard deviations for the execution times were less than 25.41 in the tests shown in Table 3.

The number of CGM iterations of each simulation are presented in the Iterations column in Table 3. In the last column in Table 3, reduction by using a heuristic for bandwidth and profile reductions is shown in relation to the CGM runtimes without reordering the graph vertices. Computational costs are shown in the first line in a corresponding test with each instance in this table. The best CGM computational-cost reductions are highlighted in the last column in Table 3. The positive or negative percentages presented in this column indicates a computational cost reduced or increased, respectively.

For the simulations shown in Table 3, linear systems were solved by the JPCGM. By using the VNS-band heuristic, no reduction in the JPCGM computational cost was reached for instances up to 23,367 vertices and, as described, the VNS-band executable program runs only with instances up to 500,000 vertices. Simulations with the best choice in each instance with respect to the VNS-band heuristic were considered in this table. No reduction in the JPCGM computational cost was obtained in this dataset using the SOM-band heuristic. In Fig. 1, these results are summed up.

The storage costs of the SOM-band and VNS-band heuristics are shown in Table 4, whose last column shows the ratio between the memory occupation of the VNS-band and SOM-band heuristics. It can be verified that the storage costs

Table 4. Memory occupations of the VNS-band and SOM-band heuristics with instances resulting from the discretization of the heat conduction equation by finite volumes.

Vertices	10728	23367	50592	108683	232052	492853	965545
VNS-band (MB)	129.50	371.10	967.10	2293.76	5048.32	10844.16	–
SOM-band (MB)	7.30	16.10	30.70	65.70	141.40	296.40	583.10
VNS-band/SOM-band	17.74	23.05	31.50	34.91	35.70	36.59	–

of both SOM-band and VNS-band heuristics are linear in the number of graph vertices. Nevertheless, the storage cost of the SOM-band heuristic is smaller than the storage cost of the VNS-band heuristic by 17.7 to 36.6 times, in this set of instances. Figure 2 shows the memory occupations of the VNS-band and SOM-band heuristics.

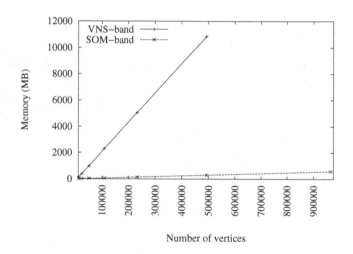

Fig. 2. Storage costs of the VNS-band and SOM-band heuristics in simulations with instances resulting from the discretization of the heat conduction equation by finite volumes.

5.3 Simulations with Linear Systems Arising from the Discretization of the Laplace Equation by Finite Volumes

This section shows simulations applied to linear systems arising from the discretization of the Laplace equation by finite volumes. Let the Dirichlet problem be represented by the Laplace equation and defined as $\nabla^2 \phi = 0$ in $\Omega \in \mathbb{R}^2$, $\phi = f \in \partial\Omega$, where ϕ is the dependent variable of the partial differential equation, Ω is a bounded domain in $\mathbb{R}^2$ and f is a smooth function defined on the boundary $\partial\Omega$. The boundary conditions were established north, south, and west of the unit square with a single value f, and the east side was set with a different value [11]. This section presents a set of simulations with linear systems in which rows of the matrices were ordered in a sequence given by a space-filling curve. More specifically, a Sierpiński-like curve was used [10,17].

Experiments were performed without reordering rows of linear systems and also reordering vertices of the corresponding graph by the VNS-band, and SOM-band heuristics. Five runs for each instance up to 192,056 vertices and 3 runs for the other instances were carried out. Executions for each heuristic in each instance finished in similar runtimes: the standard deviations for the execution times were less than 3.85 in these simulations.

The first line of each instance shown in Table 5 presents results for linear systems solved using the JPCGM without reordering the rows of linear systems. For the other lines showing each instance in Table 5, results of the resolution of linear systems are shown after reordering rows according to the SOM-band and VNS-band heuristics. The last column of Table 5 shows the percentage reduction achieved using a heuristic for bandwidth and profile reductions in relation to the JPCGM runtimes (shown in the corresponding first line for the test). In the last column of the table, the positive or negative percentages indicates that the JPCGM computational cost was reduced or increased, respectively.

For the linear system composed of 489,068 unknowns, the VNS-band heuristic increased the profile in 37.3 %. No JPCGM computational-cost reduction in any instance of this dataset was achieved when reordering rows by the VNS-band heuristic, keeping in mind that the VNS-band executable program used of this heuristic runs with instances up to 500,000 vertices.

One second was the maximum execution time used for the SOM-band and VNS-band heuristics. For the VNS-band heuristic, increasing the maximum execution time to 5 times the runtime with linear systems ranging from 1,874 to 39,716 unknowns, shown in Table 5, no better results in renumbering rows were obtained using the heuristic. For the linear systems comprised of 68,414 and 105,764 unknowns, results remained unchanged for a maximum runtime up to 3 times the runtime established to each instance shown in Table 5. For all the other instances, results of the VNS-band heuristic did not change, even when increasing the runtime to twice the value set to each instance shown in Table 5. The results of this set of instances indicated that reordering the rows in linear systems already ordered in a sequence given by a space-filling curve [11] may not be beneficial when solving linear systems using the JPCGM.

The storage costs of the SOM-band and VNS-band heuristics applied in this dataset are shown in Table 6, whose last column shows the ratio between the memory occupation of the VNS-band and SOM-band heuristics. Similarly to the other previous tests, it was verified that the storage costs of both the SOM-band and VNS-band heuristics are linear in the number of graph vertices. However, the storage cost of the SOM-band heuristic is smaller than the storage cost of the VNS-band heuristic by a factor of 15.3 to 36.9 in this dataset. Figure 3 shows the memory occupation of the VNS-band and SOM-band heuristics.

Table 5. Resolution of linear systems by the JPCGM without reordering rows and also with rows reordered by the VNS-band and SOM-band heuristics. Rows of these linear systems were originally ordered in a sequence given by a space-filling curve. The best JPCGM computational-cost reductions for each instance are highlighted in the "Reduction" column.

Vertices	Heuristic	$\bar{\beta}$	$profile$	Heuristica (s)	JPCGM (s)	Iterations	Reduction
1874	without reordering	1873	57765	-	1.0840	257	-
	VNS-band (1s)	363	105250	1.01259	1.1080	257	-95.63 %
	SOM-band	1844	1010425	0.02750	1.0680	257	-1.06 %
5882	without reordering	5881	333971	-	7.7900	454	-
	VNS-band (1s)	887	529491	1.03764	7.6920	454	-12.06 %
	VNS-band (2s)	887	529491	2.03827	7.8820	454	-27.35 %
	SOM-band	5821	10060410	0.27420	6.5260	454	**12.71%**
16922	without reordering	16921	1710910	-	46.1420	767	-
	VNS-band (1s)	4765	2328301	1.16345	47.1760	767	-4.76 %
	VNS-band (2s)	4765	2328301	2.16871	47.9080	767	-8.53 %
	SOM-band	16870	83773254	2.27590	48.0440	767	-9.05 %
39716	without reordering	39715	6309342	-	166.3220	1144	-
	VNS-band (1s)	8043	13261308	1.66940	172.3940	1144	-4.65 %
	VNS-band (2s)	8043	13261308	2.70276	172.4350	1144	-5.30 %
	SOM-band	39549	465224178	13.30850	200.5520	1144	-28.58 %
68414	without reordering	68413	14882117	-	375.8966	1514	-
	VNS-band (1s)	2542	17876119	2.78124	401.4400	1514	-7.54 %
	VNS-band (4s)	2542	17876119	5.83806	401.6720	1514	-8.41 %
	SOM-band	68267	1381895033	40.57910	470.3960	1514	-35.94 %
105764	without reordering	105763	29560801	-	705.5030	1846	-
	VNS-band (1s)	15174	22202516	5.08892	732.4900	1846	-4.55 %
	VNS-band (6s)	15174	22202516	10.16593	734.7810	1846	-5.59 %
	SOM-band	105497	3313239522	99.32100	906.4600	1846	-42.56 %
192056	without reordering	192055	80822936	-	1687.8002	2383	-
	VNS-band (1s)	2888	71935018	14.22770	1869.6033	2383	-11.61 %
	VNS-band (15s)	2888	71935018	28.41990	1868.9030	2383	-12.41 %
	SOM-band	191661	10915862166	324.80730	2117.3200	2383	-44.69 %
281954	without reordering	281953	156196014	-	2946.8490	2812	-
	VNS-band (1s)	20891	109731059	28.40410	3067.0733	2812	-5.04 %
	VNS-band (29s)	20891	109731059	56.71557	3069.0750	2812	-6.07 %
	SOM-band	281736	23513746884	702.46360	3691.4667	2812	-49.11 %
373796	without reordering	373795	256465457	-	4576.0367	3150	-
	VNS-band (1s)	25580	145246396	49.23200	5079.4266	3150	-12.08 %
	VNS-band (60s)	25580	145246396	108.62470	5080.2970	3150	-13.39 %
	SOM-band	373512	41365091373	1245.76330	5651.6631	3150	-50.73 %
489068	without reordering	489067	387216428	-	6380.9000	3519	-
	VNS-band (1s)	84473	531636561	84.66560	6441.1200	3519	-2.27 %
	VNS-band (90s)	84473	531636561	172.18467	6447.1100	3519	-3.74 %
	SOM-band	488681	70667989807	2134.44000	8052.7000	3519	-59.65 %
740288	without reordering	740287	838516571	-	11597.7233	4270	-
	SOM-band	739318	161947149989	4622.67000	14639.5000	4270	-66.09 %
1015004	without reordering	1015003	1580908606	-	17095.1000	4569	-
	SOM-band	1014083	304420871105	8071.13000	21463.7000	4569	-72.77 %
1485410	without reordering	1485409	3181425530	-	28794.2000	5270	-
	SOM-band	1483781	652546694835	19135.03330	36480.9000	5270	-93.15 %
2139050	without reordering	2139049	6473418611	-	45204.4000	5778	-
	SOM-band	2137936	1353175769523	40516.43330	57653.9000	5778	-117.17 %

Table 6. Memory occupations of the VNS-band and SOM-band heuristics with linear systems resulting from the discretization of the Laplace equation by finite volumes and ordered in a sequence given by a space-filling curve.

Vertices	VNS-band (MB)	SOM-band (MB)	VNS-band/SOM-band
1874	29.5	0.8	36.88
5882	56.6	2.6	21.77
16922	144.7	9.3	15.56
39716	327.2	20.3	16.12
68414	557.4	36.4	15.31
105764	857.2	46.9	18.28
192056	1579.1	87.1	18.13
281954	2317.1	141.7	16.35
373796	3993.6	172.9	23.10
489068	5130.2	203.8	25.17
740288	$--$	336.9	$--$
1015004	$--$	423.3	$--$
1485410	$--$	678.8	$--$
2139050	$--$	1102.2	$--$

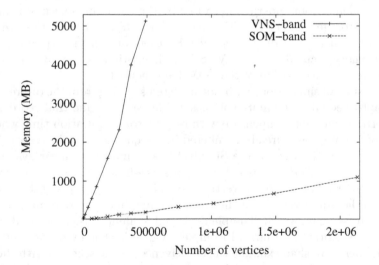

Fig. 3. Storage costs of the VNS-band and SOM-band heuristics in simulations with instances originated from the discretization of the Laplace equation by finite volumes and ordered in a sequence given by a space-filling curve.

6   Conclusions

In this work, performances of two heuristics for reordering graph vertices in order to reduce the computational cost of solving linear systems by the Jacobi-preconditioned Conjugate Gradient Method (JPCGM) were analyzed. One of the heuristics was proposed and evaluated in this work. A self-organizing map was used in the design of this new heuristic, here termed SOM-band. The SOM-band heuristic was designed in a manner that it can be run on symmetric and asymmetric matrices. Sequential sets of experiments with different instances with sizes ranging from tens to more than 2 million vertices were performed.

Experimental evaluations were performed to reduce the computational cost in solving linear systems using the Jacobi-preconditioned Conjugate Gradient Method after reordering graph vertices associated with matrices of linear systems by these heuristics for bandwidth and profile reductions. Despite having executed quickly in instances of the Harwell-Boeing sparse matrix collection, results of the new heuristic, SOM-band, were dominated by results obtained by the Variable neighborhood search for bandwidth reduction (VNS-band) heuristic. This was verified both in approximating the solution as well as comparing the computational cost to present an approximate solution in relation to these two heuristics when solving linear system by the JPCGM. In executions with instances of the Harwell-Boeing sparse matrix collection and linear systems up to approximately 11,000 vertices, the SOM-band heuristic demonstrated faster performance than the VNS-band heuristic. Nevertheless, in general, the SOM-band heuristic did not achieve competitive results in simulations involving solving linear systems by the JPCGM. In these simulations, the performance of the SOM-band heuristic may have suffered from underfitting caused by the maximum execution time set.

Verified storage costs of the SOM-band and VNS-band heuristics are linear in the number of vertices of the instance. Furthermore, in the experiments performed, memory occupation of the VNS-band heuristic was up to 36 times larger than the memory occupation of the SOM-band heuristic.

With results of simulations with linear systems arising from the discretization of the Laplace equation by finite volumes, evidence was found that reordering of graph vertices may not compensate with respect to computation time when the vertices of the mesh are already numbered in a sequence given by a space-filling curve, such as in the case of the Sierpiński-like curve, and linear systems are solved using the JPCGM. This is because a space-filling curve already provides a suitable ordering of the graph vertices; that is, it provides a suitable memory location to the graph vertices of the corresponding linear-system matrix.

In future studies, we shall implement and evaluate other preconditioners, such as the Algebraic Multigrid, Incomplete Cholesky Factorization, and ILUT preconditioners. We shall examine the effectiveness of the schemes with the computation of incomplete or approximate factorization based preconditioners as well approximate inverse preconditioners. These techniques shall be applied as preconditioners of the Conjugate Gradient Method in order to analyze their computational cost in conjunction with heuristics for reordering graph vertices.

Acknowledgements. This work was undertaken with the support of FAPEMIG - Fundação de Amparo à Pesquisa do Estado de Minas Gerais (Minas Gerais Research Support Foundation). We also thank Professor Dr. Dragan Urosevic for sending us the VNS-band executable program.

References

1. Benzi, M.: Preconditioning techniques for large linear systems: a survey. J. Comput. Phys. **182**(2), 418–477 (2002)
2. Bernardes, J.A.B., Gonzaga de Oliveira, S.L.: A systematic review of heuristics for profile reduction of symmetric matrices. Procedia Comput. Sci. **51**, 221–230 (2015). (Proceedings of the ICCS 2015 - International Conference on Computational Science, Reykjavík, Iceland)
3. Brocki, L., Korzinek, D.: Kohonen self-organizing map for the traveling salesperson problem. In: Jablonski, R., Turkowski, M., Szewczyk, R. (eds.) Recent Advances in Mechatronics, pp. 116–119. Springer, Heidelberg (2007)
4. Burgess, D.A., Giles, M.B.: Renumbering unstructured grids to improve the performance of codes on hierarchial memory machines. Adv. Eng. Softw. **28**, 189–201 (1997)
5. Chagas, G.O., Gonzaga de Oliveira, S.L.: Metaheuristic-based heuristics for symmetric-matrix bandwidth reduction: a systematic review. Procedia Comput. Sci. **51**, 211–220 (2015). (Proceedings of the ICCS 2015 - International Conference on Computational Science, Reykjavík, Iceland)
6. Das, R., Mavriplis, D.J., Saltz, J., Gupta, S., Ponnusamy, R.: The design and implementation of a parallel unstructured Euler solver using software primitives. AIAA J. **32**(3), 489–496 (1994)
7. Duff, I.S., Grimes, R.G., Lewis, J.G.: Sparse matrix test problems. ACM Trans. Math. Softw. **15**(1), 1–14 (1989)
8. Duff, I.S., Meurant, G.A.: The effect of ordering on preconditioned conjugate gradients. BIT Numer. Math. **29**(4), 635–657 (1989)
9. Gonzaga de Oliveira, S.L., Chagas, G.O.: A systematic review of heuristics for symmetric-matrix bandwidth reduction: methods not based on metaheuristics. In: The XLVII Brazilian Symposium of Operational Research (SBPO), Sobrapo, Ipojuca, Brazil, August 2015
10. Gonzaga de Oliveira, S.L., Kischinhevsky, M.: Sierpiński curve for total ordering of a graph-based adaptive simplicial-mesh refinement for finite volume discretizations. In: XXXI Brazilian National Congress in Applied and Computational Mathematics (CNMAC), Belém, Brazil, pp. 581–585. The Brazilian Society of Computational and Applied Mathematics (SBMAC) (2008)
11. Gonzaga de Oliveira, S.L., Kischinhevsky, M., Tavares, J.M.R.S.: Novel graph-based adaptive triangular mesh refinement for finite-volume discretizations. Comput. Model. Eng. Sci. CMES **95**(2), 119–141 (2013)
12. Gonzaga de Oliveira, S.L., de Oliveira, F.S., Chagas, G.O.: A novel approach to the weighted laplacian formulation applied to 2D delaunay triangulations. In: Gervasi, O., Murgante, B., Misra, S., Gavrilova, M.L., Rocha, A.M.A.C., Torre, C., Taniar, D., Apduhan, B.O. (eds.) ICCSA 2015. LNCS, vol. 9155, pp. 502–515. Springer, Heidelberg (2015)
13. Kohonen, T.: Construction of similarity diagrams for phonemes by a self-organizing algorithm. Technical Report TKK-F-A463, Helsinki University of Technology, Espoo, Finland (1981)

14. Lin, Y.X., Yuan, J.J.: Profile minimization problem for matrices and graphs. Acta Mathematicae Applicatae Sinica **10**(1), 107–122 (1994)
15. Mladenovic, N., Urosevic, D., Pérez-Brito, D., García-González, C.G.: Variable neighbourhood search for bandwidth reduction. Eur. J. Oper. Res. **200**(1), 14–27 (2010)
16. Papadimitriou, C.H.: The NP-Completeness of the bandwidth minimization problem. Computing **16**(3), 263–270 (1976)
17. Velho, L., Figueiredo, L.H., Gomes, J.: Hierarchical generalized triangle strips. Vis. Comput. **15**(1), 21–35 (1999)

Modeling Combustions: The *ab initio* Treatment of the O(³P) + CH₃OH Reaction

Leonardo Pacifici[1(✉)], Francesco Talotta[2], Nadia Balucani[1],
Noelia Faginas-Lago[1], and Antonio Laganà[1]

[1] Department of Chemistry, Biology and Biotechnologies, University of Perugia,
via Elce di Sotto, 8, 06123 Perugia, Italy
xleo@dyn.unipg.it
[2] Laboratoire de Chimie et Physique Quantiques Université Paul Sabatier - Bat.,
3R1b4 118 route de Narbonne, 31400 Toulouse, France

Abstract. In this work we tackle the problem of dealing in an *ab initio* fashion with the description of the

$$O(^3P) + CH_3OH \rightarrow OH + CH_2OH$$

reaction that is one of the most important elementary processes involved in the methanol oxidation. In particular, we carried out the following computational steps:

1. calculate the electronic structure of the $O + CH_3OH$ system
2. fit to a pseudo triatomic LEPS (London Eyring Polanyi Sato) the collinear reaction channel leading to the production of OH
3. calculate the dynamical properties of the process using quantum techniques

For the purpose of *ab initio* computing the electronic structure of the $O(^3P) + CH_3OH$ system we used various computational programs based on DFT techniques (to characterize the stationary points and work out harmonic vibrational frequencies) and CCSD(T) level of theory (to refine the energy of the stationary points, calculate the exoergicity of the considered channel and estimate the height of the barrier to reaction). For the purpose of computing quantum reactive scattering state specific probabilities on the proposed LEPS potential energy surface, the Multi Configuration Time Dependent Hartree method was used.

1 Introduction

Among bio-alcohols CH_3OH, methanol, is considered a very promising alternative fuel. In fact, it can be used in modern internal combustion engines as well as in fuel cells thanks to a catalytic electrolytic chemical process. In last years, several experiments and theoretical calculations [1–4] were carried out in order to evaluate its combustion properties, such as those important for the environment like heat release, CO_2 emission and efficiency as fuel. Moreover, thanks to the fact that methanol is the simplest alcohol, the determination of its oxidation mechanism provides a reasonable basis for improving data available

© Springer International Publishing Switzerland 2016
O. Gervasi et al. (Eds.): ICCSA 2016, Part I, LNCS 9786, pp. 71–83, 2016.
DOI: 10.1007/978-3-319-42085-1_6

for larger ones. Information derived from various kinetics studies [2,5] converge on a mechanism composed of different elementary processes, of which the most important ones are:

$$O(^3P) + CH_3OH \rightarrow OH + CH_2OH \qquad (1)$$
$$\rightarrow OH + CH_3O \qquad (2)$$
$$\rightarrow HO_2 + CH_3 \qquad (3)$$

Process (1) is exoergic and exhibits a small energy barrier, process (2) is slightly endoergic (almost thermoneutral) and process (3) is endoergic with an energy barrier so large to make its investigation irrelevant. Experimental and theoretical findings [1,2] agree on the fact that process (1) is dominant over process (2) at low temperature while the latter becomes increasingly more important as temperature rises. Yet, the study of process (1) represents a real challenge from the theoretical point of view due to the fact that $O + CH_3OH$ system is too large to apply a full quantum treatment while it is still small enough to adopt a semi empirical potential energy surface when calculating reactive scattering. As a matter of fact, *ab initio* theoretical information on this reaction are limited and, in particular, some of those needed for undertaking dynamical studies are missing. More in detail, no full quantum reactive scattering calculations have been performed and only a few dynamical studies [3], mainly addressed to the determination of rate coefficients and branching ratios via kinetics approaches, have been produced. This paper reports on the preliminary work we carried out for this reaction, within a typical GEMS (Grid Empowered Molecular Simulator) [6] based workflow scheme. In Sect. 2 the calculations of the electronic structure of the $O(^3P) + CH_3OH$ system and of the resulting shape of the reaction channels involved in the considered processes are discussed. In Sect. 3 the assemblage of the proposed model Potential Energy Surface (PES) out of the calculated *ab initio* potential energy values is illustrated. In Sect. 4 the outcomes of the preliminary dynamics calculations are presented. Finally, in Sect. 5 some conclusions are drawn.

2   *Ab Initio* Calculations of the Molecular Geometries and Energies

In order to get a global view of the energetics of the reaction in our work we have investigated first, according to the GEMS scheme, the most important geometries of the $O(^3P)+CH_3OH$ system by characterizing them using the `Gaussian 09` suite [7] at the BB1K/aug-cc-pVTZ level of theory for geometry optimization and CCSD(T)/aug-cc-pVTZ for single point calculations.

The BB1K (Becke88-Becke95 1-parameter model for kinetics) is a hybrid density functional optimized for thermochemical kinetic calculations developed by Truhlar et al. [8]. As already mentioned in the Introduction, DFT calculations allowed to characterize three different product channels:

1. channel 1: abstraction of one methyl hydrogen to form OH
2. channel 2: exchange of the hydroxyl H with the incoming O to form OH
3. channel 3: abstraction of the hydroxyl radical to form the hydroperoxyl radical

In our work, we focused our efforts on the characterization of the first two OH forming channels because of their importance at thermal energies.

The characterization of CH_3OH was carried out first by optimizing the molecule at the BB1K/aug-cc-PVTZ level of theory. Then, a single point calculation was performed at CCSD(T) level using the same basis set and the optimized geometry of CH_3OH with the addition of an oxygen at a distance of 16 and 26 Å by considering both singlet and triplet multiplicities. Results indicated that a single reference wavefunction is sufficient for describing the whole system. Related computed energy values have been used also for estimating the energy barriers and the energetic of the reactive process. The characterization of the involved transition states required three different sets of calculations: BB1K tight geometry optimization, BB1K frequency calculations and CCSD(T) single point energy refinements.

The optimized structure of the transition state (TS1) of channel (1) is shown in Fig. 1. Related values of bond lengths and angles are listed in Table 1 and agree with the ones reported in ref. [1]. For the same molecular geometry frequency calculations were performed and only one imaginary frequency was obtained as typical 1st-order saddle points. Coupled Cluster energies of singlet and triplet are shown in Table 2. The triplet state is lower in energy than the singlet one. For both cases the T_1 diagnostic, also shown in Table 2, is slightly larger than 0.02, indicating a non negligible non-dynamical electron correlation [9] because of the formation of the OH bond and the breaking of the old CH_2OH-H bond. This suggests, for future work, the use of a multi-configuration wavefunction to the end of obtaining a more accurate estimate of the energy barrier.

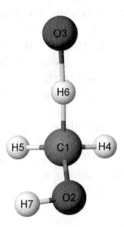

Fig. 1. Geometry of the transition state TS1 for the first reactive channel

Table 1. Characteristics of the first reactive channel transition state TS1 (bond distances in Å and angles in degree)

O3-H6	1.285
H6-C1	1.243
∠O3-H6-C1	177.83
∠O2-C1-H6	110.78
ϕO2-C1-H6-O3	−84.91

Table 2. Coupled Cluster energies (in E_h) and T_1 diagnostic for the singlet and triplet spin multiplicities.

$2S+1$	E	T_1
3	−190.52776867	0.0236
1	−190.50775793	0.0335

Following the same procedure for channel (2), we carried out DFT calculations aimed at characterizing the corresponding transition state whose optimised geometry is shown in Fig. 2. Related DFT values of bond lengths and angles are given in Table 3. For the same molecular geometry frequency calculations were performed and the estimated saddle point energies and T_1 diagnostic are given in Table 4 for the singlet and triplet spin multiplicities.

After determining the transition states TS1 and TS2 a forward minimal energy path calculation was performed at the BB1K/aug-cc-pVTZ level of theory to the end of obtaining the product's minimum energy configuration. For illustrative purposes we show in Table 5 a sample of the Minimal Energy Path (MEP) points of process (1). In the first column the difference between O3 and H6 (Δr_{O3-H6}) and in the second column the difference between II6 and C1 (Δr_{H6-C1}) internuclear distances (see Fig. 1 for the meaning of the used symbols) are given. The angle formed by the related bonds and the associated energy difference from the transition state are given in columns three and four, respectively. The values of Table 5 were used to the end of guiding the fit of the global PES (e.g. the computed overall energy gap between the transition state and the lowest product point is 0.508 eV). In order, however, to evaluate the true asymptotic energetic balance we calculated CCSD(T) single point energies for reactants, transition state and products. The corresponding correlation diagram for process (1) and (2) are given in Fig. 3. From the Figure it is apparent that the true asymptotic difference between the transition state TS1 and the product configuration of process (1) is 0.74 eV.

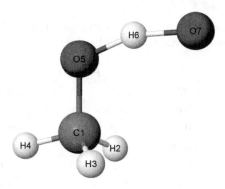

Fig. 2. Geometry of the transition state TS2 for the second reactive channel.

Table 3. Characteristics of the second reactive channel transition state TS2 (bond distances in Å and angles in degree).

O7-H6	1.170
H6-O5	1.137
∠O5-H6-O7	160.49
∠C1-O5-H6	114.60
ϕH2-C1-O5-H6	−62.72

Table 4. Coupled Cluster energies (in au) and diagnostic T_1 vector for the singlet and triplet spin multiplicities.

$2S+1$	E	T_1
3	−190.51752849	0.0224
1	−190.48659175	0.0249

3   The Fitting of the Potential Energy Surface

In order to carry out preliminary dynamical computations for the considered process (1) out of the fifteen internal degrees of freedom (DOF's) of the system we focused on those of the initial and final bond by mapping the interaction of the seven body problem into that of a three body one. This means that we freeze all the internuclear distances but H6-C1, O3-H6 and O3-C1 (or equivalently the planar angle ∠ O3-H6-C1 (see Fig. 1)). Therefore, the H7, O2, H5, H4 and C1 atoms have been considered hereinafter as compacted into a single pseudo atom (called X and coincident with C1), having a mass of ∼ 31 AMU (the sum of the atomic masses of the compacted atoms). In this way, the reactive process (1) can be formulated as:

$$O + HX \rightarrow OH + X$$

Table 5. Minimal Energy Path properties of Process (1) (see Fig. 1 for the meaning of the symbols), given as a difference from TS1 values

$\Delta r_{\text{H6-C1}}$/Å	$\Delta r_{\text{O3-H6}}$/Å	$O3 - \widehat{H6} - C1$/deg	E_{MEP}/eV
0.000	0.000	0.00	0.000
0.054	−0.055	−0.35	−0.013
0.109	−0.110	−0.29	−0.055
0.165	−0.165	−0.27	−0.121
0.222	−0.218	−0.27	−0.194
0.273	−0.262	−0.40	−0.246
0.306	−0.278	−0.40	−0.275
0.332	−0.284	−0.40	−0.300
0.356	−0.287	−0.40	−0.321
0.379	−0.290	−0.72	−0.339
0.402	−0.293	−0.72	−0.357
0.423	−0.293	−1.13	−0.373
0.445	−0.296	−1.13	−0.387
0.465	−0.296	−1.62	−0.400
0.486	−0.297	−1.62	−0.412
0.508	−0.299	−2.04	−0.423
0.528	−0.299	−2.56	−0.434
0.549	−0.301	−3.05	−0.443
0.569	−0.300	−3.61	−0.451
0.591	−0.303	−4.09	−0.459
0.610	−0.301	−4.09	−0.466
0.631	−0.303	−4.48	−0.472
0.652	−0.304	−4.92	−0.478
0.672	−0.304	−5.32	−0.483
0.693	−0.305	−5.76	−0.487
0.713	−0.305	−6.10	−0.491
0.734	−0.306	−6.56	−0.494
0.753	−0.306	−6.83	−0.497
0.775	−0.308	−7.26	−0.500
0.795	−0.308	−7.26	−0.502
0.815	−0.308	−7.67	−0.504
0.836	−0.308	−7.91	−0.505
0.856	−0.309	−8.21	−0.506
0.875	−0.309	−8.40	−0.507
0.894	−0.309	−8.40	−0.508

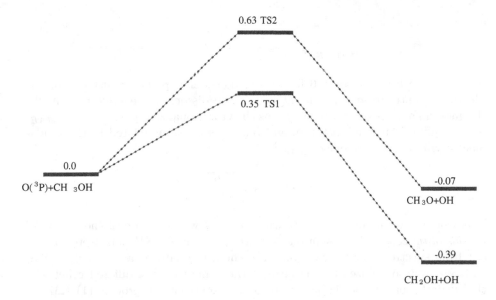

Fig. 3. CCSD(T)/aug-cc-pVTZ energy diagram in eV, without ZPE correction.

where O coincides with O2 and H with H6, while Θ (the $\widehat{XHO}$ angle) is free to vary. Furthermore, DFT optimized geometries data for TS1 show that during the reactive collision of process (1) the $\angle$ O3-H6-C1 angle is always very close to $\Theta = 180°$ (see Fig. 1), allowing us to adopt in the second step of the GEMS scheme for the related PES a LEPS formulation:

$$V(r_{\text{OH}}, r_{\text{HX}}, r_{\text{OX}}) = Q_{\text{OH}} + Q_{\text{HX}} + Q_{\text{AC}} - \frac{1}{2}\left(\Delta^2_{OH-HX} + \Delta^2_{HX-XO} + \Delta^2_{XO-OH}\right)^{1/2} \tag{4}$$

where

$$\Delta_{AB-BC} = \frac{J_{AB}}{1 + S_{AB}} - \frac{J_{BC}}{1 + S_{BC}} \tag{5}$$

leveraging on the well known bias of the LEPS PESs for collinear reactive encounters [10,11]. In Eqs. 4 and 5

$$\frac{Q_{AB}}{1 + S_{AB}} = \frac{1}{2}\left(^1\varepsilon_{AB} + \frac{1 - S_{AB}}{1 + S_{AB}}{}^3\varepsilon_{AB}\right) \tag{6}$$

$$\frac{J_{AB}}{1 + S_{AB}} = \frac{1}{2}\left(^1\varepsilon_{AB} - \frac{1 - S_{AB}}{1 + S_{AB}}{}^3\varepsilon_{AB}\right) \tag{7}$$

The parameters for the Morse

$$^1\varepsilon_{AB}(r) = D_{e(AB)}\left[e^{-2\beta_{AB}(r - r_{e(AB)})} - 2e^{-\beta_{AB}(r - r_{e(AB)})}\right] \tag{8}$$

and anti-Morse

$$^3\varepsilon_{AB}(r) = D_{e(AB)} \left[e^{-2\beta_{AB}(r-r_{e(AB)})} - 2e^{+\beta_{AB}(r-r_{e(AB)})} \right] \tag{9}$$

functions are listed in Table 6 for all the considered pseudo diatoms. For the Morse and anti-Morse functions $D_{e(AB)}$ is the dissociation energy, namely the thermodynamic dissociation energy plus the vibrational zero-point energy, $r_{e(AB)}$ is the equilibrium bond distance, and β_{AB} is a constant related to the fundamental vibrational frequency $\omega_{e(AB)}$:

$$\beta_{AB} = \omega_{e(AB)} \sqrt{\frac{\mu_{AB}}{2D_{e(AB)}}} \tag{10}$$

It is worth pointing out here that while for XO we made use of the results of the *ab initio* calculations mentioned above, for OH and XH data were worked out from spectroscopic information [12]. Then, the Sato parameters S_{OH}, S_{HX} and S_{XO} were optimized to minimize at the same time the difference between the *ab initio* barrier height ($E_a = 0.35$ eV, exoergicity of process (1) ($\Delta H = -0.39$ eV), the geometry of the transition state and the LEPS ones. As a matter of fact the collinear barrier to reaction calculated on the LEPS PES differs from the *ab initio* one of 0.11 eV. The isoenergetic contours for the collinear geometry are plotted in Fig. 4. The plot singles out the *early character* of the barrier to products with the saddle point located at $r_{OH} = 1.598$ Å and $r_{HX} = 1.142$ Å. The exit channel exhibits a small well (V_{wd}) deep 0.66 eV. Further information on the structure of the PES are given in Fig. 5 by plotting the fixed angle MEPs for different values of Θ. As shown by the figure, in agreement with *ab initio* calculations, the most favoured MEP is collinear ($\Theta = 180°$) and as Θ deviates from collinearity (i.e., decreases) the barrier gets larger. The MEPs show also a well located in the collinear exit channel, that smooths down as Θ decreases. This well disappears at $\Theta \simeq 120°$ and the MEP becomes completely repulsive at even smaller values of Θ.

Table 6. Parameters for LEPS's Morse functions.

	OH	HX	XO
μ (amu)	0.948	0.976	10.560
ω_e (cm^{-1})	3737.8	2858.5	1033.0
r_e (Å)	0.97	1.09	1.32
D_e (eV)	4.64	4.25	3.97
β (Å^{-1})	2.29	1.86	2.29
S	−0.127	0.923	−0.559

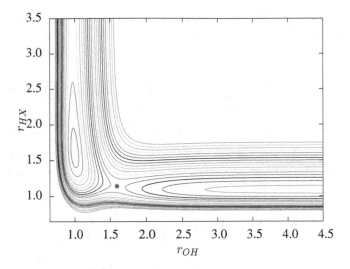

Fig. 4. The collinear isoenergetic contours of the OHX LEPS PES (spacing is 0.13 eV) plotted as a function of the OH and HX internuclear distances (given in Å). The red dot locates the saddle point. (Color figure online)

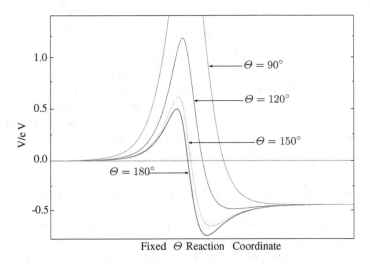

Fig. 5. Minimum Energy Paths for different approaching angles

4   Dynamical Calculations and Probabilities

In order to single out the main features of the dynamics of process (1) we carried out on the assembled LEPS, according to the third step of the GEMS scheme, some quantum dynamics calculations. To this end, we used the MCTDH (Multi Configuration Time Dependent Hartree) method [13,14], that represents the system wavefunction Ψ as a sum of multi configuration Hartree products (allowing to handle highly correlated degrees of freedom) as follows:

$$\Psi(Q_1,\ldots,Q_f,t) = \sum_{j_1}^{n_1} \cdots \sum_{j_f}^{n_f} A_{j_1\ldots j_f}(t) \prod_{k=1}^{f} \varphi_{j_k}^{(\kappa)}(Q_\kappa,t) = \sum_J A_J \Phi_J \qquad (11)$$

where $Q_1 \ldots Q_f$ are mass weighted Jacobi coordinates, $A_{j_1\ldots j_f}$ are the MCTDH expansion coefficients, and the $\varphi_{j_k}^{(\kappa)}$ are the n_κ single-particle functions (spf) for the κ degree of freedom, usually represented by a finite number N_κ of DVR or FFT primitives. By varying the coefficients one obtains:

$$i\dot{A}_J = \sum_L \langle \Phi_J| H |\Phi_L\rangle A_L - \sum_{\kappa=1}^{f} \sum_{l=1}^{n_\kappa} g_{j_\kappa l}^{(\kappa)} A_{J_l^\kappa} \qquad (12)$$

while by varying the single particle functions one obtains:

$$i\dot{\boldsymbol{\varphi}}^{(\kappa)} = g^{(\kappa)} \mathbf{1}_{n_k} \boldsymbol{\varphi}^{(\kappa)} + (1 - P^{(\kappa)})[(\boldsymbol{\rho}^{(\kappa)})^{-1}\langle \boldsymbol{H}\rangle^{(\kappa)} - g^{(\kappa)} \mathbf{1}_{n_k}]\boldsymbol{\varphi}^{(\kappa)} \qquad (13)$$

where $P^{(\kappa)} = \sum_{j=1}^{n_{(\kappa)}} |\varphi_j^{(\kappa)}\rangle\langle\varphi_j^{(\kappa)}|$ is the projection operator for the κ degree of freedom, $\rho_{jl}^{(k)}$ is the density matrix element $\sum_{\bar{J}}^{\kappa} = A_{J_j^\kappa}^* A_{J_l^\kappa}$, $g_{jl}^{(k)}$ is the constraint operator $\langle\varphi_j^{(\kappa)}|g^{(\kappa)}|\varphi_l^{(\kappa)}\rangle$ and $\langle\hat{\boldsymbol{H}}\rangle$ the Hamiltonian operator. The number n_κ is a very important computational parameter because, as we can see from Eq. 11, it determines the size of the basis function used to expand the system wavefunction.

In the collinear case, as is our calculations, the Hamiltonian operator is the sum of a kinetic operator for the translational mass scaled Jacobi coordinate Q $(\frac{\partial^2}{\partial Q^2})$, a kinetic operator for the internal Jacobi coordinate q $(\frac{\partial^2}{\partial q^2})$ of the reactant HX, plus two complex absorption potentials, $W_\alpha(Q)$ and $W_\beta(q)$, respectively:

$$\hat{H} = -\frac{1}{2\mu}\frac{\partial^2}{\partial Q^2} - \frac{1}{2\mu}\frac{\partial^2}{\partial q^2} + V(Q,q) - iW_\alpha(Q) - iW_\beta(q) \qquad (14)$$

The integration of the above given MCTDH propagation equations is then carried out in time after placing the reactant system wavepacket in the near asymptotic region (large Q values) of the interaction, as shown in Fig. 6. In the Figure the LEPS PES is skewed by an angle $\beta \sim 17°$ due to the Heavy-Light-Heavy nature of the OHX system. The small value of the skewing angle β also indicates that the two degrees freedom Q and q are highly correlated. The initial wavepacket was propagated for 1040 fs, with a Δt of 4.0 fs, and an energy interval ranging between 0.9 and 1.4 eV. The calculations reported here were performed using the Heidelberg MCTDH Package [13], in which concurrent computations are performed by distributing different initial states runs on the Grid while the single initial state runs were restructured for running on GPUs.

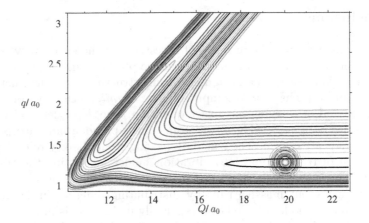

Fig. 6. Example of an initial wavepacket used in MCTDH, placed in the asymptotic region of the LEPS surface, at $Q_0 = 20\ a_0$ and $\Delta Q = 0.55$.

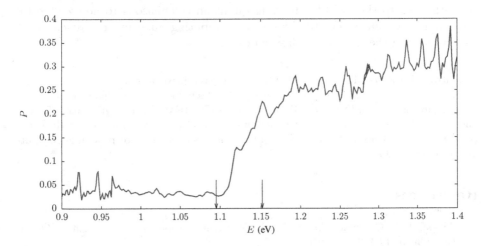

Fig. 7. State specific reaction probabilities $P_0(E)$. Left arrow is the 3rd vibrational energy of the product fragment OH, right arrow is the 3rd vibrational energy of the reagent XH.

MCTDH propagation, are shown in Fig. 7, where the $v = j = 0$ zero null total angular momentum state specific reaction probability P_0 is plotted against total energy. The value of the calculated probability is of the order of 10^{-2} for total energy lower than 1.1 eV. There, it exhibits a sudden increase (1.1 eV is about the energy of the 3rd vibrational level of the OH product), suggesting that the threshold energy is associated with a conversion of the reactant translational energy into the vibration of the product.

5   Conclusions

The present paper tackles the problem of investigating the elementary combustion process of methanol reacting with oxygen to produce OH. In order to carry out the calculations the geometry and electronic structure of the fragments and of the transition states have been computed using *ab initio* techniques of different levels. To the end of carrying out exact quantum time dependent calculations the problem has been modeled as a two atoms plus a pseudoatom (incorporating the CH_2OH radical) system. The calculations single out the key role played by the conversion of translational into vibration modes for setting the threshold to reaction. The calculations show also the possibility of exploiting concurrent platforms to carry out massive runs by distributing different initial reactant states runs over the nodes of the Grid, especially if the code is implemented on GPUs. In order to better characterize the dynamics of the methanol combustion process further work will be made with the purpose of building a more reliable global PES taking into account all the atoms of the system. Further work will be performed to the end of extending quantum calculations to non zero total angular momentum values to the end of computing quantum rate coefficients and comparing them with the classical ones.

Acknowledgments. LP thanks "Fondazione Cassa di Risparmio di Perugia (Codice Progetto: 2014.0253.021 Ricerca Scientifica e Tecnologica" for financial support. Thanks are also due to INSTM, IGI and the COMPCHEM virtual organization for the allocation of computing time. The Supercomputing Center for Education & Research (OSCER) at the University of Oklahoma (OU) is acknowledged for providing computing resources and services.

References

1. Alves, M.M., Carvalho, E.F.V., Machado, F.B.C., Roberto-Neto, O.: Int. J. Quantum Chem. **110**, 2037 (2010)
2. Lu, C.-W., Chou, S.-L., Lee, Y.-P., Xu, S., Xu, Z.F., Lin, M.C.: J. Chem. Phys. **122**, 244314 (2005)
3. Carr, S.A., Blitz, M.A., Seakins, P.W.: Chem. Phys. Lett. **511**, 207 (2011)
4. Mani Sarathy, S., Oßwald, P., Hansen, N., Kohse-Höinghaus, K.: Prog. Energy Combust. Sci. **44**, 40 (2014)
5. Aronowitz, D., Naegeli, D.W., Glassman, I.: J. Phys. Chem. **81**, 2555–2559 (1977)
6. Laganà, A.: Towards a grid based universal molecular simulator. In: Laganà, A., Lendvay, G. (eds.) Theory of Chemical Reaction Dynamics. NATO Science Series II: Mathematics, Physics and Chemistry, vol. 145, pp. 363–380. Springer Netherlands (2005). ISBN 978-1-4020-2055-1

7. Frisch, M.J., Trucks, G.W., Schlegel, H.B., Scuseria, G.E., Robb, M.A., Cheese-
man, J.R., Scalmani, G., Barone, V., Mennucci, B., Petersson, G.A., Nakatsuji,
H., Caricato, M., Li, X., Hratchian, H.P., Izmaylov, A.F., Bloino, J., Zheng, G.,
Sonnenberg, J.L., Hada, M., Ehara, M., Toyota, K., Fukuda, R., Hasegawa,
J., Ishida, M., Nakajima, T., Honda, Y., Kitao, O., Nakai, H., Vreven, T.,
Montgomery Jr., J.A., Peralta, J.E., Ogliaro, F., Bearpark, M., Heyd, J.J., Brothers,
E., Kudin, K.N., Staroverov, V.N., Kobayashi, R., Normand, J., Raghavachari, K.,
Rendell, A., Burant, J.C., Iyengar, S.S., Tomasi, J., Cossi, M., Rega, N., Millam, J.M.,
Klene, M., Knox, J.E., Cross, J.B., Bakken, V., Adamo, C., Jaramillo, J., Gomperts,
R., Stratmann, R.E., Yazyev, O., Austin, A.J., Cammi, R., Pomelli, C., Ochterski,
J.W., Martin, R.L., Morokuma, K., Zakrzewski, V.G., Voth, G.A., Salvador, P., Dannenberg, J.J., Dapprich, S., Daniels, A.D., Farkas, O., Foresman, J.B., Ortiz, J.V.,
Cioslowski, J., Fox, D.J.: Gaussian09 Revision D.01 Gaussian Inc., Wallingford CT
2009
8. Zhao, Y., Lynch, B.J., Truhlar, D.G.: J. Phys. Chem. A **108**, 2715 (2004)
9. Lee, T.J., Taylor, P.R.: Int. J. Quantum Chem. Quant. Chem. Symp. **S23**, 199
(1989)
10. Sato, S.: J. Chem. Phys. **23**, 592 (1955)
11. Sato, S.: J. Chem. Phys. **23**, 2465 (1955)
12. Huber, K.P., Herzberg, G.: Molecular spectra and molecular structure IV. In: Constants of Diatomic Molecules, vol. 4. Springer Science+Business media, LLC (1979)
13. Beck, M.H., Jäckle, A., Worth, G.A., Meyer, H.D.: Phys. Rep. **324**, 1 (2000)
14. Meyer, H.D., Gatti, F., Worth, G.: Multidimensional Quantum Dynamics,
MCTDH Theory and applications. Wiley-VCH, Weinheim (2009)

Point Placement in an Inexact Model
with Applications

Kishore Kumar V. Kannan, Pijus K. Sarker,
Amangeldy Turdaliev, and Asish Mukhopadhyay[(✉)]

School of Computer Science, University of Windsor, Windsor, ON N9B 3P4, Canada
{varadhak,sarkerp,turdalia,asishm}@uwindsor.ca

Abstract. The point placement problem is to determine the locations
of n distinct points on a line uniquely (up to translation and reflection)
by making the fewest possible pairwise distance queries of an adversary.
A number of deterministic and randomized algorithms are available when
distances are known exactly. In this paper, we discuss the problem in an
inexact model. This is when distances returned by the adversary are
not exact; instead, only upper and lower bounds on the distances are
provided. We propose an algorithm called DGPL for this problem that
is based on a distance geometry approach that Havel [8] used to solve
the molecular conformation problem. Our algorithm does not address the
problems of query choices and their minimization; these remain open. We
have used our DGPL algorithm for the probe location problem in DNA
mapping, where upper and lower bounds on distance between some pairs
of probes are known. Experiments show the superior performance of our
algorithm compared to that of an algorithm by Mumey [9] for the same
problem.

Keywords: Probe location problem · Distance geometry · Point
placement · Bound smoothing · Embed algorithm · Eigenvalue decomposition

1   Introduction

Retrieving the coordinates of n points $P = \{p_0, p_2, \ldots, p_{n-1}\}$ from their mutual
distances is a problem that is interesting both from a theoretical as well as a
practical point of view. Young and Householder [10] showed that a necessary and
sufficient condition is that the matrix $B = [b_{ij}] = [(d_{i0}^2 + d_{j0}^2 - d_{ij}^2)/2]$ is positive
semi-definite, where d_{ij} is the Euclidean distance between p_i and p_j. The rank of
B also gives the minimum dimension of the Euclidean space in which the point
set P can be embedded.

In another variation of the problem, given a set of mutual distances and an
embedding dimension k, can P be embedded in Euclidean space E^k to realize the
given distances [11]? This problem is strongly NP-complete even when $k = 1$ and
the distances are restricted to the set $\{1, 2, 3, 4\}$. It is, however, possible to solve

© Springer International Publishing Switzerland 2016
O. Gervasi et al. (Eds.): ICCSA 2016, Part I, LNCS 9786, pp. 84–96, 2016.
DOI: 10.1007/978-3-319-42085-1_7

this problem for special classes of graphs and the embedding dimension is 1 [12]. The edges of such graphs join pairs of points for which the distances are known.

We, and other researchers, extensively studied **the point placement problem** [2,3,6,13,14], which is to determine an unique (up to translation and reflection) one-dimensional embedding of n distinct points by making the fewest possible pairwise distance queries of an adversary. The queries, spread over one or more rounds, are modeled as a graph whose vertices represent the points and there is an edge connecting two vertices, if the distance between the corresponding points is queried. The goal is to keep the number of queries linear, with a constant factor as small as possible. This requires meeting a large number of constraints on the pairwise distances, making the resulting algorithms very intricate [15].

The simplest of all, the 3-cycle algorithm, has the following query graph (Fig. 1).

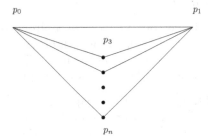

Fig. 1. Query graph using triangles

If $G = (V,E)$ is a query graph, an assignment l of lengths to the edges of G is said to be valid if there is a placement of the vertices V on a line such that the distances between adjacent vertices are consistent with l. Here in this problem, the distance between a pair of vertices returned by the adversary are exact. The algorithm designer tries to construct a graph over fixed number of rounds to minimize the number of edge queries and also make sure that there is a unique placement of the vertices. The construction of such a graph is the heart of different algorithms for this problem.

In this paper, we study a one-round version of the point placement problem in the adversarial model when the query distances returned by the adversary are not necessarily exact. A formal definition of the problem goes as follows.

Problem Statement: Let $P = \{p_0, p_1, ..., p_{n-1}\}$ be a set of n distinct points on a line. For a pair of points p_i and p_j, the query distance $D(p_i, p_j)$ lies in an interval $[l_{ij}, u_{ij}]$, where $l_{ij} \leq u_{ij}$ denote the lower and upper bound respectively. The objective is to find a set of locations of the points in P that are consistent with the given distance bounds.

The distance bounds are collectively represented by an upper distance matrix, U, for the upper bounds and a lower distance matrix, L, for the lower bounds.

The corresponding entries in the upper and lower bound matrices for pairs of points for which $l_{ij} = u_{ij}$, that is, the pairwise distances are exactly known, are identical. When the distance between a pair of points is unknown, the distance interval bound is set to $[-\infty, \infty]$.

For example with three points ($n = 3$), a possible pair of upper and lower bound distance matrices are shown below:

$$U(p_0, p_1, p_2) = \begin{pmatrix} 0 & 60 & \infty \\ 60 & 0 & 3 \\ \infty & 3 & 0 \end{pmatrix}$$

$$L(p_0, p_1, p_2) = \begin{pmatrix} 0 & 60 & -\infty \\ 60 & 0 & 3 \\ -\infty & 3 & 0 \end{pmatrix}$$

Thus, $l_{01} = u_{01} = 60$ for the pair of points p_0 and p_1, whereas $l_{02} = -\infty$ and $u_{02} = \infty$ for the pair p_0 and p_2.

Motivation: This study is motivated by a problem from computational biology, where probe locations are to be mapped on a chromosome, given distance estimates between probe pairs that are obtained from FISH experiments [16–18]. This is known in the literature as the probe location problem.

Overview of Contents: The rest of the paper is organized thus. In the next section, we show how the point placement problem in the inexact model can be formulated in the distance geometry framework. Based on this formulation, we propose an algorithm, called DGPL (short for Distance Geometry based Probe Location). We follow up with an analysis of some experimental results (1-dimensional layouts) obtained from an implementation of our DGPL algorithm. In the third section, we briefly review an algorithm by Mumey [9] for the probe location problem. In the next section, experimental results are given, comparing the performance of the DGPL algorithm with that of Mumey's. In the final section, we conclude with some observations and directions for further research.

2   The Distance Geometry Approach

The fundamental problem of distance geometry is this: "Given a set of m distance constraints in the form of lower and upper bounds on a (sparse) set of pairwise distance measures, and chirality constraints on quadruples of points, find all possible embeddings of the points in a suitable k-dimensional Euclidean space" [1,19].

Crippen and Havel's EMBED algorithm for the molecular conformation problem is based on a solution to the above fundamental problem. Let d_{ij} denote the distance between a pair of points p_i and p_j; then each of the m constraints above specifies an upper and lower bound on some d_{ij}. For our 1-dimensional point placement problem, chirality constraints do not come into the picture. Determining a feasible set of locations of the set of points in P is equivalent to finding the coordinates of these locations, with reference to some origin of coordinates. Thus the approach used for the EMBED algorithm can be directly adapted to the solution of our problem. Due to the non-availability of an implementation of the EMBED algorithm, we wrote our own.

Below, we describe the DGPL algorithm, on which our implementation is based. Before this, the so-called bound-smoothing algorithm used in the EMBED algorithm deserves some mention. The main underlying idea is to refine distance bounds into distance limits. Thus if $[l'_{ij}, u'_{ij}]$ are the distance limits corresponding to distance bounds $[l_{ij}, u_{ij}]$, then $[l'_{ij}, u'_{ij}] \subset [l_{ij}, u_{ij}]$. The distance limits are defined as follows.

Let l_{ij} and u_{ij} be the upper and lower distance bounds on d_{ij}. Then

$$l'_{ij} = \inf_e \{d_{ij} | l_{ij} \leq d_{ij} \leq u_{ij}\}$$

and

$$u'_{ij} = \sup_e \{d_{ij} | l_{ij} \leq d_{ij} \leq u_{ij}\},$$

where the inf and sup are taken over all possible embeddings $e : P \times P \to R$ of the points in P on a line.

When this is done, assuming that the triangle inequality holds for the triplet of distances among any three points p_i, p_j, p_k, then these distance limits are called triangle limits.

In [20], a characterization of the triangle limits was established. It was also shown how this could be exploited to compute the triangle limits, using a modified version of Floyd's shortest path algorithm [15].

2.1   Algorithm DGPL

An input to the DGPL program for a set of n points in its most general form consists of the following: (a) exact distances between some pairs of points; (b) finite upper and lower bounds for some other pairs; (c) for the rest of the pairs, the distance bounds are not specified. However, as we have conceived of DGPL as a solution to the point placement problem in the inexact model, the input is assumed to be adversarial and is simulated by generating an initial adversarial layout of the points. This layout is used by the algorithm to generate input data of types (a) and (c); however, DGPL will also work if some of the input data are of type (b).

The DGPL algorithm works in three main phases, as below.

Phase 1: [Synthetic Data Generation] Based on the user-input for n, the algorithm simulates an adversary to create a layout of n points; further, based on the user-input for the number of pairs for which mutual distances are not known, the algorithm assigns the distance bounds $[-\infty, \infty]$ for as many pairs and uses the layout created to assign suitable distance bounds for the rest of the pairs; the output from this stage are the lower and upper bound distance matrices, L and U.

Phase 2: [Point location or Coordinate generation] A set of coordinates of the n points, consistent with the bounds in the matrices L and U, are computed, following the approach in [8] for the molecular conformation problem.

Phase 3: [Layout generation] Finally, a layout of the embedding is generated, which allows us to verify how good the computed layout is vis-a-vis the adversarial layout.

A more detailed description of the algorithm is given below.

Algorithm 1. DGPL

Data: 1. The size n of P.
         2. The number of pairs for which $l = -\infty$ and $u = \infty$.
         3. Embedding dimension

Result: Locations of the n points in P, consistent with the distance bounds

(1) Create a random valid layout of the points p_i in $P = \{p_0, p_1, ..., p_{n-1}\}$.

(2) Create distance bound matrices L and U.

    (2.1) Assign $-\infty$ and ∞ respectively to the corresponding values in the L and U matrices for as many pairs as the the number unknown distances (user-specified).

    (2.2) Assign the distances determined from the layout of Step 1 to the remaining entries of both L and U

(3) Apply a modified version of Floyd's shortest path algorithm [8] to compute triangle limit matrices, LL and UL, from the distance bound matrices, L and U.

(4) If all triangle limit intervals have been collapsed to a single value (which is now the distance between the corresponding pair of points) go to Step 5, else collapse any remaining pair of triangle limit interval to a randomly chosen value in this interval and go to Step 3 (this step is called metrization)

(5) Compute matrix $B = [b_{ij}] = [(d_{i0}^2 + d_{j0}^2 - d_{ij}^2)/2]$, where d_{ij} is the distance between points p_i and p_j, $1 \leq i, j \leq n - 1$ [10].

(6) Compute the eigenvalue decomposition of the matrix B; the product of the largest eigenvalue with its corresponding normalized eigenvector gives the one-dimensional coordinates of all the points.

(7) The computed coordinates are plotted on a line, placing p_0 at the origin of coordinates, and are compared with the adversarial layout for the extent of agreement.

In the next section, we discuss the simulation of DGPL for a small value of n.

2.2 Simulating DGPL

1. Let the input to the program be as follows:
Number of points: 4
Number of unknown distances: 1
Embedding dimension: 1.

2. The following random one-dimensional layout of the four points is created:

$$p_0 = 0, p_1 = 59, p_2 = 48, p_3 = 74.$$

3. The input lower and upper bound distance matrices, L and U, are set to:

$$L = \begin{pmatrix} 0 & 59 & 48 & 74 \\ 59 & 0 & 11 & -\infty \\ 48 & 11 & 0 & 26 \\ 74 & -\infty & 26 & 0 \end{pmatrix}, U = \begin{pmatrix} 0 & 59 & 48 & 74 \\ 59 & 0 & 11 & \infty \\ 48 & 11 & 0 & 26 \\ 74 & \infty & 26 & 0 \end{pmatrix},$$

where the distance d_{13} between the points p_1 and p_3 is contained in the interval $[-\infty, \infty]$.

4. The lower and upper triangle limit matrices, LL and UL, obtained by bound smoothing from the lower and upper distance bound matrices are:

$$LL = \begin{pmatrix} 0 & 59 & 48 & 74 \\ 59 & 0 & 11 & 15 \\ 48 & 11 & 0 & 26 \\ 74 & 15 & 26 & 0 \end{pmatrix}, UL = \begin{pmatrix} 0 & 59 & 48 & 74 \\ 59 & 0 & 11 & 37 \\ 48 & 11 & 0 & 26 \\ 74 & 37 & 26 & 0 \end{pmatrix}.$$

5. The distance limits are then metrized, as discussed in Step 4 of the DGPL algorithm, to fixed distances as:

$$D = \begin{pmatrix} 0 & 59 & 48 & 74 \\ 59 & 0 & 11 & 35 \\ 48 & 11 & 0 & 26 \\ 74 & 35 & 26 & 0 \end{pmatrix}.$$

6. Relative to the point p_0 as the origin of coordinates, the matrix B computes to:

$$B = \begin{pmatrix} 3481 & 2832 & 3866 \\ 2832 & 2304 & 3552 \\ 3866 & 3552 & 5476 \end{pmatrix}.$$

7. From the eigenvalue decomposition of the matrix B, we obtain the following coordinate matrix as the product of the square root of largest eigenvalue with its corresponding normalized eigenvector:

$$X = \begin{pmatrix} 58.677 \\ 48.828 \\ 73.671 \end{pmatrix}.$$

The 1-dimensional embedding of the computed coordinates, after rounding to integer values, is shown in Fig. 2; the agreement with the adversarial layout is seen to be near-perfect.

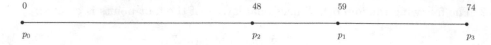

0   48   59   74

p_0 p_2 p_1 p_3

Fig. 2. Final embedding of the four input points

3   Mumey's Approach to the Probe Location Problem

Mumey [9] considered the problem of mapping probes along a chromosome based on separation or distance intervals between probe pairs, estimated from *fluorescence in-situ hybridization (FISH)* experiments. He named this as the probe location problem. The problem is challenging as the distance intervals are known only with some confidence level, some may be error-prone and these need to be identified to find a consistent map. Previous algorithmic approaches based on seriation [16], simulated annealing [17] and branch and bound algorithm [18], relying as these did on exhaustive search, could handle inputs of only up to 20 or fewer probes efficiently. However, Mumey's algorithm claims to be able solve the problem for up to 100 probes in a matter of few minutes. We briefly review Mumey's algorithm next.

From the above discussion it is clear that Mumey's problem can also be conveniently cast in the framework of distance geometry. Indeed, we have done this and the results are discussed in the next section.

3.1   Overview of Mumey's Algorithm

Let $P = \{p_0, p_1, ..., p_{n-1}\}$ be the list of probes in a chromosome, it being given that the distance interval between a pair of probes p_i and p_j lie in an interval $[l_{ij}, u_{ij}]$, where l_{ij} and u_{ij} are respectively the lower and upper bounds on the distance. The probe location problem is to identify a feasible set of probe locations $\{x_0, x_1, ..., x_{n-1}\}$ from the given distance intervals such that $|x_i - x_j| \in [l_{ij}, u_{ij}]$.

The distance bounds on probe pairs leads to a special kind of linear program, which is a system of difference constraints. A directed graph derived from these constraints is input to Bellman-Ford's single source shortest path algorithm. The shortest paths from the specially-added source vertex to all the destination vertices is a solution to the system of difference constraints [15].

For the probe location problem, the directed graph in question is called an edge orientation graph, whose construction is described next.

Edge Orientation Graph. The first step in the construction of an edge orientation graph is to set the orientation of each edge by choosing one placement. If x_i and x_j are the positions of probes i and j then one of the following is true:

$$x_j - x_i \in [l_{ij}, u_{ij}] \tag{1}$$

$$x_j - x_i \in [l_{ij}, u_{ij}] \tag{2}$$

If (1) holds then x_i is to the left of x_j whereas if (2) holds, then x_j is to the left of x_i. Assume that (1) holds. We can express this equivalently as.

$$l_{ij} \le x_j - x_i \le u_{ij}$$

or as:

$$x_j - x_i \le u_{ij} \tag{3}$$
$$x_i - x_j \le -l_{ij} \tag{4}$$

Corresponding to the two inequalities above, we have two edges in the edge orientation graph, one going from x_i to x_j, with weight u_{ij} and the other from x_j to x_i with weight $-l_{ij}$.

Similarly, if (2) holds, we can express this equivalently as:

$$l_{ij} \le x_i - x_j \le u_{ij}$$

or as:

$$x_i - x_j \le u_{ij} \tag{5}$$
$$x_j - x_i \le -l_{ij} \tag{6}$$

with two directed edges in the edge orientation graph of weights u_{ij} and $-l_{ij}$ respectively.

When $l_{ij} = u_{ij}$ for a pair of probes p_i and p_j, there is exactly one edge connecting the corresponding nodes in the edge orientation graph, its orientation determined by the relative linear order of the probes.

Finding Feasible Probe Positions. Once all the edge weights are fixed, a special source vertex is added whose distance to all other vertices is initialized to 0 and Bellman-Ford's algorithm is run on this modified edge orientation graph. If there is no negative cycle in the graph, Bellman-Ford's algorithm outputs a set of feasible solutions $(x_0, x_1, ..., x_{n-1})$. Otherwise, the algorithm resets the edge weights by changing the relative order of a probe pair and re-runs the Bellman-Ford algorithm. These two steps are repeated till a feasible solution is found.

4   Experimental Results

We implemented both the DGPL algorithm and Mumey's, discussed in the previous sections in Python 2.7. The programs were run on a computer with the following configuration: Intel(R) Xeon(R) CPU, X7460@2.66GHz OS: Ubuntu 12.04.5, Architecture:i686. We used the mathematical package numpy.linAlg to calculate eigenvalue decompositions and also for solving linear equations; the package matplotlib.pyplot was used to plot an embedding of final coordinates obtained from the programs in a space of specified dimensionality.

The chart below compares the running time of Mumey's algorithm with that of DGPL. Each of these algorithms were run on point sets of different sizes, up to 101 points. We also recorded the effect on both algorithms of the number of pairs of points (probes) for which the distances are not known or are unspecified. In these cases, we set $l_{ij} = -\infty$ and $u_{ij} = \infty$ (Table 1).

Table 1. Performance comparison of Mumey's and DGPL algorithm

No. of points	No. of unknown distances	Mumey's approach running time (hrs:mins:secs)	DGPL algorithm running time (hrs:mins:secs)
3	1	0:00:00.000184	0:00:00.001514
10	2	0:00:00.001339	0:00:00.006938
10	5	0:00:00.024560	0:00:00.006816
10	8	0:00:00.060520	0:00:00.017163
20	2	0:00:00.001369	0:00:00.007464
20	5	0:00:00.001336	0:00:00.007743
20	10	0:00:01.164363	0:00:00.007436
40	5	0:00:00.947250	0:00:00.328563
40	8	0:00:07.369925	0:00:00.315001
40	10	0:00:30.857658	0:00:00.312674
80	5	0:00:10.609233	0:00:02.503798
80	10	0:06:15.443501	0:00:02.496285
80	15	5:00:00.000000+	0:00:02.687672
101	5	0:00:14.256343	0:00:05.020695
101	10	0:10:32.299084	0:00:05.282747
101	15	5:00:00.000000+	0:00:05.192594

Interestingly, the chart (Fig. 3) shows that Mumey's approach takes a longer time when the number of unknown distances increases. Each of these algorithms were run on point sets of different sizes, up to 100 points.

Clearly, the DGPL algorithm is consistently faster; as we can see from the graph, irrespective of the number of unknown distances in the fixed number of points, DGPL's run-time is linear in the number of points. However, this is not true of Mumey's approach. This can be explained by the fact that when a negative cycle is detected Bellmann-Ford's algorithm has to be run again. This is more likely to happen as the number of unknown distances increase. Thus the running time increases rapidly. Furthermore, after each detection of a negative cycle, a new distance matrix will have to be plugged into the Bellmann-Ford algorithm to find the feasible solutions. In addition, the cost of keeping track of the distances chosen for unknown distance pairs increases with the number of such pairs.

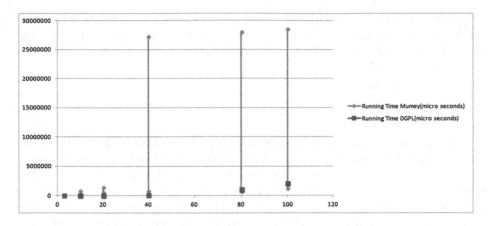

Fig. 3. Time complexity graph Mumey's vs DGPL algorithm - Increasing number of unknown distances between fixed number of points

5 Three Dimensional Embedding

The steps applied to generate the coordinates in three-dimensions using the DGPL program are slightly different from the generation of the coordinates in one-dimension. The input to DGPL are a set of upper and lower distance intervals for each pair of points, with three as the embedding dimension. Steps 1 to 5 are the same in this case also. Finally, the eigenvalue decomposition of the

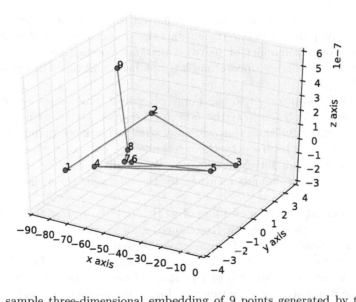

Fig. 4. A sample three-dimensional embedding of 9 points generated by the DGPL program

B matrix is obtained and the product of the three largest eigenvalues with their corresponding normalized eigenvectors yield the three-dimensional coordinates of all the points. A sample plot of a three-dimensional embedding generated by DGPL is shown in Fig. 4.

6   Conclusions

Summary of work done: The probe-location problem is an important one in Computational Genomics. In this paper we have proposed an efficient algorithm for its solution. The novelty of our contribution lies in the use of the distance geometry framework. The core of this work took the form of experiments. A representative set of results from these experiments have been presented in an earlier section. Both Mumey's algorithm (see Sect. 3) and ours were tested with different sets of points, ranging in size from 3 to 100, each with a different set of unknown distances. As the results show, the run-time of Mumey's algorithm is highly sensitive to the number of unknown distances, while that of DGPL is almost impervious to this.

Some interesting conclusions can be drawn from the experiments of the previous section. For example, with 15 unknown distances for a set 80 points, the DGPL algorithm shows a vastly superior performance as compared to that of Mumey, since the latter has to keep track of the distances chosen between each pair of unknown distances. The results have been shown in a graph with the coordinates being plotted in one-dimension. This final graph can also be used as an aid to validate the correctness of the placement of points by verifying all coordinate values against the initial layout generated by the program.

Future directions: Further work can be done on several fronts. Except for the point placement problem in one dimension, the challenging problem of finding lower bounds on the number of pairwise distances needed to embed a set of points in three and higher dimensions has not been addressed in the surveyed literature.

In our earlier approaches to the point placement problem, when the basic point-placement graph (such as the 5-cycle, 5:5 jewel etc.) was not rigid, we formulated rigidity conditions that were met over two or more rounds of queries. Now, if the distances returned by the adversary are not exact, there arises the challenging problem of satisfying the rigidity conditions, using exact lengths as well as the distance bounds returned by the adversary, over multiple query rounds.

Another important and useful line of work would be to test the DGPL program on NMR (Nuclear Magnetic Resonance) data for macromolecules, particularly protein molecules. The NMR technology is used to obtain pairwise distance data of the atoms of molecules that are not available in crystalline form. The process of obtaining coordinates of the atoms from the distance data is well-known as the molecular conformation problem. This would be a good test of the robustness of our implementation of the DGPL algorithm.

References

1. Blumenthal, L.M.: Theory and Applications of Distance Geometry. Chelsea, New York (1970)
2. Chin, F.Y.L., Leung, H.C.M., Sung, W.K., Yiu, S.M.: The point placement problem on a line – improved bounds for pairwise distance queries. In: Giancarlo, R., Hannenhalli, S. (eds.) WABI 2007. LNCS (LNBI), vol. 4645, pp. 372–382. Springer, Heidelberg (2007)
3. Damaschke, P.: Point placement on the line by distance data. Discrete Appl. Math. **127**(1), 53–62 (2003)
4. Eckart, C., Young, G.: The approximation of one matrix by another of lower rank. Psychometrika **1**(3), 211–218 (1936)
5. Malliavin, T.E., Mucherino, A., Nilges, M.: Distance geometry in structural biology: new perspectives. In: Distance Geometry, pp. 329–350. Springer, New York (2013)
6. Mukhopadhyay, A., Sarker, P.K., Kannan, K.K.V.: Randomized versus deterministic point placement algorithms: an experimental study. In: Gervasi, O., Murgante, B., Misra, S., Gavrilova, M.L., Rocha, A.M.A.C., Torre, C., Taniar, D., Apduhan, B.O. (eds.) ICCSA 2015. LNCS, vol. 9156, pp. 185–196. Springer, Heidelberg (2015)
7. Roy, K., Panigrahi, S.C., Mukhopadhyay, A.: Multiple alignment of structures using center of proteins. In: Harrison, R., Li, Y., Măndoiu, I. (eds.) ISBRA 2015. LNCS, vol. 9096, pp. 284–296. Springer, Heidelberg (2015)
8. Havel, T.F.: Distance geometry: theory, algorithms, and chemical applications. Encycl. Comput. Chem. **120**, 723–742 (1998)
9. Mumey, B.: Probe location in the presence of errors: a problem from DNA mapping. Discrete Appl. Math. **104**(1), 187–201 (2000)
10. Young, G., Householder, A.S.: Discussion of a set of points in terms of their mutual distances. Psychometrika **3**(1), 19–22 (1938)
11. Saxe, J.B.: Embeddability of weighted graphs in k-space is strongly NP-hard. In: 17th Allerton Conference on Communication, Control and Computing, pp. 480–489 (1979)
12. Mukhopadhyay, A., Rao, S.V., Pardeshi, S., Gundlapalli, S.: Linear layouts of weakly triangulated graphs. In: Pal, S.P., Sadakane, K. (eds.) WALCOM 2014. LNCS, vol. 8344, pp. 322–336. Springer, Heidelberg (2014)
13. Alam, M.S., Mukhopadhyay, A.: More on generalized jewels and the point placement problem. J. Graph Algorithms Appl. **18**(1), 133–173 (2014)
14. Alam, M.S., Mukhopadhyay, A.: Three paths to point placement. In: Ganguly, S., Krishnamurti, R. (eds.) CALDAM 2015. LNCS, vol. 8959, pp. 33–44. Springer, Heidelberg (2015)
15. Cormen, T.H., Leiserson, C.E., Rivest, R.L.: Introduction to Algorithms. The MIT Press and McGraw-Hill Book Company (1989)
16. Buetow, K.H., Chakravarti, A.: Multipoint gene mapping using seriation. i. general methods. Am. J. Hum. Genet. **41**(2), 180 (1987)
17. Pinkerton, B.: Results of a simulated annealing algorithm for fish mapping. Communicated by Dr. Larry Ruzzo, University of Washington (1993)

18. Redstone, J., Ruzzo, W.L.: Algorithms for a simple point placement problem. In: Bongiovanni, G., Petreschi, R., Gambosi, G. (eds.) CIAC 2000. LNCS, vol. 1767, pp. 32–43. Springer, Heidelberg (2000)
19. Crippen, G.M., Havel, T.F.: Distance Geometry and Molecular Conformation, vol. 74. Research Studies Press Somerset, England (1988)
20. Dress, A.W.M., Havel, T.F.: Shortest-path problems and molecular conformation. Discrete Appl. Math. **19**(1–3), 129–144 (1988)

Development and Validation of a Logistic Regression Model to Estimate the Risk of WMSDs in Portuguese Home Care Nurses

Ana C. Braga[✉] and Paula Carneiro

ALGORITMI Centre, University of Minho, 4710-057 Braga, Portugal
{acb,pcarneiro}@dps.uminho.pt

Abstract. Multivariable prediction models are being developed, validated, updated, and implemented in different domains, with the aim of assisting professionals in estimating probabilities and potentially influence their decision making. The main goal of this work was to develop and validate a logistic regression model that would predict the risk of musculoskeletal complaints involving the lumbar region in nurses working at the Health Centres of Northern Portugal and providing home-based care. The main methodology used was a questionnaire developed in electronic format, which was based on the Standardized Nordic Questionnaire for the analysis of musculoskeletal symptoms.

Internal and external validation methods were used, and their performances were compared with the ROC (Receiver Operating Characteristic) methodology.

Keywords: Logistic regression · Development · Validation · ROC curve

1 Introduction

Work-related musculoskeletal disorders (WMSDs) have been described by the scientific community as the most important occupational health problem affecting nursing professionals [2,19]. The high prevalence of symptoms and injuries associated with the musculoskeletal system of these professionals constitutes an evidence of the referred problem [1,3,10]. The high-risk level of WMSDs results from physical requirements and also from the various risk factors in the working context. Musculoskeletal disorders negatively influence many aspects of the nurses' lives, such as their productivity level, absenteeism rate, well-being, and quality of life [2,4,7].

Most studies focus on professionals who develop their activity in the hospital setting, thus excluding the group of nurses who provide home health care. This group has unique characteristics when compared with hospital professionals, as they work in a poorly controlled/standardized working environment. At the patient's home, nurses experience an increased physical demand, and consequently experience more musculoskeletal symptoms [5,14,18,20]. Several studies have revealed that WMSDs are a serious problem for nurses and nursing assistants who provide home care, affecting mainly their back [8,12,20].

© Springer International Publishing Switzerland 2016
O. Gervasi et al. (Eds.): ICCSA 2016, Part I, LNCS 9786, pp. 97–109, 2016.
DOI: 10.1007/978-3-319-42085-1_8

Although back problems have a multi-factorial etiology, including physical, psychosocial, and individual factors, the manual handling of patients (physical risk factor) is considered one of its main causes [14].

2 Problem Description

Taking into account the scarcity of studies addressing musculoskeletal problems in home care nurses, Carneiro [4] developed a study to characterize the musculoskeletal complaints of these professionals. That author found out that providing home care increases the likelihood of having lower back complaints. Based on that result, that author proposed to identify the main risk factors for lower back complaints. Thus Carneiro [4] developed a logistic regression model to predict which risk factors could contribute to the occurrence of complaints involving the lumbar region in home care nurses.

After an exhaustive univariate analysis, the final multivariate model contained seven variables that, when acting together, could contribute, negatively or positively, to the risk of having lower back complaints (Table 1).

The variables that integrate the proposed statistical model are usually associated with back complaints, some more than others, and thus contribute to the credibility of the statistical model. For example, the use of assistive devices by

Table 1. Variables of the logistic regression model and their contribution to lower back complaints, according to Carneiro [4].

Variable	Description	Contribution
X_1	Forearm posture	The adoption of a posture different from the reference posture may contribute to the absence of complaints
X_2	Static postures	The maintenance of static postures for more than 1 min may contribute to the occurrence of complaints
X_3	Arm posture	The adoption of a posture different from the reference posture may contribute to the occurrence of complaints
X_4	Arm supported	Working with the arm supported may contribute to the absence of complaints
X_5	Height of the bed	Working with a bed at an inadequate height may contribute to the occurrence of complaints
X_6	Job satisfaction	Job satisfaction may contribute to the absence of complaints
X_7	Assistive devices for moving or transferring patients	Using assistive devices to move patients may contribute to the absence of complaints

nurses for lifting/transferring patients (X_7) whenever possible, has been encouraged to decrease the risk of musculoskeletal complaints [11,13]. Job satisfaction (X_6) has also been referred in the literature as a factor that might contribute to the occurrence of WMSDs. Namely, Daraiseh et al. [9], reported that dissatisfaction with the working conditions may lead to the occurrence of musculoskeletal symptoms. Moreover, in a study involving Japanese nurses, Smith et al. [19] concluded that more importance should be given to job satisfaction, work organization, and occupational stress, in parallel with the more traditional risk reduction strategies that emphasize manual handling tasks and other occupational factors. Maintaining the same posture for more than a minute (X_2) has also been recognized as a WMSDs risk factor, and this is particularly relevant in the home care context, where the maintenance of static postures may occur in several situations, such as dressing in a limited workspace [5].

According to the model, working with patients' beds at inadequate heights may contribute to the onset of complaints involving the lumbar region. The inadequate height of the bed contributes to the adoption of inadequate postures and, therefore, to the occurrence of musculoskeletal problems. In general, the beds at patients' homes are typically low. Accordingly, Owen and Staehler [17] found that the major sources of back problems for home care workers were the height and width of the beds, their location, and the impossibility of adjusting them.

The other variables of the statistical model are less emphasized in the scientific literature as potential factors for the occurrence of musculoskeletal complaints. These Variables are the "arm's posture" (X_3), the "forearm's posture" (X_1), and the "arm supported" (X_4). For both the arm and the forearm, some postures may be favorable to reduce the moment generated on the lumbar spine, hence resulting in decreased stress on it and a consequent reduced likelihood of complaints. According to the final statistical model, and in order to minimize the complaints involving the lumbar region, nurses should avoid working with the forearm in the reference posture, which is between 60° and 100° of flexion (Fig. 1(a)), and should work with the arm in the reference posture, i.e., between 20° of extension and 20° of flexion (Fig. 1(b)). Working with the arm supported is also favorable for reducing the moment generated on the lumbar spine.

3 Model Development

Model development studies aim to derive a prediction model by selecting predictors and combining them into a multivariable model. Logistic regression is a technique commonly used in cross-sectional (diagnostic) and short-term studies [6].

Using a binary dependent variable and p independent predictors, $\mathbf{x}$, the logistic regression model, in terms of expected probability, $\pi(\mathbf{x})$, can be written as shown below in Eq. 1:

$$\pi(\mathbf{x}) = \frac{\exp(\beta_0 + \beta_1 x_1 + \beta_2 x_2 + \cdots + \beta_p x_p)}{1 + \exp(\beta_0 + \beta_1 x_1 + \beta_2 x_2 + \cdots + \beta_p x_p)} \tag{1}$$

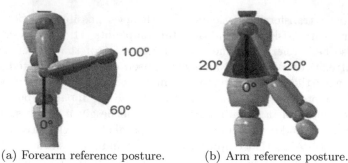

(a) Forearm reference posture.   (b) Arm reference posture.

Fig. 1. Reference postures for X_1 and X_4.

The transformation of $\pi(\mathbf{x})$ is usually called *logit* transformation and is defined in terms of $\pi(\mathbf{x})$, as shown below (Eq. 2):

$$g(\mathbf{x}) = \ln\left(\frac{\pi(\mathbf{x})}{1 - \pi(\mathbf{x})}\right) = \beta_0 + \beta_1 x_1 + \beta_2 x_2 + \cdots + \beta_p x_p \qquad (2)$$

The *logit* transformation is important due to being a linear function of the parameters β.

The construction of a model is an art; it depends not only on gathering a set of requirements and assumptions that should be statistically evaluated but also on the intuition and knowledge of the researcher.

Generally, the researcher must evaluate the association of each variable with the binary outcome. This work is exhaustive, and some researchers opt for selection techniques based on error criteria, like stepwise procedures.

3.1 Sample Size Considerations

The estimation of a sample size for logistic regression is a complex problem. Nevertheless, based on the work of Peduzzi et al. [16] a guideline for a minimum number of cases to include in this kind of study can be suggested. That guideline is described below.

Let p be the smallest of the proportions of negative or positive cases in the population and k the number of covariates (the number of independent variables or predictors), then the minimum number of cases to include is:

$$N = 10 * \frac{k}{p}$$

Thus, for example, if we have seven covariates to include in the model and the proportion of positive cases in the population is 0.60 (60 %), the minimum number of cases required is

$$N = 10 * \frac{7}{0.6} = 117$$

If the resulting number is inferior to 100 Peduzzi et al. [16] suggest increasing it to 100.

4   Model Validation

According to Terrin et al. [21] the utility of predictive models depends on their generality, which can be separated into two components: internal validity (reproducibility) and external validity (transportability).

After selecting the best set of predictors to include in the model, the researcher must evaluate the statistical significance of the estimated parameters and the diagnostic value of the resulting model. Measures such as significant p values for the parameters test and no significant p value for the Hosmer-Lemeshow goodness-of-fit statistical test suggest that the model fits the data reasonably well.

The term *"validation"*, although widely used, is misleading, because it indicates that model validation studies would result in a "yes" (good validation) or "no" (poor validation) answer to the evaluation of the model's performance. Instead, the aim of model validation is to quantitatively evaluate the model's predictive performance either on the resampled participant data of the developed dataset (often referred to as internal validation) or on the independent participant data that were not used to develop the model (often referred to as external validation) [6].

According to Collins et al. [6], the quantification of a model's predictability with the same data used to develop the model (often referred as apparent performance, Fig. 2) tends to give an optimistic estimate of performance, due to overfitting (too few outcome events in relation to the number of candidate predictors) and to the use of predictor selection strategies [6]. Studies developing new prediction models should therefore always include some form of internal validation to quantify any optimism in the predictive performance (for example, calibration and discrimination) of the developed model and adjust the model for overfitting. Internal validation techniques use only the original sample of the study and include methods such as bootstrapping and cross-validation.

Therefore, after developing a prediction model, the evaluation of its performance with participant data different from that used in the model development – external validation – is strongly recommended (Fig. 2) [6]. This type of model validation requires that, for each individual in the new participant data set, outcome predictions be made using the original model (i.e., the published model or regression formula) and compared with the observed outcomes.

The external validation of the model may be performed using participant data collected by the same investigators, typically using the same predictor and outcome definitions and measurements, but obtained later (in time or with narrow validation).

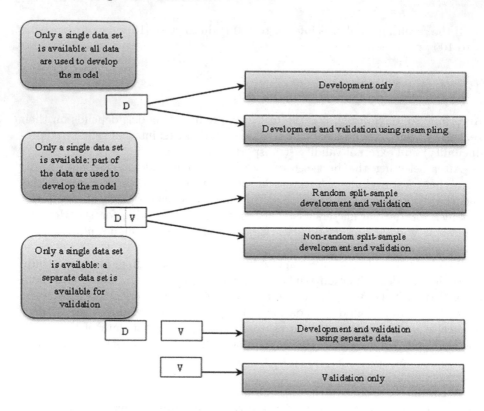

Fig. 2. Types of prediction model studies (adapted from [6]).

5   Materials and Methods

While conducting this study, we faced some difficulties, particularly in the data collection, which resulted in a small sample with many variables. We chose to develop a model starting with a univariate analysis in order to exclude variables that had little information, and/or that did not show statistical significance for the dependent variable (lower back complaints by nurses working in health centers and providing home care). This part of the study is described in more detail in Carneiro [4].

For this study the model with seven variables proposed by Carneiro [4] was considered. Using this same set of data, we restarted the development process, taking into account the missing values and dealing with them firstly.

A questionnaire based on the one developed by Carneiro [4] was used to collect the information necessary for the external validation of the model. However, changes were made to simplify and shorten the questionnaire.

That electronic questionnaire was sent to nurses during the year of 2014, and the posterior methodology was based on respondent-driven sampling. This methodology combines snowball sampling with a mathematical model that comprises a sampling method to compensate the sample being randomly collected. Thus, it allows easier access to specific groups, thereby leading to a more accurate representation of the professional group, in this case, nurses working in Health Centres in Northern Portugal.

The development of the model and its process validation were carried out in the STATA® 13.1 software using the logistic regression command .logistic with post estimation evaluation.

6 Results

Two datasets were used in this study, one from Carneiro [4] - *Dataset 1* - and other from the study conducted in 2014 - *Dataset 2*.

Table 2 summarizes the descriptive data of some variables of interest in the two datasets. For qualitative variables, the values are expressed in percentages and/or counts, for quantitative variables, the mean value and the standard deviation value (between brackets) are presented.

Table 2. Summary of the two datasets.

n		*Dataset 1*	*Dataset 2*
		147	83
Home health care		85 % (125)	100 % (83)
Region	North	125	50
Female		88 %	80 %
Age		34.7 years (8.01)	40.4 years (7.28)
Seniority in the profession Complaints:		11.9 years (7.60)	17.1 years (6.70)
	Cervical	73.6 %	74.0 %
	Lower back	68.8 %	86.0 %
	Dorsal	50.4 %	18.0 %
	Shoulders	48.0 %	44.0 %
	Elbows	8.8 %	8.4 %
	Wrists/hands	31.2 %	24.0 %
	Thighs	16.8 %	4.0 %
	Knees	20.8 %	20.0 %
	Ankles/feet	13.6 %	12.0 %

The dependent variable is musculoskeletal complaints involving the lower back reported by nursing professionals. This variable has two levels: 0 = absence of musculoskeletal complaints involving the lower back and 1 = musculoskeletal complaints involving the lower back. The considered predictor variables are also binary.

Table 3 presents the distribution for the reference level $(X_i = 0)$ of each predictor in the two datasets.

Table 3. Distribution of the predictors by dataset.

Variable	Dataset 1 (Total = 79)		Dataset 2 (Total = 50)	
	%	(n)	%	(n)
X_1	20.0 %	(25)	56.0 %	(28)
X_2	9.6 %	(12)	18.0 %	(9)
X_3	6.4 %	(8)	12.0 %	(6)
X_4	56.8 %	(71)	92.0 %	(46)
X_5	16.5 %	(13)	96.0 %	(48)
X_6	20.3 %	(16)	38.0 %	(19)
X_7	97.5 %	(77)	98.0 %	(49)

We carried out the study using the following procedure:

1. developing the logistic models based on Dataset 1, using all of the seven predictors (model 1);
2. developing alternative models based on Dataset 1, using more restrictive statistical criteria for testing the parameters' significance (p values < 0.1);
3. validating the models generated in the development process, using Dataset 2;
4. evaluating the apparent and external validation using ROC analysis.

The first developed model included all seven candidate predictors. The corresponding results of the estimated coefficients of the logit model, the standard errors, and the significance test for the coefficients (z-statistic and corresponding bilateral p values) are listed in Table 4. Then, the variable X_7 was excluded from the model due to the low percentage of associated cases, and $p > 0.2$, thus generating the Model 2 (Table 5). Considering the values of Model 2, we tried to assess the model's behaviour without the variable with a higher p value in the statistical significance tests for the coefficients. Thus, we obtained Model 3 (Table 6).

The measures of the goodness-of-fit test for each model are evaluated and resumed in Table 7.

In Table 7, X^2 represents the Pearson's goodness-of-fit test or the Hosmer-Lemeshow's goodness-of-fit test. Values of $p > 0.05$ indicate that the model seems to fit quite well. The **LL** values correspond to the log likelihood of the final model. The individual value has no meaning in and of itself; regarding models comparison, the values are of the same order of magnitude.

The pseudo R^2 is a qualitative measure of the quality of fit of the models. It is difficult to find a rule to establish the limits of McFadden R^2 values. Nevertheless, Louviere et al. [15], on page 55 of his book, indicates the following: "*Values of rho-squared between 0.2–0.4 are considered to be indicative of extremely*

Table 4. RL model: Model 1.

Predictor	Coef., B	Std. Err	z	$P > \lvert z \rvert$	[95 % CI for B]	
X_1	−2.1630	1.0691	−2.02	0.043	−4.2585	−0.0675
X_2	1.4273	0.8709	1.64	0.101	−0.2796	3.1342
X_3	3.6061	1.3750	2.62	0.009	0.9111	6.3011
X_4	−3.1618	1.0787	−2.93	0.003	−5.2760	−1.0475
X_5	−1.1283	0.8785	−1.28	0.199	−2.8501	0.5936
X_6	−2.1931	1.2081	−1.82	0.069	−4.5608	0.1746
X_7	−2.2730	1.7873	−1.27	0.203	−5.7761	1.2301
Constant	0.7105	1.6611	0.43	0.669	−2.5452	3.9661

Table 5. RL model: Model 2.

Predictor	Coef., B	Std. Err	z	$P > \lvert z \rvert$	[95 % CI for B]	
X_1	−1.5914	0.8636	−1.84	0.065	−3.2841	0.1013
X_2	1.4259	0.8792	1.62	0.105	−0.2973	3.1492
X_3	3.1884	1.2129	2.63	0.009	0.8112	5.5657
X_4	−2.9448	1.0294	−2.86	0.004	−4.9625	−0.9271
X_5	−1.3600	0.8836	−1.54	0.124	−3.0917	0.3717
X_6	−2.2053	1.1777	−1.87	0.061	−4.5135	0.1029
Constant	0.5793	1.6162	0.36	0.720	−2.5884	3.7470

Table 6. RL model: Model 3.

Predictor	Coef., B	Std. Err	z	$P > \lvert z \rvert$	[95 % CI for B]	
X_1	−1.8993	0.8707	−2.18	0.029	−3.6062	−0.1924
X_2	1.74488	0.8265	2.11	0.035	0.1250	3.3648
X_3	3.3086	1.2214	2.71	0.007	0.9146	5.7025
X_4	−2.9934	1.0114	−2.96	0.003	−4.9757	−1.0111
X_6	−2.3202	1.1944	−1.94	0.052	−4.6612	0.0207
Constant	0.3406	1.5853	0.21	0.830	−2.7664	3.4479

good model fits. Simulations by Domenich and McFadden (1975) equivalence this range to 0.7 to 0.9 for a linear function". Taking this into account, the values obtained for pseudo R^2 for the three models reveal good model fits.

Finally, the measures of the information criteria - AIC and BIC - allow comparing the models in terms of maximum likelihood. These two measures only differ in the complexity term. So, given that two models fit in the same data, the model with the smaller value of the information criterion is considered to be better. The values of AIC listed in Table 7 are similar, but Model 3 has the smaller value for BIC.

Table 7. Summary of the goodness-of-fit test.

	Model 1	Model 2	Model 3
# predictors	7	6	5
X^2	15.06	17.60	16.09
$p$value	0.4470	0.2843	0.0650
LL	-30.3319	-31.1240	-32.2996
Pseudo R^2	0.3634	0.3468	0.3221
Correctly classified	83.5%	84.8%	84.8%
Sensitivity	92.9%	94.6%	94.6%
Specificity	60.9%	60.9%	60.9%
AIC	76.6638	76.2481	76.5999
BIC	95.6936	92.8342	90.8166

Table 8. Summary of the ROC analysis for each model.

		External validation	Apparent validation
Model 1	AUC	0.7492	0.8649
	SE(AUC)	0.0842	0.0460
	chi2(1)	1.46	
	p value	0.2276	
Model 2	AUC	0.7658	0.8505
	SE(AUC)	0.0807	0.0493
	chi2(1)	0.8	
	p value	0.3700	
Model 3	AUC	0.7641	0.8412
	SE(AUC)	0.0816	0.0513
	chi2(1)	0.64	
	p value	0.4237	

When using the Dataset 2 for external validation, we compared the models in terms of area under the ROC curve (AUC). we evaluated the apparent validation in terms of AUC, and, considering the Dataset 2, we calculated the predicted probabilities for each model and calculated the AUC index for external validation. The summary results are listed in Table 8.

In Table 8, "AUC" stands for the index of the area under the empirical ROC curve, "SE(AUC)" represents the standard error associated with this estimative, "chi2" is the test statistic that corresponds to testing the null hypothesis - $H_0 : AUC(external) = AUC(apparent)$ - and the p value is the probability,

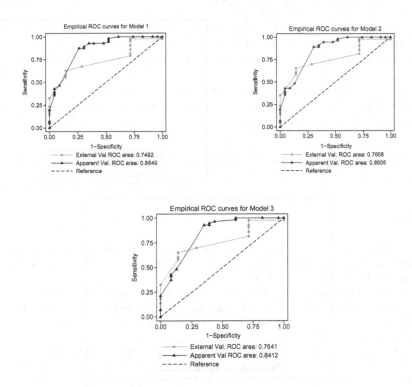

Fig. 3. Empirical ROC curves comparison.

considering this hypothesis, of obtaining a result equal to or higher than what was actually observed $Pr(\chi^2 \geq chi2)$ (Fig. 3).

7   Conclusion

According to the results obtained in the developed model, the "best" model to fit the data in the initial dataset was Model 3. The dimension of the sample, the measures evaluated in the goodness-of-fit test, and the values of the area under the ROC curves for the apparent internal validation justify this choice. In the validation process, we could conclude that the internal validation in the same dataset - the apparent validation - led to more "optimistic" values of AUC and better performance, but with no statistical differences from the corresponding results obtained in the external validation.

Therefore, taking into account these results, the model chosen to predict the risk of musculoskeletal complaints involving the lumbar region in nurses working at the Health Centres of Northern Portugal and providing home-based care, was Model 3, which contained five variables: X_1, X_2, X_3, X_4 and X_6. The contribution of each variable to this model was the same encountered by Carneiro [4] regarding sign in the logit model. However, we tried to minimize the overfitting and to maximize the information contained in the complete sample, by discarding the missing values in data used to develop the model.

The dimension of the sample represented a limitation of this study. Nevertheless, we concluded that Model 3 can be used as a prediction model for the proposed goal, because its performance with participant data different from those used for the model development was evaluated in same conditions, and showed a good performance.

Acknowledgments. This work was supported by FCT - (Fundação para a Ciência e Tecnologia) within the Project Scope: UID/CEC/00319/2013.

References

1. Alexopoulos, E.C., Burdorf, A., Kalokerinou, A.: A comparative analysis on musculoskeletal disorders between Greek and Dutch nursing personnel. Int. Arch. Occup. Environ. Health **79**, 82–88 (2006)
2. Barroso, M., Carneiro, P., Braga, A.C.: Characterization of ergonomic issues and musculoskeletal complaints. In a Portuguese District Hospital. In: Proceedings of the International Symposium Risks for Health Care Workers: Prevention Challenges, Athens (2007)
3. Barroso, M.P., Martins, J.: Assessment and analysis of musculoskeletal risk perception among nurses. In: Proceedings of the International Conference HEPS 2008 C Healthcare Systems Ergonomics and Patient Safety, Strasbourg, France, 25–27 June (2008)
4. Carneiro, P.: LME na prestação de cuidados de saúde ao domicílio: avaliação do risco e construção de modelos estatísticos de previsão. Tese de Doutoramento, Universidade do Minho (2012) (in Portuguese)
5. Cheung, K., Gillen, M., Faucett, J., Krause, N.: The prevalence of and risk factors for back pain among home care nursing personnel in Hong Kong. Am. J. Ind. Med. **49**(1), 14–22 (2006)
6. Collins, G.S., Reitsma, J.B., Altman, D.G., Moons, K.G.: Transparent Reporting of a multivariable prediction model for Individual Prognosis Or Diagnosis (TRIPOD): the TRIPOD statement. Ann. Intern. Med. **162**, 55–63 (2015). http://dx.doi.org/10.7326/M14-0697
7. Coelho, M.S.R.: Estudo da Frequência das Lesões Músculo-Esqueléticas Relacionadas com o Trabalho (LMERT) em Profissionais de Enfermagem, Proposta de um Programa de Ginástica Laboral. Dissertação de licenciatura da Faculdade de Desporto da Universidade do Porto, Porto (2009). (in Portuguese)
8. Czuba, L.R., Sommerich, C.M., Lavender, S.A.: Ergonomic and safety risk factors in home health care: exploration and assessment of alternative interventions. Work **42**, 341–353 (2012)
9. Daraiseh, N., Genaidy, A.M., Karwowski, W., Davis, L.S., Stambough, J., Huston, R.L.: Musculoskeletal outcomes in multiple body regions and work effects among nurses: the effects of stressful and simulating working conditions. Ergonomics **46**(12), 1178–1199 (2003)
10. Daraiseh, N., Cronin, S., Davis, L., Shell, R., Karwowski, W.: Low back symptoms among hospital nurses, associations to individual factors and pain in multiple body regions. Int. J. Ind. Ergon. **40**, 19–24 (2010)

11. Evanoff, B., Wolf, L., Aton, E., Canos, J., Collins, J.: Reduction in injury rates in nursing personnel through introduction of mechanical lifts in the workplace. Am. J. Ind. Med. **44**(5), 451–457 (2003)
12. Faucett, J., Kang, T., Newcomer, R.: Personal service assistance: musculoskeletal disorders and injuries in consumer-directed home care. Am. J. Ind. Med. **56**, 454–468 (2013)
13. Kromark, K., Dulon, M., Beck, B.-B., Nienhaus, A.: Back disorders and lumbar load in nursing staff in geriatric care: a comparison of home-based care and nursing homes. J. Occup. Med. Toxicol. **4**, 33, 9. http://www.occup-med.com/content/4/1/33. Accessed 15 Mar 2011
14. Lee, S., Faucett, J., Gillen, M., Krause, N.: Musculoskeletal pain among critical-care nurses by availability and use of patient lifting equipment: an analysis of cross-sectional survey data. Int. J. Nurs. Stud. **50**(12), 1648–1657 (2013)
15. Louviere, J.J., Hensher, A.D., Swait, D.J.: Stated Choice Methods: Analysis and Applications. Cambridge University Press, New York (2000)
16. Peduzzi, P., Concato, J., Kemper, E., Holford, T.R., Feinstein, A.R.: A simulation study of the number of events per variable in logistic regression analysis. J. Clin. Epidemiol. **49**, 1373–1379 (1996)
17. Owen, B., Staehler, K.: Decreasing back stress in home care. Home Healthc. Nurse **21**(3), 180–186 (2003)
18. Simon, M., Tackenberg, P., Nienhaus, A., Estryn-Behar, M., Conway, P.M., Hasselhorn, H.M.: Back or neck-pain-related disability of nursing staff in hospitals, nursing homes and home care in seven countries C results from the European NEXT-Study. Int. J. Nurs. Stud. **45**, 24–34 (2008)
19. Smith, D.R., Mihashi, M., Adachi, Y., Koga, H., Ishitake, T.: A detailed analysis of musculoskeletal disorder risk factors among Japanese nurses. J. Saf. Res. **37**, 195–200 (2006)
20. Szeto, G.P.Y., Wong, K.T., Law, K.Y., Lee, E.W.C.: A study of spinal kinematics in community nurses performing nursing tasks. Int. J. Ind. Ergon. **43**, 203–209 (2013)
21. Terrin, N., Schmid, C.H., Griffith, J.L., D'Agostino Sr., R.B., Selker, H.P.: External validity of predictive models: a comparison of logistic regression, classification trees, and neural networks. J. Clin. Epidemiol. **56**(8), 721–729 (2003)

Analytical Spatial-Angular Structure of Uniform Slab Radiation Fields for Strongly Elongated Phase Functions

Oleg I. Smokty$^{(\boxtimes)}$

St. Petersburg Institute for Informatics and
Automation of Russian Academy of Sciences, St. Petersburg, Russia
`soi@iias.spb.su`

Abstract. New nonlinear integral equations of the radiative transfer theory for determination of unified function combining the intensities of upgoing and downgoing radiation at arbitrary symmetrical levels of a uniform slab has been obtained using the Ambarzumian-Chandrasekhar classical invariance principle. By applying the mirror reflection principle proposed by the author, the unified function and photometrical invariants concept is considered. Based on fitting parametrization of the obtained nonlinear integral equation in the case of strongly elongated phase functions, the spatial-angular structure of inner radiation fields is represented as primary scattering radiation and adaptive fitting factors conditioned by multiple radiation scattering in a uniform slab. Numerical modelling has been carried out for the obtained approximate solutions of the nonlinear integral equations for the Henyey-Greenstein phase function with parameter $g \geq 0.9$ and representative models of optical parameters of Earth's aerosol atmosphere. The accuracy of obtained approximate relations has been evaluated also. The obtained fitting analytical expressions have clear physical meaning and can be used for the statement and numerical modelling of direct and inverse problems solutions in the radiative transfer theory and natural environments optics.

Keywords: Non-linear integral equations · Mirror reflection principle · Unified photometrical function · Photometrical invariants · Strongly elongated phase functions · Fitting parametrization · Numerical modelling

1 Introduction

The concept of the unified function E combining the brightness coefficients ρ and σ for a uniform slab of a finite optical thickness $\tau_0 < \infty$ has been independently developed in the radiative transfer theory by Hovenier [1] and the author [2] on the grounds of different approaches. In his work Hovenier modified the Ambarzumian–Chandrasekhar invariance principle [3, 4] for the case of primary external energy sources of radiation located symmetrically in relation to the slab middle $(1/2\tau_0)$. In [2] the author used the method of linear structural transformations of the brightness coefficients ρ and σ, as well as the Ambarzumian–Chandrasekhar functions φ_i^m and ψ_i^m [5], based on the mirror reflection principle and corresponding photometrical invariants for reflected and transmitted radiation on the slab external boundaries $\tau = 0$ and $\tau = \tau_0$. Both

© Springer International Publishing Switzerland 2016
O. Gervasi et al. (Eds.): ICCSA 2016, Part I, LNCS 9786, pp. 110–128, 2016.
DOI: 10.1007/978-3-319-42085-1_9

approaches demonstrate that the unified exit function E, being an asymmetrical function by its angular variables, allows to easily and unequivocally determine the brightness coefficients ρ and σ. This paper, based on the structural approach and the Ambarzumian–Chandrasekhar invariance principle, presents the generalization of the unified function E for scalar intensities of upgoing and downgoing radiation in mirror symmetrical sighting directions at any mirror symmetrical levels τ and $\tau_0 - \tau$ at an arbitrary optical depth τ of a uniform slab. The fitting parametrization of the obtained new integral equations using the extreme properties of strongly elongated phase functions $P(\cos\gamma)$ near small scattering angles $\gamma \simeq 0°$ allows to define the approximate spatial-angular structure of the unified function E and its components – photometrical invariants of reflected and transmitted radiation intensities at arbitrary mirror symmetrical levels τ and $\tau_0 - \tau$ in mirror symmetrical sighting directions.

2   New Integral Equations for Upgoing and Downgoing Intensities of Inner Radiation Fields

Consider a uniform slab of a finite optical thickness $\tau_{0,\lambda} < \infty$ illuminated from above $(\tau_\lambda = 0)$ by parallel solar beams falling at the angle $\theta_0 = \arccos\xi$ relative to the inner normal at azimuth $\varphi_0 = 0$ and creating monochromatic illumination of the perpendicular area πS_λ. The wavelength index λ will be omitted in further consideration. Let $\theta_0 = \arccos\eta$ denote the angle between the radiation propagation direction and the inner normal to the slab (Fig. 1). The phase function $P(\cos\gamma) = P(\eta, \eta', \varphi' - \varphi'')$ and the single scattering albedo Λ are assumed independent of the optical depth τ.

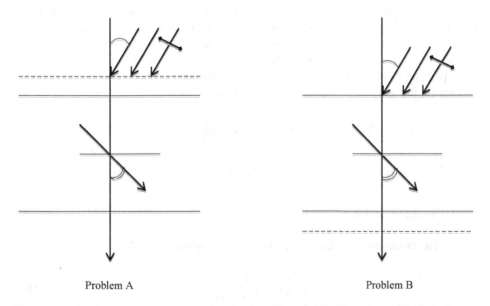

Problem A   Problem B

Fig. 1. Application of the Ambarzumian-Chandrasekhar classical invariance principle for the determination of inner radiation intensities at arbitrary optical levels τ of a uniform slab in the case of a finite optical thickness $\tau_0 < \infty$.

Let us consider the basic problem in the radiative transfer theory related to finding the intensity $I(\tau, \eta, \xi, \varphi, \tau_0)$ in the upgoing $(\eta < 0)$ and downgoing $(\eta > 0)$ sighting directions at an arbitrary optical depth τ. In order to simplify the subsequent consideration, we use the Fourier transformations of the phase function $P(\eta, \xi, \varphi)$ and radiation intensity $I(\tau, \eta, \xi, \varphi, \tau_0)$, namely:

$$P(\eta, \xi, \varphi) = P^\circ(\eta, \xi) + 2\sum_{m=1}^{M} P^m(\eta, \xi)\cos m\varphi, \ \eta \in [-1,1], \ \xi \in [0,1], \ \varphi \in [0, 2\pi], \ \tau \in [0, \tau_0], \quad (1)$$

$$I(\tau, \eta, \xi, \varphi, \tau_0) = I^\circ(\tau, \eta, \xi, \tau_0) + 2\sum_{m=1}^{M} I^m(\tau, \eta, \xi, \tau_0)\cos m\varphi,$$

$$\eta \in [-1,1], \ \xi \in [0,1], \ \varphi \in [0, 2\pi], \ \tau \in [0, \tau_0]. \tag{2}$$

Now let us apply the Ambarzumian–Chandrasekhar invariance principle in its classical interpretation [3, 4] to finding the azimuthal harmonics of intensities $I^m(\tau, \eta, \xi, \tau_0)$. According to this principle, at $\tau \geq 0$, adding (subtracting) at the upper boundary $(\tau = 0)$ or subtracting (adding) at the lower boundary $(\tau = \tau_0)$ of a uniform stab of a small optical thickness $\Delta\tau \ll 1$ will not change the values of $I^m(\tau, \eta, \xi, \tau_0)$ (Fig. 1, A, B). Thus, due to the slab's optical uniformity $(\tau_0 + \Delta\tau)$ we have the following equality relation:

$$I^m(\tau, \eta, \xi, \tau_0 + \Delta\tau)|_A = I^m(\tau + \Delta\tau, \eta, \xi, \tau_0 + \Delta\tau)|_B, m = 0, 1, \ldots, M. \tag{3}$$

For the invariance relation (3) to be fulfilled in the case of adding a small thickness $\Delta\tau \ll 1$ at the upper boundary $\tau = 0$, it is necessary to change the boundary condition at the level $\tau = 0$ in the initial boundary problem, i.e. $I^m(0, \eta, \xi, \tau_0) = 0$, at $\eta > 0$, to a new boundary condition by introducing radiation that has been directly weakened and then scattered by a small layer $\Delta\tau$ in the direction of optical depths $\tau \geq 0$. This transformation results in the following integral equations for the azimuthal harmonics of radiation intensity $I^m(\tau, \eta, \xi, \tau_0)$ in problem A:

$$I^m(\tau, \eta, \xi, \tau_0 + \Delta\tau)|_A = I^m(\tau, \eta, \xi, \tau_0)e^{-\frac{\Delta\tau}{\xi}} + \frac{\Lambda}{4}\frac{\Delta\tau}{\eta}P^m(\eta, \xi)e^{-\frac{\tau}{\eta}}\Theta(\eta) + \frac{\Lambda}{2}\Delta\tau\int_0^1 P^m(\eta', \xi)I^m(\tau, \eta, \eta', \tau_0)\frac{d\eta'}{\eta'} +$$

$$+ \frac{\Lambda}{2}\xi\frac{\Delta\tau}{\eta}e^{-\frac{\Delta\tau}{\eta}}e^{-\frac{\tau}{\xi}}\Theta(\eta)\int_0^1 P^m(-\eta', \eta)\rho^m(\eta', \xi, \tau_0)d\eta' + q\int_0^1 d\eta'\int_0^1 P^m(-\eta', \eta)\rho^m(\eta', \xi, \tau_0)I^m(\tau, \eta, \eta'', \tau_0)e^{-\frac{\Delta\tau}{\xi}}\frac{d\eta''}{\eta''}, \tag{4}$$

where q is $\Lambda\xi\Delta\tau$ and $\eta \in [-1, 1], \ \xi \in [0, 1], \ \tau \in [0, \tau_0], \ m = \overline{0, M}$.

The Heaviside function $\Theta(\eta)$ is defined in a standard manner:

$$\Theta(\eta) = \begin{cases} 1, \ \eta > 0 \\ 0, \ \eta < 0 \end{cases} \tag{5}$$

The azimuthal harmonics of brightness coefficients $\rho^m(\eta, \xi, \tau_0)$ and $\sigma^m(\eta, \xi, \tau_0)$ are defined at the upper $(\tau = 0)$ and lower $(\tau = \tau_0)$ boundaries of the slab under consideration according to the following Fourier expansions of $\rho(\eta, \xi, \varphi, \tau_0)$ and $\sigma(\eta, \xi, \varphi, \tau_0)$:

$$\rho(\eta, \xi, \varphi, \tau_0) = \rho^\circ(\eta, \xi, \tau_0) + 2\sum_{m-1}^{M} \rho^m(\eta, \xi, \tau_0)\cos m\varphi, \ (\eta, \xi) > 0, \ \varphi \in [0, 2\pi], \quad (6)$$

$$\sigma(\eta, \xi, \varphi, \tau_0) = \sigma^\circ(\eta, \xi, \tau_0) + 2\sum_{m-1}^{M} \sigma^m(\eta, \xi, \tau_0)\cos m\varphi, \ (\eta, \xi) > 0, \ \varphi \in [0, 2\pi]. \quad (7)$$

Similarly to problem A, for problem B we have:

$$I^m(\tau + \Delta\tau, \eta, \xi, \tau_0 + \Delta\tau)|_B = I^m(\tau + \Delta\tau, \eta, \xi, \tau_0) +$$

$$+ \Lambda\Delta\tau\xi \int_0^1 d\eta' \int_0^1 P^m(\eta', -\eta'')\sigma^m(\eta', \xi, \tau_0)I^m(\tau_0 - \tau - \Delta\tau, -\eta, \eta'', \tau_0)\frac{d\eta''}{\eta''} -$$

$$- \frac{\Lambda}{2}\xi\frac{\Delta\tau}{\eta}e^{-\frac{\tau_0 - \tau - \Delta\tau}{\eta}}\Theta(-\eta)\int_0^1 P^m(\eta', \eta)\sigma^m(\eta', \xi, \tau_0)d\eta' + \frac{\Lambda}{2}\Delta\tau e^{-\tau_0/\xi}\int_0^1 P^m(-\eta', \xi)I^m(\tau_0 - \tau - \Delta\tau, -\eta, \eta', \tau_0)\frac{d\eta'}{\eta'} -$$

$$- \frac{\Lambda}{4}\frac{\Delta\tau}{\eta}e^{-\tau_0/\xi}e^{-\frac{\tau - \tau_0 + \Delta\tau}{\eta}}P^m(\eta, \xi)\Theta(-\eta), \ \eta \in [-1, 1], \ \xi \in [0, 1], \ \tau \in [0, \tau_0], \ m = \overline{0, M}.$$

$$(8)$$

Further, taking into account the smallness of $\Delta\tau \ll 1$, the following relation is valid:

$$I^m(\tau + \Delta\tau, \eta, \xi, \tau_0) = I^m(\tau, \eta, \xi, \tau_0) + \Delta\tau\frac{\partial I^m(\tau, \eta, \xi, \tau_0)}{\partial\tau} + o\left[(\Delta\tau)^2\right], \quad (9)$$

$$\eta \in [-1, 1], \ \xi \in [0, 1], \ \tau \in [0, \tau_0], \ m = \overline{0, M}.$$

The expression for derivative in (9) can be easily found using the initial radiative transfer equation:

$$\frac{\partial I^m(\tau, \eta, \xi, \tau_0)}{\partial\tau} = \frac{1}{\eta}\left[-I^m(\tau, \eta, \xi, \tau_0) + \frac{\Lambda}{2}\int_{-1}^{+1} P^m(\eta, \eta')I^m(\tau, \eta', \xi, \tau_0)d\eta' + \frac{\Lambda}{4}P^m(\eta, \xi)e^{-\tau/\xi}\right], \quad (10)$$

$$\eta \in [-1, 1], \ \xi \in [0, 1], \ \tau \in [0, \tau_0], \ m = \overline{0, M}.$$

Further, according to (3), we equate (4) and (8), then by using Eqs. (9) and (10), after some simple but cumbersome operations, find the required integral equations for $I^m(\tau, \eta, \xi, \tau_0)$:

$$(\xi - \eta)I^m(\tau, \eta, \xi, \tau_0) = \frac{\Lambda}{2}\xi \int_{-1}^{+1} P^m(\eta, \eta')I^m(\tau, \eta', \xi, \tau_0)d\eta' - \frac{\Lambda}{2}\eta\xi \int_0^1 P^m(\eta', \xi)I^m(\tau, \eta', \tau_0)\frac{d\eta'}{\eta'} -$$

$$- \Lambda\eta\xi^2 \int_0^1 d\eta' \int_0^1 \rho^m(\eta', \xi, \tau_0)P^m(-\eta', \eta'')I^m(\tau, \eta, \eta'', \tau_0)\frac{d\eta''}{\eta''} +$$

$$+ \frac{\Lambda}{2}\eta\xi \int_0^1 P^m(-\eta', \xi)I^m(\tau_0 - \tau, -\eta, \eta', \tau_0)e^{-\tau_0/\xi}\frac{d\eta'}{\eta'} + \tag{11}$$

$$+ \Lambda\eta\xi^2 \int_0^1 d\eta' \int_0^1 \sigma^m(\eta', \xi, \tau_0)P^m(-\eta', \eta'')I^m(\tau_0 - \tau, -\eta, \eta'', \tau_0)\frac{d\eta''}{\eta''} + f^m(\tau, \eta, \xi, \tau_0),$$

$$\eta \in [-1, 1], \ \xi \in [0, 1], \ \tau \in [0, \tau_0], \ m = \overline{0, M},$$

where functions $f^m(\tau, \eta, \xi, \tau_0)$ are defined as follows:

$$f^m(\tau, \eta, \xi, \tau_0) = \frac{\Lambda}{4}\xi P^m(\eta, \xi)e^{-\tau/\xi} - \frac{\Lambda}{4}\xi P^m(\eta, \xi)e^{-\tau_0/\xi}e^{\frac{\tau_0 - \tau}{\eta}}\Theta(-\eta) -$$

$$- \frac{\Lambda}{4}\xi P^m(\eta, \xi)e^{-\tau/\eta}\Theta(\eta) - \frac{\Lambda}{2}\xi^2 e^{\frac{\tau_0 - \tau}{\eta}}\Theta(-\eta) \int_0^1 \sigma^m(\eta', \xi, \tau_0)P^m(\eta', \eta)d\eta' - \tag{12}$$

$$- \frac{\Lambda}{2}\xi^2 e^{-\tau/\eta}\Theta(\eta) \int_0^1 \rho^m(\eta', \xi, \tau_0)P^m(-\eta', \eta)d\eta'.$$

Thus, if the azimuthal harmonics of brightness coefficients $\rho^m(\eta, \xi, \tau_0)$ and $\sigma^m(\eta, \xi, \tau_0)$ are known, then (11) are linear Fredholm integral equations of the second kind with respect to two unknown values $I^m(\tau, \eta, \xi, \tau_0)$ and $I^m(\tau_0 - \tau, -\eta, \xi, \tau_0)$. In order to find them, carry out a spatial shift of $\tau \Rightarrow \tau_0 - \tau$ and angular rotation of $\eta \Rightarrow -\eta$ simultaneously in Eqs. (11). As a result, we obtain a needed second integral equations for finding $I^m(\tau_0 - \tau, -\eta, \xi, \tau_0)$:

$$(\xi + \eta)I^m(\tau_0 - \tau, -\eta, \xi, \tau_0) = \frac{\Lambda}{2}\xi \int_{-1}^{+1} P^m(\eta, \eta')I^m(\tau_0 - \tau, -\eta, \xi, \tau_0)d\eta' + \frac{\Lambda}{2}\eta\xi \int_0^1 P^m(\eta', \xi)I^m(\tau_0 - \tau, -\eta, \eta', \tau_0)\frac{d\eta'}{\eta'} +$$

$$+ \Lambda\eta\xi^2 \int_0^1 d\eta' \int_0^1 \rho^m(\eta', \xi, \tau_0)P^m(-\eta', \eta'')I^m(\tau_0 - \tau, -\eta, \eta'', \tau_0)\frac{d\eta''}{\eta''} - \frac{\Lambda}{2}\eta\xi \int_0^1 P^m(-\eta', \xi)I^m(\tau, \eta, \eta', \tau_0)e^{-\tau_0/\xi}\frac{d\eta'}{\eta'} - \tag{13}$$

$$- \Lambda\eta\xi^2 \int_0^1 d\eta' \int_0^1 \sigma^m(\eta', \xi, \tau_0)P^m(-\eta', \eta'')I^m(\tau, \eta, \eta'', \tau_0)\frac{d\eta''}{\eta''} + f^m(\tau_0 - \tau, -\eta, \xi, \tau_0),$$

$$\eta \in [-1, 1], \ \xi \in [0, 1], \ \tau \in [0, \tau_0], \ m = \overline{0, M}.$$

If the values of the brightness coefficients azimuthal harmonics $I^m(0, -\eta, \xi, \tau_0) = S\xi\rho^m(\eta, \xi, \tau_0), \ \eta > 0$ and $I^m(\tau_0, \eta, \xi, \tau_0) = S\xi\sigma^m(\eta, \xi, \tau_0), \ \eta > 0$ are known, then Eqs. (11)–(13) form a system of two exact linear Fredholm integral equations in respect of unknown functions $I^m(\tau, \eta, \xi, \tau_0)$ and $I^m(\tau_0 - \tau, -\eta, \xi, \tau_0)$ within the above defined variation range of variables τ, η and ξ. From Eqs. (13), at $\tau \to 0$ and $\tau_0 \to \infty$, accounting for the boundary conditions $I^m(0, \eta, \xi, \tau_0) = 0, \ \eta > 0$ and $I^m(\tau_0, -\eta, \xi, \tau_0) = 0, \ \eta > 0$, follows the known Ambarzumian nonlinear integral equations [3] for the diffuse reflection coefficients $\rho_\infty^m(\eta, \xi)$. Assuming $\tau = 0$ in (13) and then $\tau = \tau_0$ in (11), and once again using the mentioned boundary conditions, we obtain the classical system of Chandrasekhar nonlinear integral equations [4] for the azimuthal harmonics of brightness coefficients $\rho^m(\eta, \xi, \tau_0)$ and $\sigma^m(\eta, \xi, \tau_0)$ in a uniform slab of a finite optical thickness $\tau_0 < \infty$. Thus, the system of nonlinear integral equations (11)–(13) generalizes the above mentioned classical system of integral equations to an arbitrary optical depth τ which were obtained for the particular case of the slab's outer boundaries $\tau = 0$ and $\tau = \tau_0$. It should be emphasized that the boundary conditions for upgoing ($\eta < 0$) and downgoing ($\eta > 0$) radiation intensities, at $\tau = \tau_0$ and $\tau = 0$ respectively, were not used to obtain Eqs. (11) and (13). Therefore, these equations should be considered as new integral relations for radiation intensities which must be satisfied by any regular solution of the basic problem of the radiative transfer theory [5].

By adding up and subtracting (11) and (13), and then using the definition of photometrical invariants of intensities $I_\pm^m(\tau, \eta, \xi, \tau_0)$ for azimuthal harmonics [6]:

$$I_\pm^m(\tau, \eta, \xi, \tau_0) = I^m(\tau_0 - \tau, -\eta, \xi, \tau_0) \pm I^m(\tau, \eta, \xi, \tau_0), \ \eta \in [-1, 1], \ \xi \in [0, 1], \ \tau$$
$$\in [0, \tau_0], \ m = \overline{0, M}, \tag{14}$$

$$I_\pm^m(\tau, \eta, \xi, \tau_0) = \pm I^m(\tau_0 - \tau, -\eta, \xi, \tau_0), \ \eta \in [-1, 1], \ \xi \in [0, 1], \ \tau \in [0, \tau_0], \ m = \overline{0, M}, \tag{15}$$

we find new integral equations for linear combinations of (14) and (15) that will be mathematically equivalent to (11) and (13):

$$\xi I_+^m(\tau, \eta, \xi, \tau_0) + \eta I_-^m(\tau, \eta, \xi, \tau_0) = \frac{\Lambda}{2} \xi \int\limits_{-1}^{+1} P^m(\eta, \eta') I_+^m(\tau, \eta, \xi, \tau_0) d\eta' +$$

$$+ \Lambda \eta \xi^2 \int\limits_0^1 d\eta' \int\limits_0^1 R_+^m(\eta', \xi, \tau_0) P^m(-\eta', \eta'') I_-^m(\tau, \eta, \eta'', \tau_0) \frac{d\eta''}{\eta''} + \tag{16}$$

$$+ \frac{\Lambda}{2} \eta \xi \int\limits_0^1 \left[P^m(\eta', \xi) + P^m(-\eta', \xi) e^{-\tau_0/\xi} \right] I_-^m(\tau, \eta, \eta'', \tau_0) \frac{d\eta'}{\eta'} + f_+^m(\tau, \eta, \xi, \tau_0),$$

$$\eta \in [-1, 1], \ \xi \in [0, 1], \ \tau \in [0, \tau_0], \ m = \overline{0, M},$$

where $f_+^m(\tau, \eta, \xi, \tau_0) = f^m(\tau_0 - \tau, -\eta, \xi, \tau_0) + f^m(\tau, \eta, \xi, \tau_0)$, photometrical invariants $R_\pm^m(\eta, \xi, \tau_0)$ are defined according to [7]:

$$R_\pm^m(\eta, \xi, \tau_0) = \rho^m(\eta, \xi, \tau_0) \pm \sigma^m(\eta, \xi, \tau_0), \ (\eta, \xi) > 0, \ m = \overline{0, M}. \tag{17}$$

Note that, because of invariance property (15), we can use a half of the variation range of optical depth τ in equations (16), i.e. $\tau \in [0, 1/2\tau_0]$.

Let us now look at the left-hand members of Eqs. (16) which include the specific linear combinations of photometrical invariants $I_+^m(\tau, \eta, \xi, \tau_0)$ and $I_-^m(\tau, \eta, \xi, \tau_0)$, namely:

$$I_{\Sigma}^{m}(\tau, \eta, \xi, \tau_0) = \xi I_{+}^{m}(\tau, \eta, \xi, \tau_0) + \eta I_{-}^{m}(\tau, \eta, \xi, \tau_0), \ \eta \in [-1, 1], \ \xi \in [0, 1], \ \tau$$
$$\in [0, \tau_0], \ m = \overline{0, M}, \tag{18}$$

$$I_{\Sigma}^{m}(\tau, \eta, \xi, \tau_0) = (\eta + \xi)I(\tau_0 - \tau, -\eta, \xi, \tau_0) + (\eta - \xi)I(\tau, \eta, \xi, \tau_0),$$
$$\eta \in [-1, 1], \ \xi \in [0, \tau_0], \ \tau \in [0, \tau_0], \ m = \overline{0, M}. \tag{19}$$

Linear combinations (18) and (19) form new asymmetrical functions $I_{\Sigma}^{m}(\tau, \eta, \xi, \tau_0)$ by their angular variables η and ξ which we call unified photometrical functions for upgoing $(\eta < 0)$ and downgoing $(\eta > 0)$ radiation intensities at arbitrary mirror-symmetrical levels τ and $(\tau_0 - \tau)$ in in mirror sighting directions η and $-\eta$. At the same time, the following invariance property is fulfilled for (18)–(19) in the case of simultaneous translation of optical and angular variables $\tau \Leftrightarrow (\tau_0 - \tau)$ and $\eta \Leftrightarrow -\eta$:

$$I_{\Sigma}^{m}(\tau, \eta, \xi, \tau_0) = I_{\Sigma}^{m}(\tau_0 - \tau, -\eta, \xi, \tau_0), \ \eta \in [-1, 1], \ \xi \in [0, 1], \ \tau \in [0, \tau_0], \ m = \overline{0, M}. \tag{20}$$

According to relation (20), the azimuthal harmonics of the unified photometrical function $I_{\Sigma}^{m}(\tau, \eta, \xi, \tau_0)$ and its constituent photometrical invariants $I_{\pm}^{m}(\tau, \eta, \xi, \tau_0)$ can be considered not within the full change range of variables $(\tau, \eta, \xi)D = \{\eta \in [-1, 1], \ \xi \in [0, 1], \ \tau \in [0, \tau_0]\}$, as it is the case in the initial non-symmetrized boundary-value problem in the classical radiative transfer theory [3–5], but within a narrower range $D_1 = \{\eta \in [0, 1], \ \xi \in [0, 1], \ \tau \in [0, \tau_0]\}$ or alternatively $D_2 = \{\eta \in [-1, 1], \ \xi \in [0, 1], \ \tau \in [0, 1/2\tau_0]\}$. Then, as shown by the author [8], the use of known methods for solving the initial boundary-value problem in case of its invariant treatment (15) and (20) leads to additional economy of computing resources by 2–3 times.

Let us select the range D_1 in order to consistently describe the azimuthal harmonics of classical brightness coefficients $\rho^m(\eta, \xi, \tau_0)$ and $\sigma^m(\eta, \xi, \tau_0)$ at the slab's outer boundaries $\tau = 0$ and $\tau = \tau_0$ together with the azimuthal harmonics of generalized brightness coefficients $R^m(\tau, \eta, \xi, \tau_0)$ and $S^m(\tau, \eta, \xi, \tau_0)$ inside the slab $0 < \tau < \tau_0$, which are defined as

$$I^{m}(\tau, \eta, \xi, \tau_0) = S\xi T^{m}(\tau, \eta, \xi, \tau_0), \ (\eta, \xi) > 0, \ \tau \in [0, \tau_0], \ m = \overline{0, M}, \tag{21}$$

$$I^{m}(\tau, -\eta, \xi, \tau_0) = S\xi R^{m}(\tau, \eta, \xi, \tau_0), \ (\eta, \xi) > 0, \ \tau \in [0, \tau_0], \ m = \overline{0, M}. \tag{22}$$

Is should be also noted that nonlinear integral equations (16) and linear combinations (18)–(19) represent generalization of unified photometrical functions $E^m(\eta, \xi, \tau_0)$, previously introduced in [1] and [2] independently, at arbitrary optical depths τ, including the slab's outer boundaries $\tau = 0$ and $\tau = \tau_0$, namely:

$$I_{\Sigma}^{m}(\tau, \eta, \xi, \tau_0) = S\xi E^{m}(\tau, \eta, \xi, \tau_0), \ (\eta, \xi) > 0, \ \tau \in [0, \tau_0], \ m = \overline{0, M}, \tag{23}$$

$$I_{\Sigma}^{m}(\tau_0, \eta, \xi, \tau_0) = S\xi E^{m}(\eta, \xi, \tau_0), \ I_{\Sigma}^{m}(0, \eta, \xi, \tau_0) \equiv 0, (\eta, \xi) > 0, \ m = \overline{0, M}. \tag{24}$$

Furthermore, it is important to remark that the procedure of outer primary energy sources symmetrization relative to the mid-plane of a uniform slab $(1/2\tau_0)$ and modification of the classical Ambarzumian-Chandrasekhar invariance principle [3, 4] in the form proposed by Hovenier [1] unambiguously leads to the above unified photometrical function $I_{\pm}^m(\tau, \eta, \xi, \tau_0)$ coinciding with (18)–(19) and thereby eliminates its nonuniqueness from a physical point of view.

Thus, from a physical point of view, in the general case of multiple anisotropic light scattering, the concept of the unified photometrical function $I_{\Sigma}^m(\tau, \eta, \xi, \tau_0)$ and corresponding photometrical invariants $I_{\pm}^m(\tau, \eta, \xi, \tau_0)$ is a consequence of the mirror reflection principle [6] for radiation fields in a uniform slab of a finite optical thickness $\tau_0 < \infty$. In that case, the outer energy sources symmetrization relative to the mid-plane of the uniform slab $(1/2\tau_0)$ being considered allows to explicitly demonstrate the invariant properties of the above-mentioned values at the mirror-symmetric sighting directions η and $-\eta$. Form a mathematical point of view, the new constructions of unified photometrical function $I_{\Sigma}^m(\tau, \eta, \xi, \tau_0)$ and photometrical invariants $I_{\pm}^m(\tau, \eta, \xi, \tau_0)$ are derived from linear structural transformations of formal solutions of the basic boundary-value problem in the radiative transfer theory, these simultaneous transformations being related to a group of spatial shifts along $\tau \Rightarrow \tau_0 - \tau$ and to sighting line rotations $\eta \Rightarrow -\eta$. These linear transformations of intensities $I^m(\tau, \eta, \xi, \tau_0)$, as it is possible to show, will lead to structural linear transformations of the generalized brightness coefficients of the kind $R^m(\tau, \eta, \xi, \tau_0) \pm T^m(\tau, \eta, \xi, \tau_0)$ on the basis of the appropriate linear transformation of Ambarzumian-Chandrasekhar generalized functions of the kind $\varphi_i^m(\tau, \eta, \xi, \tau_0) \pm (-1)^{i+m} \psi_i^m(\tau, \eta, \xi, \tau_0)$.

3   Fitting Analytical Approximations for Upgoing and Downgoing Intensities of Radiation Fields

Note that a main problem of efficient application of the unified photometrical function $I_{\Sigma}^m(\tau, \eta, \xi, \tau_0)$ and photometrical invariants $I_{\pm}^m(\tau, \eta, \xi, \tau_0)$, based on numerical solutions of non-linear integral equations (16), is the necessity to use their spatial-angular symmetry properties at arbitrary optical levels $0 < \tau < \tau_0$ inside the slab. A similar problem naturally arises for upgoing $I^m(\tau_0 - \tau, -\eta, \xi, \tau_0)$, $\eta > 0$ and downgoing $I^m(\tau, \eta, \xi, \tau_0)$, $\eta > 0$ radiation intensities by using the real slab's phase functions. Unfortunately, unlike the slab's outer boundaries $\tau = 0$ and $\tau = \tau_0$, in the case of arbitrary optical depths $0 < \tau < \tau_0$, neither the spatial-angular symmetry properties of generalized brightness coefficients $R(\tau, \eta, \xi, \varphi, \tau_0)$ and $T(\tau, \eta, \xi, \varphi, \tau_0)$ nor their structures are known yet. For this test computations of azimuthal harmonics $I^m(\tau, \pm\eta, \xi, \tau_0)$, $\eta > 0$ even in the simple case of using exact formulae of primary radiation scattering

$$I_1^m(\tau, \eta, \xi, \tau_0) = \frac{\Lambda S}{4} \xi P^m(\eta, \xi) \frac{e^{-\tau/\xi} - e^{-\tau/\eta}}{\xi - \eta}, \ (\eta, \xi) \in [0, 1], \ \tau \in [0, \tau_0], \ m = \overline{0, M}, \quad (25)$$

$$I_1^m(\tau, -\eta, \xi, \tau_0) = \frac{\Lambda S}{4} \xi P^m(-\eta, \xi) \frac{1 - e^{-(\tau_0 - \tau)\left(\frac{1}{\eta} + \frac{1}{\xi}\right)}}{\xi + \eta} e^{-\tau/\xi}, \ (\eta, \xi) \in [0, 1], \ \tau \in [0, \tau_0], \ m = \overline{0, M}, \quad (26)$$

have shown that there are no spatial-angular symmetry properties in them. Furthermore, finding highly accurate solutions of Eqs. (11) and (16) in the case of real phase functions $P(\cos\gamma)$ is connected with overcoming their poor stability and slow grid $\{\tau_i, \eta_j, \xi_k\}$ convergence due to the need of accounting for strong maximums of the functions $P(\cos\gamma)$ in the proximity of small scattering angles $\gamma \simeq 0°$. Therefore, in order to overcome the above difficulties, it is feasible, instead of exact Eqs. (11), (13) and (16), to obtain adequate simple fitting relations having a sufficient accuracy and accounting for the main features of the spatial-angular radiation fields distribution in the case of real phase functions $P(\cos\gamma)$ for natural environments. Further it will be shown that the above mentioned approximate analytical models for the upgoing $I^m(\tau, \eta, \xi, \tau_0)$, $\eta < 0$ and downgoing $I^m(\tau, -\eta, \xi, \tau_0)$, $\eta > 0$ radiation intensities can be obtained on the basis of fitting parametrization of nonlinear integral equations (11)–(13). Indeed, real aerosol and cloud phase functions $P(\cos\gamma)$ for upgoing $P(-\eta', \eta'', \varphi' - \varphi'')$ and downgoing $P(\eta', \eta'', \varphi' - \varphi'')$ vision directions have a sharp maximum at $\eta' = \eta'' = 0$, $\varphi' = \varphi''$ and $\eta' = \eta''$, $\varphi' = \varphi''$ respectively. Let us use these properties of the function $P(\cos\gamma)$ in considering Eqs. (11) and (13). Taking the unknown values of intensities $I^m(\tau_0 - \tau, -\eta, \xi, \tau_0)$ and $I^m(\tau, \eta, \xi, \tau_0)$ in the maximum points of the given phase function $P(\cos\gamma)$ out of the integrals, we will ignore the nonlinear members in (11) and (13). An obvious physical justification of such approximation lies in the fact that the nonlinear members in Eqs. (11) and (13) correspond to photons reradiated during their multiple scattering from the "forward" $(\gamma \sim 0°)$ to "backward" $(\gamma > 90°)$ direction, which is a highly improbable event for strongly elongated real phase functions $P(\cos\gamma)$ having their sharp maximums in the proximity to angles $\gamma \simeq 0°$.

Taking into account the above considerations, after some simple transformations and rather cumbersome fitting operations connected with the parametrization of Eqs. (11) and (13) on the basis of the mentioned phase function $P(\cos\gamma)$ properties, near small scattering angles $\gamma \simeq 0°$ we obtain the desired fitting analytical approximations for the components $I^m(\tau_0 - \tau, -\eta, \xi, \tau_0)$, $\eta > 0$ and $I^m(\tau, \eta, \xi, \tau_0)$, $\eta > 0$ of the azimuthal harmonics of unified photometrical functions $I_\Sigma^m(\tau, \eta, \xi, \tau_0)$, $\eta \in [-1, 1]$ in the following form:

$$I^m(\tau_0 - \tau, -\eta, \xi, \tau_0) \simeq \frac{\Lambda S}{4} \xi P^m(-\eta, \xi) \frac{1 - e^{-\tau\left(\frac{1}{\eta} + \frac{1}{\xi}\right)}}{\xi\left[1 - \frac{\Lambda}{2}a(\eta)\right] + \eta\left[1 - \frac{\Lambda}{2}a(\xi)\right]} e^{-(\tau_0 - \tau)/\xi}, \quad (27)$$

$\eta \in [0, 1]$, $\xi \in [0, 1]$, $\tau \in [0, \tau_0]$, $m = \overline{0, M}$,

$$I^m(\tau, \eta, \xi, \tau_0) \simeq \frac{\Lambda S}{4} \xi P^m(\eta, \xi) \frac{e^{-\frac{\tau}{\eta}} - e^{-\frac{\tau}{\xi}}}{\xi\left[1 - \frac{\Lambda}{2}a(\eta)\right] - \eta\left[1 - \frac{\Lambda}{2}a(\xi)\right]}, \quad (28)$$

$\eta \in [0, 1]$, $\xi \in [0, 1]$, $\tau \in [0, \tau_0]$, $m = \overline{0, M}$,

where the fitting auxiliary function $a(\eta)$ can be found according to the following exact formula:

$$a(\eta) = \frac{1}{2\pi} \int\limits_0^{2\pi} d\varphi' \int\limits_0^1 P(\eta', \eta, \varphi') d\eta'. \tag{29}$$

4 Results of Auxiliary Functions and Slab's Brightness Coefficients Numerical Modelling

Taking into account the numerical application, the evaluation of the accuracy of the above approximate formulae for the intensities $I^m(\tau, \eta, \xi, \varphi, \tau_0)$ and $I^n(\tau_0 - \tau, -\eta, \xi, \tau_0)$ requires more detailed consideration of the auxiliary functions $a(\xi)$. For that purpose, we use formula (29) together with another expression of the function $a(\xi)$ in which the integration is performed over the new argument $\mu(\cos \gamma)$. To use this argument, we must move from the old integration variables $\theta = \arccos \eta$, $\xi = \arccos \xi$ and φ to other variables (μ, ξ, ψ), i.e. the angle μ and a new azimuth ψ. In this case, the new integration domain of (μ, ξ, ψ) is determined by the following relations:

$$0 \le \psi \le 2\pi \text{ at } 1 \ge \mu \ge \sqrt{1 - \xi^2}, \; \xi \in [0, 1], \tag{30}$$

$$\beta(\mu, \xi) \le \psi \le 2\pi - \beta(\mu, \xi) \text{ at } \sqrt{1 - \xi^2} \ge \mu \ge -\sqrt{1 - \xi^2}, \; \xi \in [0, 1], \tag{31}$$

where the function $\beta(\mu, \xi)$ is equal to

$$\beta(\mu, \xi) = \arccos \frac{\xi\mu}{\sqrt{(1 - \xi^2)(1 - \mu^2)}}, \; \xi \in [0, 1]. \tag{32}$$

The resulting expression for the function $a(\xi)$ is the following:

$$a(\xi) = \int\limits_0^1 P(\mu) d\mu - \frac{1}{\pi} \int\limits_0^{\sqrt{1-\xi^2}} [P(\mu) - P(-\mu)]\beta(\mu, \xi) d\mu. \tag{33}$$

Let us now focus on some above mentioned properties of the function $a(\xi)$. First, from the condition of normalizing the given phase function $P(\cos \gamma)$:

$$\frac{1}{2} \int\limits_0^\pi P(\cos \gamma) \sin \gamma d\gamma = 1, \tag{34}$$

it follows, as stated above, that the sum of auxiliary functions $a(\xi)$ and $a(-\xi)$ is equal to 2. In particular, at $\xi = 0$ we get $a(0) = 1$. For a strongly elongated phase function $P(\eta, \xi, \varphi)$, as evident from (34), we have $a(\xi) > a(-\xi)$ at $\xi > 0$. On account of (33), this means that $1 \leq a(\xi) < 2$. It should be also noted that at very small η and ξ, as follows from Eqs. (11)–(13) and correlations (14), the values of the intensities $I^m(\tau, \eta, \xi, \varphi, \tau_0)$ and $I^m(\tau_0 - \tau, -\eta, \xi, \tau_0)$ are mainly determined by the first order light scattering intensities $I_1^m(\tau, \eta, \xi, \varphi, \tau_0)$ and $I_1^m(\tau_0 - \tau, -\eta, \xi, \tau_0)$ described by free terms in (11) and (13) outside the integral in the right part of the considered equations.

Note that by using the well-known Henyey-Greenstein phase function

$$P(\cos \gamma) = \frac{1 - g^2}{(1 + g^2 - 2g \cos \gamma)^{3/2}}, \quad |g| < 1 \tag{35}$$

we can find explicit expression for the auxiliary function $a(\xi)$. After inserting the phase function (35) into (33) and integrating by parts, we obtain:

$$a(\xi) = \frac{1 + g}{g} - \frac{1 - g^2}{g} \frac{\xi}{\sqrt{1 + g^2 + 2g\omega}} \left[\frac{1}{1 - \omega} \prod \left(\frac{\pi}{2}, \frac{2\omega}{1 - \omega}, \Omega \right) \right] + \left[\frac{1}{1 + \omega} \prod \left(\frac{\pi}{2}, -\frac{2\omega}{1 - \omega}, \Omega \right) \right], \tag{36}$$

where the functions Ω, ω, and parameters g are defined as follows:

$$g = \frac{1}{3}x_1, \quad \Omega^2 = \frac{4g\omega}{1 + g^2 + 2g\omega}, \quad \omega = \sqrt{1 - \xi^2}. \tag{37}$$

The complete elliptic integrals $\prod$ of the third kind in formula (36) are expressed through elliptic integrals of the first and third kind [9]. The values of the auxiliary function $a(\xi)$ calculated using this formula are shown in Table 1. As follows from it, as the parameter g increases, i.e. aa the elongation of the phase function $P(\cos \gamma)$ grows, the values of the function $a(\xi)$ at $\xi > 0$ tend to 2.

Note that in the practice of numerical radiation modelling the phase function $P(\cos \gamma)$ is often used, which takes another form in comparison with (35), namely:

$$P(\cos \gamma) = \frac{2l}{(1 + l \cos \gamma) ln \left| \frac{1+l}{1-l} \right|}, \tag{38}$$

where the basic fitting parameter $|l| < 1$.

By substituting (38) into (29) and subsequently integrating, we obtain the following expression for the auxiliary function $a(\xi)$:

Table 1. Values of the auxiliary function $a(\xi)$ for Hanyey-Greenstein phase function

g	ξ							
	0	0.05	0.10	0.20	0.40	0.60	0.80	1
0.9500	1.000	1.5600	1.7100	1.8600	1.9300	1.9600	1.9700	1.9800
0.9900	1.000	1.8780	1.7400	1.9720	1.9870	1.9920	1.9940	1.9960
0.9990	1.000	1.9879	1.9940	1.9972	1.9987	1.9992	1.9994	1.9996

$$a(\xi) = \frac{2l}{ln\left|\frac{1+l}{1-l}\right|} \ln \frac{(1+l)(1+\xi)}{\xi + \sqrt{1 - l^2 + l^2\xi^2}}. \tag{39}$$

In order to estimate the actual effectiveness of approximate formula (39) as compared with (36), we need to use numerical solutions of non-linear equations (11)–(13) for the strongly elongated ($g \simeq 1$) phase function (38).

In fact, Table 2 demonstrates the angular dependence (η, ξ) comparison between exact [10] and approximate values (27)–(28) of the function $I^0(0, -\eta, \xi, \tau_0)$ in the case of $\tau_0 = \infty$ for the Henyey-Greenstein phase function (35) at $g = 0.990$ and $g = 0.999$. From the data obtained it follows that values of function $I^0(\tau, \eta, \xi, \tau_0)$ found by approximate formula (27) at $\tau_0 \rightarrow \infty$ asymptotically tend to their exact values [10] as values of optical parameter $(1 - \Lambda)/(1 - g)$ increases. In this case we have $g \rightarrow 1$ and $\Lambda \ll 1$, which stipulates a small contribution of multiple light scattering to the general diffuse radiation field of the considered uniform slab. It is this condition that allows to substantially simplify initial non-linear integral equations (11)–(13) making use of the extreme angular properties of the given phase function near scattering angles $\gamma \simeq 0°$ and, as a result, to obtain fitting analytical expressions for the intensities of inner radiation fields. Note that for the considered case of strongly elongated phase functions ($g \simeq 1$), as follows from Table 1, the function $a(\xi)$ at $\xi > 0$ is practically equal to 2. In this case, based on (27)–(28), the functions $I^m(\tau, \eta, \xi, \varphi, \tau_0)$ and $I^m(\tau_0 - \tau, -\eta, \xi, \varphi, \tau_0)$ will be equal to the functions $I_1^m(\tau, \eta, \xi, \varphi, \tau_0)$ and $I_1^m(\tau_0 - \tau, -\eta, \xi, \varphi, \tau_0)$ correspondingly multiplied by $(1 - \Lambda)^{-1}$ as will be shown below. It should be emphasized that the given estimations of the corresponding approximate formula for the intensity $I^0(0, -\eta, \xi, \tau_0)$ at $\tau_0 = \infty$ depend on the correlation of angular variables (η, ξ), parameter g and single scattering albedo Λ.

Taking into account exact relations (25)–(26), instead of (27)–(28) we obtain:

$$I^m(\tau_0 - \tau, -\eta, \xi, \tau_0) \simeq M(-\eta, \xi)I_1^m(\tau_0 - \tau, -\eta, \xi, \tau_0),\ \eta \in [0,1],\ \xi \in [0,1],\ \tau \in [0, \tau_0],\ m = \overline{0, M}, \tag{40}$$

$$I^m(\tau, \eta, \xi, \tau_0) \simeq M(\eta, \xi)I_1^m(\tau, \eta, \xi, \tau_0),\ \eta \in [0,1],\ \xi \in [0,1],\ \tau \in [0, \tau_0],\ m = \overline{0, M}. \tag{41}$$

Formulae (40) and (41) demonstrate an important role of primary light scattering in the formation of a uniform slab's radiation field in the case of strongly elongated phase functions $P(\cos\gamma)$. The adaptive fitting function $M(\pm\eta, \xi)$ in (40)–(41) which is determined according to the following formula:

$$M(-\eta, \xi) = \frac{\eta + \xi}{\xi\left[1 - \frac{\Lambda}{2}a(\eta)\right] + \eta\left[1 - \frac{\Lambda}{2}a(\xi)\right]},\ (\eta, \xi) > 0, \tag{42}$$

$$M(\eta, \xi) = \frac{\xi - \eta}{\xi\left[1 - \frac{\Lambda}{2}a(\eta)\right] - \eta\left[1 - \frac{\Lambda}{2}a(\xi)\right]},\ (\eta, \xi) > 0, \tag{43}$$

represents corrective multipliers accounting for the influence of photons multiple scattering inside the uniform slab. It should be noted that expressions (27)–(28) could be presented in a more compact form, namely:

$$I^m(\tau, \eta, \xi, \tau_0) \simeq \frac{\Lambda S}{4} \xi P^m(\eta, \xi) \frac{e^{-\tau/\xi} - e^{-\tau/\eta}\Theta(\eta) - e^{-\tau/\xi}e^{-(\tau_0-\tau)(\frac{1}{\xi}-\frac{1}{\eta})}\Theta(-\eta)}{\xi[1 - \frac{\Lambda}{2}a(|\eta|)] - \eta[1 - \frac{\Lambda}{2}a(\xi)]}, \qquad (44)$$

$\eta \in [-1, 1], \ \xi \in [0, 1], \ \tau \in [0, \tau_0], \ m = \overline{0, M}.$

Similarly to (40)–(41), using relations (44) and (42)–(43), we obtain:

$$I^m(\tau, \eta, \xi, \tau_0) \simeq M(\eta, \xi)I_1^m(\tau, \eta, \xi, \tau_0), \ \eta \in [-1, 1], \ \xi \in [0, 1], \ \tau \in [0, \tau_0], \ m = \overline{0, M}, \quad (45)$$

where the adaptive fitting function $M(\eta, \xi)$, symmetrical by its angular variables (η, ξ), is equal to

$$M(\eta, \xi) = \frac{\xi - \eta}{\xi[1 - \frac{\Lambda}{2}a(|\eta|)] - \eta[1 - \frac{\Lambda}{2}a(\xi)]}, \ \eta \in [-1, 1], \ \xi \in [0, 1], \ \tau \in [0, \tau_0], \ m = \overline{0, M}. \quad (46)$$

Formula (46) allows to carry out further simplification in the case of an arbitrary optical thickness $\tau_0 \leq \infty$ if we use the above property of auxiliary function $a(\eta)$ for strongly elongated phase functions $P(\cos \gamma)$, i.e. on condition $a(-\eta) = 2 - a(\eta) \ll 1$, which is fulfilled, as can be easily shown, for $\eta \geq \eta_0 \geq 0.9$ when $a(\eta) \simeq 2$:

$$M(\eta, \xi) \simeq \frac{1}{1 - \Lambda}, \ \eta \in [-1, 1], \ \xi \in [0, 1], \ \tau \in [0, \tau_0], \ m = \overline{0, M}. \quad (47)$$

Taking into account (47), instead of (45) we have:

$$I^m(\tau, \eta, \xi, \tau_0) \simeq \frac{1}{1 - \Lambda} I_1^m(\tau, \eta, \xi, \tau_0), \ \eta \in [-1, 1], \ \xi \in [0, 1], \ \tau \in [0, \tau_0], \ m = \overline{0, M}. \quad (48)$$

The structural similarity of approximate formulae (40)–(48) with (25)–(26) in the applied form of fitting adaptive multipliers $M(\eta, \xi)$ and $M(-\eta, \xi)$ for downgoing and upgoing radiation intensities in the case of photons single scattering becomes understandable if we take notice of the fact that initial nonlinear integral equations (11)–(13) and (16) are a generalization of the above mentioned Chandrasekhar system of non-linear integral equations [4] at the finite optical thickness $\tau_0 < \infty$ of the considered uniform slab.

It should be noted however that the inner radiation field described by approximate formula (44) can be described by alternative adaptive structural functions $j^m(\tau, \eta, \xi, \tau_0)$ different from the fitting function $M(\eta, \xi)$ that is symmetrical by its angular variables (η, ξ). Indeed, let us represent the unknown radiation intensity $I^m(\tau, \eta, \xi, \tau_0)$ as follows:

$$I^m(\tau, \eta, \xi, \tau_0) = S\xi\left[\Theta(\eta) + \Theta(-\eta)e^{-\frac{\tau}{\xi}} - e^{-\frac{\tau}{\xi}}e^{-(\tau_0-\tau)(\frac{1}{\xi}-\frac{1}{\eta})}\Theta(-\eta)\right]j^m(\tau, \eta, \xi, \tau_0),$$

$\eta \in [-1, 1], \ \xi \in [0, 1], \ \tau \in [0, \tau_0], \ m = \overline{0, M}.$

$$(49)$$

Table 2. Comparison of angular dependencies of exact and approximate values of function $\rho^0(\eta, \xi, \tau_0)$ in case of semi-infinite atmosphere depending on parameters g and Λ

Λ	0.80				0.90				0.99			
g	**0.990**		**0.999**		**0.990**		**0.999**		**0.990**		**0.999**	
$\rho^0(\eta, \xi, \tau_0)$ / η	Exact	Approx.	Exact	Approx.	Exact	Approx.	Exact	Approx.	Exact	Approx.	Exact	Approx.
ξ = 0.1												
0	2.0	2.1	0.2	0.2	2.5	2.6	0.2	0.2	3.9	3.1	0.33	0.31
0.2	0.24	0.22	0.024	0.024	0.46	0.45	0.054	0.054	2.0	1.8	0.40	0.48
0.4	0.052	0.50	0.005	0.005	0.11	0.10	0.012	0.012	0.51	0.39	0.12	0.11
0.6	0.020	0.019	0.002	0.002	0.042	0.037	0.0045	0.0045	0.22	0.14	0.048	0.040
0.8	0.0095	0.0090	0.0010	0.0010	0.020	0.018	0.0022	0.0022	0.12	0.067	0.024	0.019
1	0.00053	0.00051	0.00056	0.00056	0.012	0.010	0.0012	0.0012	0.073	0.038	0.014	0.011
$(1-\Lambda)/(1-g)$	20		200		10		100		1		10	
ξ = 1												
0	0.0028	0.0024	0.0024	0.00023	0.0043	0.0029	0.00029	0.00028	0.015	0.0035	0.0006	0.00034
0.2	0.0046	0.0043	0.00045	0.00045	0.0100	0.0092	0.00100	0.00100	0.089	0.051	0.0120	0.0099
0.4	0.0032	0.0030	0.00031	0.00031	0.0075	0.0066	0.0068	0.00058	0.092	0.050	0.0083	0.0072
0.6	0.0023	0.0022	0.00022	0.00022	0.0055	0.0048	0.00049	0.00049	0.080	0.041	0.0600	0.0052
0.8	0.0017	0.0160	0.00016	0.00016	0.0041	0.0035	0.00036	0.00036	0.067	0.033	0.0045	0.0039
1	0.0013	0.0013	0.00012	0.00012	0.0031	0.0028	0.00028	0.00028	0.055	0.026	0.0034	0.0030
$(1-\Lambda)/(1-g)$	20		200		10		100		1		10	

Then from approximate relation (48) it follows that the new adaptive structural functions $\tilde{j}^m(\tau, \eta, \xi, \tau_0)$ at $\eta \in [-1, 1]$, $\xi \in [0, 1]$, $\tau \in [0, \tau_0]$ and $m = \overline{0, M}$ is equal to

$$\tilde{j}^m(\tau, \eta, \xi, \tau_0) = \frac{\Lambda}{4} P^m(\eta, \xi) \frac{e^{-\frac{\tau}{\xi}} - e^{-\frac{\tau}{\eta}}}{\xi[1 - \frac{\Lambda}{2}a(\eta)] - \eta[1 - \frac{\Lambda}{2}a(\xi)]}, \ (\eta, \xi) > 0, \ \tau \in [0, \tau_0], \ m$$
$$= \overline{0, M},$$

(50)

$$\tilde{j}^m(\tau, -\eta, \xi, \tau_0) = \frac{\Lambda}{4} P^m(-\eta, \xi) \frac{1}{\xi[1 - \frac{\Lambda}{2}a(\eta)] + \eta[1 - \frac{\Lambda}{2}a(\xi)]}, \ (\eta, \xi) > 0, \ \tau$$
$$\in [0, \tau_0], \ m = \overline{0, M}.$$

(51)

Using the angular symmetry of phase function $P^m(\pm\eta, \xi) = P^m(\pm\xi, \eta)$, from formulae (50)–(51) we obtain:

$$\tilde{j}^m(\tau, \eta, \xi, \tau_0) = \tilde{j}^m(\tau, \xi, \eta, \tau_0), \tilde{j}^m(\tau, -\eta, \xi, \tau_0) = \tilde{j}^m(\tau, -\xi, \eta, \tau_0), (\eta, \xi) > 0, \ \tau$$
$$\in [0, \tau_0], \ m = \overline{0, M},$$

(52)

Thus, for downgoing ($\eta > 0$) and upwngoing ($\eta < 0$) vision directions of the uniform slab's radiation fields that the structural functions $\tilde{j}^m(\tau, \eta, \xi, \tau_0)$ are symmetrical functions of its angular arguments η and ξ.

Making use of the above mentioned condition $a(\eta) \simeq 2$ for strongly elongated phase functions $P(\cos\gamma)$, we have analogously to (47)–(48) simpler structural forms for representation of relations (50)–(51):

$$\tilde{j}^m(\tau, \eta, \xi, \tau_0) \simeq \frac{\Lambda P^m(\eta, \xi)}{4(1 - \Lambda)(\xi - \eta)} \left[e^{-\tau/\xi} - e^{-\tau/\eta} \right], \ (\eta, \xi) > 0, \ \tau \in [0, \tau_0], \ m = \overline{0, M},$$

(53)

$$\tilde{j}^m(\tau, -\eta, \xi, \tau_0) \simeq \frac{\Lambda P^m(-\eta, \xi)}{4(1 - \Lambda)(\xi + \eta)}, \ (\eta, \xi) > 0, \ \tau \in [0, \tau_0], \ m = \overline{0, M}.$$

(54)

The above obtained approximate analytical expressions (27) and (28) are basic for construction of the photometrical intensities invariants $I_{\pm}^m(\tau, \eta, \xi, \tau_0)$ and unified photometrical function $I_{\Sigma}^m(\tau, \eta, \xi, \tau_0)$, $\eta \in [-1, 1]$, $\xi \in [0, 1]$, $\tau \in [0, \tau_0]$, $m = \overline{0, M}$. Using (14)–(15) and (40)–(41), for the photometrical intensities invariants $I_{\pm}^m(\tau, \eta, \xi, \tau_0)$ in the case of strongly elongated phase functions $P(\cos\gamma)$, we find the following fitting relations:

$$I_{\pm}^m(\tau, \eta, \xi, \tau_0) \simeq M(-\eta, \xi)I_1^m(\tau_0 - \tau, -\eta, \xi, \tau_0) \pm M(\eta, \xi)I_1^m(\tau, \eta, \xi, \tau_0),$$
$$\eta \in [-1, 1], \ \xi \in [0, 1], \ \tau \in [0, \tau_0], \ m = \overline{0, M} \cdot$$

(55)

Similarly, for the unified photometrical function $I_{\Sigma}^m(\tau, \eta, \xi, \tau_0)$, using (18) and (55) we have:

$$I_\Sigma^m(\tau, \eta, \xi, \tau_0) \simeq (\xi + \eta)M(-\eta, \xi)I_1^m(\tau_0 - \tau, -\eta, \xi, \tau_0) + (\xi - \eta)M(\eta, \xi)I_1^m(\tau, \eta, \xi, \tau_0),$$
$$\eta \in [-1, 1], \ \xi \in [0, 1], \ \tau \in [0, \tau_0], \ m = \overline{0, M}.$$

$$(56)$$

Taking into account relations (47)–(48), instead of (55)–(56) we obtain simpler structural representations for azimuthal harmonics of photometrical invariants $I_\pm^m(\tau, \eta, \xi, \tau_0)$ and unified photometrical functions $I_\Sigma^m(\tau, \eta, \xi, \tau_0)$, namely:

$$I_\pm^m(\tau, \eta, \xi, \tau_0) \simeq \frac{1}{1 - \Lambda} I_{1,\pm}^m(\tau, \eta, \xi, \tau_0), \ \eta \in [-1, 1], \ \xi \in [0, 1], \ \tau \in [0, \tau_0], \ m = \overline{0, M},$$

$$(57)$$

$$I_\Sigma^m(\tau, \eta, \xi, \tau_0) \simeq \frac{1}{1 - \Lambda} I_{1,\Sigma}^m(\tau, \eta, \xi, \tau_0), \ \eta \in [-1, 1], \ \xi \in [0, 1], \ \tau \in [0, \tau_0], \ m = \overline{0, M},$$

$$(58)$$

where the values $I_{1,\pm}^m(\tau, \eta, \xi, \tau_0)$ and $I_{1,\Sigma}^m(\tau, \eta, \xi, \tau_0)$ are received from (14) and (19) correspondingly, on the basis of exact formulae (25) and (26).

The spatial-angular structure of the above obtained adaptive fitting formulae allows for a clear physical interpretation of the transition from exact integral equations (11) and (13) to approximate relations (40)–(46) and (55)–(56). Namely, this operation imitates selective "freezing" the primary scattering radiation field into the common field of multiple photons scattering with subsequent multiplicative correction with the help of the adaptive symmetrical function $M(\eta, \xi)$ and $M(-\eta, \xi)$ due to the strong elongation of the given phase function $P(\cos \gamma)$ in the proximity of angles $\gamma \simeq 0°$.

For the case of a semi-infinite uniform slab ($\tau_0 = \infty$), Table 2 demonstrates the comparison between exact values of diffuse reflection coefficient $\rho_\infty^m(\eta, \xi)$ calculated in [10] for m = 0 and the Henyey-Greenstein phase function at $g = 0.990$ and $g = 0.999$, and respective approximate values of $\rho_\infty^0(\eta, \xi)$ calculated according to formula (27) at $\tau = 0$ and $\tau_0 = \infty$ for similar optical parameters. From the given data it is evident that in the case of strongly elongated phase function $P(\cos \gamma)$, approximate values of $\rho_\infty^0(\eta, \xi)$ tend asymptotically to their exact values as the basic parameter $(1 - \Lambda)/(1 - g)$ increases.

In conclusion it should be noted that nonlinear integral equations (16) for azimuthal harmonics of the unified photometrical function $I_\Sigma^m(\tau, \eta, \xi, \tau_0)$ was not used directly in solving the above considered problem. The obvious reason is that in order to carry out the fitting adaptive parametrization of Eqs. (16), as noted above, we need to understand the angular symmetry properties of the functions $I_\pm^m(\tau, \eta, \xi, \tau_0)$ which still remain unknown. This difficulty can be avoided, however, if the strong elongation of a given phase function $P(\cos \gamma)$ is accounted for by using the known similarity ratios between the initial environment optical parameters τ_0 and Λ, on the one hand, and the same optical parameters after cutting the maximum of the functions $P(\cos \gamma)$ in the proximity to small scattering angles $\gamma \simeq 0°$, on the other hand [2]. Indeed, let us represent the strongly elongated phase function $P(\cos \gamma)$ as a sum of the Dirac δ-function and the smoothed phase function $P(\cos\gamma)$ having no maximum in the proximity to small scattering angles $\gamma \sim 0°$, namely:

$$P(\cos\gamma) = L\delta(\cos\gamma - \cos\gamma_0) + (1-L)\tilde{P}(\cos\gamma), \qquad (59)$$

where $\tilde{P}(\cos\gamma) = P(\cos\gamma_0)$ at $\gamma \leq \gamma_0$ and $\tilde{P}(\cos\gamma) = P(\cos\gamma)$ at $\gamma > \gamma_0$. The "cutting" angle $\gamma_0 \simeq 0°$ values of the function $P(\cos\gamma)$ maximum are defined by normalization of the residual phase function $\tilde{P}(\cos\gamma)$. Using (59) and taking into account the above considerations, the photometrical invariants $I_\pm^m(\tau,\eta,\xi,\tau_0)$ can be approximated in the following form:

$$I_\pm^m(\tau,\eta,\xi,\tau_0) \simeq \tilde{I}_\pm^m(\tau,\eta,\xi,\tau_0) + f_1^\pm(\tau,\eta,\xi,\tau_0)P^m(\eta,\xi) + f_2^\mp(\tau,\eta,\xi,\tau_0)P^m(-\eta,\xi),$$
$$\eta \in [-1,1],\ \xi \in [0,1],\ \tau \in [0,\tau_0],\ m = \overline{0,M}, \qquad (60)$$

where the adaptive fitting functions $f_1^\pm$ and $f_2^\mp$ correlate with the intensities of upgoing $(\eta < 0)$ and downgoing $(\eta > 0)$ radiation in the primary scattering approximation and can be found by using the main property of δ-function:

$$\int_{-\infty}^{+\infty} \delta(x-x')f_{1,2}^\pm(x')dx' = f_{1,2}^\pm(x). \qquad (61)$$

Applying relations (60)–(61) to Eqs. (16) and neglecting their nonlinear term (for reasons mentioned above), we obtain the following approximate relation for azimuthal harmonics $I_\Sigma^m(\tau,\eta,\xi,\tau_0)$:

$$I_\Sigma^m(\tau,\eta,\xi,\tau_0) \simeq \tilde{I}_\Sigma^m(\tau,\eta,\xi,\tilde{\tau}_0) + \frac{1}{1-\Lambda}I_{1,\Sigma}^m(\tau,\eta,\xi,\tau_0),\ \eta \in [-1,1],\ \xi \in [0,1],\ \tau$$
$$\in [0,\tau_0],\ m = \overline{0,M}, \qquad (62)$$

where $\tilde{I}_\Sigma^m(\tau,\eta,\xi,\tau_0)$ are unified photometrical functions for the smoothed phase functions $\tilde{P}(\cos\gamma)$, and the functions $I_{1,\Sigma}^m(\tau,\eta,\xi,\tau_0)$ are defined on the basis of primary scattering exact formulae (25)–(26) with the use of representation (18):

$$I_{1,\Sigma}^m(\tau,\eta,\xi,\tau_0) = \xi I_{1,+}^m(\tau,\eta,\xi,\tau_0) + \eta I_{1,-}^m(\tau,\eta,\xi,\tau_0),\ \eta \in [-1,1],\ \xi \in [0,1],\ \tau \in [0,\tau_0],\ m = \overline{0,M}, \quad (63)$$

where the azimuthal harmonics of photometrical invariants $I_{1,\pm}^m(\tau,\eta,\xi,\tau_0)$ are as follows:

$$I_{1,\pm}^m(\tau,\eta,\xi,\tau_0) = \frac{\Lambda S}{4}P^m(-\eta,\xi)\frac{1-e^{-\tau(\frac{1}{\eta}+\frac{1}{\xi})}}{\xi+\eta}e^{-\frac{\tau_0-\tau}{\xi}} \pm P^m(\eta,\xi)\frac{e^{-\tau/\xi}-e^{-\tau/\eta}}{\xi-\eta}, \qquad (64)$$
$$\eta \in [0,1],\ \xi \in [0,1],\ \tau \in [0,\tau_0],\ m = \overline{0,M}.$$

The fitting values of the optical parameters $\tilde{\tau}_0$ and $\tilde{\Lambda}$ resulting from cutting the maximum of the initial strongly elongated phase functions $P(\cos\gamma)$ in the proximity to $\gamma_0 \simeq 0°$ are defined by the following similarity ratios:

$$\tilde{\tau}_0 = \tau_0\left(1 - \frac{L}{2}\Lambda\right), \ \tilde{\Lambda} = \frac{(2-L)\Lambda}{2-L\Lambda}. \tag{65}$$

With such approach, the values of the adaptive fitting optical parameters $\tilde{\tau}_0$ and $\tilde{\Lambda}$ can be found from the condition of normalization of the smoothed phase function $\tilde{P}(\cos\gamma)$:

$$L = 2 - \int\limits_0^\pi \tilde{P}(\cos\gamma)\sin\gamma d\gamma. \tag{66}$$

Substituting (63)–(64) into (62), after some simple operations we obtain the approximate representation of the unified photometrical function $I_\Sigma^m(\tau, \eta, \xi, \tau_0)$ azimuthal harmonics in the case of strongly elongated phase function $P(\cos\gamma)$:

$$I_\Sigma^m(\tau, \eta, \xi, \tau_0) \simeq \tilde{I}_\Sigma^m(\tau, \eta, \xi, \tilde{\tau}_0) + P^m(\eta, \xi)\tilde{f}_1^\pm(\tau, \eta, \xi, \tau_0) + P^m(-\eta, \xi)\tilde{f}_2^\mp(\tau, \eta, \xi, \tau_0),$$
$$\eta \in [-1, 1], \ \xi \in [0, 1], \ \tau \in [0, \tilde{\tau}_0], \ m = \overline{0, M}, \tag{67}$$

where the adaptive fitting functions $\tilde{f}_1^\pm$ and $\tilde{f}_2^\mp$ are defined by the following expressions:

$$\tilde{f}_1^\pm(\tau, \eta, \xi, \tau_0) = \frac{\Lambda S}{4(1-\Lambda)}\left(e^{-\tau/\xi} - e^{-\tau/\eta}\right), \ (\eta, \xi) > 0, \ \tau \in [0, \tau_0], \tag{68}$$

$$\tilde{f}_2^\mp(\tau, \eta, \xi, \tau_0) = \frac{\Lambda S}{4(1-\Lambda)}e^{-\frac{\tau_0-\tau}{\xi}}\left[1 - e^{-\tau\left(\frac{1}{\eta}+\frac{1}{\xi}\right)}\right], \ (\eta, \xi) > 0, \ \tau \in [0, \tau_0]. \tag{69}$$

Note that the structure of formula (67) for the unified photometrical functions $I_\Sigma^m(\tau, \eta, \xi, \tau_0)$ is very close to that of formula (60) for the photometrical invariants $I_\pm^m(\tau, \eta, \xi, \tau_0)$. Apparently, such closeness could be explained by the great influence of primary scattering on the formation of reflected and transmitted radiation in a uniform slab of a finite optical thickness τ_0 in the case of strongly elongated phase functions $\tilde{P}(\cos\gamma)$ in the proximity to small scattering angles $\gamma \simeq 0°$.

5   Conclusion

The angular-spatial structure of the obtained fitting relations is based on the exact formula for primary light scattering and taking into account the approximate calculation of multiple light scattering contribution in a uniform slab of an arbitrary optical thickness τ_0. These approximate relations generalize similar formulae of the classical radiative transfer theory applied to calculations of multiple light scattering at the outside boundaries of a uniform slab of an arbitrary optical thickness. The performed parametrization of exact nonlinear integral equations (11) and (13) allows to construct simple adaptive fitting relations for radiation field intensities both inside and at the boundaries of a plane-parallel uniform slab in the case of real strongly elongated phase functions $P(\cos\gamma)$ with maximums in the proximity of angles $\gamma \simeq 0°$. Such parametrization of the initial non-linear integral equations provides a sufficiently high accuracy of multiple light scattering calculations based on a much simpler structure of computational formulae. Besides, the received fitting relations have a clear physical meaning and demonstrate how primary light scattering transforms into multiple scattering in the case of strongly elongated phase functions. It should be also noted that the obtained approximate expressions for radiation field intensities of a free atmosphere can obviously be applied to the case of the "atmosphere – underlying surface" system. Furthermore, the received fitting analytical relations can be used for the statement and numerical modelling of various direct and inverse problem solutions in the radiative transfer theory and applied environment optics propositions.

References

1. Hovenier, J.W.: A unified treatment of reflected and transmitted intensities of homogeneous plane-parallel atmospheres. Astrom. Astrophys. **68**, 239–250 (1978)
2. Smokty, O.I.: Modeling of radiation fields in problems of space spectrophotometry. Nauka, Leningrad (1986). (in Russian)
3. Ambarzumian, V.A.: Nauchnye Trudy, vol. 1. Erevan (1960). (in Russian)
4. Chandrasekhar, C.: Radiative Transfer. Oxford Univ. Press, London (1950)
5. Sobolev, V.V.: Light Scattering in Planetary Atmospheres. Pergamon Press, Oxford (1975)
6. Smokty, O.I.: Photometrical invariants in the radiative transfer theory. In: Proceedings of the IGARSS 1993, pp. 1960–1961. Kogakuin Univ., Tokyo (1993)
7. Smokty, O.I.: Development of radiative transfer theory on the basis of mirror symmetry principle. In: Proceedings of the IRS 2001, pp. 341–342. A. Deepak Publ. Co., Hampton (2001)
8. Smokty, O.I.: Improvements of the methods of radiation fields numerical modeling on the basis of mirror reflection principle. In: Murgante, B., Misra, S., Carlini, M., Torre, C.M., Nguyen, H.-Q., Taniar, D., Apduhan, B.O., Gervasi, O. (eds.) ICCSA 2013, Part V. LNCS, vol. 7975, pp. 1–16. Springer, Heidelberg (2013)
9. Press, W., Flanery, B., Teukolsky, S., Vittering, W.: Numerical Recipes (Fortran Version). Cambridge Univ. Press, Cambridge (1989)
10. Smokty, O.I., Anikonov, A.S.: Light Scattering in the Large Optical Thickness Media. Nauka, St. Petersburg (2008). (in Russian)

Acetone Clusters Molecular Dynamics Using a Semiempirical Intermolecular Potential

Noelia Faginas-Lago[1]($\boxtimes$), Margarita Albertí[2], and Andrea Lombardi[1]

[1] Dipartimento di Chimica, Biologia e Biotecnologie,
Università di Perugia, Perugia, Italy
noelia.faginaslago@unipg.it
[2] IQTCUB, Departament de Ciència de Materials i Química Física,
Universitat de Barcelona, Barcelona, Spain

Abstract. A semiempirical force field for the intermolecular acetone (CH_3-CO-CH_3) small clusters interaction has been build and applied to characterize the acetone behaviour of some small clusters by Molecular Dynamics. Preliminary theoretical calculations of the structural and dynamical properties of the dimer and the trimer acetone have been investigated at atomistic level of detail by molecular dynamics simulations considering a microcanonical ensemble (NVE). Predictions of the $(CH_3$-CO-$CH_3)_{2-3}$ binding energies have been performed by extrapolating to $0\,K$ the mean potential energy values obtained at low temperatures. The probability of isomerization processes has been also analyzed. The extent to which a classical molecular simulation accurately predicts properties depends on the quality of the force field used to model the interactions in the fluid. The intermolecular interactions involved have been modelled by a recently developed approach targeted to accuracy and low computational cost adopting the Improved Lennard-Jones (ILJ) function to describe the long-range interaction of small clusters systems.

Keywords: Molecular dynamics · Empirical potential energy surface · Acetone clusters · DL_POLY

1 Introduction

Acetone (systematically named propanone) is the organic compound with the formula CH_3-CO-CH_3. It is a colorless, volatile, flammable liquid, and is the simplest ketone. Acetone is miscible with water and serves as an important solvent in its own right, typically for cleaning purposes in the laboratory. About 6.7 million tonnes were produced worldwide in 2010. It is a common building block in organic chemistry. Familiar household uses of acetone are as the active ingredient in nail polish remover and as paint thinner. Acetone is produced and disposed of in the human body through normal metabolic processes. It is normally present in blood and urine. People with diabetes produce it in larger amounts. Reproductive toxicity tests show that it has low potential to cause reproductive problems. Pregnant women, nursing mothers and children have

© Springer International Publishing Switzerland 2016
O. Gervasi et al. (Eds.): ICCSA 2016, Part I, LNCS 9786, pp. 129–140, 2016.
DOI: 10.1007/978-3-319-42085-1_10

higher levels of acetone [1]. Ketogenic diets that increase acetone in the body are used to counter epileptic attacks in infants and children who suffer from recalcitrant refractory epilepsy.

There are two main ways of manufacturing propanone, both produced from propene. One of them is via cumene and the other one via isopropanol. By far the most important route is the cumene process [2–4]. Acetone is able to dissolve a wide range of chemical compounds, including polar, non-polar and polymeric materials [5,6] and is very soluble in water. Accordingly, most of the acetone production is used as a solvent, the properties as a solvent being known for a long time [7]. Due to the capability to dissolve different chemical compounds, acetone is often the primary component of cleaning agents, which present specific benefits because being effective and not expensive, they have a very low toxicity. Moreover, acetone is also important in the chemical industry, being widely used in the production of methyl methacrylate [8] and bisphenol A [9].

Owing to the central role of water in everyday life, water is the most investigated solvent. However, the understanding of the solvation mechanisms in solutions of organic solvents is crucial to improve the efficiency of processes of technological interest [10–13]. From an experimental point of view, to obtain detailed dynamical information on simple and complex liquids, spectroscopic methods based on a second rank molecular property, such as the polarizability anisotropy, can be suitably applied [12]. On an other hand, Molecular Dynamics (MD) simulations is a theoretical powerful tool to analyze extensively the structural rearrangement of pure solvents, solution mixtures and the combustion processes. Combustion is a complex of chemical and physical nature of the process that leads to the conversion of chemical energy into heat energy stored in fuels. From a molecular point of view, the combustion is the result of numerous elementary processes, chemical reactions including mono-, bi- and ter-molecular and energy-transfer processes (which ensure the distribution of the energy released by the exothermic reactions to everything the system). The characterization of the elementary processes (like organic compounds, acetone) is therefore essential for a detailed understanding and for the optimization of combustion systems. Given the complexity of the problem, we resort to a multidisciplinary approach based on the use of computer models that are considered all possible elementary processes. However, the accuracy of the results depends largely on the reliability of the interaction potential. In the last years some of us have formulated a potential energy function [14] to describe the non permanent charge interactions (V_{nel}), which combined with the electrostatic contributions (V_{el}), when present, is able to describe accurately ionic and neutral systems. In such model, the relevant parameters of V_{nel} are derived from the values of the molecular polarizability. On another hand, polarization effects are indirectly considered by increasing the value of the dipole moment of the molecules in respect with those corresponding to gas phase. Such formulation has been proved to be adequate not only to describe systems on gas phase (see for instance Refs. [15–17]) but also the liquid behavior of water, [18–21] acetonitrile, [13] and ammonia [22].

All the previous results encouraged us the use of the same methodology to construct the acetone-acetone force field.

The original potential model used here is based on a formulation of the non electrostatic approach to the intermolecular interaction that exploits the decomposition of the molecular polarisability [23] into effective components associated with atoms, bonds or groups of atoms of the involved molecules. This type of contribution to the intermolecular energy was already applied in the past to the investigation of several neutral [24–32] and ionic [33–35] systems, often involving weak interactions [24,36], difficult to calculate. The adequacy of such potential energy functions to describe several intermolecular systems was proved by comparing energy and geometry predictions at several configurations with ab initio calculations.

In the present paper, we present the preliminary theoretical results of the potential model, based in a modification of the Lennard Jones (LJ) function which is applied to investigate the small clusters of acetone, $(CH_3\text{-}CO\text{-}CH_3)_2$ and $(CH_3\text{-}CO\text{-}CH_3)_3$. The paper is organized as follows: in Sect. 2, we outline the construction of the potential energy function. We give in Sect. 3 the details of the Molecular Dynamics simulations. Results are presented in Sect. 4 and concluding remarks are given in Sect. 5.

2   Potential Energy Surface

As it has been indicated in the Introduction, the formulation of the potential energy interaction has been performed by assuming the separability of non permanent (V_{nel}) and permanent (V_{el}) charge interaction contributions. V_{nel} is described by considering four interaction centers localized on the $CH_3\text{-}CO\text{-}CH_3$ molecule, considered as a rigid body. The representation of the acetone molecule is given in Fig. 1.

Two of the interaction centers are placed on the C atoms of the methyl groups, represented by CM, and the other two placed on the C and O atoms of the C=O group, described by C and O respectively. This means that the acetone molecule is given by a rigid set of four interaction sites, located on oxygen

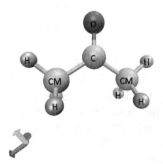

Fig. 1. Graphical representation of the labeled atoms in the acetone molecule.

and carbon atoms. Accordingly, V_{nel} is represented as a sum of pair contributions between interaction centers on different molecules. Each pair contribution described by means of an ILJ function, $V_{ILJ}(r)$, depending on the distance r between the considered centers. Specifically,

$$V_{ILJ}(r) = \varepsilon \left[\frac{m}{n(r) - m} \left(\frac{r_0}{r}\right)^{n(r)} - \frac{n(r)}{n(r) - m} \left(\frac{r_0}{r}\right)^m \right] \tag{1}$$

with,

$$n(r) = \beta + 4 \left(\frac{r}{r_0}\right)^2 \tag{2}$$

The ILJ function defined in Eq. 1, thanks to the additional parameter β, is more flexible than the traditional Lennard Jones (LJ) one. For the acetone-acetone interactions, independently of the involved interaction centers, β has been taken equal to 11, while m, as correspond to neutral-neutral interactions, is equal to 6. For neutral systems usually β take values between 6 and 11. Due to the relative role played by β and the dipole moment (μ), the specific value depends on the selected value of μ [21]. Small modifications of β can be compensated by varying μ. As it has been indicated in the Introduction section, the relevant parameters, ε describing for each pair the depth of the potential well and its location r_0 are evaluated using the values of the polarizability associated to each interaction center [37]. The values of such parameters are given in Table 1.

Table 1. ε (well depth), r_0 (equilibrium distance) parameters used to define V_{nel} for the acetone-acetone intermolecular interactions using ILJ functions. The atoms are named as in Fig. 1.

Interaction partners	ε/meV	r_0/Å
CM-CM	11.362	4.053
C-C	6.520	3.628
O-O	5.160	3.398
CM-C	8.819	3.866
CM-O	6.688	3.789
C-O	5.640	3.521

The electrostatic contribution, V_{el}, is calculated from a ten point charge distribution by applying the Coulomb law. Such distribution is compatible with the value of the experimental dipole moment of acetone, equal to 2.88 D [38] and charges of -0.4290 a.u and 0.4290 a.u. placed on the C and O atoms, while on the methyl groups charges of 0.406 a.u., -0.406 and 0.0702 are placed on the CM and H centers, respectively.

It is interesting to note that using the ILJ function, the consideration of different charge distributions can be compensated by variations of the β parameter. The modulation of β can also partially compensate the case of different representations of the charge distributions and indirectly include in the contribution of other less important components, at intermediate and short distances [13,18,19,21,22,39].

3   The Simulation Protocol of the Molecular Dynamics

Classical molecular dynamics simulations were performed using the DL_POLY [40] molecular dynamics simulation package. We performed classical MD simulations of $(CH_3\text{-}CO\text{-}CH_3)_n$, $n = 2 - 3$. For each optimized structure, we performed simulations with increasing temperature. A microcanonical ensemble (NVE) of particles, where the number of particles, N, volume, V, and total energy, E, are conserved, has been considered. The total energy, E, is expressed as a sum of potential and kinetic energies. The first one is decomposed in non electrostatic and electrostatic contributions and its mean value at the end of the trajectory is represented by the average configuration energy E_{cfg} ($E_{cfg} = E_{nel} + E_{el}$). The kinetic energy at each step, E_{kin}, allows to determine the instantaneous temperature, T, whose mean values can be calculated at the end of the simulation. The total time interval for each simulation trajectory was set equal to 5 ns after equilibration (1 ns) at each temperature. We implementated the Improved Lennard-Jones potential function and used it to treat all the intermolecular interactions in $(CH_3\text{-}CO\text{-}CH_3)_2$ and $(CH_3\text{-}CO\text{-}CH_3)_3$.

4   The $(CH_3\text{-}CO\text{-}CH_3)_2$ and $(CH_3\text{-}CO\text{-}CH_3)_3$ Small Aggregates

The binding energies for $(CH_3\text{-}CO\text{-}CH_3)_2$ and $(CH_3\text{-}CO\text{-}CH_3)_3$ have been estimated by performing MD calculations at low temperatures T. As it has been explained in previous papers [16,22,41], when no isomerization occurs, the extrapolation of the configuration (potential) energy values, E_{cfg}, to 0 K provides a good estimation of the binding energy (as E is conserved along the trajectory, at very low temperatures, the configuration energy, tends to become almost constant).

The E_{cfg} values at different (low) temperatures for $(CH_3\text{-}CO\text{-}CH_3)_2$ and $(CH_3\text{-}CO\text{-}CH_3)_3$ are given in the Table 2 and represented on the Fig. 2, respectively. As it can be seen, the results at temperatures lower than 40 K (values reported in Table 2) shown a linear trend, indicating the absence of isomerization processes. In the chosen temperature conditions, the linear dependence is easily reached for T in the range of 5 K–40 K which gives a value of −4.943 kcal mol^{-1} when it was extrapolated to 0 K. At higher temperatures, when isomerization processes are likely to occur, the different steps of the simulation probe several configurations with different potential energy values, and, according with

Table 2. E_{cfg} (Kcal mol^{-1}) at different (low) temperatures (K) for the acetone-acetone dimer.

Temp. (K)	E_{cfg} (Kcal mol^{-1})
5.1400	−4.911
8.4487	−4.891
10.523	−4.877
13.310	−4.861
18.024	−4.831
21.041	−4.811
22.574	−4.802
24.162	−4.791
26.358	−4.778
28.464	−4.762
29.089	−4.760
32.543	−4.733
35.283	−4.722
36.149	−4.712
38.143	−4.698
41.766	−4.671
42.045	−4.671
43.120	−4.661
43.889	−4.655
48.273	−4.625
50.697	−4.609
51.025	−4.604
51.434	−4.601
51.717	−4.598
54.285	−4.581
59.947	−4.535
60.117	−4.536
68.913	−4.466

the fact that total energy must be conserved, the kinetic energy (and T) also varies. For this reason, the difficulty to achieve a chosen value of T increases.

The same range of temperatures have been used to analyze the acetone trimer. In this case, results suggest the presence of more than one isomer. As a matter of fact, the simulations, beginning from a random configuration for the trimer, show that the increase of T causes an initial linear decrease of the absolute value of E_{cfg}, as it was observed for the dimer. However, for T approaching to

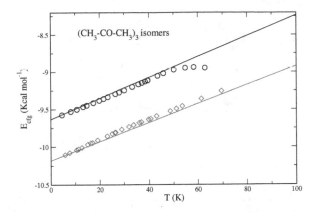

Fig. 2. Configuration (potential) energy values, E_{cfg}, for the acetone trimer as a function of T. The markers correspond to the MD results.

70 K, more negative values of E_{cfg} can be obtained, pointing out the presence of a stabler isomer. In order to characterize both isomers, we have performed additional MD simulations at decreasing temperatures, starting from the last configuration computed at 70 K (see red markers in Fig. 2). This procedure allows to obtain two sets of independent E_{cfg} data (black and red markers) as a function of T that, by their extrapolation to 0 K, provide binding energy predictions for both isomers. For acetone the linear extrapolation has been performed in the 5–20 K temperature range. However, MD simulations have been performed from 5 K to 70 K. These results point out that, in spite of the fact that only low temperatures are used in the linear extrapolation, a study including a wider range of T is mandatory to check the presence of more isomers. The process of running consecutive MD simulations at decreasing temperatures in the low T range can be seen as a simulated annealing minimization. [42] The extrapolated values of E_{cfg} to 0 K, are equal to -10.182 and -9.632 kcal mol^{-1}.

The analysis of the configuration energy at each step of the trajectory calculated at the lowest temperature investigated shows that, for the dimer, the lowest energy value attained along the trajectory is equal to -4.931 kcal mol^{-1}, very similar to the extrapolated value, representing the binding energy. For the trimer, the same analysis has been performed at the lowest temperature associated to each data set. Results indicate that the lowest energy values attained for the two data sets amount to -9.628 and -10.177 kcal mol^{-1}, which are also very similar to the extrapolated values. Accordingly, the geometries associated to the energies of -4.931, -9.628 and -10.177 kcal mol^{-1} should be very similar to the actual, for dimer and trimer respectively minima. Such equilibrium-like configurations are shown in Fig. 3.

On the panel (a) of Fig. 3 is shown the equilibrium-like geometry for the dimer, which corresponds to an antiparallel structure in which the $C = O$ bond of one molecule points toward the other one in such a way that the O of each molecule participates in the formation of two hydrogen(H)-bonds. This means that,

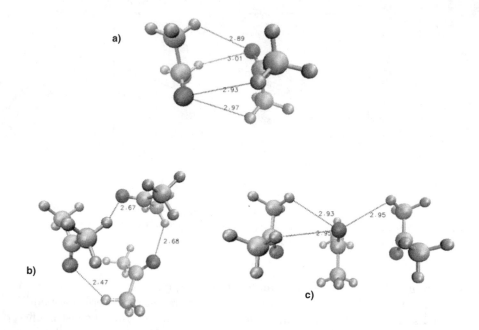

Fig. 3. Equilibrium-like structures for the acetone dimer (panel a) and for the trimer (panels b and c).

in agreement with other theoretical results, [43,44] four H-bonds are involved in the most stable configuration predicted by the potential model. All four O-H distances are very similar around to 2.9 Å. This value is in very good agreement with that of 3.0 Åreported by Frurip et al. [45]. Moreover, the same authors give a C-C distance equal to 3.84 Åsimilar to the value of 3.78 Åpredicted by own potential model.

The equilibrium-like structures for the two isomers of the trimer are shown in panels (b) and (c) of Fig. 3. In agreement with the results of Tamenori et al. [43] the cyclic structure shown in panel (b) is the most stable one and the O-H distances are about a 15 % shorter than those found for the dimer, while values of about 2.95 Å are found for the O-H distances of structure (c). The existence of two isomers and its relative stability is in agreement with the experimental results of Guan et al. [46]. The authors use size-selected IR-VUV spectroscopy to detect vibrational characteristics of neutral acetone and some of its clusters. In spite of the presence of two isomers, in the range of temperatures investigated, isomerization between them is quite unlikely. In particular, the less stable isomer can transform into the most stable one only at the high T.

5   Conclusions

In this paper we have done the preliminary theoretical calculation to study the properties of acetone small clusters by classical MD simulations by using a semiempirical intermolecular potential energy surface based on the ILJ pair interaction function. The underlying interaction model is targeted to an accurate description of the weak interactions, whose reliability has been validated here by comparison of our results with those available in the literature from theory and experiments. The predicted binding energies for the acetone dimer has been found in good agreement with both experimental and ab initio results. No isomerization processes have been observed for the dimer at temperatures lower than 40 K, instead for the trimer system some evidence of the presence of more than one isomer has been observed. Perspectives regard the application to extensive studies of clusters with more acetone molecules and the liquid acetone at several conditions of pressure and temperature.

Acknowledgement. M. Noelia Faginas Lago acknowledges financial support from Fondazione Cassa di Risparmio di Perugia (P 2014/1255, ACT 2014/6167), 2014.0253.021 Ricerca Scientifica e Tecnologica and the OU Supercomputing Center for Education & Research (OSCER) at the University of Oklahoma (OU) for the computing time. M. Albertí acknowledges financial support from the Ministerio de Educación y Ciencia (Spain, Projects CTQ2013-41307-P) and to the Comissionat per a Universitats i Recerca del DIUE (Generalitat de Catalunya, Project 201456R25). The Centre de Serveis Científics i Acadèmics de Catalunya CESCA and Fundació Catalana per a la Recerca are also acknowledged for the allocated supercomputing time. A. Lombardi also acknowledges financial support to MIUR-PRIN 2010-2011 (contract 2010ERFKXL 002).

References

1. American Chemistry Council: Acetone vccep submission (2013)
2. Andrigo, P., Caimi, A., d'Oro, P.C., Fait, A., Roberti, L., Tampieri, M., Tartari, V.: Phenol-acetone process: cumene oxidation kinetics and industrial plant simulation. Chem. Eng. Sci. **47**, 2511–2516 (1992)
3. Sifniades, S., Levy, A.: "Acetone" in Ullmann's Encyclopedia of Industrial Chemistry. Willey-VCH, Weinheim (2005)
4. Zakoshansky, V.: The cumene process for phenol-acetone product. Pet. Chem. **47**, 301–313 (2007)
5. Evchuk, Y., Musii, R., Makitra, R., Pristanskii, R.: Solubility of polymethyl methacrylate in organic solvents. Russ. J. Appl. Chem. **78**, 1576–1580 (2005)
6. Hadi Ghatee, M., Taslimian, S.: Investigation of temperature and pressure dependent equilibrium and transport properties of liquid acetone by molecular dynamics. Fluid Phase Equilib. **358**, 226–232 (2013)
7. Remler, R.: The solvent properties of acetone. Eng. Chem. **15**, 717–720 (1923)
8. Nagai, K.: New developments in the production of methyl methacrylate. Appl. Catal. A: Gen. **221**, 367–377 (2001)
9. Uglea, C., Negulescu, I.: Synthesis and Characterization of Oligomers. CRC Press, New York (1991)
10. Asada, M., Fujimori, T., Fujii, K., Umebayashi, Y., Ishiguro, S.I.: Solvation structure of magnesium, zinc, and alkaline earth metal ions in N, N-dimethylformamide, N, N-dimethylacetamide, and their mixtures studied by means of raman spectroscopy and dftcalculations - ionic size and electronic effects on steric congestion. J. Raman Spectrosc. **38**, 417–426 (2007)

11. Mollner, A., Brooksby, J., Loring, I., Palinkas, G., Ronald, W.: Ion-solvent inter-actions in acetonitrile solutions of lithium iodide and tetrabutylammonium iodide. J. Phys. Chem. A **108**, 3344–3349 (2004)

12. Palombo, F., Paolantoni, M., Sassi, P., Morresi, A., Giorgini, M.: Molecular dynam-ics of liquid acetone determined by depolarized rayleigh and low-frequency raman scattering spectroscopy. Phys. Chem. Chem. Phys. **13**, 16197–16207 (2011)

13. Albertí, M., Amat, A., De Angelis, F., Pirani, F.: A model potential for acetonitrile: from small clusters to liquid. J. Phys. Chem. B **117**, 7065–7076 (2013)

14. Pirani, F., Albertí, M., Castro, A., Moix, M., Cappelletti, D.: Atom-bond pairwise additive representation for intermolecular potential energy surfaces. Chem. Phys. Lett. **394**, 37–44 (2004)

15. Faginas-Lago, N., Lombardi, A., Albertí, M., Grossi, G.: Accurate analytic inter-molecular potential for the simulation of Na^+ and K^+ ion hydration in liquid water. J. Mol. Liq. **204**, 192–197 (2015)

16. Albertí, M., Faginas-Lago, N., Laganà, A., Pirani, F.: A portable intermolecular potential for molecular dynamics studies of NMA-NMA and NMA-H_2O aggregates. Phys. Chem. Chem. Phys. **13**, 8422–8432 (2011)

17. Albertí, M.: Rare gas-benzene-rare gas interactions: structural properties and dynamic behavior. J. Phys. Chem. A **114**, 2266–2274 (2010)

18. Albertí, M., Aguilar, A., Bartolomei, M., Cappelletti, D., Laganà, A., Lucas, J., Pirani, F.: A study to improve the van der waals component of the interaction in water clusters. Phys. Scr. **78**, 058108 (2008)

19. Albertí, M., Aguilar, A., Bartolomei, M., Cappelletti, D., Laganà, A., Lucas, J.M., Pirani, F.: Small water clusters: the cases of rare gas-water, alkali ion-water and water dimer. In: Gervasi, O., Murgante, B., Laganà, A., Taniar, D., Mun, Y., Gavrilova, M.L. (eds.) ICCSA 2008, Part I. LNCS, vol. 5072, pp. 1026–1035. Springer, Heidelberg (2008)

20. Albertí, M., Aguilar, A., Cappelletti, D., Laganà, A., Pirani, F.: On the develop-ment of an effective model potential to describe water interaction in neutral and ionic clusters. Int. J. Mass Spectrom. **280**, 50–56 (2009)

21. Faginas-Lago, N., Huarte-Larrañaga, F., Albertí, M.: On the suitability of the ilj function to match different formulations of the electrostatic potential for water-water interactions. Eur. Phys. J. D **55**(1), 75 (2009)

22. Albertí, M., Amat, A., Farrera, L., Pirani, F.: From the $(NH_3)_{2-5}$ clusters to liquid ammonia: molecular dynamics simulations using the nve and npt ensembles. J. Mol. Liq. **212**, 307–315 (2015)

23. Pirani, F., Cappelletti, D., Liuti, G.: Range, strength and anisotropy of intermole-cular forces in atom-molecule systems: an atom-bond pairwise additivity approach. Chem. Phys. Lett. **350**(3–4), 286–296 (2001)

24. Albertí, M., Castro, A., Laganà, A., Pirani, F., Porrini, M., Cappelletti, D.: Prop-erties of an atom-bond additive representation of the interaction for benzene-argon clusters. Chem. Phys. Lett. **392**(4–6), 514–520 (2004)

25. Bartolomei, M., Pirani, F., Laganà, A., Lombardi, A.: A full dimensional grid empowered simulation of the $CO_2 + CO_2$ processes. J. Comput. Chem. **33**(22), 1806–1819 (2012)

26. Lombardi, A., Faginas-Lago, N., Pacifici, L., Costantini, A.: Modeling of energy transfer from vibrationally excited CO_2 molecules: cross sections and probabilities for kinetic modeling of atmospheres, flows, and plasmas. J. Phys. Chem. A **117**(45), 11430–11440 (2013)

27. Faginas-Lago, N., Albertí, M., Laganà, A., Lombardi, A., Pacifici, L., Costantini, A.: The molecular stirrer catalytic effect in methane ice formation. In: Murgante, B., Misra, S., Rocha, A.M.A.C., Torre, C., Rocha, J.G., Falcão, M.I., Taniar, D., Apduhan, B.O., Gervasi, O. (eds.) ICCSA 2014, Part I. LNCS, vol. 8579, pp. 585–600. Springer, Heidelberg (2014)

28. Lombardi, A., Laganà, A., Pirani, F., Palazzetti, F., Faginas-Lago, N.: Carbon oxides in gas flows and earth and planetary atmospheres: state-to-state simulations of energy transfer and dissociation reactions. In: Murgante, B., Misra, S., Carlini, M., Torre, C.M., Nguyen, H.-Q., Taniar, D., Apduhan, B.O., Gervasi, O. (eds.) ICCSA 2013, Part II. LNCS, vol. 7972, pp. 17–31. Springer, Heidelberg (2013)

29. Falcinelli, S., Rosi, M., Candori, P., Vecchiocattivi, F., Bartocci, A., Lombardi, A., Faginas-Lago, N., Pirani, F.: Modeling the intermolecular interactions and characterization of the dynamics of collisional autoionization processes. In: Murgante, B., Misra, S., Carlini, M., Torre, C.M., Nguyen, H.-Q., Taniar, D., Apduhan, B.O., Gervasi, O. (eds.) ICCSA 2013, Part I. LNCS, vol. 7971, pp. 69–83. Springer, Heidelberg (2013)

30. Lombardi, A., Faginas-Lago, N., Laganà, A., Pirani, F., Falcinelli, S.: A bond-bond portable approach to intermolecular interactions: simulations for n-methylacetamide and carbon dioxide dimers. In: Murgante, B., Gervasi, O., Misra, S., Nedjah, N., Rocha, A.M.A.C., Taniar, D., Apduhan, B.O. (eds.) ICCSA 2012, Part I. LNCS, vol. 7333, pp. 387–400. Springer, Heidelberg (2012)

31. Laganà, A., Lombardi, A., Pirani, F., Gamallo, P., Sayos, R., Armenise, I., Cacciatore, M., Esposito, F., Rutigliano, M.: Molecular physics of elementary processes relevant to hypersonics: atom-molecule, molecule-molecule and atom-surface processes. Open Plasma Phys. J. **7**, 48 (2014)

32. Faginas-Lago, N., Albertí, M., Costantini, A., Laganà, A., Lombardi, A., Pacifici, L.: An innovative synergistic grid approach to the computational study of protein aggregation mechanisms. J. Mol. Model. **20**(7), 2226 (2014)

33. Albertí, M., Castro, A., Laganá, A., Moix, M., Pirani, F., Cappelletti, D., Liuti, G.: A molecular dynamics investigation of rare-gas solvated cation-benzene clusters using a new model potential. J. Phys. Chem. A **109**(12), 2906–2911 (2005)

34. Albertí, M., Aguilar, A., Lucas, J., Pirani, F.: Static and dynamic properties of anionic intermolecular aggregates: the I-benzene-Ar_n case. Theoret. Chem. Acc. **123**(1–2), 21–27 (2009)

35. Faginas-Lago, N., Albertí, M., Laganà, A., Lombardi, A.: Water $(H_2O)_m$ or benzene $(C_6H_6)_n$ aggregates to solvate the K^+? In: Murgante, B., Misra, S., Carlini, M., Torre, C.M., Nguyen, H.-Q., Taniar, D., Apduhan, B.O., Gervasi, O. (eds.) ICCSA 2013, Part I. LNCS, vol. 7971, pp. 1–15. Springer, Heidelberg (2013)

36. Albertí, M.: Rare gas-benzene-rare gas interactions: structural properties and dynamic behavior. J. Phys. Chem. A **114**(6), 2266–2274 (2010)

37. Cambi, R., Cappelletti, D., Liuti, G., Pirani, F.: Generalized correlations in terms of polarizability for van der waals interaction potential parameter calculations. J. Chem. Phys. **95**(3), 1852–1861 (1991)

38. Manion, J.A., Huie, R.E., Levin Jr., R.D., D.R.B., Orkin, V.L., Tsang, W., McGivern, W.S., Hudgens, J.W., Knyazev, V.D., Atkinson, D.B., Chai, E., Tereza, A.M., Lin, C.Y., Allison, T.C., Mallard, W.G., Westley, F., Herron, J.T., Hampson, R.F., Frizzell, D.H.: Nist chemical kinetics database, nist standard reference database 17 (2013)

39. Albertí, M., Aguilar, A., Cappelletti, D., Laganà, A., Pirani, F.: On the development of an effective model potential to describe ater interaction in neutral and ionic clusters. Int. J. Mass Spectrom. **280**, 50–56 (2009)

40. Smith, W., Yong, C., Rodger, P.: DL_POLY: application to molecular simulation. Mol. Simul. **28**(5), 385–471 (2002)
41. Albertí, M., Amat, A., Aguilar, A., Huarte-Larrañaga, F., Lucas, J., Pirani, F.: A molecular dynamics study of the evolution from the formation of the C_6F_6-$(H_2O)_n$ small aggregates to the C_6F_6 solvation. Theor. Chem. Acc. **134**, 61(1)–61(12) (2015)
42. Kirkpatrick, S., Gelatt Jr., C., Vecchi, M.: Optimization by simulated annealing. Science **220**, 671–680 (1983)
43. Tamenori, Y., Takahashi, O., Yamashita, K., Yamaguchi, T., Okada, K., Tabayashi, K., Gejo, T., Honma, K.: Hydrogen bonding in acetone clusters probed by near-edge x-ray absorption fine structure spectroscopy in the carbon and oxygen k -edge regions. J. Chem. Phys. **131**, 174311(1)–174311(9) (2009)
44. Hermida-Ramón, J., Ríos, M.: The energy of interaction between two acetone molecules: a potential function constructed from ab initio data. J. Phys. Chem. A **102**, 2594–2602 (1998)
45. Frurip, D.J., Curtiss, L.A., Blander, M.: Characterization of molecular association in acetone vapor. thermal conductivity measurements and molecular orbital calculations. J. Phys. Chem. **82**(24), 2555–2561 (1978)
46. Guan, J., Hu, Y., Xie, M., Bernstein, E.: Weak carbonyl-methyl intermolecular interactions in acetone clusters explored by IR plus VUV spectroscopy. Chem. Phys. **405**, 117–123 (2012)

Cumulative Updating of Network Reliability with Diameter Constraint and Network Topology Optimization

Denis A. Migov$^{(\boxtimes)}$, Kseniya A. Nechunaeva,
Sergei N. Nesterov, and Alexey S. Rodionov

Institute of Computational Mathematics and Mathematical Geophysics SB RAS,
prospect Akademika Lavrentjeva 6, 630090 Novosibirsk, Russia
mdinka@rav.sscc.ru, ksu.nech@gmail.com,
cepera_666@inbox.ru, alrod@sscc.ru
http://www.sscc.ru/

Abstract. Reliability-based optimization of a network topology is to maximize the network reliability within certain constraints. For modeling of unrelaible networks we use random graphs due to their good applicability, wide facilities and profound elaborating. However, graph optimization problems in conditions of different constraints are NP-hard problems mostly. These problems can be effectively solved by optimization methods based on biological processes, such as genetic algorithms or clonal selection algorithms. As a rule, these techiques can provide an applicable solution for network topology optimization within an acceptable time. In order to speed up fitness function calculation, we improve operators of a genetic algorithm and a clonal selection algorithm by using the method of cumulative updating of lower and upper bounds of network reliability with diameter constraint. This method allows us to make a decision about the network reliability (or unreliability) with respect to a given threshold without performing the exhaustive calculation. Based on this method, we obtain the genetic algorithm and the clonal selection algorithm for network topology optimization. Some computational results are also presented for demonstration of an applicability of the proposed approach.

Keywords: Network reliability · Network topology optimization · Genetic algorithm · Clonal selection algorithm · Random graph · Diameter constraint · Factoring method · Cumulative updating

1 Introduction

Genetic algorithms (GAs) and clonal selection algorithms (CSAs) [1–3] are widely used for network topology optimization [4–6]. These approaches are based on selection and recombination of promising solutions. GAs and CSAs

Supported by Russian Foundation for Basic Research under 16-37-00345 grant.

O. Gervasi et al. (Eds.): ICCSA 2016, Part I, LNCS 9786, pp. 141–152, 2016.
DOI: 10.1007/978-3-319-42085-1_11

have achieved great success in solving numerous graph optimization problems. However, the performance of these algorithms depends on the right choice of operators (such as selection, crossover, cloning, mutation etc.), and method for calculation of fitness function. Otherwise, computational time of GAs and CSAs can be increased on an enormous scale. On the other hand, we can significantly improve the algorithms performance by using various techniques for rejecting inapplicable chromosomes.

In present paper we deal with the problem of obtaining the most reliable network topology within a given budget.

It is assumed that network has unreliable elements which are subject to random fault that occur mutually independently. Random graphs are commonly used for modeling such networks. We consider the case of absolutely reliable nodes and unreliable edges which corresponds to real networks where the reliability of nodes is much higher than reliability of communication links.

One of the basic reliability measures for such networks is the probabilistic connectivity, i.e. the probability of a given subset of nodes to be connected. This measure is quite well examined, various exact and approximate reliability calculation methods have been proposed [7]. Another popular measure of network reliability is the diameter constrained network reliability (Petingi and Cancela, 2001 [8,9]). Further on we use abbreviation DCNR for notation of diameter constrained network reliability. DCNR is a probability that every two nodes from a given set of terminals are connected with a path of length less or equal to a given integer. By the length of a path we understand the number of edges in this path. This reliability measure is more applicable in practice, for example, in the case of P2P networks [10]. However, the problems of computing these characteristics are known to be NP-hard. Moreover, DCNR calculation problem is NP-hard for most combinations of a diameter value and a number of terminals [11].

The new approach in the area of network reliability analysis was introduced in [12,13]: cumulative updating of lower and upper bounds of all-terminal network reliability for faster feasibility decision. This method allows to decide the feasibility of a given network without performing the exhaustive calculation. The approach was further developed with help of network decomposition [14]. In our previous research [15] we've proposed the method for network topology optimization with use of cumulative updating. The most reliable topology was sought from the point of view of all-terminal reliability. In present study we obtain the cumulative updating method for DCNR and use it for network topology optimization.

2 Problem Statement

Let us have a set of vertices $V = \{V_1, ..., V_n\}$ and a set $S = \{S_1, ..., S_t\}$ of weighted edges, $C = \{C_1, ..., C_t\}$ and $P = \{r_1, ..., r_t\}$ — weights and connection probabilities of edges from S accordingly; $K = \{k_1, ..., k_l\}$ — terminal nodes. Values of budget constraint C^* and diameter constraint d are given. We use notation $R_K^d(G)$ for reliability of G with diameter constraint d. Probabilistic connectivity of G is denoted by $R(G)$. For rigorous definitions of the described reliabilities measures we refer to [16] (for probabilistic connectivity) and [17] (for DCNR).

Need to construct connected undirected graph $G = (V, E \subset S)$ with maximum diameter constrained reliability value with following restrictions:

$$\begin{cases} R_k^d(G) \to \max; \\ Weight(G) < C^*. \end{cases} \tag{1}$$

Preference is given to the cheapest solution in case of equal probability values.

3   Brief Survey of Exact Methods for Network Reliability Calculation

Present section describes some methods of network reliability calculation, but this is not meant to be a complete summary of the work in this field.

The usually used method for calculating any network reliability measure is the factoring method. The main idea is to partition the probability space into two sets based on the success or failure of a chosen network's element which is referred to as factored element. Thus, given a graph G and a factored element e we will obtain two graphs G/e and $G \backslash e$. In the first of them the factored element is absolutely reliable and in the second one the factored element is absolutely unreliable, e.g. it could just be removed. The probability of G/e is equal to the reliability of factored element and the probability of $G \backslash e$ is equal to the failure probability of factored element. The same procedure is to be applied for the both graphs involved. Using the total probability law the following expression is obtained [7]:

$$Rel(G) = r_e Rel(G/e) + (1 - r_e) Rel(G \backslash e), \tag{2}$$

Recursions continue until either obtained network is clearly unreliable (procedure returns 0) or it is absolutely reliable (returns 1). In some cases it is possible to improve factoring process by calculating reliabilities of intermediate networks directly, i.e. without further factorization. For example, the formula 5-vertex graph reliability can be applied for $R(G)$ calculation [14].

For DCNR calculation we have a modified factoring method which is much faster than the basic factoring method (2) in the diameter constrained case [9]. The main feature of this method is operating with the list of paths instead of operating with graphs: in the preliminary step for any pair of terminals s and t the list $P_{st}(d)$ of all paths with limited length between s, t is generated. It automatically removes all the edges which don't belong to any such path from consideration. For example, all so called "attached trees" without terminals are no longer considered.

Afterwards all the operations described above are performed with the list of all paths $P = \cup_{s,t \in T} P_{st}(d)$. In these terms the success of the factored element will also make it absolutely reliable while failure of the factored element will remove all the paths which contained it from further consideration. Further on we refer to described method as CPFM (Cancela&Petingi factoring method).

Another approach for network reliability calculation is applying methods of reduction and decomposition. One of the most effective among them is the parallel-series transformation which removes chains and multiple edges from graph. For probabilistic connectivity [18,19] it is possible to apply such reduction in every call of the factoring procedure. For computing DCNR we can only use parallel-series transformation in the preliminary step, before factoring process starts [17]. Below we assume that the described reduction is performed during network reliability calculation by factoring method, both for probabilistic connectivity and DCNR.

4   Cumulative Updating of Network Reliability

Recent research [12] considered problem of determination whether a network is reliable enough in terms of network probabilistic connectivity (without diameter constraint). The idea of the proposed method is to check if a network is feasible without exact calculating a value of network reliability. For this purpose we define so called threshold R_0 which is a requirement of the network reliability. By RL and RU we will denote the lower bound and the upper bound of $R(G)$ respectively, and initialize them by 0 and 1. These bounds are updated in such a way that on i-th iteration $RL_i \geq RL_{i-1}$ and $RU_i \leq RU_{i-1}$. Decision process stops when either RL_l exceeds R_0 or R_0 exceeds RU_l. In the first case the network is supposed to be reliable and in the second one the network is unreliable.

Let us assume that during factoring procedure we obtain L final graphs $G_1, G_2, \ldots, G_L$, for which the reliability can be easily calculated. Let P_l for $1 \leq l \leq L$ be the probability to have G_l. Thus, $\sum_{l=1}^{L} P_l = 1$ and the following inequality holds for any $1 \leq k \leq L$ [12]:

$$\sum_{l=1}^{k} P_l R(G_l) \leq R(G) \leq 1 - \sum_{l=1}^{k} P_l(1 - R(G_l)). \qquad (3)$$

This inequality gives the algorithm for cumulative updating of the lower and upper bounds of $R(G)$. Every time whenever reliability of some G_l for any $1 \leq l \leq L$ is calculated, we can update RL_l and RU_l in the following way:

$$\begin{aligned} RL_l &= RL_{l-1} + P_l R(G_l) \\ RU_l &= RU_{l-1} - P_l(1 - R(G_l)). \end{aligned} \qquad (4)$$

RL_l and RU_l approach exact $G(R)$ value every time when l increases. Once either RL_l or RU_l reaches R_0, the proposed algorithm concludes the feasibility of G: if RL_l reaches R_0, G is feasible; if RU_l passes R_0, G is infeasible. Thus, we can set any acceptable value of R_0 in order to stop the method during execution without performing exact calculating of the network reliability.

We have applied this approach for DCNR bounds updating by CPFM. As it was mentioned above, CPFM doesn't oprate with graphs directly, instead it operates with the list of all paths P. In cases when either at least one pair of terminals cannot be connected by any path or all pairs of terminals are connected by absolutely reliable paths we can update our RL_l and RU_l values. In other words, these cases will play a role of the final graphs $G_1, G_2, \ldots, G_L$. We also denote by P_l the probability of the network obtained on l-th iteration. P_0 will be initialized by 1. Any time during the factoring procedure we should multiply P_l by either r_e or $1 - r_e$ depending on the factored element e status.

Parameters of the modified factoring procedure in CPFM aren't graphs. Instead we use 6 parameters, which describe the corresponding graph from the viewpoint of P_d. Listed below is the parameters of the CPFM and the pseudocode of the proposed method for DCNR bounds cumulative updating.

- np_{st}: the number of paths of length at most d between s and t in the graph being considered.
- $links_p$: the number of non-perfect edges (edges e such that $r(e) < 1$) in path p, for every $p \in P_d$.
- $feasible_p$: this is a flag, which has value False when the path is no longer feasible, i.e. it includes an edge which failed; and True otherwise.
- $connected_{st}$: this is a flag, which has value True when s and t are connected by a perfect path of length at most d and False otherwise.
- $connectedPairs$: this is the number of connected pairs of terminals (those between which there is a perfect path of length at most d).

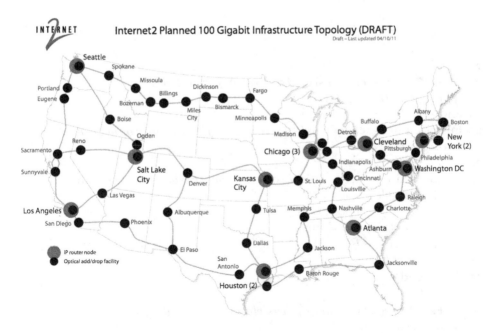

Fig. 1. Tested network (Color figure online)

Input: $G = (V, E)$, d, P_d, $P(e)$, $np(s, t)$, $links(p)$, $feasible(p)$,
        $connected(s, t)$, $connectedPairs$, $RL = 0$, $RU = 1$, $P_l = 1$

1 **Function** FACTO($np(s, t)$, $links(p)$, $feasible(p)$, $connected(s, t)$,
   $connectedPairs$, P_l)

2   **if** $nowTime - startTime > T_0$ **or** $RL > R_0$ **or** $RU < R_0$ **then**

3 | **return**

4   **end**

5    $e \leftarrow$ arbitrary edge : $0 < r_e < 1$

6   contractEdge($np(s, t)$, $links(p)$, $feasible(p)$, $connected(s, t)$,
   $connectedPairs$, P_l)

7   deleteEdge($np(s, t)$, $links(p)$, $feasible(p)$, $connected(s, t)$,
   $connectedPairs$, P_l)

8 **end**

9 **Function** contractEdge($np(s, t)$, $links(p)$, $feasible(p)$, $connected(s, t)$,
   $connectedPairs$, P_l)

10    $P_l \leftarrow P_l * r_e$

11   **foreach** $p = (s, \ldots, t)$ in $P(e)$ such that $feasible(p) = true$ **do**

12       $links(p) \leftarrow links(p) - 1$

13   **if** $connected(s, t) = false$ **and** $links(p) = 0$ **then**

14          $connected(s, t) \leftarrow true$

15          $connectedPairs \leftarrow connectedPairs + 1$

16   **if** $connectedPairs = \frac{k \times (k-1)}{2}$ **then**

17             $RL \leftarrow RL + P_l$

18   **return**

19   **end**

20   **end**

21   **end**

22   FACTO $(np(s, t)$, $links(p)$, $feasible(p)$, $connected(s, t)$,
   $connectedPairs$, P_l)

23 **end**

24 **Function** deleteEdge($np(s, t)$, $links(p)$, $feasible(p)$, $connected(s, t)$,
   $connectedPairs$, P_l)

25    $P_l \leftarrow P_l * r_e$

26   **foreach** $p = (s, \ldots, t)$ in $P(e)$ such that $feasible(p) = true$ **do**

27       $feasible(p) \leftarrow false$

28       $np(s, t) \leftarrow np(s, t) - 1$

29   **if** $np(s, t) = 0$ **then**

30          $RU \leftarrow RU - P_l$

31   **return**

32   **end**

33   **end**

34   FACTO $(np(s, t)$, $links(p)$, $feasible(p)$, $connected(s, t)$,
   $connectedPairs$, P_l)

35 **end**

   1. Pseudocode of the method for cumulative updating of DCNR bounds

Diagram (Fig. 2) shows updating of RL and RU during execution of the proposed procedure for topology of Internet2 network (Fig. 1) for the diameter value 25. Edge reliability is equal 0.9 for each edge. R_0 value was equal to exact value of DCNR. Calculation time was about 48 s.

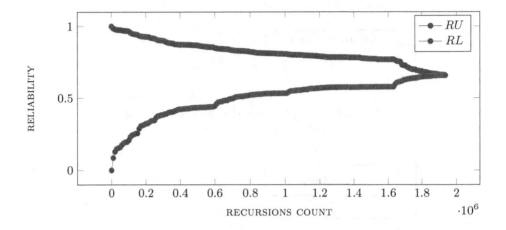

Fig. 2. D = 25, P = 0.9 (Color figure online)

5   Genetic Algorithm

Genetic Algorithms are a common probabilistic optimization method based on the model of natural evolution. Imitation of natural processes — selection, mutation, recombination, reproduction, proliferation is used as a basis, refer to Charles Darvin's theory presented in "On the Origin of Species" [20].

Algorithmic scheme (Fig. 3): At first, individual is presented as chromosomes — sequenced collection of elements (more often as a bit string). Each chromosome presents some solution. Then, the population is defined like an arbitrary subset of full set of chromosomes. The most appropriate individuals are found by a fitness function. Next step is selection of chromosomes — choosing mates (for example, wheel selection). Crossover occurs among the fittest individuals specially selected for it. After crossover, new offspring goes through mutation. New population is combined from fittest individuals from the current population and new individuals with possible addition of randomly chosen new individuals. Algorithms stops after given time constraints, given number of generations, or when given number of generations does not produce any improvement of solution.

Cumulative updating lets the GA operators work faster due to cutting down bad chromosomes (with worse fitness function value).

Mutation: let A_0 be original chromosome with known diameter constrained reliability R_0. Verification new chromosome A_1:

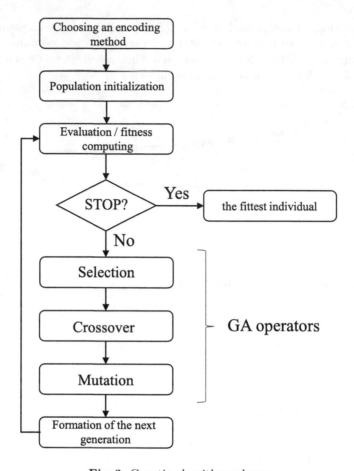

Fig. 3. Genetic algorithm scheme

$$Feas(A_1, R_0) = \begin{cases} 1, R_k^d(A_1) > R_0; \\ 0, else. \end{cases} \tag{5}$$

Crossover: let A_0 and A_1 be parents with reliability minimum of R_{min}. Verification offspring A_2:

$$Feas(A_2, R_{min}) = \begin{cases} 1, R_k^d(A_2) > R_{min}; \\ 0, else. \end{cases} \tag{6}$$

Obtained offspring is accepted only if it is better fitted than one of the parents.

6   Artificial Immune System

Artificial immune system (AIS) theory comes from theoretical immunology in the middle of 80th. First immune algorithms were used by Bersini [2] for solving different problems. The main aim of AIS is to use immunology principles for creating systems to solve different optimization problems [3].

One of the main algorithm among AIS algorithms is the Clonal Selection Algorithm (CSA) [5]. The CSA is simulating the natural B-cell response mechanism. When the antigen (virus, for instance) get to a blood, B-cells start to secret antibodies. Each cell secrets only one type of antibody specific for antigen. During the process the B-cells are cloning and mutating for achieving the best match for antigen. Those B-cells with better matching start quickly spread antibodies and become plasma cells, part of them becomes memory cells and is circulating in blood until new invasion.

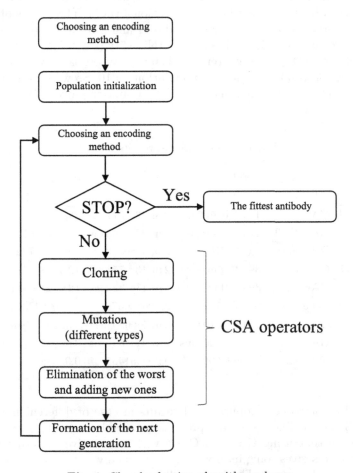

Fig. 4. Clonal selection algorithm scheme

Algorithmic scheme (Fig. 4): Like chromosomes in GA, antibodies are coded as bit strings or as collection of elements. An arbitrary set of them is defined as the population. Then we calculate affinity of each antibody in the population. The next step is to clone antibodies accordingly their affinity values. Every clone goes through mutation to obtain better solution. After elimination of some part of the worst clones, new group of new random antibodies adds to population in the same number. Next, algorithm works repeatedly with new population until stop criteria: time constraints, number of generations etc.

Same as GAs we can accelerate clonal selection operators by cutting down worst antibodies using the cumulative updating.

7   Case Studies

This section presents series of experiments aimed to demonstrate relation between given data and running time. We demonstrate this on not dense graphs with 15 vertices (cause of their fast computational time). The results of the GA and CSA performance are presented in Tables 1 and 2 respectively. Changeable parameters are diameter (5, 8, 10) and edge reliability (0.1, 0.5, 0.9), which is the same for all edges. Other parameters fixed: mutation probability for GA is 0.1, size of population is 50, a number of populations is 10. So we have to calculate exactly reliabilities for 500 graphs for mutation and crossover. The number of terminals is equal to 3.

Table 1. Computational results for GA

Graph $|V| = 15, |E| = 22$		d = 5	d = 8	d = 10
r = 0,1	GA	1 m 55 s 789 ms	2 m 43 s 671 ms	3 m 50 s 336 ms
	GAwithCU	1 m 56 s 70 ms	2 m 43 s 563 ms	3 m 50 s 374 ms
	Reliability	0,053377031	0,035480363	0,042657732
r = 0,5	GA	39 s 247 ms	2 m 13 s 754 ms	12 m 18 s 777 ms
	GAwithCU	39 s 321 ms	2 m 11 s 322 ms	12 m 21 s 311 ms
	Reliability	0,801047325	0,796380579	0,830094337
r = 0,9	GA	1 m 59 s 10 ms	3 m 5 s 699 ms	4 m 1 s 559 ms
	GAwithCU	1 m 54 s 263 ms	2 m 41 s 897 ms	3 m 27 s 677 ms
	Reliability	0,999966901	0,999951435	0,999883982

We've also compared computational results in case of different numbers of terminals (3 and 7). The results are presented in Table 3.

We can see that using GAs and CSAs with cumulative updating for such types of problems gives some improvements, especially in the case of not dense and highly reliable graphs. The best results were obtained for problems where amount of terminals is large.

Table 2. Computational results for CSA

Graph $|V| = 15$, $|E| = 22$		d = 5	d = 8	d = 10
r = 0,1	CSA	42 s 939 ms	1 m 18 s 686 ms	1 m 20 s 586 ms
	CASwithCU	43 s 484 ms	1 m 19 s 501 ms	1 m 20 s 227 ms
	Reliability	0,033715333	0,035687813	0,035632029
r = 0,5	CSA	46 s 870 ms	1 m 46 s 51 ms	2 m 14 s 788 ms
	CSAwithCU	46 s 965 ms	1 m 44 s 935 ms	2 m 5 s 659 ms
	Reliability	0,771484375	0,756820679	0,652519226
r = 0,9	CSA	44 s 178 ms	1 m 52 s 241 ms	2 m 50 s 238 ms
	CSAwithCU	43 s 616 ms	1 m 45 s 308 ms	2 m 31 s 871 ms
	Reliability	0,999604889	0,999946036	0,999672344

Table 3. Computational results for $T = 3$ and $T = 7$

	$|T| = 3$	$|T| = 7$
GA	3 m 5 s 699 ms	39 m 20 s 844 ms
GAwithCU	2 m 41 s 897 ms	33 m 29 s 853 ms
Reliability	0,999951434	0,998523183
CSA	1 m 52 s 241 ms	13 m 34 s 306 ms
CSAwithCU	1 m 45 s 308 ms	9 m 53 s 614 ms
Reliability	0,999946036	0,992177548

8   Conclusion

Proposed optimization approach allows to reduce computational time for obtaining an appropriate solution, i.e. a reliable enough network topology. Nevertheless, network topology optimization problems still show a great level of complexity. Our ongoing research involves studying of new improvements of cumulative updating method for further speeding up the optimization process.

References

1. Holland, J.H.: Adaptation in Natural and Artificial Systems. University of Michigan Press, Ann. Arbor (1975)
2. Bersini, C., Varela, F.J.: Hints for adaptive problem solving gleaned from immune networks. In: Schwefel, H.-P., Männer, R. (eds.) PPSN 1990. LNCS, vol. 496, pp. 343–354. Springer, Heidelberg (1991)
3. Ishida, Y.: The immune system as a self-identification process: a survey and a proposal. In: Proceedings of the ICMAS International Workshop on Immunity-Based Systems, pp. 2–12 (1996)

4. Landwehr, B.: A genetic algorithm based approach for multi-objective data-flow graph optimization. In: Asia and South Pacific Design Automation Conference 1999 (ASP-DAC 1999), vol. 1, pp. 355–358 (1999)
5. De Castro, L.N., Von Zuben, F.J.: The clonal selection algorithm with engineering applications. In: Workshop Proceedings of GECCO. Workshop on Artificial Immune Systems and Their Applications, Las Vegas, USA, pp. 36–37 (2000)
6. Kumar, A., Mishra, K.K., Misra, A.K.: Optimizing the reliability of communication network using specially designed genetic algorithm. In: World Congress on Nature & Biologically Inspired Computing, NaBIC 2009, pp. 499–502 (2009)
7. Ball, M.O.: Computational complexity of network reliability analysis: an overview. IEEE Trans. Reliab. **35**, 230–239 (1986)
8. Petingi, L., Rodriguez, J.: Reliability of networks with delay constraints. Congressus Numerantium. **152**, 117–123 (2001)
9. Cancela, H., Petingi, L.: Diameter constrained network reliability: exact evaluation by factorization and bounds. In: International Conference on Industrial Logistics, Okinawa, Japan, pp. 359–356 (2001)
10. Pandurangan, G., Raghavan, P., Upfal, E.: Building low-diameter peer-to-peer networks. IEEE J. Sel. Areas Commun. (JSAC) **21**(6), 995–1002 (2003)
11. Canale, E., Cancela, H., Robledo, F., Romero, P., Sartor, P.: Full complexityanalysis of the diameter-constrained reliability. Int. Trans. Oper. Res. **22**(5), 811–821 (2015)
12. Won, J.-M., Karray, F.: Cumulative update of all-terminal reliability for faster feasibility decision. IEEE Trans. Reliab. **59**(3), 551–562 (2010)
13. Won, J.-M., Karray, F.: A greedy algorithm for faster feasibility evaluation of all-terminal-reliable networks. IEEE Trans. Syst. Man Cybern. Part B Cybern. **41**(6), 1600–1611 (2011)
14. Rodionov, A.S., Migov, D.A., Rodionova, O.K.: Improvements in the efficiency of cumulative updating of all-terminal network reliability. IEEE Trans. Reliab. **61**(2), 460–465 (2012)
15. Nechunaeva, K.A., Migov, D.A.: Speeding up of genetic algorithm for network topology optimization with use of cumulative updating of network reliability. In: ACM IMCOM2015, article 42. ACM New York (2015)
16. Migov, D.A., Rodionov, A.S.: Parallel implementation of the factoring method for network reliability calculation. In: Murgante, B., et al. (eds.) ICCSA 2014, Part VI. LNCS, vol. 8584, pp. 654–664. Springer, Heidelberg (2014)
17. Migov, D.A., Nesterov, S.N.: Methods of speeding up of diameter constrained network reliability calculation. In: Gervasi, O., Murgante, B., Misra, S., Gavrilova, M.L., Rocha, A.M.A.C., Torre, C., Taniar, D., Apduhan, B.O. (eds.) ICCSA 2015, Part II. LNCS, vol. 9156, pp. 121–133. Springer, Heidelberg (2015)
18. Shooman, A.M., Kershenbaum, A.: Exact graph-reduction algorithms for network reliability analysis. In: IEEE Global Telecommunications Conference GLOBECOM 1991, pp. 1412–1420. IEEEP Press, New York (1991)
19. Rodionova, O.K., Rodionov, A.S., Choo, H.: Network probabilistic connectivity: exact calculation with use of chains. In: Laganá, A., Gavrilova, M.L., Kumar, V., Mun, Y., Tan, C.J.K., Gervasi, O. (eds.) ICCSA 2004. LNCS, vol. 3045, pp. 315–324. Springer, Heidelberg (2004)
20. Darwin, C.R.: On the Origin of Species by Means of Natural Selection, or the Preservation of Favoured Races in the Struggle for Life. John Murray, London (1859)

Set Covering Problem Resolution
by Biogeography-Based Optimization Algorithm

Broderick Crawford[1,2,3], Ricardo Soto[1,4,5], Luis Riquelme[1(✉)],
Eduardo Olguín[3], and Sanjay Misra[6]

[1] Pontificia Universidad Católica de Valparaíso, Valparaíso, Chile
lriquelme@outlook.com
[2] Universidad Central de Chile, Santiago, Chile
[3] Facultad de Ingeniería y Tecnología, Universidad San Sebastián, Santiago, Chile
[4] Universidad Autónoma de Chile, Santiago Metropolitan, Chile
[5] Universidad Cientifica del Sur, Lima, Peru
[6] Covenant University, Ota, Nigeria

Abstract. The research on Artificial Intelligence and Operational
Research has provided models and techniques to solve many industrial
problems. For instance, many real life problems can be formulated as a
Set Covering Problem (SCP). The SCP is a classic NP-hard combinatorial problem consisting in find a set of solutions that cover a range of
needs at the lowest possible cost following certain constraints. In this
work, we use a recent metaheuristic called Biogeography-Based Optimization Algorithm (BBOA) inspired by biogeography, which mimics
the migration behavior of animals in nature to solve optimization and
engineering problems. In this paper, BBOA for the SCP is proposed.
In addition, to improve performance we provide a new feature for the
BBOA, which improve stagnation in local optimum. Finally, the experiment results show that BBOA is a excellent method for solving such
problems.

Keywords: Biogeography-Based Optimization Algorithm · Set Covering Problem · Metaheuristics

1 Introduction

Different solving methods have been proposed in the literature to solve Combinatorial Optimization Problems. Exact algorithms are mostly based on Branch-
and-Bound and Branch-and-Cut techniques, Linear Programing and Heuristic
methods [2,16]. However, these algorithms are rather time consuming and can
only solve instances of very limited size. For this reason, many research efforts
have been focused on the development of heuristics to find good results or near-
optimal solutions within a reasonable period of time.

In Artificial Intelligence (AI) heuristics are used, that in a very generic way,
is a set of techniques or methods for solve a problem more quickly, finding an
approximate solution when classic methods fail to find any exact solution. From

© Springer International Publishing Switzerland 2016
O. Gervasi et al. (Eds.): ICCSA 2016, Part I, LNCS 9786, pp. 153–165, 2016.
DOI: 10.1007/978-3-319-42085-1_12

this point of view, the heuristic is a procedure which tries to give good solutions, with quality and good performance.

Metaheuristics, as the prefix says, are upper level heuristics. They are intelligent strategies to design or improve general heuristic procedures with high performance. In their original definition, metaheuristics are general purpose approximated optimization algorithms; they find a good solution for the problem in a reasonable time (not necessarily the optimal solution). They are iterative procedures that smartly guide a subordinate heuristic, combining different concepts to suitably explore and operate the search space. Over time, these methods have also come to include any procedures that employ strategies for overcoming the trap of local optimum in complex solution spaces, especially those procedures that utilize one or more neighborhood structures as a means of defining admissible moves to transition from one solution to another, or to build or destroy solutions in constructive and destructive processes.

To get good solutions, any search algorithm must establish an adequate balance between two overlayed process characteristics:

- Intensity: Effort put on local space search (space exploitation).
- Diversity: Effort put on distant space search (space exploration).

This balance is needed to quickly identify regions with good quality solutions and to not spend time in promising or visited regions.

The metaheuristics are categorized based on the procedures types which it refers. Some of the fundamental types of metaheuristics are:

- Constructive heuristics: Start from an empty solution and go adding components until a solution is built.
- Trajectory methods: Start from an initial solution and then, iteratively, try to replace it with a better one from their neighborhood.
- Population-based methods: Iteratively evolve a solution-population.

One of the fairly new and existing metaheuristics is the Biogeography-Based Optimization Algorithm (BBOA). It is based on the behavior of natural migration of animals, considering emigration, immigration and mutation factors. This is a population algorithm for binary and real problems, and it's useful for maximizing and minimizing problems [28]. In general, BBOA is based on the concept of Habitat Suitability Index (HSI) which it is generated from the characteristics of an habitat, where the habitat that has better characteristics have a higher HSI and worst features, lower HSI. It is also considered that the more HSI have an habitat, more species inhabit it, contrary to lower HSI [28,36]. Each habitat also has a single rate of immigration, emigration and mutation probabilities, which come from the habitat number of species.

This metaheuristic is applied for solving the Set Covering Problem (SCP), whose aim is to cover a range of needs at the lowest cost, following certain restrictions on the context of the problem where the needs are constraints. SCP can be applied for location services, selection of files in a database, simplifying boolean expressions, slot allocation, among others [3]. Currently, there is extensive literature on methods for SCP resolutions. They are the exact methods as mentioned in [2,16], and heuristic methods to solve a range of problems such in [21]. In case of SCP, this is solved by a variety of heuristics, so there is considerable literature. Among the metaheuristics that has tried to solve the SCP, they are: hybrid algorithms [15], hybrid ant algorithm [13], binary cat swarm optimization [7], bat algorithm [10], cuckoo search [30], artificial bee colony algorithm [8], binary firefly algorithm [11], among others.

BBOA has been used to solve other problems of optimization, among them are the classic and one of the most important optimization problems: The Traveling Salesman Problem of NP-hard class, which it is to find the shortest route between a set of points, visiting them all at once and returning to the starting point [26]. This was solved by using BBOA in [25], demonstrating that behaves very effectively for some combinations of optimization and even outperforms other traditional methods inspired by nature. Also, BBOA has been used to solve constraint optimization problems such as in [23], where indicate that BBO generally performs better than Genetic Algorithm (GA) and Particle Swarm Optimization (PSO) in handling constrained single-objective optimization problems. Undoubtedly, the BBOA is a method that may have great potential to solve the SCP.

The remaining of this document is structured as follows: a description of the SCP, then the technique (BBOA) used to solve SCP. Then, the changes to the algorithm to relate and integrated at problem. Subsequently, results of experiments comparing with known global optimums and, finally, the corresponding conclusions.

2 Set Covering Problem

The Set Covering Problem is a popular NP-hard problem [19] that has been used to a wide range of airlines and buses crew scheduling [29], location of emergency facilities [32], railway crew management [5], steel production [33], vehicle scheduling [18], ship scheduling [17], etc.

The SCP consists of finding a set of solutions which covers a range of needs at the lowest cost. In a zero-one matrix view (a_{ij}), the needs correspond to m-rows (constraints), while the whole solution is the selection of n-columns that optimally cover the rows. Among the real-world applications in which it applies are: location of emergency facilities, steel production, vehicle routing, network attack or defense, information retrieval, services location, among others [3].

The SCP was also successfully solved with metaheuristics such as taboo search [6], simulated annealing [4,31], genetic algorithm [20,22,24], ant colony optimization [1,24], swarm optimization particles [9], artificial bee colony [12,35] and firefly algorithms [14].

2.1   Formal Definition

The SCP is mathematically modeled as follows.

$$Minimize\, Z = \sum_{j=1}^{n} c_j x_j.\tag{1}$$

Subject to:

$$\sum_{j=1}^{n} a_{ij} x_j \geq 1 \qquad \forall i \in \{1,2,3,...,m\}$$

$$x_j \in \{0,1\} \qquad \forall j \in \{1,2,3,...,n\}.\tag{2}$$

where Eq. (1) minimizes the number of sets, analogous to obtain the minimum cost (c_j). Subject to Eq. (2), ensuring that each m-row is covered by at least one n-column. Where the domain constraint x_j is 1 if the column belongs to solution and 0 otherwise.

3   Biogeography-Based Optimization Algorithm

Biogeography studies the migration between habitats, speciation and extinction of species. Simon (2008) proposes the BBOA by mathematical models of biogeography made in the 1960s [28]. This says that areas that are well adapted as a residence for biological species have a high HSI. Some features are related to this index; precipitation, vegetation diversity, diversity of topography, land surface and temperature. Variables that characterize the habitability are called Suitability Index Variables (SIV). SIVs can be considered the independent variables of the habitat, and HSI can be considered the dependent variable [28].

Then, based on the species number, it is possible to predict the rate of immigration and emigration: habitats that are more HSI have higher rate of emigration, since the big population causes that species migrate to neighboring habitats. They also have a low inmigration rate because they are already nearly saturated with species. Furthermore, habitats with a low HSI have a high species immigration rate because of their sparse populations and a high rate of emigration, as conditions cause rapid way or species extinction. This behavior is shown in Fig. 1.

Where I and E are the highest rates of immigration and emigration, the same for simplicity. S_{max}, the maximum amount of species and S_0 the equilibrium number of species. Finally, λ is the inmigration rate and μ is the emigration rate. k is the habitat species number.

3.1   Migration Operator

As mentioned in biogeography, species can migrate between habitats. In BBO, the characteristics of the solutions may affect others and themselves, using immigration and emigration rates to share information between them probabilistically.

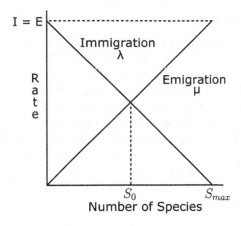

Fig. 1. Species model of a single habitat

In BBOA and based on Fig. 1, immigration curve is used to probabilistically decide whether or not to immigrate each feature in each solution. If a characteristic of solution is selected to immigrate, a solution to migrate one of its features are probabilistically selected randomly. Based on above description the main steps in the BBOA are detailed in Algorithm 1. Also, note that "probability λ_i" and "probability μ_j" are a random number (0 to 1) compared with the respective rate.

Algorithm 1. Migration operator

1: {N the size of the population}
2: **for** i=1 to N **do**
3:    Select H_i with probability λ_i
4:    **if** H_i is selected **then**
5:       {D the solution length}
6:       **for** k=1 to D **do**
7:          Select H_j with probability μ_j
8:          **if** H_j is selected **then**
9:             Select random $k \in [1, D]$
10:            Set $H_{ik} = H_{jk}$
11:         **end if**
12:      **end for**
13:   **end if**
14: **end for**

We note that the best solutions are the least likely to immigrate characteristics, since immigration rates are lower. Opposite of this the solutions with lower fitness are more likely to immigrate, given their high rates of immigration. However, the solutions they provide to their emigration to these worst solutions are those having good fitness for its high rate of emigration [36].

3.2 Mutation Operator

A natural habitat may be affected by cataclysmic events drastically changing its HSI. This could cause a count of species that is different from its equilibrium value (species arriving from neighboring habitats, diseases, natural disasters and others). Thus, the HSI of habitat could suddenly change due to random events. In BBOA likely number of species (P_s) is used to determine mutation rates. These are determined by the balance between immigration and emigration rates (Fig. 1) as a balance between these rates, the probability that S number of species is greater: immigrating species at a rate that is similar to the number of species that migrate in the same habitat. Given that, the best and worst habitats are less likely to have S number of species. This is mentioned in detail in [28]. Then, the mutation rate is calculated as Eq. (3)

$$m(s) = m_{max} \left(\frac{1 - P_s}{P_{max}} \right), \tag{3}$$

where m_{max} is a maximum probability of mutation given by parameter, and P_{max} the probability of S maximum existing. Then, the algorithm 2 explain this operator: where for each habitat the probability of S species is calculated, and then for each feature if selected to be mutated by this probability, it is replaced with another SIV random.

Algorithm 2. Mutation Operator

1: {M the size of the solution}
2: **for** j=1 to M **do**
3:    Calculate probability of mutation P_i based on (3)
4:    Select SIV $H_i(j)$ with probability $P_i(j)$
5:    **if** $H_i(j)$ is selected **then**
6:       Replace $H_i(j)$ with a randomly generated SIV.
7:    **end if**
8: **end for**

Note that in binary problems, the mutation operator to exchange a SIV does so that $H_i(j) = 1 - H_i(j)$ [36]

3.3 Algorithm Description

The features and steps are described in general terms of the BBOA:

1. Initialize parameters. Mapping SVI and habitats according to problem solutions. Initialize a maximum of species S_{max} (for simplicity, matching with the size of the population); immigration, emigration and mutation maximum rates. An elitist parameter to save the best solutions.
2. Initialize set of habitats, where each habitat corresponds to a possible solution of the problem.

3. For each habitat, calculate the HSI and accordingly, the number of species (A greater HSI, the greater the number of species). Then calculate rates of immigration and emigration.
4. Probabilistically using rates of immigration and emigration to modify habitats (Migration operator).
5. For each habitat, update the probability of number of species. Then mutate based on their probability of mutation (Mutation Operator).
6. Back to step 3 and finish until a stopping criterion is satisfied.

Note that after each habitat is modified (steps 2, 4, and 5), its feasibility as a problem solution should be verified. If it does not represent a feasible solution, then some method needs to be implemented in order to map it to the set of feasible solutions [28].

4 Biogeography-Based Optimization Algorithm for the SCP

After the description of the problem and the technique to use, finally it continues with the implementation and adaptation of BBOA to obtain acceptable results for the SCP.

4.1 General Considerations

As general considerations of the algorithm implementation, we can highlight:

- The long (SIVs) of each solution is the same size as the amount of costs instance of SCP.
- Repair function for infeasible solutions is used.
- A parameter of elitism, which stores the 2 solutions with the lowest cost over the generations is used.
- The stop criterion is a maximum number of generations.
- We use an optimized stagnation in local optimum by a created method for this purpose.

4.2 Fitness

An important point of the implemented algorithm is the calculation of the HSI, also called fitness in other population optimization algorithms. BBOA indicates that greater HSI solutions are best; and lower HSI, the worst. In addition, these estimates based on the costs of the problem being optimized. This contradicts the SCP, since this is minimization. It must find a solution with the lowest cost; therefore, lower cost solution is the best. Given that the fitness is calculated as:

$$HSI = \frac{1}{total\ cost\ solution} \tag{4}$$

Thus, at lower cost, the greater the value of the HSI. And higher cost, lower it.

4.3 Repair Infeasible Solutions

Due to changes in the solution features, some of these may be unfeasible for the instance of SCP; this means that the solution generated not comply with constraints. To resolve this problem, repair function is used. The repair is based on analyzing the solution in each constraint (row) verifying the feasibility; i.e., occurs at least one active column covering the restriction. If not exists a column that covers the row, then it is considered infeasible. To fix this, sought and activated columns from unfeasible rows with lower cost that will make the solution becomes feasible.

4.4 Delete Redundant Columns

Other technique for improving solutions is delete redundant columns [27]. A column is considered as redundant, w.r.t a given solution, if after deleting it the solution remains feasible. Therefore, we check the columns of the solution to find possible removals. With this, the solutions costs is reduced without losing feasibility.

4.5 Optimize Stagnation in Local Optimum

In BBOA, a very high maximum mutation rate allowed varied solutions, affecting the cost of these. For this, the value in parameter tends to lower numbers (0.0005 to 0.004 approximately). In the convergence of the BBOA, solutions generally stagnate in a local optimum, losing valuable iterations. When this happens, we created and applied a method that increase the maximum mutation rate, adding diversity and avoiding long stagnation.

The maximum rate of mutation should be increased to allow for new solutions when there is stagnation. For this, we calculated a percentage of 10 % of deadlock over the missing iterations. If this is true, the maximum mutation rate is increased in a 0.0009 over the rate (the latter parameter value subject to more experimentation). Then, if the percentage of stagnation continues to increase up to 20 %, the rate increases again and so that the local optimum change. By applying this method, the maximum mutation rate, which is a fixed parameter of BBOA, becomes variable.

We created this method over experimentation, nothing improvements in results. The values mentioned in this section, are very subject to more experimentation, since they could improve much more the solutions and avoiding the stagnation.

5 Experiments and Results

For the experiments, the optimization algorithm was implemented in Java programming language. In addition, they were carried out on a laptop with Windows 8.1 operating system, Intel Core i3 2.50 GHz with 6 GB of RAM. Moreover, we

Table 1. Results of preprocessed instances experiments - Problem Set 4

Instance	Optimal	Best R	Worst R	Average
mscp41	429	430	433	430,83
mscp410	514	514	519	516,53
mscp42	512	512	512	512,00
mscp43	516	516	521	516,53
mscp44	494	495	495	495,00
mscp45	512	514	517	516,50
mscp46	560	560	570	561,47
mscp47	430	430	433	431,73
mscp48	492	493	499	498,20
mscp49	641	641	656	646,07

Table 2. Results of preprocessed instances experiments - Problem Set 5

Instance	Optimal	Best R	Worst R	Average
mscp51	253	253	263	255,70
mscp510	265	265	267	265,87
mscp52	302	305	307	305,70
mscp53	226	226	230	228,07
mscp54	242	242	243	242,37
mscp55	211	211	212	211,50
mscp56	213	213	216	213,57
mscp57	293	293	301	294,53
mscp58	288	288	299	289,13
mscp59	279	279	287	280,27

used preprocessed [34] instances for SCP, obtained from OR-Library [3]. The table columns are formatted following: the first, for instance executed; the second, the global optimum known; the third best result obtained; fourth worst result and in the fifth the average of the results obtained.

The next parameters, obtained through experimentation was used: Population size = 15, maximum mutation probability = 0.004, maximum immigration probability = 1, maximum emigration probability = 1 and a maximum number of iterations = 6000. Each instance was executed 30 times. We divide the results on instances set. This can be seen in Tables 1, 2, 3, 4, 5, 6, 7 and 8.

Given the above results, we can see an excellent performance with the preprocessed instances. Getting the global optimum in 41 of 48 instances, and low cost average. Furthermore, thanks to the preprocessed method, this instances allow numerous experiments because to the speed of execution.

Table 3. Results of preprocessed instances experiments - Problem Set 6

Instance	Optimal	Best R	Worst R	Average
mscp61	138	138	148	142,57
mscp62	146	146	151	149,90
mscp63	145	145	148	146,60
mscp64	131	131	134	131,10
mscp65	161	161	169	164,83

Table 4. Results of preprocessed instances experiments - Problem Set A

Instance	Optimal	Best R	Worst R	Average
mscpa1	253	253	258	255,33
mscpa2	252	252	261	255,73
mscpa3	232	232	239	234,00
mscpa4	234	234	235	234,60
mscpa5	236	236	238	236,70

Table 5. Results of preprocessed instances experiments - Problem Set B

Instance	Optimal	Best R	Worst R	Average
mscpb1	69	69	75	70,37
mscpb2	76	76	80	76,50
mscpb3	80	80	82	80,77
mscpb4	79	79	83	80,53
mscpb5	72	72	74	72,13

Table 6. Results of preprocessed instances experiments - Problem Set C

Instance	Optimal	Best R	Worst R	Average
mscpc1	227	227	233	229,93
mscpc2	219	219	225	221,13
mscpc3	243	248	255	250,40
mscpc4	219	219	227	221,20
mscpc5	215	215	218	216,83

Table 7. Results of preprocessed instances experiments - Problem Set D

Instance	Optimal	Best R	Worst R	Average
mscpd1	60	60	62	60,27
mscpd2	66	66	69	67,43
mscpd3	72	72	76	73,83
mscpd4	62	62	65	63,37
mscpd5	61	61	64	61,57

Table 8. Results of preprocessed instances experiments - Problem Set NR

Instance	Optimal	Best R	Worst R	Average
mscpnre1	29	29	32	29, 63
mscpnrf1	14	14	15	14, 47
mscpnrg1	176	177	190	181, 77

6   Conclusion

After analyzed the problem and the technique to solve it, the algorithm is implemented, showing good results with full experiments; finding some low-cost solutions and low average cost. We created and implemented a technique that occur very good behavior in the algorithm, adding diversity and avoiding long stagnation. This, together with delete redundant columns and a simple repair method, allowed improve the results and algorithm performance. This type of algorithm modifications were made in order to obtain better quality results, shown results with 41 optimum solutions of 48 instances, including big instances.

Undoubtedly, new methods applied had great impact on the quality of results, due to the native algorithm not shown as good behavior. We could carry out more experiments, with new good repair methods, since even a basic repair method is used; as to find more precise parameters in the change of maximum mutation rate or BBOA input. This could generate a full optimum table. Finally, we can say that BBOA is very good to solve the SCP.

Acknowledgements. The author Broderick Crawford is supported by grant CONICYT/FONDE-CYT/REGULAR/1140897 and Ricardo Soto is supported by grant CONICYT/FONDECYT/REGULAR/1160455.

References

1. Amini, F., Ghaderi, P.: Hybridization of harmony search and ant colony optimization for optimal locating of structural dampers. Appl. Soft Comput. **13**(5), 2272–2280 (2013)
2. Balas, E., Carrera, M.C.: A dynamic subgradient-based branch-and-bound procedure for set covering. Oper. Res. **44**(6), 875–890 (1996)

3. Beasley, J.E., Jornsten, K.: Enhancing an algorithm for set covering problems. Eur. J. Oper. Res. **58**(2), 293–300 (1992). http://ideas.repec.org/a/eee/ejores/v58y1992i2p293-300.html

4. Brusco, M.J., Jacobs, L.W., Thompson, G.M.: A morphing procedure to supplement a simulated annealing heuristic for cost- and coverage-correlated set-covering problems. Ann. Oper. Res. **86**, 611–627 (1999)

5. Caprara, A., Fischetti, M., Toth, P., Vigo, D., Guida, P.L.: Algorithms for railway crew management. Math. Program. **79**(1–3), 125–141 (1997). http://dx.doi.org/10.1007/BF02614314

6. Caserta, M.: Tabu search-based metaheuristic algorithm for large-scale set covering problems. In: Doerner, K.F., Gendreau, M., Greistorfer, P., Gutjahr, W., Hartl, R.F., Reimann, M. (eds.) Metaheuristics. OR/CSIS, vol. 39, pp. 43–63. Springer, Heidelberg (2007). http://dx.doi.org/10.1007/978-0-387-71921-4_3

7. Crawford, B., Soto, R., Berríos, N., Johnson, F., Paredes, F.: Solving the set covering problem with binary cat swarm optimization. In: Tan, Y., Shi, Y., Buarque, F., Gelbukh, A., Das, S., Engelbrecht, A. (eds.) ICSI-CCI 2015. LNCS, vol. 9140, pp. 41–48. Springer, Heidelberg (2015)

8. Crawford, B., Soto, R., Cuesta, R., Paredes, F.: Application of the artificial bee colony algorithm for solving the set covering problem. Sci. World J. **2014**(189164), 1–8 (2014)

9. Crawford, B., Soto, R., Monfroy, E., Palma, W., Castro, C., Paredes, F.: Parameter tuning of a choice-function based hyperheuristic using particle swarm optimization. Expert Syst. Appl. **40**(5), 1690–1695 (2013)

10. Crawford, B., Soto, R., Olea, C., Johnson, F., Paredes, F.: Binary bat algorithms for the set covering problem. In: 2015 10th Iberian Conference on Information Systems and Technologies (CISTI), pp. 1–4, June 2015

11. Crawford, B., Soto, R., Olivares Suarez, M., Paredes, F., Johnson, F.: Binary firefly algorithm for the set covering problem. In: 2014 9th Iberian Conference on Information Systems and Technologies (CISTI), pp. 1–5, June 2014

12. Crawford, B., Soto, R., Cuesta, R., Paredes, F.: Application of the artificial bee colony algorithm for solving the set covering problem. Sci. World J. **2014**, 1–8 (2014)

13. Crawford, B., Soto, R., Monfroy, E., Paredes, F., Palma, W.: A hybrid ant algorithm for the set covering problem. Int. J. Phys. Sci. **6**, 4667–4673 (2011)

14. Crawford, B., Soto, R., Olivares-Suárez, M., Paredes, F.: A binary firefly algorithm for the set covering problem. In: Silhavy, R., Senkerik, R., Oplatkova, Z.K., Silhavy, P., Prokopova, Z. (eds.) Modern Trends and Techniques in Computer Science. AISC, vol. 285, pp. 65–73. Springer, Heidelberg (2014). http://dx.doi.org/10.1007/978-3-319-06740-7_6

15. Eremeev, A.V., Kolokolov, A.A., Zaozerskaya, L.A.: A hybrid algorithm for set covering problem. In: Proceedings of International Workshop Discrete Optimization Methods in Scheduling and Computer-Aided Design, pp. 123–129 (2000)

16. Fisher, M.L., Kedia, P.: Optimal solution of set covering/partitioning problems using dual heuristics. Manage. Sci. **36**(6), 674–688 (1990)

17. Fisher, M.L., Rosenwein, M.B.: An interactive optimization system for bulk-cargo ship scheduling. Naval Res. Logistics (NRL) **36**(1), 27–42 (1989)

18. Foster, B.A., Ryan, D.: An integer programming approach to the vehicle scheduling problem. Oper. Res. **27**, 367–384 (1976)

19. Garey, M.R., Johnson, D.S.: Computers and Intractability: A Guide to the Theory of NP-Completeness. W. H. Freeman & Co., New York, NY, USA (1990)

20. Goldberg, D.E.: Real-coded genetic algorithms, virtual alphabets, and blocking. Complex Syst. **5**, 139–167 (1990)
21. Guanghui Lan, A., Depuy, G.W.B., G.E.W.C.: Discrete optimization an effective-and simple heuristic for the set covering problem abstract (2005)
22. Han, L., Kendall, G., Cowling, P.: An adaptive length chromosome hyperheuristic genetic algorithm for a trainer scheduling problem. In: Proceedings of the fourth Asia-Pacific Conference on Simulated Evolution And Learning, (SEAL 2002), Orchid Country Club, Singapore, pp. 267–271 (2002)
23. Ma, H., Simon, D.: Biogeography-based optimization with blended migration for constrained optimization problems. In: Pelikan, M., Branke, J. (eds.) Genetic and Evolutionary Computation Conference, GECCO 2010, Proceedings, Portland, Oregon, USA, July 7–11, 2010. pp. 417–418. ACM (2010). http://doi.acm.org/10.1145/1830483.1830561
24. Michalewicz, Z.: Genetic algorithms + data structures = evolution programs, 3rd edn. Springer-Verlag, London, UK (1996)
25. Mo, H., Xu, L.: Biogeography migration algorithm for traveling salesman problem. In: Tan, Y., Shi, Y., Tan, K.C. (eds.) ICSI 2010, Part I. LNCS, vol. 6145, pp. 405–414. Springer, Heidelberg (2010). http://dx.doi.org/10.1007/978-3-642-13495-1_50
26. Mudaliar, D., Modi, N.: Unraveling travelling salesman problem by genetic algorithm using m-crossover operator. In: 2013 International Conference on Signal Processing Image Processing Pattern Recognition (ICSIPR), pp. 127–130, February 2013
27. Naji-Azimi, Z., Toth, P., Galli, L.: An electromagnetism metaheuristic for the unicost set covering problem. Eur. J. Oper. Res. **205**(2), 290–300 (2010). http://EconPapers.repec.org/RePEc:eee:ejores:v:205:y:2010:i:2
28. Simon, D.: Biogeography-based optimization. Evol. Comput. IEEE Trans. **12**(6), 702–713 (2008)
29. Smith, B.M.: Impacs - a bus crew scheduling system using integer programming. Math. Program. **42**(1), 181–187 (1988). http://dx.doi.org/10.1007/BF01589402
30. Soto, R., Crawford, B., Olivares, R., Barraza, J., Johnson, F., Paredes, F.: A binary cuckoo search algorithm for solving the set covering problem. In: Vicente, J.M.F., Álvarez-Sánchez, J.R., López, F.P., Toledo-Moreo, F.J., Adeli, H. (eds.) Bioinspired Computation in Artificial Systems. LNCS, vol. 9108, pp. 88–97. Springer, Heidelberg (2015). http://dx.doi.org/10.1007/978-3-319-18833-1_10
31. Thomson, G.: A simulated annealing heuristic for shift-scheduling using non-continuously available employees. Comput. Oper. Res. **23**, 275–288 (1996)
32. Toregas, C., Swain, R., ReVelle, C., Bergman, L.: The location of emergency service facilities. Oper. Res. **19**(6), 1363–1373 (1971). http://dx.doi.org/10.1287/opre.19.6.1363
33. Vasko, F.J., Wolf, F.E., Stott, K.L.: A set covering approach to metallurgical grade assignment. Eur. J. Oper. Res. **38**(1), 27–34 (1989). http://EconPapers.repec.org/RePEc:eee:ejores:v:38:y:1989:i:1:p:27-34
34. Xu, Y., Kochenberger, G., Wang, H.: Pre-processing method with surrogate constraint algorithm for the set covering problem (2008)
35. Zhang, Y., Wu, L., Wang, S., Huo, Y.: Chaotic artificial bee colony used for cluster analysis. In: Chen, R. (ed.) ICICIS 2011 Part I. CCIS, vol. 134, pp. 205–211. Springer, Heidelberg (2011). http://dx.doi.org/10.1007/978-3-642-18129-0_33
36. Zhao, B.B., Deng, C., Yang, Y., Peng, H.: Novel binary biogeography-based optimization algorithm for the knapsack problem. In: Tan, Y., Shi, Y., Ji, Z. (eds.) ICSI 2012, Part I. LNCS, vol. 7331, pp. 217–224. Springer, Heidelberg (2012). http://dx.doi.org/10.1007/978-3-642-30976-2_26

Finding Solutions of the Set Covering Problem with an Artificial Fish Swarm Algorithm Optimization

Broderick Crawford[1,2,5], Ricardo Soto[1,3,4], Eduardo Olguín[5], Sanjay Misra[6], Sebastián Mansilla Villablanca[1(✉)], Álvaro Gómez Rubio[1], Adrián Jaramillo[1], and Juan Salas[1]

[1] Pontificia Universidad Católica de Valparaíso, 2362807 Valparaíso, Chile
{broderick.crawford,ricardo.soto}@pucv.cl,
{sebastian.mansilla.v,alvaro.gomez.r,adrian.jaramillo.s,
juan.salas.f}@mail.pucv.cl
[2] Universidad Central de Chile, 8370178 Santiago, Chile
[3] Universidad Autónoma de Chile, 7500138 Santiago, Chile
[4] Universidad Científica del Sur, Lima 18, Peru
[5] Facultad de Ingeniería y Tecnología, Universidad San Sebastián, Bellavista 7,
8420524 Santiago, Chile
eduardo.olguin@uss.cl
[6] Covenant University, Ogun 110001, Nigeria
sanjay.misra@covenantuniversity.edu.ng

Abstract. The Set Covering Problem (SCP) is a matrix that is composed of zeros and ones and consists in finding a subset of zeros and ones also, in order to obtain the maximum coverage of necessities with a minimal possible cost. In this world, it is possible to find many practical applications of this problem such as installation of emergency services, communications, bus stops, railways, airline crew scheduling, logical analysis of data or rolling production lines. SCP has been solved before with different nature inspired algorithms like fruit fly optimization algorithm. Therefore, as many other nature inspired metaheuristics which imitate the behavior of population of animals or insects, Artificial Fish Swarm Algorithm (AFSA) is not the exception. Although, it has been tested on knapsack problem before, the objective of this paper is to show the performance and test the binary version of AFSA applied to SCP, with its main steps in order to obtain good solutions. As AFSA imitates a behavior of a population, the main purpose of this algorithm is to make a simulation of the behavior of fish shoal inside water and it uses the population as points in space to represent the position of fish in the shoal.

Keywords: Set Covering Problem · Artificial Fish Swarm Optimization Algorithm · Metaheuristics · Combinatorial optimization

O. Gervasi et al. (Eds.): ICCSA 2016, Part I, LNCS 9786, pp. 166–181, 2016.
DOI: 10.1007/978-3-319-42085-1_13

1   Introduction

Metaheuristics provide "acceptable" solutions in a reasonable time for solving hard and complex problems in science and engineering when it is expensive to find the best solution especially with a computing power limited.

One of the classical problems that Metaheuristics try to solve is Set Covering Problem (SCP) which consists in finding a set of solutions at the lowest possible cost, accomplishing with the constraints of a matrix that has zeros and ones, where each row must be covered of at least one column. In the past, SCP has been solved with different algorithms such as cultural algorithm [5], fruit fly optimization algorithm [13] or teaching-learning-based optimization algorithm [14]. There are many applications of this problem such as optimal selection of ingot sizes [25] or assign fire companies to fire houses [26].

The main goal of this paper is to show the performance of Artificial Fish Swarm Algorithm (AFSA) applied to SCP, previously, it was tested on the knapsack problem [19–21]. This algorithm, simulates the behavior of a fish inside the water which belongs to a shoal and it uses a population of points in space to represent the position of fish in the shoal. In the original version of AFSA, there are five main behaviors such as random, chasing, swarming, searching and leaping. In the following work, it will be showed its simplified version of AFSA in order to solve SCP. The proposed algorithm has the following steps, initialization of its population, generation of trial points, the effect-based crossover, dealing with SCP constraints, selection of a new population, reinitialization of the population, local search and termination conditions. This method will be tested on each one of the 70 SCP benchmarks obtained from OR-Library website. These 70 files are formatted as: number of rows n, number of columns m, the cost of each column $c_j, j \in \{1, \ldots, n\}$, and for each row $i, i \in \{1, ..., m\}$ the number of columns which cover row i followed by a list of the columns which cover rows i. These 70 files were chosen in order to solve SCP in an academic and theoretical way.

The remainder paper is organized as follows: In Sect. 2, it will be explained the set covering problem. In Sect. 3, it is going to be presented the artificial fish swarm algorithm in general terms. In Sect. 4, It will be showed the AFSA method and its simplified version in order to solve the SCP with its main steps for solving this problem and, it is going to be illustrated the proposed algorithm. In Sect. 5, it will be exposed the experimental results. Finally, in Sect. 6, it is going to be presented the conclusions of this paper.

2   Set Covering Problem

This is a classical and well-known NP-hard problem. It is a representation of a sort of combinatorial optimization problem which has many practical applications in the world such as construction of firemen stations in different places or installation of network cell phones in order to obtain the maximum coverage with a minimal possible cost. The SCP can be formulated as follows [1]:

$$\text{minimize} \quad Z = \sum_{j=1}^{n} c_j x_j \tag{1}$$

Subject to:

$$\sum_{j=1}^{n} a_{ij} x_j \geq 1 \quad \forall i \in I \tag{2}$$

$$x_j \in \{0, 1\} \quad \forall j \in J \tag{3}$$

where $A = (a_{ij})$ be a $m \times n$ 0–1 matrix with $I = \{1, \ldots, m\}$ and $J = \{1, \ldots, n\}$ be the row and column sets respectively. Column j can be covered a row i if $a_{ij} = 1$. Where c_j is a nonnegative value that represents the cost of selecting the column j and x_j is a decision variable, that can be 1 if column j is selected or 0 otherwise. The objective is to find a minimum cost subset $S \subseteq J$, such that each row $i \in I$ is covered by at least one column $j \in S$.

2.1   How Has It Been Solved Before?

There are two kinds of methods that have been used to solve the SCP. First, the methods which produce optimal solutions but sometimes need a lot of time and/or high computational cost. Those methods could be constraint programming, branch and bound or integer linear programming. On the other hand, there are the metaheuristics that provide "acceptable or good" solutions in a reasonable time. Even, it is possible to find optimal solutions with many metaheuristics.

2.2   Metaheuristics that Have Solved SCP

In the past, the SCP was successfully solved with many metaheuristics such as artificial bee colony [2–4], cultural algorithm [5], swarm optimization particles [6], ant colony optimization [7], firefly algorithm [8–10], shuffled frog leaping algorithm [11,12], fruit fly optimization algorithm [13], teaching-learning-based optimization algorithm [14], or genetic algorithm [15,16,27]. In others works there are comparisons among different kind of metaheuristics [17], or comparisons among different kinds of nature-inspired metaheuristics [18] in order to solve the SCP.

2.3   Metaheuristics, Bio-Inspired and AFSA

As it mentioned before, one of the main advantage of metaheuristics is to provide good solutions in a "reasonable" time, especially when the computing resources are not infinite. On the other hand, metaheuristics not always provide the optimal results. However, in real life not always is required the best known solution because this world moves quickly and it is necessary to obtain responses in little time, for instance, in companies when they try to increase benefits or decrease costs. Moreover, metaheuristics can be helped by other techniques, according to [23], such as handling of constraints that problem impose and/or the dimension reduction of the problem.

Bio-inspired metaheuristics are a sort of metaheuristics and they are methods that simulate the behavior of a swarm or group of animals. Also, they try to solve problems of optimization such as maximization or minimization. In this case, AFSA was created with the intention of solving the knapsack problem which is a kind of maximization problem. Hence, the proposed work has the objective of transforming the application of a maximization problem into a minimization problem and observe its results because many others nature inspired metaheuristics have been good results on SCP, as it mentioned above.

3   Artificial Fish Swarm Algorithm

As many other nature inspired metaheuristics which imitate the behavior of population of animals or insects, AFSA is not the exception. According to [19], this algorithm was proposed and applied in order to solve optimization problems and it simulates the behavior of a fish swarm inside the water where a fish represents a point or a fictitious entity of a true fish in a population and the swarm movements are randomly.

3.1   Main Behaviors of AFSA

The fish swarm behavior is summarized as follows [19]:

1. **Random Behavior:** In order to find companion and food, a fish swims randomly inside water.
2. **Chasing Behavior:** If food is discovered by a fish, the others in the neighborhood go quickly after it.
3. **Swarming Behavior:** In order to guarantee the survival of the swarm and avoid dangers from predatory, Fish come together in groups.
4. **Searching Behavior:** Fish goes directly and quickly to a region, when that region is discovered with more food by it. That can be by vision or sense.
5. **Leaping Behavior:** Fish leaps to look for food in other regions, when it stagnates in a region.

These five behaviors are the responsible that the artificial fish swarm tries to search good results. Also, AFSA works with feasible solutions.

3.2 Another Description of AFSA

Additionally to the explanation showed above there is another description of AFSA which is proposed with more details in [20]. The concept of "visual scope" is the main concept utilized in that version of AFSA, and it represents how close is the neighborhood in comparison to a point/fish.

Depending on the position of a point related to the population, there could occur three situations [20]:

(a) The "visual scope" is empty, and the current point with no other points in its neighborhood, moves randomly looking for a better region.
(b) When the "visual scope" is not crowded, the current point can move towards the best point inside the "visual scope", or, if this best point does not improve the objective function value it moves towards the central point of the "visual scope".
(c) When the "visual scope" is crowded, the current point has some difficulty in following any particular point, and searches for a better region by choosing randomly another point (from the "visual scope") and moving towards it.

The condition that decides when the "visual scope" of the current point is not crowded and the central point inside the "visual scope" are explained in [20].

3.3 Proposed Algorithm of AFSA Binary Version

Before of applying AFSA on SCP, it is necessary to show the binary version of this algorithm which was proposed by [19], but it was applied on the knapsack problem. That algorithm is the following:

Algorithm 1. Binary version of AFSA

```
 1: Set parameter values
 2: Set t = 1 and randomly initialize x^{i,t}, i = 1, 2, ..., N
 3: Perform decoding and evaluate z. Identify x^{max} and z_{max}
 4: if Termination condition is met then
 5:     Stop
 6: end if
 7: for all x^{i,t} do
 8:     Calculate"visual scope" and "crowding factor"
 9:     Perform fish behavior to create trial point y^{i,t}
10:      Perform decoding to make the trial point feasible
11: end for
12: Perform selection according to step 4 to create new current points
13: Evaluate z and identify x^{max} and z_{max}
14: if t%L = 0 then
15:     Perform leaping
16: end if
17: Set t = t + 1 and go to step 4
```
---

4   Artificial Fish Swarm Algorithm and Its Simplified Binary Version

In this section, it is going to be showed the main steps of AFSA and its simplified binary version in order to solve SCP. According to the authors [21] AFSA converges to a non-optimal solution in previous versions like [20]. Therefore, there were proposed some modifications of AFSA and they were slightly modified in order to solve SCP.

4.1   Features that Were Modified in AFSA

In [21] the main modifications were: the "visual scope" concept was rejected; The behavior depends on two probability values; Swarming behavior is never utilized; An effect-based crossover is used instead of an uniform crossover in different behaviors to create trial points; A local search with two steps was implemented; Among other modifications that are explained with more details in [21]. Also, it was introduced a repair function for handling the SCP constaraints.

Next, it will be explained the steps of AFSA in order to obtain SCP results.

4.2   Initialization of Population

As many other metaheuristics, it is necessary to initialize the population with objective to find good solutions. Therefore, the best representation of a population is N current points, x^i, where $i \in \{1, 2, ..., N\}$ each one represented by a binary string of 0/1 bits of length n and are randomly generated.

4.3   Generation of Trial Population

This metaheuristic works with a trial population at each iteration. So, in order to create trial points in consecutive iterations based on behaviors of random, chasing, and searching is necessary utilize crossover and mutation after the initialization of population. In [21] probabilities of $0 \leq \tau_1 \leq \tau_2 \leq 1$ were introduced and they are the responsible to reach this objective.

Random Behavior: If a fish does not have companion in its neighborhood, then it moves randomly looking for food in another region [21]. This happens when a random number $rand(0, 1)$ is less than or equal to τ_1. The trial point y^i is created randomly setting 0/1 bits of length n [21].

Chasing Behavior: When a fish, or a group of fish in the swarm, discover food, and the others go quickly after it [21]. This happens when $rand(0, 1) \geq \tau_2$ and it is related to the movement towards the best point found so far in the population, x^{min}. The trial point y^i is created using an effect-based crossover (see *Algorithm 2*) between x^i and x^{min} [21].

Searching Behavior: When fish discovers a region with more food, by vision or sense, it goes directly and quickly to that region [21]. This behavior is related to the movement towards a point x^{rand} where "rand" is an index randomly chosen from the set $\{i = 1, 2, ..., N\}$. When $\tau_1 < rand(0, 1) < \tau_2$ it is implemented. An effect-based crossover between x^{rand} and x^i is utilized to create the trial point y^i [21].

Trial Point Corresponding to the Best Point: In [21], the 3 behaviors explained above are implemented to create $N - 1$ trial points; the best point x^{min} uses a 4 flip-bit mutation. It is performed on the point x^{min} to create the corresponding trial point y^i. In this operation 4 positions are randomly selected, and the bits of the corresponding positions are changed from 0 to 1 or vice versa [21].

4.4   The Effect-Based Crossover in Simplified Binary Version of AFSA

In order to obtain the trial point in chasing and searching behavior, it necessary to calculate the *effect ratio* ER_{u,x^i} of u on the current point x^i, according to [21]. It can obtain with the following two formulas:

$$ER_{u,x^i} = \frac{q(u)}{q(u) + q(x^i)} \tag{4}$$

$$q(x^i) = \exp\left[\frac{-(z(x^{min}) - z(x^i))}{(z(x^{min}) - z(x^{max}))}\right] \tag{5}$$

$u = x^{min}$ is used with chasing behavior, $u = x^{rand}$ is used with searching behavior and x^{max} is the worst point of the population. The effect-based crossover to obtain the trial point y^i is showed in Algorithm 2 according to [21].

Algorithm 2. Effect-based crossover

```
Require: current point xⁱ, u and ER_{u,xⁱ}
1: for j = 1 to n do
2:     if rand(0,1) < ER_{u,xⁱ} then
3:         yʲⁱ = uⱼ
4:     else
5:         yʲⁱ = xⱼ
6:     end if
7: end for
8: return trial point yⁱ
```

4.5   Deal with Constraints of SCP

In order to obtain good results, it is necessary to introduce an appropriate method that helps with SCP constraints. Therefore, it is going to be showed a repair function for handling the SCP constraints in the Algorithm 3. According to [23], Algorithm 3 shows a repair method where all rows not covered are identified and the columns required are added. Hence, in this way all the constraints will be covered. The search of these columns are based in the relationship showed in the next equation.

$$\frac{cost\ of\ one\ column}{amount\ of\ columns\ not\ covered} \tag{6}$$

Once the columns are added and the solution is feasible, a method is applied to remove redundant columns of the solution. The redundant columns are those that are removed, the solution remains a feasible solution. The algorithm of this repair method is detailed in the Algorithm 3. Where:

(a) I is the set of all rows
(b) J is the set of all columns
(c) J_i is the set of columns that cover the row $i, i \in I$
(d) I_j is the set of rows covered by the column $j, j \in J$
(e) S is the set of columns of the solution
(f) U is the set of columns not covered
(g) w_i is the number of columns that cover the row $i, \forall i \in I$ in S

Algorithm 3. Repair Operator for Dealing with SCP Constraints

---

```
1:  w_i ← | S ∩ J_i | ∀i ∈ I;
2:  U ← {i | w_i = 0}, ∀i ∈ I;
3:  for i ∈ U do
4:       find the first column j in J_i that minimize  c_j/|U∩I_j| S ← S ∩ j;
5:       w_i ← w_i + 1, ∀i ∈ I_j;
6:       U ← U − I_j;
7:  end for
8:  for j ∈ S do
9:       if w_i ≥ 2, ∀i ∈ I_j then
10:           S ← S − j;
11:           w_i ← w_i − 1, ∀i ∈ I_j;
12:       end if
13: end for
```

4.6 Selection of New Population

The new population is selected between trial population and current population. Each trial point contends against the current point, therefore, if $z(y^i) \le z(x^i)$, then the trial point becomes a member of the new population to the next iteration; otherwise, the current point is maintained to the next iteration, according to [21].

$$x^{i,t+1} = \begin{cases} y^{i,t}\ if\ z(y^{i,t}) \le z(x^{i,t}),\ i = 1,2,...,N \\ \\ x^{i,t}\ \text{otherwise} \end{cases} \tag{7}$$

4.7 Reinitialization of Current Population

In [21], Every certain iterations, this metaheuristic replaces the population of the last iteration with a new population to the next iteration. Therefore, utilizing the same values, it will be done a randomly reinitialization of 50 % of the population at every R iterations, where R is a positive integer parameter.

4.8 Exploitation or Local Search

Exploitation is related to leaping behavior of AFSA. So, according to the authors [21], in oder to obtain better solutions and improve old versions of AFSA, the concept of exploitation/local search was utilized and its purpose is to find better solutions when the method obtains the same solution as the iterations pass. This is based on a flip-bit mutation which N_{loc} points are selected randomly from the population, where $N_{loc} = \tau_3 N$ with $\tau_3 \in (0,1)$. This mutation changes the bit values of those points from 0 to 1 and vice versa according to p_m probability. After that, those new points are made feasible by using the repair function of SCP explained in Sect. 4.5. Then they become members of the population, if they improve z_{min} at that moment. Afterwards, the best point of the population is identified and another mutation is operated on N_{ref} positions, with $N_{ref} = \tau_3 n$, those positions are randomly selected from the point [21]. Then it becomes a member of the population, if it improves z_{min} at that moment. This mutation is used L times, where L is s a positive integer parameter.

4.9 Conditions to Finish the Algorithm

As many other metaheuristics, it is necessary to stop this metaheuristic when it is almost impossible to find a better solution and it is unnecessary continuing wasting the computational resource. Therefore, in accordance with [21], AFSA terminates when the known optimal solution is reached or a maximum number of iterations, T_{max}, is exceeded.

$$t > T_{max} \ or \ z_{min} \le z_{opt} \qquad (8)$$

where z_{min} is the best objective function value attained at iteration t and z_{opt} is the known optimal value available in the literature.

4.10 AFSA-SCP Proposed Algorithm

After showing the main steps of this algorithm, it is going to show the proposed algorithm by mean of two tools. First, the flow chart of this algorithm in Fig. 1. Then it will be showed the pseudocode of the proposed method which is shown in Algorithm 4 and it has the appropriate modifications in order to solve SCP.

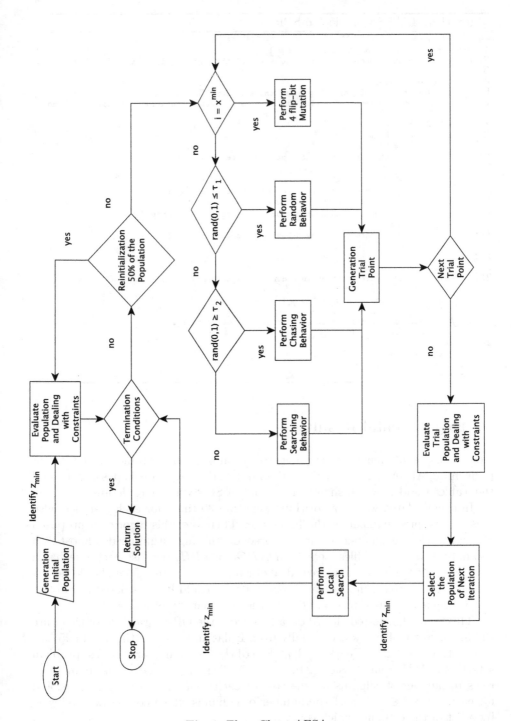

Fig. 1. Flow Chart AFSA

Algorithm 4. AFSA applied to SCP

Require: T_{max} and z_{opt} and other values of parameters
 1: Set $t = 1$ Initialize population $x^{i,t}$, $i = 1, 2, ..., N$
 2: Execute SCP repair function in order to evaluate the population, identify x^{min} and z_{min}
 3: while $t \leq T_{max}$ or $z_{min} \geq z_{opt}$ do
 4:     if $t\%R = 0$ then
 5:         Reinitialize 50% of the population, keeping x^{min} and z_{min}
 6:         Execute SCP repair function in order to evaluate population, identify x^{min} and z_{min}
 7:     end if
 8:     for $i = 1$ to N do
 9:         if $i = x^{min}$ then
10:             Execute 4 flip-bit mutation to create trial point $y^{i,t}$
11:         else
12:             if $rand(0,1) \leq \tau_1$ then
13:                 Execute random behavior to create trial point $y^{i,t}$
14:             else if $rand(0,1) \geq \tau_2$ then
15:                 Execute chasing behavior to create trial point $y^{i,t}$
16:             else
17:                 Execute searching behavior to create trial point $y^{i,t}$
18:             end if
19:         end if
20:     end for
21:     Execute SCP repair function in order to evaluate and get $y^{i,t}$, $i = 1, 2, ..., N$ and evaluate them
22:     Select new population $x^{i,t+1}$, $i = 1, 2, ..., N$
23:     if $t\%L = 0$ then
24:         Execute exploitation/local search - leaping behavior
25:         Identify x^{min} and z_{min}
26:     end if
27:     Set $t = t + 1$
28: end while
29: return x^{min} and z_{min}

---

5   Experimental Results

After many experiments, it is going to be showed the obtained results after performing AFSA to solve SCP. At the final of this section it is possible to find the Tables 1 and 2, which shows the results of SCP with more details.

In other related works, algorithms were run 30 times for each instance, also it is an accepted number in the literature. Therefore, this algorithm proposed was run with that number of times. Moreover, this algorithm needs almost 20 h to analyze the last 10 files, sets from NRG to NRH, because each one has a matrix with a big dimension, great deal of rows and columns. That should not be considered a problem in academic area but that could be considered a problem if it is applied in a real problem due to the big quantity of hours.

This algorithm tested the 70 data files from the OR-Library, 25 of them are the instance sets $4, 5, 6$ was originally from Balas and Ho [22], the others 25, the sets A, B, C, D, E from Beasley [23] and 20 of these data files are the test problem sets E, F, G, H from Beasley [24]. These 70 files are formatted as: number of rows n, number of columns m, the cost of each column $c_j, j \in \{1, ..., n\}$, and for each row $i, i \in \{1, ..., m\}$ the number of columns which cover row i followed by a list of the columns which cover rows i.

Table 1. Experimental results of SCP benchmark sets (4, 5, 6, A, B, C, D, E, NRE, NRF, NRG and NRH)

Number	Number of Files	Instance	ARPD
1	10	4	0,84
2	10	5	0,83
3	5	6	0,78
4	5	A	1,83
5	5	B	2,92
6	5	C	2,29
7	5	D	3,65
8	5	E	0,0
9	5	NRE	4,93
10	5	NRF	7,16
11	5	NRG	5,82
12	5	NRH	6,72

This algorithm was implemented in Java programming language, using Eclipse IDE, with the following hardware: Intel core i5 dual core 2.60 GHz processor, 8 GB RAM and it was run under OSX Yosemite.

Finally, the program was executed only with feasible solutions, with a population of $N = 20$ fish, probability $\tau_1 = 0.1$, probability $\tau_2 = 0.9$, probability $\tau_3 = 0.1$, probability $p_m = 0.1$, $L = 50$, reinitialization of population $R = 10$ and each trial was run 1000 iterations, and 30 times each one. After obtained all results, it was obtained the averages values from these 30 times for each one of the files.

Table 1 shows an overview of all sets. It contains the instance number, set number and $ARPD$ which is an average of the deviation of the objective value (best known solution). With these results, it is possible to show that the best results which have an $ARPD$ minor to 1 %, are the groups $4, 5, 6$ and E, and due to that reason they were selected because they have better results in comparison with other benchmark groups. Hence, Table 2 shows the best results obtained where the first column is the number of experiment of each instance, the second column $Instance$ indicates each benchmark evaluated, and Z_{opt} shows the best known solution value of each instance. The next columns Z_{min}, Z_{max}, Z_{avg} represents the minimum, maximum among minimums, and average of minimums solutions obtained. The last column reports the relative percentage deviation RPD which represents the deviation of the objective value or best known solution f_{opt} from f_{min} which is the minimum value obtained for each instance. RPD was calculated as follows:

$$RPD = \frac{100(f_{min} - f_{opt})}{f_{opt}} \tag{9}$$

Table 2. Experimental results of SCP benchmark sets (4, 5, 6 and E)

Number	Instance	Z_{opt}	Z_{min}	Z_{max}	Z_{avg}	RPD
1	4.1	429	430	445	437,4	0,23
2	4.2	512	515	546	530,83	0,59
3	4.3	516	519	543	528,27	0,58
4	4.4	494	495	532	514,83	0,20
5	4.5	512	514	536	521,73	0,39
6	4.6	560	565	597	580,9	0,89
7	4.7	430	432	447	437,37	0,47
8	4.8	492	492	514	501,73	0,0
9	4.9	641	658	688	669,8	2,65
10	4.10	514	525	559	539,6	2,14
11	5.1	253	254	271	263,03	0,40
12	5.2	302	310	318	314,27	2,65
13	5.3	226	228	244	232,77	0,88
14	5.4	242	242	247	244,77	0,0
15	5.5	211	212	215	212,6	0,47
16	5.6	213	214	242	227,77	0,47
17	5.7	293	299	315	307,9	2,05
18	5.8	288	291	313	298,97	1,04
19	5.9	279	279	296	285,73	0,0
20	5.10	265	266	276	272,07	0,38
21	6.1	138	138	153	146,37	0,0
22	6.2	146	149	156	151,97	2,05
23	6.3	145	145	161	149,63	0,0
24	6.4	131	131	137	134,17	0,0
25	6.5	161	164	181	172,67	1,86
26	E.1	5	5	6	5,87	0,0
27	E.2	5	5	6	5,5	0,0
28	E.3	5	5	6	5,2	0,0
29	E.4	5	5	6	5,7	0,0
30	E.5	5	5	6	5,57	0,0

6   Conclusions

AFSA optimization has a good performance with some instances. It has been observed that sets $4, 5, 6$ and E obtained each one an $ARPD$ minor to $1\,\%$, and sets A, B, C, D obtained an $ARPD$ minor to $4\,\%$. Then with the last 20 instances the results of $ARPD$ start to decrease. Due to that reason sets $4, 5, 6$ and E were selected because they have better results in comparison with other benchmark groups and their results were showed in Table 2. Moreover, in comparison to other metaheuristics such as Artificial Bee Colony Algorithm [4], Binary Firefly Algorithm [8] or Genetic Algorithm [27], the results of AFSA are similar with sets $4, 5, 6$ and E and worse than the other groups. For this reason, this technique requires more study.

It was observed that in the first 200 iterations, this algorithm converges quickly to very good solutions or optimal solutions in some cases. Between 200 and 1000 iterations, sometimes, the algorithm can obtain a slightly better results if it does not obtain the optimal result before but in other cases it maintain the same result reached at first 200 iterations. Therefore, It could be possible that it is necessary more than 1000 iterations and/or find a better configuration of parameters to obtain the optimal results or closest to the optimal in all instances. Thus, reduce the variability in its results.

Another characteristic that was observed is that AFSA requires a lot of processing time with some benchmarks, especially the last 10 files, sets from NRG to NRH need almost 20 h. Therefore, It could be possible that it is necessary to introduce a the technique of the dimension reduction of the problem in order to reduce the processing time.

Although, this paper showed that this algorithm is a good way to solve SCP with feasible solutions, in a binary domain. An interesting future work is to apply other versions of AFSA on SCP or making a comparison among the different versions of AFSA.

References

1. Garey, M.R., Johnson, D.S.: Computers and Intractability: A Guide to the Theory of NP-Completeness. W. H. Freeman & Co, New York (1990)
2. Crawford, B., Soto, R., Aguilar, R.C., Paredes, F.: A new artificial bee colony algorithm for set covering problems. Electr. Eng. Inf. Technol. **63**, 31 (2014)
3. Crawford, B., Soto, R., Aguilar, R.C., Paredes, F.: Application of the Artificial Bee Colony Algorithm for Solving the Set Covering Problem. Sci. World J. 2014 (2014)
4. Cuesta, R., Crawford, B., Soto, R., Paredes, F.: An artificial bee colony algorithm for the set covering problem. In: Silhavy, R., Senkerik, R., Oplatkova, Z.K., Silhavy, P., Prokopova, Z. (eds.) CSOC 2014. AISC, vol. 285, pp. 53–63. Springer, Switzerland (2014)
5. Crawford, B., Soto, R., Monfroy, E.: Cultural algorithms for the set covering problem. In: Tan, Y., Shi, Y., Mo, H. (eds.) ICSI 2013, Part II. LNCS, vol. 7929, pp. 27–34. Springer, Heidelberg (2013)
6. Crawford, B., Soto, R., Monfroy, E., Palma, W., Castro, C., Paredes, F.: Parameter tuning of a choice-function based hyperheuristic using Particle Swarm Optimization. Expert Syst. Appl. **40**(5), 1690–1695 (2013)
7. Crawford, B., Soto, R., Monfroy, E., Paredes, F., Palma, W.: A hybrid Ant algorithm for the set covering problem (2014)
8. Crawford, B., Soto, R., Olivares-Suárez, M., Paredes, F.: A binary firefly algorithm for the set covering problem. Modern Trends Tech. Comput. Sci. **285**, 65–73 (2014)

9. Crawford, B., Soto, R., Riquelme-Leiva, M., Peña, C., Torres-Rojas, C., Johnson, F., Paredes, F.: Modified binary firefly algorithms with different transfer functions for solving set covering problems. In: Silhavy, R., Senkerik, R., Oplatkova, Z.K., Prokopova, Z., Silhavy, P. (eds.) CSOC 2015. AISC, vol. 349, pp. 307–315. Springer, Switzerland (2015)

10. Crawford, B., Soto, R., Olivares-Suárez, M., Paredes, F.: A new approach using a binary firefly algorithm for the set covering problem. WIT Trans. Inf. Commun. Technol. **63**, 51–56 (2014)

11. Crawford, B., Soto, R., Peña, C., Palma, W., Johnson, F., Paredes, F.: Solving the set covering problem with a shuffled frog leaping algorithm. In: Nguyen, N.T., Trawiński, B., Kosala, R. (eds.) ACIIDS 2015. LNCS, vol. 9012, pp. 41–50. Springer, Heidelberg (2015)

12. Crawford, B., Soto, R., Peña, C., Riquelme-Leiva, M., Torres-Rojas, C., Johnson, F., Paredes, F.: Binarization methods for shuffled frog leaping algorithms that solve set covering problems. Software Engineering in Intelligent Systems, pp. 317–326 (2015)

13. Crawford, B., Soto, R., Torres-Rojas, C., Peña, C., Riquelme-Leiva, M., Misra, S., Johnson, F., Paredes, F.: A binary fruit fly optimization algorithm to solve the set covering problem. In: Gervasi, O., Murgante, B., Misra, S., Gavrilova, M.L., Rocha, A.M.A.C., Torre, C., Taniar, D., Apduhan, B.O. (eds.) ICCSA 2015. LNCS, vol. 9158, pp. 411–420. Springer, Heidelberg (2015)

14. Crawford, B., Soto, R., Aballay, F., Misra, S., Johnson, F., Paredes, F.: A teaching-learning-based optimization algorithm for solving set covering problems. In: Gervasi, O., Murgante, B., Misra, S., Gavrilova, M.L., Rocha, A.M.A.C., Torre, C., Taniar, D., Apduhan, B.O. (eds.) ICCSA 2015. LNCS, vol. 9158, pp. 421–430. Springer, Heidelberg (2015)

15. Michalewicz, Z.: Genetic Algorithms + Data Structures = Evolution Programs, 3rd edn. Springer, Heidelberg (1996)

16. Michalewicz, Z.: Genetic Algorithms + Data Structures = Evolution Programs. Springer Science & Business Media, Heidelberg (2013)

17. Soto, R., Crawford, B., Galleguillos, C., Barraza, J., Lizama, S., Muñoz, A., Vilches, J., Misra, S., Paredes, F.: Comparing cuckoo search, bee colony, firefly optimization, and electromagnetism-like algorithms for solving the set covering problem. In: Gervasi, O., Murgante, B., Misra, S., Gavrilova, M.L., Rocha, A.M.A.C., Torre, C., Taniar, D., Apduhan, B.O. (eds.) ICCSA 2015. LNCS, vol. 9155, pp. 187–202. Springer, Heidelberg (2015)

18. Crawford, B., Soto, R., Peña, C., Riquelme-Leiva, M., Torres-Rojas, C., Misra, S., Johnson, F., Paredes, F.: A comparison of three recent nature-inspired meta-heuristics for the set covering problem. In: Gervasi, O., Murgante, B., Misra, S., Gavrilova, M.L., Rocha, A.M.A.C., Torre, C., Taniar, D., Apduhan, B.O. (eds.) ICCSA 2015. LNCS, vol. 9158, pp. 431–443. Springer, Heidelberg (2015)

19. Azad, M.A.K., Rocha, A.M.A.C., Fernandes, E.M.G.P.: Solving multidimensional 0–1 knapsack problem with an artificial fish swarm algorithm. In: Murgante, B., Gervasi, O., Misra, S., Nedjah, N., Rocha, A.M.A.C., Taniar, D., Apduhan, B.O. (eds.) ICCSA 2012, Part III. LNCS, vol. 7335, pp. 72–86. Springer, Heidelberg (2012)

20. Azad, M.A.K., Rocha, A.M.A., Fernandes, E.M.: Improved binary artificial fish swarm algorithm for the 0–1 multidimensional knapsack problems. Swarm Evol. Comput. **14**, 66–75 (2014)

21. Azad, M.A.K., Rocha, A.M.A., Fernandes, E.M.: Solving large 0–1 multidimensional knapsack problems by a new simplified binary artificial fish swarm algorithm. J. Math. Model. Algorithms Oper. Res. **14**(3), 1–18 (2015)
22. Balas, E., Ho, A.: Set covering algorithms using cutting planes, heuristics, and subgradient optimization: a computational study. In: Padberg, M.W. (ed.) Combinatorial Optimization, vol. 12, pp. 37–60. Springer, Heidelberg (1980)
23. Beasley, J.E.: An algorithm for set covering problem. Eur. J. Oper. Res. **31**(1), 85–93 (1987)
24. Beasley, J.E.: A lagrangian heuristic for set-covering problems. Naval Res. Logistics (NRL) **37**(1), 151–164 (1990)
25. Vasko, F.J., Wolf, F.E., Stott, K.L.: Optimal selection of ingot sizes via set covering. Oper. Res. **35**(3), 346–353 (1987)
26. Walker, W.: Using the set-covering problem to assign fire companies to fire houses. Oper. Res. **22**(2), 275–277 (1974)
27. Beasley, J.E., Chu, P.C.: A genetic algorithm for the set covering problem. Eur. J. Oper. Res. **94**(2), 392–404 (1996)

Linear Programming in a Multi-Criteria Model for Real Estate Appraisal

Benedetto Manganelli[1]([⊠]), Pierfrancesco De Paola[2],
and Vincenzo Del Giudice[2]

[1] University of Basilicata, Viale dell'Ateneo Lucano, 85100 Potenza, Italy
benedetto.manganelli@unibas.it
[2] University of Naples "Federico II", Piazzale Vincenzo Tecchio,
80125 Naples, Italy
pfdepaola@libero.it, vincenzo.delgiudice@unina.it

Abstract. In real estate appraisal, research has long been addressed to the experimentation of multi-parametric models able to reduce the margin of error of the estimate and to overcome or to limit, as far as possible, the problems and difficulties that the use of these models often involves. On the one hand, researchers are trying to overcome the essentially deductive approach that has characterized the traditional discipline, and on the other, to minimize the problems arising from a merely inductive approach. The real estate market is characterized by an inelastic supply and by properties whose complexity and differentiation often involve, also and especially on the demand side, subjective and psychological elements that could distort the results of an inductive investigation. This problem can be overcome by increasing the size of the survey sample, and by using statistical analysis. Statistical analyses, however, are often based on very strong assumptions. A multi-criteria valuation model that uses linear programming is applied to the real estate market. The model, integrated with the inductive and deductive approach, exceeds many of the assumptions of the best known statistical approaches.

Keywords: Linear programming · Multi-criteria valuation model · Real estate market

1 Introduction

The multi-parametric models in real estate appraisal that use quantitative data analysis, can be divided into two groups: (a) those based on statistical techniques, such as multiple regression analysis [1–5], neural network [6–11], genetic algorithms [12, 13], and (b) those using only mathematical processing, such as structural equation systems [14, 15] and rough set theory [16], UTA [17, 18]. The statistical and mathematical approaches are different, of course, not so much for the content but for the way of thinking of Statistics and the way of thinking of Mathematics. The fundamental nature that distinguishes the two approaches is that Statistics is an inductive discipline, while Mathematics is rather deductive.

© Springer International Publishing Switzerland 2016
O. Gervasi et al. (Eds.): ICCSA 2016, Part I, LNCS 9786, pp. 182–192, 2016.
DOI: 10.1007/978-3-319-42085-1_14

The model applied in this study is of the second type [19]; it uses linear programming techniques [20–22] and derives its theoretical basis from tools developed as part of decision theory and operational research.

The decision making process, approached with multi-criteria analysis, formally can be summarized in an evaluation matrix. The columns of the matrix represent the alternatives, while the rows describe the evaluation criteria. The values in each cell of the matrix are the attributes, namely the qualitative or quantitative level reached by each alternative for each criterion. In turn, the appraisal process involves the comparison of the property to be estimated with a sample of which the selling prices are known. This comparison is done by measuring the difference between the property characteristics taken into consideration among those that contribute to the formation of the value. The estimate may thus derive from the solution of a system of equations that has the following expression:

$$s = D - 1 \cdot p,$$

where s is the vector of unknowns, namely, the property value and marginal prices of the characteristics, p is the price vector and D is the matrix of the differences between the characteristics. Therefore, it is evident the analogy with a decision-making process where, in this case, the alternatives are the property of the sample and the criteria are the characteristics used for the comparison.

The search for a model able to reproduce the actual decision-making of market participants (supply and demand) as best as possible, has led many researchers to propose the application, in the field of real estate appraisal, of procedures borrowed from decision theory. These analysis techniques approximate the utility functions of the players in the housing market. The utility function must be understood as that which describes the marginal price of the property characteristic.

Researchers, in an attempt to retrace the decision-making process in a real way, have tried to increase the flexibility of the models using non-linear functions. On the other hand, efforts have been directed to the development of models able to limit the negative impact on the results due to the presence of strongly correlated explanatory variables.

This model interprets the process of the price's formation in the same way as a multi-criteria choice [23], with a multi-objective approach, where the features of the properties that the market considers to be essential represent the selection criteria.

It is therefore a multi-equation model, but unlike other models of the same type, the equations have no endogenous variables except for those that each equation tries to explain. This is why no problem arises for the identification and simultaneous estimation of the parameters. The model equations describe the contributions of the features taken into account in the estimation process of the market value.

The contributions are integrated into a single function-price additive [24]. Because the price variable is expressed as the sum of univariate functions, one for each feature of the property, the model is able to obviate the typical statistical approach problem, relative to the size of the sample data, which is a function of the number of variables taken into consideration. The problem, known as curse of dimensionality requires, for example, that in a multivariate regression model, the amount of data required to maintain the same statistical accuracy, grows more rapidly than the number of variables taken into consideration. Hence, compared with multiple linear regression models or other models using multivariate analysis, its basic assumptions appear much weaker. Moreover, the model allows to impose limitations on the functions that describe the individual contributions. The operator can deductively impose whether the contribution is positive or negative.

In deductive logic, it is also possible to assign piecewise-defined functions. In this way, preserving the simplicity of linear forms, the marginal contribution of individual features can better adapt to the economic logic or to the very special conditions of the housing market. You can thus take into account some economic principles as the law of diminishing marginal utility (with the increase of the internal floor area of the units, a reduction in the marginal price is expected); or you can adapt the model to the actual trend of some observable phenomena (for example the change in the sign function of the price based on the floor level, given that the intermediate levels generally have the highest values).

This paper is structured as follow. The following section provides the model description, focusing specially on the way in which to set the constraints, able to impose the shape of utility functions that describe marginal prices. Section 3 presents the case study, the real estate properties used for comparison and the constraints imposed. The last section illustrates and discusses the results also comparing them to those obtained using other approaches.

2   The Mathematical Formalism of the Model

In the mathematical formalism of the model $A = \{i, 1 \leq i \leq m\}$ is the set of m units of the sample, $C = \{j, 1 \leq j \leq n\}$ describes the n criteria (features) that identify the units, chosen from among those that the market considers to be most significant. Given these two sets, for each criterion j, V_{ij} is the score, that is, the numerical value of the generic element of the set A (housing unit). A prerequisite for the development of the analytical model is for the scores to always be greater than zero ($V_{ij} > 0$). Scores assigned to the units V_{ij}, for each of the selected criteria j, contribute to the formation of the sale price of a property. This contribution is represented under the symbol U_{ji}; it can be positive or negative and is expressed as a linear piecewise-defined function that binds it to V_{ij} score based on criterion j. For this purpose, the range of non-null values of the j criterion should be split into subintervals T_j (integer), constructed so that the elements of each are not present in another and that the set of elements of all the subintervals contains all measured values of the j criterion. Once the subintervals have been defined, for each of those relative to the j-th criterion, the upper D_{tj}^+ limit and the lower D_{tj}^- limit shall be indicated; where t is the generic subinterval between all the subintervals T_j. The symbols α e W indicate, respectively, the constant and the angular coefficient of a generic linear function. The marginal evaluation function of the j criterion then assumes the following expression:

$$U_{ij}^t = f_j^t(V_{ij}) = \begin{cases} 0, & \text{if } V_{ij} = 0 \\ \alpha_{tj} + V_{ij}W_{tj} & \text{if } V_{ij} \neq 0 \text{ and } D_{tj}^- \leq V_{ij} \leq D_{tj}^+ \end{cases}$$

The piecewise linear representation of the function expressing the contribution of the j criterion, provides an approximation of the probable non-linear function that could possibly represent the relationship between price and scoring of different characteristics. The subdivision into sub-intervals or sections therefore depends on the nature of the criterion adopted, and should be an expression of real elasticity of market prices towards changes in the values of the measured scores for the j-th criterion. The indispensable condition for the division into sections is that for every defined section there correspond a number of observations, sufficient to provide a representation of the function. If due to the peculiarities of the survey sample, the number of discrete values of the scores relative to a given criterion is poor, the problem can be resolved by considering each value as a section. The price function of the i-th property is constructed as an additive sum of the individual contributions which are obtained with respect to each j criterion (property feature). Its analytical form is:

$$f(U_{ij}, \ldots \ldots, U_{in}) = U_o + \sum_{j=1}^{n} U_{ij}$$

The d_i^- and d_i^+ symbols respectively indicate the negative and positive residual. One of them of course will be null. These residues are expressed in absolute value as the difference between the observed price pi and the estimated value for the i-th property:

$$Pr_i - \left[U_o + \sum_{j=1}^{n} U_{ij} \right] = d_i = \begin{cases} d_i^+ & \text{if } d_i \geq 0 \\ d_i^- & \text{if } d_i < 0 \end{cases}$$

DA is the sum of the residues, weighted on relative prices observed.

$$\sum_{i \in A} \frac{1}{Pr_i} (d_i^- + d_i^+)$$

The model is developed on the minimum calculation of DA function and respecting the constraints imposed.

$$Min(DA)$$

with the following constraints

$$U_o + \sum_{j=1}^{n} U_{ij} - d_i^+ + d_i^- = Pr_i, \quad d_i^+ \geq 0, \quad d_i^- \geq 0 \quad \text{with } i \in A$$

$$f\left(D_{tj}^+\right) \leq f\left(D_{t+1j}^-\right) \quad \text{for} \quad 1 \leq t \leq T_j - 1, \quad j \in C^+$$
$$W_{tj} \geq 0 \text{for} \quad \text{for} \quad 1 \leq t \leq T_j, \quad j \in C^+$$
$$f\left(D_{tj}^+\right) \geq f\left(D_{t+1j}^-\right) \quad \text{for} \quad 1 \leq t \leq T_j - 1, \quad j \in C^-$$
$$W_{tj} \leq 0 \text{for} \quad \text{for} \quad 1 \leq t \leq T_j, \quad j \in C^-$$

3 The Case Study

The data used for the study are those already elaborated in a previous research [25], specifically 148 sales of residential property units located in a central district of a city of the Campania region, i.e. in a homogeneous market area with identical extrinsic characteristics, over a period of eight years (Tables 1, 2).

Table 1. Variable description

Variable	Description
Price observed (pr)	expressed in thousands of Euros
Age of the property (age)	expressed retrospectively in no. of years
Date of sale (date)	expressed retrospectively in no. of months
Internal area (int)	expressed in sqm
Balconies area (balc)	expressed in sqm
Connected area (conn)	expressed in sqm (lofts, cellars, etc.)
Number of services (serv)	no. of services in residential unit
Number of views (views)	no. of views on the street
Maintenance (main)	expressed with dichotomous scale(1 or 0, respectively, for the presence or absence of optimal maintenancestate)
Floor level (f_lev)	no. of floor levels of residential unit

Table 2. Statistical description of variables

Variable	Std. Dev.	Median	Mean	Min	Max
Pr	32.134,11	114.585	113.502	41.316	250.000
age	7,40	24	23,51	10	35
date	23,77	33	37,30	2	96
int	25,88	117	118,80	48	210
balc	10,19	16	17,56	0	59
conn	15,78	0	9,53	0	62
serv	0,52	2	1,61	1	3
views	0,60	2	2,27	1	4
man	0,33	0	0,12	0	1
f_lev	1,63	3	3,16	1	7

Table 3. Sections of variation of the scores

criterion	number of sections	Range of variation
age	3	[10, 20], [21, 30]; [31,35];
date	5	[0, 12]; [13, 36]; [37, 60]; [61, 74]; [75, 96]
int	5	[48, 60]; [61, 100]; [101, 130; [131, 170]; [171, 210]
bal	3	[0,15]; [16, 30]; [31, 60];
conn	3	[0, 20]; [21, 40]; [41, 62];
serv	3	[1,1]; [2, 2]; [3, 3]
views	3	[1,1]; [2, 2]; [3, 3]
man	2	[0,1];
f_level	3	[1, 2]; [3, 5]; [6, 7];

Table 3 describes the sub-intervals of variation of the scores assigned to each criterion.

The empirical knowledge of the likely contribution to the price of the selected features has enabled the following choices: for the age variable the function is constrained to be decreasing; for the variables related to the surfaces, the facilities, the

Table 4. Coefficients of the piecewise functions

$U_o =$	-12.339,38					
section (t)	1		2		3	
$U = a + VW$	a	W	a	W	a	W
age	16221,39	-811,07	0,00	0,00	0,00	0,00
date	31294,90	0,00	39826,11	-656,25	19300,76	-83,77
int	-35436,66	738,26	-45779,67	823,57	43214,95	0,00
bal	3525,93	895,61	16960,02	0,00	-3825,78	670,51
conn	929,83	775,20	15443,66	51,14	4886,67	307,38
serv	0,00	0,00	1770,63	0,00	1770,63	0,00
views	0,00	0,00	2762,04	0,00	5330,64	0,00
man	0,00	0,00	2402,50	0,00		
f_level	30455,22	1173,27	26901,20	1475,65	46107,88	-2750,55

section (t)	4		5	
$U = a + VW$	a	W	a	W
date	56297,18	-771,19	6781,24	0,00
int	-53335,69	791,25	104973,30	0,00

views and the maintenance, the functions are assumed to increase, while the functions related to the floor level and the date of sale are free from constraints.

4 The Results

The Lp_solve software is used to analyse the real estate data. Lp_solve is a free Mixed Integer Linear Programming (MILP) solver.

Table 4 shows the coefficients defining the piecewise functions of the individual criteria.

The residue analysis indicates that the model has good predictive ability. The average percentage error is 7.13 %. The predictive power of the model is therefore higher than the multiple regression analysis (MRA) 7.84 % but slightly lower compared to a semi-parametric regression method based on Penalized Spline Smoothing, 6.47 %. The comparison between the residuals of these three models, however, shows that the one proposed in this study has the best performance on a percentage of 85 % of the sample in comparison with the MRA and on a percentage of 68 % compared to the non-linear regression model (Fig. 1).

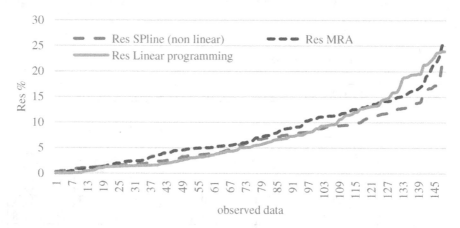

Fig. 1. Residue analysis (Color figure online)

The graphs in Figs. 2, 3, 4 and 5 show the marginal value piecewise functions of some features.

For a better understanding of the graphs, when the piecewise functions have the steps, the moving average is calculated.

The results are consistent with the expectations, especially with regard to those parameters that are not constrained. The law of diminishing marginal utility is respected (Figs. 2, 4), and the variation of the marginal contribution of the floor level is consistent, first increasing then decreasing (Fig. 5). Even the marginal contribution on the date of sale reflects the dynamics of the market observed by official Observatories of the real estate market (Fig. 3).

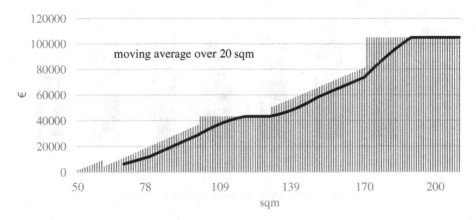

Fig. 2. Value Piecewise functions of *internal area*

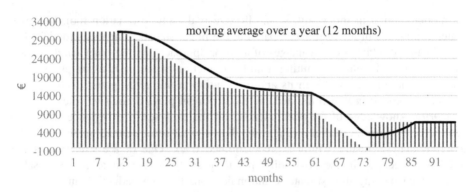

Fig. 3. Value Piecewise functions of *date of sale*

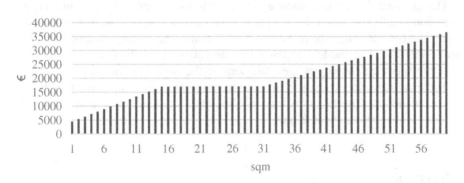

Fig. 4. Value Piecewise functions of *balconies area*

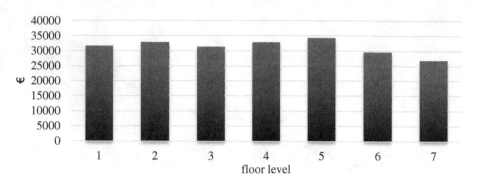

Fig. 5. Value Piecewise functions of *floor level*

5   Conclusion

The model applied in this study, is an effective real estate estimation tool. It shows greater confidence in the results compared to the linear statistical analysis models. On the one hand, it retains the advantages of a linear approach; on the other hand, it adds some operating advantages resulting from the reduction of the basic assumptions. The improved reliability comes from its ability to interpret the investigated phenomenon, due to an approach in which the deductive component plays a decisive role.

A *priori* knowledge of the phenomenon and the evaluator's experience with adequate perception of the market mechanism, allows to set some constraints that lead the inductive analysis results within a predetermined track. One of the strengths of the model lies precisely in the possibility, not necessarily conceived as an exercisable option, of connecting the estimate to the indications that the deductive analysis provides about the shape and/or direction of the individual functions of the marginal contributions.

The essential feature in inductive reasoning is its generalizability, however, generalization is effective when the sample is very large and representative of the population. The complexity of the real estate market makes the construction of this type of sample difficult. The model proposed by integrating the inductive analysis with a deductive approach overcomes the limitations of the statistical analysis.

The marginal prices that easily derived from the piecewise functions may be used as benchmarks even in the estimates, very frequent in practice, which are based on a reduced number of comparable units.

References

1. Isakson, H.R.: Using multiple regression analysis in real estate appraisal. Appraisal J. **69**(4), 424–430 (2001)
2. Manganelli, B., Pontrandolfi, P., Azzato, A., Murgante, B.: Using geographically weighted regression for housing market segmentation. Int. J. Bus. Intell. Data Min. **9**(2), 161–177 (2014)

3. Manganelli, B., Pontrandolfi, P., Azzato, A., Murgante, B.: Urban residential land value analysis: the case of Potenza. In: Murgante, B., Misra, S., Carlini, M., Torre, C.M., Nguyen, H.-Q., Taniar, D., Apduhan, B.O., Gervasi, O. (eds.) ICCSA 2013, Part IV. LNCS, vol. 7974, pp. 304–314. Springer, Heidelberg (2013)
4. Newsome, B.A., Zietz, J.: Adjusting comparable sales using multiple regression analysis - the need for segmentation. Appraisal J. **60**, 129–135 (1992)
5. Simonotti, M., Salvo, F.: Ciuna, M,: Multilevel methodology approach for the construction of real estate monthly index numbers. J. Real Estate Lit. **22**(2), 281–302 (2015)
6. Do, Q., Grudnitski, G.: A neural network approach to residential property appraisal. The Real Estate Appraiser, pp. 38–45, December 1992
7. Evans, A., James, H., Collins, A.: Artificial neural networks: an application to residential valuation in the UK. J. Property Valuation Investment **11**(2), 195–203 (1992)
8. Tay, D., Ho, D.: Artificial intelligence and the mass appraisal of residential apartments. J. Property Valuation Investment **10**(2), 525–540 (1993)
9. Worzala, E.M., Lenk, M.M., Silva, A.: An exploration of neural networks and its application to real estate valuation. J. Real Estate Res. **10**(2), 185–201 (1995)
10. Peterson, S., Flanagan, A.: Neural network hedonic pricing models in mass real estate appraisal. J. Real Estate Res. **31**(2), 147–164 (2009)
11. Tajani, F., Morano, P., Locurcio, M., D'Addabbo, N.: Property valuations in times of crisis: artificial neural networks and evolutionary algorithms in comparison. In: Gervasi, O., Murgante, B., Misra, S., Gavrilova, M.L., Rocha, A.M.A.C., Torre, C., Taniar, D., Apduhan, B.O. (eds.) ICCSA 2015. LNCS, vol. 9157, pp. 194–209. Springer, Heidelberg (2015)
12. Morano, P., Tajani, F., Locurcio, M.: Land use, economic welfare and property values: an analysis of the interdependencies of the real estate market with zonal and socio-economic variables in the municipalities of the Region of Puglia. Int. J. Agric. Environ. Inf. Syst. **6**(4), 16–39 (2015)
13. Manganelli, B., De Mare, G., Nesticò, A.: Using genetic algorithms in the housing market analysis. In: Gervasi, O., Murgante, B., Misra, S., Gavrilova, M.L., Rocha, A.M.A.C., Torre, C., Taniar, D., Apduhan, B.O. (eds.) ICCSA 2015. LNCS, vol. 9157, pp. 36–45. Springer, Heidelberg (2015)
14. Manganelli, B.: Un sistema di equazioni strutturali per la stima di masse di immobili. Genio Rurale, 2 (2001)
15. Bravi, M., Fregonara, E.: Structural equations models in real estate appraisal. In: International Real Estate Conference - American Real Estate and Urban Economics Association, Orlando, 23–25 May 1996
16. D'Amato, M.: Comparing rough set theory with multiple regression analysis as automated valuation methodologies. Int. Real Estate Rev. **10**(2), 42–65 (2007)
17. Jacquet-Lagreze, E., Siskos, Y.: Assessing a set of additive utility functions for multicriteria decision-making, the UTA Method. Eur. J. Oper. Res. **10**(2), 151–164 (1982)
18. Siskos, Y., Yannacopoulos, D.: UTASTAR: an ordinal regression method for building additive value functions. Investigacao Operacional **5**(1), 39–53 (1985)
19. Kettani, O., Oral, M., Siskos, Y.: A multiple criteria analysis model for real estate evaluation. J. Global Optim. **12**(2), 197–214 (1998)
20. Oral, M., Kettani, O.: A mathematical programming model for market share prediction. Int. J. Forecast. **5**(1), 59–68 (1989)
21. Pekelman, D., Subrata, K.S.: Mathematical programming models for the determination of the attribute weights. Manage. Sci. **20**, 1217–1229 (1974)
22. Sivrinisan, V., Shocker, A.D.: Linear programming techniques for multidimensional analysis of preferences. Psychometrica **38**(3), 337–369 (1973)

23. Oral, M., Kettani, O.: Modeling the process of multiattribute choice. J. Oper. Res. Soc. **40**(3), 281–291 (1989)
24. Pace, K.: Appraisal using generalized additive models. J. Real Estate Res. **15**(1), 77–99 (1998)
25. Del Giudice, V., Manganelli, B., De Paola, P.: Spline smoothing for estimating hedonic housing price models. In: Gervasi, O., Murgante, B., Misra, S., Gavrilova, M.L., Rocha, A. M.A.C., Torre, C., Taniar, D., Apduhan, B.O. (eds.) ICCSA 2015. LNCS, vol. 9157, pp. 210–219. Springer, Heidelberg (2015)

A Prioritisation Model Aiding
for the Solution of Illegal Buildings Problem

Fabiana Forte[1], Maria Fiorella Granata[2(✉)], and Antonio Nesticò[3]

[1] Department of Architecture and Industrial Design "Luigi Vanvitelli",
Second University of Naples,
Abazia di S.Lorenzo Ad Septimum, 81031 Aversa, CE, Italy
fabiana.forte@unina2.it
[2] Department of Architecture, University of Palermo,
Viale delle Scienze - building 14, 90128 Palermo, Italy
maria.granata@unipa.it
[3] Department of Civil Engineering, University of Salerno,
Via Giovanni Paolo II, 132, 84084 Fisciano, SA, Italy
anestico@unisa.it

Abstract. In Italy, especially in the southern areas, illegal buildings and unauthorized settlements are serious and very old problems that remain unsolved in spite of three building amnesties. As a consequence, local authorities have to face the demolition of an enormous number of buildings, as prescribed by the Italian law. Being possible the execution of only a certain number of demolition orders during a short period due to financial and social reasons, the problem of how to select the buildings arises. This worrying situation had become the object of a special bill giving elements to establish priorities for demolition of illegal buildings. In this work we propose a multicriteria model aimed at giving a rational ranking of buildings on which a demolition order lays on. It is based on TOPSIS method and, given the only database with information on buildings; it is able to provide the competent authority with priority for building work demolitions according to the bill.

Keywords: Illegal buildings problem · Prioritisation model · Multicriteria decision making · TOPSIS method

1 Introduction

Illegal developments are widespread in almost all parts of the world. They are sometimes the result of the natural need to have a protection from the elements, as it is the case in the cities of the Third World countries and even of European and United States countries, but they are also a way to produce and accumulate capital. Unauthorized buildings an settlements do not have all of the same nature, since an illegal building activity by necessity due to the need of low income families and individuals can be

---

The authors equally contributed to this paper.

© Springer International Publishing Switzerland 2016
O. Gervasi et al. (Eds.): ICCSA 2016, Part I, LNCS 9786, pp. 193–206, 2016.
DOI: 10.1007/978-3-319-42085-1_15

distinguished from a speculative one [1]. In this paper we focus on the phenomenon of illegal buildings and settlements in Italy.

Urbanization in Europe is an ongoing phenomenon both in terms of urban land expansion and increasing population share, along with its impacts extending beyond city borders [2]. A particular and worrying form of urbanization is representing by unplanned urban development, frequently characterized by informal settlements or unauthorized buildings, especially in the east and south of Europe [3]. Informal settlements are residential areas where a certain number of housing units has been built on lands on which the occupants have no legal claim or are entirely unplanned settlements [3]. Unauthorized buildings are not in compliance with current planning and building regulations or have been constructed without a regular planning permission.

In Italy, informal settlements and unauthorized buildings are serious and very old problems that remain unsolved in spite of three building amnesties. The aforesaid phenomena are particularly accentuated in the southern Italian areas, especially in the Region of Campania [4] where has reached a worrying extension. As a consequence local authorities have to face the demolition of a huge number of buildings. The execution of all demolition orders in the same time is not possible for financial and social constrains, thus the problem of how to select the buildings arises. This critical situation had become the object of a special bill (No. 580-A/2013) providing elements to establish priorities for demolition of illegal buildings.

Dealing with public decision is a hard problem due the complexity of interests concerned with. The assessment tools typically used belong to the family of cost-benefit analysis [5, 6], financial analysis, environmental analysis [7], and multi-criteria analysis [8–11]. In this work we propose a multicriteria model aimed at giving a rational ranking of buildings on which a demolition order lays on. It is based on the Technique for Order Preference by Similarity to Ideal Solution (TOPSIS) and, given the only database with information on buildings; it is able to provide the competent authority with priority for building work demolitions according to the bill. Our objective was to develop a practical and completely automatic computer-based tool for ranking the concerned buildings according the law priority. Using the proposed tool, the competent public authority should only implement a database with the required information on buildings and automatically would get the ranking. The ranking will be continually updated when new buildings will added to the database.

After the present Introduction, this work is organized as follows. The Sects. 2 and 3 briefly describe respectively the illegal building phenomenon in Italy and the instruments by which public authorities can contrast illegal building activity. The Sect. 4 deals with the decision problem about assigning an order of precedence on demolition of illegal building works, while the Sects. 5, 6, 7 and 8 describe the proposed multi-criteria decision model. Then, in Sect. 9, the evaluation model is tested through a simulation on fictitious data. The paper ends with some concluding remarks.

2 The Problem of Illegal Buildings in Italy

The informal settlements problem, together with the dynamics of authorized buildings, has a strong territorial impact with relevant effects on the individual and collective welfare (urban sustainability, quality of life, safety, etc.). In the South of Italy and the Islands, according to the BES – Benessere Equo e Solidale – Report [12], the index of illegal building (the ratio of the number of unauthorized buildings to the number of building permits issued by the Municipalities) has recently risen over 35 %. In particular, in the Region of Campania, where the phenomenon has recorded an intensity that is second only to that of the Region of Calabria, it has been estimated that the share of illegal buildings is almost equal to half of those legally built (Table 1).

Table 1. Illegal building rates in the Italian Regions

Italian Regions	Illegal building rate 2013
Northern Italy	**5.3**
Piemonte	4.4
Valle d'Aosta	4.4
Ligury	15.3
Lombardy	5
Trentino-Alto Adige	1.6
Veneto	6.8
Friuli-Venezia Giulia	4.4
Central Italy	**11.6**
Emilia-Romagna	5.4
Tuscany	10.8
Umbria	14.9
Marche (The Marches)	8.7
Lazio (Latium)	12.2
Abruzzo	27.4
Molise	49.4
Southern Italy and Islands	**35.9**
Campania	62.1
Puglia (Apulia)	21.7
Basilicata	29.5
Calabria	69.3
Sicilia (Sicily)	47.7
Sardegna (Sardinia)	21.2

Source: our elaboration on data in BES – Benessere
Equo e Solidale – Report [12].

Most infringements concern illegal buildings constructed in extra-urban areas by low or middle class families and housing construction of modest (but sometimes good) quality, often on legally owned land. The illegal nature of these buildings derives from the lack of formal urban plans and/or planning permissions [13]. Other types of informal settlements occur in Italy, like illegal single-family houses or multi-family buildings within the city, illegal extensions in violation of building regulations, illegal buildings in protected areas with landscape and environmental constraints or in archeological areas [14].

3 The Management of Illegal Building Activity in Italy

As in other European countries, in Italy actions set out by current regulations to address illegal settlements follow several approaches: the repressive approach that concerns the demolition of illegal structures, the mitigatory approach, that is the confiscation of properties and their regeneration by means of a detailed urban plan [15], and a comprehensive approach providing the legalization of informal settlements or buildings on the basis of a pecuniary sanctions to obtain a "retrospective building permit". This last approach, considering the consistency of the phenomenon, for the cases in which is possible to legalize the settlements or buildings, should not be underestimated, especially in light of the disastrous effects of the current economic recession on the budges of local authorities.

In fact, the Presidential Decree No. 380/2001 (the so-called "Testo Unico dell'Edilizia") provides building activities regulations and establishes the system of administrative and penal sanctions for their infringements. According to the Italian legal system, Local Authorities (or Municipalities) – on the basis of territorial government rules – are concerned with planning and building matters and, above all, have the faculty to transform urban areas. The right or permission to build is given under the payment of a contribution of construction or "development charge".

The revenues from the issuing of "building permits" are constrained funds in the budget of the local authority to be used for financing the "public city", that is for the construction of primary and secondary urbanization works, urban facilities as well as the maintenance of the assets, green urban areas, etc. In addition, the revenues deriving from retrospective building permit or permit in sanatoria for the illegal buildings are also constrained funds, intended, among the other things, for the demolition of works that are not subject to the amnesty or sanatoria and for interventions of urban and environment regeneration [16].

According to the Presidential Decree No. 380/2001, capo II, sections of statute No. 30-51, the record of case of building work abuses for which the demolition penalty is inflicted, together with other administrative and penal sanctions, comprehends: illegal lotting out; building constructed without the required public permission ("permesso di costruire") or without the required declaration of works opening ("denuncia di inizio attività"); building constructed on the base of a legal public permission or declaration of works opening, but in full or essential difference from them; illegal building on public property and on protected areas; illegal building works on buildings subjected to artistic, historical, archaeological, environmental, hydro-geological, landscape constrains.

4   Assigning an Order of Precedence on Demolition of Illegal Building Works

Over the time the failure of the Local Authorities to contrast the effects of illegal building activity through the repressive approach has become more and more evident. For example, considering the Campania Region, despite the "promotion of the sustainable development of urban and extra-urban territory trough the minimum soil consumption" being one of the main goals of the Regional Law of Campania on Territory Government (Law No. 16/2004), the intense urbanization process, frequently chaotic and unplanned, which has involved for a long time the regional territory and, above all, the urban area of Naples seems to contradict this attempt. In particular, in the Region of Campania the repressive approach has been frequently disregarded for political, social and financial reasons.

As stated in the Relation on the bill No. 580-A containing Provisions for the rationalization of competences on demolition of illegal buildings ("Disposizioni per la razionalizzazione delle competenze in materia di demolizione dei manufatti abusivi"), the severe situation in the Region of Campania due to the stratification of informal settlements and unauthorized buildings has produced since the 2013 the pronouncement of about 70,000 demolition orders, while a triple number of administrative proceedings have also been started (Relation of the Second Permanent Committee of Justice presented to the Presidency of the Senate of the Italian Republic, April 29, 2013).

The huge size of the phenomenon poses two fundamental problems. On the one hand, if all the ordered demolition would be fulfilled, in the Regions highly affected by illegal building, local authorities would be unable to accommodate the enormous number of families living there and to dispose of the relative debris. For example, in Campania the houses that would be knocked down would be equivalent to the city of Naples. On the other hand, if only a few buildings would be demolished every year public competent authorities would expose themselves to a risk of arbitrariness of the choices. Furthermore, the financial issue is not the sole matter. There are also the social and the environmental questions to be considered. The aforementioned Relation claims the opportunity of making a distinction between infringements due to a state of necessity and speculative illegal initiatives.

In order to avoid the discretionary choice of demolition provisions, the bill No. 580-A (2013) suggests to refer to some factors of priority for the demolition executions of illegal building works. Illegal building works have different levels of severity in different profiles, due to their economic value and speculative aspect or, on the contrary, to their use for necessity. Other factors to be considered are the availability of other housing solution for those who are housed; the possible risks to public and private safety that have many buildings, both for the way in which the work had been made, either because of the characteristics of the area where it was made; the use for illegal activities by organized crime; the environmental and esthetical impacts of the building; the date of declaratory judgment; and the gravity of the inflicted penalty by the sentence.

The bill gives a clear elucidation of DM preferences in terms of single demolition priority, but do not contemplate the case in which a plurality of factors concerns the same illegal building. This is the reason why a comprehensive evaluation model is required.

5 The Assessment Points of View

The following set of criteria (Table 2) was identified according to the bill No. 580-A, whose contents are described in the previous section.

Table 2. The value tree

Goal	Criteria	Codes	Indicators
Overall index of demolition priority	*Dangerousness for public and private safety*	DS	Dangerousness for safety
	Work progress level	WPL	Work progress level
	Social issues	STE	Social state of emergency
		LU	Lawfulness of the use
	Environmental issues	EI	Environmental impacts
		PETC	Presence of effective territorial constrains
	Crime declaration date	PC	Persistence of the crime

- *Dangerousness for Public and Private Safety*. It depends on the static conditions of buildings and may result from deterioration of structural elements or from the same construction modality of the carrying structure. We consider the following levels of dangerousness: impending danger due to a structural or geological deficiency; building realized without a structural design; danger of collapse of non-structural parts; no danger.
- *Work Progress Level*. Buildings that are still under construction or however are not completed must be demolished ahead of the finished ones. We consider four levels of work progress: completed building; completed functional components; rustic or raw state; bearing structure under construction.
- *Social Issues*. The criterion is divided into two sub-criteria. The first one, concerning the social state of emergency related to the satisfaction of the primary need of occupancy, contemplates three possible situations (unique dwelling of a family in a state of poverty, house for letting, second home or holiday home). The second one, regarding the lawfulness of the use and the belonging to owners sentenced for organized crime, considers five use conditions (use for illegal economic activities,

residential use of owners convicted of illegal activities, use for commercial and industrial productive activities, temporary residence, permanent residence).
- *Environmental Issues.* They regard the existence of important environmental impacts due to illegal buildings and the presence of effective archaeological, hydro-geological, landscape or environmental constrains. Buildings in non-authorized settlements are considered of important environmental impact.
- *Crime Declaration Date.* Older dates will take precedence over the others. The considered priority levels are: more than 10 years; between 5 and 10 years; less than 5 years.

Table 2 presents the value tree including goal, criteria and indicators with their codes.

All the indicators are measured on a nominal scale and can be valued, for example, on a 0–10 scale (Table 3). With the exception of WPL and STE, they must be maximized. In this way the assignment of values to indicators will be natural and then more simple for the functionaries of the competent public authority engaged in the attribution of information on buildings.

Table 3. The directions and value scales of indicators

Codes	Directions	Levels	Values
DS	To be maximized	Impending danger due to a structural or geological deficiency Building realized without a structural design Danger of collapse of non-structural parts No danger	10 7.33 3.67 0
WPL	To be minimized	Completed building Completed functional components Rustic or raw state Bearing structure under construction	10 7.33 3.67 0
STE	To be minimized	Unique dwelling of a family in a state of poverty House for letting Second home or holiday home	10 5 0
LU	To be maximized	Use for illegal economic activities Residential use of owners convicted of illegal activities Use for commercial and industrial productive activities Temporary residence Permanent residence	10 7.5 5 2.5 0
EI	To be maximized	Existence of important environmental impacts or building in non-authorized settlements Nonexistence	10 0
PETC	To be maximized	Presence of effective archaeological, hydro-geological, landscape or environmental constrains Nonexistence	10 0
PC	To be maximized	More than 10 years Between 5 and 10 years Less than 5 years	10 5 0

6 Choice of the Evaluation Aggregation Procedure

The choice of a suitable aggregation procedure for a given decision making situation is a crucial task on which depends the quality of the same decision. We referred to the choice framework proposed by Guitouni and Martel [17].

The decision making problem under consideration is characterized by an only decision maker, the competent public authority, who wants to get an alternatives ranking. Given the great complexity of concerned social issues, frequently involving the sphere of the satisfaction of primary needs, and of environmental implications in the areas most hit by the illegal building phenomenon, we assumed that compensations among criteria could be accepted [18].

The choice of the methodology was also based on the need to do not give too much space to interpretations of the bill that often give rise to misunderstand in application of laws. Since this aspect is considered a strategic one [19] we chose a model that uses prior information on preferences of decision makers, that is the lawmaker.

The Technique for Order Preference by Similarity to Ideal Solution (TOPSIS) [20] do not presuppose an interaction with the decision maker, and we regard this feature as useful in the decision problem under consideration, where the aim is the application of decision preferences given by the bill, avoiding its possible subjective interpretation. The TOPSIS method seem suitable for the here considered decision making also for the clearness with which the ranking solution is obtained. We consider this feature of fundamental importance for our decision situation in order to prevent petitions against the demolition orders.

7 The TOPSIS Method

7.1 Literary Review

The Technique for Order Preference by Similarity to Ideal Solution (TOPSIS) is a widespread multicriteria decision making method. A review of TOPSIS application is given in [21] that identifies the main areas on: supply chain management and logistics; design, engineering and manufacturing systems; business and marketing management; health, safety, and environment management; human resources management; energy management; chemical engineering; and water resources management. Recently, either in its original version or in its several versions used alone or in conjunction with other multicriteria techniques, the TOPSIS method has been applied mainly in decision making on technical and industrial issues, for example in [22–24], nevertheless there are some applications on financial issues [25, 26], on sociological themes [27], and rarely on planning decisions [28, 29].

7.2 Basic Concepts and Features

The TOPSIS method [20] is based on the idea that a preferable alternative has a shorter geometrical distance from the ideal solution than from the negative-ideal solution.

Preferable alternatives are closer to the ideal solution, and then the preference order of the alternatives can be derived through the comparisons of geometrical distances from the ideal and negative-ideal solutions. These distances can be measured by the Euclidean approach, although alternative distance measures could be used.

The method deals with numerical information on the attributes and assumes that each of them is monotonically increasing or decreasing. It is a compensatory aggregation procedure, assuming that a poor performance on an attribute criterion can be offset by a good performance on another one. The method provides a cardinal ranking of alternatives and does not require the criteria to be independent [30].

7.3   The Aggregative Procedure

Given a set of alternatives $A = \{A_1, A_2, \ldots, A_m\}$ and a set of criteria $G = \{g_1, g_2, \ldots, g_n\}$ with the relative set of normalized weights $W = \{w_1, w_2, \ldots, w_n\}$, let x_{ij} is the performance measure of the i-th alternative against the j-th criterion, with $i = 1, 2, \ldots, m$ and $j = 1, 2, \ldots, n$. Since, in general, the various criteria have different dimensions, the x_{ij} performances are converted in non-dimensional measures by the following normalization:

$$r_{ij} = \frac{x_{ij}}{\sqrt{\sum_{h=1}^{m} x_{hj}^2}}. \tag{1}$$

Called v_{ij} the weighted normalized performance of i-th alternative on j-th criterion, the ideal alternative, A^+, and the negative-ideal alternative, A^-, are respectively defined as follows:

$$A^+ = \left\{ \left(\max_i vij \middle| j \in J \right), \left(\min_i vij \middle| j \in J' \right), i = 1, 2, \ldots, m \right\} \tag{2}$$
$$= \{v_1^+, v_2^+, \ldots, v_n^+\}$$

$$A^- = \left\{ \left(\min_i vij \middle| j \in J \right), \left(\max_i vij \middle| j \in J' \right), i = 1, 2, \ldots, m \right\} \tag{3}$$
$$= \{v_1^-, v_2^-, \ldots, v_n^-\}$$

where:

$$J = \{j = 1, 2, \ldots, n | j \text{ is associated with benefit criteria}\} \tag{4}$$

$$J' = \{j = 1, 2, \ldots, n | j \text{ is associated with cost/loss criteria}\}. \tag{5}$$

According to Euclidean distance approach the distance of alternatives from the ideal solution A^+ is:

$$D_i^+ = \sqrt{\sum_{j=1}^{n} \left(v_{ij} - v_j^+ \right)^2} \quad for \ i = 1, 2, \ldots, m \tag{6}$$

while the distance of alternatives from the negative-ideal solution A^- is:

$$D_i^- = \sqrt{\sum_{j=1}^{n} \left(v_{ij} - v_j^- \right)^2} \quad for \ i = 1, 2, \ldots, m. \tag{7}$$

Then, the relative closeness of the alternative A_i with respect to the ideal solution is given by:

$$C_i^+ = \frac{D_i^-}{D_i^+ + D_i^-} \quad i = 1, 2, \ldots, m, \tag{8}$$

where $0 \leq C_i^+ \leq 1$.

The preference ranking of alternatives follows the rank order of C_i^+.

8   Weights Elicitation

It is known that weights assume values highly depending on the eliciting technique [31].

The Rank Order Centroid (ROC) [32] method seems to fit well the eliciting weights problem under consideration. This method give weight (w_j) to a number of items ranked according to their importance through the following formula:

$$w_j = \frac{1}{n} \sum_{r=1}^{n} \frac{1}{r} \quad j = 1, 2, \ldots, n \tag{9}$$

where n is the number of items and r is the rank position.

9   Verification of the Assessment Model

The proposed assessment model was tested by a simulation on hypothetical data. Supposing that is given the following evaluation table (Table 4) concerning ten buildings on which exist a demolition order. It, small as it is, corresponds to the information database of a competent public authority. As our aim is testing the model, the data were conceived for giving a clear response on the validity of the assessment procedure.

Weights (Table 5) were obtained through the ROC method, according to the priority order for the considered item that can be derived from the bill.

The ideal alternative, A^+, and the negative-ideal alternative, A^-, are:

Table 4. The evaluation table

Buildings	Criteria						
	DS	WPL	STE	LU	EI	PETC	PC
B1	10	10	10	0	0	0	0
B2	0	0	10	0	0	0	0
B3	0	10	0	0	0	0	0
B4	0	10	10	10	0	0	0
B5	0	10	10	0	10	0	0
B6	0	10	10	0	0	10	0
B7	0	10	10	0	0	0	10
B8	3.67	3.67	5	10	10	10	0
B9	7.33	0	0	2.5	10	0	5
B10	0	7.33	10	5	0	10	10

Table 5. The system of weights

Criteria						
DS	WPL	STE	LU	EI	PETC	PC
0.3704	0.2276	0.1561	0.1085	0.0728	0.0442	0.0204

$$A^+ = \{3.7041, 0.0000, 0.0000, 1.0850, 0.7279, 0.4422, 0.2041\}$$

$$A^- = \{0.0000, 2.2755, 1.5612, 0.0000, 0.0000, 0.0000, 0.0000\}$$

The distances of alternatives from the ideal solution, D^+, and from the negative-ideal solution, D^-, are given in Table 6, together with the relative closeness of alternatives with respect to the ideal solution, C^+.

Table 6. Distances of alternatives from ideal solutions and their relative closeness to the ideal solution

Buildings	D^+	D^-	C^+
B1	3.0919	3.7041	0.5450
B2	4.2546	2.2755	0.3485
B3	4.5654	1.5612	0.2548
B4	4.7013	1.0850	0.1875
B5	4.7697	0.7279	0.1324
B6	4.8046	0.4422	0.0843
B7	4.8206	0.2041	0.0406
B8	2.6165	2.5367	0.4923
B9	1.3588	3.9498	0.7440
B10	4.4457	0.9490	0.1759

Then, the preference ranking of alternatives is B9, B1, B8, B2, B3, B4, B10, B5, B6, B7. This result is clearly consistent with the data. In particular it is coherent with expected outcomes about the first seven buildings, characterized by elementary data, for which was attended the same priority relating the criteria.

10 Concluding Remarks

The present work points out the usefulness of multicriteria evaluations in helping public authorities on decisions concerning the management of illegal buildings, when it is not possible to legalize them.

The assessment simulation on fictitious data showed the validity of the model that could be used as an automatic tool for prioritization the demolitions of illegal buildings. It is able to give outcomes easily explainable and, then, representing a guarantee against possible impugnments.

Moreover, the assessment model is flexible, as weights and even indicators can be easily modified in order to adapt itself to local specificity, supposing that the bill will become a law admitting regional adaptations. One or more criteria could be easily excluded, for example, for the areas in which grave reasons might suggest the preservation of buildings as social houses.

Further developments of this work could be a deeper test on real data and a fuzzy version of the assessment model, to be used if deterministic input information could not be available.

Finally we recall that all over the world, illegal buildings and settlements produce unwelcome effects, mainly of environmental and financial nature, and are concerned with social issues. The model presented in this paper is a first attempt to provide public authorities involved in the management of this kind of problem with a tool able to give an objective response, far from individual interpretation, to the need of apply the law. The decision model here proposed fits exactly the Italian bill, but can be easily modified to take into account local values and priorities typical of other countries. Then, it could be usefully applied in international contexts affected by analogous problems, making the suitable modifications in performance values, criteria, and weights to better represent the local requirements.

References

1. Patton, C.V. (ed.): Spontaneous Shelter: International Perspectives and Prospects. Temple University Press, Philadelphia (1988)
2. European Environment Agency: SOER 2015 – The European environment state and outlook 2015 (2015). www.eea.europa.eu/soer
3. Potsiou C.: Informal Urban Development in Europe. Experiences from Albania and Greece. Summary version. UN-HABITAT (2010)
4. Forte, F.: The management of informal settlements for urban sustainability: experiences from the Campania Region (Italy). In: Brebbia, C.A., Florez-Escobar, W.F. (eds.) The Sustainable City X, pp. 153–164. WIT Press, Southampton (2015)

5. Nesticò, A., Macchiaroli, M., Pipolo, O.: Costs and benefits in the recovery of historic buildings: the application of an economic model. Sustainability 7(11), 14661–14676 (2015)
6. Bellia, C., Granata, M.F., Scavone, V.: Aree dismesse ed orti urbani: un "valore sociale complesso" nelle città - Abandoned areas and urban gardens: a "complex social value" in the cities. Agribusiness Paesaggio & Ambiente vol. 17, pp. 61–70 (2014). Special Issue 2
7. Nesticò, A., Pipolo, O.: A protocol for sustainable building interventions: financial analysis and environmental effects. Int. J. Bus. Intell. Data Min. 10(3), 199–212 (2015)
8. Fusco Girard, L., Cerreta, M., De Toro, P., Forte, F.: The human sustainable city: values, approaches and evaluative tools. In: Deakin M., Mitchell G., Nijkamp P., Vreeker R. (eds.) Sustainable Urban Development. The Environmental Assessment Methods, vol. 2, pp. 65–93, Routledge (2007)
9. Granata, M.F.: Economia eco-sistemica ed efficienza bio-architettonica della città. FrancoAngeli, Milano (2008)
10. Cilona, T., Granata, M.F.: Multicriteria prioritization for multistage implementation of complex urban renewal projects. In: Gervasi, O., Murgante, B., Misra, S., Gavrilova, M.L., Rocha, A.M.A.C., Torre, C., Taniar, D., Apduhan, B.O. (eds.) ICCSA 2015. LNCS, vol. 9157, pp. 3–19. Springer, Heidelberg (2015)
11. Cilona, T., Granata, M.F.: A Choquet integral based assessment model of projects of urban neglected areas: a case of study. In: Murgante, B., et al. (eds.) ICCSA 2014, Part III. LNCS, vol. 8581, pp. 90–105. Springer, Heidelberg (2014)
12. ISTAT-CNEL: Il benessere equo e sostenibile in Italia. ISTAT (2014)
13. Economic Commission for Europe, In Search for Sustainable Solutions for Informal Settlements in the ECE Region: Challenges and Policy Responses, unedited draft, Geneva (2008)
14. De Mare, G., Fasolino, I., Ferrara, B.: Abusivismo edilizio: ipotesi metodologica di rifunzionalizzazione urbana, E-STIMO, pp. 1–15 (2010). online journal
15. De Biase, C., Forte, F., Unauthorised building and financial recovery of urban areas: evidences from Caserta Area. In: 6th EuroMed Conference book of proceedings, pp. 816–831. EuroMed Press, Estoril (2013)
16. Forte, F.: Illegal buildings and local finance in new metropolitan perspectives. In: Bevilacqua, C., Calabrò F., Della Spina, L. (eds) New Metropolitan Perspectives. Advanced Engineering Forum, vol. 11, pp. 600–606 (2014). www.scientific.net
17. Guitouni, A., Martel, J.-M.: Tentative guidelines to help choosing an appropriate MCDA method. Eur. J. Oper. Res. 109, 501–521 (1998)
18. De Mare, G., Granata, M.F., Nesticò, A.: Weak and strong compensation for the prioritization of public investments: multidimensional analysis for pools. Sustainability 7, 16022–16038 (2015)
19. Livingston, M.A., Monateri, P.G., Parisi, F.: The Italian Legal System. An Introduction, 2nd edn. Stanford University Press, Stanford (2015)
20. Hwang, C.-L., Youn, K.: Multiple Attribute Decision Making - Methods and Application: A State of the Art Survey. Springer, New York (1981)
21. Behzadian, M., Otaghsara, S.K., Yazdani, M., et al.: A state-of the-art survey of TOPSIS applications. Expert Syst. Appl. 39, 13051–13069 (2012)
22. Abedi, M., Norouzi, G.-H.: A general framework of TOPSIS method for integration of airborne geophysics, satellite imagery, geochemical and geological data. Int. J. Appl. Earth Obs. Geoinf. 46, 31–44 (2016)
23. Wang, P., Zhu, Z., Wang, Y.: A novel hybrid MCDM model combining the SAW, TOPSIS and GRA methods based on experimental design. Inf. Sci. 345, 27–45 (2016)

24. Onat, N.C., Gumus, S., Kucukvar, M., Tatari, O.: Application of the TOPSIS and intuitionistic fuzzy set approaches for ranking the life cycle sustainability performance of alternative vehicle technologies. Sustain. Prod. Consumption **6**, 12–25 (2016)
25. Mandic, K., Delibasic, B., Knezevic, S., Benkovic, S.: Analysis of the financial parameters of Serbian banks through the application of the fuzzy AHP and TOPSIS methods. Econ. Model. **43**, 30–37 (2014)
26. Bilbao-Terol, A., Arenas-Parra, M., Cañal-Fernández, V., Antomil-Ibias, J.: Using TOPSIS for assessing the sustainability of government bond funds. Omega **49**, 1–17 (2014)
27. Sekhar, C., Patwardhan, M., Vyas, V.: A Delphi-AHP-TOPSIS based framework for the prioritization of intellectual capital indicators: a SMEs perspective. Procedia – Soc. Behav. Sci. **189**, 275–284 (2015)
28. Zhao, J., Fang, Z.: Research on campus bike path planning scheme evaluation based on TOPSIS method: Wei'shui campus bike path planning as an example. Procedia Eng. **137**, 858–866 (2016)
29. Assari, A., Mahesh, T., Assari, E.: Role of public participation in sustainability of historical city: usage of TOPSIS method. Indian J. Sci. Technol. **5**(3), 2289–2294 (2012)
30. Yoon, K.P., Hwang, C.L.: Multiple attribute decision making. Sage Publication, Thousand Oaks (1995)
31. Barron, F., Barrett, B.E.: Decision quality using ranked attribute weights. Manage. Sci. **42**, 1515–1523 (1996)
32. Solymosi, T., Dompi, J.: Method for determining the weights of criteria: the centralized weights. Eur. J. Oper. Res. **26**, 35–41 (1985)

Solving the Set Covering Problem with a Binary Black Hole Inspired Algorithm

Álvaro Gómez Rubio[1(✉)], Broderick Crawford[1,2,5(✉)], Ricardo Soto[1,3,4],
Eduardo Olguín[5], Sanjay Misra[6], Adrián Jaramillo[1],
Sebastián Mansilla Villablanca[1], and Juan Salas[1]

[1] Pontificia Universidad Católica de Valparaíso, 2362807 Valparaíso, Chile
{alvaro.gomez.r,adrian.jaramillo.s,
sebastian.mansilla.v,juan.salas.f}@mail.pucv.cl,
{broderick.crawford,ricardo.soto}@pucv.cl
[2] Universidad Central de Chile, 8370178 Santiago, Chile
[3] Universidad Autónoma de Chile, 7500138 Santiago, Chile
[4] Universidad Espíritu Santo, Guayaquil, Ecuador
[5] Facultad de Ingeniería y Tecnología, Universidad San Sebastián,
Bellavista 7, 8420524 Santiago, Chile
eduardo.olguin@uss.cl
[6] Covenant University, 110001 Ogun, Nigeria
sanjay.misra@covenantuniversity.edu.ng

Abstract. There are multiple problems in several industries that can be solved with combinatorial optimization. In this sense, the Set Covering Problem is one of the most representative of them, being used in various branches of engineering and science, allowing find a set of solutions that meet the needs identified in the restrictions that have the lowest possible cost. This paper presents an algorithm inspired by binary black holes (BBH) to resolve known instances of SPC from the OR-Library. Also, it reproduces the behavior of black holes, using various operators to bring good solutions.

Keywords: Set Covering Problem · Binary black hole · Meta heuristics · Combinatorial optimization problem

1 Introduction

The SCP is one of 21 NP-Hard problems, representing a variety of optimization strategies in various fields and realities. Since its formulation in the 1970s has been used, for example, in minimization of loss of materials for metallurgical industry [1], preparing crews for urban transportation planning [2], safety and robustness of data networks [3], focus of public policies [4], construction structural calculations [5]. This problem was introduced in 1972 by Karp [6] and it is used to optimize problems of elements locations that provide spatial coverage, such as community services, telecommunications antennas and others.

© Springer International Publishing Switzerland 2016
O. Gervasi et al. (Eds.): ICCSA 2016, Part I, LNCS 9786, pp. 207–219, 2016.
DOI: 10.1007/978-3-319-42085-1_16

The present work applied a strategy based on a binary algorithm inspired by black holes to solve the SCP, developing some operators that allow to implement an analog version of some characteristics of these celestial bodies to support the behavior of the algorithm and improve the processes of searching for the optimum. This type of algorithm was presented for the first time by Abdolreza Hatamlou in September 2012 [7], registering some later publications dealing with some applications and improvements. In this paper it will be detailed methodology, developed operators, experimental results and execution parameters and handed out some brief conclusions about them, the original version for both the proposed improvements.

The following section explains in detail the Set Covering Problem (SPC) and will be defined and briefly explain the black holes and their behavior. Then, in Sect. 3 the algorithm structure and behavior will be reviewed. Sections 4 and 5 will explain the experimental results and the conclusions drawn from these results.

2   Conceptual Context

This section describes the necessary concepts to understand the operation of the SPC and the basic nature of black holes, elements necessary to understand the subsequent coupling of both concepts.

2.1   SCP Explanation and Detail

Considering a binary numbers array A, of m rows and n columns (a_{ij}), and a C vector (c_j) of n columns containing the costs assigned to each one, then we can then define the SCP such as:

$$\text{Minimize} \sum_{j=1}^{m} c_j x_j \tag{1}$$

Where a:

$$\sum_{j=1}^{n} a_{ij}x_j \geq 1 \; \forall \; i \in \{1, ..., m\}$$

$$x_j \in \{0,1\}; \; j \in \{1, ..., n\}$$

This ensures that each row is covered by at least one column and that there is a cost associated with it [8].

This problem was introduced in 1972 by Karp [6] and is used to optimize problems of elements locations that provide spatial coverage, such as community services, telecommunications antennas and others. In practical terms it can be explained by the following example:

Imagine a floor that is required to work on a series of pipes (represented in red) that are under 20 tiles, rising the least amount possible of them. The diagram of the situation would be as follows (Fig. 1).

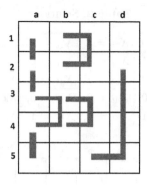

Fig. 1. SCP Example (Color figure online)

We define a variable X_{ij} which represents each tile by its coordinates in row i and column j. It will store a 1 if it is necessary to lift the back plate and a 0 if it is not. Then we will have:

$$i = \{1, 2, 3, 4, 5\}; \quad j = \{a, b, c, d\}$$

We will define the "objective function" as:

Min z= $X_{1a} + X_{2a} + X_{3a} + X_{4a} + X_{5a} + X_{1b} + X_{2b} + X_{3b} + X_{4b} + X_{5b} + X_{1c} + X_{2c} + X_{3c} + X_{4c} + X_{5c} + X_{1d} + X_{2d} + X_{3d} + X_{4d} + X_{5d}$

Then the following system of equations represent the constraints of the problem (Table 1):

Table 1. Equations system

$X_{2d} +$	$X_{3d} +$	$X_{4d} +$	X_{5d}	≥ 1
$X_{1a} +$	X_{2a}			≥ 1
$X_{2a} +$	X_{3a}			≥ 1
$X_{3a} +$	$X_{3b} +$	$X_{4a} +$	X_{4b}	≥ 1
$X_{4a} +$	X_{5a}			≥ 1
$X_{1b} +$	$X_{1c} +$	$X_{2c} +$	X_{2b}	≥ 1
$X_{3b} +$	$X_{3c} +$	$X_{4c} +$	X_{4b}	≥ 1

Then, each restriction corresponds to the tile on top of a main. It is only necessary to pick one per tube. The solution of the system can be seen on Table 2.

Finally, it is only necessary to lift 5 tiles (Fig. 2):

This type of strategy has been widely used in aerospace turbine design [9], timetabling design [10], probabilistic queuing [11], geographic analysis [12], services location [13], scheduling [14] and many others [15].

Table 2. Solutions

$X_{1a} = 0,$	$X_{1b} = 1,$	$X_{1c} = 0,$	$X_{1d} = 0$
$X_{2a} = 1,$	$X_{2b} = 0,$	$X_{2c} = 0,$	$X_{2d} = 0$
$X_{3a} = 0,$	$X_{3b} = 3,$	$X_{4b} = 0,$	$X_{5b} = 0$
$X_{4a} = 0,$	$X_{4b} = 0,$	$X_{4c} = 0,$	$X_{4d} = 0$
$X_{5a} = 1,$	$X_{5b} = 0,$	$X_{5c} = 1,$	$X_{5d} = 0$

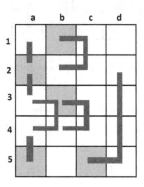

Fig. 2. SCP example solution

A variety of algorithms "bio-inspired" that mimic the behavior of some living beings [16] to solve problems, as well as others who are inspired by elements of nature [15], cultural and other types.

The present work applied a strategy based on a binary algorithm inspired by black holes to solve the SCP, developing some operators that allow to implement an analog version of some characteristics of these celestial bodies to support the behavior of the algorithm and improve the processes of searching for the optimum. This type of algorithm was presented for the first time by Abdolreza Hatamlou in September 2012 [7], registering some later publications dealing with some applications and improvements. In this paper it will be detailed methodology, developed operators, experimental results and execution parameters and handed out some brief conclusions about them, the original version for both the proposed improvements.

2.2   Black Holes

Black holes are the result of the collapse of a big star's mass that after passing through several intermediate stages is transformed in a so massively dense body that manages to bend the surrounding space because of its immense gravity. They are called "black holes" due to even light does not escape their attraction and therefore is undetectable in the visible spectrum, knowing also by "singularities", since inside traditional physics loses meaning. Because of its immense gravity, they tend to be orbited by other stars in binary or multiple systems consuming a little mass of bodies in its orbit [17] (Fig. 3).

Fig. 3. Celestial bodies orbiting a black hole

When a star or any other body is approaching the black hole through what is called "event horizon", collapses in its interior and is completely absorbed without any possibility to escape, since all its mass and energy become part of singularity (Fig. 4). This is because at that point the exhaust speed is the light one [17].

Fig. 4. Event horizon in a black hole

On the other hand, black holes also generate a type of radiation called "Hawking radiation", in honor of its discoverer. This radiation have a quantum origin and implies transfer of energy from the event horizon of the black hole to its immediate surroundings, causing a slight loss of mass of the dark body and an emission of additional energy to the nearby objects [18].

3   Algorithm

The algorithm presented by Hatamlou [7] faces the problem of determination of solutions through the development of a set of stars called "universe", using an algorithm type population similar to those used by genetic techniques or particles swarm. It proposes the rotation of the universe around the star that has the best fitness, i.e., which has the lowest value of a defined function, called "objective function". This rotation is applied by an operator of rotation that moves all stars in each iteration of the algorithm and determines in each cycle if there is a new black hole, that will replace the previous one. This operation is repeated until find the detention criteria, being the last black holes founded the proposed solution and the last of the black holes found corresponds to the final solution proposed. Eventually, a star can ever exceed the defined by the radius of the event horizon [17]. In this case, the star collapses into the black hole and is removed from the whole universe being taken instead by a new star. Thus, stimulates the exploration of the space of solutions. The following is the proposed flow and the corresponding operators according to the initial version of the method (Fig. 5):

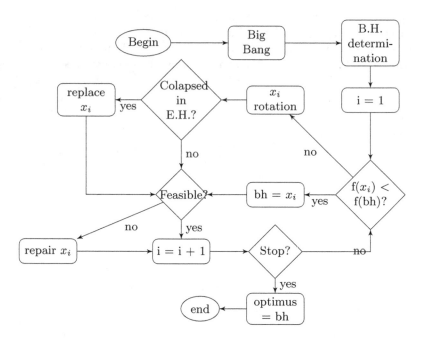

Fig. 5. Black Hole algorithm

3.1   Big Bang

It is the initial random creation of the universe. Corresponds to the creation of a fixed amount of feasible binary vectors, i.e., that comply with all the restrictions defined in the problem.

3.2   Fitness Evaluation

For each star x_i fitness is calculated by evaluating the objective function, according to the initial definition of the problem. In others terms described in the following way:

$$\sum_{j=1}^{n} c_j x_j \tag{2}$$

It should be remembered that c_j corresponds to the cost of that column in the matrix of costs. In other words, the fitness of a star is the sum of the product of the value of each column covered with a star in particular, multiplied by the corresponding cost. The black hole will be those who have minor fitness among all existing stars at the time of the evaluation.

3.3   Rotation Operator

The rotation operation occurs above all the universe of x_i stars of t iteration, with the exception of the black hole, which is fixed in its position. The operation sets the new t+1 position follows:

$$X_i(t+1) = X_i(t) + random(X_{BH} - X_i(t)), where\ i = 1, 2, ..., N \tag{3}$$

where random $\in \{0,1\}$ and change in each iteration, $x_i(t)$ and $x_i(t+1)$ are the positions of the star x_i at t and t+1 iterations respectively, x_{BH} is the black hole location in the search space and N is the number of stars that make up the universe (candidate solution). It should be noted that the only exception in the rotation is designated as black hole star, which retains the position.

3.4   Collapse into the Black Hole

When a star is approaching a black hole at a distance called event horizon is captured and permanently absorbed by the black hole, being replaced by a new randomly generated one. In other words, it is considered when the collapse of a star exceeds the radius of Schawarzchild (R) defined as:

$$R = \frac{f_{BH}}{\sum_{i=1}^{n} f_i} \tag{4}$$

where f_{BH} is the value of the fitness of the black hole and f_i is the ith star fitness. N is the number of stars in the universe.

3.5   Implementation

The algorithm implementation was carried out with a I-CASE tool, generating Java programs and using a relational database as a repository of the entry information and gathered during executions. The parameters finally selected are the result of the needs of the original design of the algorithm improvements made product of the tests performed. In particular, attempted to improve the capacity of exploration of the metha heuristics [19]. Is contrast findings with tables of known optimal values [8], in order to quantitatively estimate the degree of effectiveness of the presented metha heuristics.

The process begins with the random generation of a population of binaries vectors (Stars) in a step that we will call "big bang". With a universe of m stars formed by vectors of n binary digits, the algorithm must identify the star with better fitness value, i.e., the which one the objective function obtains a lower result. The next step is to rotate the other stars around the black hole detected until some other presents a better fitness and take its place.

The number of star generated will remain fixed during the iterations, notwithstanding that many vectors (or star) will be replaced by one of the operators. The binarization best results have been achieved with that was the standard to be applied in the subsequent benchmarks.

3.6 Feasibility and Unfeasibility

The feasibility of a star is given by the condition if it meets each of the constraints defined in the matrix A. In those cases which unfeasibility was detected, opted for repair of the vector to make it comply with the constraints. We implemented a repair function in two phases, ADD and DROP, as way to optimize the vector in terms of coverage and costs. The first phase changes the vector in the column that provides the coverage at the lowest cost, while the second one removes those columns which only added cost and do not provide coverage.

3.7 Event Horizon

One of the main problems for the implementation of this operator is that the authors refer to vectorial distances determinations or some other method. However, in a 2015 publication, Farahmandian, and Hatamlouy [20] intend to determine the distance of a star x_i to the radius R as:

$$|f(x_{BH}) - f(x_i)| \tag{5}$$

I.e. a star x_i will collapse if the absolute value of the black hole and his fitness subtraction is less than the value of the radius R:

$$|f(x_{BH}) - f(x_i)| < R \tag{6}$$

4   Experimental Results

The original algorithm was subjected to a test by running the benchmark 4, 5, 6, A, B, C, D, NRE, NRF, NRG and NRH from OR library [21]. Each of these data sets ran 30 times with same parameters [22], presenting the following results:

Where Z_{BKS} is the optimal for instance, Z_{min} is the minimum value found, Z_{min} is the maximum value found, Z_{avg} is the average value and Z_{RPD} is the percentage of deviation from the optimum (Table 3).

Table 3. Experimental results

Instance	Z_{BKS}	Z_{min}	Z_{max}	Z_{avg}	RPD	Instance	Z_{BKS}	Z_{min}	Z_{max}	Z_{avg}	RPD
4.1	429	435	603	519,00	1,39	C.1	227	252	287	269,5	9,92
4.2	512	544	633	588,50	6,25	C.2	219	245	289	267	10,01
4.3	516	551	696	623,50	6,78	C.3	243	266	399	332,5	8,65
4.4	494	527	749	638,00	6,68	C.4	219	252	301	276,5	13,10
4.5	512	520	730	625,00	1,56	C.5	215	247	295	271	12,96
4.6	560	566	674	620,00	1,07	D.1	60	71	146	108,5	15,49
4.7	430	461	514	487,50	7,21	D.2	66	73	177	125	9,59
4.8	492	528	613	570,50	7,32	D.3	72	81	120	100,5	11,11
4.9	641	688	767	727,50	7,33	D.4	62	70	135	102,5	11,43
4.10	514	547	660	603,50	6,42	D.5	61	72	208	140	15,28
5.1	253	269	398	333,50	6,32	E.1	5	9	53	31	44,44
5.2	302	302	329	315,50	0,00	E.2	5	12	61	36,5	58,33
5.3	226	246	275	281,50	8,85	E.3	5	10	112	61	50,00
5.4	242	249	261	255,00	2,89	E.4	5	11	76	43,5	54,55
5.5	211	228	258	243,00	8,06	E.5	5	13	71	42	61,54
5.6	213	230	359	294,50	7,98	NRE1	29	81	169	125	64,20
5.7	293	322	372	347,00	9,90	NRE2	30	44	152	98	31,82
5.8	288	308	459	383,50	6,94	NRE3	27	435	522	478,5	93,79
5.9	279	296	449	372,50	6,09	NRE4	28	44	62	53	36,36
5.10	265	283	412	347,50	6,79	NRE5	28	213	346	279,5	86,85
6.1	138	151	201	176,00	9,42	NRF1	14	658	711	684,5	97,87
6.2	146	157	281	219,00	7,53	NRF2	15	18	163	90,5	16,67
6.3	145	153	195	175,50	7,59	NRF3	14	69	116	92,5	79,71
6.4	131	144	233	188,50	9,92	NRF4	14	45	147	96	68,89
6.5	161	177	258	217,50	9,94	NRF5	13	222	362	292	94,14
A.1	253	298	414	356,00	17,79	NRG1	176*	770	797	783,5	77,14
A.2	252	301	430	365,50	19,44	NRG2	151*	876	1006	941	82,76
A.3	232	256	390	323,00	10,34	NRG3	166*	1012	1046	1029	83,60
A.4	234	268	316	292	14,53	NRG4	168*	289	398	343,5	41,87
A.5	236	266	369	317,50	12,71	NRG5	168*	1211	1339	1275	620,83
B.1	69	82	149	115,50	18,84	NRH1	63*	2143	2242	2192,5	3301,59
B.2	76	99	184	133,50	30,26	NRH2	63*	701	810	755,5	1012,70
B.3	80	89	145	117	11,25	NRH3	59*	893	915	904	1413,56
B4	79	88	104	96	11,39	NRH4	59*	329	464	396,5	457,63
B.5	72	88	119	99,50	22,22	NRH5	55*	715	845	780	1200

* = Best results found in literature [23]

5   Analysis and Conclusions

Comparing the results of experiments with the best reported in the literature
[24], we can see that results are acceptably close to the best known fitness for 5
and 6 benchmarks, and far away from them in the case of final ones. It is relevant
to the case of series 4 and 5, which reached the optimum in a couple of instances.
In the case of the first ones are deviations between 0 % and 61,54 %, while in the
case of the latter ones reach 3.301,59 % of deviation. In both cases more than 30
times algorithm is executed, considering 20.000 iterations in each one. The rapid
initial convergence is achieved, striking finding very significant improvements in
the first iteration, to find very significant improvements in early iterations, being

much more gradual subsequent and requiring the execution of those operators that stimulate exploration, such as collapse and Hawking radiation. This suggests that the algorithm has a tendency to fall in optimal locations, where cannot leave without the help of scanning components. In order to illustrate these trends, some graphics performance benchmarks are presented (Fig. 6):

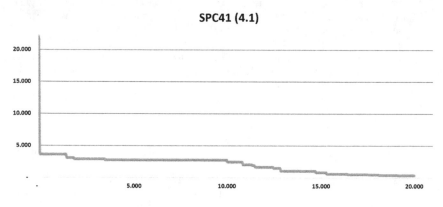

Fig. 6. SPC41 results

While in the benchmark results which threw poor results have significant percentages of deviation from the known optimal, in absolute terms the differences are low considering the values from which it departed iterating algorithm. It is probably why these tests require greater amount of iterations to improve its results, since the values clearly indicate a consistent downward trend, the number of variables is higher and the difference between the optimum and the start values is broader. An interesting analysis element is that the gap between the best and the worst outcome is small and relatively constant in practically all benchmarks, indicating the algorithm tends continuously towards an improvement of results and the minimums are not just a product of suitable random values. The following chart explains this element (Fig. 7):

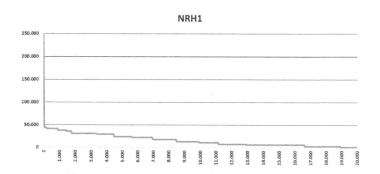

Fig. 7. NRH1 results

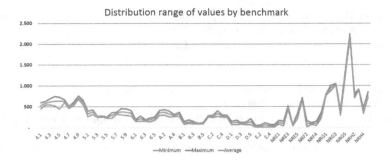

Fig. 8. Evolution of maxs and mins (Color figure online)

On the other hand, it is also important to note that in those initial tests in which the stochastic component was greater than that has been postulated as the optimal, the algorithm presented lower performance, determining optimal much higher probably by the inability to exploit areas with better potential solutions. All this is what it can be noted that the associated parameters to define the roulette for decision-making are quite small ranges in order that the random component be moderate. Other notorious elements are the large differences in results obtained with different methods of transfer and binarization, some ones simply conspired against acceptable results. Various possibilities already exposed to find a satisfactory combination were explored (Fig. 8). Some investigation lines that can be interesting approach for a possible improvement of results may be designed to develop a better way to determine the concept of distance, with better tailored criteria to the nature of the algorithm, as well as a more sophisticated method of mutation for those stars subjected to Hawking radiation. Additionally, some authors treat the rotation operator by adding additional elements such as mass and electric charge of the black holes [25], what was not considered in this work becouse the little existing documentation.

Acknowledgements. Broderick Crawford is supported by Grant CONICYT / FONDECYT / REGULAR / 1140897. Ricardo Soto is supported by Grant CONICYT / FONDECYT / REGULAR / 1160455. Sebastiásn Mansilla, Álvaro Gómez and Juan Salas are supported by Postgraduate Grant Pontificia Universidad Católica de Valparaiso 2015 (INF-PUCV 2015).

References

1. Vasko, F., Wolf, F., Stott, K.: Optimal selection of ingot sizes via set covering. Oper. Res. **35**(3), 346–353 (1987)
2. Desrochers, M., Soumis, F.: A column generation approach to the urban transit crew scheduling problem. Transp. Sci. **23**(1), 1–13 (1989)
3. Bellmore, M., Ratliff, H.D.: Optimal defense of multi-commodity networks. Manage. Sci. **18**(4–part–i), B-174 (1971)

4. Garfinkel, R.S., Nemhauser, G.L.: Optimal political districting byimplicit enumeration techniques. Manage. Sci. **347**, 267–276 (2015)
5. Amini, F., Ghaderi, P.: Hybridization of harmony search and ant colony optimization for optimal locating of structural dampers. Appl. Soft Comput. **13**(5), 2272–2280 (2013)
6. Karp, R.: Reducibility among combinatorial problems (1972). http://www.cs.berkeley.edu/~luca/cs172/karp.pdf
7. Hatamlou, A.: Black hole: a new heuristic optimization approach for data clustering. Inf. Sci. **222**, 175–184 (2013)
8. Ataim, P.: Resolución del problema de set-covering usando un algoritmo genético (2005)
9. Wang, B.S.: Caro and Crawford, "Multi-objective robust optimization using a postoptimality sensitivity analysis technique: application to a wind turbine design". J. Mech. Design **137**(1), 11 (2015)
10. Crawford, B., Soto, R., Johnson, F., Paredes, F.: A timetabling applied case solved with ant colony optimization. In: Silhavy, R., Senkerik, R., Oplatkova, Z.K., Prokopova, Z., Silhavy, P. (eds.) Artificial Intelligence Perspectives and Applications. AISC, vol. 347, pp. 267–276. Springer, Heidelberg (2015)
11. Marianov, R.: The queuing probabilistic location set coveringproblem and some extensions. Department of Electrical Engineering, PontificiaUniversidad Católica de Chile (2002)
12. ReVelle, C., Toregas, C., Falkson, L.: Applications of the location set-covering problem. Geog. Anal. **8**(1), 65–76 (1976)
13. Walker, W.: Using the set-covering problem to assign fire companies to fire houses. The New York City-Rand Institute (1974)
14. Crawford, B., Soto, R., Johnson, F., Misra, S., Paredes, F.: The use of metaheuristics to software project scheduling problem. In: Murgante, B., Misra, S., Rocha, A.M.A.C., Torre, C., Rocha, J.G., Falcão, M.I., Taniar, D., Apduhan, B.O., Gervasi, O. (eds.) ICCSA 2014, Part V. LNCS, vol. 8583, pp. 215–226. Springer, Heidelberg (2014)
15. Soto, R., Crawford, B., Galleguillos, C., Barraza, J., Lizama, S., Muñoz, A., Vilches, J., Misra, S., Paredes, F.: Comparing cuckoo search, bee colony, firefly optimization, and electromagnetism-like algorithms for solving the set covering problem. In: Gervasi, O., Murgante, B., Misra, S., Gavrilova, M.L., Rocha, A.M.A.C., Torre, C., Taniar, D., Apduhan, B.O. (eds.) ICCSA 2015. LNCS, vol. 9155, pp. 187–202. Springer, Heidelberg (2015)
16. Crawford, B., Soto, R., Peña, C., Riquelme-Leiva, M., Torres-Rojas, C., Misra, S., Johnson, F., Paredes, F.: A comparison of three recent nature-inspired metaheuristics for the set covering problem. In: Gervasi, O., Murgante, B., Misra, S., Gavrilova, M.L., Rocha, A.M.A.C., Torre, C., Taniar, D., Apduhan, B.O. (eds.) ICCSA 2015. LNCS, vol. 9158, pp. 431–443. Springer, Heidelberg (2015)
17. Hawking, S.: Agujeros negros y pequeños universos. Planeta (1994)
18. S. Hawking and M. Jackson, A brief history of time. Dove Audio, 1993
19. Crawford, B., Soto, R., Riquelme-Leiva, M., Peña, C., Torres-Rojas, C., Johnson, F., Paredes, F.: Modified binary firefly algorithms with different transfer functions for solving set covering problems. In: Silhavy, R., Senkerik, R., Oplatkova, Z.K., Prokopova, Z., Silhavy, P. (eds.) Software Engineering in Intelligent Systems. AISC, vol. 349, pp. 307–315. Springer, Heidelberg (2015)
20. Farahmandian, M., Hatamlou, A.: Solving optimization problems using black hole algorithm. J. Adv. Comput. Sci. Technol. **4**(1), 68–74 (2015)

21. Beasley, J.: Or-library (1990). http://people.brunel.ac.uk/~mastjjb/jeb/orlib/scpinfo.html
22. Beasley, J.E.: An algorithm for set covering problem. Euro. J. Oper. Res. **31**(1), 85–93 (1987)
23. Gervasi, O., Murgante, B., Misra, S., Gavrilova, M.L., Rocha, A.M.A.C., Torre, C., Taniar, D., Apduhan, B.O. (eds.): ICCSA 2015. LNCS, vol. 9155. Springer, Heidelberg (2015)
24. Beasley, J.E.: A Lagrangian heuristic for set-covering problems. Naval Res. Logistics **1**(37), 151–164 (1990)
25. Nemati, M., Salimi, R., Bazrkar, N.: Black holes algorithm: a swarm algorithm inspired of black holes for optimization problems. IAES Int. J. Artif. Intell. (IJ-AI) **2**(3), 143–150 (2013)

Solving Biobjective Set Covering Problem Using Binary Cat Swarm Optimization Algorithm

Broderick Crawford[1,2,3], Ricardo Soto[1,4,5], Hugo Caballero[1(✉)],
Eduardo Olguín[3], and Sanjay Misra[6]

[1] Pontificia Universidad Católica de Valparaíso, 2362807 Valparaíso, Chile
hcaballec@gmail.com
[2] Universidad Central de Chile, 8370178 Santiago, Chile
[3] Facultad de Ingeniera y Tecnología, Universidad San Sebastián,
Bellavista 7, 8420524 Santiago, Chile
[4] Universidad Autónoma de Chile, 7500138 Santiago, Chile
[5] Universidad Científica del Sur, Lima 18, Peru
[6] Covenant University, Ogun 110001, Nigeria

Abstract. The set cover problem is a classical question in combinatorics, computer science and complexity theory. It is one of Karp's 21 NP-complete problems shown to be NP-complete in 1972. Several algorithms have been proposed to solve this problem, based on genetic algorithms (GA), Particle Swarm Optimizer (PSO) and in recent years algorithms based in behavior algorithms based groups or herds of animals, such as frogs, bats, bees and domestic cats. This work presents the basic features of the algorithm based on the behavior of domestic cats and results to solve the SCP bi-objective, experimental results and opportunities to improve results using adaptive techniques applied to Cat Swarm Optimization. For this purpose we will use instances of SCP OR-Library of Beasley by adding an extra function fitness to transform the Beasly instance to Bi-Objective problem.

Keywords: Multiobjective problems · Evolutionary algorithm · Swarm optimization · Cat swarm optimization · Multiobjective cat swarm optimization · Pareto dominance

1 Introduction

Optimization problems require complex and optimal solutions because they relate to distribute limited basic resources. To resolve these problems it means improving the lives of poor people directly and enabling the growth of businesses, for example: resources related to social welfare, reaction by natural disasters, medical distribution capabilities. For these reasons the optimization generates a wide area of research in the sciences of Operations Research and Computer.

In the last decades bio-inspired algorithms have called the attention of researchers, in particular the heuristic Particle Swarm Optimization, which is

© Springer International Publishing Switzerland 2016
O. Gervasi et al. (Eds.): ICCSA 2016, Part I, LNCS 9786, pp. 220–231, 2016.
DOI: 10.1007/978-3-319-42085-1_17

based the behavior of some species: Bugs, fish, felines. These species use the collective intelligence to reach specific objectives guided by some community member. This paper is focused on studying the heuristic based on the behavior of domestic cats to solve a classical problem the Set Covering Problem (SCP).

2   Basic Concepts

2.1   Swarm Intelligent

Swarm intelligence (SI) is the collective behavior of independent individuals, that generate self-organizing, natural or artificial systems. Algorithms based on this principle are generally composed of simple agents that interact directly, locally, with simple rules, without centralized control, with interactions with stochastic components. This interaction between different autonomous agents generates an "intelligent" behavior, which gives rise to a pattern of global functioning that is used for optimization of complex mathematical functions. These techniques are inspired by nature (Bio Inspired), in processes such as ant colonies, schools, flocks or herds [1–3].

2.2   Multi Objective

Decision problems involves multiple evaluation criteria and generally they are in conflict [4]. To resolve a multi objective problem it required to optimize multiple criteria simultaneously. Exists a wide variety of cases in our society, for example: vehicle route optimization, environmental problems, allocation of medical resources [5]. The solution to multi-objective optimization problem it is presented by a set of feasible solutions, and the best of them define a set of non-dominated solutions, this set we will call Front. Formally the multi objective problem is defined as:

$$min \quad z(x) \;=\; [z_1(x), z_2(x), z_3(x), z_4(x), ..., z_M(x)] \qquad (1)$$

The goal consists in minimizing a function z with M components with a vector variable $x = (x_1, \ldots, x_n)$ in a universe U, i.e., A solution u dominates v if u performs at least as well as v across all the objectives and performs better than v in at least one objective.

The dimensions of the target area corresponding to the number of functions to optimize. In this single-objective problem is one-dimensional space, since each decision vector corresponds to only a scalar number. In multi-objective problems, this is multi-dimensional space, where each dimension corresponds to each objective function to be optimized [6].

2.3   Pareto Dominance

If we have two candidate solutions u and v from U, vector z(u) is said to dominate vector $z(v)$ denoted by: z(u)$\prec$ z(v), if and only if:

$$z_i(u) \;\leq\; z_i(v), \qquad \forall\, i \,\in\, \{1, ..., M\} \qquad (2)$$

$$z_i(u) \leq z_i(v), \qquad \exists\, i \in \{1, ..., M\} \tag{3}$$

If solution u is not dominated by any other solution, then u is declared as a Non Dominated or Pareto Optimal Solution. There are no superior solutions to the problem than u, although there may be other equally good solutions [7,8].

2.4   Hypervolume

When measuring the quality of multi-objective algorithms we consider two aspects: minimizing the distance of Pareto Front obtained by the algorithm to Front exact Pareto problem and maximize the spread of solutions on the front so that the distribution is as uniform possible [6]. Hypervolume is designed to measure both aspects: convergence and diversity - in a given front. This metric calculates the volume (in the objective space) covered by members of a given set, Q, non-dominated solutions to problems where all the objectives are to be minimized. Mathematically, for each $i \in Q$ a hypercube v_i is built with a reference point W and the solution i that define the diagonal thereof. The point W can be obtained simply with the worst values of the objective functions. Then the union of all hypercubes is what defines the hypervolume (HV):

$$HV = \bigcup_{i=1}^{|Q|} v_i \tag{4}$$

3   Set Covering Problem

SCP is defined as a fundamental problem in Operations Research and often described as a problem of coverage of *m-rows n-columns* of a binary matrix by a subset of columns to a minimum cost [9]. It is one of Karp's 21 NP-complete problems. This is the problem of covering the rows of an m-row, n column, zero-one $m \ x \ n$ matrix a_{ij} by a subset of the columns at minimal cost. Formally, the problem can be defined as:

Defining $x_j = 1$ if column j with cost c_j is in the solution and $x_j = 0$ otherwise

$$Minimize \ Z = \sum_{j=1}^{n} c_j x_j \qquad j \in \{1, 2, 3, ..., n\} \tag{5}$$

Subject to:

$$\sum_{j=1}^{n} a_{ij} x_j \geq 1 \qquad i \in \{1, 2, 3, ..., m\} \tag{6}$$

$$x_j = \{0, 1\} \tag{7}$$

This definition contains a one fitness function, there is just one objective to be optimized. We study the case for two objective functions, using meta heuristic Cat Swarm Optimization (CSO) and using position vector of ones and zeros. A complete case study of SCP using CSO was done **Pontificia Universidad**

Cátolica de Valparaíso [10]. This work focuses on solving the SCP with two fitness functions, i.e., textit M = 2. To ensure the fitness functions have opposed criteria the second cost vector will be transposed the first, therefore the definition will be:

$$c_2 = (c_1)^t \tag{8}$$

$$min \quad z(x) = [z_1(x), z_2(x)] \tag{9}$$

$$Minimize \ Z_1 = \sum_{j=1}^{n} c_j^1 x_j \quad j \in \{1, 2, 3, ..., n\} \tag{10}$$

$$Minimize \ Z_2 = \sum_{j=1}^{n} c_j^2 x_j \quad j \in \{1, 2, 3, ..., n\} \tag{11}$$

Subject to:

$$\sum_{j=1}^{n} a_{ij} x_j \geq 1 \quad i \in \{1, 2, 3, ..., m\} \tag{12}$$

4   Cat Swarm Optimization CSO

4.1   Basic Concepts

A detailed description of the behavior of cats and especially domestic cats may be revised in [10, 11]. This work indicates the specific concepts that control the behavior of the algorithm. There are importants features in their behavior and they employ to achieve their goals:

- Seeking mode. Resting but being alert - looking around its environment for its next move.
- Tracing mode. The cat is moving after its prey.
- The presentation of solution sets. How many cats we would like to use in the iteration and we must to define a mixture ratio (MR) which dictates the joining of seeking mode with tracing mode. According to observations, the cats spend a lot more time resting therefore MR should take a low value to ensure this feature.

The CSO was originally developed for continuous valued spaces. But there exist a number of optimization problems, as the SCP, in which the values are not continuous numbers but rather discrete binary integers. Sharafi et al. introduced a discrete binary version of CSO for discrete optimization problems: Binary Cat Swarm Optimization (BCSO) [14]. BCSO is based on CSO algorithm proposed by Chu, Tsai ann Pan in 2006 [11]. The difference is that in BCSO the vector position consists of ones and zeros, instead of the real numbers of CSO.

4.2   Parameters Important

In this paper, we considered the following BCSO parameters as relevant for our experimentation. We have considered the impact on the results in previous experiments

- NC: Number of population or pack cats
- MR: Mixture Rate that defines number of cats mode, this parameter must be chosen between 0 and 1. Define what percentage of cats are in seeking mode and tracing mode
- Termination condition. Normally is used a number of iterations.
- Adaptative criteria. To get better results usually we choose a parameter modified to perform the process again. In our work we choose the MR parameter especially to calculate the front not dominated ranging from 0.1 to 0.9 on increasing 0.1

4.3   Description of Cat Swarm Optimization - Main Algorithm

The main mechanism used in our experiments was consider as criteria of adaptive change the mixing ratio. In our experiment It was modified from 0.6 until 0.9 and determined non dominated solution

(a) Create N cats
(b) Initiate MR_p in min value ($= 0.5$)
(c) Create the cat swam, N cats working to solve the problem
(d) Define, randomly the position and velocity for each cat
(e) Distribute the swarm in tracing and seeking mode based on MR_p
(f) Check the cat mode, if cat is in Seeking Mode, applay Seekin Mode, else apply Tracing Mode
(g) Check if the cat is feasible solution (Ec. 11). if the cat satisfies the restriction compute the fitness (Ec. 9 and Ec.10) and compare with the non dominated solutions in the archive
(h) Update de solution file
(i) If number iteration less than the max iteration continue work, goto step c
(j) If MR_p les than max value MR, increment MR_p and go to step b
(k) Calculate the pareto front from non domination file

Below the two main modes are described

4.4   Seeking Mode

The seeking mode corresponds to a global search technique in the search space of the optimization problem. A term used in this mode is seeking memory pool (SMP). It is the number of copies of a cat produced in seeking mode.

There are four essential factors in this mode: seeking memory pool (SMP), seeking range of the selected dimension (SRD), counts of dimension to change (CDC), and self-position considering (SPC).

- SMP is used to define the size of seeking memory for each cat. SMP indicates the points explored by the cat. This parameter can be different for different cats.
- SRD declares the mutation ratio for the selected dimensions.
- CDC indicates how many dimensions will be varied.
- SPC is a Boolean flag, which decides whether current position of cat

The steps involved in this mode are:

(a) Create T (=SMP) copies of jth cat i.e. Y_{kd} where $(1 \leq k \leq T)$ and $(1 \leq d \leq D)$. D is the total number of dimensions.
(b) Apply a mutation operator to Y_k.
(c) Evaluate the fitness of all mutated copies.
(d) Update the contents of the archive with the position of those mutated copies which represent non dominated solutions.
(e) Pick a candidate randomly from T copies and place it at the position of jth cat.

4.5   Tracing Mode

The tracing mode corresponds to a local search technique for the optimization problem. In this mode, the cat traces the target while spending high energy. The rapid chase of the cat is mathematically modeled as a large change in its position. Define position and velocity of ith cat in the D-*dimensional* space as $X_i = (X_{i1}, X_{i2}, X_{i3} \ldots X_{iD})$ and $V_i = (V_{i1}, V_{i2}, V_{i3} \ldots V_{iD})$ where $(1 \leq d \leq D)$ represents the dimension. The global best position of the cat swarm is represented as $X_g = (X_{g1}, X_{g2}, X_{g3} \ldots X_{gD})$. The steps involved in tracing mode are:

(a) Compute the new velocity of ith cat using (13)

$$V_{id} = w * V_{id} + c * r * (X_{gd} - X_{id}) \tag{13}$$

where
$\mathbf{w} = $ is the inertia weight
$\mathbf{c} = $ is the acceleration constant
$\mathbf{r} = $ is a random number uniformly distributed in the range $[0, 1]$
(b) Compute the new position of ith cat using

$$V_{id} = X_{gd} - X_{id} \tag{14}$$

(c) If the new position of ith cat corresponding to any dimension goes beyond the search space, then the corresponding boundary value is assigned to that dimension and the velocity corresponding to that dimension is multiplied by -1 to continue the search in the opposite direction.
(d) Evaluate the fitness of the cats.
(e) Update the contents of the archive with the position of those cats which represent no dominated vectors.

5 The Execution of the Algorithm

1 Load Instance SCP
2 Initiate phase
   (a) For to obtein the best Pareto Front: $MR = 0, 1...0.99$
   (b) For Analysis CSO, $MR = 0.5$
   (c) Initiate: $SRD, CDC, SPC, w, c1$
   (d) Define the size file Pareto
   (e) Define Iteration number
   (f) Define, randomly the position and velocity for each cat
3 Distribute the swarm in tracing and seeking mode based on MR
4 Repeat until iteration number reached
   (a) If *cat* is seeking mode, apply seeking process and return a solution candidate $\vec{X}$
   (b) If *cat* is tracing mode, apply seeking process and return a solution candidate $\vec{X}$
   (c) Check if the $\vec{X}$ satisfies the restriction problem: $\sum_{j=1}^{n} a_{ij} X_j \geq 1$
   (d) Compute
   $$f1 = \sum_{j=1}^{n} c_{1j} X_j$$
   $$f2 = \sum_{j=1}^{n} c_{2j} X_j$$
   (e) Store the position of the cats representing non-dominated solutions in the archive
5 Calculate the pareto front from non domination file

This algorithm was executed 30 times for each SCP Instance, and the pareto front was obtained $ParetoFront = \bigcup_{i=1}^{30} (pf)_i$

6 Experimental Results

The CSO was evaluated using the next features:

Table 1. Parameter values CSO

Name	Parameter	Value	Obs
Number of cats	C	30	
Mixture ratio	MR	0.5	
Seeking memory pool	SMP	20	-
Probability of Mutation	PMO	1	-
Counts of dimensions to change	CDC	0,001	-
Inertia weight	w	1	-
Factor c_1	c_1	1	-

Table 2. Experimentals results

INST	HYPER	MAX	MIN	PROM	DESV
scp41	0,6223	132,441	109,345	118,84	7,48
scp42	0,6845	156,556	121,211	143,1	10,03
scp43	0,7261	135,3	115,309	125,22	6,53
scp44	0,5804	154,609	129,51	140,53	6,38
scp45	0,7426	134,763	105,963	119,49	9,28
scp46	0,5435	140,833	114,68	134,02	7,41
scp47	0,5172	147,812	126,058	136,42	7,26
scp48	0,7319	135,586	114,344	120,57	7,09
scp49	0,6029	159,194	135,4	148,2	7,13
scp51	0,6156	270,516	247,489	256,56	7,63
scp52	0,6378	282,612	259,742	270,77	7,24
scp53	0,6613	257,966	203,538	229,88	17,42
scp54	0,8511	259,181	212,809	241,01	15,83
scp55	0,5872	234,38	205,496	225,25	9,034
scp56	0,7223	265,601	218,673	238,11	14,736
scp57	0,6036	259,252	234,426	245,85	8,5
scp58	0,6242	270,754	242,436	254,9	9,502
scp59	0,5338	243,13	209,511	227,58	11,928
scp61	0,5992	103,339	81,946	94,31	7,73
scp62	0,6673	100,748	79,064	91,99	8,472
scp63	0,6873	96,555	77,817	86,94	6,077
scp64	0,6363	103,206	78,183	90,4	9,285
scp65	0,6696	101,83	78,088	90,66	7,244
scpa1	0,7834	506,95	463,377	482,7	14,943
scpa2	0,5462	618,465	513,501	559,59	34,738
scpa3	0,5631	517,718	474,45	496,63	13,896
scpa4	0,6269	526,053	469,132	502,76	17,687
scpa5	0,7679	529,614	445,58	488,48	27,499

MAX, MIN, PROM, DESV in sec

1 Using 65 Instances for set covering from OR-Library of Beasley [12]
2 MacBook Pro (13-inch, Mid 2012), CPU MacBook Pro (13-inch, Mid 2012),
   16 GB 1333 MHz DDR3, OS X Yosemite, version 10.10.5
3 IDE: BlueJ version 3.1.5 (Java version 1.8.0_31)

The Optimal Pareto Front was estimated individually for each instance varying the value of MR from 0.1 until 0.9 using an increment 0.1, and we choose
use the hypervolume because is the only one indicator of performance that is

compatible unary with Pareto dominance and it has been able to demonstrate that its maximization is equivalent to achieve convergence to the true Pareto front for a specific value of MR (fig. a) and collectively by varying the value of MR from 0.5 until 0.9 using an increment 0.1 (fig b). According to the results, the best result is obtained collectively. The working conditions of the process were:

1 1.500 iterations for each MR value
2 30 times each Beasly instance
3 Varying MR from 0.5 until 0.9 using an increment 0.1
4 The parameters BCO was obteined from [10,13] and show in Table 1.

Table 2 shows the experimental results for each Beasly Instance. The Optimal Pareto Front was obtained varying MR from 0.1 until 0.99 and determined by the union of fronts obtained MR.

In our experiments the best HV was with SCP54 instance, however the worst HV was with SCP47. If we observe the charts there are zones of the solution space that they need a better exploration strategy by changing the calculation (Fig. 1).

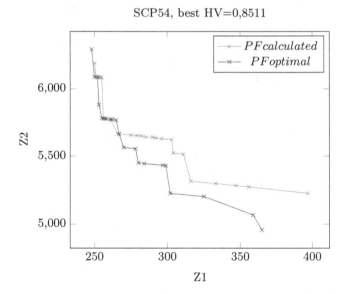

Fig. 1. Best HV was with SCP54 instance (Color figure online)

SCP47, worst HV=0,5172

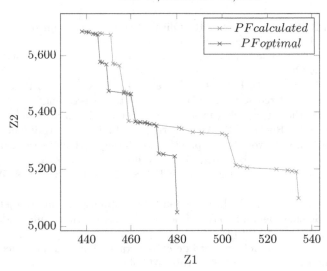

Fig. 2. Worst HV was with SCP54 instance (Color figure online)

7 Conclusion and Future Work

There are not published results on a SCP Bi Objective using BCSO and we think the Pareto Front is quite promising considering just we varied only MR. We also think that applying adaptive mechanisms in other parameters of the CSO metaheuristic we can improve results. We also believe that this particular CSO should be compared with genetic and evolutionary algorithms. Within the same field with metaheuristics highlighted with ants, bees and frogs. The proposed BCSO is implemented and tested using 65 SCP test instances from the OR-Library of Beasley. This is first phase of our research. We only work with MR parameter, however. We think that using adaptive techniques for parameter we wil improve our results (Fig. 2). The next step:

1 To use adaptive techniques for BCSO parameters to improve HV
2 To obtain a Pareto optimal front using genetic, evolutionary algorithms. Within the same field with metaheuristics highlighted with ants, bees and frogs and

Acknowledgements. The author Broderick Crawford is supported by grant CONICYT/FONDECYT/REGULAR/1140897 and Ricardo Soto is supported by Grant CONICYT/FONDECYT/REGULAR/1160455.

References

1. Crawford, B., Soto, R., Cuesta, R., Olivares-Suárez, M., Johnson, F., Olguin, E.: Two swarm intelligence algorithms for the set covering problem. In: 2014 9th International Conference on Software Engineering and Applications (ICSOFT-EA), pp. 60–69. IEEE (2014)
2. Crawford, B., Soto, R., Peña, C., Riquelme-Leiva, M., Torres-Rojas, C., Misra, S., Johnson, F., Paredes, F.: A comparison of three recent nature-inspired metaheuristics for the set covering problem. In: Gervasi, O., Murgante, B., Misra, S., Gavrilova, M.L., Rocha, A.M.A.C., Torre, C., Taniar, D., Apduhan, B.O. (eds.) ICCSA 2015. LNCS, vol. 9158, pp. 431–443. Springer, Heidelberg (2015)
3. Gervasi, O., Murgante, B., Misra, S., Gavrilova, M.L., Rocha, A.M.A.C., Torre, C., Taniar, D., Apduhan, B.O. (eds.): ICCSA 2015. LNCS, vol. 9158. Springer, Heidelberg (2015)
4. Lopez, J., Lanzarini, L.C., Leguizamón, G.: Optimización multiobjetivo: aplicaciones a problemas del mundo real. Buenos Aires, Argentina, Universidad Nacional de la Plata, pp. 66–90 (2013)
5. Wikipedia, Problema del conjunto de cobertura – wikipedia, la enciclopedia libre (2014). [Internet; descargado 29-octubre-2015]
6. Zitzler, E., Thiele, L.: Multiobjective optimization using evolutionary algorithms– a comparative case study. In: Eiben, A.E., Bäck, T., Schoenauer, M., Schwefel, H.-P. (eds.) Parallel Problem Solving from Nature—PPSN V. LNCS, vol. 1498, pp. 292–301. Springer, Heidelberg (1998)
7. Wikipedia, Pareto efficiency – wikipedia, the free encyclopedia (2015). Accessed 29 Oct 2015
8. Knowles, J., Corne, D.: The pareto archived evolution strategy: a new baseline algorithm for pareto multiobjective optimisation. In: Proceedings of the 1999 Congress on Evolutionary Computation, 1999, CEC 99, vol. 1, IEEE (1999)
9. Caprara, A., Toth, P., Fischetti, M.: Algorithms for the set covering problem. Ann. Oper. Res. **98**(1–4), 353–371 (2000)
10. Crawford, B., Soto, R., Berrios, N., Johnson, F., Paredes, F.: Binary cat swarm optimization for the set covering problem pp. 1–4 (2015)
11. Chu, S.-C., Tsai, P.-W.: Computational intelligence based on the behavior of cats. Int. J. Innovative Comput. Inf. Control **3**(1), 163–173 (2007)
12. Beasley, J.E.: A lagrangian heuristic for set-covering problems. Nav. Res. Logistics (NRL) **37**(1), 151–164 (1990)
13. Pradhan, P.M., Panda, G.: Solving multiobjective problems using cat swarm optimization. Expert Syst. Appl. **39**(3), 2956–2964 (2012)
14. Chu, S.-C., Tsai, P., Pan, J.-S.: Cat swarm optimization. In: Yang, Q., Webb, G. (eds.) PRICAI 2006. LNCS (LNAI), vol. 4099, pp. 854–858. Springer, Heidelberg (2006)
15. Panda, G., Pradhan, P.M., Majhi, B.: IIR system identification using cat swarm optimization. Expert Syst. Appl. **38**(10), 12671–12683 (2011)
16. Lust, T., Tuyttens, D.: Two-phase pareto local search to solve the biobjective set covering problem. In: 2013 Conference on Technologies and Applications of Artificial Intelligence (TAAI), pp. 397–402. IEEE (2013)
17. Durillo, J.J., Nebro, A.J.: jmetal: A java framework for multi-objective optimization. Ad. Eng. Softw. **42**(10), 760–771 (2011)
18. Wikipedia, Optimización multiobjetivo – wikipedia, la enciclopedia libre (2013). [Internet; descargado 29-octubre-2015]

19. Friedrich, T., He, J., Hebbinghaus, N., Neumann, F., Witt, C.: Approximating covering problems by randomized search heuristics using multi-objective models*. Evol. Comput. **18**(4), 617–633 (2010)
20. Vazirani, V.V.: Approximation Algorithms. Springer Science & Business Media, New York (2013)
21. Bouzidi, A., Riffi, M.E.: Cat swarm optimization to solve flow shop scheduling problem. J. Theor. Appl. Inf. Technol. 72(2) (2015)
22. Hadi, I., Sabah, M.: Improvement cat swarm optimization for efficient motion estimation. Int. J. Hybrid Inf. Technol. **8**(1), 279–294 (2015)
23. Musliu, N.: Local search algorithm for unicost set covering problem. In: Ali, M., Dapoigny, R. (eds.) IEA/AIE 2006. LNCS (LNAI), vol. 4031, pp. 302–311. Springer, Heidelberg (2006)
24. Zhang, L.-B., Zhou, C.-G., Liu, X., Ma, Z., Ma, M., Liang, Y.: Solving multi objective optimization problems using particle swarm optimization. In: Proceedings of IEEE congress on evolutionary computation, pp. 2400–2405 (2003)
25. Crawford, B., Soto, R., Aballay Leiva, F., Johnson, F., Paredes, F.: The set covering problem solved by the binary teaching-learning-based optimization algorithm. In: 2015 10th Iberian Conference on Information Systems and Technologies (CISTI), pp. 1–4. IEEE (2015)
26. Crawford, B., Soto, R., Peña, C., Riquelme-Leiva, M., Torres-Rojas, C., Johnson, F., Paredes, F.: Binarization methods for shuffled frog leaping algorithms that solve set covering problems. In: Silhavy, R., Senkerik, R., Oplatkova, Z.K., Prokopova, Z., Silhavy, P. (eds.) Software Engineering in Intelligent Systems. AISC, vol. 349, pp. 317–326. Springer, Heidelberg (2013)
27. Crawford, B., Soto, R., Aballay, F., Misra, S., Johnson, F., Paredes, F.: A teaching-learning-based optimization algorithm for solving set covering problems. In: Gervasi, O., Murgante, B., Misra, S., Gavrilova, M.L., Rocha, A.M.A.C., Torre, C., Taniar, D., Apduhan, B.O. (eds.) ICCSA 2015. LNCS, vol. 9158, pp. 421–430. Springer, Heidelberg (2015)
28. Crawford, B., Soto, R., Torres-Rojas, C., Peña, C., Riquelme-Leiva, M., Misra, S., Johnson, F., Paredes, F.: A binary fruit fly optimization algorithm to solve the set covering problem. In: Gervasi, O., Murgante, B., Misra, S., Gavrilova, M.L., Rocha, A.M.A.C., Torre, C., Taniar, D., Apduhan, B.O. (eds.) ICCSA 2015. LNCS, vol. 9158, pp. 411–420. Springer, Heidelberg (2015)
29. Cuesta, R., Crawford, B., Soto, R., Paredes, F.: An artificial bee colony algorithm for the set covering problem. In: Silhavy, R., Senkerik, R., Oplatkova, Z.K., Silhavy, P., Prokopova, Z. (eds.) Modern Trends and Techniques in Computer Science. AISC, vol. 285, pp. 53–63. Springer, Heidelberg (2014)

An Accelerated Multistart Derivative-Free Framework for the Beam Angle Optimization Problem in IMRT

Humberto Rocha[1,2(✉)], Joana M. Dias[1,2], Tiago Ventura[3],
Brígida C. Ferreira[4], and Maria do Carmo Lopes[3]

[1] Faculdade de Economia, Universidade de Coimbra, 3004-512 Coimbra, Portugal
hrocha@mat.uc.pt
[2] INESC-Coimbra, 3030-290 Coimbra, Portugal
joana@fe.uc.pt
[3] Serviço de Física Médica, IPOC-FG, EPE, 3000-075 Coimbra, Portugal
{tiagoventura,mclopes}@ipocoimbra.min-saude.pt
[4] School for Allied Health Technologies, 4400-330 Porto, Portugal
bcf@estsp.ipp.pt

Abstract. Radiation therapy, either alone or combined with surgery or chemotherapy, is one of the main treatment modalities for cancer. Intensity-modulated radiation therapy (IMRT) is an advanced form of radiation therapy, where the patient is irradiated using non-uniform radiation fields from selected beam angle directions. The goal of IMRT is to eradicate all cancer cells by delivering a radiation dose to the tumor volume, while attempting to spare, simultaneously, the surrounding organs and tissues. Although the use of non-uniform radiation fields can favor organ sparing, the selection of appropriate irradiation beam angle directions – beam angle optimization – is the best way to enhance organ sparing. The beam angle optimization (BAO) problem is an extremely challenging continuous non-convex multi-modal optimization problem. In this study, we present a novel approach for the resolution of the BAO problem, using a multistart derivative-free framework for a more thoroughly exploration of the search space of the highly non-convex BAO problem. As the objective function that drives the BAO problem is expensive in terms of computational time, and a multistart approach typically implies a large number of function evaluations, an accelerated framework is explored. A clinical case of an intra-cranial tumor treated at the Portuguese Institute of Oncology of Coimbra is used to discuss the benefits of the accelerated multistart approach proposed for the BAO problem.

Keywords: IMRT · Beam angle optimization · Multistart · Derivative-free optimization

1 Introduction

Radiation therapy, either alone or combined with surgery or chemotherapy, is one of the main treatment modalities for cancer. The goal of radiation therapy is

© Springer International Publishing Switzerland 2016
O. Gervasi et al. (Eds.): ICCSA 2016, Part I, LNCS 9786, pp. 232–245, 2016.
DOI: 10.1007/978-3-319-42085-1_18

to eradicate all cancer cells by delivering a radiation dose to the tumor volume, while attempting to spare, simultaneously, the surrounding organs and tissues. Radiation therapy is used with curative intent or to palliate symptoms giving important symptom relief. Intensity-modulated radiation therapy (IMRT) is an advanced form of radiation therapy, where the radiation beam is modulated by a multileaf collimator allowing its discretization into small beamlets of different intensities. This discretization of the radiation beam enables the irradiation of the patient using non-uniform radiation fields from selected beam angle directions. The use of non-uniform radiation fields in IMRT provides an accurate control of the different doses to be optimized which can favor organ sparing. However, appropriate selection of individualized irradiation beam directions – beam angle optimization (BAO) – depositing in an additive way the total radiation dose in the tumor while attempting to spare the surrounding organs and tissues that only receive radiation from a small subset of radiation beams, is the best way to enhance a proper organ sparing.

Despite the fact that for some treatment sites, in particular for intra-cranial tumors, BAO substantially improves plan quality [4], in clinical practice, coplanar equispaced beam directions, i.e. evenly spaced beam directions that lay on the plane of rotation of the linear accelerator's gantry, are still commonly used. Alternatively, beam directions are manually selected on a long trial-and-error procedure by the treatment planner, as commercial treatment planning systems have limited resources available for BAO. One of the reasons for the scarce commercial offer for beam angle directions optimal selection is the difficulty of solving the BAO problem, a highly non-convex multi-modal optimization problem on a large search space [12].

The BAO approaches can be separated into two different classes. The first class considers a discrete sample of all possible beam angle directions and addresses the BAO problem as a combinatorial optimization problem. As the best ensemble of beam angles cannot be obtained through exhaustive searches, in a reasonable computational time, different methods are commonly used to guide the searches including genetic algorithms [13], simulated annealing [14], particle swarm optimization [15], gradient search [12], neighborhood search [1], response surface [2], branch-and-prune [16] or hybrid approaches [5]. The combinatorial formulation of the BAO problem leads to an NP hard problem. Thus, there is no algorithm known capable of finding, in a polynomial run time, the optimal solution of the combinatorial BAO problem [3]. Another common and successful combinatorial approach is iterative BAO [3,8], where beams are sequentially added, one at a time, to a treatment plan, significantly reducing the number of beam combinations. The second class of BAO approaches considers a completely different methodological approach by exploring the continuous search space of the highly non-convex BAO problem [18–20].

In this study, we present a novel approach that belong to the second class of approaches for the resolution of the BAO problem. A multistart derivative-free framework is sketched for a more thoroughly exploration of the continuous search space of the highly non-convex BAO problem. As the objective function that drives the BAO problem is expensive in terms of computational time, and a multistart approach typically implies a large number of function evaluations, an accelerated multistart framework is tested using a clinical case of an intra-cranial tumor treated at the Portuguese Institute of Oncology of Coimbra (IPOC). The paper is organized as follows. In the next section we describe an accelerated multistart framework for the BAO problem. Computational tests are presented in Sect. 3. In the last section we have the conclusions.

2   Multistart Derivative-Free Framework for BAO

2.1   BAO Formulation

The formulation of the BAO problem as a continuous optimization problem was proposed in our previous works [18–20]. In order to model the BAO problem as a mathematical optimization problem, a measure of the quality of the beam angles ensemble is required. The straightforward measure for driving the BAO problem is the optimal solution of the fluence map optimization (FMO) problem [1,3,8, 12–14,16,18–20], the problem of finding the optimal radiation intensities. Let n be the fixed number of (coplanar) beam directions, a continuous formulation for the BAO problem is obtained by selecting an objective function f such that the best set of beam angles is obtained for the function's minimum:

$$\min f(\theta_1, \ldots, \theta_n)$$
$$s.t. \ (\theta_1, \ldots, \theta_n) \in \mathbb{R}^n, \tag{1}$$

where $\theta_i, i = 1, \ldots, n$ are beam angles selected from all continuous beam irradiation directions. For this study, the objective function $f(\theta_1, \ldots, \theta_n)$ that measures the quality of the beam angle ensemble $\theta_1, \ldots, \theta_n$ was the FMO problem objective function, modeled as a multicriterial optimization problem. Nevertheless, as the FMO model is used as a black-box function, the conclusions drawn using this particular formulation of the FMO problem can be extended to different FMO formulations.

2.2   FMO for an Intra-cranial Tumor Case

Typically, the FMO problem is modeled as a weighted sum function with conflicting objectives which difficult their trade-off. Thus, treatment plan optimization is inherently a multicriteria procedure. Different multicriteria approaches have been proposed for the FMO problem where solutions can be selected from a set of Pareto-optimal treatment plans *a posteriori* [11,17] or a set of criteria (objectives and constraints) that have to be met is defined *a priori* [6–8]. The latter approach is more suitable for a fully automated BAO procedure. Thus, a multicriterial optimization procedure based on a prescription called wish-list [6–8] is used to address the FMO problem.

Table 1 depicts the wish-list constructed for the clinical intra-cranial tumor case treated at IPOC. Intra-cranial cases are complex tumors to treat with radiation therapy due to the large number of sensitive organs in this region. Beyond the planning target volume (PTV), i.e. the tumor to be irradiated, a large number of organs at risk (OARs) are included in the wish-list. Different dose levels were defined for the tumor (PTV-T) and for the lymph nodes (PTV-N), according to the IPOC protocols defined for this pathology. Several auxiliary structures (PTV-T Ring, PTV-N Ring, PTV-N shell and External Ring) were constructed by computerized volume expansions to support the dose optimization.

The wish-list contains 11 hard constraints, all maximum-dose constraints, that have to be strictly met. It also contains 28 prioritized objectives that are sequentially optimized following the priorities defined in the wish-list. For the target dose optimization, the logarithmic tumor control probability $(LTCP)$ was considered [8], $LTCP = \frac{1}{N_T} \sum_{l=1}^{N_T} e^{-\alpha(D_l - T_l)}$, where N_T is the number of voxels (small volume elements) in the target structure, D_l is the dose in voxel l, T_l is the prescribed dose, and α is the cell sensitivity parameter. For most OARs maximum-dose constraints were considered. For some OARs, a generalized Equivalent Uniform Dose $(gEUD)$ objective was considered [8], $gEUD = k\left(\frac{1}{N_S} \sum_l D_l^a\right)^{\frac{1}{a}}$, where k is the number of treatment fractions, N_S the number of voxels of the discretized structure, D_l the dose in voxel l and a is the tissue-specific parameter that describes the volume effect.

A primal-dual interior-point algorithm tailored for multicriteria IMRT treatment planning, $2p\epsilon c$ [6], was used for optimization of the FMO problem using the described wish-list. The $2p\epsilon c$ algorithm automatically generates a single Pareto optimal IMRT plan for a given number of beams. For a detailed description of the $2p\epsilon c$ algorithm see Breedveld et al. [6].

2.3 Multistart Approach for the Continuous BAO Problem

Multistart approaches have two phases that can be designated as global and local phases [10]. In the global phase, a number of starting points is selected for which the objective function is evaluated. Then, local search procedures are used to improve each of the starting points outcome. In previous works, we have shown that a beam angle set can be locally improved in a continuous manner using Pattern Search Methods (PSM) [18–20]. An important feature of PSM is its ability to converge globally, i.e., from arbitrary points to local minimizers [21]. Furthermore, PSM have the ability to avoid local entrapment and require few function evaluations to converge. Thus, PSM were selected for the local search procedure to be embedded in the multistart framework.

In the global phase, typically, starting points (beam ensembles) are randomly selected. However, for search spaces with peculiar characteristics as the BAO continuous search space, different strategies need to be adopted. As the order of the beam directions of a beam angle ensemble is irrelevant, the BAO continuous search space has symmetry properties. This means that different solutions can in reality correspond to the same solution, the only difference being that the same

Table 1. Wish-list for the intra-cranial tumor case.

	Structure	Type	Limit	
Constraints	PTV-T	max.	74.9 Gy	(= 107 % of prescribed dose)
	PTV-N	max.	63.6 Gy	(= 107 % of prescribed dose)
	PTV-N shell	max.	63.6 Gy	(= 107 % of prescribed dose)
	Brainstem	max.	54 Gy	
	Spinal cord	max.	45 Gy	
	Retinas	max.	45 Gy	
	Optics	max.	55 Gy	
	PTV-T Ring	max.	59.5 Gy	(= 85 % of prescribed dose)
	PTV-N Ring	max.	50.5 Gy	(= 85 % of prescribed dose)
	External Ring	max.	45 Gy	
	Body	max.	70 Gy	

	Structure	Type	Priority	Goal	Sufficient	Parameters
Objectives	PTV-N	LTCP	1	1	0.5	$T_l = 59.4$ Gy; $\alpha = 0.75$
	PTV-T	LTCP	2	1	0.5	$T_l = 70$ Gy; $\alpha = 0.75$
	PTV-N shell	LTCP	3	1	0.5	$T_l = 59.4$ Gy; $\alpha = 0.75$
	External ring	max.	4	42.75 Gy	–	–
	Spinal cord	max.	5	42.75 Gy	–	–
	Brainstem	max.	6	51.3 Gy	–	–
	Optics	max.	7	52.25 Gy	–	–
	Retinas	max.	8	42.75 Gy	–	–
	Lens	gEUD	9	12 Gy	–	$a = 12$
	Ears	mean	10	50 Gy	–	–
	Parotids	mean	11	50 Gy	–	–
	Oral cavity	mean	12	45 Gy	–	–
	TMJ	max.	13	66 Gy	–	–
	Mandible	max.	14	66 Gy	–	–
	Esophagus	mean	15	45 Gy	–	–
	Larynx	mean	16	45 Gy	–	–
	Optics	gEUD	17	48 Gy	–	$a = 12$
	Retinas	gEUD	18	22 Gy	–	$a = 12$
	Lens	gEUD	19	6 Gy	–	$a = 12$
	Ears	mean	20	45 Gy	–	–
	Parotids	mean	21	26 Gy	–	–
	Oral cavity	mean	22	35 Gy	–	–
	Oesophagus	mean	23	35 Gy	–	–
	Larynx	mean	24	35 Gy	–	–
	Brain	gEUD	25	54v	–	$a = 12$
	Pituitary gland	gEUD	26	60 Gy	–	$a = 12$
	Thyroid	mean	27	27.5 Gy	–	–
	Lungs	mean	28	5 Gy	–	–

angles appear in different positions. This simple observation allows for a drastic reduction of the search space by simply ordering the angles in each solution. For n-beam directions, by keeping the beam angles sorted, the search space is reduced by 2^n. For the 7-beam angle search space, e.g., the reduced search space is only 0.78 % of $[0, 360]^7$. However, the reduced search space may take a peculiar shape. Thus, a strategy to sample the reduced search space is required.

The strategy sketched for selecting the starting points considers all the possible distributions of the sorted beam angle directions by quadrants. For illustration purposes, all possible distributions of 3-beam angle directions by the four quadrants are depicted in Fig. 1. Examples of 3-beam angle directions for each one of the 20 possible distributions by the four quadrants are displayed in Fig. 1(a) while Fig. 1(b) displays the corresponding painted cubes of the reduced search space. In general, for the n-beam angle direction search space, the total number of (hyper)cubes of the entire search space is 4^n while the number of (hyper)cubes of the reduced search space is the combination with repetition of $\binom{n+4-1}{4} = \frac{(n+4-1)!}{4!(n-1)!}$. For the 7-beam angle optimization problem, e.g., the reduced search space has 120 (hyper)cubes while the entire search space has 16384. A good strategy for sampling the reduced search space consists in selecting one starting point for each one of the (hyper)cubes of the reduced search space. Such strategy guarantees that the starting points belong to the reduced search space, they are well spread and most importantly they cover well all the reduced search space.

A major drawback of multistart methods, particularly for a parallel setting, is that the same region of the search space may be simultaneously searched by local procedures originated from starting points of different regions. Thus, the same local minima can be found more than once wasting precious computational time. A generalization of the notion of region of attraction of a local minimum can be used to avoid overlap of local searches and simultaneously accelerate the optimization procedure. One can define each (hyper)cube of the reduced search space as a region of attraction of a local minimum and allow a single "active" local search for each region of attraction. If two local searches end up simultaneously searching the same (hyper)cube during the optimization procedure, only the local search with the current best solution, i.e. with the solution corresponding to the lowest objective function value, will remain active which accelerates the overall optimization process.

At the end of the global phase of the multistart method, the objective function value is evaluated at the $N = \frac{(n+4-1)!}{4!(n-1)!}$ initial beam ensembles selected, $\mathbf{x}_i^0 \in [0, 360]^n, i = 1, \ldots, N$. For many of the regions, the starting points will lead to poor objective function results. Beam ensembles with many (or all) beams in the same quadrant will have beams too close to produce good results. Thus, another strategy to accelerate the multistart framework is to explore only the most promising regions at each iteration. In order to do so, a boolean vector, $\mathbf{Active}_{N \times 1}$, that stores the information of which regions have active local search procedures is updated in every iteration and the overall best objective function

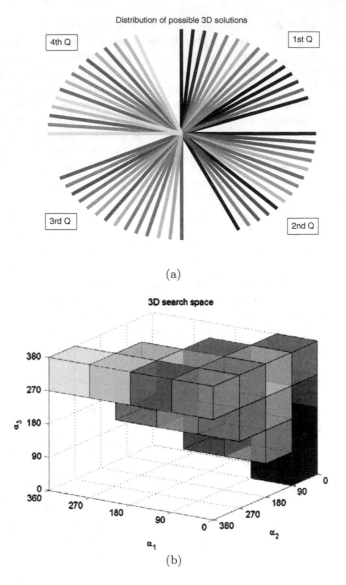

Fig. 1. Distribution of 3-beam directions by the four quadrants – 1(a) and the corresponding cubes in the search space $[0, 360]^3$ – 1(b).

value $f_*^k = f(\mathbf{x}_*^k) = \min\{f(\mathbf{x}_1^k), f(\mathbf{x}_2^k), \ldots, f(\mathbf{x}_N^k)\}$ is determined at each iteration k. The regions i where local searches remain active, **Active**$_i = 1$, correspond to the regions whose best objective function value is not worst than the overall best objective function value f_* within a defined threshold $p \geq 0$, i.e. $f_i \leq (1 + p)f_*$. The different local search procedures can always progress towards regions whose local procedures are not currently active. The accelerated multistart derivative-free algorithm is described in Algorithm 1.

Algorithm 1. Accelerated multistart derivative-free algorithm framework

---

Initialization:

  - Choose $\mathbf{x}_i^0 \in [0, 360]^n, i = 1, \ldots, N$;
  - Evaluate, in parallel, the objective function value at $\mathbf{x}_i^0 \in [0, 360]^n, i = 1, \ldots, N$;
  - Set $\mathbf{x}_i^{best} \leftarrow \mathbf{x}_i^0, i = 1, \ldots, N$ and $f_i^{best} \leftarrow f(\mathbf{x}_i^0), i = 1, \ldots, N$;
  - Determine the best initial beam ensemble $\mathbf{x}_*^0$ and the corresponding best initial objective function value $f_* \leftarrow f(\mathbf{x}_*^0)$;
  - Choose $p \geq 0$. Set **Active**$_i \leftarrow 1$ if $f_i^0 \leq (1 + p)f_*^0$, **Active**$_i \leftarrow 0$ otherwise;
  - Set $k \leftarrow 1$;
  - Choose a positive spanning set, a step size $s_i^1 > 0, i = 1, \ldots, N$ and a minimum step size s_{min} for PSM algorithm;

Iteration:

1. Using PSM, perform local search, in parallel, for the active regions;
2. For each active region i do
   If local search is successful, i.e. $f(\mathbf{x}_i^{k+1}) < f(\mathbf{x}_i^k)$ then
       If $\mathbf{x}_i^{k+1}$ remains in region i then
           $\mathbf{x}_i^{best} \leftarrow \mathbf{x}_i^{k+1}$;
           $f_i^{best} \leftarrow f(\mathbf{x}_i^{k+1})$;
       Else
           Active$_i \leftarrow 0$;
           Find $j \neq i$ where $\mathbf{x}_i^{k+1}$ is;
           If $f(\mathbf{x}_i^{k+1}) < f(\mathbf{x}_j^{best})$ then
               $\mathbf{x}_j^{best} \leftarrow \mathbf{x}_i^{k+1}$;
               $f_j^{best} \leftarrow f(\mathbf{x}_i^{k+1})$;
               Active$_j \leftarrow 1$;
   Else
       $s_i^{k+1} \leftarrow \frac{s_i^k}{2}$;
       If $s_i^{k+1} < s_{min}$ then
           Active$_i \leftarrow 0$;
3. Determine the overall best beam ensemble $\mathbf{x}_*^k$ and the corresponding overall best objective function value $f_* \leftarrow f(\mathbf{x}_*^k)$;
4. Set **Active**$_i \leftarrow 1$ if $f_i^k \leq (1 + p)f_*^k$, **Active**$_i \leftarrow 0$ otherwise;
5. Set $k \leftarrow k+1$. If any region is still active then return to step 1 for a new iteration.

---

3   Computational Results

A modern 8-core workstation was used to perform the computational tests. An in-house optimization platform written in MATLAB, named YARTOS, developed at Erasmus MC Cancer Institute in Rotterdam [6–8], was used to import DICOM images, create new structures, compute dosimetric input, compute the optimal fluence dose maps and compute/visualize dosimetric output. The YARTOS fluence map optimizer, 2pεc, was used to obtain the optimal value of the FMO problem required to drive our multistart derivative-free BAO framework. As the FMO model is treated as a black-box function, other FMO models can be easily coupled with this multistart BAO framework.

The PSM algorithm implemented for the local search procedure of the multi-start derivative-free framework used the maximal positive basis ($[I \quad -I]$), where I is the n-dimensional identity matrix. These directions correspond to the rotation of each beam direction clockwise and counter-clockwise for a certain amount (step-size) at each iteration. The initial step-size considered was $s^1 = 2^5 = 32$ and the minimal value allowed was one $s_{min} = 1$, defining the stopping criteria. As the step-size is halved at unsuccessful iterations, this choice of initial step-size implies that all beam directions will be integer until the termination criteria, when the step-size becomes inferior to one. No trial point was computed in the search step to avoid increasing the number of FMO evaluations.

Treatment plans with 7-beam angle directions, obtained using the multistart derivative-free framework and denoted *Multistart BAO*, were compared against treatment plans with 7-beam angle directions, obtained using iterative BAO and denoted *Iterative BAO*. These treatment plans were compared against treatment plans with 7-beam angle equispaced coplanar ensembles, denoted *Equi*, commonly used at IPOC and in clinical practice to treat intra-cranial tumor cases [1] and used here as benchmark.

The performance of a BAO algorithm should be evaluated using two criteria. While the main goal is to obtain the best objective function value possible, another important goal is to obtain a good solution as fast as possible. For the BAO problem, the computation of the optimal value of the FMO problem dominates the computational time, consuming more than 95 % of the overall computational time. In our accelerated multistart framework, the total number of FMO evaluations depends on the threshold p that defines which regions have active local search procedures. Obviously, by decreasing p the number of function evaluations will decrease. The main goal of this study is to acknowledge how much can we decrease that parameter, accelerating the multistart strategy, without significantly deteriorating the objective function value. We tested $p = 1, p = 0.5, p = 0.1, p = 0.05$ and $p = 0.01$ corresponding to consider, at each iteration, all regions active, regions for which the best objective function value is not 50 % (10 %, 5 % or 1 %, respectively) worst than the overall best objective function value. The results of these tests are displayed in Fig. 2. In Fig. 2(a) the relative FMO improvement comparing the benchmark beam ensemble, *Equi*, (0 % improvement) and the best treatment plan, *Multistart BAO* with $p = 1$, (100 % improvement) is displayed. The number of FMO evaluations required to obtain the corresponding solutions are displayed in Fig. 2(b). It can be seen that the results in terms of objective function value slowly deteriorate until $p = 0.1$. For smaller values of p the deterioration of the optimal FMO value is more accentuated. A steep decrease on the number of FMO evaluations is seen as the value of p decreases. The number of function evaluations for $p = 0.1$, e.g., is inferior to half of the number of function evaluations for $p = 1$ at a very small cost of objective function deterioration.

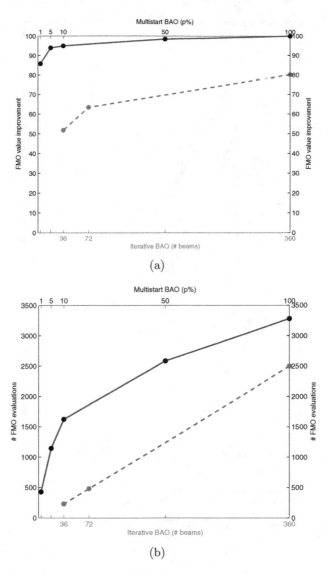

Fig. 2. Relative FMO value improvement – 2(a) and corresponding number of function evaluations – 2(b).

Iterative BAO results for different discrete samples of all possible beam irradiation directions are also depicted in Fig. 2. Gantry angles are typically discretized into equally spaced beam directions with a given angle increment. We considered angle increments of 10, 5 and 1 degrees, originating 3 discrete samples with 36, 72 and 360 beam angles. The iterative BAO greedy strategy drastically reduces the number of FMO problem evaluations compared to the combinatorial BAO, particularly for larger angle increments. However, since the search space is trun-

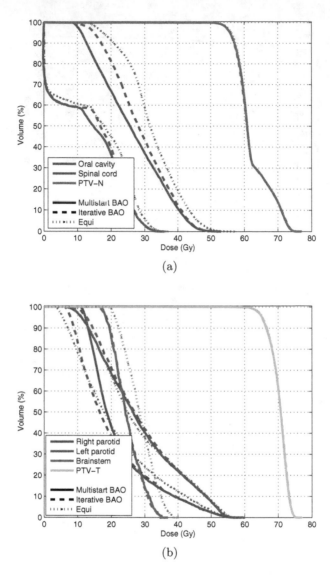

Fig. 3. Cumulative dose volume histogram comparing the results obtained by *Multistart BAO* with $p = 0.1$, *Iterative BAO* with 360 angles and *Equi* treatment plans.

cated at each iteration, at the very first iteration this strategy can disregard the best ensembles with n-beam directions. In Fig. 2(a) the consequences of that greedy strategy are clearly seen with worse FMO improvements even for larger numbers of FMO evaluations (for 360 beam angles).

In clinical practice, the quality of the results is typically assessed by cumulative dose-volume histograms (DVHs). The DVH displays the fraction of a structure's volume that receives at least a given dose. DVH results for *Multistart BAO* with $p = 0.1$, *Iterative BAO* with 360 angles and *Equi* treatment plans are displayed in Fig. 3. For clarity, DVH curves were divided into two figures and only spinal cord, brainstem, oral cavity, parotids and tumor volumes are displayed. The DVH curves show that for a similar tumor coverage, a better organ sparing is generically obtained by the treatment plans using the optimized beam directions, in particular by *Multistart BAO* treatment plans. In Fig. 3(a), *Multistart BAO* DVH curves show a better sparing of spinal cord and oral cavity compared to *Iterative BAO* DVH curves, corresponding to better maximum and mean doses for spinal cord and oral cavity (26.1 Gy vs 28.7 Gy and 33.6 Gy vs 34.3 Gy, respectively). At IPOC, two of the salivary glands, submandibular glands and sublingual glands, are included in the oral cavity structure. These two salivary glands, along with parotids, are very important in saliva production. Thus, the enhanced oral cavity sparing is of the utmost interest to prevent xerostomia caused by over-irradiation of salivary glands. In Fig. 3(b), both *Multistart BAO* and *Iterative BAO* DVH curves show a better sparing of brainstem compared to *Equi* DVH curves, corresponding to an improvement of 3 Gy on the brainstem maximum dose. The significant sparing obtained for brainstem is very important, e.g. for re-irradiation cases. Parotids show similar results for all treatment plans.

4   Conclusions and Future Work

Multistart methods with local search procedures are globally convergent [9] and their interest and application fields continue to rise [10]. For an extremely challenging non-convex optimization problem as the BAO problem, a multistart derivative-free framework is a suitable approach. This approach combines a global strategy for sampling the search space with a local strategy for improving the sampled solutions. While the local strategy using PSM proved, in previous works, to be successful in improving locally beam angle ensembles, requiring few function evaluations and avoiding local entrapment, the global strategy inherently requires a larger number of function evaluations. Several strategies were embedded in this methodology to reduce, as much as possible, the number of function evaluations, including the non-random strategy sketched to take advantage of a reduced search space and the use of a generalization of regions of attraction of a local minimum. Knowing in advance that some regions would most probably produce poor solutions, a strategy to explore locally the most promising regions at each iteration was also drafted and tested in this study.

The accelerated multistart derivative-free framework was tested using a clinical intra-cranial tumor treated at IPOC to acknowledge how much can we speed up the optimization process before significant deterioration of results occur. For the intra-cranial clinical case retrospectively tested, FMO optimal values of treatment plans obtained using *Multistart BAO* beam directions suffer small deterioration if up to 10 % of the most promising regions where locally explored at each iteration corresponding to a decrease superior to 50 % on the number of function evaluations. Thus, for the accelerated multistart derivative-free framework, the results indicate that a good compromise between the number of function evaluations and the quality of the solutions is the threshold $p = 0.1$. Furthermore, treatment plans obtained using *Multistart BAO* beam directions clearly outperform results obtained by *Iterative BAO* treatment plans even when the number of function evaluations was large (360 beam angle set).

As future work, different strategies for accelerating the proposed multistart derivative-free framework will be tested including the truncation of the number of iterations of the FMO. The multistart framework presented is tailored for coplanar beam directions and in future work this framework will be extended to include noncoplanar beam directions as well.

Acknowledgements. This work has been supported by the Fundação para a Ciência e a Tecnologia (FCT) under project grant UID/MULTI/00308/2013. We would like to show gratitude to Ben Heijmen and Sebastiaan Breedveld for giving us permission to install Yartos.

References

1. Aleman, D.M., Kumar, A., Ahuja, R.K., Romeijn, H.E., Dempsey, J.F.: Neighborhood search approaches to beam orientation optimization in intensity modulated radiation therapy treatment planning. J. Global Optim. **42**, 587–607 (2008)
2. Aleman, D.M., Romeijn, H.E., Dempsey, J.F.: A response surface approach to beam orientation optimization in intensity modulated radiation therapy treatment planning. INFORMS J. Comput. Computat. Biol. Med. Appl. **21**, 62–76 (2009)
3. Bangert, M., Ziegenhein, P., Oelfke, U.: Characterizing the combinatorial beam angle selection problem. Phys. Med. Biol. **57**, 6707–6723 (2012)
4. Bangert, M., Ziegenhein, P., Oelfke, U.: Comparison of beam angle selection strategies for intracranial imrt. Med. Phys. **40**, 011716 (2013)
5. Bertsimas, D., Cacchiani, V., Craft, D., Nohadani, O.: A hybrid approach to beam angle optimization in intensity-modulated radiation therapy. Comput. Oper. Res. **40**, 2187–2197 (2013)
6. Breedveld, S., Storchi, P., Keijzer, M., Heemink, A.W., Heijmen, B.: A novel approach to multi-criteria inverse planning for IMRT. Phys. Med. Biol. **52**, 6339–6353 (2007)
7. Breedveld, S., Storchi, P., Heijmen, B.: The equivalence of multicriteria methods for radiotherapy plan optimization. Phys. Med. Biol. **54**, 7199–7209 (2009)
8. Breedveld, S., Storchi, P., Voet, P., Heijmen, B.: iCycle: integrated, multicriterial beam angle, and profile optimization for generation of coplanar and noncoplanar IMRT plans. Med. Phys. **39**, 951–963 (2012)
9. Brimberg, J., Hansen, P., Mladenovic, N.: Convergence of variable neighborhood search. Les Chaiers du Gerard (2004)
10. Mart, R., Resende, M.G.C., Ribeiro, C.C.: Multistart methods for combinatorial optimization. Eur. J. Oper. Res. **226**, 1–8 (2013)
11. Craft, D., Halabi, T., Shih, H., Bortfeld, T.: Approximating convex Pareto surfaces in multiobjective radiotherapy planning. Med. Phys. **33**, 3399–3407 (2006)
12. Craft, D.: Local beam angle optimization with linear programming and gradient search. Phys. Med. Biol. **52**, 127–135 (2007)

13. Dias, J., Rocha, H., Ferreira, B.C., Lopes, M.C.: A genetic algorithm with neural network fitness function evaluation for IMRT beam angle optimization. Cent. Eur. J. Oper. Res. **22**, 431–455 (2014)
14. Dias, J., Rocha, H., Ferreira, B.C., Lopes, M.C.: Simulated annealing applied to IMRT beam angle optimization: A computational study. Physica Medica **31**, 747–756 (2015)
15. Li, Y., Yao, D., Yao, J., Chen, W.: A particle swarm optimization algorithm for beam angle selection in intensity modulated radiotherapy planning. Phys. Med. Biol. **50**, 3491–3514 (2005)
16. Lim, G.J., Cao, W.: A two-phase method for selecting IMRT treatment beam angles: Branch-and-Prune and local neighborhood search. Eur. J. Oper. Res. **217**, 609–618 (2012)
17. Monz, M., Kufer, K.H., Bortfeld, T.R., Thieke, C.: Pareto navigation Algorithmic foundation of interactive multi-criteria IMRT planning. Phys. Med. Biol. **53**, 985–998 (2008)
18. Rocha, H., Dias, J., Ferreira, B.C., Lopes, M.C.: Selection of intensity modulated radiation therapy treatment beam directions using radial basis functions within a pattern search methods framework. J. Glob. Optim. **57**, 1065–1089 (2013)
19. Rocha, H., Dias, J., Ferreira, B.C., Lopes, M.C.: Beam angle optimization for intensity-modulated radiation therapy using a guided pattern search method. Phys. Med. Biol. **58**, 2939–2953 (2013)
20. Rocha, H., Dias, J., Ferreira, B.C., Lopes, M.C.: Pattern search methods framework for beam angle optimization in radiotherapy design. Appl. Math. Comput. **219**, 10853–10865 (2013)
21. Torczon, V.: On the convergence of pattern search algorithms. SIAM J. Optim. **7**, 1–25 (1997)

Collisional Energy Exchange in CO_2–N_2 Gaseous Mixtures

Andrea Lombardi[✉], Noelia Faginas-Lago, Grossi Gaia, Palazzetti Federico,
and Vincenzo Aquilanti

Dipartimento di Chimica, Biologia e Biotecnologie,
Università di Perugia, Perugia, Italy
ebiu2005@gmail.com, piovro@gmail.com,
{gaia.grossi,federico.palazzetti,vincenzo.aquilanti}@unipg.it
http://www.chm.unipg.it/gruppi?q=node/48

Abstract. The calculation of series of vibrational state-specific collision cross sections and rate coefficients over extended energy and temperature ranges is required to set up accurate kinetic models of gaseous
mixtures under nonequilibium conditions, such as those encountered in
hypersonic flows. Processes involving carbon oxides (particularly CO_2)
and N_2 have to be included of models designed to simulate and interpret important aspects of the Earth and planetary atmospheres, also in
connection to spacecraft's entry problems, where hypersonics flows are
involved. Here we summarize a theoretical approach to the calculation
of energy transfer cross section and rates for CO_2-N_2 mixtures, based
on classical trajectory simulations of the collision dynamics and on a
bond-bond semi-empirical description of the intermolecular interaction,
which includes the dependence of the intermolecular interaction on the
monomer deformations.

Keywords: Intermolecular interactions · Molecular dynamics · Carbon
oxides · Gas flows · Earth and planetary atmospheres

1 Introduction

Molecular collisions in gas phase determine the redistribution of energy among
the molecular degrees of freedom, permitting the establishment of equilibrium
conditions and the relaxation of systems towards dynamics and thermodynamics
favourite states. In turn, relaxation and state population, affected by the energy
transfer, influence the chemical processes which are sensitive to the internal
energy content of the molecules and to the collision energies. In general, what-
ever the conditions, the modeling of gas phase systems requires an accurate
description of the energy transfer processes, which have to be carefully represented within the kinetic models. This applies to many fields, such as combustion
chemistry, chemistry of plasmas, gases under flow conditions (e.g. in aircraft and
spacecraft design [1–4]).

© Springer International Publishing Switzerland 2016
O. Gervasi et al. (Eds.): ICCSA 2016, Part I, LNCS 9786, pp. 246–257, 2016.
DOI: 10.1007/978-3-319-42085-1_19

The energy transfer is strongly dependent on the features of the intermolecular potentials [5,6] and part of the efforts to improve the accuracy of models must be spent to obtain a realistic description of the interactions. A further key point is that energy transfer is state-specific in its nature, since it depends on the initial quantum states of the colliding molecules, and this dependency is more pronounced when excited states are involved, as for hypersonic flows and high temperature plasmas [4]. This is due to the distortion of the molecular geometry generated by rotations and vibrations, that modifies polarizabilities and charge distributions, inducing variations in the intermolecular interaction. It is therefore preferable that a suitable intermolecular potential energy surface (PES), that we want to be realistic, represents the interaction energy as a function of the internal coordinates of the molecules. Important attempts are currently being carried out in the gas dynamics community to develop kinetic models at a state-specific level [7,8].

Concerning the simulation of the collision processes, a pure quantum mechanical approach (e.g. close-coupling representation of the scattering equations) is limited to three-atom systems, since for larger systems the dimension of the vibro-rotational basis sets would be too large for the calculation to be feasible (see [9] for an example of semiclassical approach). Work has be done to improve the feasibility of pure quantum calculations, relying upon convenient choices of the coordinate system and of the related internal and rotational state basis set [10–14], however these calculations remain confined to few-atom systems and cannot be performed on wide ranges of conditions. On the other hand, semiclassical approaches [15–18] are also demanding with respect to computing resources. As a consequence, calculations based on classical trajectories are the only viable approach to simulate collisions, in the wide range of conditions required to set up a data base of state-to-state vibrational exchange cross sections and thermal rate coefficients, to be exploited in state-to-state kinetic models.

In the present paper, we focus on gaseous systems containing carbon dioxide and nitrogen molecules, which are important components of the Earth and other planetary atmospheres (in Mars and Venus CO_2 and N_2 are both major components) [19,20]. Here we take care of the interactions adopting an effective formulation of the intermolecular interaction potential, based on the so called bond-bond approach (see [21–24] and references therein), which expresses the intermolecular interaction in terms of bond properties and parameters characterizing the internal molecular structure, such as charge distributions and polarizabilities [25–29].

The bond-bond based potential energy surfaces are versatile and usable in different environments and under different conditions, being also extensible to the description of large molecular systems (see for instance Refs. [25,30–40] and Refs. [27,28,41–46]).

The paper is organized as follows. Background information and details of the PES are given in Sect. 2. In Sect. 3 a description of the method adopted for the simulation of the collisions is provided along with examples of typical results. Conclusions are discussed in Sect. 4.

2 Intermolecular Interactions by the Bond-Bond Approach

The intermolecular potential energy, V_{inter}, for a given pair of molecules, such as CO_2 and N_2, is conveniently represented as a sum of two contributions, as follows:

$$V_{inter} = V_{vdW} + V_{elect}, \qquad (1)$$

where V_{vdW} is the van der Waals (size repulsion plus dispersion-attraction) inter-action contribution and V_{elect} is the electrostatic potential. V_{elect} is a consequence of anisotropies in the charge distributions of the molecules, and tends, at the asymptote, to the permanent quadrupole-permanent quadrupole interaction. The two energy terms depend on the distance R between the centers of mass of the two molecules (say a for CO_2 and b for N_2), and on a set of angular coordinates Θ_a, Θ_b and Φ (Jacobi type) which define the mutual orientation of them. To obtain a realistic representation of the intermolecular interactions, the accurate knowledge (as a function of R) of the potential energy for a sufficient number of fixed orientations is necessary. For the CO_2–N_2 system it is convenient to consider the following configurations, expressed in terms of the Jacobi angles: $(\Theta_a, \Theta_b, \Phi) = (90°, 90°, 0°), (90°, 90°, 90°), (90°, 0°, 0°), (0°, 90°, 0°), (0°, 0°, 0°), (45°, 45°, 0°)$ and $(60°, 60°, 0°)$, referred to as H, X, T_a, T_a, I, S^{45} and S^{60}, respectively (see Ref. [29]).

According to the bond-bond approach, the van der Waals term, V_{vdW} of Eq. 1, is obtained as a sum of bond pair contributions (all those obtained pairing the molecular bonds) denoted here as V_{vdW}^i:

$$V_{vdW}(R, \Theta_a, \Theta_b, \Phi) = \sum_i^K V_{vdW}^i (R_i, \theta_{ai}, \theta_{bi}, \phi_i), \qquad (2)$$

where $K = 2$ corresponds to the number of all possible CO_2-N_2 bond pairs, the distance R_i is that between the reference points of the two bonds of the i-th pair and θ_{ai}, θ_{bi} and ϕ_i are the bond mutual orientation angles. The interaction contributions so formulated can be added by virtue of the additivity of the bond polarizability components (whose sum gives the molecular polarizability) the fundamental physical quantity to which the van der Waals interaction can be connected. It is worth to remark that this representation, although indirectly, accounts for three body effects [21], as indicated by the fact that each bond polarizability component differs from the sum of those of the free atoms.

The V_{vdW}^i terms are an extension of atom-bond terms, previously defined for the case of the intermolecular interactions of atom-diatom systems, discussed in Refs. [21,22] for a case of atom-molecule interactions. A key point that has to be mentioned, is that each of the bonds is considered as an independent diatomic sub-unit having characteristic polarizability and electronic charge distribution of cylindrical symmetry. For the N_2 molecule bond, the reference point has been placed at its geometric center, while for each CO bond it has been displaced of 0.1278 Å towards the oxygen atom because in this case the dispersion and the bond centers do not coincide [47].

The V_{vdW}^i terms are represented by means of the the Pirani potential function developed by Pirani et al. (see Refs. [22,48]):

$$\frac{V_{vdW}^i(R_i,\gamma_i)}{\varepsilon_i(\gamma_i)} = f(x_i) = \left[\frac{6}{n(x_i)-6}\left(\frac{1}{x_i}\right)^{n(x_i)} - \frac{n(x_i)}{n(x_i)-6}\left(\frac{1}{x_i}\right)^6\right] \quad (3)$$

where the reduced distance x_i is defined as

$$x_i = \frac{R_i}{R_{mi}(\gamma_i)} \quad (4)$$

γ_i denotes collectively the angles $(\theta_{ai},\theta_{bi},\phi_i)$, which define the mutual orientation of the bonds, ε_i and R_{mi} are respectively the well depth and position of the pair interaction. The above function has been proved to be much more realistic than the original Lennard-Jones(12, 6) one, with a more accurate size repulsion profile and long range dispersion attraction tail [22–24]. The exponent n in Eq. 3, contains both R_i and γ_i being defined by the following empirical equation [22]:

$$n(x_i) = \beta + 4.0x_i^2 \quad (5)$$

where the parameter β represents the hardness of the pair of the interacting particles. In the present case the value chosen for β is 9 for all the bond pairs.

The Pirani potential function (see Eq. 3) must depend on the orientation of the interacting bonds. This is achieved in the formulation by an expansion of the parameters ε_i and R_{mi} in terms of bipolar spherical harmonics $A^{L_1L_2L}(\gamma)$. Consequently, the $f(x_i)$ function, the reduced i-th bond-bond pair potential, has the same form for all the orientations (see Refs. [49–52]). The CO_2–N_2 system required an expansion to the fifth order:

$$\varepsilon_i(\gamma) = \varepsilon_i^{000} + \varepsilon_i^{202}A^{202}(\gamma) + \varepsilon_i^{022}A^{022}(\gamma) + \varepsilon_i^{220}A^{220}(\gamma) + \varepsilon_i^{222}A^{222}(\gamma) \quad (6)$$

$$R_{mi}(\gamma) = R_{mi}^{000} + R_{mi}^{202}A^{202}(\gamma) + R_{mi}^{022}A^{022}(\gamma) + R_{mi}^{220}A^{220}(\gamma) + R_{mi}^{222}A^{222}(\gamma). \quad (7)$$

A method to estimate, at fixed γ_i values (defining selected limiting configurations), the ε_i and R_{mi} parameters from diatomic (or molecular bond) polarizability values is illustrated in Appendix A of Ref. [53].

The five limiting configurations (mutual orientations) considered for each i-th bond pair are the following: $H_i(\theta_{ai}=90°,\theta_{bi}=90°,\phi_i=0°)$, $X_i(\theta_{ai}=90°,\theta_{bi}=90°,\phi_i=90°)$, $T_{ai}(\theta_{ai}=90°,\theta_{bi}=0°,\phi_i=0°)$, $T_{bi}(\theta_{ai}=0°,\theta_{bi}=90°,\phi_i=0°)$, and $I_i(\theta_{ai}=0°,\theta_{bi}=0°,\phi_i=0°)$ [54]. A simple inversion of the Eqs. 6 and 7 [53] gives the expansion coefficients $\varepsilon_i^{L_1L_2L}$ and $R_{mi}^{L_1L_2L}$ for the five selected geometries of each i-th bond-bond pair, if ε_i and R_{mi} have been previously determined.

Once the above coefficients are known, a tentative full dimensional PES can be promptly generated and subsequently refined by fine tuning the ε_i and R_{mi}

parameters by means of fitting of experimental data and/or comparison with accurate ab initio electronic structure calculations.

The electrostatic potential, the V_{elect} term of Eq. 1, is conveniently defined as a sum of Coulomb terms, corresponding to point charge interactions, as follows:

$$V_{elect}(R, \Theta_a, \Theta_b, \Phi) = \sum_{jk} \frac{q_{ja} q_{kb}}{r_{jk}} \tag{8}$$

where q_{ja} and q_{kb} are the point charges for monomers a and b, respectively, and r_{jk} the distances between them. Note that the molecular charge distributions are such that the corresponding calculated molecular electric quadrupoles are reproduced. For both molecules, a linear distribution of charges has been adopted. In the case of CO_2 we have chosen the charge distribution previously used in Refs. [25, 26], made up by five charges, one for each atom and each bond of the molecule. The charge distribution of N_2 has been modelled by three charges placed on the N atoms and at the center of the N–N bond.

For charge values q_i (as well as for the corresponding position r_i) we took for CO_2 the values reported in Ref. [25] while for N_2 we have obtained them from the corresponding quadrupole moment and equilibrium distance reported in Refs. [29, 55].

2.1   Extension of the Bond-Bond Model

Removing the rigid constraint to the molecules from the PES so far described, the flexible PES, namely containing the dependence of V_{vdW} and V_{elec} (see Eq. 1) on the molecular deformations, is achieved making explicit the change of the bond polarizability due to the bond deformations, which can be used to derive the modulation of the V_{vdW} potential parameters. As seen above, the definition of the charge point positions determining the electrostatic contribution (see Eq. 8) is already dependent on bond length and angles. For the CO_2 deformations the empirical radial dependence of the C-O bond polarizability α from the corresponding bond length as well as that of the point charges distribution on the molecular symmetric (and asymmetric) stretching and bending, is used, as obtained in Ref. [25] (see Appendices A, B and C therein for more details).

For the N_2 molecule the empirical bond length dependence of the polarizablity α reported in Appendix B of Ref. [53], has been used to introduce the dependence on the N_2 bond length. Details can be found in Ref. [29].

3   Molecular Dynamics Simulations

The intermolecular interaction PES obtained, permits to run simulations of the CO_2–N_2 collision processes, by classical trajectories, from which collision cross sections, probabilities and rate coefficients for the energy transfer involving specific molecular vibrational states can be calculated. As anticipated above, the distorting effect of the vibrations on the molecular geometry of the system and on the related charge distribution, is taken into account by the PES improving the realism of the model.

To calculate state-to-state energy transfer cross sections, rates and probabilities, the intermolecular potential energy surface described in previous sections (see Eq. 1) must be coupled to an intramolecular potential energy surface, say V_{intra}, to give a full dimensional total potential energy:

$$V(R, \Theta_a, \Theta_b, \Phi, \mathbf{q}) = V_{intra}(\mathbf{q}) + V_{inter}(R, \Theta_a, \Theta_b, \Phi; \mathbf{q}) \tag{9}$$

where $\mathbf{q}$ denotes collectively the internal coordinates of CO_2 and N_2 and V_{inter} depends parametrically on $\mathbf{q}$. In the present work, a many-body expansion type potential function has been used for CO_2 (see [56]), while for N_2 a Morse potential has been adopted.

In the trajectories, the initial vibrational states of the molecules are explicitly selected, while the initial rotational states and the relative velocity of the colliding partners are assigned by sampling the related Boltzmann distributions. In order to obtain energy transfer probabilities specific for vibrational states, but averaged over rotation and translation (a choice based on the different time scales of energy exchange processes involving vibrations, rotations and translations, see Ref. [29] for a discussion of this point) a rotational temperature T_{rot} is introduced and set equal to the translational temperature T. These two values are then used to generate trajectory initial conditions as follows: a collision energy is chosen within the Boltzmann distribution at temperature T; the initial rotational angular momenta of the molecules are sampled from a Boltzmann distribution at the chosen rotational temperature T_{rot}, and the corresponding vectors randomly oriented; the initial vibrational states of the two molecules are generated to agree to the selected vibrational quantum numbers (for the two molecules a and b); initial coordinates and momenta for the relative motion are generated and the impact parameter b is given a value from 0 to b_{max}, the maximum impact parameter; the initial separation of the two molecules is set at a value large enough to make the intermolecular interaction negligible; finally, the molecules are given a random mutual orientation. The vibrational state of the linear CO_2 molecule is defined by three quantum numbers, neglecting the vibrational angular momentum quantum number, on the basis of assumptions discussed in previous works reporting calculations involving CO_2 [25, 27, 28].

According to this scheme, each bunch is labelled by the temperatures T and T_{rot} ($=T$), and by the initial vibrational quantum numbers of the colliding CO_2 and N_2 molecules. All the observables estimated by such calculations are therefore state specific for vibration and thermalized with respect to rotations and translations. The typical size of a bunch must be ~50000 trajectories to ensure acceptable accuracy, for the most common observables, such as energy distributions and state-to-state probabilities greater than 0.1%. The maximum impact parameter b_{max} is set equal to 18 Å, to account the long range of the intermolecular interactions.

Following the quasiclassical trajectory method, the thermal state to state probabilities P_{vw}, linking initial and final vibrational states v and w, can be obtained by the ratio $\frac{N_{vw}}{N_t}$ between the number of the related transition trajectories, N_{vw} and the total number of trajectories N_t. The cross section corresponding to the given vibrational transition, σ_{vw}, is defined as an integral of the probability P_{vw} over the impact parameter interval $[0, b_{max}]$

$$\sigma_{vw} = \int_0^{b_{max}} 2\pi b P_{vw}(b) db \tag{10}$$

where the opacity function, the dependence of P_{vw} on the impact parameter b, contains the information on the probability dependency on the short and long range parts of the intermolecular interactions. Due to the sampling of the initial conditions, the thermal probability $P_{vw}(T)$ for the same transition, which are averaged over translational and rotational Boltzmann distributions at temperatures T and T_{rot}, is simply the fraction of trajectories that lead to the transition $v \to w$, while the corresponding thermal cross section $\sigma_{vw}(T)$ is obtained by multiplying it by the available circular target area of the collisions (a function of b_{max}) in which the intermolecular potential is effective, as follows:

$$P_{vw}(T) = \frac{N_{vw}(T)}{N_t} \tag{11}$$

$$\sigma_{vw}(T) = P_{vw}(T)\pi b_{max}^2$$

where $N_{vw}(T)$ is the number of trajectories terminating with the given transition and N the total number of trajectories that are analyzed.

Moreover, being the cross section thermal, it is proportional to the corresponding state-to-state rate coefficient, and the following relationship holds:

$$k(T)_{vw} = \left(\frac{8kT}{\mu\pi}\right)^{\frac{1}{2}} \sigma_{vw}(T) \tag{12}$$

μ being the reduced mass of the $CO_2 + N_2$ system.

The classical final states (particle coordinates and velocities) obtained from trajectory calculations have to be expressed in a quantum fashion, meaning that the vibrational energies have to be expressed as quantum numbers, three and one for CO_2 and N_2 respectively. As a first approximation, the final vibrational energies of each vibrational mode of CO_2 can be calculated by projecting the final phase-space vectors into the CO_2 normal mode basis vectors (as obtained from the intramolecular potential energy function V_{intra}), assuming harmonic motion and separable vibrational modes. The harmonic vibration can be obtained from the Hessian eigenvalues of V_{intra} calculated at the minimum energy configuration of CO_2.

For the N_2 molecule, a vibrational quantum number n is deduced from the Einstein-Brillouin-Keller (EBK) quantization rule (see e.g. [57]). Then, the classical normal mode energies of CO_2 have to be discretized. Accordingly, the output of any bunch of trajectories is subject to a data-binning procedure in order to get corresponding quantum numbers. A uniform binning procedure is the simplest method, although not so refined, and consists in rounding to the closest integer the ratio between the energy content of each vibrational mode and the corresponding energy quantum. The trajectories terminating with molecular energy values lower than the zero point energy, which are physically meaningless, were discarded.

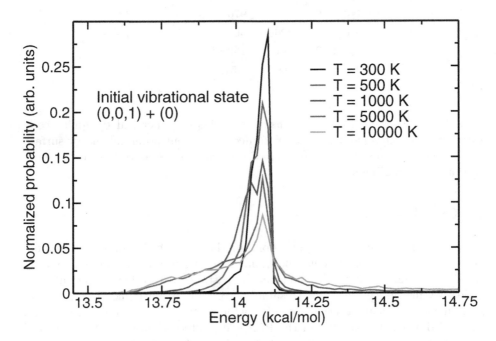

Fig. 1. Vibrational energy distributions of CO_2 molecules, as a result of $CO_2 + N_2$ collisions at different translational and rotational temperatures T = 300, 500, 1000, 5000 and 10000 K, with N_2 initially in its ground vibrational state and CO_2 with a quantum of energy in the symmetric stretching mode. (Color figure online)

As an example of collision observables that can be calculated by our PES, by running classical trajectories, Fig. 1 shows the vibrational energy distribution of the CO_2 molecules after $CO_2 + N_2$ collision events involving CO_2 initially excited with a quantum of energy in the asymmetric stretching mode and N_2 in its vibrational ground state, at various temperatures, ranging from T = 300 K to T = 10000 K, with equal values of rotational temperatures. As it can be seen, the model correctly reproduces the expected behaviour of an increasingly widespread vibrational energy distribution, which indicates that a greater translational and rotational energy content (corresponding to increasing temperatures) populates an increasing number of CO_2 excited vibrational states. Table 1 shows the main vibrational transition probabilities and cross sections for $CO_2 + N_2$ collisions at 5000 K, and processes listed indicate that only four quantum states are significantly populated, involving bending and symmetric stretching mode, initially not excited.

Table 1. Probabilities and cross sections for $CO_2 + N_2$ collisions at a translational and rotational temperature of 5000 K. The final vibrational CO_2 quantum numbers (v'_1, v'_2, v'_3, corresponding to stretching, bending and asymmetric stretching modes, respectively) and the m final vibrational quantum number of N_2, are indicated.

v'_1 v'_2 v'_3 m	Prob	Cross section (Å^2)
1 0 1 0	0.71185	970.56852
0 0 1 0	0.26890	367.88674
1 1 1 0	0.01133	15.38742
2 0 1 0	0.00238	0.51108

Calculations were carried out using the classical chemical dynamics code VENUS96 [58,59], modified in order to incorporate our potential energy surface and to generate the initial conditions of the trajectories according to the most appropriate sampling scheme.

4   Conclusions

This paper illustrates how the bond-bond method and the Pirani potential function to build up intermolecular potential energy surfaces can be successfully applied to simulate molecular collisions at a state-to-state level of detail. The trajectory method permits one to sample large sets of initial conditions, corresponding to physical situations covering wide temperatures and energy ranges. The full dimension character of the PES is expected to improve the realism of simulations at high temperatures, when excited vibrational states are involved and strong coupling of internal and intermolecular degrees of freedom takes place. The inherent parallelism of the trajectory method, is optimal for implementation in distributed calculation environments (e.g.Grid environments) for massive calculations of cross sections and rate constants of use in modeling of atmospheres and gaseous flows.

Acknowledgments. Andrea Lombardi acknowledges financial support from the Dipartimento di Chimica, Biologia e Biotecnologie dell'Universitá di Perugia (FRB grant), from MIUR PRIN 2010-2011 (contract 2010ERFKXL_002) and from "Fondazione Cassa Risparmio Perugia (Codice Progetto: 2015.0331.021 Ricerca Scientifica e Tecnologica)". Federico Palazzetti, Andrea Lombardi and Vincenzo Aquilanti acknowledge the Italian Ministry for Education, University and Research, MIUR, for financial support through SIR 2014 Scientific Independence for young Researchers (RBSI14U3VF) and FIRB 2013 Futuro in Ricerca (RBFR132WSM_003). Vincenzo Aquilanti thanks CAPES for the appointment as Professor Visitante Especial at Instituto de Fìsica, Universidade Federal de Bahia, Salvador (Brazil). Calculations were performed thanks to the support of the Virtual Organization COMPCHEM and the allocated computing time from the OU Supercomputing Center for Education &Research (OSCER) at the University of Oklahoma(OU).

References

1. Capitelli, M., Bruno, D., Colonna, G., D'Ammando, G., Esposito, F., Laricchiuta, A., Pietanza, L.D.: Rendiconti Lincei **22**, 201–210 (2011)
2. Panesi, M., Jaffe, R.L., Schwenke, D.W., Magin, T.E.: J. Chem. Phys. **138**, 044312 (2013)
3. Laganà, A., Lombardi, A., Pirani, F., Belmonte, P.G., Ortega, R.S., Armenise, I., Cacciatore, M., Esposito, F., Rutigliano, M.: Open Plasma Phys. J. **7**, 48–59 (2014)
4. Celiberto, R., et al.: Plasma Sources Sci. Tech. (2016)
5. Hirschfelder, J.O.: Advances in Chemical Physics: Intermolecular Forces, p. 12. Wiley, New York (1967)
6. Maitland, G.C., Rigby, M., Smith, E.B., Wakeham, W.A.: Intermolecular Forces. Clarendon Press, Oxford (1987)
7. Kustova, E.V., Nagnibeda, E.A.: Chem. Phys. Lett. **398**, 111–117 (2012)
8. Kustova, E.V., Kremer, G.M.: Chem. Phys. **445**, 82–94 (2014)
9. Faginas-Lago, N., Lombardi, A., Pacifici, L., Costantini, A.: Comput. Theor. Chem. **2013**, 103–107 (1022)
10. Lombardi, A., Ragni, M., De Fernandes, I.F.: Proceedings - 12th International Conference on Computational Science and Its Applications (ICCSA 2012), art. no. 6257613, pp. 77–82 (2012)
11. Palazzetti, F., Munusamy, E., Lombardi, A., Grossi, G., Aquilanti, V.: Int. J. Quantum Chem. **111**, 318–332 (2011)
12. Palacio, J.C.C., Abad, L.V., Lombardi, A., Aquilanti, V., Soneira, J.R.: Chem. Phys. **126**, 174701 (2008)
13. Aquilanti, V., Lombardi, A., Littlejohn, R.G.: Theor. Chem. Acc. **111**, 406 (2004)
14. Aquilanti, V., Novillo, E.C., Garcia, E., Lombardi, A., Sevryuk, M.B., Yurtsever, E.: Comput. Mater. Sci. **35**, 187–191 (2006)
15. Gargano, R., Barreto, P.R.P., Faginas Lago, N., Laganaà, A.: J. Chem. Phys. **125**, 114311–114316 (2006)
16. Laganà, A., Lago, N.F., Rampino, S., Huarte-Larrañaga, F., Garcia, E.: Phys. Scripta **78**, 058113 (2008)
17. Faginas-Lago, N., Huarte-Larrañaga, F., Laganà, A.: Chem. Phys. Lett. **464**, 249–255 (2008)
18. Rampino, S., Lago, N.F., Naga, F.H.-L., Lagan'a, A.: J. Comp. Chem. **33**, 708–714 (2012)
19. Khalil, M.A., Rasmussen, R.A.: Nature **332**, 242 (1988)
20. Palazzetti, F., Maciel, G.S., Lombardi, A., Grossi, G., Aquilanti, V.J.: Chin. Chem. Soc-Taip. **59**, 1045–52 (2012)
21. Pirani, F., Cappelletti, D., Liuti, G.: Chem. Phys. Lett. **350**, 286 (2001)
22. Pirani, F., Albertí, M., Castro, A., Teixidor, M.M., Cappelletti, D.: Chem. Phys. Lett. **37**, 394 (2004)
23. Pirani, F., Brizi, S., Roncaratti, L., Casavecchia, P., Cappelletti, D., Vecchiocattivi, F.: Phys. Chem. Chem. Phys. **10**, 5489 (2008)
24. Lombardi, A., Palazzetti, F.: J. Mol. Struct. (THEOCHEM) **852**, 22 (2008)
25. Bartolomei, M., Pirani, F., Laganà, A., Lombardi, A.: J. Comp. Chem. **33**, 1806 (2012)
26. Albertí, M., Pirani, F., Laganà, A.: Carbon dioxide clathrate hydrates: selective role of intermolecular interactions and action of the SDS catalyst. J. Phys. Chem. A. **117**, 6991–7000 (2013)

27. Lombardi, A., Faginas-Lago, N., Pacifici, L., Costantini, A.: J. Phys. Chem. A **117**, 11430–11440 (2013)
28. Lombardi, A., Faginas-Lago, N., Pacifici, L., Grossi, G.: J. Chem. Phys. **143**, 034307 (2015)
29. Lombardi, A., Pirani, F., Laganà, A., Bartolomei, M.: J. Comp. Chem. **37**, 1463–1475 (2016)
30. Lago, F.N., Albertì, M.: Eur. Phys. J. D **67**, 73 (2013)
31. Albertì, M., Lago, N.F., Pirani, F.: J. Phys. Chem. A **115**, 10871–10879 (2011)
32. Bruno, D., Catalfamo, C., Capitelli, M., Colonna, G., De Pascale, O., Diomede, P., Gorse, C., Laricchiuta, A., Longo, S., Giordano, D., Pirani, F.: Phys. Plasmas **17**, 112315 (2010)
33. Albertí, M., Huarte-Larrañaga, F., Aguilar, A., Lucas, J.M., Pirani, F.: Phys. Chem. Chem. Phys. **13**, 8251 (2011)
34. Lombardi, A., Lago, N.F., Laganà, A., Pirani, F., Falcinelli, S.: A bond-bond portable approach to intermolecular interactions: simulations for n-methylacetamide and carbon dioxide dimers. In: Murgante, B., Gervasi, O., Misra, S., Nedjah, N., Rocha, A.M.A.C., Taniar, D., Apduhan, B.O. (eds.) ICCSA 2012, Part I. LNCS, vol. 7333, pp. 387–400. Springer, Heidelberg (2012)
35. Lago, N.F., Albertí, M., Laganà, A., Lombardi, A.: Water $(H_2O)_m$ or Benzene $(C_6H_6)_n$ aggregates to solvate the K^+? In: Murgante, B., Misra, S., Carlini, M., Torre, C.M., Nguyen, H.-Q., Taniar, D., Apduhan, B.O., Gervasi, O. (eds.) ICCSA 2013, Part I. LNCS, vol. 7971, pp. 1–15. Springer, Heidelberg (2013)
36. Falcinelli, S., et al.: Modeling the intermolecular interactions and characterization of the dynamics of collisional autoionization processes. In: Murgante, B., Misra, S., Carlini, M., Torre, C.M., Nguyen, H.-Q., Taniar, D., Apduhan, B.O., Gervasi, O. (eds.) ICCSA 2013, Part I. LNCS, vol. 7971, pp. 69–83. Springer, Heidelberg (2013)
37. Lombardi, A., Laganà, A., Pirani, F., Palazzetti, F., Lago, N.F.: Carbon oxides in gas flows and earth and planetary atmospheres: state-to-state simulations of energy transfer and dissociation reactions. In: Murgante, B., Misra, S., Carlini, M., Torre, C.M., Nguyen, H.-Q., Taniar, D., Apduhan, B.O., Gervasi, O. (eds.) ICCSA 2013, Part II. LNCS, vol. 7972, pp. 17–31. Springer, Heidelberg (2013)
38. Faginas-Lago, N., Lombardi, A., Albertí, M., Grossi, G.J.: Mol. Liquids **204**, 192–197 (2015)
39. Faginas-Lago, N., Albertí, M., Costantini, A., Laganà, A., Lombardi, A., Pacifici, L.: J. Mol. Model. **20**, 2226 (2014)
40. Pacifici, L., Verdicchio, M., Faginas-Lago, N., Lombardi, A., Costantini, A.J.: Comput. Chem. **34**, 2668 (2013)
41. Faginas-Lago, N., Albertí, M., Laganà, A., Lombardi, A.: Ion-water cluster molecular dynamics using a semiempirical intermolecular potential. In: Gervasi, O., Murgante, B., Misra, S., Gavrilova, M.L., Rocha, A.M.A.C., Torre, C., Taniar, D., Apduhan, B.O. (eds.) ICCSA 2015. LNCS, vol. 9156, pp. 355–370. Springer, Heidelberg (2015)
42. Faginas Lago, N., Huarte-Larra ñaga, F., Albertí, M.: Eur. Phys. J. D **55**, 75–85 (2009)
43. Albertí, M., Faginas Lago, N., Pirani, F.: Chem. Phys. **399**, 232 (2012)
44. Albertí, M., Faginas Lago, N.: Ion size influence on the Ar solvation shells of Mi^+-C_6F_6 Clusters (M = Na, K, Rb, Cs). J. Phys. Chem. A **116**, 3094 (2012)
45. Albertí, M., Faginas Lago, N., Laganà, A., Pirani, F.: Phys. Chem. Chem. Phys. **13**, 8422–8432 (2011)

46. Lombardi, A., Faginas-Lago, N., Laganà, A.: Grid calculation tools for massive applications of collision dynamics simulations: carbon dioxide energy transfer. In: Murgante, B., Misra, S., Rocha, A.M.A.C., Torre, C., Rocha, J.G., Falcão, M.I., Taniar, D., Apduhan, B.O., Gervasi, O. (eds.) ICCSA 2014, Part I. LNCS, vol. 8579, pp. 627–639. Springer, Heidelberg (2014)

47. Schatz, G.C.: Fitting potential energy surfaces. In: Laganà, A., Riganelli, A. (eds.) Reaction and Molecular Dynamics. Lecture Notes in Chemistry, vol. 75, pp. 15–32. Springe, Heidelberg (2000)

48. Albernaz, A., Aquilanti, V., Barreto, P., Caglioti, C., Cruz, A.C., Grossi, G., Lombardi, A., Palazzetti, F.: J. Phys. Chem. A (2016). doi:10.1021/acs.jpca.6b01718

49. Pack, R.T.: Chem. Phys. Lett. **55**, 197 (1978)

50. Candori, R., Pirani, F., Vecchiocattivi, F.: Chem. Phys. Lett. **102**, 412 (1983)

51. Beneventi, L., Casavecchia, P., Volpi, G.G.: J. Chem. Phys. **85**, 7011 (1986)

52. Beneventi, L., Casavecchia, P., Pirani, F., Vecchiocattivi, F., Volpi, G.G., Brocks, G., van der Avoird, A., Heijmen, B., Reuss, J.: J. Chem. Phys. **95**, 195 (1991)

53. Cappelletti, D., Pirani, F., Bussery-Honvault, B., Gomez, L., Bartolomei, M.: Phys. Chem. Chem. Phys. **10**, 4281 (2008)

54. Gomez, L., Bussery-Honvault, B., Cauchy, T., Bartolomei, M., Cappelletti, D., Pirani, F.: Chem. Phys. Lett. **445**, 99 (2007)

55. Bartolomei, M., Carmona-Novillo, E., Hernández, M.I., Campos-Martínez, J., Hernández-Lamoneda, R.: J. Comput. Chem. **32**, 279 (2011)

56. Carter, S., Murrell, J.N.: Croat. Chem. Acta **57**, 355 (1984)

57. Gutzwiller, M.C.: Chaos in Classical and Quantum Mechanics. Springer, New York (1990)

58. Hase, W.L., Duchovic, R.J., Hu, X., Komornicki, A., Lim, K.F., Lu, D.-H., Peslherbe, G.H., Swamy, K.N., Linde, S.R.V., Zhu, L., Varandas, A.M., Wang, H., Wolf, R.J.: J. Quantum Chem. Prog. Exch. Bull. **16**, 671 (1996)

59. Hu, X., Hase, W.L., Pirraglia, T.: J. Comput. Chem. **1014**, 12 (1991)

A Theoretical and Computational Approach to a Semi-classical Model for Electron Spectroscopy Calculations in Collisional Autoionization Processes

Stefano Falcinelli[1](✉), Marzio Rosi[1], Fernando Pirani[2],
Noelia Faginas Lago[2], Andrea Nicoziani[2], and Franco Vecchiocattivi[1]

[1] Department of Civil and Environmental Engineering,
University of Perugia, Via G. Duranti 93, 06125 Perugia, Italy
{stefano.falcinelli,marzio.rosi}@unipg.it,
franco@vecchio.it
[2] Department of Chemistry, Biology and Biotechnologies,
University of Perugia, Via Elce di Sotto, 8, 06123 Perugia, Italy
{fernando.pirani,noelia.faginaslago,
andrea.nicoziani}@unipg.it

Abstract. The analysis of energy spectra of emitted electrons is of great relevance to understand the main characteristics of the potential energy surfaces and of the stereodynamics of the collisional autoionization processes. In this work we analyze the electron kinetic energy spectra obtained in our laboratory in high resolution crossed beam experiments. For such an analysis, a novel semi-classical method is proposed, that assumes ionization events as mostly occurring in the vicinities of the collision turning points. The potential energy driving the system in the relevant configurations of the entrance and exit channels, used in the spectrum simulation, has been formulated by the use of a semi-empirical method. The analysis puts clearly in evidence how different approaches of the metastable atom to the target molecule lead to ions in different electronic states.

Keywords: Semi-classical model · Collisional autoionization · Penning ionization electron spectroscopy · Metastable atoms · Molecular beam technique

1 Introduction

In general, we can represent collisional autoionization processes as follows:

$$A + B \rightarrow [A \cdots B]^*$$

where A and B are atoms or molecules, and $[A \cdots B]^*$ the intermediate excited collision complex in an autoionizing state evolving towards the electron ejection

$$[A \cdots B]^* \rightarrow [A \cdots B]^+ + e^-.$$

© Springer International Publishing Switzerland 2016
O. Gervasi et al. (Eds.): ICCSA 2016, Part I, LNCS 9786, pp. 258–272, 2016.
DOI: 10.1007/978-3-319-42085-1_20

The formed $[A \cdots B]^+$ ionic complex can then continue its dynamical evolution towards the final ionic species

$$[A \cdots B]^+ \rightarrow \text{ion products.}$$

Such collisional autoionization reactions are also commonly called as Penning ionization reactions after the early observation in 1927 by F.M. Penning [1].

In general, the autoionization processes are interesting both from a fundamental and applied point of view. For example, many applications are available to important fields: (i) radiation chemistry; (ii) development of excimer laser sources; (iii) combustion processes; (iv) plasma physics and chemistry [2–9]. In addition to low energy ionized plasmas and electric discharges, Penning ionization can play an important role also in planetary atmospheres and interstellar environments [10–13]. Relevance and the possible role played by ionic species in planetary atmospheres and in the interstellar medium are discussed in a number of recent papers [14–19]. For example, ions formed by UV photo-ionization or by collision with excited species, are extremely important for the transmission of radio and satellite signals, and they govern the chemistry of planetary ionospheres [12, 20, 21]. In fact, since more than twenty years, our studies have shown that, when atoms and simple molecules (like for example H, O, rare gas atoms, and N_2 molecules) are subjected to electrical discharges, they can produce a large amount of species in highly excited electronic metastable states (see the following Table 1) [22–26].

Table 1. The main characteristics of the most important metastable atomic and molecular species [4, 5, 12, 26].

Excited species	Excitation energy (eV)	Lifetime (s)
$H^*(2s^2S_{1/2})$	10.1988	0.14
$He^*(2^1S_0)$	20.6158	0.0196
$He^*(2^3S_1)$	19.8196	9000
$Ne^*(^3P_0)$	16.7154	430
$Ne^*(^3P_2)$	16.6191	24.4
$Ar^*(^3P_0)$	11.7232	44.9
$Ar^*(^3P_2)$	11.5484	55.9
$Kr^*(^3P_0)$	10.5624	0.49
$Kr^*(^3P_2)$	9.9152	85.1
$Xe^*(^3P_0)$	9.4472	0.078
$Xe^*(^3P_2)$	8.3153	150
$O^*(2p^33s^5S_2)$	9.146	1.85×10^{-4}
$H_2^*(c^3\Pi_u)$	11.764	1.02×10^{-3}
$N_2^*(A^3\Sigma_u^+)$	6.169	1.9
$N_2^*(A^1\Pi_g)$	8.549	1.20×10^{-4}
$N_2^*(E^3\Sigma_g^+)$	11.875	1.90×10^{-4}
$CO^*(a^3\Pi)$	6.010	$\geq 3 \times 10^{-3}$

Looking at data of Table 1, it is evident that metastable rare gas atoms can produce autoionization reactions by collisions with neutral atomic or molecular targets with high efficiency [12, 26, 27]. In particular, the presence in our terrestrial atmosphere of argon (the third component of air with its 0.93%), and of a minor amount of neon and helium (with a concentration of about 18 and 5 ppm, respectively), makes highly probable the production of ionized species by Penning ionization induced by gas metastable atoms. It has to be noted, that these processes are still now not fully considered in kinetic models used to describe the physical chemistry of such environments [12]. We are confident that our research should be able to clarify the possible role played by Penning ionization processes in Earth's atmosphere, giving a contribution in the attempt to explain some unclear phenomena, as for example the "sprites" formation, i.e. phenomena similar to lightning, showing a big influence on the radio waves propagation. Because these phenomena happens in certain meteorological conditions and in the presence of electrical discharges, we suggest that they could be related to the formation in the atmosphere of a strong local increase in the ionic concentration due to collisional autoionization processes induced by argon, neon and helium metastable atoms, that would facilitate the transmission of the radio waves by reflection.

Several experimental techniques are used to study the microscopic dynamics of these collisional autoionization processes. It is well established that the most valuable information about the dynamics of a collisional process is provided by molecular beam scattering experiments. In such cases one can study the process by detecting the metastable atoms, the electrons or the product ions. It has to be noted that the intermediate excited complex $[A \cdots B]^*$, having a lifetime so short ($\sim 10^{-15}$ s), can be considered as very close to the true "transition state" of the process, and consequently, the measurement of energy and momentum of its emitted electron provides a real spectroscopy of the transition state for such reactions [2, 3]. In Fig. 1 the main possible molecular beam experiments are schematically shown.

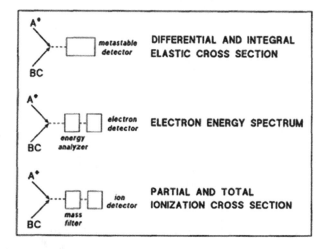

Fig. 1. Possible molecular beam experiments studying collisional autoionization reactions.

In our laboratory we are able to carry out all the three types of experiments sketched in Fig. 1. In particular, we performed integral elastic cross section experiments characterizing the interaction between hydrogen metastastable atoms $H^*(2 s)$ and simple atomic (Ar, Xe, Kr) and molecular species (O_2, Cl_2, CCl_4) [28, 29]. On the other hand, by using the crossed molecular beam technique coupled with mass spectrometry we were able to measure the total and partial ionization cross sections for many autoionization reactions as those induced by metastable rare gas atoms colliding with H_2, HD, and D_2 [30], HCl [31, 32], Cl_2 [33], CH_3Cl [24], and C_{60} [34, 35] molecules. More recently the insertion in our experimental apparatus (see next section) of the Penning Ionization Electron Spectroscopy (PIES) technique allowed to obtain important information concerning electronic structure, geometry and intermolecular interactions inside the intermediate collision complex in various autoionizing systems such as Ne^*-Kr [25], Ne^*-N_2O [6, 7], Ne^*-H_2O [36, 37], He^*, Ne^*-H_2S [38, 39], and NH_3 [40, 41].

2 Experimental

Our high resolution molecular beam scattering experiments have been done by using a crossed molecular beam apparatus able to perform: (i) integral elastic cross section experiments; (ii) mass spectrometry determinations (for the measurements of the total and partial ionization cross sections); (iii) electron kinetic energy analysis of emitted electrons (in the case of PIES spectra). The used experimental set up, described in detail in previous papers [25, 30, 34] and showed in Fig. 2, basically consists of a metastable hydrogen or rare gas atom beam crossing at right angles an effusive molecular beam.

The metastable atoms beam can be produced by two sources that can be used alternatively. The first one is a standard effusive source at room temperature, while the second one is a supersonic device (60 μm nozzle diameter; stagnation pressure 4–5 atm, producing a supersonic atomic beam with a rather well-defined velocity $\Delta v/v \approx 0.1$) that can be heated to different temperatures in the range 300–700 K. In both cases the metastable atoms are produced by electron bombardment at ~ 150 eV. The metastable atom velocity can be analyzed by a time-of-flight (TOF) technique. In general, the two metastable sources are used to carry out both the ionization cross section measurements as well as the PIES spectra. The first method allowed us to obtain the collision energy dependence for the ionization cross sections in a single experiment, while the second one, which requires an experiment for each single collision energy, permits to work with higher intensity signals and to record PIES spectra with relatively short accumulation time. In our apparatus we are able to perform in the same experimental conditions both collisional autoionization (by using metastable rare gas atoms as excited species inducing ionization) and photoionization (by using photons having almost the same energy of the relative metastable atoms) studies. For such a purpose, instead of the electron bombardment metastable atom beam source a microwave discharge in rare gas is exploited. In general, such a source was employed by using a ≈ 6 or ≈ 10 Torr inlet pressure of pure Helium or Neon, respectively. When we use such a device performing Penning ionization experiments with Ne^* metastable atoms, we are producing both metastable atoms and a high intensity of $Ne(I_{\alpha,\beta})$

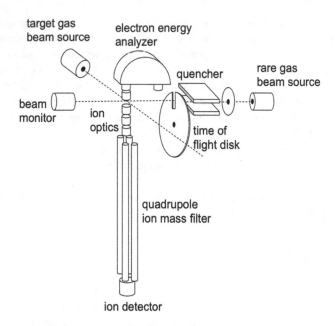

Fig. 2. A schematic sketch of the experimental set up used for the high resolution crossed molecular beam measurement of: (i) integral elastic cross sections; (ii) ionization cross sections; (iii) Penning ionization electron kinetic energy spectra.

photons, essentially of 73.6 and 74.4 nm wavelength respectively, in a α:β ratio of about 5.3, checked by photoelectron spectrometric measurements of Kr atoms [22, 25]. The use of the microwave discharge beam source allows us also to calibrate the electron transmission efficiency and the electron energy scale in our PIES measurements by recording photoionization spectra for the investigated systems and comparing them with data reported in literature [42].

In mass spectrometric determinations the ions produced in the beam crossing volume are extracted, focused, mass analyzed by a quadrupole filter and then detected by a channel electron multiplier.

The use of the TOF technique easily allows the separation of photo ions and Penning ions, the former being detected at practically zero delay time. Absolute values of the total ionization cross section for different species can be obtained by the measurement of relative ion intensities in the crossing volume with the same conditions of both metastable atom and target gas densities. This allows the various systems to be put on a relative scale which can be then normalized by reference to a known cross section such as, in the present case, the Ne^*–Ar absolute total ionization cross-section by West *et al.* [43].

For electron energy spectra determinations, a hemispherical electrostatic selector has been used [25, 36]. This is located above the beam crossing volume in opposite direction with respect to the quadrupole mass spectrometer. During electron spectroscopy experiments, the ion extracting field of the mass filter is maintained off, in order to do not perturb the electron energy. An appropriate electron optics is used to focus the electrons from the scattering volume to the hemispherical analyzer and, after energy selection, the electrons are detected by a channel electron multiplier. The resolution of our electron spectrometer is ~ 45 meV at a transmission energy of 3 eV, as determined by measuring the photoelectron spectra of Ar, O_2, and N_2 by He(I) radiation with the procedure described in a previous paper [6]. Spurious effects due to the geomagnetic field have been reduced to ≤ 20 mG by an appropriate μ-metal shielding.

3 A New Semi-classical Computational Approach for the Electron Kinetic Energy Spectra Calculation

The start point of the adopted theoretical model considers that a collisional autoionizing system formed by a metastable atom interacting with a triatomic molecule, as H_2O or H_2S, dynamically evolves on a multidimensional potential energy surface (PES). Under the assumption of rigid water and hydrogen sulfide molecules, the optical potential W, controlling ionization processes, can be written as the combination of a real V and an imaginary Γ part [4, 5]:

$$W(R,\vartheta, \varphi) = V(R,\vartheta, \varphi) - \frac{i}{2} \Gamma(R,\vartheta, \varphi) \tag{1}$$

In Eq. (1), R represents the distance of the metastable atom from the center of mass of the molecule, ϑ is the polar angle between the C_{2v} axis of H_2O or H_2S and R ($\vartheta = 0$ refers to the oxygen or sulfur side), and φ is the azimuthal angle ($\varphi = \pi/2$ describes the atom located on the H_2O or H_2S molecular plane). As we have recently demonstrated in the case of water [37] and hydrogen sulfide [38], most of the phenomena are expected to be controlled by the real part of the intermolecular potential, $V(R,\vartheta,\varphi)$, in six limiting configurations, whose interactions are indicated as V_{C2v-O} (or V_{C2v-S}) and V_{C2v-H} (with $\vartheta = 0$ and $\vartheta = \pi$, respectively), $V_\perp$ ($\vartheta = \pi/2$, $\varphi = 0$) and $V_{planar}(\vartheta = \pi/2, \varphi = \pi/2)$, where $V_\perp$ and V_{planar} are doubly degenerate.

In our semi-classical model describing the microscopic dynamics of collisional autoionization reactions, we assume that the ionization probability, at a certain R, is given by the ratio between the residence time of the system at that point and its lifetime, with respect to the ionization event, defined as the $\hbar/\Gamma$ ratio. Usual semi-classical treatments of the Penning ionization, giving total cross section and electron spectrum, have been discussed elsewhere [4, 5]. It has to be noted that such models are suitable for atom-atom systems only, and difficulties arise when we consider a polyatomic molecule colliding with an excited metastable atom. In the case of He^*-H_2O autoionizing system, a quasiclassical trajectory calculation has been used successfully by Ishida in modeling the PIES spectrum [44]. In general, quasiclassical trajectory calculations describing total cross sections for autoionizing collisions are quite well reliable assuming that the maximum probability for the electron ejection occurs essentially in the proximity of the turning point R_0 [4, 5]. Such an assumption is strongly reliable, because the local collision velocity at such a distance region tends to vanish, producing a very high autoionization probability. For the investigated Ne^*-H_2O, and H_2S systems, we applied this assumption concerning the turning point in the computation of related electron kinetic energy spectra.

Assuming that the ionization occurs dominantly at the turning point $R_0(E_{coll}, \vartheta, \varphi)$, where E_{coll} is the collision energy, the probability that an electron is emitted at R_0 with an energy E_{electr} is given by

$$P_{E_{electr}} = \frac{\Gamma(R_0)}{\hbar} \frac{\Delta t}{\Delta E_{electr}} \tag{2}$$

where Δt is the time spent by the system in a short range, ρ, around R_0. ΔE_{electr} accounts for the energy change of emitted electrons in such ρ range. Consequently, the time interval Δt can be computed by the following integral

$$\Delta t = \frac{2}{g} \int_{R_0}^{\rho} \left[1 - \frac{V(R)}{E_{coll}} - \frac{b^2}{R^2} \right]^{\frac{1}{2}} dR \tag{3}$$

where g indicates the relative collision asymptotic velocity, and it is assumed that ϑ and φ angles remain fixed around the turning point R_0. In our calculations the ρ value was chosen to be 0.12 Å, which provides a Δt longer than the typical electron emission time (of about 10^{-14}–10^{-15} s) and comparable with the rotational period of H_2O (typically $\sim 10^{-13}$ s). Therefore PIES spectra, for a certain orientation of the target molecule and for a given collision energy E_{coll}, is obtained by summing the probability $P_{Eelectr}$ for all impact parameters leading to the same E_{electr}, which is the difference between neutral (entrance channel) and ionic (exit channel) potentials at that distance. For a comparison with the measured electron spectra, such a difference has to be added of the asymptotic energy separation of the two neutral and ionic channels. The used potential energy curves for Ne^*-H_2O and Ne-H_2O^+ interactions, as calculated by applying a semi-empirical method developed in our laboratory and already calculated for such investigated system [37], are reported on the left panel of Fig. 3. In the Figure only the main $V(R, \vartheta, \varphi)$ configurations, V_{C2v-O}, V_{C2v-H}, and $V_\perp$, as defined above, are reported.

In Fig. 3 (on the right panel) are also shown some characteristic features of the investigated collision dynamics in the case of Ne^*-H_2O, as a meaningful example clarifying the computational procedure followed for electron kinetic energy spectra calculations. In the lower panel of the Figure, the turning point, at 300 meV collision energy, as a function of the impact parameter is reported, while in the panel just above, at those R_0 distances, is reported the difference between the neutral potential (in this case for the perpendicular orientation) and that of the ionic potential. When we calculate such a difference, and correct it for the asymptotic separation, we can obtain the energy of emitted electrons, with the relevant probability shown in the second panel from the top. In the upper panel is reported the total ionization probability in the case of the orientation here considered.

In Fig. 4 is shown the electron energy (without the correction for the asymptotic value) as a function of the turning point R_0 distance, for the three basic orientations of the H_2O molecule with respect to the approaching direction of the metastable neon atom (see Fig. 3 left-upper panel). It appears well evident, that a strong stereo-dynamic effect is present and controls the dependence of the emitted electron energy on the molecular orientation. A further interesting characteristic concerning the shape and

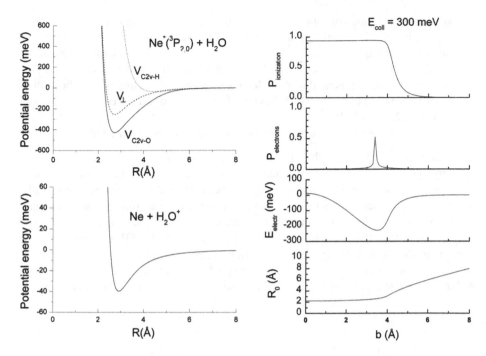

Fig. 3. Left side: (Upper panel) Plots of the $V_{C2v\text{-}H}$, $V_{C2v\text{-}O}$ and $V_\perp$ potential energy curves (see text) in the Ne^*-H_2O entrance neutral channels [37]; (lower panel) the interaction potential in the Ne^+-H_2O exit ionic channel [37]. Right side: Leading dynamics features for Ne^*-H_2O at a 300 meV relative collision energy. From the bottom to the top of the Figure, are plotted as a function of the classical impact parameter b the following panels reporting: (i) the turning point distance R_0; (ii) the ejected electron energy E_{electr}; (iii) the probability $P_{electrons}$ for the electron emission in the vicinities of R_0; (iv) and the probability $P_{ionization}$ for total ionization.

position of the electron energy peaks is appreciable in Fig. 5: here are reported the energy peaks of electrons as obtained for the $V_\perp$ and $V_{C2v\text{-}O}$ interaction potentials, that is for those orientations leading to the formation of the ground $X(^2B_1)$ and first excited $A(^2A_1)$ electronic states of the H_2O^+ ion, respectively.

Such an electron kinetic energy is not yet added with the asymptotic value of the neutral-ionic channel energy difference, therefore a zero value for that energy is what should be obtained when the intermolecular potential vanishes in the entrance and in the exit channel.

The peaks in the Figure are computed for two collision energy values (40 and 300 meV, being the minimum and maximum values investigated in our experiments) but the results appear to be indistinguishable. This observation indicates the following important features of the calculation:

(i) Peaks are very narrow and shifted to negative values (and more negative in the case of $V_{C2v\text{-}O}$ interaction); this is a direct consequence of the strong attractive behavior of the Ne^*-H_2O potential energy surface when the Neon metastable atom is approaching water molecules along the direction of the C_{2v} molecular symmetry axis towards the oxygen atom side [37] (see also Fig. 3 left-upper panel); in fact, it is well known that in PIES studies, when the position of the maximum of a recorded peak exhibits a sizeable negative shift with respect to the nominal energy (defined as the difference between the excitation energy of Ne[*] and the ionization potential of the relevant molecular state, this is an indication of a phenomenological attractive behavior of the interaction between the collisional partners [2, 37, 45, 46];

(ii) The peak shape do not change significantly with the collision energy.

A complete simulation of the PIES spectrum requires also to consider the width of the energy peaks. They are somewhat larger respect to those calculated, and reported in Fig. 5, because of several effects reducing the electron kinetic energy resolution, like the profile and velocity spread of two crossed beams, the real dimension of the scattering volume, and the instrumental resolution of the hemispherical electron energy analyzer. After that, the electron kinetic energy so calculated for PIES spectra has to be added with the asymptotic energy, as mentioned above, which is determined by the metastable neon atom electronic energy, (16.619 eV for the Ne[*](^{3}P$_2$) state, and 16.716 eV for the Ne[*](^{3}P$_0$) state, with these two states in 5:1 statistical ratio).

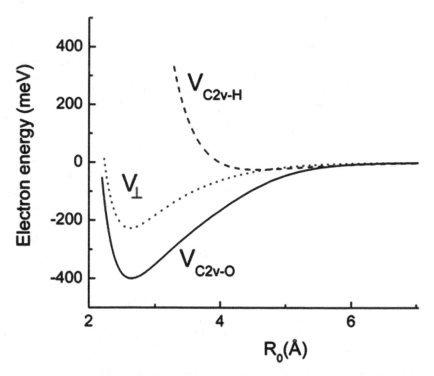

Fig. 4. The energy of emitted electrons for three basic configurations of the Ne[*]-H$_2$O ionizing complex (see text and Fig. 3 left-upper panel) as a function of turning point distance. It is assumed that the potential energy curves, in the entrance and in the exit channels, are asymptotically at the same level, to make more evident the role of the intermolecular interaction.

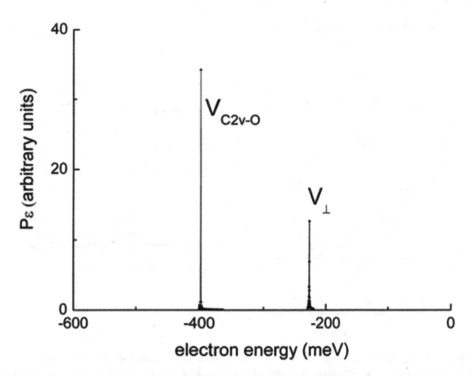

Fig. 5. The computed probability of emitting electrons, P_ε, as a function of the emission electron kinetic energy. The two peaks are related to the two potentials, V_{C2v-O} and $V_\perp$, associated to the relevant different orientations (as defined in the text), and for two collision energies 0.040 and 0.300 eV (maximum and minimum values experimentally investigated. The two peaks are very narrow and do not change their shape with the collision energy.

Necessary for such determination are also the energy levels of the formed H_2O^+ ion, which can be provided by previous accurate photoionization experiments performed in our laboratory (see data reported in Fig. 6 on the left-upper panel), and in previous experiments by Reutt, *et al.* [47]. To test the reliability of PIES spectra simulation procedure, we have applied our method to the calculation of the photoionization spectrum obtained by using Ne(I) photons, and we have compared it with our photoionization measurements reported in Fig. 6 (left-upper panel). For such a purpose, we considered that: (i) our used Ne(I) lamp produces photons at 73.6 and 74.4 nm in a 5.3 to 1 ratio (as discussed in the Experimental Section); a peak width of 50 meV full width at half maximum was reasonable for our simulations. In such a way the result of our calculation, shown in the left-upper panel of Fig. 6, appears to be very reliable.

Such a computational procedure has been adopted for the simulation of the PIES spectrum, with a peak width of 80 meV full width at half maximum, reported (dotted line) in the lower part of the left panel of Fig. 6. In the Figure such a simulated PIES spectrum (on the bottom) is compared with the experimental data (reported on the top). Also shown in the Figure is a simulation where the potential does not affect the shifting of the peaks (dashed line). It can be noted that the one "without potential" is shifted towards larger energies, while, when the interaction potential effect is included in the

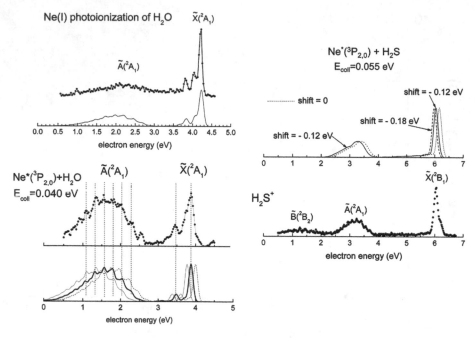

Fig. 6. Left panel (upper part): Photoionization electron spectrum of H_2O by Ne(I) photons, at 73.6 and 74.4 nm. The continuous line shifted below is our performed simulation and taking into account of our experimental conditions (see text). Left panel (lower part): The PIES spectrum measured for the Ne*-H_2O ionization at the collision energy of 40 meV (on the top) compared with a simulation (on the bottom) as obtained with the computational procedure proposed in this work. The dashed line, the one shifted towards higher energies, is the simulation obtained without the effect of the intermolecular potential and considering only the asymptotical energy difference between entrance and exit channels. The dotted line, the one shifted towards lower energies, is the one calculated by assuming a fixed orientation of the collision complex. The bold line is obtained by shifting the calculated line until a best reproduction of the bending structure for the $A(^2A_1)$ excited electronic state, and of the stretching structure for the ground $X(^2A_1)$ ion state. Right panel: on the bottom is reported the experimental PIES spectrum for Ne*+H_2S system at 55 meV collision energy; on the top is shown the simulation performed by using same procedure applied for Ne*-H_2O case (see text). The best fit of the data for the $X(^2B_1)$ state is given by -0.12 eV.

calculus, the PIES spectrum is shifted towards lower energies. Tentatively we found that a satisfactory fit is obtained with negative shifts of 110 meV for the $X(^2B_1)$ ground electronic state and of 180 meV for the $A(^2A_1)$ first excited state. Such energy shifts allow for a better reproduction of the stretching levels in the $X(^2B_1)$ peaks and the bending levels in the $A(^2A_1)$ band, and their difference is a clear indication of a strong anisotropic nature of the potential energy surface describing the Ne*-H_2O interaction [37]. In our previous investigations, we have found that the autoionization of hydrogenated simple molecules, like water, hydrogen sulfide, and ammonia by metastable helium and neon atoms strongly depends on how the molecule M is oriented with respect the approaching excited rare gas atom, determining the electronic state of the intermediate ionic complex [Ne···M]$^+$ and therefore its following reactivity. In fact, as discussed above, a direct probe of the anisotropy of the potential energy surface for Ne$^+$-H$_2$O system is provided by the comparative analysis of the features of measured electron energy spectra both in photoionization (Fig. 6 – left upper panel) and in Penning ionization (see Fig. 6 – left-lower panel) experiments. All the electron spectra reported of Fig. 6 appeared to be composed by two bands: one (at higher electron kinetic energies) for the formation of H$_2$O$^+$ ion (in case of photoionization measurements) and of [Ne···H$_2$O]$^+$ intermediate ionic complex (in Penning ionization determinations) in their ground X(^{2}B$_1$) electronic state, and another one (at lower electron energy) for the formation of H$_2$O$^+$ and [Ne···H$_2$O]$^+$ in their first excited A(^{2}A$_1$) electronic state. The production of these two electronic states corresponds to the removal of one electron from one of the two non bonding orbitals of the neutral molecule. The ground state X(^{2}B$_1$) H$_2$O$^+$ ion is formed by the removal of one electron from the 2pπ lone pair orbital, while for the formation of the excited A(^{2}A$_1$) H$_2$O$^+$ ion, the electron must be removed from the 3a$_1$ sp^2 lone pair orbital [36, 37]. Considering the "exchange mechanism" for Penning ionization (one outer shell electron of the molecule to be ionized is transferred into the inner shell hole of the excited atom; then the excess of energy produces the ejection of one electron from the collision complex [4, 5]), we can assert that the two bands in the spectra of Fig. 6 are related to ionization events occurring with two different geometries of the system: the X(^{2}B$_1$) band is originated from a collision complex with the metastable neon atom perpendicular to the plane of the molecule, while for the A(^{2}A$_1$) band, at the instant of the ionization, the metastable atom approaches water in the direction of the C$_{2v}$ symmetry axis towards the oxygen atom side. The negative energy shifts obtained by our analysis of the electron kinetic energy spectra (shift = −110 meV for the formation of H$_2$O$^+$ in its X(^{2}B$_1$) ground electronic state, and shift = −180 meV for the A(^{2}A$_1$) first excited state of H$_2$O$^+$) are in agreement with the different attractive behavior of the of the related V$_\perp$ and V$_{C2v-O}$ potential energy curves (reported in the left side and upper panel of Fig. 3), having the latter a well depth almost twice respect to the former.

The simulation of the PIES spectrum for Ne*+H$_2$S collisional autoionization has been carried out by using the same computational procedure already applied to Ne*+ H$_2$O, and discussed above. The spectrum is therefore calculated up by using the energy levels for H$_2$S$^+$ ionic states available in the literature, appropriately shifted because the effect of the intermolecular interaction driving the collision complex in the specific orientations promoting the selective formation of product ions in the final elec-tronic states. A Gaussian shape, related to the 45 meV energy resolution of our experiment, is then given to the levels. The H$_2$S$^+$ion levels are taken from the photoionization experiment by Hochlaf et al. [48], for the X(^{2}B$_1$) electronic ground state, and by Baltzer et al. [49] for the A(^{2}A$_1$) first excited state. Unfortunately we did not find suitable data available for the B(^{2}B$_2$) second excited electronic state of H$_2$S$^+$.

The experimental and simulated electron energy spectra for Ne*–H$_2$S are reported in the right panel of Fig. 6. In the lower part are shown the experimental data, while in the upper part of such a panel are displayed results of our calculations, where the dotted line concerns the spectrum without any energy shift (where shift = 0 it means to assume no effects due to intermolecular interaction). The calculation for a Ne* approach along the perpendicular direction respect to the H$_2$S molecular plane with the formation of

$H_2S^+(X^2B_1)$ provides a negative shift of -0.18 eV. On the other hand, the calculus performed for the $H_2S^+(A^2A_1)$ ionic excited state formation (related to Ne^* atom approaching H_2S along the C_{2v} molecular axis direction) provides a negative shift of -0.12 eV. By looking carefully at the PIES spectra of Fig. 6 (right panel), it appears that our simulation satisfactorily reproduces the position and the shape of the experimental $A(^2A_1)$ state band. On the contrary, the $X(^2B_1)$ state is too shifted with respect the experimental band. In such a case an energy shift of -0.12 eV appears to be more appropriate, and in reasonable agreement with the results already published that we have estimated on the basis of empirical methods [38, 50]. This is a clear confirmation of our preliminary estimates concerning the nature of the Ne^*-H_2S potential energy surface being less attractive and anisotropic respect to the Ne^*-H_2O case [38, 51].

4   Conclusions

In this paper is proposed a novel semi-classical method for the analysis and computation of kinetic energy spectra of emitted electrons in autoionizing collisions involving metastable atoms and simple hydrogenated molecules, as H_2O and H_2S. Such a theoretical method is based on the microscopic dynamical assumption that ionization events are mostly occurring in the vicinities of the collision turning points.

This computational procedure allowed to cast light on the basic features of the involved potential energy surfaces that control the stereo-selectivity in the collision dynamics [52], and to rationalize the changes observed between Ne^*+H_2O and Ne^*+H_2S autoionizing collisions. Further efforts will be done to characterize in a deeper detail the potential energy surface controlling the microscopic dynamics of Penning ionization collisions in order to clarify the role played in such reactions by the halogen and hydrogen bonds [53–55], and to extend the application of our semi-classical method calculating PIES spectra to other similar autoionizing systems.

Acknowledgments.   Financial contributions from the MIUR (Ministero dell'Istruzione, dell'Università e della Ricerca) through PRIN 2009 (Grant 2009W2W4YF_002) project is gratefully acknowledged. The authors thank "Fondazione Cassa di Risparmio di Perugia" for a partial support (Project code: 2014.0255.021).

References

1. Penning, F.M.: Naturwissenschaflen **15**, 818 (1927)
2. Hotop, H., Niehaus, A.A.: Z. Phys. **228**, 68 (1969)
3. Benz, A., Morgner, H.: Mol. Phys. **57**, 319–336 (1986)
4. Siska, P.E.: Rev. Mod. Phys. **65**, 337 (1993)
5. Brunetti, B., Vecchiocattivi, F.: Cluster Ions, pp. 359–445. Wiley & Sons Ltd., New York (1993). Ng, C.Y., Baer, T., Powis, I. (eds.)
6. Biondini, F., Brunetti, B.G., Candori, P., et al.: J. Chem. Phys. **122**, 164307 (2005)
7. Biondini, F., Brunetti, B.G., Candori, P., et al.: J. Chem. Phys. **122**, 164308 (2005)
8. Leonori, F., Balucani, N., Nevrly, V., Bergeat, A., et al.: J. Phys. Chem. Lett. **5**, 4213–4218 (2014)

9. Vanuzzo, G., Balucani, N., Leonori, F., Stranges, D., et al.: J. Phys. Chem. Lett. **7**, 1010–1015 (2016)
10. Rosi, M., Falcinelli, S., Balucani, N., Casavecchia, P., Leonori, F., Skouteris, D.: Theoretical study of reactions relevant for atmospheric models of titan: interaction of excited nitrogen atoms with small hydrocarbons. In: Murgante, B., Gervasi, O., Misra, S., Nedjah, N., Rocha, A.M.A., Taniar, D., Apduhan, B.O. (eds.) ICCSA 2012, Part I. LNCS, vol. 7333, pp. 331–344. Springer, Heidelberg (2012)
11. Alagia, M., Balucani, N., Candori, P., Falcinelli, S., Richter, R., Rosi, M., Pirani, F., Stranges, S., Vecchiocattivi, F.: Rendiconti lincei scienze fisiche e naturali **24**, 53–65 (2013)
12. Falcinelli, S., Pirani, F., Vecchiocattivi, F.: Atmosphere **6**, 299–317 (2015)
13. Falcinelli, S.: Penning ionization of simple molecules and their possible role in planetary atmospheres. In: Batzias, F. et al. (eds.): Recent Advances in Energy, Environment and Financial Planning – Mathematics and Computers in Science and Engineering Series **35**, pp. 84–92 (2014) © WSEAS press 2014. ISSN: 2227-4588; ISBN 978-960-474-400-8
14. Alagia, M., Bodo, E., Decleva, P., et al.: Physi. Chem. Chem. Phys. **15**, 1310–1318 (2013)
15. Falcinelli, S., Rosi, M., Candori, P., Vecchiocattivi, F., Farrar, J.M., Pirani, F., Balucani, N., Alagia, M., Richter, R., Stranges, S.: Plan. Space Sci. **99**, 149–157 (2014)
16. Falcinelli, S., et al.: The escape probability of some ions from mars and titan ionospheres. In: Murgante, B., Misra, S., Rocha, A.M.A., Torre, C., Rocha, J.G., Falcão, M.I., Taniar, D., Apduhan, B.O., Gervasi, O. (eds.) ICCSA 2014, Part I. LNCS, vol. 8579, pp. 554–570. Springer, Heidelberg (2014)
17. Pei, L., Carrascosa, E., Yang, N., Falcinelli, S., Farrar, J.M.: J. Phys. Chem. Lett. **6**, 1684–1689 (2015)
18. Schio, L., Li, C., Monti, S., Salén, P., Yatsyna, V.: Feifel, et al. Phys. Chem. Chem. Phys. **17**, 9040–9048 (2015)
19. Falcinelli, S., Rosi, M., et al.: Angular distributions of fragment ions produced by coulomb explosion of simple molecular dications of astrochemical interest. In: Gervasi, O., Murgante, B., Misra, S., Gavrilova, M.L., Rocha, A.M.A.C., Torre, C., Taniar, D., Apduhan, B.O. (eds.) ICCSA 2015. LNCS, vol. 9156, pp. 291–307. Springer, Heidelberg (2015)
20. Rosi, M., Falcinelli, S., Balucani, N., Casavecchia, P., Skouteris, D.: A theoretical study of formation routes and dimerization of methanimine and implications for the aerosols formation in the upper atmosphere of titan. In: Murgante, B., Misra, S., Carlini, M., Torre, C.M., Nguyen, H.-Q., Taniar, D., Apduhan, B.O., Gervasi, O. (eds.) ICCSA 2013, Part I. LNCS, vol. 7971, pp. 47–56. Springer, Heidelberg (2013)
21. Skouteris, D., Balucani, N., Faginas-Lago, N., et al.: Astron. Astrophys. **584**, A76 (2015)
22. Kraft, T., Bregel, T., Ganz, J., Harth, K., Ruf, M.-W., Hotop, H.: Z. Phys. D **10**, 473–481 (1988)
23. Ben Arfa, M., Lescop, B., Cherid, M., et al.: Chem. Phys. Lett. **308**, 71–77 (1999)
24. Brunetti, B.G., Candori, P., Falcinelli, S., Kasai, T., Ohoyama, H., Vecchiocattivi, F.: Phys. Chem. Chem. Phys. **3**, 807–810 (2001)
25. Brunetti, B.G., Candori, P., Falcinelli, S., Lescop, B., Liuti, G., Pirani, F., Vecchiocattivi, F.: Eur. Phys. J. D **38**, 21–27 (2006)
26. Hotop, H.: In atomic, molecular, and optical physics: atoms and molecules. In: Dunning, F. B., Hulet, R.G. (eds.) Academic Press, Inc.: San Diego CA, USA (1996); vol. 29B Experimental Methods in the Physical Sciences, 191–216 (1996) ISBN 0-12-475976-9
27. Falcinelli, S., Rosi, M., Candori, P., Vecchiocattivi, F., Bartocci, A., Lombardi, A., Lago, N. F., Pirani, F.: Modeling the intermolecular interactions and characterization of the dynamics of collisional autoionization processes. In: Murgante, B., Misra, S., Carlini, M., Torre, C.M., Nguyen, H.-Q., Taniar, D., Apduhan, B.O., Gervasi, O. (eds.) ICCSA 2013, Part I. LNCS, vol. 7971, pp. 69–83. Springer, Heidelberg (2013)

28. Brunetti, B.G., Falcinelli, S., Giaquinto, E., Sassara, A., Prieto-Manzanares, M., Vecchiocattivi, F.: Phys. Rev. A **52**, 855–858 (1995)
29. Brunetti, B.G., Candori, P., De Andres, J., Falcinelli, S., Stramaccia, M., Vecchiocattivi, F.: Chem. Phys. Lett. **290**, 17–23 (1998)
30. Brunetti, B., Falcinelli, S., Sassara, A., De Andres, J., Vecchiocattivi, F.: Chem. Phys. **209**, 205–216 (1996)
31. Aguilar Navarro, A., Brunetti, B., Falcinelli, S., Gonzalez, M., Vecchiocattivi, F.: J. Phys. **96**, 433–439 (1992)
32. Brunetti, B., Cambi, R., Falcinelli, S., Farrar, J.M., Vecchiocattivi, F.: J. Phys. Chem. **97**, 11877–11882 (1993)
33. Brunetti, B., Falcinelli, S., Paul, S., Vecchiocattivi, F., Volpi, G.G.: J. Chem. Soc., Faraday Trans. **89**, 1505–1509 (1993)
34. Brunetti, B., Candori, P., Ferramosche, R., et al.: Chem. Phys. Lett. **294**, 584–592 (1998)
35. Brunetti, B., Candori, P., Falcinelli, S., et al.: J. Phys. Chem. **104**, 5942–5945 (2000)
36. Brunetti, B., Candori, P., Cappelletti, D., et al.: Chem. Phys. Lett. **539–540**, 19–23 (2012)
37. Balucani, N., Bartocci, A., Brunetti, B., Candori, P., et al.: Chem. Phys. Lett. **546**, 34–39 (2012)
38. Falcinelli, S., Candori, P., Bettoni, M., Pirani, F., Vecchiocattivi, F.: J. Phys. Chem. A **118**, 6501–6506 (2014)
39. Falcinelli, S., Bartocci, A., Candori, P., Pirani, F., Vecchiocattivi, F.: Chem. Phys. Lett. **614**, 171–175 (2014)
40. Falcinelli, S., Bartocci, A., Cavalli, S., Pirani, F., Vecchiocattivi, F.: J. Chem. Phys. **143**, 164306 (2015)
41. Falcinelli, S., Bartocci, A., Cavalli, S., Pirani, F., Vecchiocattivi, F.: Chem. Eur. J. **22**, 764–771 (2016)
42. Kimura, K., Katsumata, S., Achiba, Y., Yamazaky, T., Iwata, S.: *Handbook of HeI photoelectron spectra of fundamental organic molecules.* Japan Scientific Societies Press, Tokyo (1981)
43. West, W.P., Cook, T.B., Dunning, F.B., Rundel, R.D., Stebbings, R.F.: J. Chem. Phys. **6**, 1237–1242 (1975)
44. Ishida, T.: J. Chem. Phys. **105**, 1392 (1996)
45. Cermák, V., Yencha, A.J.: J. Electron Spectr. Rel. Phenom. **11**, 67 (1977)
46. Sanders, R.H., Muschlitz, E.E.: Int. J. Mass Spectrom. Ion Phys. **23**, 99 (1977)
47. Reutt, J.E., Wang, L.S., Lee, Y.T., Shirley, D.A.: J. Chem. Phys. **85**, 6928 (1986)
48. Hochlaf, M., Weitzel, K.-M., Ng, C.Y.: J. Chem. Phys. **120**, 6944 (2004)
49. Baltzer, P., Karlsson, L., Lundqvist, M., Wannberg, B., Holland, D.M.P., Mac-Donald, M. A.: J. Chem. Phys. **195**, 403 (1995)
50. Candori, P., Falcinelli, S., Pirani, F., Tarantelli, F., Vecchiocattivi, F.: Chem. Phys. Lett. **436**, 322–326 (2007)
51. Falcinelli, S., Rosi, M., Stranges, D., Pirani, F., Vecchiocattivi, F.: J. Phys. Chem A., in press (2016) doi:10.1021/acs.jpca.6b00795
52. Brunetti, B.G., Candori, P., Falcinelli, S., Pirani, F., Vecchiocattivi, F.: J. Chem. Phys. **139**, 164305 (2013)
53. Cappelletti, D., Bartocci, A., Grandinetti, F., et al.: Chem. Eur. J. **21**, 6234–6240 (2015)
54. Bartocci, A., Belpassi, L., Cappelletti, D., et al.: J. Chem Phys. **142**, 184304 (2015)
55. Cappelletti, D., Candori, P., Falcinelli, S., Albertì, M., Pirani, F.: Chem. Phys. Lett. **545**, 14–20 (2012)

Solving Set Covering Problem
with Fireworks Explosion

Broderick Crawford[1,2,3], Ricardo Soto[1,4,5], Gonzalo Astudillo[1(✉)],
Eduardo Olguín[3], and Sanjay Misra[6]

[1] Pontificia Universidad Católica de Valparaíso, Valparaíso, Chile
{broderick.crawford,ricardo.soto}@ucv.cl,
gonzalo.astudillo.sepu@gmail.com
[2] Universidad Central de Chile, Santiago, Chile
[3] Facultad de Ingeniería y Tecnología, Universidad San Sebastián, Bellavista 7,
8420524 Santiago, Chile
[4] Universidad Autónoma de Chile, Santiago, Chile
[5] Universidad Cientifica del Sur, Lima, Peru
[6] Covenant University, Ota, Nigeria
sanjay.misra@covenantuniversity.edu.ng

Abstract. To solve the Set Covering Problem we will use a metaheuristic Fireworks Algorithm inspired by the fireworks explosion. Through the observation of the way that fireworks explode is much similar to the way that an individual searches the optimal solution in swarm. Fireworks algorithm (FWA) consists of four parts, i.e., the explosion operator, the mutation operator, the mapping rule and selection strategy. The Set Covering Problem is a formal model for many practical optimization problems. It consists in finding a subset of columns in a zero/one matrix such that they cover all the rows of the matrix at a minimum cost.

Keywords: Firework algorithm · Set covering problem · Metaheuristic

1 Introduction

The Set Covering Problem (SCP) is a classic problem that consists in finding a set of solutions which allow to cover a set of needs at the lowest cost possible. There are many applications of these kind of problems, the main ones are: location of services, files selection in a data bank, simplification of boolean expressions, balancing production lines, among others [1].

In the field of optimization, many algorithms have been developed to solve the SCP. Examples of these optimization algorithms include: Genetic Algorithm (GA) [2], Ant Colony Optimization (ACO) [3], Particle Swarm Optimization (PSO) [4], Firefly Algorithm [5,6], Shuffled Frog Leaping [7], and Cultural Algorithms [1] have been also successfully applied to solve the SCP.

Our proposal of algorithm uses fireworks behavior to solve optimization problems, it is called Fireworks Algorithm (FWA) [8].

© Springer International Publishing Switzerland 2016
O. Gervasi et al. (Eds.): ICCSA 2016, Part I, LNCS 9786, pp. 273–283, 2016.
DOI: 10.1007/978-3-319-42085-1_21

When a firework explodes, a shower of sparks is shown in the adjacent area. Those sparks will explode again and generate other shows of sparks in a smaller area. Gradually, the sparks will search (almost) the entire search space ending (eventually) good enough solutions. This work in addition to search a solution to the problem SCP, seeks to get better results for each instance of OR-Library. We use a new method of setting parameters, where we choose different parameters for each instances set. Moreover, in order to binarize we use eight transfer functions, 5 discretization techniques. These were combined with each other and be selected to deliver the best solution.

This document consists of five principal sections. In the Sect. 2, a brief description of Set Covering Problem. In the second section FWA is explained, highlighting the parts of this technique. In the third it is explained the algorithm used and implemented to solve the SCP. The penultimate section disclosed the results of the 65 instances, so in the last section conclusions from these results are presented.

2   Set Covering Problem

The SCP [9] can be formally defined as follows. Let $A = (a_{ij})$ be an m-row, n-column, zero-one matrix. We say that a column j can cover a row if $a_{ij} = 1$. Each column j is associated with a nonnegative real cost c_j. Let $I = \{i,...,m\}$ and $J = \{j,...,n\}$ be the row set and column set, respectively. The SCP calls for a minimum cost subset $S \subseteq J$, such that each row $i \in I$ is covered by at least one column $j \in S$. A mathematical model for the SCP is

$$v(\text{SCP}) = \min \sum_{j \in J} c_j x_j \tag{1}$$

subject to

$$\sum_{j \in J} a_{ij} x_j \geq 1, \quad \forall i \in I, \tag{2}$$

$$x_j \in \{0,1\}, \forall j \in J \tag{3}$$

The objective is to minimize the sum of the costs of the selected columns, where $x_j = 1$ if column j is in the solution, 0 otherwise. The constraints ensure that each row i is covered by at least one column.

The SCP has been applied to many real world problems such as crew scheduling [10], location of emergency facilities [11], production planning in industry [12], vehicle routing [13], ship scheduling [14], network attack or defense , assembly line balancing [15], traffic assignment in satellite communication systems [16], simplifying boolean expressions [17], the calculation of bounds in integer programs [18], information retrieval, political districting [19], stock cutting, crew scheduling problems in airlines [20] and other important real life situations. Because it has wide applicability, we deposit our interest in solving the SCP.

3   Fireworks Algorithm

When a firework is set off, a shower of sparks will fill the local space around the firework. In our opinion, the explosion process of a firework can be viewed as a search in the local space around a specific point where the firework is set off through the sparks generated in the explosion. Mimicking the process of setting fireworks.

After a firework exploded, the sparks are appeared around a location. The process of exploding can be treated as searching the neighbor area around a specific location. Inspired by fireworks in real world, fireworks algorithm is proposed. Fireworks algorithm utilizes N D-dimensional parameter vectors X_i^g as a basic population in each generation. Parameter i varied from 1 to N and parameter G stands for the index of generations. Every individual in the population explodes and generates sparks around him/her. The number of sparks and the amplitude of each individual are determined by certain strategies. Furthermore, a Gaussian explosion is used to generate sparks to keep the diversity of the population. Finally, the algorithm keeps the best individual in the population and selects the rest (N−1) individuals based on distance for next generation.

3.1   Components of FWA

Operator Explosion. To initialize the algorithm is necessary to generate N fireworks, thus generating sparks fireworks explosion. In FWA, the operator explosion is key and it plays an important role. The explosion operator including explosion strength, explosion amplitude and displacement operation. The explosion strength is a core operation in explosion operator. It simulates the way of explosion of fireworks in real life. When a firework blasts, the firework vanished in one second and then many small bursts appear around it. Fireworks algorithm first determines the number of sparks, then calculates the amplitudes of each explosion. Through the observations on the curves of some typical optimization functions, it can be seen that there are more points with good fitness values around the optima than that away from the optima. Therefore, the fireworks with better fitness values produce more sparks, avoiding swing around the optima but fail to locate it. For the fireworks with worse fitness values, their generated sparks are less in number and sparse in distribution, avoiding unnecessary computing. The fireworks with worse fitness values are used to explore the feasible space, preventing the algorithm from premature convergence. Fireworks algorithm determines the number and amplitude of the fireworks according to their fitness values, letting the fireworks with better fitness values produce more sparks within a smaller amplitude and vice versa. The Explosion Amplitude through the observation on the curves of some typical optimization functions, the points around the local optima and global optima always have better fitness values. Therefore, by controlling the explosion amplitude, the amplitude of the fireworks with better fitness values gradually reducing, leading fireworks algorithm find the local optima and global optima. On the contrary, the fireworks with worse fitness values will explore the optima through a large amplitude.

This is how the FWA controls the magnitude of the explosion amplitude. After the calculation of explosion amplitude, it is necessary to determine the displacement within the explosion amplitude. FWA uses the random displacement. In this way, each firework has its own specific explosion number and amplitude of sparks. FWA generates different random displacements within each amplitude to ensure the diversity of population. Through the explosion operator, each firework generates a shower of sparks, helping finding the global optimal of an optimization function, this is called displacement operation

Explosion Strength. In the explosion strength, i.e., the number of sparks, is determined as follows.

$$S_i = m\frac{Y_{max} - f(x_i)}{\sum_{j=1}(Y_{max} - f(x_i))} \tag{4}$$

where S_i is the number of sparks for each individual or firework, m is a constant stands for the total number of sparks and $Y_{max}s$ means the fitness value of the worst individual among the N individuals in the population. Function $f(x_i)$ represents the fitness for an individual x_i.

Explosion Amplitude. The explosion amplitude is defined below.

$$A_i = A\frac{f(x_i) - Y_{min}}{\sum_{j=1}(f(x_i) - Y_{min})} \tag{5}$$

where A_i denotes the amplitude of each individual, A is a constant as the sum of all amplitudes where initially the value of A is the difference between Ymax $-Y_{min}$., while Y_{min} means the fitness value of the best individual among the N individuals. The meaning of function $f(x_i)$ is the same as aforementioned in Eq. (4).

Displacement Operation. Displacement operation is to make displacement on each dimension of a firework and can be defined as

$$\triangle x_i^k = x_i^k + U(-A_i, A_i), \tag{6}$$

where $U(-A_i, A_i)$ denotes the uniform random number within the intervals of the amplitude A_i.

Mutation Operator. To further improve the diversity of a population, the Gaussian mutation is introduced to FWA. The way of producing sparks by Gaussian mutation is as follows: choose a firework from the current population, then apply Gaussian mutation to the firework in randomly selected dimensions. For Gaussian mutation, the new sparks are generated between the best firework and the selected fireworks. Still, Gaussian mutation may produce sparks exceed the feasible space. When a spark lies beyond the upper or lower boundary, the mapping rule will be carried out to map the spark to a new location within the feasible space.

Suppose the position of current individual be stated as x_i^k, where i varies from 1 to N and k denotes the current dimension. The sparks of Gaussian explosion are calculated by

$$x_i^k = x_i^k * g, \tag{7}$$

where g is a random number in Gaussian distribution with mean 1 and variance 1 such as

$$g = N(1; 1). \tag{8}$$

Mapping Ruler. If a firework is near the boundary of the feasible space, while its explosion amplitude covers both the feasible and infeasible space, the generated sparks may lie out of the feasible space. As such, the spark beyond the feasible space is useless. Therefore, it needs to be getting back into the feasible space. The mapping rule is used to deal with this situation. The mapping rule ensures that all sparks are in the feasible space. If there is any spark that is generated by a firework beyond the feasible space, it will be mapped back to the feasible space.

Selection Strategy. After applying the explosion operator, the mutation operator and the mapping rule, some of the generated sparks need to be selected and passed down to the next generation. The distance-based strategy is used in fireworks algorithm. In order to select the sparks for next generation, first of all, the best spark is always kept for next generation. And then, the other $(N-1)$ individuals are selected based on distance maintaining the diversity of the population. The individual that is farther from other individuals has more chance to be selected than those individuals near the other individuals.

4   Binary Firework Algorithm

In this section it is presented the functioning of the algorithm.

Step 1 Initialization the Firework parameters (initial amount of fireworks, mutation rate, number of iterations).

Step 2 Generate fireworks (at first, the number of fireworks will be given by the initial parameter).

Step 3 Calculate the amount and breadth of fireworks for each firework and also its fitness.

Step 4 Generate new solutions (fireworks) with the displacement operator equation.

Step 5 However, the operation of step 4, provides solutions to real numbers, and in this case (SCP) our solution must be in terms of 0 and 1. It is for this reason that the binarization of the solution is necessary. To fix this, we use the transfer functions (Table 1) that helps us define a chance to change an element of the solution from 1 to 0 , or vice versa.

Table 1. Transfer functions [21].

S-shape	V-shape
S1 $T(V_i^d) = \frac{1}{1+e^{-2V_i^d}}$	**V1** $T(V_i^d) = \left\| erf\left(\frac{\sqrt{\pi}}{2}V_i^d\right) \right\|$
S2 $T(V_i^d) = \frac{1}{1+e^{-V_i^d}}$	**V2** $T(V_i^d) = \left\| tanh(V_i^d) \right\|$
S3 $T(V_i^d) = \frac{1}{1+e^{\frac{-V_i^d}{2}}}$	**V3** $T(V_i^d) = \left\| \frac{V_i^d}{\sqrt{1+(V_i^d)^2}} \right\|$
S4 $T(V_i^d) = \frac{1}{1+e^{\frac{-V_i^d}{3}}}$	**V4** $T(V_i^d) = \left\| \frac{2}{\pi}arctan\left(\frac{\pi}{2}V_i^d\right) \right\|$

Besides the Transfer functions, 5 discretization methods were used, Roulette wheel (12), Complement (9), Set the Best (10), Standard (13), Statics probability (11), these are showed below:

Complement.

$$
x_i^d(t+1) = \begin{cases} complement(x_i^k) & \text{if } rand \le V_i^d(t+1) \\ 0 & \text{otherwise} \end{cases} \tag{9}
$$

Set the Best.

$$
x_i^d(t+1) = \begin{cases} x_{best}^k & \text{if } rand \le V_i^d(t+1) \\ 0 & \text{otherwise} \end{cases} \tag{10}
$$

Statics Probability.

$$
x_i^d(t+1) = \begin{cases} x_i^d & \text{if } V_i^d(t+1) \le \alpha \\ x_{best}^d & \text{if } \alpha \le V_i^d(t+1) \le \frac{1}{2}(1+\alpha) \\ x_1^d & \text{if } \frac{1}{2}(1+\alpha) \le V_i^d(t+1) \end{cases} \tag{11}
$$

Roulette.

$$p_i = \frac{f_i}{\sum_{j=1}^{k} f_j} \tag{12}$$

Standard.

$$x_i^d(t+1) = \begin{cases} 1 \text{ if } rand \leq V_i^d(t+1) \\ \\ 0 \text{ otherwise} \end{cases} \tag{13}$$

Step 6 Fireworks mutate randomly selected (the quantity index indicating initialized in step 1).

Step 7 Again with new fireworks (generated and mutated) is calculated the fitness and is necessary to keep the minimum to be compared in a next iteration. The number of iterations is given in the initialization parameters.

5 Solving the Set Covering Problem

Next is described the Solving SCP pseudocode:

Algorithm 1. $FWA()$

1: Generate N Fireworks
2: Establish number of iterations I
3: Calculate Fitness of each firework X_i with $i = (1, ..., N)$
4: Calculate maximum value of fitness
5: Calculate minimum value of fitness
6: **For** all iterations I **do**
7: Calculate number of spark of each firework and sum total of sparks s
8: calculate amplitude of each firework X_i
9: **For** $k = 1 \rightarrow s$ **do**
10: Generate sparks S_i from quantity and amplitude firework X_i
11: **end for;**
12: Binarize sparks S_i
13: // m is the number of sparks generated by Gaussian mutation
14: **for** $k = 1 \rightarrow m$ **do**
15: Select the best firework
16: Randomly select a firework X_i and generate a spark
17: **end for;**
18: $X_i = S_i$
19: Calculate the New Fitness of each firework X_i
20: save minimum of all fireworks
21: **end while;**

Table 2. Experimental results over the first 35 instances of SCP.

Instance	Method of discretization	Transfer functions	Opt.	Min.	Man.	Avg.	RPD
4.1	Standard	S1	429	436	437	436.7	1,61
4.2	Standard	S3	512	533	536	534.6	3,94
4.3	Standard	S1	516	526	526	526	1,90
4.4	The Best	V3	494	505	532	523.85	2,18
4.5	The Best	V2	512	517	527	525.7	0,97
4.6	The Best	V2	560	564	607	598.55	0,71
4.7	The Best	V2	430	434	447	444.2	0,92
4.8	Standard	V2	492	499	509	505,8	1,42
4.9	The Best	V3	641	670	697	691.9	4,33
4.10	The Best	V3	514	538	572	560.85	4,46
5.1	Roulette	V4	253	274	280	279.65	7,66
5.2	Standard	V2	302	312	317	314,4	3,31
5.3	Standard	V2	226	233	247	236,7	3,10
5.4	Standard	V2	242	246	251	248,5	1,65
5.5	Roulette	V2	211	219	225	224.7	3,65
5.6	Standard	V2	213	230	247	237,1	7,98
5.7	Standard	V2	293	311	315	314,9	6,14
5.8	Roulette	V1	288	302	316	314.8	4,64
5.9	Roulette	V1	279	292	315	312.65	4,45
5.10	Roulette	S1	265	275	280	279.05	3,64
6.1	Roulette	S2	138	147	152	151.45	6,12
6.2	Standard	V2	146	151	155	153,9	3,42
6.3	Standard	V2	145	150	160	156	3,45
6.4	Roulette	S1	131	134	140	139.5	2,24
6.5	Standard	V2	161	175	184	180,1	8,70
A.1	Standard	V2	253	257	261	260,4	1,58
A.2	Standard	V2	252	269	277	274	6,75
A.3	Roulette	S1	232	249	252	205,8	7,33
A.4	Roulette	S2	234	242	294	259.5	3,31
A.5	Standard	V2	236	239	241	240,3	1,27

6 Result

The FWA performance was evaluated experimentally using 65 SCP test instances from the OR-Library of Beasley [22]. The optimization algorithm was coded in Java 1.8 in Eclipse Luna 4.4.2 and executed on a Computer with 2.1 GHz AMD A10-5745M APU CPU and 8.0 GB of RAM under Windows 8 Operating System.

The Tables 2 and 3 shows the results of the 65 instances. The Transfer and Discretization columns reports the technique which the best results were obtained, that is, shows the best transfer function and the best discretization technique respectively. The Z_{Opt} column reports the optimal value or the best known solution for each instance. The Z_{Min} and Z_{Avg} columns report the lowest cost and the average of the best solutions obtained in 30 runs respectively. The quality of a solution is evaluated in terms of the percentage deviation relative

Table 3. Experimental results over the 30 instances of SCP.

Instance	Methods of discretization	Transfer functions	Opt.	Min.	Man.	Avg.	RPD
B.1	Standard	V2	69	79	86	83,7	14,49
B.2	Standard	V2	70	83	88	87,03	9,21
B.3	Roulette	S1	80	84	100	85.8	4,76
B.4	Standard	V2	79	83	84	83,9	5,06
B.5	Standard	V2	72	72	78	77,23	0,00
C.1	Standard	V2	227	234	235	234,8	3,08
C.2	Standard	V2	219	231	236	235,1	5,48
C.3	Standard	V2	243	264	270	269,2	8,64
C.4	Standard	V2	219	239	246	244,6	9,13
C.5	Standard	V2	215	219	223	221,5	1,86
D.1	Roulette	S1	60	61	92	63.95	1,64
D.2	Standard	V2	66	71	73	72,5	7,58
D.3	Roulette	V4	72	78	79	78.9	7,69
D.4	Standard	V2	62	65	68	63,3	4,84
D.5	Roulette	V2	61	64	66	65.55	4,69
NRE.1	The Best	V3	29	30	30	30	3,33
NRE.2	Roulette	V3	30	34	35	34.95	11,76
NRE.3	Standard	V2	27	30	34	32	11,11
NRE.4	Standard	V2	28	32	33	32,8	14,29
NRE.5	Standard	V2	28	29	30	29,9	3,57
NRF.1	Roulette	S1	29	30	112	36.5	3,33
NRF.2	Standard	V2	15	17	18	17,9	13,33
NRF.3	Roulette	S1	14	17	180	33.45	17,65
NRF.4	Roulette	S1	14	16	18	17.75	12,50
NRF.5	Roulette	S1	13	16	16	16	18,75
NRG.1	Standard	V2	176	193	196	194,6	9,66
NRG.2	Standard	V2	154	166	168	167,3	7,79
NRG.3	Standard	V2	166	170	180	179,4	2,41
NRG.4	Standard	V2	168	180	184	182,1	7,14
NRG.5	Standard	V2	168	185	188	186,9	10,12
NRH.1	Standard	V2	63	71	73	72,4	12,70
NRH.2	Standard	V2	63	66	67	66,9	4,76
NRH.3	Roulette	S2	59	66	69	68.85	10,61
NRH.4	Roulette	S2	58	66	68	67.8	12,12
NRH.5	Standard	V2	55	60	61	60,9	9,09

(RPD) of the solution reached Z_b and Z_{opt} (which can be either the optimal or the best known objective value), to compute RPD we use $Z = Min$, calculate as follows:

$$RPD = \frac{(Z - Z_{opt})}{Z_{opt}} * 100 \tag{14}$$

About the solutions obtained is reached only five B.5 optimal in the instance. Discretization methods with best results were the "Standard", "Roulette" and "Set The Best", on the other hand, the transfer functions with better results were family of V-shape (V1, V2, V3, V4). Discretization methods "Static Probability" and "Complement" not achieved a better result than the others.

7   Conclusions

From the experimental results it is concluded that the metaheuristic behaves good in almost all instances, highlighting , finding the best solution known (B.5) and in many other instancias it was a point of getting the best optimal known. We can also see that the RPD average of all instances is 6.11 %.

Considering the experiments, the initial parameters depend on the instance to solve the SCP, as we advance in the instances, it is necessary to increase the number of iterations increase the percentage of mutated sparks and the number of fireworks. This is because the firework must travel or generate more sparks to find a lower value, that is, we need to explore more on solutions. The effectiveness of the proposed approach was tested on benchmark problems and the obtained results sow that Binary Firework Algorithms is a good alternative to solve the SCP, being the main use of this metaheuristic for continous domains.

Acknowledgements. The author Broderick Crawford is supported by grant CON-ICYT/FONDECYT/REGULAR/1140897 and Ricardo Soto is supported by grant CONICYT/FONDECYT/INICIACION/1160455.

References

1. Crawford, B., Soto, R., Monfroy, E.: Cultural algorithms for the set covering problem. In: Tan, Y., Shi, Y., Mo, H. (eds.) ICSI 2013, Part II. LNCS, vol. 7929, pp. 27–34. Springer, Heidelberg (2013)
2. Goldberg, D.: Real-coded genetic algorithms, virtual alphabets, and blocking. Complex Syst. **5**, 139–167 (1990)
3. Amini, F., Ghaderi, P.: Hybridization of harmony search and ant colony optimization for optimal locating of structural dampers. Appl. Soft Comput. **13**(5), 2272–2280 (2013)
4. Crawford, B., Soto, R., Monfroy, E., Palma, W., Castro, C., Paredes, F.: Parameter tuning of a choice-a function based hyperheuristic using particle swarm optimization. Expert Syst. Appl. **40**(5), 1690–1695 (2013)
5. Crawford, B., Soto, R., Olivares-Suárez, M., Palma, W., Paredes, F., Olguin, E., Norero, E.: A binary coded firefly algorithm that solves the set covering problem. Rom. J. Inf. Sci. Technol. **17**, 252–264 (2014)
6. Crawford, B., Soto, R., Olivares-Suárez, M., Paredes, F.: A binary firefly algorithm for the set covering problem. In: Silhavy, R., Senkerik, R., Oplatkova, Z.K., Silhavy, P., Prokopova, Z. (eds.) Modern Trends and Techniques in Computer Science. Advances in Intelligent Systems and Computing, vol. 285, pp. 65–73. Springer, Heidelberg (2014)

7. Crawford, B., Soto, R., Peña, C., Palma, W., Johnson, F., Paredes, F.: Solving the set covering problem with a shuffled frog leaping algorithm. In: Nguyen, N.T., Trawiński, B., Kosala, R. (eds.) ACIIDS 2015. LNCS, vol. 9012, pp. 41–50. Springer, Heidelberg (2015)

8. Tan, Y.: Fireworks Algorithm: A Novel Swarm Intelligence Optimization Method. Springer, Heidelberg (2015)

9. Caprara, A., Fischetti, M., Toth, P.: Algorithms for the set covering problem. Ann. Oper. Res. **98**, 353–371 (2000)

10. Ali, A.I., Thiagarajan, H.: A network relaxation based enumeration algorithm for set partitioning. Eur. J. Oper. Res. **38**(1), 76–85 (1989)

11. Walker, W.: Using the set-covering problem to assign fire companies to fire houses. Oper. Res. **22**, 275–277 (1974)

12. Vasko, F.J., Wolf, F.E., Stott, K.L.: Optimal selection of ingot sizes via set covering. Oper. Res. **35**, 346–353 (1987)

13. Balinski, M.L., Quandt, R.E.: On an integer program for a delivery problem. Oper. Res. **12**(2), 300–304 (1964)

14. Fisher, M.L., Rosenwein, M.B.: An interactive optimization system for bulk-cargo ship scheduling. Naval Res. Logist. **36**(1), 27–42 (1989)

15. Freeman, B.A., Jucker, J.V.: The line balancing problem. J. Ind. Eng. **18**, 361–364 (1967)

16. Ribeiro, C.C., Minoux, M., Penna, M.C.: An optimal column-generation-with-ranking algorithm for very large scale set partitioning problems in traffic assignment. Eur. J. Oper. Res. **41**(2), 232–239 (1989)

17. Breuer, M.A.: Simplification of the covering problem with application to boolean expressions. J. Assoc. Comput. Mach. **17**, 166–181 (1970)

18. Christofides, N.: Zero-one programming using non-binary tree-search. Comput. J. **14**(4), 418–421 (1971)

19. Garfinkel, R.S., Nemhauser, G.L.: Optimal political districting by implicit enumeration techniques. Manag. Sci. **16**(8), B495–B508 (1970)

20. Housos, E., Elmroth, T.: Automatic optimization of subproblems in scheduling airline crews. Interfaces **27**(5), 68–77 (1997)

21. Mirjalili, S., Lewis, A.: S-shaped versus v-shaped transfer functions for binary particle swarm optimization. Swarm Evol. Comput. **9**, 1–14 (2013)

22. Beasley, J.: A lagrangian heuristic for set covering problems. Naval Res. Logist. **37**, 151–164 (1990)

Simulation of Space Charge Dynamics in High Intensive Beams on Hybrid Systems

Nataliia Kulabukhova$^{(\boxtimes)}$, Serge N. Andrianov, Alexander Bogdanov, and Alexander Degtyarev

Saint-Petersburg State University, Saint Petersburg, Russia
n.kulabukhova@spbu.ru

Abstract. The method for construction of analytical expressions for electric and magnetic fields for some set of the distributions of the charge density is described. These expressions are used for symbolic computation of the corresponding electric and magnetic fields generated by the beam during the evolution in accelerators. Here we focus on the use of the matrix form for Lie algebraic methods for calculating the beam dynamics in the presence of self-field of the beam. In particular, the corresponding calculations are based on the predictor-corrector method. The suggested approach allows not only to carry out numerical experiments, but also to provide accurate analytical analysis of the impact of different effects with the use of ready-made modules in accordance with the concept of Virtual Accelerator Laboratory. To simulate the large number of particle distributed resources for computations are used. Pros and cons of using described approach on hybrid systems are discussed. In particular, the investigation of overall performance of the predictor-corrector method is made.

1 Introduction

It is not necessary to say how popular and important accelerators are presently. There are a lot of facilities all over the world created for different purposes. The number of various software packages based on different approaches is even bigger that the accelerators we have. It is better to give a brief survey of methods that are used today to calculate the dynamics of beam with space charge.

High intensive beams are interesting from both side – the theoretical and practical point of view. More particles we have, more information about the beam we will get. On the other hand, intensive beams play a great role in medicine, when needed to irradiate only diseased cells, but not the healthy ones. But it is obvious that with intensity different effects of the beam that can not be denied occur. On of this is the forces of the self field of the particles. In the works [1–4] pay attention on the impact of space charge forces especially in the case that it can lead to the so called the filamentation effect or to the Halo (e.g. see Fig. 1). And for that purposes it is important to consider the space charge forces.

The work is supported by SPbSU 0.37.155.2014 and RFBR 16-07-01113A.

O. Gervasi et al. (Eds.): ICCSA 2016, Part I, LNCS 9786, pp. 284–295, 2016.
DOI: 10.1007/978-3-319-42085-1_22

Fig. 1. (a) Filamentation, (b) Halo

The method which is commonly used [5–7] is Particle-in-Cell method (PIC). The methods popularity caused its conceptual simplicity, the relative ease at which simulations may be implemented. Often, PIC simulations are implemented from first-principles (without the need for an approximate equation of state). However, these simulations often are computationally expensive with restrictive time step and mesh spacing limitations [8].

The Fortran-based environment COSY INFINITY [9] is also well known and used. The main use of the code lies in the field of nonlinear dynamics, where it is used for the computation of perturbation expansions of Poincare maps to high orders as well as their analysis based on normal forms and other methods.

Another approach is given by Alex Dragt and his team. In [10] is said that Lie algebraic methods may be used for particle tracking around or through a lattice and for analysis of linear and nonlinear lattice properties. When used for tracking, they are both exactly symplectic and extremely fast. Tracking can be performed element by element, lump by lump, or any mixture of the two. (A lump is a set of elements combined together and treated by a single transfer map.)

In addition to single-particle tracking, Lie algebraic methods may also be used to determine how particle phase-space distribution functions evolve under transport through both linear and nonlinear elements. These methods are useful for the self-consistent treatment of space-charge effects and for the study of how moments and emittances evolve.

MARYLIE [11] is a FORTRAN program for beam transport and tracking based on a Lie algebraic formulation of charged particle trajectory calculations. This software is useful for the design and evaluation of both linear transport systems and circulating storage rings. The program is able to compute transfer maps and trace rays through single or multiple beamline elements for the full six-dimensional phase space without the use of numerical integration or traditional matrix methods. The effects of high-order aberrations are computed as an integral part of the Lie algebra approach. All non-linearities, including chromatic effects, through third order are included [12].

The number of methods and software is not restricted by these once. There are TRANSPORT, BEAMBEAM3D, IMPACT-Z, IMPACT-T, MAD [13] and some others. But in all these methods the trajectory of one particle is calculated. And in case of intensive beams the number of particles to count is bigger then 1 billion. Though, the computer resources allow us to calculate large amount of data, the practice shows that it is better to have a parallel algorithm at the beginning than a good machine. That was our goal to make the algorithm that can be parallelized easily.

2   Lie Algebra in Accelerator Physics

The approach about which we will speak is very similar to the one on what MARYLIE is based. The evolution operator for dynamic systems is used in theoretical physics for a long time. This operator can be written in general form as Lie operator (see, for example, [10]):

$$\frac{d\mathcal{M}(\boldsymbol{U}(t), t|t_0)}{dt} = \mathcal{L}(\boldsymbol{U}(t), t) \circ \mathcal{M}(\boldsymbol{U}(t), t|t_0), \mathcal{M}(t_0|t_0) = \mathcal{I}d \ \forall \ t_0 \in [T, t_0]. \quad (1)$$

These operators define the Lie transformations $\mathcal{M}(\boldsymbol{U}(t), t|t_0)$, generated by an infinitesimal Lie operator $\mathcal{L}(\boldsymbol{U}(t), t)$ (the vector field of the dynamical system), where $\boldsymbol{U}(t)$ is a vector of control functions (for simplicity, we will omit the argument of $\boldsymbol{U}(t)$). Note that the Eq. (1) generally has the form of a nonautonomous linear operator equation. The equality 1 in the integral form has the following form

$$\mathcal{M}(t|t_0) = \mathcal{I}d + \int_{t_0}^{t} \mathcal{L}(\tau) \circ \mathcal{M}(t|\tau) d\tau. \quad (2)$$

The general solution can be written in the form of a chronological series of Volterra [14]

$$\mathcal{M}(t|t_0) = \mathcal{I}d + \sum_{k=1}^{\infty} \int_{t_0}^{t} \int_{t_0}^{\tau_1} \dots \int_{t_0}^{\tau_{k-1}} \mathcal{L}(\tau_k) \circ \mathcal{L}(\tau_{k-1}) \circ \dots \circ \mathcal{L}(\tau_1) \, d\tau_k \dots d\tau_1. \quad (3)$$

or using the Magnus presentation [15] this equality can be written as

$$\mathcal{M}(t|t_0) = \exp \mathcal{W}(t|t_0; \mathcal{L}). \quad (4)$$

Here $\mathcal{W}(t|t_0; \mathcal{L})$ – a new vector field, generated by the "old" vector field $\mathcal{L}$. In Ref. [14] analytical expressions (for step-by-step calculations) for the new operator are presented $\mathcal{W}(t|t_0; \mathcal{L})$ using "nested" series

$$\mathcal{W}(t|t_0) - \int\limits_{t_0}^{t} \mathcal{V}(\tau)d\tau + \alpha_1 \int\limits_{t_0}^{t} \left\{ \mathcal{L}(\tau), \int\limits_{t_0}^{\tau} \mathcal{L}(\tau') \right\} d\tau +$$

$$+ \alpha_1^2 \int\limits_{t_0}^{t} \left\{ \mathcal{L}(\tau), \int\limits_{t_0}^{\tau} \left\{ \mathcal{L}(\tau'), \int\limits_{t_0}^{\tau'} \mathcal{L}(\tau'')d\tau'' \right\} d\tau' \right\} d\tau +$$

$$+ \alpha_1\alpha_2 \int\limits_{t_0}^{t} \left\{ \left\{ \mathcal{L}(\tau), \int\limits_{t_0}^{\tau} \mathcal{L}(\tau')d\tau' \right\}, \int\limits_{t_0}^{\tau} \mathcal{L}(\tau')d\tau' \right\} d\tau + \dots \quad (5)$$

In the work [14] the necessary conditions for the convergence of the corresponding series as well as the convergence rate are described. Thus, these relations allow us to find correct solutions of nonlinear operator equations in the form of convergent series (under some relatively simple assumptions). However, the operator form for practical solutions cannot be used in computations, so we have to choose some functional basis (in our case we use the well-known Poincare-Witt basis), which can provide appropriate solutions in the form of the following equality:

$$\mathcal{M}(t|t_0) \circ \boldsymbol{X}_0 = \sum_{k=1}^{\infty} \mathbb{M}^{[1k]}(t|t_0) \boldsymbol{X}_0^{[k]}, \quad \boldsymbol{X}_0 = \boldsymbol{X}(t_0), \quad (6)$$

where $\boldsymbol{X}$, $\boldsymbol{X}_0$ are vectors of current and initial phase coordinates of a particle, $\mathbb{M}^{[1k]}(t|t_0)$, $k \geq 1$ are matrices (two dimensional arrays) responsible for the nonlinearity of k-th order in the solution of the equation of evolution. Thus, the task of investigating the evolution of a nonlinear system is reduced to the computation of the matrices $\mathbb{M}^{[1k]}(t|t_0)$ up to the necessary order of nonlinearity with the corresponding estimates of accuracy [14]. So in the absence of effect of space-charge, we can compute corresponding matrices in the nonlinear approximation step by step up to the desired order of nonlinearity. However, taking the space charge into account, the matrices $\mathbb{M}^{[1k]}(t|t_0)$ can be computed using the method of successive approximations, if necessary [14,16].

It should be noted, that the knowledge of the matrix $\mathbb{M}^{[1k]}$ up to the required order of nonlinearity allows to calculate the dynamics of the beam as an ensemble of particles with the given accuracy. Indeed, the equality (6) allows to describe the particle beam using various forms of its description. Let's consider an ensemble of particles $\mathfrak{M}_0$ consisting of N particles at some initial time. Then the beam evolution may be described by different methods:

- with the help of a matrix phase states beam $\mathbb{M}_0^N = \left\{ \boldsymbol{X}_0^1, \boldsymbol{X}_0^2, \dots, \boldsymbol{X}_0^N \right\}$,
- an envelopes matrix σ_0,
- or a particle distribution function $f(\boldsymbol{X}, 0) = f_0(\boldsymbol{X})$, $\boldsymbol{X}_0^k \in \mathfrak{M}_0$, $\forall k = \overline{1, N}$.

All these methods of particle beam description can be considered a base for forming information objects that characterize the state of the beam at the initial and current moments. About two last methods we will speak later.

On the next step we should introduce information objects that are responsible for evolution of the initial state. In our approach we use the matrices $M^{[1k]}$, $\forall k \geq 1$. These matrices can be calculated according to the Lie formalism [10, 14]. We should note that the group property for the evolution operator allows us to calculate the operator successively (step-by-step) for each control element of the lattice (here we refer to the control elements, such as dipoles, multipole lenses, free spaces and etc.) and for the accelerator system in the whole.

Moreover, the map that describes the impact of a control element can in turn be represented as a series of maps that reflect the impact of electromagnetic fields [10] up to nonlinearities of necessary order. These properties of the evolution operator and its consequent effect on control systems allows us to introduce information objects, each of which is responsible for mapping generated by a particular control element as a set of particular units in the defined sequence. So, for each control element (multipoles, drifts and etc.) we can calculate the necessary matrices $M^{[1k]}$ and then to construct these matrices for some lattices using some concatenation procedure (see, e.g. [17]).

Naturally, beside the corresponding objects, we have a set of mathematical rules with which they act on the data objects, characterizing the state of the beam in the initial or current moments. However, these objects themselves consist of a set of virtual subagents (for example, responsible for the fringing fields or some other characteristics of control fields).

In other words, we have an additional internal subset of subagents [18] designed to study the effects of additional characteristics of the transport system on beam behaviour. It should be noted that these objects themselves do not have autonomy from the physical point of view. Introduction of such objects is justified by the fact that their use provides the necessary degree of flexibility and performance from a computational point of view. Thus, physical objects that are responsible for the impact on the ensemble of particles are in turn divided into a set of subagents, which are responsible for certain properties, but do not have an independent physical interpretation. An assembly of these sub-objects helps to secure the necessary variability and implement the optimization of control system as a whole. In other words we can change the necessary subagent without distortion of physical sense and computational sequence. Moreover, you must also monitor the state of the beam itself (values of beam envelopes, polarization, etc.), because these characteristics are very important for realization of the optimal working regime. Besides, we should carry out necessary additional computational procedures to analyze the impact of the effects of control errors on the beam characteristics. These additional computing can be also realized using additional subagents. All necessary properties of agents and subagents should be divided on physical properties (derived from physical characteristics of a primary physical object) and information properties in according to general concept of forming of information agents.

3   Self-consistent Particle Dynamics with Space Charge Forces

In this part the predictor-corrector method based on matrix formalism will be discussed. By saying predictor-corrector we mean a multiple step method. This one can be used as an alternative to the well known Runge-Kutta method. The scheme of this method is simple. First, the extrapolation method is used to predict the value of some y_{j+1} by known previous y_j, y_{j-1}, etc. Then the obtained value is estimated and is correcting to get the better approximation of y_{j+1}. If the difference between the correcting and the predicting values exceeds a certain value, then the next iteration is running [19].

In our case the main idea of this method is to predict the distribution of the particles, which we want to get at the end, by correcting the intermediate results during the calculations. First, let's talk about the solution algorithm of self-consistent dynamic of the beam in general. Here the distribution functions will be discussed.

At the beginning we set an interval $T = [t_0, t_1]$, on which the solution is looking for, and $\Delta t = t_1 - t_0$. The transportation system is given on this interval, that means that the external fields $\mathbf{B}^{ext}$, $\mathbf{E}^{ext}$ and the function $\mathbf{F}^{ext}$ can be obtained.

Besides, the distribution function is selected from the set of initial distributions. In the simple case it can be the ideal Kapchinskij-Vladimirslij distribution. Or it can be the modifications of KV, which are given in [20]. It is a base distribution in such cases, because K-V distribution is a four dimensional distribution in phase space and has properties:

- the space charge forces are linear;
- it transforms into a K-V distribution under a linear mapping.

With other distributions this one forms something like the class of initial distributions:

- linear: $\rho_1(x, y) = \rho_0(1 - 4\varkappa^2/9)\Theta(1 - 4\varkappa^2/9))$;
- uniform: $\rho_2(x, y) = \rho_0\Theta(1 - \varkappa^2)$;
- normal: $\rho_3(x, y) = \rho_0 exp(-\alpha_3^2\varkappa^2)$, $\alpha_3 = -\frac{\pi}{2\,i\,erf(i)}$.

Where $erf(x) = \int_0^x exp(-t^2/2)dt$ - probability integral.
- quadratic: $\rho_4(x, y) = \rho_0(1 - (4/5)^4\varkappa^4)\Theta(1 - (4/5)^4\varkappa^4)$;
- co-sinusoidal: $\rho_5(x, y) = \rho_0\cos^2(\pi\alpha_5^2\varkappa^2)\Theta(1 - \alpha_5^2\varkappa^2)$.

Where α_5 calculates with Frenel integral.

On Fig. 2 the density function ρ_i, $i = \overline{1,5}$ as the function of scalar variable R is shown. See the [16] for more details. This list can be supplemented by other distributions depending on the task.

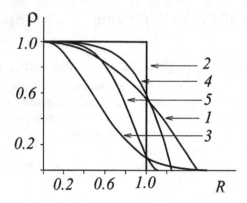

Fig. 2. Distribution of charge density: 1 - linear, 2 - uniform, 3 - normal, 4 - quadratic, 5 - co-sinusoidal.

Now we are turning back to the algorithm. After all the initial parameters are set, the evolution operator is calculated using the Eq. 3 and the technology as described above.

After that the distribution function is obtained with the help of evolution operator applying on the previous value of distribution:

$$f^1(\mathbf{X}, t) = f_0((\mathcal{M}^0)^{-1} \circ \mathbf{X_0}).$$

Substituting the obtained function to the field equations, we get $(\mathbf{B}^{self})^1$, $(\mathbf{E}^{self})^1$.

Now we are ready to calculate the function

$$(\mathbf{F}^{self})^1 = \mathbf{F}^{self}(\mathbf{B}^{self})^1, (\mathbf{E}^{self})^1, \mathbf{X}, t)).$$

Or the self-consistent Hamiltonian $(\mathcal{H}^{self})^1 = \mathcal{H}^{self}((\mathbf{B}^{self})^1, (\mathbf{E}^{self})^1, \mathbf{X}, t)$ is found.

Thereafter, the evolution operator $\mathcal{M}^1 = \mathcal{A} \circ \mathcal{M}^0$ is evolved by equation

$$\mathcal{M}(t|t_0; \mathcal{V}^{ext} + \mathcal{V}^{self}) = \mathcal{I}d + \int_{t_0}^{t} (\mathcal{V}^{ext}(\tau) + \mathcal{V}^{self}(\tau)) \circ \mathcal{M}(\tau|t_0; \mathcal{V}^{ext}(\tau) + \mathcal{V}^{self}(\tau)) d\tau.$$

$$(7)$$

The new average value of distribution function is found by the following equation:

$$\langle f(\mathbf{X}, t_0) \rangle^1_{\mathfrak{M}_1} = (1 - \alpha)\langle f_0((\mathcal{M}^0)^{-1} \circ \mathbf{X_0}) \rangle_{\mathfrak{M}_0} + \alpha \langle f_0((\mathcal{M}^0)^{-1} \circ \mathbf{X_0}) \rangle_{\mathfrak{M}_0}, \quad (8)$$

The final step is to verify the criteria, e.g.:

$$\|\mathcal{M}^k - \mathcal{A} \circ \mathcal{M}^{k-1}\| < \epsilon, k \geq 1. \quad (9)$$

It is obvious that if the condition 9 is carried out, the solution is obtained. Otherwise, the $(\mathbf{B}^{self})^i$, $(\mathbf{E}^{self})^i$ are calculating again.

This algorithm is suitable for analytical analysis, but on practice we can not measure the distribution function. To get the result that can be proved by experiment the algorithm based on envelope dynamic was designed.

Similarly to the previous algorithm we set an interval $T = [t_0, t_1]$, $\Delta t = t_1 - t_0$, and $f(\mathbf{X}, t_0) = f_0(X)$ is the initial value of the distribution function consisting of the set of phase points when $t = t_0$, and N is the order of approximation.

First step. We calculate matrices $\mathfrak{S}_0^{ik}$, $i, k = \overline{1, N}$ by the following equation:

$$\mathfrak{S}_0^{ik} = \int_{\mathfrak{M}_0} f_0(\mathbf{X}) \mathbf{X}^{[i]} (\mathbf{X}^{[k]})^* d\mathbf{X}$$

As a form-matrix A_0 $(\mathfrak{S}_0^{11})^{-1}$ can be chosen, or $\mathfrak{S}_0^{-1}$ – if the initial set $\mathfrak{M}_0$ is an ellipsoid with the border

$$X_0^* \mathfrak{S}_0^{-1} \mathbf{X}_0 = \varepsilon.$$

Next we built approximant $\varphi_0(\varkappa_0^2) \approx f_0(\mathbf{X}_0)$, where $\varkappa_0^2 = \mathbf{X}_0^* A_0 \mathbf{X}_0$ and turn to the next step.

Second step. Here we get the block-matrices $\mathbb{P}^{1k}(B^{ext}, E^{ext}, t)$ and $\mathbb{N}_1^{1k} = \mathbb{P}^{1k}(\mathbf{B}^{ext}, \mathbf{E}^{ext}, t)$ [14] for external fields. The (ij) element of matrix $\mathbb{P}^{1k}$, for example, can be found by the following form:

$$\left\{ \mathbb{P}^{1k}(t) \right\}_{ij} = \frac{1}{k_1! \ldots k_n!} \frac{\partial^k \mathbf{F}_i(X_j, t)}{\partial x_1^{k_1} \ldots \partial x_n^{k_n}} \bigg|_{x_1 = \cdots = x_n = 1}$$

Third step. It is necessary to find $\mathbf{E}^{self} = \mathbf{E}(\varphi_0(\varkappa_0^2))$ depending of the distribution function we have chosen (e.g. uniform, normal, etc.).

On the fourth step we calculate block-matrices $\mathbb{P}^{1k}(\mathbf{E}^{self}, t)$ with space charge effect: $\mathbb{N}_2^{1k} = \mathbb{P}^{1k}(\mathbf{E}^{ext}, t)$

Fifth step. Then comes the calculations of block-matrices $\mathbb{M}^{ik}$ where $i \leq k \leq N$,

$$\mathbb{M}_1^{ik} = \mathbb{M}^{ik}(t|t_0; \{\mathbb{N}_1^{1l}\}), l = \overline{1, k},$$

$$\mathbb{M}_2^{ik} = \mathbb{M}^{ik}(t|t_0; \{\mathbb{N}_2^{1l}\}),$$

$$\mathbb{M}_0^{ik} = \mathbb{M}_1^{ik} + \mathbb{M}_2^{ik}.$$

Block-matrices $\mathbb{M}^{ik}$ are the matrix form of the evolution operator.

On the next step, after all necessary matrices have been obtained, we can substitute them into block-matrices $\mathfrak{S}_0^{ik}$

$$\mathfrak{S}_0^{ik} = \sum_{l=i}^{\infty} \sum_{j=k}^{\infty} \mathbb{M}_0^{il} \mathfrak{S}_0^{lj} (\mathbb{M}_0^{jk})^T.$$

Step seven. Before the conditions will be checked the virtual changes of settings while beam evolution must be made:

$$\mathfrak{S}_1^{ik} = \alpha \mathfrak{S}_0^{ik} + (1 - \alpha)\mathfrak{S}_0^{ik}, 0 < \alpha < 1.$$

Virtual change implies changes of settings that are necessary to built a map. Envelope matrices, functions of distributions, etc., are not changed.

Step eight. Now we can check the conditions:

$$\frac{2\|\mathfrak{S}_1^{ik} - \mathfrak{S}_0^{ik}\|}{\|\mathfrak{S}_1^{ik}\| + \|\mathfrak{S}_0^{ik}\|} < \varepsilon^{ik}. \tag{10}$$

Different equivalent rules can be used as 10. If the condition is right, the process stops. Otherwise:

$$\mathfrak{S}_0^{ik} = \mathfrak{S}_1^{ik},$$

and before turning back to the algorithm the final step is to find the approximate value $\varphi(\varkappa^2)$ for function $f(\mathbf{X}, t)$:

$$\varphi(\varkappa^2) \approx f_0(\mu_0^{-1} \circ \mathbf{X}_0) = f_0(\sum_{i=1}^{\infty} \mathbb{T}_0^{1l} X_0^{[i]}).$$

Assuming that $\varphi_0(\varkappa^2) = \varphi(\varkappa^2)$ return to the step three.

Considered algorithm can be simplified by using as approximate value the function that is constant on the ellipsoid and zero outside it. After that step three is modified to be easier and the final changes are not needed at all. That significantly accelerates the process. Moreover, choosing the approximate value from the class of polynomials allows us to use pre-computed block-matrices from special database. So this approach can significantly reduce the computations instead of numerical simulations.

4   Parallelization of Predictor-Corrector Method

Practically for every one it is obvious that the number of particles needed to compute on practice can not be provided by any known approach. The natural parallelization and distributed structures of beam physics problems allow using parallel and distributed computer systems (see works [21–23]. Analyzing the situation, it became clear that it is no matter how much resources we have, if the algorithm is not suitable for parallelization there will be not great benefit of it. As we can see from the algorithm shown above matrix formalism is a high-performance mapping approach for ODE solving. It allows to present the intermediate results and the solution of the system in the form of matrices. That makes the approach to be easy implemented in parallel code.

Due to the fact that only matrix multiplication and addition are used, we think of a GPU programming [24]. The present research is shown that there is no great benefit via parallelization of computational code for one particle by using OpenMP library (see Fig. 3 and Tables 1 and 2). In this case overhead

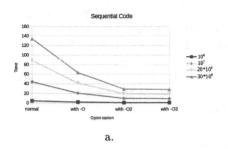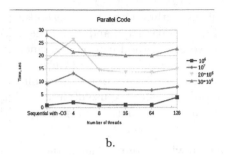

a.   b.

Fig. 3. a – Sequential code; b – Parallel code

Table 1. Time (sec.) in sequential code for different number of particles

Optimization/Num. of particles	10^6	10^7	$20 * 10^6$	$30 * 10^6$
normal	4,3942	44.0785	88.42	133.955
-O	2.0449	20.5504	41.6161	63.3152
-O2	0.9780	9.7588	19.4914	29.2429
-O3	0.9114	9.14508	18.3444	28.1367

on data sending is significant and take the greatest part of time. On the other hand, matrix formalism allows to process a set of initial points, where parallelization is more preferable on GPUs. But using only GPUs is not justified. In our experiment we have the system that can provide the power to compute only the number of particles nearly 10^7. It's less that required, but our goal in this part of research was to test the algorithm before using in on the real system. The results has shown good parallelization of the described approach.

Table 2. Time (sec.) in parallel code for different number of particles

Num. threads/Num. of particles	10^6	10^7	$20 * 10^6$	$30 * 10^6$
Sequential with -O3	0,911436	9,14508	18,3444	28,1367
4	2,00863	13,1787	26,3788	21,5358
8	1,06923	7,16977	14,5777	20,8549
16	1,06208	6,8548	13,6071	20,15
64	1,00906	6,70448	13,5593	20,0794
128	3,86119	7,92809	14,8894	22,748

5   Conclusion

Our challenge is to provide computer simulation for developed algorithm for solving the problem of accounting space-charge forces in general and compare this algorithm with other methods. It allows simulate both long-term evolution of a set of particles, and evaluating based on envelope description. As it was said above the method can be implemented in parallel codes on GPU+CPU hybrid Cluster. That is why the future development of the research also can be based on writing software to compare different parallel techniques for Hybrid Systems, in order to effective use of described approach to compute the required number of particles in long-term evolution of the beam.

Acknowledgments. The authors would like to express gratitude to Vladimir Korkhov for valuable help. Scientific research were performed using the equipment of the Research Park of St.Petersburg State University. The work was sponsored by the Russian Foundation for Basic Research under the projects: 16-07-01113 "Virtual supercomputer as a tool for solving complex problems" and by the Saint-Petersburg State University under the project 0.37.155.2014 "Research in the field of designing and implementing effective computational simulation for hydrophisical and hydro-meteorological processes of Baltic Sea (and the open Ocean and offshores of Russia)".

References

1. Batygin, Y.K., Scheinker, A.: Suppresion of halo formation in fodo channel with nonlinear focusing. In: Proceedings of IPAC 2013. JACOW (2015)
2. Batygin, Y.K., Scheinker, A., Kurennoy, S.: Nonlinear Optics for Suppresion of Halo Formation in Space Charge Dominated Beams (2015)
3. Batygin, Y.K.: Space-charge neutralization of 750-keV proton beam in lansce injector LIN. In: Proceedings of IPAC 2015. JACOW (2015)
4. Ryne, R.D., Habib, S., Wangle, T.P.: Halos of Intense Proton Beams. IEEE (1996)
5. Paret, S., Qiang, J.: Collisional effects in particle-in-cell beam-beam simulation. In: Proceedings of IPAC 2013. JACOW (2013)
6. Wolfheimer, F., Gjonaj, E., Weiland, T.: Parallel particle-in-cell (PIC) codes. In: Proceedings of ICAP 2006. JACOW (2006)
7. Stancari, G., Redaelli, S., Moens, V.: Beam dynamics in an electron lens with the warp particle-in-cell code. In: Proceedings of IPAC 2014. JACOW (2014)
8. Bowers, K.J.: Accelerating a paticle-in-cell simulation using a hybrid counting sort. J. Comput. Phys. **173**, 393–411 (2001). Academic Press
9. Makino, K., Berz, M.: COSY INFINITY Version 9. Nuclear Instruments and Methods A558 (2005)
10. Dragt, A.J.: Lie methods for nonlinear dynamics with applications to accelerator physics. University of Maryland (2015)
11. Dragt, A.J., Ryne, R.D., et al.: MARYLIE 3.0 Users Manual: A Program for Charged Particle Beam Transport Based on Lie Algebraic Methods. University of Maryland (2003)
12. Dragt, A.J., Ryne, R.D., et al.: Numerical computation of transfer maps using lie algebraic methods. In: Proceedings of PAC 1987 (1987)
13. Ryne, R.D.: Advanced computing tools and models for accelerator physics. In: Proceedings of EPAC 2008 (2008)
14. Andrianov, S.N.: Dynamical Modeling of Control Systems for Particle Beams'. Saint Petersburg State University, SPb (2004)

15. Magnuss, W.: On the exponential solution of differential equations for a linear operator. Comm. Pure Appl. Math. **7**(4), 649–673 (1954)
16. Kulabukhova, N., Degtyatev, A., Bogdanov, A., Andrianov, S.: Simulation of space charge dynamics on HPC. In: Proceedings of IPAC 2014. JACOW (2014)
17. Healy, L.M., Dragt, A.J.: Concatenation of Lie algebraic maps. Lie Methods in Optics II. Lect. Notes in Physics, vol. 352 (1989)
18. Andrianov, S., Kulabukhova, N.: Lie algebraic methods as mathematical models for high performance computing using the multi-agent approach. In: Gervasi, O., et al. (ed.) ICCSA 2016, Part I. LNCS, vol. 9786, pp. 418–430. Springer, Heidelberg (2016)
19. Szilagui, M.: Electron and ion optics (in russian). Mir, Moscow (1990)
20. Venturini, M.: Lie methods, exact map computation, and the problem of dispertion in space charge dominated beams. Ph.D. thesis (1998)
21. Giovannozzi, M.: Space-Charge Simulation Using Parallel Algorithms
22. Bowers, K.J.: Accelerating a particle-in-cell simulation using a hybrid counting sort. J. Comput. Phys. **173**, 393–411 (2001)
23. Qiang, J., Ryne, R.D., Habib, S., Decy, V.: An object-oriented parallel particle-in-cell Code for beam dynamics simulation in linear accelerators. J. Comput. Phys. **163**, 434–451 (2000)
24. Kulabukhova, N.: GPGPU implementation of matrix formalism for beam dynamics simulation. In: Proceedings of ICAP 2012. JACOW (2012)

A Theoretical Study on the Relevance of Protonated and Ionized Species of Methanimine and Methanol in Astrochemistry

Marzio Rosi[1,2(✉)], Stefano Falcinelli[1], Nadia Balucani[3], Noelia Faginas-Lago[3], Cecilia Ceccarelli[4], and Dimitrios Skouteris[5]

[1] Dipartimento di Ingegneria Civile ed Ambientale, University of Perugia, Via Duranti 93, 06125 Perugia, Italy
{marzio.rosi,stefano.falcinelli}@unipg.it
[2] ISTM-CNR, 06123 Perugia, Italy
[3] Dipartimento di Chimica, Biologia e Biotecnologie, University of Perugia, 06123 Perugia, Italy
{nadia.balucani,noelia.faginaslago}@unipg.it
[4] Institute de Planétologie et d'Astrophysique de Grenoble, 38041 Grenoble, France
Cecilia.Ceccarelli@obs.ujf-grenoble.fr
[5] Scuola Nornale Superiore, 56126 Pisa, Italy
dimitrios.skouteris@sns.it

Abstract. Under the low T conditions of the interstellar medium or planetary atmospheres like that of Titan, which presents a surface temperature of 94 K and a temperature of ca. 180 K in the upper layers of the atmosphere, reactions involving neutral species, which usually show relatively high or even huge energy barriers, are not efficient processes unless a significant external energy source is provided and transient species, like atomic or molecular radicals, are involved. A completely different picture holds when ionized or protonated species are involved, since most reactions in this case are barrierless. In order to show this point we will present two case studies: the dimerization of methanimine and the production of dimethyl ether from methanol. Methanimine is a molecule of interest in astrobiology, as it is considered an abiotic precursor of the simplest amino acid glycine. Methanimine has been observed in the interstellar medium and in the upper atmosphere of Titan. In particular, it has been speculated that its polymerization can contribute to the formation of the haze aerosols that surround the massive moon of Saturn. To assess its potential role in the formation of Titan's aerosol, we have performed a theoretical investigation of a possible dimerization process. The aim of this study is to understand whether dimerization of methanimine and, eventually, its polymerization, is possible under the conditions of the atmosphere of Titan. The second case study presented is the reaction of methanol with protonated methanol which has been considered for a long time to be an important step towards the formation of gaseous interstellar dimethyl ether. Results of high-level electronic structure calculations are reported.

Keywords: Ab initio calculations · Astrochemistry · Atmospheric models

© Springer International Publishing Switzerland 2016
O. Gervasi et al. (Eds.): ICCSA 2016, Part I, LNCS 9786, pp. 296–308, 2016.
DOI: 10.1007/978-3-319-42085-1_23

1 Introduction

In the interstellar medium (ISM) as well as in the upper atmosphere of Titan (the massive moon of Saturn [1], particularly interesting for its reminiscence of the primeval atmosphere of Earth [2, 3]) the low number density and low temperature conditions prevent reactivity among neutral closed-shell molecules because of the presence of relatively high activation energy barriers. Nevertheless, a rich chemistry leading to some molecular complexity has been observed in those environments. We have recently performed a theoretical investigation on the dimerization of methanimine [4, 5], which is an important molecule in prebiotic chemistry because it is a possible precursor of glycine, via its reactions with HCN and H_2O [2, 6, 7], and an intermediate in the Strecker synthesis of aminoacetonitrile [8] and glycine [9–13]. Methanimine was observed in the ISM [14] and its presence has been inferred in the upper atmosphere of Titan from the analysis of the ion spectra recorded by Cassini Ion Neutral Mass Spectrometer (INMS) [1]. Methanimine can be produced in the atmosphere of Titan by the reactions of N (^2D) with both methane and ethane [15, 16], as well as by other simple processes, including the reaction between NH and CH_3 [17] or reactions involving ionic species [18]. The recent photochemical model by Lavvas et al. [19, 20] derived a larger quantity of methanimine than that derived by the analysis of the ion spectra recorded by Cassini INMS [1]. The reason for this discrepancy could be the severe lack of knowledge on the possible chemical loss pathways of methanimine [19, 20]. Since imines are well-known for their capability of polymerizing, CH_2NH could polymerize and copolymerize with other unsaturated nitriles or unsaturated hydrocarbons. Lavvas et al. [19, 20] suggested that polymerization of methanimine provides an important contribution to the formation of the nitrogen-rich aerosols, but a quantitative inclusion of this process in the model could not be obtained as there is no information (either experimental or theoretical) on methanimine polymerization. Since the first step of polymerization is dimerization, we will review our theoretical investigations [4, 5] on methanimine dimerization comparing the differences in reactivity when neutral or ionic or protonated species are involved. A second reaction involving protonated species, which will be reported in this contribution, is the reaction between methanol and protonated methanol. The reaction $CH_3OH_2^+ + CH_3OH$ has been considered for a long time an important step towards the formation of gaseous interstellar dimethyl ether. Nevertheless, after some experimental evidence that protonated dimethyl ether does not produce significant amount of neutral CH_3OCH_3 via electron recombination [21], this reaction has been disregarded from new astrochemical models. Recent work, however, suggested that the proton transfer reaction with ammonia can be a process efficient enough to account for interstellar CH_3OCH_3 [22]. This mechanism could be important also for other reactions relevant for astrochemical models.

2   Computational Details

The stationary points, both minima and saddle points, were optimized performing density functional (DFT) calculations using the Becke 3-parameter exchange and Lee −Yang−Parr correlation (B3LYP) [23, 24] hybrid functional and the correlation consistent valence polarized set aug-cc-pVTZ [25–27]. This level of theory has been used also to compute the harmonic vibrational frequencies necessary to check the nature of the stationary points, *i.e.* minimum if all the frequencies are real, saddle point if there is one, and only one, imaginary frequency. The assignment of the saddle points was performed using the intrinsic reaction coordinate (IRC) method [28, 29]. For comparison purposes and for selected cases, the energy of the stationary points was computed also using the more accurate coupled cluster theory with the inclusion of single and double excitations and a perturbative estimate of connected triples (CCSD(T)) [30–32] in connection with the same basis set aug-cc-pVTZ [25–27]. Both the B3LYP and the CCSD(T) energies were corrected to 0 K by adding the zero point energy correction computed using the scaled harmonic vibrational frequencies evaluated at B3LYP/aug-cc-pVTZ level. All calculations were done using Gaussian 09 [33] while the analysis of the vibrational frequencies was performed using Molekel [34, 35].

3   Results and Discussion

3.1   Dimerization of Methanimine in the Atmosphere of Titan or in the Interstellar Space

The interaction of two molecules of methanimine leads, without any barrier, to a van der Waals adduct (species **1** in Fig. 1) which is only 2.1 kcal/mol more stable than the reactants at B3LYP/aug-cc-pVTZ level. This adduct is characterized by a weak hydrogen bond, being the N—H bond distance as long as 2.223 Å, as we can see in Fig. 1 where the optimized structures of the stationary points relevant for the dimerization of methanimine are reported. Adduct **1** can isomerize to species **2**, characterized by a new C—N bond, or species **3**, which show a new N—N bond. Species **2** is more stable than species **1** by 12.2 kcal/mol at B3LYP/aug-cc-pVTZ level, but the reaction leading to its formation shows an activation energy of 44.2 kcal/mol at B3LYP/aug-cc-pVTZ level. Species **3** is even less stable than adduct **1** (although by only 1.9 kcal/mol at B3LYP/aug-cc-pVTZ level), but its formation requires the overcoming of a barrier as high as 82.1 kcal/mol at B3LYP/aug-cc-pVTZ level. These reactions are reported in a schematic representation in Fig. 2. Under the low T ($\leq$ 180 K) conditions of Titan, there is not enough energy to compensate for the high barriers shown by the dimerization of the methanimine process. Being these barriers so high (larger than 40 kcal/mol) also in the ISM the probability for this process is very low. In order to check the accuracy of our results, we performed the same calculations also at the more accurate, but more expensive, CCSD(T)/aug-cc-pVTZ level. We compute the energies of the stationary points, both minima and transition states, at the CCSD(T)/aug-cc-pVTZ level considering the B3LYP optimized geometries. The results are shown in Figs. 1 and 2 in parentheses. The energy differences between the two methods are within ±3 kcal/mol and therefore the CCSD(T)/aug-cc-pVTZ results confirm the description obtained at B3LYP/aug-cc-pVTZ level. This is an important result since the B3LYP method can be applied also to systems too big to be described at CCSD(T) level. For this reason in the following discussion we will consider only B3LYP calculations.

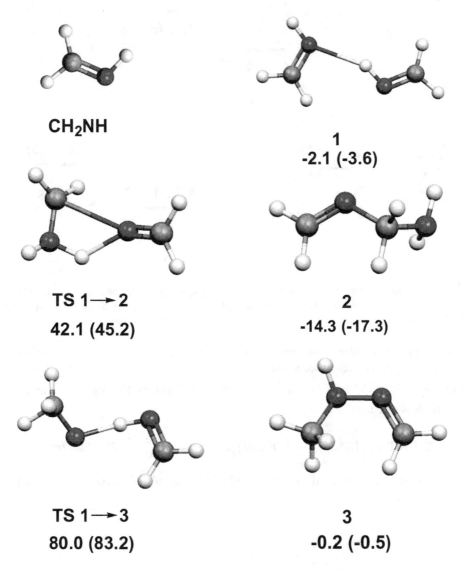

CH$_2$NH

1
-2.1 (-3.6)

TS 1→2
42.1 (45.2)

2
-14.3 (-17.3)

TS 1→3
80.0 (83.2)

3
-0.2 (-0.5)

Fig. 1. B3LYP optimized structures and relative energies with respect to two molecules of methanimine not interacting (kcal/mol) at 0 K of minima and saddle points localized on the PES for the dimerization of methanimine; CCSD(T) relative energies are reported in parentheses. Carbon atoms in green, nitrogen atoms in blue and hydrogen atoms in white. (Color figure online)

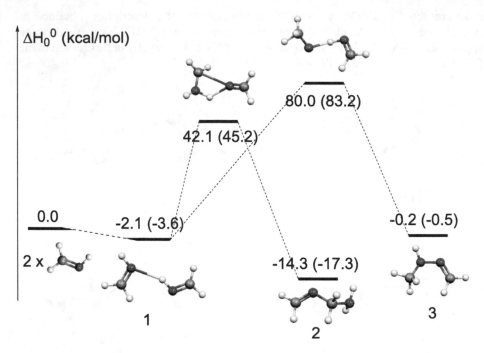

Fig. 2. Schematic representation of the two main reaction paths leading to dimethanimine. Relative energies (kcal/mol, 0 K) computed at B3LYP/aug-cc-pVTZ (CCSD(T)/aug-cc-pVTZ) level with respect to the energy of two molecules of methanimine not interacting.

A complete different picture holds when we consider the reaction of a methanimine molecule with a protonated second molecule of methanimine. This interaction gives the protonated dimer $CH_2NHCH_2NH_2^+$, whose structure (species **4**) is shown in Fig. 3, in a barrierless reaction:

$$CH_2NH + CH_2NH_2^+ \rightarrow (CH_2NH)_2H^+ \qquad \Delta H_0^0 = -26.6 \text{ kcal/mol}$$

However, the protonated dimer $(CH_2NH)_2H^+$ can only dissociate in very endothermic processes:

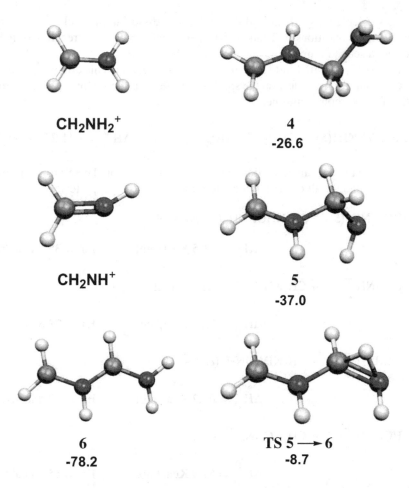

Fig. 3. B3LYP optimized structure and relative energies with respect to the reactants (kcal/mol) at 0 K of minima and saddle points localized on the PES for the reaction of methanimine with a protonated molecule of methanimine and methanimine with an ionized molecule of methanimine. Carbon atoms in green, nitrogen atoms in blue and hydrogen atoms in white. (Color figure online)

$$(CH_2NH)_2H^+ \rightarrow CH_2NHCHNH_2^+ + H \qquad \Delta H_0^0 = 68.6 \text{ kcal/mol}$$

$$(CH_2NH)_2H^+ \rightarrow CH_2NHCH_2^+ + NH_2 \qquad \Delta H_0^0 = 76.1 \text{ kcal/mol}$$

$$(CH_2NH)_2H^+ \rightarrow CH_2NHCH_2NH^+ + H \qquad \Delta H_0^0 = 99.8 \text{ kcal/mol}$$

$$(CH_2NH)_2H^+ \rightarrow CH_2NCH_2NH_2^+ + H \qquad \Delta H_0^0 = 102.1 \text{ kcal/mol}$$

$$(CH_2NH)_2H^+ \rightarrow CHNHCH_2NH_2^+ + H \qquad \Delta H_0^0 = 104.7 \text{ kcal/mol}$$

Kinetics calculations [5] showed that only back-dissociation is relevant under the low temperature conditions of Titan or ISM. More interesting is the reaction of methanimine with a second ionized molecule. Also in this case this process leads to a dimer, whose structure is reported in Fig. 3, in a barrierless reaction. This dimer (species **5** in Fig. 3) can isomerize to a more stable species (species **6** in Fig. 3) through an hydrogen migration from carbon to nitrogen:

$$CH_2NHCH_2NH^+(\textbf{5}) \rightarrow CH_2NHCHNH_2^+ (\textbf{6}) \qquad \Delta H_0^0 = -41.2 \text{ kcal/mol}$$

This reaction shows an activation energy of 32.7 kcal/mol. There are also several dissociation channels, all of them lying under the reactants asymptote:

$$CH_2NHCH_2NH^+(\textbf{5}) \rightarrow CH_2NHCHNH^+(trans) + H$$

$$\Delta H_0^0 = 31.5 \text{ Kcal/mol} \qquad Ea = 35.9 \text{ kcal/mol}$$

$$CH_2NHCHNH_2^+ (\textbf{6}) \rightarrow CH_2NHCHNH^+(trans) + H$$

$$\Delta H_0^0 = 72.7 \text{ Kcal/mol} \qquad Ea = 73.3 \text{ kcal/mol}$$

$$CH_2NHCHNH_2^+ (\textbf{6}) \rightarrow CH_2NHCHNH^+(cis) + H$$

$$\Delta H_0^0 = 66.2 \text{ Kcal/mol} \qquad Ea = 67.4 \text{ kcal/mol}$$

$$CH_2NHCHNH_2^+ (\textbf{6}) \rightarrow CH_2NCHNH_2^+ + H$$

$$\Delta H_0^0 = 51.9 \text{ Kcal/mol} \qquad Ea = 55.9 \text{ kcal/mol}$$

However, the most interesting process is the formation of dimer **5** in an exothermic and barrierless reaction:

$$CH_2NH + CH_2NH^+ \rightarrow (CH_2NH)_2^+ \qquad \Delta H_0^0 = -37.0 \text{ kcal/mol}$$

The dimerization of methanimine, which is not feasible under the low T conditions of Titan, is possible when an ionized species is involved. We have investigated also the addition of a third molecule of methanimine to species **5**, considering the following reaction:

$$(CH_2NH)_2^+ + CH_2NH \rightarrow (CH_2NH)_3^+ \qquad \Delta H_0^0 = -28.2 \text{ kcal/mol}$$

Also this reaction is barrierless and its exothermicity suggests that the polymerization of methanime, provided that one molecule is in an ionized form, is possible also under the conditions of Titan. Work is in progress to investigate further this point. The optimized structure of $\left(CH_2NH\right)_3^+$ is shown in Fig. 4.

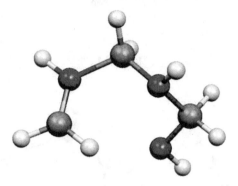

Fig. 4. B3LYP optimized structure of $\left(CH_2NH\right)_3^+$. Carbon atoms in green, nitrogen atoms in blue and hydrogen atoms in white. (Color figure online)

3.2   Formation of Dimethyl Ether in the Interstellar Medium

The reaction $CH_3OH + CH_3OH_2^+$ has been considered for a long time to be an important step towards the formation of gaseous dimethyl ether. For this reason we investigated this reaction which leads to protonated dimethyl ether and water:

$$CH_3OH + CH_3OH_2^+ \rightarrow CH_3OHCH_3^+ + H_2O$$

The B3LYP/aug-cc-pVTZ optimized structures of minima and saddle points involved in this process are shown in Fig. 5. The reaction between methanol and protonated methanol leads, in a barrierless process, to the adduct 1^+ where a new C—O interaction is formed.

$$CH_3OH + CH_3OH_2^+ \rightarrow CH_3OH \cdots CH_3OH_2^+ \left(1^+\right) \qquad \Delta H_0^0 = -9.6 \, kcal/mol$$

The C—O bond length, 2.639 Å, is very long suggesting that this could be considered an electrostatic interaction rather than a chemical bond. Species 1^+ through a transition state which lies under the reactants isomerizes to species 2^+ where the C—O interaction becomes a true chemical bond (bond length 1.509 Å) while the terminal C—O bond becomes an electrostatic interaction, as we can notice from the C—O bond length which changes from 1.554 Å to 3.438 Å.

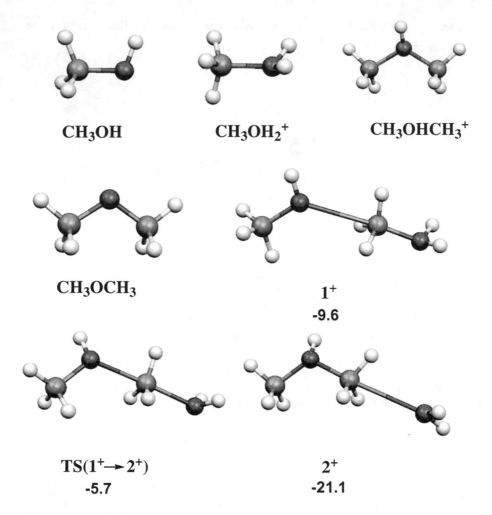

Fig. 5. B3LYP optimized structure and relative energies with respect to the reactants (kcal/mol) at 0 K of minima and saddle points localized on the PES for the reaction of methanol with a protonated molecule of methanol. Carbon atoms in green, oxygen atoms in red and hydrogen atoms in white. (Color figure online)

$$CH_3OH\cdots CH_3OH_2^+ \ (\mathbf{1^+}) \rightarrow CH_3OHCH_3\cdots OH_2^+ \ (\mathbf{2^+})$$

$$\Delta H_0^0 = -11.5 \ \text{Kcal/mol} \qquad Ea = 3.9 \ \text{kcal/mol}$$

Species $\mathbf{2^+}$ easily dissociates to protonated dimethyl ether and water.

$$CH_3OHCH_3\cdots OH_2^+ \ (\mathbf{2^+}) \rightarrow CH_3OHCH_3^+ + H_2O \qquad \Delta H_0^0 = -7.5 \ \text{kcal/mol}$$

The interesting thing of this process involving protonated species is that all the involved intermediates and transition states lie under the energy of the reactants asymptote; this process is, therefore, an efficient way to produce protonated dimethyl ether. Notably, after that some experiments revealed that protonated dimethyl ether does not produce significant amount of neutral CH_3OCH_3 via electron recombination [21], the

$$CH_3OH_2^+ \rightarrow CH_3OHCH_3^+ \rightarrow CH_3OCH_3$$ pathway has been disregarded as a route of

dimethyl ether formation from new astrochemical models. Recent work by Taquet et al. [22], however, suggested that the proton transfer reaction with species which show a larger proton affinity than dimethyl ether like ammonia could be a process efficient enough to account for interstellar CH_3OCH_3.

For this reason we plan to theoretically investigate also the reaction

$$CH_3OHCH_3^+ + NH_3 \rightarrow CH_3OCH_3 + NH_4^+$$

Kinetics calculations are planned to provide quantitative rate coefficients to be included in astrochemical models. If proved to be efficient, this set of two reactions will be an additional gas phase route to dimethyl ether formation to complement the radiative association reaction between CH_3 and CH_3O [36].

4   Conclusions

Polymerization of methanimine in the gas phase has been invoked to explain the difference between model predictions and Cassini measurements [19, 20]. According to the present results, dimerization of methanimine (the first step toward polymerization) is a process characterized by very high energy barriers and is difficult to believe it can be important under the conditions of Titan. A complete different situation holds when we consider ionized species: dimerization and also trimerization are barrierless processes which can be possible also under very low T conditions. In this respect it should be noted that methanimine can react with many radicals which are relatively abundant in the upper atmosphere of Titan, such as CN, HNO, CH_2NCH etc. [37–40]. Recent studies have indeed pointed out that these reactions can be very fast under those conditions [37]. Protonated species are important also for the production of interstellar CH_3OCH_3 which can be formed by the reaction of methanol with protonated methanol. This process, which implies intermediates and transition states which lie under the energy of the reactants, leads to protonated dimethyl ether, which can trasfer easily the proton to a molecule of ammonia [22].

Acknowledgment.   The authors gratefully thank "Fondazione Cassa di Risparmio di Perugia" for financial support (Project code: 2014.0255.021). M. Noelia Faginas Lago acknowledges financial support from Fondazione Cassa di Risparmio di Perugia (P 2014/1255, ACT 2014/6167).

References

1. Vuitton, V., Yelle, R.V., Anicich, V.G.: The nitrogen chemistry of titan's upper atmosphere revealed. Astrophys. J. **647**, L175 (2006)

2. Balucani, N.: Elementary reactions of N atoms with hydrocarbons: first steps towards the formation of prebiotic N-containing molecules in planetary atmospheres. Chem. Soc. Rev. **41**, 5473 (2012)

3. Vuitton, V., Dutuit, O., Smith, M.A., Balucani, N.: Chemistry of Titan's atmosphere. In: Mueller-Wodarg, I., Griffith, C., Lellouch, E., Cravens, T. (eds.) Titan: Surface, Atmosphere and Magnetosphere. University Press, Cambridge (2013). ISBN: 9780521199926

4. Rosi, M., Falcinelli, S., Balucani, N., Casavecchia, P., Skouteris, D.: A theoretical study of formation routes and dimerization of methanimine and implications for the aerosols formation in the upper atmosphere of Titan. In: Murgante, B., Misra, S., Carlini, M., Torre, C.M., Nguyen, H.-Q., Taniar, D., Apduhan, B.O., Gervasi, O. (eds.) ICCSA 2013, Part I. LNCS, vol. 7971, pp. 47–56. Springer, Heidelberg (2013)

5. Skouteris, D., Balucani, N., Faginas-Lago, N., Falcinelli, S., Rosi, M.: Dimerization of methanimine and its charged species in the atmosphere of Titan and interstellar/cometary ice analogs. A&A **584**, A76 (2015)

6. Balucani, N.: Elementary reactions and their role in gas-phase prebiotic chemistry. Int. J. Mol. Sci. **10**, 2304 (2009)

7. Woon, D.E.: Ab initio quantum chemical studies of reactions in astrophysical ices. 4. Reactions in ices involving HCOOH, CH$_2$NH, HCN, HNC, NH$_3$, and H$_2$O. Int. J. Quant. Chem. **88**, 226 (2002)

8. Danger, G., Borget, F., Chomat, M., Duvernay, F., Theulé, P., Guillemin, J.-C., Le Sergeant d'Hendecourt, L., Chiavassa, T.: Experimental investigation of aminoacetonitrile formation through the Strecker synthesis in astrophysical-like conditions: reactivity of methanimine (CH$_2$NH), ammonia (NH$_3$), and hydrogen cyanide (HCN). A&A **535**, A47 (2011)

9. Rimola, A., Sodupe, M., Ugliengo, P.: Deep-space glycine formation via Strecker-type reactions activated by ice water dust mantles. A computational approach. Phys. Chem. Chem. Phys. **12**, 5285 (2010)

10. Ugliengo, P., Rimola, A., Sodupe, M.: In silico study of the interstellar prebiotic formation and delivery of glycine. Rendiconti Lincei **22**, 137 (2011)

11. He, C., Smith, M.A.: Identification of nitrogenous organic species in Titan aerosols analogs: implication for prebiotic chemistry on Titan and early Earth. Icarus **238**, 86 (2014)

12. Holtom, P.D., Bennet, C.J., Osamura, Y., Mason, N.J., Kaiser, R.I.: A combined experimental and theoretical study on the formation of the amino acid glycine (NH$_2$CH$_2$COOH) and its isomer (CH$_3$NHCOOH) in extraterrestrial ices. Astrophys. J. **626**, 940 (2005)

13. Arnaud, R., Adamo, C., Cossi, M., Milet, A., Vallée, Y., Barone, V.: Addition of hydrogen cyanide to methanimine in the gas phase and in aqueous solution. J. Am. Chem. Soc. **122**, 324 (2000)

14. Godfrey, P.D., Brown, R.D., Robinson, B.J., Sinclair, M.W.: Discovery of interstellar methanimine (formaldimine). Astrophys. Lett. **13**, 119 (1973)

15. Balucani, N., Bergeat, A., Cartechini, L., Volpi, G.G., Casavecchia, P., Skouteris, D., Rosi, M.: Combined crossed molecular beam and theoretical studies of the N(^{2}D) + CH$_4$ reaction and implications for atmospheric models of Titan. J. Phys. Chem. **113**, 11138 (2009)

16. Balucani, N., Leonori, F., Petrucci, R., Stazi, M., Skouteris, D., Rosi, M., Casavecchia, P.: Formation of nitriles and imines in the atmosphere of Titan: combined crossed-beam and theoretical studies on the reaction dynamics of excited nitrogen atoms N(^{2}D) with ethane. Faraday Discuss. **147**, 189 (2010)

17. Redondo, P., Pauzat, F., Ellinger, Y.: Theoretical survey of the NH+CH$_3$ potential energy surface in relation to Titan atmospheric chemistry. Planet. Space Sci. **54**, 181 (2006)

18. Yelle, R.V., Vuitton, V., Lavvas, P., Klippenstein, S.J., Smith, M.A., Hörst, S.M., Cui, J.: Formation of NH$_3$ and CH$_2$NH in Titan's upper atmosphere. Faraday Discuss. **147**, 31 (2010)

19. Lavvas, P., Coustenis, A., Vardavas, I.: Coupling photochemistry with haze formation in Titan's atmosphere, Part I: model description. Planet. Space Sci. **56**, 27 (2008)
20. Lavvas, P., Coustenis, A., Vardavas, I.: Coupling photochemistry with haze formation in Titan's atmosphere, Part II: results and validation with Cassini/Huygens data. Planet. Space Sci. **56**, 67 (2008)
21. Hamberg, M., Österdahl, F., Thomas, R.D., Zhaunerchyk, V., Vigren, E., Kaminska, M., af Ugglas, M., Källberg, A., Simonsson, A., Paál, A., Larsson, M., Geppert, W.D.: Experimental studies of the dissociative recombination processes for the dimethyl ether ions $CD_3OCD_2^+$ and $(CD_3)_2OD^+$. A&A **514**, A83 (2010)
22. Taquet, V., Wirstrom, E.S., Charnley, S.B.: Formation and recondensation of complex organic molecules during protostellar luminosity outbursts. http://arxiv.org/pdf/1602.05364.pdf
23. Becke, A.D.: Density-functional thermochemistry. III. The role of exact exchange. J. Chem. Phys. **98**, 5648 (1993)
24. Stephens, P.J., Devlin, F.J., Chabalowski, C.F., Frisch, M.J.: Ab Initio calculation of vibrational absorption and circular dichroism spectra using density functional force fields. J. Phys. Chem. **98**, 11623 (1994)
25. Dunning Jr., T.H.: Gaussian basis sets for use in correlated molecular calculations. I. The atoms boron through neon and hydrogen. J. Chem. Phys. **90**, 1007 (1989)
26. Woon, D.E., Dunning Jr., T.H.: Gaussian basis sets for use in correlated molecular calculations. III. The atoms aluminum through argon. J. Chem. Phys. **98**, 1358 (1993)
27. Kendall, R.A., Dunning Jr., T.H., Harrison, R.J.: Electron affinities of the first-row atoms revisited. Systematic basis sets and wave functions. J. Chem. Phys. **96**, 6796 (1992)
28. Gonzales, C., Schlegel, H.B.: An improved algorithm for reaction path following. J. Chem. Phys. **90**, 2154 (1989)
29. Gonzales, C., Schlegel, H.B.: Reaction path following in mass-weighted internal coordinates. J. Phys. Chem. **94**, 5523 (1990)
30. Bartlett, R.J.: Many-body perturbation theory and coupled cluster theory for electron correlation in molecules. Annu. Rev. Phys. Chem. **32**, 359 (1981)
31. Raghavachari, K., Trucks, G.W., Pople, J.A., Head-Gordon, M.: A fifth-order perturbation comparison of electron correlation theories. Chem. Phys. Lett. **157**, 479 (1989)
32. Olsen, J., Jorgensen, P., Koch, H., Balkova, A., Bartlett, R.J.: Full configuration–interaction and state of the art correlation calculations on water in a valence double-zeta basis with polarization functions. J. Chem. Phys **104**, 8007 (1996)
33. Frisch, M.J., Trucks, G.W., Schlegel, H.B., Scuseria, G.E., Robb, M.A., Cheeseman, J.R., Montgomery Jr., J.A., Vreven, T., Kudin, K.N., Burant, J.C., Millam, J.M., Iyengar, S.S., Tomasi, J., Barone, V., Mennucci, B., Cossi, M., Scalmani, G., Rega, N., Petersson, G.A., Nakatsuji, H., Hada, M., Ehara, M., Toyota, K., Fukuda, R., Hasegawa, J., Ishida, M., Nakajima, T., Honda, Y., Kitao, O., Nakai, H., Klene, M., Li, X., Knox, J.E. Hratchian, H.P., Cross, J.B., Adamo, C., Jaramillo, J., Gomperts, R., Stratmann, R.E., Yazyev, O., Austin, A.J., Cammi, R., Pomelli, C., Ochterski, J.W., Ayala, P.Y., Morokuma, K., Voth, G.A., Salvador, P., Dannenberg, J.J., Zakrzewski, V.G., Dapprich, S., Daniels, A.D., Strain, M.C., Farkas, O., Malick, D.K., Rabuck, A.D., Raghavachari, K., Foresman, J.B., Ortiz, J.V., Cui, Q., Baboul, A.G., Clifford, S., Cioslowski, J., Stefanov, B., Liu, B.G., Liashenko, A., Piskorz, P., Komaromi, I., Martin, R.L., Fox, D.J., Keith, T., Al-Laham, M.A., Peng, C.Y., Nanayakkara, A., Challacombe, M., Gill, P.M.W., Johnson, B., Chen, W., Wong, M.W., Gonzalez, C., Pople, J.A.: Gaussian 09, Revision A.02. Gaussian, Inc., Wallingford (2004)
34. Flükiger, P., Lüthi, H.P., Portmann, S., Weber, J.: MOLEKEL 4.3, Swiss Center for Scientific Computing, Manno (Switzerland) (2000–2002)

35. Portmann, S., Lüthi, H.P.: MOLEKEL: an interactive molecular graphics tool. Chimia **54**, 766 (2000)
36. Balucani, N., Ceccarelli, C., Taquet, V.: A combined crossed molecular beams and theoretical study of the reaction CN + C_2H_4. MNRAS **449**, L16 (2015)
37. Vazart, F., Latouche, C., Skouteris, D., Balucani, N., Barone, V.: Cyanomethanimine isomers in cold interstellar clouds: Insights from electronic structure and kinetic calculations. Astrophys. J. **810**, 111 (2015)
38. Balucani, N., Cartechini, L., Casavecchia, P., Homayoon, Z., Bowman, J.M.: A combined crossed molecular beam and quasiclassical trajectory study of the Titan-relevant $N(^2D) + D_2O$ reaction. Mol. Phys. **113**, 2296 (2015)
39. Homayoon, Z., Bowman, J.M., Balucani, N., Casavecchia, P.: Quasiclassical trajectory calculations of the $N(^2D) + H_2O$ reaction elucidating the formation mechanism of HNO and HON seen in molecular beam experiments. J. Phys. Chem. Lett. **5**, 3508 (2014)
40. Balucani, N., Skouteris, D., Leonori, F., Petrucci, R., Hamberg, M., Geppert, W.D., Casavecchia, P., Rosi, M.: Combined crossed beam and theoretical studies of the $N(^2D)+ C_2H_4$ reaction and implications for atmospheric models of Titan. J. Phys. Chem. A **116**, 10467 (2012)

An Algorithm for Smallest Enclosing Circle Problem of Planar Point Sets

Xiang Li[1(✉)] and M. Fikret Ercan[2]

[1] Department of Physics and Electronic Engineering,
Hanshan Normal University, Chaozhou 521041, China
xl_huse@126.com
[2] School of Electrical and Electronic Engineering,
Singapore Polytechnic, 500 Dover Road, Singapore 139651, Singapore
mfercan@sp.edu.sg

Abstract. In this paper, a new computational algorithm is proposed for accurately determining the Smallest Enclosing Circle (SEC) of a finite point set (P) in plane. The set P that we are concerned here contains more than two non-collinear points which is a typical case. The algorithm basically searches for three particular points from P that forms the desired SEC of the set P. The SEC solution space of arbitrary P is uniform under this algorithm. The algorithmic mechanism is simple and it can be easily programmed. Our analysis proved that algorithm is robust and our empirical study verified its effectiveness. The computational complexity of the algorithm is found to be O($n$log$n$).

Keywords: Computational geometry · Planar point sets · Smallest enclosing circle · Convex hull

1 Introduction

The smallest enclosing circle (SEC) is defined as the circle with minimum radius that encloses all the given points in plane. SEC feature plays an important role in modern manufacturing industry and in solving many practical location problems. In manufacturing, the derived SEC is essential to the functionality oriented quality evaluation of manufactured shaft or ball features that are specified with the maximum material condition [1]. This is because the functional suitability of a part for an assembly or the proper functional suitability of a part within a mechanical system is largely influenced by the geometric envelope surfaces that are defined by the least or maximum material condition. To evaluate the geometric errors on manufactured parts, features such as planes, circles and spheres are utilized and the analysis often requires SEC features to be obtained [2]. The center of SEC is usually taken as the optimal position for a typical location problem. Typical applications in practice include identifying the location of a single facility used in an emergency facility model, locating a head office or an industrial plant, and setting up a radar station or radio transmitter to cover a particular region [3]. In these cases, it is desirable to minimize the maximum travel distance rather than the average or total distance. In addition, some geometric problems in the area of robotics and image analysis involve processing a simple geometric model that

© Springer International Publishing Switzerland 2016
O. Gervasi et al. (Eds.): ICCSA 2016, Part I, LNCS 9786, pp. 309–318, 2016.
DOI: 10.1007/978-3-319-42085-1_24

represents a set of points [4]. It is easier in these situations to use a non polygonal SEC model that may be expressed algebraically, though the convex hull of the point set is also a workable option.

A number of algorithms for computing the SEC have been reported in the literature. Earlier algorithms constructed an initial covering circle through a pair of points and then obtained the SEC by iteratively updating the pair of points [5–7]. Recently, more iterative algorithms sought the SEC solution that passes through two or three points [8, 9]. The Voronoi diagram [10, 11] and polar coordinate transformation [12, 13] based algorithms were developed effectively. The optimization based approaches were also reported including quadratic programming [14], stochastic optimization based on simulated annealing and Hooke-Jeeves pattern search [15], and Tschebyscheff approximation [16]. However, as pointed out by Anthony et al. [17], there is only a few research outcome reported on the solution configurations of these problems. The possible reason is that the solution configurations vary in different situations. For instance, the SEC may be defined by three points forming an acute triangle. It may also be defined by two points forming the diameter of the SEC. However, in this paper we demonstrate that there exists uniform solution configuration Minimal Angle (MA) subset in P that covers all the possible solution configurations for the SEC of the set P. The subset consists of three particular points from the P. The three points follow a particular geometric property, called the Minimal Angle (MA) in this work. The SEC of the MA subset searched out from P by the proposed algorithm will also be the SEC of the P. The presented solution theory guarantees that the MA subset searched out by our algorithm is a sufficient condition for determining the SEC of P. Therefore, a new SEC algorithm is developed to first identify a MA point subset in P and then solve its SEC taken as the SEC of the P. The proposed algorithm is simple and has a low computational complexity equivalent to $O(n\log n)$.

Remainder of the paper is organized as following. We first give the related notations and definitions, including the MA subset, and then present our SEC solution theory in Sect. 2. Section 3 describes the SEC solution algorithm in detail and also presents its computational complexity. In Sect. 4, the numerical simulating results of the proposed algorithm are demonstrated. Section 5 concludes the paper.

2 SEC Solution Theory

Unless otherwise mentioned, here P represents a finite set of $n(n > 2)$ points with no collinear in plane. The set P is the commonly encountered case in practice.

Definition 1 (MA subset): If the three points q_1, q_2, q_3 in set P simultaneously satisfy the equations of $\angle q_2 q_1 q_3 = min(\angle q_2 q_i q_3)$ (i ≠ 2, 3), $\angle q_1 q_2 q_3 = min(\angle q_1 q_i q_3)$ (i ≠ 1, 3) and $\angle q_1 q_3 q_2 = min(\angle q_1 q_i q_2)$ (i ≠ 1, 2), where $q_i \in P$, the subset consisting of q_1, q_2, q_3 are defined as a *MA subset* of P, denoted as $\{q_1, q_2, q_3\}_P$. $\triangle q_1 q_2 q_3$ is called a *MA triangle* of P.

Definition 2 (Convex hull [18]): The minimal convex polygon enclosing all points in P is defined as the *convex hull* of P, denoted as $CH(P)$.

It is known from Definition 2, *CH*(P) is a *convex polygon*. An example of *convex polygon CH*(P) is shown in Fig. 1 with dashed lines.

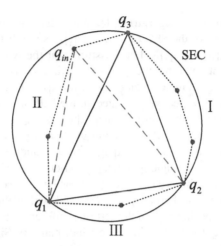

Fig. 1. $\triangle q1q2q3$ – acute triangle.

Lemma 1: Let Q be the set of points in P at the vertex positions of *CH*(P), and then the SEC of Q must be the SCE of P.

Proof: Any point in P at the vertex position of *CH*(P) is denoted by q_j. Q is equal to set $\{q_j\}$. Evidently, $Q \subseteq P$ holds. Let SEC_Q denote the smallest enclosing circle of Q. Since $Q \subseteq P$, thus we only need to interpret that SEC_Q must enclose all points in P\Q. Since SEC_Q enclose all points in Q, it must enclose the convex polygon, i.e. *CH*(P), determined by Q. According to *Definition* 2, *CH*(P) encloses all points in P\Q. Therefore SEC_Q must enclose all points in P\Q. That is to say that SEC_Q is the SEC of P. The proof of *Lemma* 1 is completed.   □

Lemma 2: Let Q be the set of points at the vertex of a convex polygon in plane, $\{q_1, q_2, q_3\}_Q$ is a MA set of Q, and then the SEC of $\{q_1, q_2, q_3\}_Q$ must be the SEC of Q.

Proof: The proposition is proven for two possible cases below:

(1) $\triangle q_1q_2q_3$ is an acute triangle. In this case, the SEC of $\{q_1, q_2, q_3\}_Q$ is the same as the circumscribed circle of $\triangle q_1q_2q_3$ and passes through the three points. We only need to interpret that all points in $Q\backslash\{q_1, q_2, q_3\}_Q$ are also within the SEC. As shown in Fig. 1, the three edges of $\triangle q_1q_2q_3$ separate the set $Q\backslash\{q_1, q_2, q_3\}_Q$ into three subsets which are respectively in the half planes I, II and III. Since $\{q_1, q_2, q_3\}_Q$ is a MA subset of Q, there is $\angle q_1q_3q_2 = \min(\angle q_1q_iq_2)$, $\forall\, q_i \in Q\backslash\{q_1, q_2, q_3\}_Q$. Therefore, for any point q_{in} of $Q\backslash\{q_1, q_2, q_3\}_Q$ in the half planes I and II, there is $\angle q_1q_{in}q_2 > \angle q_1q_3q_2$. Based on triangle geometry, the circumference angles of same arcs are equal, hence it can be derived that the point q_{in} is definitely within the SEC. Due to the arbitrary nature of q_{in}, the part of points in $Q\backslash\{q_1, q_2, q_3\}_Q$ locating in the half planes I and II are all

within the SEC. Similarly, it can be derived that the part of points in $Q\backslash\{q_1, q_2, q_3\}_Q$ locating in the half plane III are all within the SEC too. In another words, when $\triangle q_1 q_2 q_3$ is an acute triangle, all the points in $Q\backslash\{q_1, q_2, q_3\}_Q$ must be within the SEC of $\{q_1, q_2, q_3\}_Q$.

(2) $\triangle q_1 q_2 q_3$ is an obtuse (or right) triangle. When $\triangle q_1 q_2 q_3$ is an obtuse triangle, it is known that the diameter of the SEC of $\{q_1, q_2, q_3\}_Q$ is the distance between the two farthest apart points in $\{q_1, q_2, q_3\}_Q$. Let's assume that the two points are q_1 and q_3. Then, we only need to interpret that the points in $Q\backslash\{q_1, q_2, q_3\}_Q$ are definitely within the SEC. As shown in Fig. 2, the diameter segment $q_1 q_3$ separates set $Q\backslash\{q_1, q_2, q_3\}_Q$ into the two subsets which respectively located in the half planes I and II. Choosing a reference position q' on the right circumference, there is $\angle q_1 q' q_3 = 90°$. Let q_i denote any one of the points in $Q\backslash\{q_1, q_2, q_3\}_Q$ locating in the half plane I. Since $\{q_1, q_2, q_3\}_Q$ is a MA subset of Q, there is $\angle q_1 q_i q_3 > \angle q_1 q_2 q_3 > 90°$, and then $\angle q_1 q_i q_3 > \angle q_1 q' q_3$ holds. Also known from the triangle geometry mentioned above, point q_i is certainly within the SEC. Due to the arbitrary nature of q_i, it can be derived that the part of points in $Q\backslash\{q_1, q_2, q_3\}_Q$ locating in the half plane I are all within the SEC. Similarly, we choose another reference position q'' on the left circumference, let q_j denote any one of the points in $Q\backslash\{q_1, q_2, q_3\}_Q$ locating in the left half plane II. Since $\{q_1, q_2, q_3\}_Q$ is a MA point subset of Q, so there is $\angle q_1 q_j q_3 > \angle q_1 q_2 q_3 > \angle q_1 q' q_3$, yet $\angle q_1 q'' q_3 = \angle q_1 q' q_3 = 90°$, then $\angle q_1 q_j q_3 > \angle q_1 q'' q_3$ holds. So q_j is definitely within the SEC. It is derived that the part of points in $Q\backslash\{q_1, q_2, q_3\}_Q$ locating in the half plane II are all within the SEC too. In summary, when $\triangle q_1 q_2 q_3$ is an obtuse triangle, all the points in $Q\backslash\{q_1, q_2, q_3\}_Q$ are certainly within the SEC. When $\triangle q_1 q_2 q_3$ is exactly a right triangle, additionally the SEC of $\{q_1, q_2, q_3\}_Q$ passes another point q_2. Just taking q_2 as q', similar analysis can be conducted and the same result that is all points in $Q\backslash\{q_1, q_2, q_3\}_Q$ are within the SEC can be obtained.

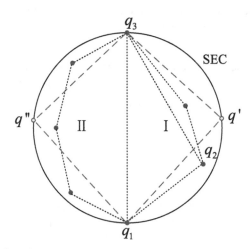

Fig. 2. $\triangle q_1 q_2 q_3$ – obtuse triangle.

Synthesizing the above discussion of the two cases, the proof of *Lemma* 2 is completed.   □

Theorem 1: Let Q be the set of points in P at the vertex positions of $CH(P)$, $\{q_1, q_2, q_3\}_Q$ is a MA subset of Q, and then the SEC of $\{q_1, q_2, q_3\}_Q$ must be the SEC of P.

Proof: From *Lemma* 2 it is known that the SEC of $\{q_1, q_2, q_3\}_Q$ must be the SEC of Q. Further that the SEC of Q must be the SEC of P is drawn from *Lemma* 1. So the SEC of $\{q_1, q_2, q_3\}_Q$ must be the SEC of P. This completes the proof of *Theorem* 1.   □

3   Algorithm Description

According to Theorem 1, we present the following specific computational procedure to accurately determine the SEC of P. Input to the algorithm is P which is a finite set of n ($n > 0$) points in plane.

STEP 1: Obtain the set of points at edges of $CH(P)$ using Graham method [19] according to substep (1) and (2).

(1) Sort points in P. The point with minimal value of y coordinate in P is denoted by p_1. If there are more than one point found, further sort them and consider the point with minimal value of x coordinate as p_1. The p_1 is usually called *base point*. Let p_i denote any one of the points in $P\backslash\{p_1\}$. Then sort points in $P\backslash\{p_1\}$ from small to large based on the angle size between p_ip_1 and x coordinate direction. If there are multiple points having the same angle size, further sort this part of points from near to far based on their distances from p_1. Finally, we have a point sequence $p_1, p_2, ..., p_n$, where p_1 and p_n must be at the vertex positions of $CH(P)$, p_2 must be at a edge of $CH(P)$. Let $p_{n+1} = p_1$.

(2) Delete the points not at the edges of $CH(P)$.

```
BEGIN
<1> k=4;
<2> j=2;
<3> IF p_1 and p_k are at both sides of segment p_{k-j+1}p_{k-j}, then delete p_{k-1}, let the indexes of
follow-up points minus one, and k=k-1, j=j-1; ELSE temporarily took p_{k-1} as a point at
the edges of CH(P).
<4> j=j+1, goto <3>, till j=k-1
<5> k=k+1, goto <2>, till k=n+1 (Note: n will decrease successively due to deleting
points)
END
```

After substep (2), let P_1 denote the set of rest points in P.

The feature of Graham method is that almost all middle points at edges of $CH(P)$ are retained except that at the left edge connecting *base point* p_1. For instance, a point set P in plane is distributed as shown in Fig. 3(a), the dashed convex polygon is the $CH(P)$. For this particular distribution, each edge contains multiple points. After using Graham method, the output set P_1, $\{p_1, p_2, ..., p_{12}\}$, is shown in Fig. 3(b).

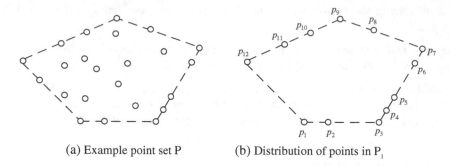

(a) Example point set P   (b) Distribution of points in P_1

Fig. 3. The result of Graham method applied to P.

STEP 2: Delete all the middle points at edges of $CH(P)$ in P_1, if they exist. When delete a point each time, decrement the indexes of following point by one. The set of remaining points in P_1 is denoted by P_2. In fact P_2 is the minimal set related to $CH(P)$.

IF $|P_2| = 1$, the SEC of P doesn't exist, EXIT. (Note: $|S|$ is the number of elements in set S)

IF $|P_2| = 2$, the SEC of P exists. Its diameter is the distance between the two points in P_2, EXIT.

IF $|P_2| = 3$, the three points in P_2 are denoted as p_1', p_2' and p_3' respectively, and GOTO **STEP 4**.

For instance, after applying deleting operation to P_1 in Fig. 3(b), P_2 would be as shown in Fig. 4.

STEP 3: Search out a MA subset $\{p_1', p_2', p_3'\}$ in P_2. Specifically, use iteration to obtain the three points p_1', p_2' and p_3' which simultaneously satisfy $min(\angle p_2' p_1' p_3')$ ($p_1' \in P_2 \backslash \{p_2', p_3'\}$), $min(\angle p_1' p_2' p_3')$ ($p_2' \in P_2 \backslash \{p_1', p_3'\}$) and $min(\angle p_1' p_3' p_2')$ ($p_3' \in P_2 \backslash \{p_1', p_2'\}$). When beginning to iterate, we consider the front three points as initial p_1', p_2' and p_3'. Alternatively, the initial three points can be selected randomly at start. If p_1' is not satisfying $min(\angle p_2' p_1' p_3')$, then consider the point p_i that satisfy $min(\angle p_2' p_i p_3')$ as p_1'. The same operation applies to other two points p_2' and p_3' in turn, continue until two of the three points are not updating anymore.

STEP 4: Solve the SEC of $\{p_1', p_2', p_3'\}$ as the desired SEC of P, EXIT.

The SEC of $\{p_1', p_2', p_3'\}$ is namely the SEC of $\Delta p_1', p_2', p_3'$. It is notable that the SEC of an acute triangle is the same as the circumscribed circle of the triangle. But the SEC of an obtuse (or right) triangle is the circle whose diameter is the distance between the two farthest apart points in $\{p_1', p_2', p_3'\}$

Next, we make a brief analysis for computational time complexity of the proposed algorithm. As set P contains n points, looking at *STEP 1*, the number of transferring to <2> from <5> is no more than n in sub step (2), each point is possibly deleted one time at most, and the number of deleted points can be no larger than n, so sub step

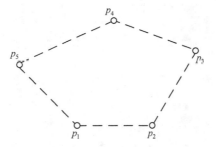

Fig. 4. Distribution of points in P_2.

(2) of *STEP 1* need linear time complexity. Sub step (1) of *STEP 1* is clearly related to the sort problem of angles. The $n-1$ included angles need to compute and be classified according to their sizes. The computation of each included angle only used constant time, and then the computation of $n-1$ included angles costs linear time. However, the classification of the angles need time $O(n\log n)$. Therefore, the total time complexity of Graham method part, i.e. *STEP1*, is $O(n\log n)$. It can be proved that Graham method is the best method for solving convex hull of a finite planar point set [20]. The size of the sets conducted in *STEP 2* and *STEP 3* is much smaller in comparison with initial set P, and needs linear time cost. *STEP 4* costs constant time. So, the overall computational time complexity of the algorithm can be approximated as $O(n\log n)$.

4 Numerical Simulation

We have implemented the above computational procedure of SEC on Breve platform [21] and verified its effectiveness considering three typical distributions of planar points. For each simulation, eighteen points from a random distribution are selected as input. In the first distribution, the MA subset searched out by our algorithm constitutes an acute triangle. The MA subset of the second distribution constitutes an obtuse triangle. But, in the third distribution, the situation of multi points locating at the left and right edges of convex hull is experimented. Using each point set as input, our algorithm calculate the center and diameter of corresponding SEC. Results are consistent with that of other complex or approximate algorithms [10, 15]. The specific distributions of each input point set and the calculated results are listed below for comparison. To facilitate comparative observation, the calculated SEC and corresponding point distribution are demonstrated in each snapshot. The three red points in each snapshot are marked as the MA subset of the related P searched out by our algorithm.

(1) MA points of this distribution constitute an acute triangle. Input point set A: (14.988556, −24.409619), (16.662191, 29.948729), (−23.631397, −17.392804), (−12.214423, −26.020997), (14.506973, −27.447432), (−15.812555, 21.681265), (−21.327860, 11.850032), (23.845637, 23.263344), (−5.734123, −24.039735), (9.458602, −25.348979), (25.806757, 16.413160), (20.060732, −29.774773), (7.999207, 22.992340), (7.713553, −22.827540), (−4.860683, 15.298013), (17.310404, 22.437513), (−9.290139, −6.517838), (5.878780, −5.673696). The calculated SEC:

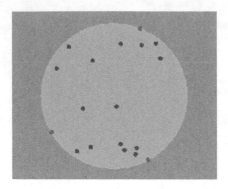

Fig. 5. The calculated SEC for point set A.

center = (4.70249, −0.69028); radius = 32.890475. The relative positions of A and its SEC are demonstrated in Fig. 5.

(2) MA points of this distribution constitute an obtuse triangle. Input point set B: (−11.872005, −1.324808), (21.494491, −28.066347), (−1.584826, −11.080966), (−24.638508, 15.030671), (0.779138, 5.223243), (−24.180731, 4.102603), (−10.714743, 1.211280), (−0.960417, −22.792749), (8.413038, 14.847560), (10.588397, −26.350597), (13.888058, 5.382550), (−11.829890, −19.458296), (17.433088, −19.566332), (9.205908, −5.805536), (−6.124149, −17.147435), (−19.692679, 10.200201), (5.129856, −1.103244), (12.245552, −23.556322). The calculated SEC: center = (−1.57201, −6.51784), radius = 31.565831. The relative positions of point set B and its SEC are demonstrated in Fig. 6.

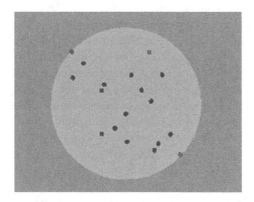

Fig. 6. The calculated SEC for point set B.

(3) Multi points locate at the left and right edges of convex hull. Input point set C: (−30.00000, −4.232612), (23.129673, 8.466140), (19.906919, −14.924467), (−23.177282, −20.386670), (−30.00000, 26.608783), (15.257729, 22.953887), (−11.403241, 7.188025), (−30.00000, −17.588733), (30.00000, 3.921323), (30.00000,

−16.129337), (8.947722, 11.086459), (−25.847041, −27.560961), (30.00000, −26.475845), (1.689199, 10.894192), (−6.202887, −23.680837), (30.00000, 25.966063), (−14.171880, −15.768609), (−4.419385, 1.061129). The calculated SEC: center = (0.00000, 0.06647), radius = 40.056141. The relative positions of point set C and its SEC are demonstrated in Fig. 7.

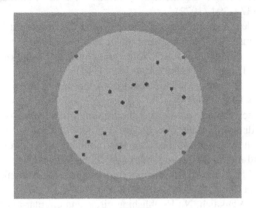

Fig. 7. The calculated SEC for point set C.

5 Conclusions

This paper described a computational algorithm for determining the smallest enclosing circle (SEC) of a finite point set in plane. Compared to the optimization based methods, the proposed algorithm is able to determine the accurate SEC of a finite set of points in plane. Accurate SEC is generally required in the context of needing quantitative analysis. Compared to other algorithms, the algorithm presented here is simple and easy to program. When the input data is a general set of $n(n > 2)$ points with no collinear in plane, the solution configuration (i.e. a MA subset) is invariant for the proposed algorithm. Specifically, three particular points which constitute a MA subset are searched out in the input set with the algorithm. The SEC of the three points must enclose all other points in the input set. The correctness of the proposed algorithm is proved theoretically. The result of numerical simulation also demonstrates its effectiveness. The computational complexity of the proposed algorithm approximates to O $(n\log n)$.

Acknowledgements. The authors appreciate the financial support from Doctor Scientific Research Startup Project of Hanshan Normal University (No. QD20140116). This work is also partly supported by the 2013 Comprehensive Specialty (Electronic Information Science and Technology) Reform Pilot Projects for Colleges and Universities granted by the Chinese Ministry of Education (No. ZG0411) and the Education Department of Guangdong Province in China (No. [2013]322).

References

1. Weckenmann, A., Eitzert, H., Garmer, M., Weber, H.: Functionality-oriented evaluation and sampling strategy in coordinate metrology. Precis. Eng. **17**(4), 244–252 (1995)
2. Huang, X., Gu, P.: CAD-model based inspection of sculptured surfaces with datum. Int. J. Prod. Res. **36**(5), 1351–1367 (1998)
3. Elzinga, D.J., Hearn, D.W.: Geometrical solutions for some minimax location problems. Transp. Sci. **6**(4), 379–394 (1972)
4. Oommen, B.J.: A learning automaton solution to the stochastic minimum spanning circle problem. IEEE Trans. Syst. Man Cybern. **16**(4), 598–603 (1986)
5. Chakraborty, R.K., Chaudhuri, P.K.: Note on geometrical solution for some minimax location problems. Transp. Sci. **15**(2), 164–166 (1981)
6. Oommen, B.J.: An efficient geometric solution to the minimum spanning circle problem. Oper. Res. **35**(1), 80–86 (1987)
7. Chrystal, G.: On the problem to construct the minimum circle enclosing N given points in the plane. Proc. Edinburgh Math. Soc. **3**, 30–33 (1885)
8. Li, X., Shi, Z.: The relationship between the minimum zone circle and the maximum inscribed circle and the minimum circumscribed circle. Precis. Eng. **33**(3), 284–290 (2009)
9. Jywe, W.Y., Liu, C.H., Chen, C.K.: The min-max problem for evaluating the form error of a circle. Measurement **26**(4), 777–795 (1999)
10. Shunmugam, M.S., Venkaiah, N.: Establishing circle and circular-cylinder references using computational geometric techniques. Int. J. Adv. Manuf. Technol. **51**(1), 261–275 (2010)
11. Gadelmawla, E.S.: Simple and efficient algorithms for roundness evaluation from the coordinate measurement date. Measurement **43**(2), 223–235 (2010)
12. Li, X., Shi, Z.: Development and application of convex hull in the assessment of roundness error. Int. J. Mach. Tools Manuf. **48**(1), 135–139 (2008)
13. Lei, X., Zhang, C., Xue, Y., Li, J.: Roundness error evaluation algorithm based on polar coordinate transform. Measurement **44**(2), 345–350 (2011)
14. Nair, K.P.K., Chandrasekaran, R.: Optimal location of a single service center of certain types. Naval Res. Logist. Q. **18**, 503–510 (1971)
15. Chen, M.C., Tsai, D.M., Tseng, H.Y.: A stochastic optimization approach for roundness measurements. Pattern Recogn. Lett. **20**(7), 707–719 (1999)
16. Goch, G., Lübke, K.: Tschebyscheff approximation for the calculation of maximum inscribed/minimum circumscribed geometry elements and form deviations. CIRP Ann. Manuf. Technol. **57**(1), 517–520 (2008)
17. Anthony, G.T., Anthony, H.M., Bittner, B., Butler, B.P., Cox, M.G., Drieschner, R., et al.: Reference software for finding Chebyshev best-fit geometric elements. Precis. Eng. **19**(1), 28–36 (1996)
18. Zhou, P.: An algorithm for determining the vertex of the convex hull. J. Beijing Inst. Technol. **13**(1), 69–72 (1993)
19. Graham, R.L.: An efficient algorithm for determining the convex hull of a finite planar set. Info. Proc. Lett. **1**, 132–133 (1972)
20. Zhou, P.: Computational geometry algorithm design and analysis, 4th edn, pp. 79–81. Tsinghua University Press, Beijing (2011)
21. Klein, J.: Breve: a 3D environment for the simulation of decentralized systems and artificial life. In: Proceedings of Artificial Life VIII, the 8th International Conference on the Simulation and Synthesis of Living Systems. The MIT Press (2002)

Simulation of Methane Production from Carbon Dioxide on a Collaborative Research Infrastructure

Carles Martí[1(✉)], Leonardo Pacifici[1],
Andrea Capriccioli[2], and Antonio Laganà[1]

[1] Dipartimento di Chimica, Biologia e Biotecnologie,
Università di Perugia, 06123 Perugia, Italy
carles.martialiod@studenti.unipg.it, leonardo.pacifici@unipg.it,
lagana05@gmail.com
[2] ENEA, Via Fermi 45, Frascati, Italy
andrea.capriccioli@enea.it

Abstract. We have implemented the numerical simulation of the kinetics of the Ni-catalyzed $H_2 + CO_2$ process to assist the development of a prototype experimental apparatus producing methane. To this end the simulation program has been ported onto the segment of the distributed platform available to the Virtual Organization COMPCHEM as part of a set of use cases gathered to the end of establishing a European Research Infrastructure. The model adopted, the structure of the software and its parallel reorganization are discussed by taking as a reference the present working conditions of the apparatus in its location at the University of Perugia.

Keywords: Sabatier reaction · Kinetic model · Concurrent computing

1 Introduction

The present trend of chemical research and, as a consequence, of computational chemistry activities is increasingly moving out from purely *per se* theoretical investigations into service oriented ones meeting societal challenges. In this perspective our research laboratory has taken part into a joint private-public effort to build PROGEO 20 kW, a prototype apparatus based on a Power-to-Gas technology for electricity storage and CO_2 valorization. PROGEO 20 kW is an innovative electricity storage apparatus, based on the conversion of low cost (or not to produce in order to avoid instability in the electric network) electricity into methane through a high efficient process made of:

- an electrolyzation stage in which electricity is converted into high purity hydrogen (H_2);
- a methanation stage in which hydrogen reacts with carbon dioxide (CO_2) to produce methane and steam water.

© Springer International Publishing Switzerland 2016
O. Gervasi et al. (Eds.): ICCSA 2016, Part I, LNCS 9786, pp. 319–333, 2016.
DOI: 10.1007/978-3-319-42085-1_25

Due to the exothermic nature of the reaction, the setup requires a negligible additional energetic cost to be self-sustained. Moreover, CO_2 is a by-product of several processes and, consequently, it is interesting to consider it for recycling. The PROGEO apparatus has also the following positive additional characteristics:

- a fast start-up thanks to a highly flexible solution;
- the possibility of being used for feeding both electrical and methane distribution nets;
- a great safety level thanks to an intrinsically passive control system;
- an extremely large number of charge/discharge cycles;
- a modular integrability in stacks.

The apparatus, designed by Andrea Capriccioli, has been assembled at the PLC System s.r.l. an international company operating as supplier of electrical power. In particular, the company is specialized in designing, installing and monitoring energy supplying apparatuses (both from fossil and renewable sources). PROGEO is likely to become, in the near future, a product to offer to its customers. PROGEO 20 kW, in fact, targets a low level user who wishes to combine cheap energy (saving on its cost) with CO_2 recycling (saving on carbon tax) for its own business. However, the next planned PROGEO apparatus is PROGEO 0.5 MW more targeted to industrial applications (though still being a prototype experimental one) suitable for coupling with small power plants [1].

The increasing involvement of different expertise in the scale up of the applications (as is the mentioned increase in power from 20 to 500 kW and the already planned further upgrades) has originated from the synergy between research and infrastructural activities fostered by the forming of Virtual Organizations (VO) [2] and Virtual Research Communities [3] promoted by the EGEE (Enabling Grids for E-Science in Europe [4]) and EGI (European Grid Infrastructure) [5] calls (in particular, the COMPCHEM VO [6] and of the Chemistry, Molecular & Materials Science and Technology (CMMST) VRC [7] in the field of Molecular Sciences and Technologies (MST). The COMPCHEM VO, first, and (with a wider participation) the CMMST VRC, later, were in fact established with the specific purpose of supporting collaborative computational chemistry applications by both sharing grid resources and workflowing different competences and codes (more details will be given in Sect. 2). Along this line of action in order to create a more solid ground for collaborative research and innovation, jointly with the computer centres of Westminster (UK) and Tubingen (DE) and the cooperation of other institutions and companies, we assembled a proposal for the "INFRAIA-02-2017: Integrating Activities for Starting Communities" call aimed at establishing an MST Research Infrastructure (RI). The proposal [8], named Supporting Research in Computational and Experimental Chemistry via Research Infrastructure (SUMO-CHEM), addresses several societal challenges among which the storage of renewable energies as chemical energy and, in particular, as CH_4 (Power-to-gas) that is the goal of the PROGEO apparatus.

In this paper we report, indeed, on the first implementation of the numerical simulation (developed in parallel with the assemblage of the experimental apparatus) designed for the rationalization of the methanation process of PRO-GEO. Accordingly, in Sect. 2 we discuss the frame of the proposed MST RI; in Sect. 3 we describe in some detail the structure of the simulator of the Sabatier Ni-catalyzed carbon dioxide reduction to methane of PROGEO; in Sect. 4 we analyse the outcome of the simulations and we infer some features of the reactive mechanism; in Sect. 5 we present some indications on future work.

2   The Molecular Science and Technology Research Infrastructure

The main aim of the proposed MST RI is to integrate Molecular Science research facilities, Technology infrastructures and e-infrastructure resources to the end of enabling synergetic Computational and Experimental Chemistry joint endeavours targeting some primary societal challenges. This implies, in fact, not only the provision of the know how for an intuitive and seamless access to advanced experimental and compute facilities but also for the utilization of high level of theory multiscale computer simulations and their coordination with experimental outcomes. Crucial in both respects is the shared and efficient management of data including its creation, publishing, accessing, curation and preservation using metadata and ontologies based on advanced services. For this purpose, significant efforts need to be spent to articulate the produced knowledge for an optimal use by both academia and industry.

Experimental raw data is expected to be generated by different experimental apparatuses. The wider range of MST experimental apparatuses is that related to measurements of properties related to light-matter interaction. The top ranking type of experimental apparatuses involved in the proposal are Synchrotrons and Lasers (in particular those of the Trieste European facility: the Elettra synchrotron and the Fermi free-electron laser). At Elettra all of the most important X-ray based techniques are available, as well as facilities for infrared microscopy, spectroscopy or ultraviolet scattering. FERMI is a single pass seeded FEL light source delivering ultra-high brilliance, fully coherent, ultra-short UV and X-ray pulses from the UV down to ~4 nm. The two light sources allow to characterize material properties and functions with sensitivity down to molecular and atomic levels, to pattern and nanofabricate new structures and devices, and to develop new processes. The light source and the available experimental stations offer unique opportunities for exploring the structure and transient states of condensed, soft and low-density matter using a variety of diffraction, scattering and spectroscopy techniques. In addition the ultrafast lasers of the University of Madrid CLUR partner of the project provides the members of the proposal with high power lasing combined with multi-photon ionization spectroscopy and time-of-flight mass spectrometry. These research facilities also allow manipulation of materials (laser micro-fabrication, laser modification of materials, laser ablation) and the synthesis of new materials. They also allow to investigate laser-induced plasma and phenomena under non-equilibrium conditions. In addition FLASH (the Free-Electron LASer for VUV and soft X-ray radiation in Hamburg operated in the "self-amplified spontaneous emission" (SASE) mode and currently covering a wavelength range from 4.2 nm to about 45 nm in the first harmonic with GW peak power and pulse durations between 50 fs and 200 fs), PETRA (the accelerator of the DESY site in Hamburg now operating as the most brilliant storage ring based X-ray source worldwide enabling Pump-probe X-ray absorption experiments with high photon flux in the energy range of 50–150 keV that can be performed on it in a 40-bunch mode) and LENS (the European Laboratory for Non-linear Spectroscopy, providing short-pulse lasers as experimental facilities for spectroscopic and non-linear optics research with frequency domain from the far IR to the extreme UV with the highest available resolution, with a time domain ranging from few femto to nanoseconds, allowing running time-resolved spectroscopic experiments such as degenerate and non-degenerate four wave mixing experiments (photon echo, optical Kerr effect, transient gratings, etc.), time resolved fluorescence, two-dimensional infrared spectroscopy, transient absorption and stimulated emission and multiphoton spectroscopy) are also other examples of the invaluable light-matter experimental machinery of the proposed European MST RI.

The second set of MST experimental apparatuses involved in the proposal are those producing raw data related to matter-matter interaction. Typical instrumentation of this type are the molecular beam-molecular beam or -gas collision apparatuses. Molecular Beams Laboratories can provide velocity, internal state, angular momenta dependence (in addition to mass spectrometric and time of flight) selected for reactants, products and transient analysis for single collision processes. This allows to estimate the value of the transition matrices, evaluate the role played by the reaction intermediate, validate the formulation of the molecular interaction and propose a process mechanism for different collision and internal energy values. More averaged, though spanning larger ranges of energy, are the experiments carried out for gas phase plasmas, combustion and biomasses. Related facilities support research on the formation and exploitation of oil, the design and implementation of advanced engines, the thermochemical conversion of biomasses, the combustion through chemical looping, the processes for transforming and storing energy. Moreover, laser-induced plasma and phenomena under non-equilibrium conditions provide information on the efficient use of energy in technological applications, such as negative ion sources for nuclear fusion, material science for aerospace and microelectronics applications, plasma-based energy recovery devices as well as micro-discharges for active flow control, and material synthesis and characterization.

The third set of raw data is obtained by more commercial apparatuses of large use by scientific laboratories and institutions like NMR, HPLC, X-ray, spectroscopes, etc. or assemblages of various types of this kind of equipments like burners, reactors, heat exchangers, etc. plus some *ad hoc* designed pieces as is the case of the already mentioned PROGEO object of our investigations.

The resulting layer of raw data resources will have to be made available in a proper (even if only *de facto*) standard format to researchers access through a science gateway. The science gateway will not only enable researchers to access research facilities, computing and data resources allowing experimental chemists to run experiments on remotely available research facilities but it will support also training activities, community building, and especially designed data services to run simulations through the submission service using different types of data resources (such as data archives, databases, data collections, data storages) and workflows.

The above given picture of the experimental facilities belonging to the MST RI will allow to manage extended libraries of computational applications covering the real Virtual Laboratories programs. Invaluable to this end is the extensive use of the computational resources and innovative on-line services of the EGI cloud and grid resources and of some partners like ENEA (CRESCO). This platform provides a unified user environment and a seamless user-friendly access method to combine computing resources and experimental facilities. It provides a unified user environment and a seamless user-friendly access method to combine computing resources and experimental facilities irrespective of their location. Similar support is planned to be given by MoSGrid the compute infrastructure configuring and providing Grid services for molecular simulations leveraging on an extensive use of D-Grid Infrastructure for high-performance computing. It includes metadata and their provision for data mining and knowledge generation. MoSGrid aims at supporting the users in all fields of simulation calculations. Via a portlet, the user can access data repositories where information on molecular properties as well as on "recipes" - standard methods for the provided applications are stored. By means of these recipes simulation jobs can be automatically generated and submitted into the Grid (Pre-processing and Job Submission). Moreover, the users are supported at the analysis of their calculation results. Through the cross-referencing of different result data sets, new insights can be achieved. The data repository additionally allows external referencing of simulation results. Further compute support will be provided from ReCaS (a cluster of 4 national Italian data centres (Napoli, Bari, Catania, Cosenza) of the national Italian Grid Infrastructure (IGI) that is part of the European Grid Infrastructure (EGI) and INFN. The key users are the research institutions in southern Italy regions and CERN (ATLAS, CMS, ALICE, LHCb). In addition to the mentioned compute infrastructures the following centers (EGI Federated Cloud and Grid, CINECA, PRACE, CRESCO, ZIH, some national and regional computer centres, openstack clouds, computer clusters, etc.) provide also compute support. As to the libraries of computational applications, the most popular of them are the *ab initio* simulation software packages such as ADF [9], DALTON [10], MCTDH [11], NWCHEM [12], etc. running on high performance computing resources, *ab initio* and density functional electronic molecular structure calculations using electronic structure calculation software (MOLPRO [13], MOLCAS [14], GAUSSIAN [15]) and dynamical calculation software (ABC [16], HIBRIDON [17], VENUS [18], QCTUCM [19]). Other in-house, open source or commercial codes available for use in the MST RI are: BTE (Boltzmann Transport Equations), DSMC (Direct Simulation Monte Carlo), PIC (Particle-in-Cell), PLASMA-FLU (plasma simulation), EPDA (Elementary Processes Data Aggregator), ORCA, Jaguar, MOPAC, DFTB+, MNDO99, QM/MM, CRECK, Pope,

ANOVA, EVB (Empirical Valence Bond), MOLARIS, Q depending on the needs of different experiments (we refer here to the list given).

3   The PROGEO Simulation

As mentioned above, our approach to the PROGEO simulations has been already designed along the guidelines of the CMMST VRC on which also the MST RI is being built using its compute platform.

In particular, the simulation of the methanation process of PROGEO was assembled starting from the already established Grid Empowered Molecular Simulator (GEMS) [20] used in recent times to carry out the OH + CO reaction virtual experiment whose products had been measured in a crossed beam experiment [21]. In the latter case the workflow was articulated in the following modules:

(a) INTERACTION: Devoted to the calculation of the electronic energies of atomic and molecular aggregates of the considered system using high level *ab initio* methods (was not used by us due to the fact that potential energy routines were already available from the literature [22–24]);

(b) FITTING: Devoted to the fitting to a suitable functional form of the electronic energy values available for the system of interest and to the assemblage of an analytical formulation of the Potential Energy Surface (PES) (was not used by us due to what has been mentioned above);

(c) DYNAMICS: Devoted to the integration of the equations of motion for the nuclei of the system with initial conditions adequate to reproduce the experimental situation (was used by us to perform extended quasiclassical trajectory calculations mimicking the internal and collision energy distributions of the experiment);

(d) OBSERVABLES: Devoted to the averaging over the unobserved parameters of the product pulses counted by the detector and properly transformed, thanks to the knowledge of the geometry of the apparatus, into the value of the cross section and of the angular and translational distributions.

To the end of simulating the processes occurring inside the PROGEO methanator, however, we have further articulated the OBSERVABLES module so as to incorporate the time evolution of the kinetics of the interleaved elementary processes. For this purpose, we extended the above described GEMS schema to the simulation of the PROGEO methanator both

– by introducing the integration in time of the system of the kinetic equations intervening in the process of forming CH_4 from H_2 and CO_2 over Ni (111)

and

– by considering as part of the kinetic system all the processes related to the interaction of the involved atoms and molecules with the surface of the catalyzer (see the leftmost column of Table 1).

Table 1. Elementary processes intervening in the $H_2 + CO_2$ and related activation energies (with the source reference in the rhs) for the forward and reverse process. Species with an asterisk (*) aside refer to adsorbed ones, meanwhile asterisk by their own refer to free adsorption sites.

Step	E_a forward (kJ/mol)	E_a reverse (kJ/mol)
$CO_2 + * \leftrightarrow CO_2*$	0.0 [25]	8.3 [26]
$H_2 + 2* \leftrightarrow 2H*$	4.0 [25]	77.1 [26]
$CO + * \leftrightarrow CO*$	0.0 [25]	127.7 [26]
$H_2O + * \leftrightarrow H_2O*$	0.0 [25]	49.0 [26]
$CO_2* + H* \leftrightarrow COOH* + *$	113.1 [26]	155.6 [26]
$CO_2* + 2H* \leftrightarrow C(OH)_2* + 2*$	292.3 [26]	217.8 [26]
$CO_2* + * \leftrightarrow CO* + O*$	93.7 [26]	169.3 [26]
$COOH* + * \leftrightarrow CO* + OH*$	306.8 [26]	308.7 [26]
$C(OH)_2* + H* \leftrightarrow CH_2O* + OH*$	98.7 [26]	125.7 [26]
$CH_2O* + H* \leftrightarrow CH_2* + OH*$	163.7 [26]	154.1 [26]
$CO* + * \leftrightarrow C* + O*$	237.4 [26]	111.8 [26]
$CO* + 2H* \leftrightarrow CH* + OH* + *$	221.4 [26]	146.1 [26]
$2CO* \leftrightarrow CO_2* + C*$	339.6 [27]	109.0 [27]
$C* + H* \leftrightarrow CH* + *$	69.2 [26]	154.1 [26]
$CH* + H* \leftrightarrow CH_2* + *$	68.2 [26]	61.9 [26]
$CH_2* + H* \leftrightarrow CH_3* + *$	71.4 [26]	105.6 [26]
$CH_3* + H* \rightarrow CH_4 + 2*$	137.4 [26]	178.7 [26]
$O* + H* \leftrightarrow OH* + *$	137.9 [26]	116.0 [26]
$OH* + H* \leftrightarrow H_2O* + *$	137.9 [26]	99.9 [26]
$H* + * \leftrightarrow * + H*$	13.0 [25]	13.0 [25]
$CO* + * \leftrightarrow * + CO*$	10.0 [25]	10.0 [25]
$O* + * \leftrightarrow * + O*$	48.0 [25]	48.0 [25]
$OH* + * \leftrightarrow * + OH*$	21.0 [25]	21.0 [25]

It is worth noticing here that the kinetic system of equations does not include at present pure gas phase reactions because, at the energetic conditions considered for our runs, they are not competitive with those occurring through adsorption on the surface of the catalyzer. The second important point to make is that, as common to many kinetic simulation packages, the rate coefficients k_p of the intervening elementary processes p are expressed in an Arrhenius like form (i.e. as the product of a pre-exponential factor (A) and an exponential term (X)). The two terms refer to an intermediate molecular geometry of the system assumed to be a transition state separating reactants from products along a properly chosen reaction coordinate. Moreover, once crossed, the transition state is assumed not to be recrossed back. Empirical corrections to this formulation are sometimes introduced by mitigating the no-recrossing condition and by introducing a steric factor accounting for the different efficiency of the process as the attack angle is varied. The exponential term is usually expressed as $X = e^{-E_p/k_B T}$, where k_B is the usual Boltzmann constant and T is the temperature of the system. E_p is a parameter quantifying the energetic barrier to reaction associated with the PES of the related elementary process p (for the direct or forward "f" and the reverse or backward "b") i.e. E_p is the difference between the energy associated with the stationary point of the potential Minimum Energy Path (MEP) at the transition state and the one associated with the original asymptote of the process. The values of E_p collected from the quoted references for both forward and reverse processes are reported in the central left and the central right columns, respectively, of Table 1. In the transition state approach adopted in our investigation, for entirely surface processes the pre-exponential factor can be formulated as $A = k_B T/h$ (where h is the Plank constant) while for adsorption processes, reaction rate (r_i) can be obtained from the well-known Hertz-Knudsen equation:

$$r_i = \frac{p_i A_{site}}{\sqrt{2\pi m_i k_B T}},$$

where p_i is the i species' partial pressure, A_{site} the surface area of adsorption site and m_i the mass of species i.

On this ground, we implemented a version of ZACROS [28, 29], a kinetic Monte Carlo (kMC) [30, 31] software package written in Fortran 2003, for simulating the kinetic behaviour of the $H_2 + CO_2$ system. ZACROS properly deals with molecular phenomena on catalytic surfaces [39] leveraging on the Graph-Theoretical kMC methodology coupled with cluster expansion Hamiltonians for the ad-layer energetics and Brønsted-Evans-Polanyi relations for the activation energies of elementary events. The rates of these elementary processes are expected to be computed in an *ab initio* fashion so as to enable the prediction of catalytic performances (such as activity and selectivity) from first principles. The package can also perform simulations of desorption/reaction spectra at a given temperature providing so far a rationale for designing kinetic mechanisms and supporting them when carrying out a comparison with experimental data. The ZACROS framework can naturally capture:

- steric exclusion effects for species that bind in more than one catalytic sites;
- complex reaction patterns involving adsorbates in specific binding configurations and neighbouring patterns;
- spatial correlations and ordering arising from adsorbate lateral interactions that involve many-body contributions;
- changes in the activation energies of elementary events, influenced by the energetic interactions of reactants with neighbouring spectator species;
- the elementary processes considered for our simulations are given in the leftmost column of Table 1.

4   The Simulation and Its Results

In spite of being the Sabatier reaction a well established textbook example, the mechanism of the Ni-catalyzed CO_2 methanation is still not fully understood as we shall discuss later in this section. The already mentioned preliminary measurements on PROGEO performed by A. Capriccioli suggest that the production of CH_4 with a H_2/CO_2 ratio of 5:1 (hydrogen excess) is effective in producing CH_4 with a good yield. Two main mechanisms have been proposed: schema (A) considers the reaction happening through adsorbed CO as intermediate [33,34], while schema (B) considers a direct hydrogenation of adsorbed CO_2 [35,36]. Furthermore, even for the subsequent steps of the first schema (the CO hydrogenation to methane) there is no consensus on the further steps. One hypothesis maintains that reaction evolves towards a CO dissociation to C + O, another considers the CO to be directly hydrogenated while a third one proposed by Martin et al. [37] considers the mechanisms mainly occurs through the disproportionation of CO, being, in all three cases, the proposed reaction the rate limiting one.

The kinetic Monte Carlo (kMC) simulations performed by us include the four aforementioned schemas as illustrated in Table 1. By using the activation energy values given in the table and the appropriate preexponential factor, the time evolution of the system components was computed at the desired conditions of temperature, pressure and initial gas phase molar fractions.

As a first step, an optimization of the most important parameters of the kMC simulation has been performed. When taking original reaction rates, simulation times are really small. This happens because the kinetic Monte Carlo timestep (Δt) is taken as

$$\Delta t = \frac{1}{\sum_{i=1}^{n} \sum_{j=1}^{k} r_{ij}},$$

where r_{ij} is the ith reaction on the jth site, n is the total number of considered steps and k the number of sites. Since diffusion rates are far larger than reactive ones, the system spends most of the time simulating the first ones. In order to accelerate the convergence of the calculations, prexponential factors for diffusive processes were multiplied by a α scaling factor ($\in (0,1]$). When this scaling factor is applied a huge increase in timestep is obtained, enabling to simulate longer times. It is always important to take care that no change on the final results would occur when applying this scaling factor. The final adopted values of α are given in Table 2 which led to six orders of magnitude increase in the simulated time.

Another parameter that was also optimized in the simulation is the size of the grid (given as the number of copies of the unit cell in two dimensions). It is to be clarified that in our case the unit cell used is not the primitive but one containing two sites. In this case too, a compromise between the length of the simulation time and the convergence of the results was to be found. A value of 25×25 unit cells was found to reproduce same results as larger lattices while significantly reducing computation time.

Table 2. Elementary processes which optimized scaling factor α was applied. Species with an asterisk (*) aside refer to adsorbed ones, meanwhile asterisk by their own refer to free adsorption sites.

Process	original preexponential factor	α coefficient applied
H* + * ↔ * + H*	$2.027 \cdot 10^{13}$	10^{-4}
CO* + * ↔ * + CO*	$2.027 \cdot 10^{13}$	10^{-5}
O* + * ↔ * + O*	$2.027 \cdot 10^{13}$	10^{-2}
OH* + * ↔ * + OH*	$2.027 \cdot 10^{13}$	10^{-3}

Using the optimized parameters several runs were performed by varying the temperature (T=300, 500, 600, 700, 800, 900, 1000 and 1100°C) to show which mechanism was the one that occurred the most. The rate limiting steps of each one were the CO decomposition (CO* + * ↔ C* + O*), the $C(OH)_2$ formation (CO_2* + 2H* ↔ $C(OH)_2$* + 2*), the CO hydrogenation (CO* + 2H* ↔ CH* + OH* + *) and the CO disproportionation (2CO* ↔ CO_2* + C*), where species with an asterisk (*) aside refer to adsorbed ones, meanwhile asterisk by their own refer to free adsorption sites. Computed results show that while at 300 and 500 °C, no CH_4 is formed, from 600 to 1100 methane is indeed produced and, as shown in Fig. 1 the largest fraction of produced CH_4 (well above 90 %) is obtained from mechanism A (the hydrogenation of the adsorbed CO concretely through the process CO* + 2H* ↔ CH* + OH* + *).

Accordingly, the CO decomposition plays always a residual (even more marginal at low temperatures) role. This finding is in contrast with the dominance of mechanism B proposed by the authors of ref [26] that postulates a CO decomposition into C + O on the ground of some *ab initio* electronic structure considerations,

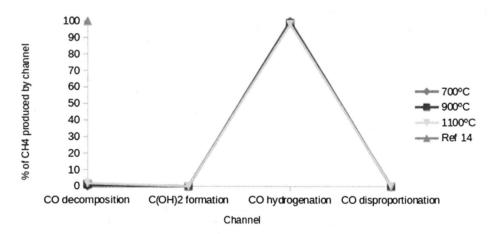

Fig. 1. Percent of formation of CH_4 from the different candidate rate determining steps (reaction mechanism). Green value is not quantitative but only to indicate which is the rate limiting step proposed. (Color figure online)

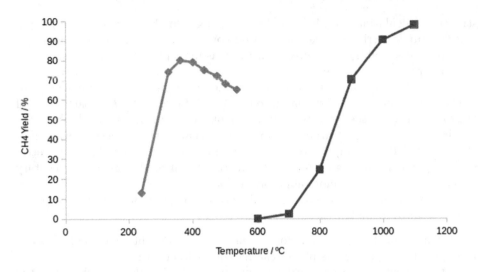

Fig. 2. Plot of the yield of CH_4 as a percent of the reactant CO_2.

see the green triangle shown in 1 (upper lhs corner) without resorting into a true kinetic study.

The analysis of the computed CH_4 yield at a pressure of 10 atm and a H_2/CO_2 ratio of 5:1 as a function of temperature (T) (see Fig. 2) exhibits a threshold at about 700°C. The yield grows to about 70 % at 900°C and reaches almost 100 % at 1100°C. This behavior is in line with the fact that the rate determining step of the CH_4 production was found in our calculations to be the hydrogenation of CO^* that has a higher energy barrier than that of the CO production but also a higher pre-exponential factor leading to higher production at high temperature. As is apparent from the Figure there is a clear difference between the measured and the computed threshold temperature [38], possibly due to the fact that we were simulating only Ni(111) surface instead of the whole catalyst, which is far more complex [39]. As a matter of fact, the computed threshold temperature is 600°C higher than that of the experiment and the computed rate of increase of the yield as a function of T is one half of that of the experiment. However, being our simulation the first attempt to obtain a computational estimate of the yield of the Ni catalyzed process of PROGEO, it is not suprprising that though not quantitative the agreement between the experiment and the simulation is satisfactory.

5 Conclusions and Possible Developments

In this paper we discuss the problems met when implementing a simulation of the PROGEO methanator.

The general qualitative agreement of the simulation with the experimental results shows that not only the problem is a suitable problem to investigate and improve on the proposed European Research infrastructure but it already confirms that mechanisms suggested on the pure ground of *ab initio* electronic structure calculations can be completely wrong. On the contrary even simplified (assuming Arrhenius like formulations of the rate coefficients) calculations lead to results qualitatively agreeing with measurements performed on an experimental apparatus. On the computational side, however, the quantitative differences between computed and measured quantities confirm that there is need for improvement obtainable for a more complex collaborative RI approach. In a RI approach, in fact, the expected improvements are provided not only from the possibility of getting accurate *ab initio* data (thanks to the better availability of high level HPC compute resources) but also from the possibility of better handling first raw data and then related metadata, of combining complementary expertise, of picking up more appropriate codes.

Moreover, the discussion of the results obtained clearly indicates the need for further improvements in the:

– evaluation of the performance of the simulation under different conditions like different pressure or gas phase reactive molar fractions;
– improvement of the chemical mechanisms with the incorporation into the formulation of the rate coefficients corrective terms associated with the partition functions of reactants and transition states, tunneling effects, sticking coefficients, etc.;
– investigation of the possibility of obtaining the evaluation of the rate coefficients from accurate dynamical calculations;
– performing a sensitivity analysis to evaluate the effect of different catalysts.
– better integration of conditions with output statistical analysis and further graphical tools supporting a better understanding and a multimedia dissemination of the results.

Acknowledgments. Thanks are due to the PLC SYSTEM srl and to MASTER-UP srl for making the PROGEO experimental apparatus available at the University of Perugia.

Thanks are also due to COST CMST (action D37), EGI (projects EGEE III and EGI Inspire) and the Fondazione Cassa di Risparmio of Perugia (project 2014.0253.021 scientific and technological research) and the European Joint Doctorate on TCCM project ITN-EJD-642294. Theoretical Chemistry and Computational Modeling.

References

1. Capriccioli, A.: Report on PROGEO progress. VIRT&L-COMM. **8**, 1–2 (2015)
2. EGI Virtual Organisations definition. http://www.egi.eu/community/vos/. Accessed 30 Mar 2016
3. EGI Virtual Research Communities definition. http://www.egi.eu/community/vos/vrcs. Accessed 30 Mar 2016
4. EGEE Main website. http://cern.ch/egee. Accessed 30 Mar 2016
5. EGI Main website. http://www.egi.eu. Accessed 30 Mar 2016
6. Laganá, A., Riganelli, A., Gervasi, O.: On the structuring of the computational chemistry virtual organization COMPCHEM. In: Gavrilova, M.L., Gervasi, O., Kumar, V., Tan, C.J.K., Taniar, D., Laganá, A., Mun, Y., Choo, H. (eds.) ICCSA 2006. LNCS, vol. 3980, pp. 665–674. Springer, Heidelberg (2006)

7. VT Towards a CMMST VRC definition. https://wiki.egi.eu/wiki/Towards_a_ CMMST_VRC

8. Sumo-Chem, INFRAIA-02-2017, Proposal number: 731010-1

9. te Velde, G., Bickelhaupt, F.M., Baerends, E.J., Fonseca Guerra, C., van Gisbergen, S.J.A., Snijders, J.G., Ziegler, T.: Chemistry with ADF. J. Comput. Chem. **22**, 931–967 (2001)

10. Aidas, K., Angeli, C., Bak, K.L., Bakken, V., Bast, R., Boman, L., Christiansen, O., Cimiraglia, R., Coriani, S., Dahle, P., Dalskov, E.K., Ekström, U., Enevoldsen, T., Eriksen, J.J., Ettenhuber, P., Fernández, B., Ferrighi, L., Fliegl, H., Frediani, L., Hald, K., Halkier, A., Hättig, C., Heiberg, H., Helgaker, T., Hennum, A.C., Hettema, H., Hjertenæs, E., Høst, S., Høyvik, I.M., Iozzi, M.F., Jansik, B., Jensen, H.J.Aa., Jonsson, D., Jørgensen, P., Kauczor, J., Kirpekar, S., Kjærgaard, T., Klopper, W., Knecht, S., Kobayashi, R., Koch, H., Kongsted, J., Krapp, A., Kristensen, K., Ligabue, A., Lutnæs, O.B., Melo, J.I., Mikkelsen, K.V., Myhre, R.H., Neiss, C., Nielsen, C.B., Norman, P., Olsen, J., Olsen, J.M.H., Osted, A., Packer, M.J., Pawlowski, F., Pedersen, T.B., Provasi, P.F., Reine, S., Rinkevicius, Z., Ruden, T.A., Ruud, K., Rybkin, K., Salek, P., Samson, C.C.M., de Merás, A.S., Saue, T., Sauer, S.P.A., Schimmelpfennig, B., Sneskov, K., Steindal, A.H., Sylvester-Hvid, K.O.P., Taylor, R., Teale, A.M., Tellgren, E.I., Tew, D.P., Thorvaldsen, A.J., Thøgersen, L., Vahtras, O., Watson, M.A., Wilson, D.J.D., Ziolkowski, M., Ågren, H.: The dalton quantum chemistry program system. WIREs Comput. Mol. Sci. **4**, 269–284 (2014)

11. Beck, M.H., Jäckle, A., Worth, G.A., Meyer, H.D.: The multi-configuration time-dependent Hartree (MCTDH) method: a highly efficient algorithm for propagating wave packets. Phys. Rep. **324**, 1–105 (2000)

12. Valiev, M., Bylaska, E.J., Govind, N., Kowalski, K., Straatsma, T.P., Van Dam, H.J.J., Wang, D., Nieplocha, J., Apr, E., Windus, T.L., De Jong, W.A.: NWChem: a comprehensive and scalable open-source solution for large scale molecular simulations. Comput. Phys. Commun. **181**, 1477–1489 (2010)

13. Werner, H., Knowles, P.J., Knizia, G., Manby, F.R., Schutz, M.: Molpro: a general-purpose quantum chemistry program package. Wiley Interdiscip. Rev. Comput. Mol. Sci. **2**, 242–253 (2011)

14. Karlström, G., Lindh, R., Malmqvist, P.Å., Roos, B.O., Ryde, U., Veryazov, V., Widmark, P.-O., Cossi, M., Schimmelpfennig, B., Neogrády, P., Seijo, L.: MOLCAS: a program package for computational chemistry. Comp. Mater. Sci. **28**, 222–239 (2003)

15. Frisch, M.J., Trucks, G.W., Schlegel, H.B., Scuseria, G.E., Robb, M.A., Cheeseman, J.R., Scalmani, G., Barone, V., Mennucci, B., Petersson, G.A., Nakatsuji, H., Caricato, M., Li, X., Hratchian, H.P., Izmaylov, A.F., Bloino, J., Zheng, G., Sonnenberg, J.L., Hada, M., Ehara, M., Toyota, K., Fukuda, R., Hasegawa, J., Ishida, M., Nakajima, T., Honda, Y., Kitao, O., Nakai, H., Vreven, T., Montgomery Jr., J.A., Peralta, J.E., Ogliaro, F., Bearpark, M., Heyd, J.J., Brothers, E., Kudin, K.N., Staroverov, V.N., Kobayashi, R., Normand, J., Raghavachari, K., Rendell, A., Burant, J.C., Iyengar, S.S., Tomasi, J., Cossi, M., Rega, N., Millam, J.M., Klene, M., Knox, J.E., Cross, J.B., Bakken, V., Adamo, C., Jaramillo, J., Gomperts, R., Stratmann, R.E., Yazyev, O., Austin, A.J., Cammi, R., Pomelli, C., Ochterski, J.W., Martin, R.L., Morokuma, K., Zakrzewski, V.G., Voth, G.A., Salvador, P., Dannenberg, J.J., Dapprich, S., Daniels, A.D., Farkas, F., Foresman, J.B., Ortiz, J. V., Cioslowski, J., Fox, D.: J. Gaussian Inc., Wallingford CT (2009)

16. Skouteris, D., Castillo, J.F., Manolopulos, D.E.: Abc: a quantum reactive scattering program. Comp. Phys. Comm. **133**, 128–135 (2000)
17. Alexander, M.H., Manolopoulos, D.E.: A stable linear reference potential algorithm for solution of the quantum closecoupled equations in molecular scattering theory. J. Chem. Phys. **86**, 2044–2050 (1987)
18. Hase, W.L., Duchovic, R.J., Hu, X., Komornicki, A., Lim, K.F., Lu, D.-H., Peslherbe, G.H., Swamy, K.N., Vande Linde, S.R., Varandas, A.J.C., Wang, H., Wolf, R.J.: VENUS96: a general chemical dynamics computer program. QCPE Bull. **16**, 43 (1996)
19. Inhouse Program
20. Costantini, A., Gervasi, O., Manuali, C., Faginas Lago, N., Rampino, S., Laganà, A.: COMPCHEM: progress towards GEMS a grid empowered molecular simulator and beyond. J. Grid Comp. **8**(4), 571–586 (2010)
21. Laganà, A., Garcia, E., Paladini, A., Casavecchia, P., Balucani, N.: The last mile of molecular reaction dynamics virtual experiments: the case of the OH (N=1-10) + CO (j=0-3) → H + CO₂ reaction. Faraday Discuss. **157**, 415–436 (2012)
22. Yang, M., Zhang, D.H., Collins, M.A., Lee, S.-Y.: Ab initio potential-energy surfaces for the reactions OH + H2 & ↔ H2O + H. J. Chem. Phys. **115**, 174 (2001)
23. Chen, J., Xu, X., Xu, X., Xhang, D.H.: A global potential energy surface for the H2 + OH ↔ H2O + H reaction using neural networks. J. Chem. Phys. **138**, 154301 (2013)
24. Medvedev, D.M., Harding, L.B., Gray, S.K.: Methyl radical: ab initio global potential surface, vibrational levels and partition function. Mol. Phys. **104**, 73 (2006)
25. Blaylock, D., Ogura, T., Green, W., Beran, G.: Computational investigation of thermochemistry and kinetics of steam methane reforming on Ni(111) under realistic conditions. J. Phys. Chem. C. **113**(12), 4898–4908 (2009)
26. Ren, J., Guo, H., Yang, J., Qin, Z., Lin, J., Li, Z.: Insights into the mechanisms of CO2 methanation on Ni(111) surfaces by density functional theory. Appl. Surf. Sci. **351**, 504–516 (2015)
27. Catapan, R., Oliveira, A., Chen, Y., Vlachos, D.: DFT study of the watergas shift reaction and coke formation on Ni(111) and Ni(211) surfaces. J. Phys. Chem. C. **116**, 20281–20291 (2012)
28. Stamatakis, M., Vlachos, D.G.: A graph-theoretical kinetic monte carlo framework for on-lattice chemical kinetics. J. Chem. Phys. **134**(21), 214115 (2011)
29. Nielsen, J., d'Avezac, M., Hetherington, J., Stamatakis, M.: Parallel kinetic monte carlo simulation framework incorporating accurate models of adsorbate lateral interactions. J. Chem. Phys **139**(22), 224706 (2013)
30. Mller-Krumbhaar, H., Binder, K.: Dynamic properties of the Monte Carlo method in statistical mechanics. J. Stat. Phys. **8**(1), 1–24 (1973)
31. Bortz, A.B., Kalos, M.H., Lebowitz, J.L.: A new algorithm for Monte Carlo simulation of Ising spin systems. J. Comput. Phys. **17**(1), 10–18 (1975)
32. Ziff, R.M., Gulari, E., Barshad, Y.: Kinetic phase transitions in an irreversible surface-reaction model. Phys. Rev. Lett. **56**, 2553–2556 (1986)
33. Weatherbee, G.D., Bartholomew, C.H.: Hydrogenation of CO2 on group VIII metals: II. Kinetics and mechanism of CO2 hydrogenation on nickel. J. Catal. **77**, 460–472 (1982)
34. Lapidus, A.L., Gaidai, N.A., Nekrasov, N.V., Tishkova, L.A., Agafonov, Y.A., Myshenkova, T.N.: The mechanism of carbon dioxide hydrogenation on copper and nickel catalysts. Pet. Chem. **47**, 75–82 (2007)

35. Fujita, S., Terunuma, H., Kobayashi, H., Takezawa, N.: Methanation of carbon monoxide and carbon dioxide over nickel catalyst under the transient state. React. Kinet. Catal. Lett. **33**, 179–184 (1987)
36. Schild, C., Wokaun, A., Baiker, A.: On the mechanism of CO and CO2 hydrogenation reactions on zirconia-supported catalysts: a diffuse reflectance FTIR study: Part II. Surface species on copper/zirconia catalysts: implications for methanol synthesis selectivity. J. Mol. Catal. **63**, 243–254 (1990)
37. Martin, G.A., Primet, M., Dalmon, J.A.: Reactions of CO and CO2 onNi/SiO$_2$ above 373 K as studied by infrared spectroscopic andmagnetic methods. J. Catal. **53**, 321–330 (1978)
38. Martí, C., Pacifici, L., Laganà, A.: Networked computing for abinitio modeling the chemical storage of alternative energy: first term report (September-November 2015). VIRT&L-COMM. **8**, 3–9 (2015)
39. John Matthey webpage. http://www.jmprotech.com/methanation-catalysts-for-hydrogen-production-katalco. Accessed 5 May 2016

Multi-pattern Matching Algorithm
with Wildcards Based on Euclidean Distance
and Hash Function

Ahmed Abdo Farhan Saif[(⊠)] and Liang Hu

College of Computer Science and Technology,
Jilin University, Changchun, China
saifahmedabdofarh12@mails.jlu.edu.cn, hul@jlu.edu.cn

Abstract. In this paper, we present an algorithm to solve the problem of multi-pattern matching with a fixed number of wildcards. The method requires each pattern in a pattern set P to be partitioned into l length blocks, and then, the Euclidean distance is calculated between each first block $b_{y,1}$ of patterns and every possible alignment of the text t. If the Euclidean distance at position i is 0, then the block $b_{y,1}$ of pattern p^y matches the text t at position i. The Euclidean distance values are used as hash values to check the matches of the remaining blocks of the partially matched pattern. The complexity of our algorithm is $O(k\,n\,log\,l + o + d)$ time. Where n is the length of the text, l is the length of the blocks, k is the number of patterns, o is the number of blocks that match using the hash values and d is the number of wildcard symbols in the blocks that match using the hash values. The major advantages of our algorithm are that (a) it can find the matches of long patterns efficiently, (b) if the alphabet size σ is large, the algorithm is still efficient and (c) it supports non-interrupted pattern updates.

Keywords: Multi-pattern matching · Wildcard · FFT · Euclidean distance · Hash function

1 Introduction

The multi-pattern matching algorithm with wildcards finds the occurrence of every pattern of a pattern set $P = \{p^1, ..., p^k\}$ in a text t with $p^y = p_0, p_1 \ldots p_m$ and $t = t_0, t_1 \ldots t_n$ such that $t_i \in \sum \cup \{*\}$ and $p_j^y \in \sum \cup \{*\}$, where $\sum$ is an alphabet with size σ and σ is the number of symbols in the alphabet.

Multi-pattern matching algorithms are used in many important applications, such as antivirus systems, intrusion detection/prevention systems (IDS/IPS), text compression, text retrieval, music retrieval, computational biology and data mining, some of which require more sophisticated forms of searching algorithms. For example, in the misuse detection mechanism in antiviruses systems and Intrusion Detection/Prevention Systems (IDPS), various attacks can be detected by means of predefined rules, where each rule is represented by a regular expression that defines a set of strings. Pattern matching with wildcards (PMW) has also become especially crucial in exploring valuable information from DNA sequences; the wildcards are used to substitute unimportant

© Springer International Publishing Switzerland 2016
O. Gervasi et al. (Eds.): ICCSA 2016, Part I, LNCS 9786, pp. 334–344, 2016.
DOI: 10.1007/978-3-319-42085-1_26

symbols in protein sequences. Robust pattern matching algorithms require multiple considerations, such as stringent worst case performance, required memory, preprocessing time, alphabet size, pattern length, support searching with wildcards, non-interrupted pattern updating, algorithmic attacks and pattern group size, which are the main features of efficient pattern matching algorithms.

There are two types of pattern matching algorithms that solve the problem of matching with wildcards: (1) pattern matching with a fixed number of wildcard algorithms [1–8, 11] and (2) pattern matching with gap algorithms, either by a matching with a restricted gap length [17–24] or a matching with an unbounded gap length [9, 10, 15, 16]. Recently, pattern matching with a fixed number of wildcards has increasingly drawn attention.

Single pattern matching algorithms with a fixed number of wildcards have been well studied. In 1974, Fischer and Paterson [2] presented the first pattern matching algorithm with a wildcard; the algorithm is based on Fast Fourier Transform (FFT) and runs in $O(n \ log \ m \ log \ \sigma)$ time. In 1994, Muthukrishan and Palem [3] removed the dependency of the Fischer and Paterson algorithm on alphabet size and proposed a new algorithm that runs in $O(n \ log \ n)$ time. Indyk [5] contributed a randomized algorithm that runs in $O(n \ log \ n)$ time and is also based on FFT, where FFT is used to calculate the convolution between the pattern and text. In 2002, Cole and Hariharan [6] presented the first deterministic $O(n \ log \ m)$ time solution. In 2007, Clifford and Clifford [1] proposed a simpler deterministic algorithm with the same time complexity. By preprocessing the text, Rahman and Iliopoulos [7] produced an efficient solution without using FFT and developed an algorithm that runs in $O(n + m + occ)$ time, where occ is the total number of occurrences of p in t. Based on prime number encoding, Linhart and Shamir [8] presented an efficient algorithm, where if $m^\sigma = n$, the algorithm runs in $O(n \ log \ m)$ time. In 2014, Barton and Iliopoulos [4] introduced two average-case algorithms for PMW; in the first algorithm, wildcards occur only in the text, and in the second algorithm, the wildcards occur only in the pattern. The complexities of the two algorithms are $O((n \ log\sigma \ m)/m)$ and $O(n(g + log\sigma \ m)/(m - g))$, respectively, where g is the number of wildcards in the pattern.

1.1   Related Work and Previous Results

Multi-pattern matching algorithms with wildcards have received much attention. Based on Directed Acyclic Word Graph (DAWG) [13, 14], Kucherov and Rusinowitch [9] presented an algorithm to solve the problem of multi-pattern matching with a variable length wildcard; the algorithm runs in $O((n + L(P)log \ L(P))$ time, where $L(P)$ is the total length of keywords in every pattern of the pattern set P. Zhang et al. [10] improved Kucherov and Rusinowitch's [9] algorithm and presented a faster algorithm that runs in $O((n + \|P\|) \ log \ k/log \ log \ k)$ time, where $\|P\|$ and k are the total number of keywords and the number of distinct keywords, respectively, in all the patterns of the pattern set P. Their algorithm is based on Aho–Corasick automaton, which uses the solutions of the Dynamic Marked Ancestor Problem of Chan et al. [12]. Qiang et al. [17] introduced two algorithms (BWW and FWW) based on bit-parallelism technology to solve the problem of multiple PMW using user specification under a one-off condition. The complexities of the two algorithms are $O(G_s\sigma + n + f_s g_m + S(G_s/w))$ and $O(2G_s\sigma + n + rG_m + S(G_s/w))$, respectively, where f_s is the occurrence frequency of the patterns, S is the occurrence printing phase, r is the number of the positions when the prefix NFA moves the terminal states and G_s and G_m are the gap sum and gap maximum, respectively.

Based on FFT, Zhang et al. [11] presented three algorithms to solve the problem of multi pattern matching with a fixed number of wildcards. The first algorithm finds the matches of a small set of patterns with small alphabet size and runs in $O(n \log |P| + occ \log k)$ time, where occ is the total number of occurrences of P in t. The main objective of the algorithm is to construct a composed pattern $p^1, ..., p^k$, where the constructed composed pattern shows the convolution result of every pattern in a separate bit-vector machine word. The second algorithm is based on prime number encoding and the Chinese remainder theorem and is designed to find a small set of patterns and runs in $O(n \log m + occ \log k)$ time, where m is the maximum length of a pattern in the pattern set P. The third algorithm is based on computing the Hamming distance between the patterns and the text and finds the occurrence of patterns in $O(n \log |P| \log \sigma + occ \log k)$ time.

2 Our Result

In this paper, we present an efficient algorithm to solve the problem of multiple PMW. The algorithm preprocessing time is $O(|P| - lk)$ and the complexity of the algorithm is $O(k n \log l + o + d)$. where n is the length of the text t, l is the block length, k is the number of patterns, o is number of matched blocks using hash values and d is the number of wildcard symbols in the blocks that match by hash values in both the text and patterns.

When the pattern is very long or the alphabet size is large, FFT on machines wherein the word size is 32- or 64-bit is not efficient, because the calculation of the convolution between the pattern and the text cannot be executed in a single machine word. Our algorithm is based on FFT and efficiently runs even if the patterns are long and the alphabet size is large. The algorithm also supports non-interrupted pattern updating.

3 Preliminaries

Let $P = \{p^1, ..., p^k\}$ denote a set of k patterns and $|P|$ be the sum of the length of all the patterns. Each pattern in the pattern set is partitioned into l length blocks; the blocks from the second block to the last block are called the remaining blocks $b_{y,q}$, where $q = 1, 2, ..., |p^y/l|$ is an integer number that indicates the block number, n denotes the length of the text t and $\sum$ denotes an alphabet with size σ from which the symbols in P and t are chosen. The wildcard symbol (do not care or *- matching) can match any symbol including itself. Every symbol in the pattern p and text t is encoded with a unique non-zero integer number and the wildcard symbol is encoded with 0. The modified sum of squared differences (the modified Euclidean distance) is used as a hash function. The terms hash value and the modified sum of squared differences value are used interchangeably throughout the paper.

$$\text{Hash Function} : \sum_{j=0}^{m-1} p_j t_{i+j} \left(p_j - t_{i+j}\right)^2 \qquad (1)$$

The above sum is 0 if there is an exact match with the wildcard [1], where FFT is used to calculate the sum in $O(n \ log \ m)$ time. The numbers (hash values) calculated by the sum can be as large as $4m \ (\sigma - 1)^4/27$.

4 Multi-pattern Matching Algorithm with Wildcards Based on Euclidean Distance and Hash Function

The main plan of our algorithm is to partition the patterns into suitable l length blocks, and then, the matches of the first block of each pattern are checked. The matches of the remaining blocks of pattern p^y are checked if the first block of that pattern matches the text. The matches of the first blocks are checked by calculating the modified sum of squared differences between the first block and every possible alignment of the text. If the first block $b_{y,1}$ of the pattern p^y matches the text at position i, then the match of a remaining block $b_{y,q}$ is checked at position $i + (q-1)l$ in the text. The modified sum of squared differences values can be used as hash values to check the matches of the remaining blocks. Assume that $H[b_{y,1}, b_{y,q}]$ is the modified sum of squared differences between the first block $b_{y,1}$ and block $b_{y,q}$ and $H[b_{y,1}, t_{i+(q-1)l}]$ is the modified sum of squared differences between the first block and the text at position $i + (q-1)l$. These two sums are equal if there is an exact match of block $b_{y,q}$ at position $i + (q-1)l$ without a wildcard. If there are wildcard symbols, the lost values caused by the wildcard symbols need to be calculated.

The algorithm includes three main steps: preprocessing, checking the first block of every pattern and checking the remaining part of the patterns when the first block is matched.

4.1 Preprocessing

In the preprocessing step, each pattern is partitioned into suitable length l blocks assuming that the length of the shortest pattern is sufficient to create one block. If the last part of the pattern is not sufficient to create a complete block, the last block is overlapped with the previous block, as shown in Fig. 1.

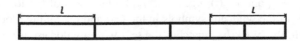

Fig. 1. Dividing patterns into small blocks.

The modified sum of squared differences between the first block of pattern p^y and the remaining blocks of that pattern are calculated. First, every symbol in the patterns and text is encoded with a unique integer number and the wildcards with $0s$; then, the modified sum of squared differences value $R_p[b_{y,1}, b_{y,q}]$ between the $b_{y,1}$ and $b_{y,q}$ blocks is calculated as follows:

$$R_p[b_{y,1}, b_{y,q}] = \sum_{j=0}^{1} (b_{y,1}(j) * b_{y,q}(j)) * (b_{y,1}(j) - b_{y,q}(j))^2 \qquad (2)$$

The $R_p[b_{y,1}, b_{y,q}]$ values are used as hash values, $H[b_{y,1}, b_{y,q}]$, to check the match of the remaining blocks.

4.2   Checking the First Blocks

To check the match of the first blocks of every pattern, the modified sum of squared differences between each first block $b_{y,1}$ of the patterns and the text t for every possible alignment are calculated as follows:

$$R_t[b_{y,1}, t_i] = \sum_{j=0}^{l} (b_{y,1}(j) * t_{i+j}) * (b_{y,1}(j) - t_{i+j})^2 \qquad (3)$$

The sum equals 0 at position i if the block $b_{y,1}$ matches the text at position i. If the block $b_{y,1}$ of the pattern p^y matches the text, the pattern p^y is a partial match and its remaining blocks need to be checked.

4.3   Checking the Remaining Blocks

To check the match of the remaining blocks of the partially matched pattern p^y, assume that block $b_{y,q}$ is a remaining block of the partially matched p^y and that the first block $b_{y,1}$ matches the text at position i; then, the match of block $b_{y,q}$ needs to be checked at the position $i + (q - 1)l$. The block $b_{y,q}$ matches the text at position $i + (q - 1)l$ without a wildcard if $H[b_{y,1}, b_{y,q}] = H[b_{y,1}, t_{i+(q - 1)l}]$, where $H[b_{y,1}, b_{y,q}]$ is the hash value between $b_{y,1}$ and $b_{y,q}$ and $H[b_{y,1}, t_{i+(q - 1)l}]$ is the hash value between $b_{y,1}$ and the text at position $i + (q - 1)l$.

If there are wildcard symbols in the matched remaining blocks of the patterns or in the text, there are lost values caused by the zero wildcard values. The lost values can be easily calculated because the positions of the wildcards can be saved during the encoding step.

Let $R_t{'}(j)[b_{y,1}, b_{y,q}]$ and $R'_p(j)[b_{y,q}, t_{i+(q - 1)l}]$ be the sums of the lost values caused by the zero values of the wildcard symbols in the text and remaining block pattern $b_{y,q}$, respectively. Then, the block $b_{y,q}$ matches the text with a wildcard at position $i + (q - 1)l$ if

$$R_p\left[b_{y,1}, b_{y,q}\right] + R'_p\left[b_{y,1}, t_{i+ql}\right] = R_t(i+ql)\left[b_{y,1}, t_{i+ql}\right] + R'_t(i+ql)\left[b_{y,1}, b_{y,q}\right] \qquad (4)$$

If the first block and all the remaining blocks of the pattern p^y match the text, the pattern p^y is a match.

Pseudo code

1. Input: text t and k patterns set $\{p^1, p^2, ..., p^k\}$.
2. Partition each pattern into l length blocks: $p = b_{y,1}, b_{y,2}, ..., b^y_{y,\lfloor p/l \rfloor}$.
3. Compute the hash values between the first block and the remaining blocks of every pattern individually, $H[b_{y,1}, b_{y,q}]$.
4. Compute the hash values between the first block of each pattern and alignment of the text, $H[b_{y,1}, t_i]$.
5. If $H[b_{y,1}, t_i] = 0$, check the match of the remaining blocks of pattern p^y.
   a. Compute the lost values caused by wildcards in both hash values.
   b. The block $b_{y,q}$ is a match if $H[b_{y,q}, t_{i+(q-1)l}]$ + pattern lost values = $H[b_{y,1}, b_{y,q}]$ + text lost values.
6. If the all blocks of pattern p^y are matched, report that the pattern p^y is matched.

5 Discussion

In the preprocessing step, the hash values between the first block and the remaining blocks of each pattern are calculated, where the total length of the remaining blocks of the patterns is $(|P| - |lk|)$. Then, the hash values of the remaining blocks can be calculated in $O(|P| - |lk|)$ time using wise-component multiplication.

When there is no match for any first block of patterns, our algorithm runs in $O(k\,n\,\log l)$ time, which is the time to convolute k first blocks against n length of text using FFT. When the first block $b_{y,1}$ of pattern p^y is matched, the match of the remaining blocks of that pattern should be checked. Let o be the number of matched remaining blocks and d be the number of wildcard symbols in the remaining blocks of the matched pattern p^y and the part of the text that matches these blocks. Then, the algorithm's overall running time is $O(k\,n\,\log l + o + d)$. Matching by hash values has a trivial probability of error; therefore, in a high accuracy device, the prime number encoding is used instead of integer number encoding. In this case, the block length will be reduced to fit the FFT function.

As shown in Table 1, algorithms are directly affected by the technique that they are based on. Algorithms based on FFT have trouble dealing with long patterns (LB) and large alphabet sizes (BA), and some have difficulties when the number of patterns (PN) is limited. While most of these algorithms are based on Euclidean distance, Humming distance and prime number encoding are also based on FFT. Algorithms that are based on structured data, such as AC automaton, have trouble supporting non-interrupted pattern updates (NPU).

FFT is thought to be the fastest algorithm for calculating the convolution between two strings, where the FFT convolution can be executed in $(n\log m)$ time. The main drawbacks of FFT are the length and alphabet size limitations of the two convoluted

Table 1. Comparison between various multi-pattern matching algorithms with wildcards.

	Algorithm based on	Running time	LP	BA	NPU	PN
1	Euclidian distance & hash function	$O(k\,n\,log\,l + o + d)$	√	√	√	√
2	Euclidian distance [11]	$O(n\,log\,|P| + occ\,log\,k)$	X	X	√	X
3	Humming distance [11]	$O(n\,log\,|P|\,log\,\sigma + occ\,log\,k)$	X	X	√	√
4	Prime number Encoding [11]	$O(n\,log\,m + occ\,log\,k)$	X	X	√	X

strings, where the string length and alphabet size need to be small enough to store their convolution result in a single machine word. Our algorithm uses FFT to calculate the modified sum of squared differences between the first blocks and every possible alignment of text, and by partitioning the patterns to the appropriate block length l, our algorithm overcomes the FFT drawbacks.

The block length l is proportional to the machine word size and the length of the patterns and is inversely proportional to the alphabet size. If the alphabet size is large, the block size should be small enough that the FFT result can be stored in a single machine word. If the block size is small, the check time for the first blocks is decreased. The block length should also be large enough to reduce the number of blocks, which directly reduces the check time of the remaining blocks. The algorithm is not based on structured data, such as AC automaton, which gives the algorithm the 'supporting non-interrupted pattern update' property.

6 Experimental Results

The algorithm performance was tested on the first 16 KB of three datasets: two of the datasets were natural language data and one was DNA sequence data. The enwik8 corpus that compromises the first 100 MB of Wikipedia in the English language was used as the first natural language dataset. The second natural language dataset was the 4.245 MB King James Version (KJV) of the Bible. The 96.7 MB hs13.chr file corresponding to the plain ASCII form of the human chromosome 13 (DNA) was used as the third dataset. The main difference between these three datasets was the dataset alphabet size, where the alphabet of the enwik8 dataset contains the entire alphabet of the English language in capital and lowercase letters and includes numbers and punctuation marks as well as some special symbols such as mathematical symbols and web script symbols. The KJV dataset has the same alphabet as enwik8 except the special symbols. Meanwhile, the alphabet size of the DNA chromosome 13 dataset is only 9 symbols.

The computer used in the tests was a Lenovo brand 3000G430 with a 2.16 GHz Pentium (R) dual 64 bit CPU T3400 with 2 GB RAM and a 32 bit Windows 8.1 enterprise operating system (OS). All code was compiled with visual studio 2005 in the DOS environment. The wildcard character was randomly selected.

The Clifford and Clifford algorithm is based on FFT, which theoretically is the fastest deterministic algorithm for single pattern matching with a wildcard with $O(n\,log\,m)$ time. In addition, the Clifford and Clifford algorithm is a simple, deterministic algorithm that directly deals with strings without preprocessing. In our experiments, we use FFT to compute the sum of squared differences with three convolutions, where the sum of squared differences between 16 KB length of text and a block of length 8 can be executed in 0.265 s.

The preprocessing time of our algorithm is the time spent calculating the remaining blocks' hash values and can be calculated in $O(|P| - |lk|)$ time using bitwise multiplication. As shown in Figs. 2, 3 and 4, the experimental results show that the algorithm preprocessing time is linear with unusual, unexplained behaviour at some points. In the preprocessing step, the algorithm can process about 350,000 patterns with 64 characters in less than 8 s.

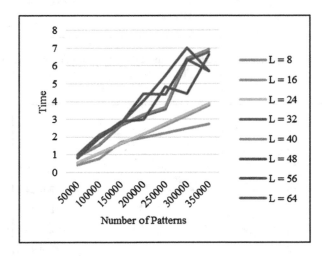

Fig. 2. Algorithm preprocessing time on the Wiki. dataset. (Color figure online)

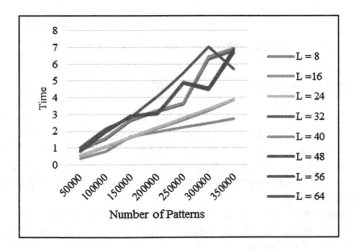

Fig. 3. Algorithm preprocessing time on the KJV dataset. (Color figure online)

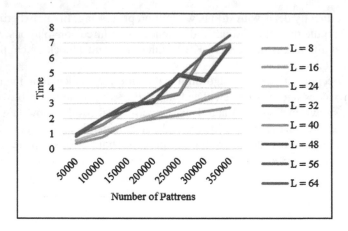

Fig. 4. Algorithm preprocessing time on the DNA dataset. (Color figure online)

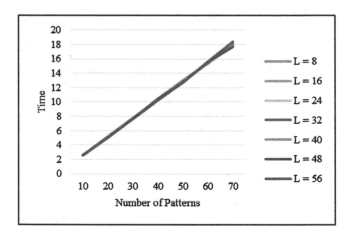

Fig. 5. Our algorithm performance. (Color figure online)

The performance of our algorithm was tested against three factors: the number of patterns, the length of patterns and the alphabet size. The patterns were randomly selected from the dataset, and every pattern had at least one match. The number of patterns were 1, 10, 20,..., 70, the length of the patterns were 8, 16,..., 56 symbols and the patterns were partitioned into fixed length blocks of $l = 8$. To show the performance gain of our algorithm, the single pattern matching algorithm of Clifford and Clifford was repeated to check the matching of the same patterns, where every block was checked individually.

The experiment results of our algorithm's performance are almost identical for the three datasets; therefore, the algorithm performance is not affected by the dataset alphabet size and the algorithm can efficiently deal with datasets that have large alphabet sizes.

The experiment result, as shown in Fig. 5, indicates that our algorithm efficiently deals with long patterns and that the effect of increasing pattern length on the algorithm performance is trivial. Conversely, the running time of the repeated single pattern of the Clifford and Clifford algorithm is doubled with doubling pattern length, as shown in Fig. 6. The algorithm time versus the number of patterns is linear.

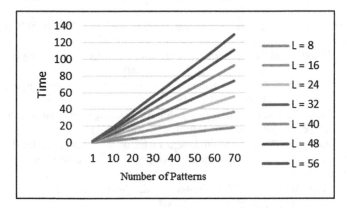

Fig. 6. Performance of the repeated single pattern matching of the Clifford and Clifford algorithm. (Color figure online)

7 Conclusions and Future Work

In this paper, we presented an algorithm that combines Euclidean distance and hash function to develop an algorithm that is more efficient than those using pure Euclidean distance. The algorithm can overcome problems of long patterns and large alphabet sizes. The algorithm performance can be improved by using a string package technique.

Acknowledgment. I would like to express my gratitude to Jilin University for providing me with a full scholarship and an excellent environment for my Ph.D. study and research. Furthermore, my appreciation extends to Dr. Meng Zhang for his continuous support and discussions.

Conflict of Interest. The author declares no conflict of interest.

References

1. Clifford, P., Clifford, R.: Simple deterministic wildcard matching. Inf. Process. Lett. **101**(2), 53–54 (2007)
2. Fischer, M., Paterson, M.: String matching and other products. In: Karp, R., (ed.) Proceedings of the 7th SIAMAMS Complexity of Computation, pp. 113–125 (1974)
3. Muthukrishan, S., Palem, K.: Non-standard stringology: algorithms and complexity. In: Proceedings of the 26th Symposium on the Theory of Computing, Canada (1994)

4. Barton, C., Iliopoulos, C.S.: On the average-case complexity of pattern matching with wildcards. CoRR, abs/1407.0950 (2014)
5. Indyk, P.: Faster algorithms for string matching problems: matching the convolution bound. In: Proceedings of the 38th Annual Symposium on Foundations of Computer Science, pp. 166–173 (1998)
6. Cole, R., Hariharan, R.: Verifying candidate matches in sparse and wildcard matching. In: [7] Proceedings of the Annual ACM Symposium on Theory of Computing, pp. 592–601 (2002)
7. Rahman, M., Iliopoulos, C.: Pattern matching algorithms with don't cares. SOFSEM 2, 116–126 (2007)
8. Linhart, C., Shamir, R.: Faster pattern matching with character classes using prime number encoding. J. Comput. Syst. Sci. 75(3), 155–162 (2009)
9. Kucherov, G., Rusinowitch, M.: Matching a set of strings with variable length don't cares. Theor. Comput. Sci. 178(1–2), 129–154 (1997)
10. Zhang, M., Zhang, Y., Hu, L.: A faster algorithm for matching a set of patterns with variable length don't cares. Inf. Process. Lett. 110(6), 216–220 (2010)
11. Zhang, M., Zhang, Y., Tang, J.: Multi-pattern matching with wildcards. In: Proceeding of Third International Symposium on Parallel Architectures, Algorithms and Programming (PAAP 2010) (2010)
12. Chan, H.-L., Hon, W.-K., Lam, T.W., Sadakane, K.: Compressed indexes for dynamic text collections. ACM Trans. Algorithms 3(2) (2007)
13. Bertossi, A.A., Logi, F.: Parallel string matching with variable length don't cares. J. Parallel Distrib. Comput. 22, 229–234 (1994)
14. Blumer, A., Blumer, J., Haussler, D., Ehrenfeucht, A., Chen, M.T., Seiferas, J.: The smallest automaton recognizing the subwords of a text. Theor. Comput. Sci. 40, 31–55 (1985)
15. Ding, B., Lo, D., Han, J., Khoo, S.: Efficient mining of closed repetitive gapped subsequences from a sequence database. In: Proceedings of the 25th IEEE International Conference on Data Engineering, pp. 1024–1035 (2009)
16. Wu, X., Zhu, X., He, Y., Arslan, A.N.: PMBC: pattern mining from biological sequences with wildcard constraints. Comput. Biol. Med. 43(5), 481–492 (2013)
17. Qiang, J., Guo, D., Fang, Y., Tian, W., Hu, X.: Multiple pattern matching with wildcards and one-off condition. J. Comput. Inf. Syst. 9, 14 (2013)
18. Guo, D., Hu, X., Xie, F., Wu, X.: Pattern matching with wildcards and gap-length constraints based on a centrality-degree graph. Appl. Intell. 39(1), 57–74 (2013)
19. Navarro, G., Raffinot, M.: Fast and simple character classes and bounded gaps pattern matching, with application to protein searching. In: Proceedings of the 5th Annual International Conference on Computational Biology, pp. 231–240 (2001)
20. Morgante, M., Policriti, A., Vitacolonna, N., Zuccolo, A.: Structured motifs search. J. Comput. Biol. 12(8), 1065–1082 (2005)
21. Cole, R., Gottlieb, L., Lewenstein, M.: Dictionary matching and indexing with errors and don't cares. In: Proceedings of the 36th Annual ACM Symposium on the Theory of Computing, pp. 91–100 (2004)
22. Haapasalo, T., Silvasti, P., Sippu, S., Soisalon-Soininen, E.: Online dictionary matching with variable-length gaps. In: Pardalos, P.M., Rebennack, S. (eds.) SEA 2011. LNCS, vol. 6630, pp. 76–87. Springer, Heidelberg (2011)
23. Arslan, A.N., He, D., He, Y., Wu, X.: Pattern matching with wildcards and length constraints using maximum network flow. J. Discrete Algorithms 35(C), 9–16 (2015)
24. Kalai, A.: Efficient pattern-matching with don't cares. In: Proceedings of the 13th Annual ACM-SIAM Symposium on Discrete Algorithms, pp. 655–656 (2002)

Improving Efficiency of a Multistart with Interrupted Hooke-and-Jeeves Filter Search for Solving MINLP Problems

Florbela P. Fernandes[1]([✉]), M. Fernanda P. Costa[2,3],
Ana Maria A.C. Rocha[4,5], and Edite M.G.P. Fernandes[5]

[1] ESTiG, Polytechnic Institute of Bragança, 5301-857 Bragança, Portugal
fflor@ipb.pt
[2] Department of Mathematics and Applications,
University of Minho, 4800-058 Guimarães, Portugal
mfc@math.uminho.pt
[3] Centre of Mathematics, Braga, Portugal
[4] Department of Production and Systems, University of Minho,
4710-057 Braga, Portugal
[5] Algoritmi Research Centre, University of Minho, 4710-057 Braga, Portugal
{arocha,emgpf}@dps.uminho.pt

Abstract. This paper addresses the problem of solving mixed-integer nonlinear programming (MINLP) problems by a multistart strategy that invokes a derivative-free local search procedure based on a filter set methodology to handle nonlinear constraints. A new concept of componentwise normalized distance aiming to discard randomly generated points that are sufficiently close to other points already used to invoke the local search is analyzed. A variant of the Hooke-and-Jeeves filter algorithm for MINLP is proposed with the goal of interrupting the iterative process if the accepted iterate falls inside an ϵ-neighborhood of an already computed minimizer. Preliminary numerical results are included.

Keywords: nonconvex MINLP · Multistart · Hooke-and-Jeeves · Filter method

1   Introduction

In this paper, we address the problem of solving mixed-integer nonlinear programming (MINLP) problems by a multistart strategy and a derivative-free local search procedure. Nonlinear inequality and equality constraints are handled by a filter set methodology. The MINLP problem has the form

$$\min f(x, y)$$
$$\text{subject to } g_j(x, y) \leq 0, \ j = 1, \ldots, m_g$$
$$h_i(x, y) = 0, \ i = 1, \ldots, m_h \tag{1}$$
$$x \in \Gamma_x, \ y \in \Gamma_y$$

© Springer International Publishing Switzerland 2016
O. Gervasi et al. (Eds.): ICCSA 2016, Part I, LNCS 9786, pp. 345–358, 2016.
DOI: 10.1007/978-3-319-42085-1_27

where $f : \mathbb{R}^n \to \mathbb{R}$ is the objective function, $g : \mathbb{R}^n \to \mathbb{R}^{m_g}$ and $h : \mathbb{R}^n \to \mathbb{R}^{m_h}$ are the constraint functions. The continuous variables are represented by the vector x, with $\Gamma_x \subset \mathbb{R}^{n_c}$ being the set of simple bounds on x:

$$\Gamma_x = \{x \in \mathbb{R}^{n_c} : l_x \le x \le u_x\},$$

with $l_x, u_x \in \mathbb{R}^{n_c}$. The integer variables are represented by the vector y, where $\Gamma_y \subset \mathbb{Z}^{n_i}$ is the set of simple bounds on y:

$$\Gamma_y = \{y \in \mathbb{Z}^{n_i} : l_y \le y \le u_y\},$$

with $l_y, u_y \in \mathbb{Z}^{n_i}$, and the parameter $n = n_c + n_i$ represents the total number of variables.

MINLP problems combine the combinatorial difficulty of optimizing over discrete variable sets with the challenge of handling nonlinear functions. When all functions involved in the problem are convex, the problem is a convex MINLP problem; otherwise it is a nonconvex one. The feasible region is in general nonconvex and multiple minima may appear. Furthermore, if some of the involved functions are nonsmooth then finding a solution to the MINLP problem is a challenging issue. For example, when the objective and constraints are provided as black-boxes, the MINLP is a nonsmooth and nonconvex problem. Convex MINLP problems are much easier to solve than nonconvex ones and therefore likely to be tractable, at least in theory. New techniques for MINLP can be found in [1–5]. The recent work presented in [6] is a detailed literature review and contains theoretical contributions, algorithmic developments, software implementations and applications for MINLP. The area of nonconvex MINLP has deserved a lot of attention from researchers working with heuristics, like genetic algorithm [7], ant colony [8], evolutionary algorithms [9], pattern search algorithms [10], multistart Hooke-and-Jeeves algorithm [11,12] and differential evolution [13].

The two major issues to be addressed when solving nonsmooth and nonconvex MINLP problems are: (i) the integrality of some variables, (ii) lack of smoothness and convexity of the involved functions. The issue of locating multiple solutions of MINLP problems by using a multistart method [14], the Hooke-and-Jeeves (HJ) algorithm [15,16] for the local search, and a filter set methodology [17] to handle inequality and equality constraints has been addressed in the past (see [11]). The HJ algorithm has been extended to define a special pattern of points spread over the search space for the integer variables and continuous variables separately. Furthermore, a filter methodology [10,17,18] to assess objective function and constraint violation function separately in order to promote convergence to feasible and optimal solutions has been incorporated.

The present study is an extension of the work presented in [11] in two aspects. First, the multistart algorithm is modified in a way that randomly generated points in the search space that are sufficiently close to other already generated and used points are discarded. This strategy aims to improve the diversification attribute of the multistart strategy, since points too close to other already used points would be starting points (to the local search) that would converge to already computed minimizers. To check closeness, the concept of componentwise normalized distance is presented. A new stopping condition for the new multistart algorithm is also proposed. The condition joins the probability of a point being generated in the area of the search space not yet covered and the probability of finding a new minimizer after the decision of invoking the local search. Second, the HJ filter based algorithm [11] is further explored by incorporating a methodology that goals the interruption of the HJ iterations when an accepted trial iterate enters an ϵ-neighborhood of an already computed minimizer.

The remaining part of the paper is organized as follows. In Sect. 2, the new multistart algorithm is presented and discussed and Sect. 3 describes the interrupted HJ filter local search methodology. Section 4 reports on the preliminary numerical experiments and the paper is concluded in Sect. 5.

2 An Efficient Multistart Method

The multistart algorithm is a stochastic strategy that repeatedly applies a local search procedure starting from sampled points, i.e., points randomly generated inside the search space, aiming to converge to a solution of the optimization problem. The goal of the local search procedure is to deeply exploit around the starting sampled point. When a problem has a multimodal objective function, the multistart algorithm is capable of converging to the multiple solutions, although some or all of them may be found over and over again.

In this section, we present a multistart algorithm that incorporates two important concepts. One is concerned with the componentwise normalized distance of the current sampled point to other previously used sampled point, and the other is related with the region of attraction of a minimizer X^*, herein represented by $A(X^*)$.

When the componentwise normalized distance of the current sampled point to other previously used sampled points is used, the algorithm avoids using a sampled point that is sufficiently close to other sampled points already used to start the local search, by discarding it and generating another one. This strategy aims to improve efficiency and increase the diversification capacity of the algorithm when points are randomly generated in the search space $\Gamma_x \times \Gamma_y$.

Using regions of attraction, the algorithm avoids invoking a local search procedure starting from a sampled point when the probability of converging to an already detected minimizer is high. This way the algorithm avoids convergence to minimizers over and over again. A region of attraction of a minimizer is a popular concept to avoid convergence to previously computed solutions by defining prohibited regions around them [14,19]. Each time a sampled point falls inside that prohibited region, it will be discarded since the implementation of a local search procedure will produce one of the previously computed solutions. Thus, the efficiency of the algorithm improves since the number of calls to the local search procedure is reduced. Clustering techniques are sophisticated variants of the multistart method that use a typical distance concept and a gradient criterion to decide which point is used to start a local search. They define a set of points that are believed to belong to the region of attraction of a particular minimizer, also denoted as a cluster [20].

Hereafter, the following notation is used: $X = (x, y)^T$ represents the vector of the n variables, $L = (l_x, l_y)^T$ and $U = (u_x, u_y)^T$ represent the lower and upper bound vectors on the variables. In a multistart context, to randomly generate a point with n_x continuous variables and n_y integer variables, in $[L, U]$, we proceed as follows:

$$x_i = (l_x)_i + \lambda_i((u_x)_i - (l_x)_i) \text{ for } i = 1, \ldots, n_x \qquad (2)$$
$$y_j = (l_y)_j + \tau_j \qquad\qquad \text{for } j = 1, \ldots, n_y$$

where the notation $(l_x)_i$ represents the component i of the vector l_x (similarly for $(u_x)_i, (l_y)_j, (u_y)_j$), λ_i is a number uniformly distributed in $[0, 1]$ and τ_j is a number randomly selected from the set $\{0, 1, \ldots, ((u_y)_j - (l_y)_j)\}$.

The componentwise normalized distance of the current sampled point (x, y) to another already used sampled point $\bar{X} = (\bar{x}, \bar{y})^T$ depends on the (componentwise) size of the search space, the number of already used sampled points, t_{rand}, and is computed for the continuous and integer variables separately as follows:

$$D_x(x, \bar{x}) = \sum_{i=1}^{n_x} \frac{(x_i - \bar{x}_i)^2}{(d_{avg}^i)^2} \text{ and } D_y(y, \bar{y}) = \sum_{i=1}^{n_y} \frac{(y_i - \bar{y}_i)^2}{(d_{avg}^{n_x+i})^2} \qquad (3)$$

where each component of the vector $d_{avg} = (d_{avg}^1, \ldots, d_{avg}^n)^T$ represents the average distance between two different points and is defined by

$$d_{avg}^i = \frac{(u_x)_i - (l_x)_i}{(t_{rand} + 1)}, i = 1, \ldots, n_x \text{ and } d_{avg}^{n_x+i} = \frac{(u_y)_i - (l_y)_i}{(t_{rand} + 1)}, i = 1, \ldots, n_y.$$

The use of the normalized distance aims to take into account different variable and search domain scaling. Furthermore, separate D_x and D_y values allow the use of different continuous and integer distance tolerances. The proposed multistart algorithm uses both distances $D_x(x, \bar{x})$ and $D_y(y, \bar{y})$ to check if the current sampled point X is sufficiently close to an already used sampled point $\bar{X}$, in which case the point is discarded since that region has been already explored and the local search procedure applied to X is likely to converge to an already detected minimizer. The herein implementation of the normalized distances aiming to discard a sampled point X that is sufficiently close to $\bar{X}$ uses the following criteria and tolerances:

$$D_x(x, \bar{x}) \leq 1 \text{ and } D_y(y, \bar{y}) \leq 1. \qquad (4)$$

This strategy also goals the diversification of the sampled points by effectively using points that are far apart.

The region of attraction of a minimizer, $X^* = (x^*, y^*)^T$, associated with a local search procedure $\mathbf{L}$, is defined as

$$A(X^*) \equiv \{X \in [L, U] : \mathbf{L}(X) = X^*\}, \qquad (5)$$

which means that if a point X is randomly selected from the set $[L, U]$ and belongs to the region of attraction $A(X^*)$, then X^* is obtained when $\mathbf{L}$ is invoked starting from X [14]. The regions of attraction are used to check if the sampled point does not belong to any of the regions of attraction of already computed minimizers or, equivalently, to the union of those regions of attraction (since they do not overlap). In this case, the local search procedure may be invoked to converge to a minimizer not yet detected. Since the region of attraction $A(X^*)$ is difficult to compute in practice, the probability, p, that the sampled point X will not belong to the union of the regions of attraction of the previously computed minimizers is used instead. This probability is easily estimated by the probability that X will not belong to the region of attraction of the nearest to X minimizer, X_o^*,

$$p = Prob[X \notin \cup_{i=1}^{s} A(X_i^*)] = \prod_{i=1}^{s} Prob[X \notin A(X_i^*)] \approx Prob[X \notin A(X_o^*)]$$

where o is the index of the nearest to X minimizer and s is the number of the already computed minimizers. Furthermore, $Prob[X \notin A(X_o^*)]$ is approximated by $Prob[X \notin B(X_o^*, R_o)]$ where $B(X_o^*, R_o)$ is the closed n-ball of radius R_o and center X_o^*. For any i, the maximum attractive radius of the minimizer X_i^* [14], denoted by R_i, is defined by

$$R_i = \max_j \left\{ \left\| \bar{X}_j - X_i^* \right\|_2 \right\}, \tag{6}$$

where $\bar{X}_j$ is the sampled point j that has already converged to the minimizer X_i^* after invoking the local search procedure, being the distance between X (the current sampled point) and the minimizer X_i^* given by $d_i = \|X - X_i^*\|_2$. Then if $d_i < R_i$, the point X is likely to be inside the region of attraction of X_i^* and the local search procedure ought not to be invoked since the probability of converging to the minimizer X_i^* is high. However, if the direction from X to X_i^* is ascent, i.e., the objective function increases if one goes from X to X_i^*, then X is likely to be outside the region of attraction of X_i^* and the local search procedure ought to be invoked, starting from X, since a new minimizer could be detected with high probability [14]. Thus, p is set to 1 if $d_i \geq R_i$ or if $d_i < R_i$ but the direction from X to X_i^* is ascent; otherwise $0 < p < 1$ and we set simply as a fraction of d_i/R_i.

Algorithm 1 displays the steps of the proposed multistart algorithm. The set Δ^*, empty at the beginning of the iterative process, contains all different computed minimizers. The condition $X^* \in \Delta^*$ is represented by the following conditions

$$|f(X^*) - f(X_l^*)| \leq \gamma^* \text{ and } \|x^* - x_l^*\|_2 \leq \gamma^* \text{ and } \|y^* - y_l^*\|_2 = 0 \tag{7}$$

for any $l \in \{1, \ldots, s\}$ and a small $\gamma^* > 0$, and means that the minimizer X^* has been previously detected.

Although the basic multistart algorithm is easily perceived and simple to code, it requires an adequate stopping rule to be effective. The goal of the stopping rule is to make the algorithm to stop when all minimizers have been located and it should not require a large number of calls to the local search procedure to decide that all minimizers have been found.

Algorithm 1. Efficient multistart algorithm

Require: L, U, $\xi > 0$, $0 < \delta < 1$, $K_{\max}$, set $\Delta^* = \emptyset$, $s = 1$, $t_{local} = 1$, $t_{rand} = 1$, $k = 1$;

1: Randomly generate $X \in [L, U]$ (sampled point), set $\bar{X}_1 \leftarrow X$;
2: Compute $X_1^* = \mathbf{L}(X)$, $R_1 = \|X - X_1^*\|_2$, set $r_1 = 1$, $\Delta^* = \Delta^* \cup X_1^*$;
3: **repeat**
4:     Randomly generate $X \in [L, U]$ (sampled point);
5:     **if** $D_x(x - \bar{x}_i) > 1 \vee D_y(y - \bar{y}_i) > 1$ for all $i \in \{1, \ldots, t_{rand}\}$ **then**
6:         Set $t_{rand} = t_{rand} + 1$, set $\bar{X}_{t_{rand}} = X$;
7:         Set $o = \arg\min_{j=1,\ldots,s} d_j \equiv \|X - X_j^*\|_2$;
8:         **if** $d_o < R_o$ **then**
9:             **if** the direction from X to X_o^* is ascent **then**
10:                 Set $p = 1$;
11:             **else**
12:                 Set $p = \delta(d_o/R_o)$;
13:             **end if**
14:         **else**
15:             Set $p = 1$;
16:         **end if**
17:         **if** $rand(0, 1) < p$ **then**
18:             Compute $X^* = \mathbf{L}(X)$, set $t_{local} = t_{local} + 1$;
19:             **if** $X^* \in \Delta^*$ **then**
20:                 Update $R_l = \max\{R_l, \|X - X_l^*\|_2\}$, $r_l = r_l + 1$;
21:             **else**
22:                 Set $s = s + 1$, $X_s^* = X^*$, $r_s = 1$, $\Delta^* = \Delta^* \cup X_s^*$, compute $R_s = \|X - X_s^*\|_2$;
23:             **end if**
24:         **else**
25:             Update $R_o = \max\{R_o, \|X - X_o^*\|_2\}$, $r_o = r_o + 1$;
26:         **end if**
27:     **end if**
28:     Set $k = k + 1$;
29: **until** $\dfrac{t_{rand}}{k} \dfrac{s}{t_{local}} \leq \xi$ or $t_{local} > K_{\max}$

A simple stopping rule [21] uses an estimate of the fraction of uncovered space, $s(s+1)/(t_{local}(t_{local}-1)) \leq \xi$ where s represents the number of (different) computed minimizers, t_{local} represents the number of local search calls and ξ is a small positive constant. Previous experiments show that this condition might not be efficient since a large number of sampled points may be required to start a local search procedure (making t_{local} to increase) although the search ends up by locating a previously identified minimizer (and s is not increased).

In this paper, we propose a different and yet simple rule to reflect a reasonable coverage of the search space. To check if the diversification procedure was able to cover all the search space and detect all minimizers, the product of the probability that a good sampled point is found (in the area of the search space not yet covered), by the probability that a different minimizer is found when the local search is invoked (noting that invocation itself follows the region of attraction concept) is required to be small:

$$\frac{t_{rand}}{k} \frac{s}{t_{local}} \leq \xi,$$

for a small $\xi > 0$. The parameter t_{rand} represents the number of effectively used sampled points and k is the total number of generated points; s and t_{local} have the same meaning as above. Alternatively, the multistart algorithm stops when the number of local search calls exceeds a maximum value $K_{\max}$.

3   The 'Interrupted HJ-filter' Method for MINLP

In this section, we address the issue related with the local search procedure invoked inside the multistart algorithm, represented by **L** in Algorithm 1. A derivative-free pattern search method that is prepared to handle inequality and equality constraints by means of a filter methodology is used. Furthermore, the algorithm has been extended to be able to handle continuous and integer variables simultaneously. Thus, the proposed algorithm relies on the HJ approach, as outlined in [15], for solving a MINLP problem. This is an extension of the work presented in [11,12] in the sense that multiple solutions of problems like (1) are computed using a multistart algorithm that invokes a modified HJ filter based local search.

The herein proposed 'Interrupted HJ-filter' local search algorithm is an iterative process that is applied to a sampled point $X = (x, y)^T$ in order to provide a trial point $X^+ = (x^+, y^+)^T$ that is a global or a local minimizer of the problem (1). At each iteration, an acceptable trial point is obtained by searching around a central point. The term acceptable point means that the point is not dominated by any other point that belongs to a filter set.

Two stopping conditions are used to terminate the 'Interrupted HJ-filter' algorithm. When it is not possible to find an acceptable trial point in a very small neighborhood of the central point of the search, the iterative process can be terminated with the central point as final solution. When the acceptable trial point falls inside an ϵ-neighborhood of a previously computed minimizer, the iterative process is interrupted since the likelihood that the process will converge to that minimizer is high. This strategy aims to improve efficiency by avoiding convergence to previously detected solutions.

Thus, the most relevant issues addressed by the 'Interrupted HJ-filter' algorithm are the following. The algorithm

- accommodates two different pattern of points to be able to handle simultaneously continuous and integer variables, when generating trial points;
- uses a projection onto the search space $[L, U]$ when generated trial points violate the bound constraints during exploratory and pattern moves;

- incorporates the filter set methodology to handle equality and inequality constraints by accepting trial points that improve feasibility or optimality;
- interrupts the iterative search process when a trial point is accepted as the new iterate and is inside an ϵ-neighborhood of a previously computed minimizer;
- terminates the iterations when the step size α_x for the continuous variables falls below a pre-specified small tolerance $\alpha_{\min}$.

Using a filter methodology [7,10,17,18], problem (1) is reformulated as a bi-objective optimization problem

$$\min [f(x,y), \theta(x,y)] \tag{8}$$
$$\text{subject to } x \in \Gamma_x, y \in \Gamma_y$$

where $\theta(x,y) = \|g(x,y)_+\|_2^2 + \|h(x,y)\|_2^2$ is the nonnegative constraint violation function and $v_+ = \max\{0, v\}$. The concept of nondominance, borrowed from the multi-objective optimization, aims to build a filter set that is able to accept trial points if they improve the constraint violation or the objective function value. The filter $\mathcal{F}$ is defined as a finite set of points (x,y), corresponding to pairs $(\theta(x,y), f(x,y))$, none of which is dominated by any of the others. A point (x,y) is said to dominate a point (x',y') if and only if $\theta(x,y) \leq \theta(x',y')$ and $f(x,y) \leq f(x',y')$.

Every time the local search procedure is invoked inside the multistart algorithm, the filter is initialized to $\mathcal{F} = \{(\theta, f) : \theta \geq \theta_{\max}\}$, where $\theta_{\max} > 0$ is an upper bound on the acceptable constraint violation.

We now discuss the implemented 'Interrupted HJ-filter' local search algorithm with pseudo-code shown in Algorithm 2. Like the classical HJ, the HJ-filter search procedure comprises both the exploratory and the pattern moves. The exploratory move starts by generating a set of n possible trial points along the unit coordinate vectors $e_i \in \mathbb{R}^n$ with a fixed step size $\alpha_x \in (0,1]$ for the continuous variables and a unitary step for the integer ones:

$$x^+ = \bar{x} \pm \alpha_x D e_i, \ i = 1, \ldots, n_x, \text{ and } y^+ = \bar{y} \pm e_i, \ i = n_x + 1, \ldots, n, \tag{9}$$

where (x^+, y^+) is a trial point, $(\bar{x}, \bar{y})$ is the central point of the search and $D \in \mathbb{R}^{n_x \times n_x}$ is a weighting diagonal matrix. The initial central point of the current iteration is the sampled point (x,y). If one of the following conditions

$$\theta(x^+, y^+) < (1 - \gamma_\theta)\,\theta(\bar{x}, \bar{y}) \text{ or } f(x^+, y^+) \leq (1 - \gamma_f)\,f(\bar{x}, \bar{y}) \tag{10}$$

holds, for fixed constants $\gamma_\theta, \gamma_f \in (0,1)$, and (x^+, y^+) is acceptable by the filter, then the point (x^+, y^+) is accepted and replaces $(\bar{x}, \bar{y})$. We note that if a sequence of trial points is feasible, the condition (10) guarantees that the trial approximation (x^+, y^+) must satisfy the second condition in order to be acceptable. This way the optimal solution is guaranteed. Whenever a point is acceptable, the point is added to the filter $\mathcal{F}$, and all dominated points are removed from the filter. The search then selects the most nearly feasible point, (x^{inf}, y^{inf}), among all the accepted trial points.

If $(x^{inf}, y^{inf}) \neq (x, y)$, the iteration is successful and a pattern move is carried out through the direction $(x^{inf}, y^{inf}) - (x, y)$. A new set of n possible trial points along the unit coordinate vectors is generated, as shown in (9), replacing $(\bar{x}, \bar{y})$ by $(x^{inf}, y^{inf}) + ((x^{inf}, y^{inf}) - (x, y))$ (as central point). If the most nearly feasible point, selected among the trial points, is acceptable, (x^{inf}, y^{inf}) is accepted as the new iterate, replaces $(\bar{x}, \bar{y})$, and the pattern move is repeated.

Algorithm 2. Interrupted HJ-filter algorithm

Require: X (sampled point), $\alpha_{\min}$ and computed minimizers $X_i^*, i \in \{1, \dots, s\}$;
1: Initialize the filter;
2: Set central point $\bar{X} = X$, set $k = 0$;
3: **repeat**
4:     Generate n possible trial points around central point;
5:     Check feasibility and optimality conditions of trial points;
6:     **if** trial points are acceptable by the filter **then**
7:         Update the filter;
8:         Choose the most nearly feasible point (x^{inf}, y^{inf});
9:         Set $\bar{X} = (x^{inf}, y^{inf})^T$;
10:         **while** (x^{inf}, y^{inf}) is a non-dominated point **do**
11:             (pattern move) Make a pattern move and define new central point;
12:             Generate n possible trial points around the central point;
13:             Check feasibility and optimality conditions of trial points;
14:             **if** trial points are acceptable by the filter **then**
15:                 Update the filter;
16:                 Choose the most nearly feasible point (x^{inf}, y^{inf});
17:                 Set $\bar{X} = (x^{inf}, y^{inf})^T$;
18:             **else**
19:                 Recuperate $\bar{X}$;
20:             **end if**
21:         **end while**
22:     **else**
23:         (restoration phase) Recuperate the most nearly feasible point in the filter as central point;
24:         Generate n possible trial points around the central point;
25:         Check feasibility and optimality conditions of trial points;
26:         **if** trial points are acceptable by the filter **then**
27:             Update the filter;
28:             Choose the most nearly feasible point (x^{inf}, y^{inf});
29:             Set $\bar{X} = (x^{inf}, y^{inf})^T$;
30:         **else**
31:             Recuperate $\bar{X}$;
32:             Reduce α_x;
33:         **end if**
34:     **end if**
35:     Set $k = k + 1$;
36: **until** $\alpha_x \leq \alpha_{\min}$ or $(\mod(k, 5) = 0 \wedge \exists i \in \{1, \dots, s\} : \|\bar{x} - x_i^*\|_2 \leq \epsilon \wedge \|\bar{y} - y_i^*\|_2 \leq 1)$

However, when $(x^{inf}, y^{inf}) = (x, y)$, the iteration is unsuccessful and a restoration phase is invoked. In this phase, the most nearly feasible point in the filter, $(x_{\mathcal{F}}^{inf}, y_{\mathcal{F}}^{inf})$, is recuperated and a set of n possible trial points along the unit coordinate vectors is generated, as shown in (9), replacing $(\bar{x}, \bar{y})$ by $(x_{\mathcal{F}}^{inf}, y_{\mathcal{F}}^{inf})$ as the central point. If a non-dominated trial point is found, then it will become the central point for the next iteration. Otherwise, the iteration remains unsuccessful, the search returns back to the current $(\bar{x}, \bar{y})$, the step size

α_x is reduced, and a new search consisting of exploratory moves and pattern moves is repeated taking $(\bar{x}, \bar{y})$ as the central point [11,12].

When α_x is reduced within an unsuccessful iteration, it may fall below a sufficiently small positive tolerance, $\alpha_{\min}$, thus indicating that no further improvement is possible and a solution is found. For the interruption issue of the HJ-filter algorithm, an acceptable trial point X is considered to be in the ϵ-neighborhood of an already computed minimizer X_i^*, for some $i \in \{1, \ldots, s\}$, if the distance to the minimizer verifies

$$||x - x_i^*||_2 < \epsilon \text{ and } ||y - y_i^*||_2 \leq 1, \tag{11}$$

where ϵ is a small positive tolerance. In this case, the likelihood that the search will converge to the minimizer X_i^* is high and the local search is interrupted. We remark that conditions (11) are tested only every five iterations.

4   Numerical Results

In this section, we aim to analyze the performance of the proposed multistart method based on the 'Interrupted HJ-filter' local search procedure. Four small problems are selected for these preliminary experiments. The results were obtained in a PC with an Intel Core i7-2600 CPU (3.4 GHz) and 8 GB of memory. The parameters of the multistart and 'Interrupted HJ-filter' algorithms are set after an empirical study as follows: $K_{\max} = 20$, $\xi = 0.1$, $\delta = 0.5$, $\gamma^* = 0.005$, $\gamma_\theta = \gamma_f = 10^{-8}$, $\theta_{\max} = 10^2 \max\{1, \theta(\bar{x}, \bar{y})\}$, $\alpha_{\min} = 10^{-4}$ and $\epsilon = 0.05$. A solution is considered feasible when $\theta(x, y) \leq 10^{-8}$. Due to the stochastic nature of the multistart algorithm, each problem was solved 30 independent times and the results correspond to average values.

This section presents the full description of the problems, the results obtained by the proposed algorithm and a comparison with the work presented in [11].

Problem 1. (Example 1 in [22]) with 2 known solutions (1 global and 1 local)

$$\min -x - y$$
$$\text{subject to } xy - 4 \leq 0,$$
$$0 \leq x \leq 4, \ y \in \{0, \ldots, 6\}$$

In [11] two solutions were detected. The global solution was found in all the 30 runs and the local in 24. The new Algorithm 1 based on the 'Interrupted HJ-filter' local search produces the global solution in all runs and the local in 18 out of 30. An average of 34 iterations were carried out and the average number of local solver calls was 9.1.

Problem 2. (Example 11 in [22]) with 2 known solutions (1 global and 1 local)

$$\min 35x_1^{0.6} + 35x_2^{0.6}$$
$$\text{subject to } 600x_1 - 50y - x_1y + 5000 = 0$$
$$600x_2 + 50y - 15000 = 0$$
$$0 \leq x_1 \leq 34, 0 \leq x_2 \leq 17, \ y \in \{100, \ldots, 300\}$$

When solving this problem, [11] detected two optimal solutions, one global and one local. The global was located in all runs and the local only in two out of the 30 runs. The herein presented Algorithm 1 detected the global in all runs and the local solution in four out of 30 runs, after an average of 21 iterations and 12.5 local search calls.

Problem 3. (Example 21 in [22]) with 2 known solutions (1 global and 1 local)

$$\min x_1^{0.6} + y_1^{0.6} + y_2^{0.4} - 4y_2 + 2x_2 + 5y_3 - y_4$$
$$\text{subject to } x_1 + 2x_2 - 4 \le 0$$
$$y_1 + y_3 - 4 \le 0$$
$$y_2 + y_4 - 6 \le 0$$
$$-3x_1 + y_1 - 3x_2 = 0$$
$$-2y_1 + y_2 - 2y_3 = 0$$
$$4x_2 - y_4 = 0$$
$$0 \le x_1 \le 3, \ 0 \le x_2 \le 2, \ y_1, y_2 \in \{0, \dots, 4\}, \ y_3 \in \{0, 1, 2\}, \ y_4 \in \{0, \dots, 6\}$$

Two optimal solutions, one global and one local, were found in [11]. The global was located in all runs and the local in 22 out of 30 runs. The new Algorithm 1 located the global in all runs and the local in four, after an average of 32 iterations and 20 local search calls.

Problem 4. (Example 13 in [22], f1 in [7]) with 2 known solutions (1 global and 1 local)

$$\min 2x + y$$
$$\text{subject to } 1.25 - x^2 - y \le 0$$
$$x + y - 1.6 \le 0$$
$$0 \le x \le 1.6, \ y \in \{0, 1\}$$

In [11] two solutions were detected, one global and one local. The global was located in all runs and the local was located only in two runs. In the present study, and after an average of 12 iterations and 2.1 local search calls, Algorithm 1 produces the global solution in all runs and the local in one out of the 30 runs.

The results presented in Table 1 aim to compare the efficiency of the Algorithm 1 based on the 'Interrupted HJ-filter' strategy with the multistart (MS) HJ-filter based algorithm in [11]. For each problem, the table shows averaged values over the 30 runs – the average overall number of function evaluations of a run, 'Nfe^{avg}'; the average overall time of a run (in seconds), 'T^{avg}'; the average number of local search calls, t_{local}^{avg}; and for each identified global and local solution, the average of the best f values obtained during the runs where the solution was located, 'f^{avg}'; the average number of function evaluations required by the HJ local search while converging to those best solutions, 'Nfe_{local}^{avg}'; and the success rate (the percentage of runs that found that particular solution at least during one local search), 'SR'.

Based on the results of the table, it can be concluded that the herein proposed multistart algorithm with the 'Interrupted HJ-filter' local search wins on efficiency when compared with the MS HJ-filter algorithm, in the sense that reductions on Nfe^{avg}, T^{avg} and t_{local}^{avg} are notorious. The algorithm was also

Table 1. MS 'HJ-filter' [11] *vs* Algorithm 1 based on the 'Interrupted HJ-filter'

Method	Solution		Problem			
			1	2	3	4
MS with 'HJ-filter' (in [11])		Nfe^{avg}	11513	13109	79892	4199
		T^{avg}	33	113	477	36
		t^{avg}_{local}	16.5	9.2	20.5	8.1
	global	f^{avg}	−6.666394	189.29356	−13.401916	2.000250
		Nfe^{avg}_{local}	590	1495	4929	458
		SR (%)	100	100	100	100
	local	f^{avg}	−5.000000	291.75820	−4.258899	2.236262
		Nfe^{avg}_{local}	519	2122	2986	527
		SR (%)	80	6.7	73.3	6.7
Algorithm 1 with 'Interrupted HJ-filter' ($\epsilon = 0.05$)		Nfe^{avg}	3110	9193	14596	628
		T^{avg}	6	42	36	4
		t^{avg}_{local}	9.1	12.5	20.0	2.1
	global	f^{avg}	−6.662584	189.29714	−13.401973	2.003090
		Nfe^{avg}_{local}	318	712	762	297
		SR (%)	100	100	100	100
	local	f^{avg}	−5.000000	291.02098	−4.258890	2.236161
		Nfe^{avg}_{local}	450	1327	480	372
		SR (%)	60	13.3	13.3	3.3
Algorithm 1 with 'Interrupted HJ-filter' ($\epsilon = 0.01$)		Nfe^{avg}	3759	12894	17479	628
		T^{avg}	8	59	44	4
		t^{avg}_{local}	9.6	15.5	21	1.9
	global	f^{avg}	−6.655440	189.29836	−13.401970	2.002982
		Nfe^{avg}_{local}	372	793	781	325
		SR (%)	100	100	100	100
	local	f^{avg}	−5.000000	291.69833	−4.258890	2.236149
		Nfe^{avg}_{local}	427	990	559	464
		SR (%)	60	10	23.3	6.7

tested with the smaller value $\epsilon = 0.01$ (the neighborhood radius on the continuous variables for the HJ interruption strategy, see (11)). From the corresponding results shown in Table 1, we may conclude that the algorithm is able to locate the local solutions more often but at a cost of more function evaluations, time and HJ local search calls.

5   Conclusions

The strategy of discarding randomly generated points that are sufficiently close to other generated and already used points, coupled with the interruption of the HJ-filter local search, for solving MINLP problems, have shown to be effective. The proposed stopping rule of the new multistart algorithm is able to detect multiple solutions without requiring a large number of local search calls. On the other hand, interrupting the local search procedure when an iterate that falls inside an ϵ-neighborhood of an already detected solution is found has also improved the overall efficiency of the algorithm.

The preliminary numerical results presented in this paper show that the combination of the componentwise normalized distance and region of attraction concepts in the multistart algorithm, as well as the interruption of the HJ-filter iterative search, make the derivative-free based multistart framework more competitive.

Future developments will be directed to the implementation of other mixed-integer point distance concepts, as well as the most popular Euclidean distance. Large-dimensional benchmark problems in the context of engineering applications will also be used to test the effectiveness of the proposed algorithm.

Acknowledgments. The authors wish to thank two anonymous referees for their comments and suggestions. This work has been supported by COMPETE: POCI-01-0145-FEDER-007043 and FCT - Fundação para a Ciência e Tecnologia, within the projects UID/CEC/00319/2013 and UID/MAT/00013/2013.

References

1. Abramson, M.A., Audet, C., Chrissis, J.W., Walston, J.G.: Mesh adaptive direct search algorithms for mixed variable optimization. Optim. Lett. **3**, 35–47 (2009)
2. Belotti, P., Kirches, C., Leyffer, S., Linderoth, J., Luedtke, J., Mahajan, A.: Mixed-Integer Nonlinear Optimization. Acta Numer. **22**, 1–131 (2013)
3. Burer, S., Letchford, A.: Non-convex mixed-integer nonlinear programming: a survey. Surv. Oper. Res. Manage. Sci. **17**, 97–106 (2012)
4. Gueddar, T., Dua, V.: Approximate multi-parametric programming based B&B algorithm for MINLPs. Comput. Chem. Eng. **42**, 288–297 (2012)
5. Liuzzi, G., Lucidi, S., Rinaldi, F.: Derivative-free methods for mixed-integer constrained optimization problems. J. Optimiz. Theory App. **164**(3), 933–965 (2015)
6. Boukouvala, F., Misener, R., Floudas, C.A.: Global optimization advances in mixed-integer nonlinear programming, MINLP, and constrained derivative-free optimization. CDFO. Eur. J. Oper. Res. **252**, 701–727 (2016)
7. Hedar, A., Fahim, A.: Filter-based genetic algorithm for mixed variable programming. Numer. Algebra Control Optim. **1**(1), 99–116 (2011)
8. Schlüter, M., Egea, J.A., Banga, J.R.: Extended ant colony optimization for non-convex mixed integer nonlinear programming. Comput. Oper. Res. **36**, 2217–2229 (2009)
9. Lin, Y.C., Hwang, K.S.: A mixed-coding scheme of evolutionary algorithms to solve mixed-integer nonlinear programming problems. Comput. Math. Appl. **47**, 1295–1307 (2004)

10. Abramson, M.A., Audet, C., Dennis Jr., J.E.: Filter pattern search algorithms for mixed variable constrained optimization problems. Pac. J. Optim. **3**(3), 477–500 (2007)
11. Costa, M.F.P., Fernandes, F.P., Fernandes, E.M.G.P., Rocha, A.M.A.C.: Multiple solutions of mixed variable optimization by multistart Hooke and Jeeves filter method. Appl. Math. Sci. **8**(44), 2163–2179 (2014)
12. Fernandes, F.P., Costa, M.F.P., Fernandes, E., Rocha, A.: Multistart Hooke and Jeeves filter method for mixed variable optimization. In: Simos, T.E., Psihoyios, G., Tsitouras, C. (eds.) ICNAAM 2013, AIP Conference Proceeding, vol. 1558, pp. 614–617 (2013)
13. Liao, T.W.: Two hybrid differential evolution algorithms for engineering design optimization. Appl. Soft Comput. **10**(4), 1188–1199 (2010)
14. Voglis, C., Lagaris, I.E.: Towards "Ideal Multistart". A stochastic approach for locating the minima of a continuous function inside a bounded domain. Appl. Math. Comput. **213**, 1404–1415 (2009)
15. Hooke, R., Jeeves, T.A.: Direct search solution of numerical and statistical problems. J. Assoc. Comput. **8**, 212–229 (1961)
16. Kolda, T.G., Lewis, R.M., Torczon, V.: Optimization by direct search: new perspectives on some classical and modern methods. SIAM Rev. **45**(3), 385–482 (2003)
17. Fletcher, R., Leyffer, S.: Nonlinear programming without a penalty function. Math. Program. **91**, 239–269 (2002)
18. Audet, C., Dennis Jr., J.E.: A pattern search filter method for nonlinear programming without derivatives. SIAM J. Optimiz. **14**(4), 980–1010 (2004)
19. Tsoulos, I.G., Stavrakoudis, A.: On locating all roots of systems of nonlinear equations inside bounded domain using global optimization methods. Nonlinear Anal. Real World Appl. **11**, 2465–2471 (2010)
20. Tsoulos, I.G., Lagaris, I.E.: MinFinder: Locating all the local minima of a function. Comput. Phys. Commun. **174**, 166–179 (2006)
21. Lagaris, I.E., Tsoulos, I.G.: Stopping rules for box-constrained stochastic global optimization. Appl. Math. Comput. **197**, 622–632 (2008)
22. Ryoo, H.S., Sahinidis, N.V.: Global optimization of nonconvex NLPs and MINLPs with applications in process design. Comput. Chem. Eng. **19**(5), 551–566 (1995)

Strengths and Weaknesses of Three Software Programs for the Comparison of Systems Based on ROC Curves

Maria Filipa Mourão[1](✉) and Ana C. Braga[2]

[1] School of Technology and Management-Polytechnic Institute of Viana do Castelo, 4900-348 Viana do Castelo, Portugal
fmourao@estg.ipvc.pt
[2] ALGORITMI Centre, University of Minho, 4710-057 Braga, Portugal
acb@dps.uminho.pt

Abstract. The Receiver Operating Characteristic (ROC) analysis is a technique that is applied in medical diagnostic testing, especially for discriminating between two health status of a patient: normal (negative) and abnormal (positive). Its ability to compare the performance of different diagnostic systems based on an empirical estimation of the area under the ROC curve (AUC) has made this technique very attractive. Thus, it is important to select an appropriate software program to carry out this comparison. However, this selection has been a difficult task, considering the operational features available in each program. In this work, we aimed to demonstrate how three of the software programs available on the market allow comparing different systems based on AUC indicator, and which tests they use. The features, functionality and performance of the three software programs were evaluated, as well as advantages and disadvantages associated with their use. For illustrative purposes, we used one dataset of the Clinical Risk Index for Babies (CRIB) from Neonatal Intensive Care Units (NICUs) in Portugal.

Keywords: ROC curve · ROCNPA · Comp2ROC · Stata · CRIB (Clinical Risk Index for Babies)

1 Introduction

The ROC analysis was developed in the 50s, based on the decision theory, and its first application was motivated by practical problems in signal detection theory [8,10]. The book by Swets and Pickett [19] marked the beginning of the vast application of this technique in biomedicine, starting in radiology. The ROC analysis theory was based on the idea that, for any signal, there is always a noisy that varies randomly around a mean value. When a stimulus is present, the activity that it creates in the image acquisition system is added to the existing noise at that time. The observer or a proper program must determine whether the level of activity in the system is caused only by noise or by an additional

© Springer International Publishing Switzerland 2016
O. Gervasi et al. (Eds.): ICCSA 2016, Part I, LNCS 9786, pp. 359–372, 2016.
DOI: 10.1007/978-3-319-42085-1_28

incentive to it. In medicine, one of the first applications of ROC analysis was published in the 1960 [13], but the ROC curve only has reached its popularity in 1970 [14].

The area under the ROC curve (AUC), is one of the indexes used to summarize the curve quality in individuals rating [11,16,19]. It represents the probability that the result of the diagnostic test for a randomly selected abnormal individual will exceed a randomly selected normal case.

A diagnostic system with higher accuracy (higher discriminating capability) is represented by a movement of the ROC curve upwards and to the left.

When ROC curves are developed for unpaired data and obtained without any distributional assumption, statistical tests to compare areas under the curves are based on the estimates of AUC for the curves corresponding to each test and on the estimates of the corresponding standard deviation [11].

This paper aims to compare the performance of the following software programs: ROCNPA (Receiver Operating Characteristic Nonparametric Analysis), Comp2ROC and Stata® 13.1, for the comparison of systems based on ROC curves.

When comparing diagnostic tests, if the corresponding ROC curves intersect at one or more points in space, causing alternation a change in the discriminating ability of these tests, it is necessary to conduct a more thorough analysis to evaluate and compare the performance of those diagnostic tests. The Comp2ROC package developed by Braga [4] may be used in that situation.

2 Empirical ROC Curve

The ROC curve is a graphical representation of pairs of values (1-specificity, sensitivity). The greatest advantage of the ROC curve lies in its simplicity. It is a direct visual representation of test performances according to all their possible answers. Consider the random variable T as representative of the result of a diagnostic test, which relies on the following decision rule based on a cutoff t_0: if $T > t_0$, the individual is classified as abnormal (positive) and if $T \leq t_0$, the individual is classified as normal (negative). Hence, for any cutoff value t_0, we have $Sensibility = TPF = P(X > t_0)$ and $Specifity = TNF = P(Y \leq t_0)$, where X e Y are random variables that represent the possible results of the diagnostic test for abnormal and normal subjects, respectively and TPF e TNF represent the True Positive Fraction and True Negative Fraction, respectively. Therefore, the ROC curve is a continuous function for different cutoffs t_0 obtained in the sample space of T, connected by straight lines. Each cutoff point is thus associated with a particular pair (1-specificity, sensitivity). In general, the smaller the value of the cutoff point, the greater its test's ability to classify patients as positive, i.e., the higher the sensitivity. However, it is inevitable that some healthy individuals might be classified as positive, which means a lower specificity. The best cutoff point is often selected based on the point where the sensitivity and specificity are simultaneously high, which is not always appropriate. According to Fletcher et al. [9], a very sensitive test will rarely diagnose individuals with the disease and a very specific test rarely categorizes an individual as not having the disease. According to Metz [16], a ROC curve is an empirical description of the test's ability to discriminate between two status (abnormal, normal) in which each point represents a different compromise between the obtained TPF and FPF (False Positive Fraction), for example, by adopting different cutoff values. The ROC curve is non-decreasing - a property that reflects the compromise between sensitivity and specificity. If the values of the diagnostic test do not allow discriminating between abnormal and normal subjects, the ROC curve is represented by a diagonal line connecting the lower left and the upper right corners of the plane. The discriminating power of the test increases as the curve moves from the diagonal to the upper left corner of the plane. If the discrimination is perfect, corresponding to 100 % sensitivity and specifity, the curve will pass this point [15]. Among the methods for obtaining a ROC curve, parametric and non-parametric, we highlight the empirical nonparametric method. This method is a graphical representation of pairs of values (FPF,TPF) for all possible cutoffs. This method as the advantage of being strong since it is free of distributional assumptions regarding the test results. Several software programs implement it. The establishment of whether the test is abble or not to discriminate between patients with and without the disease is directly related to the measurement accuracy of the ROC curve - the AUC.

2.1   Area Under the ROC Curve (AUC)

The AUC is a summary measure widely used [11,16,19] to evaluate the performance of a diagnostic test, because its estimates take into account all the sensitivities and specificities of each cutoff point t_0. Considering the random selection of a diseased individual and a non-diseased one, the AUC is interpreted as the probability of the diseased individual to have a diagnostic test result with a magnitude greater than that of the non-diseased one. A test incapable of discriminating between the two health status - diseased and non-diseased - would have an area under the ROC curve of 0.5. The higher the test's ability to discriminate individuals of these two groups, the closer the ROC curve will be to the upper left corner of the graph and the closer to 1.0 the AUC will be. The AUC may be estimated using methods such as the trapezoidal rule (TR), the Maximum Likelihood Estimation (MLE), the binormal approach (BA), and the approach to the Wilcoxon-Mann-Whitney U test (WMW).

Bamber [1], using a nonparametric approach, empirically estimated the area under the adjusted ROC curve with k values t_k, by adding up the areas of the $k - 1$ trapezoids that divide the curve in half.

Consider the random variables X and Y that, as previously defined, represent the values of T, and without loss of generality, consider them discrete random variables. The area of the kth trapezoid is given by the equation below:

$$A_k = P(X = t_k)[P(Y \leq t_{k-1}) + 0.5P(Y = t_k)] \tag{1}$$

The total area under the empirical ROC curve is given by the following equation:

$$A(X,Y) = P(Y < X) + \frac{1}{2}P(X = Y) \tag{2}$$

This latter equation is proportional to the Mann-Whitney version of the Wilcoxon statistical test for two samples. Its result can be generalized when X and Y are continuous random variables, and the AUC can be determined using the equation:

$$A(X,Y) = P(Y \leq X) \tag{3}$$

2.2   Comparing ROC Curves

To evaluate the performance of a diagnostic tests, we may choose the better of two tests for classifying individuals, or compare the results of the same test applied to different populations. For that evaluation, the AUC based on the ROC curve may be used as an appropriate measure and ordering performance tests as a function of this measure.

However, it should be noted that ROC curves may cross over each other and in a given range for FPF values one curve may perform better, while in the other FPF values the other curve is better. Since different values of FFP correspond to different cutoff values, for certain cutoff values the first curve is better than the second while for other values the second curve is better than the first.

When we know what cutoff value to use, this evaluation is not difficult. Likewise, if we know what value of FPF or TPF to use, there will be no difficulty in this evaluation. However, in all cases where the cutoff value is not predetermined, difficulties arise. The AUC has the advantage of not requiring the determination of a cutoff value. It gives the overall performance for all cutoff values, summarizing the behaviour of the ROC curve. Nonetheless, this advantage turns into a disadvantage when the curves intersect.

Four factors should be considered when comparing ROC curves. First, we must decide if we want to formulate any hypothesis about the curves or specific summary measures. Second, we may be interested in comparing two or more curves. Third, comparisons shall be based on the performance of the tests applied to data sets. The same data can be used to produce each of the ROC curves, and different sets may be used for different curves, i.e., the comparison can be for paired or unpaired data. If paired data sets are used then ROC curves and any resulting statistical summary are correlated. Lastly, the sampling process influences the variance and covariance structures.

If the ROC curves are obtained from unpaired data and without any distributional assumption, the statistical test and the confidence intervals used to compare the AUCs may be based on estimates of the AUCs corresponding to each test. For two tests T1 and T2, estimates of the AUCs and corresponding standard errors (SEs) can be obtained. Consider $\hat{A}_1$ and $\hat{A}_2$ as estimates of the AUCs associated with tests T1 and T2, respectively, and $SE(A_1)$, $SE(A_2)$ as the estimates of the corresponding SEs. The relevant hypothesis to test, H_0, is that the two data sets come from curves with the same AUC, and is represented as follow:

$$H_0 : A_2 - A_1 = 0 \text{ vs } H_1 : A_2 - A_1 \neq 0$$

One of the first proposals to compare two ROC curves for unpaired samples, using the AUC, was presented by Hanley and McNeil [11] and is based on the critical ratio (Z):

$$Z = \frac{\hat{A}_2 - \hat{A}_1}{\sqrt{SE_{A_1}^2 + SE_{A_2}^2}} \sim N(0, 1) \qquad (4)$$

If the ROC curves intersect, this Z statistic could be insufficient to assess the statistical significance when comparing ROC curves. It is necessary to make comparisons based on where these intersections occur and identify the regions of the curve where they occur. Braga et al. [2] proposed an alternative method to compare two ROC curves and developed a package to compare ROC curves that intersect [3]. When two or more empirical curves are constructed based on tests performed on the same individuals, statistical analysis on differences between curves must take into account the correlated nature of the data. DeLong et al. [6] present a nonparametric approach to the analysis of areas under correlated ROC curves, by using the theory on generalized U-statistics to generate an estimated covariance matrix.

3   Software for Comparing ROC Curves

Many software that allow us to obtain the ROC curve and the corresponding measures of performance. The first algorithm was developed by Dorfman and Alf [7]. This program determine the maximum likelihood estimates of the ROC curve parameters, considering the binormal model. Since 1980, a team led by Metz presented several programs to study the ROC curve, always considering the binormal model. However, other software programs have emerged. In this section we present a brief description of the three software programs.

3.1   ROCNPA

In 2000, Braga developed the ROCNPA (Receiver Operating Characteristic Non-Parametric Analysis) program. This software was developed in Java and can be used on any operating system. It allows assessing the performance of several tests, with the drawing of multiple ROC curves on the same plane and estimation of the corresponding measures of their performances. This software, can be found at http://pessoais.dps.uminho.pt/acb/englacb/.

This program allows viewing the distributions for positive and negative cases and displaying the empirical ROC curve in the unitary plane for one or more variables. It also allows, based on the AUC, evaluating the performance of more than two diagnostic systems, whether the samples are paired or unpaired, using the methodology proposed by DeLong et al. [6] and Hanley and McNeil [12]. The AUC index is determined by three different processes: the TR, the WMW nonparametric approach to statistics, and the approach in the binormal plane using estimated coefficients of the regression line. The values of the SEs are determined using the routine suggested by DeLong et al. [6], based on the variance and covariance matrix. When data are from paired samples, the correlation coefficient between the areas is also calculated using the procedure developed by DeLong and DeLong [6]. In a previous work [17], the developers of this work show how to introduce the data and how to obtain the histograms, the ROC curve and multiple comparison tests able to calculate using it.

3.2   Comp2ROC R® Package

R is a language and an environment for statistical computing and graphics. It provides a wide variety of statistical and graphical techniques and is highly inclusive. This language provides a great number of options for the estimation of the ROC curve and associated measures.

The Comp2ROC package developed by Frade and Braga [3], allows comparing two diagnostic systems that cross each other, based on AUC index, and use the methodology proposed by Braga et al. [2]. It consists of several functions that allow obtaining the ROC curve and precision measurements. This package has a function that allows comparing the two ROC curves and other that allows calculating some statistical measures like extension and location: "Total Area of Curve 1" (using triangles), "Total Area of Curve 2" (using triangles), "Proportion of Curve1", "Proportion of Curve2", "Proportion of ties", "Location of Curve 1", "Location of Curve 2", "Location of Ties", "Number of sampling lines", "Slopes of sampling lines", "Difference of area of triangles", "Distance of the intersection points of Curve 1 to reference point" and "Distance of the intersection points of Curve 2 to reference point". The function comp.roc.delong allows calculating the area under the curve of each curve and some statistical measures. It calculates the WMW estimate area for each modality, standard deviations, variances, and global correlations [5]. The Comp2ROC package is dependent on two other package available in R®: ROCR and boot. The ROCR package is a useful tool for two-dimensional creation of a ROC curve. Several statistical methods and various performance measures are available, including the AUC index, which is determined by nonparametric method using the WMW statistics. The function boot of the boot package, allows bootstrap resampling and generating bootstrap replicates of a statistical test applied to data. For the nonparametric bootstrap, possible resampling methods include the ordinary bootstrap, the balanced bootstrap, antithetic resampling, and permutation.

3.3   .roccomp on Stata®

Stata® is a general-purpose statistical software package created in 1985 by StataCorp. Stata's capabilities include data management, statistical analysis, graphics, simulations, regression, and custom programming. This program has

Table 1. Features of the ROCNPA, Comp2ROC and Stata® software.

	ROCNPA	Comp2ROC	Stata®
Load data file	Yes	Yes	Yes
Load data manualyy	Yes	Yes	Yes
Save to file	Yes	Yes	Yes
Help function	No	Yes	Yes
Tutorial	No	Yes	Yes
Demonstration version	No	Yes	Yes
User manual	No	Yes	Yes
ROC curve Plot	One or more	Two	One or more
Paired Samples	Yes	Yes	Yes
Unpaired Samples	Yes	Yes	Yes
		Triangle Area	
Estimatives of AUC	TR and WMW	TR and WMW	TR and WMW
	Binormal Area		Binormal Area
Comparing ROC curves	Two or more	Two	Two or more
Cross ROC curves Analysis	No	Yes	No
Statistical Tests	Yes	Yes	Yes
Confidence Intervals	No	Yes	Yes
Open source	Yes	Yes	No

always emphasized a command-line interface that facilitates replicable analyses and includes a graphical user interface that uses menus and dialog boxes to give access to nearly all built-in commands. The dataset can be viewed or edited in spreadsheet format. Stata® can import data in a variety of formats, including ASCII data formats (such as CSV or databank formats) and spreadsheet formats (including various Excel formats). Within the ROC analysis, Stata® allows nonparametric and parametric (semiparametric) methods for generating the ROC curve. It includes six commands to carry out this type of analysis. The .roctab command performs nonparametric ROC analysis for a single classifier and .roccomp extends this analysis to situations where we have multiple diagnostic tests to be compare and test. The .rocgold command also provides ROC analysis for multiple classifiers, compares each classifier's ROC curve to a *gold standard* ROC curve and makes adjustments for multiple comparisons in the analysis. Futhermore, both .rocgold and .roccomp allow parametric estimation of the ROC curve through a binormal fit. The .rocfit command is an estimation command that also estimates the ROC curve of a classifier through a binormal fit. When covariates affect ROC analysis, the .rocreg command performs ROC analysis using nonparametric and parametric (semiparametric) methods. It allows conditional and specific ROC curves. This estimation command provides many postestimation options [18]. In Table 1 we show the features of the three software.

4   Pratical Application

The clinical severity index CRIB (Clinical Risk Index for Babies), which is applied in Neonatology Intensive Care Unit (NICUs) of Portugal, was used as an illustrative example to compare the three software in terms of output interpretation. This index is an ordinal indicator in which a greater value correspond to a positive test. The study involved a total of 3115 newborns from several NICUs, whose distribution is presented in Table 2.

4.1   CRIB Index Performance with ROCNPA

Figure 1 illustrates one possible output of ROCNPA, presenting the distribution of newborns that died and survivied, based on the CRIB index of the NICUs, and the correspondent ROC curves (Fig. 1(g)). Data was entered manually according with the menu displayed in the literature [17].

The values of the AUCs may be obtained with ROCNPA through the TR and using the U statistic of WMW with the correspondent SE. Table 3 shows these values regarding each ROC curve associated with the classification obtained from the CRIB index of each NICU.

ROCNPA also allows calculating multiple comparison tests without the Bonferroni correction for unpaired samples. The resulting values for these tests are presented in Table 4.

The results obtained for the AUC of the ROC curves (Table 3) shows that the CRIB index of NICUs in the Center region ($AUC = 0.904$ and $SE = 0.023$) and Alentejo ($AUC = 0.900$ and $SE = 0.066$) showed a better performance in the classification of newborns, and hence, in assessing the risk of death in newborns, with a very low weight in this two NICUs. However, the results of multiple comparison tests (Table 4) show that there are no significant differences between the performances of the various NICUs, considering the CRIB index as an indicator for the risk severity.

4.2   CRIB Index Performance with Comp2ROC

In Comp2ROC, data can be entered manually or imported from a .csv file by following the guidelines of the help menu. As mentioned above, the Comp2ROC package compares two diagnostic systems when the corresponding ROC curves intersect at one or more points on the ROC space. The analysis between the NICUs in the North region and the Islands is presented for illustrative purpose. The functions to be used in Comp2ROC for reading data from a .csv file format for independent data samples is the following:

```
read.file(name.file.csv, header.status = TRUE, separator =";",
decimal = ",", modality1,testdirection1, modality2, testdirection2,
status1, related = FALSE, status2)
```

Table 2. Distribution of newborns by NICU and mortality rate.

NUTS II	Total of newborns	Mortality rate
North	782	9.97 %
Center	514	9.34 %
Lisbon	1288	12.34 %
Alentejo	215	6.05 %
Algarve	174	7.47 %
Islands	142	16.90 %

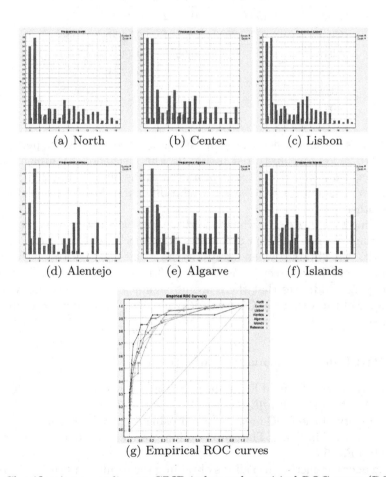

(a) North   (b) Center   (c) Lisbon

(d) Alentejo   (e) Algarve   (f) Islands

(g) Empirical ROC curves

Fig. 1. Classification according to CRIB index and empirical ROC curve (ROCNPA output).

Table 3. AUC and SE associated with the ROC curves.

NICU	Wilcoxon area	SE
North	0.882	0.022
Center	0.904	0.023
Lisbon	0.872	0.016
Alentejo	0.900	0.066
Algarve	0.887	0.044
Islands	0.879	0.034

Table 4. Multiple comparison tests for independent variables

	Center	Lisbon	Alentejo	Algarve	Islands
North	0.2543	0.96013	0.7872	0.9045	0.9522
Center		0.2543	0.9601	0.7490	0.5485
Lisbon			0.6818	0.7414	0.8572
Alentejo				0.8729	0.7795
Algarve					0.8808

The results obtained when using this function for comparing the performance of the CRIB index are illustrated in Fig. 2. The methodology implemented in this package, using the graph of area produced in the output (Fig. 2(d)), enables identifying the regions of ROC space where one curve has better performance than the other.

According to Fig. 2(d), in which the three lines are not above or below the zero line, there are no significant differences between the AUCs of the curves. The same results was obtained when analysing the results in Fig. 2(b), in which the confidence interval by percentil method is $[-0.1176149, 0.1666251]$. Like ROC-NPA, Comp2ROC indicates that there are no significant differences between the performance of the various NICUs, considering the CRIB index as an indicator for the risk severity.

4.3   CRIB Index Performance with .roccomp

After importing data from an Excel sheet, Stata® allows comparing ROC curves following the steps shown in Fig. 3.

The equality of the ROC areas is tested using the .roccomp function, whose menu is illustrated in Fig. 4. The graph option produce the ROC curves that are presented in Fig. 5.

The .roccomp function also allows obtaining the results for the AUC, the corresponding SE, and the confidence intervals for the AUC of each curves (Table 5). The theory on generalized U-statistics to generate a contrast matrix it could also be used [6].

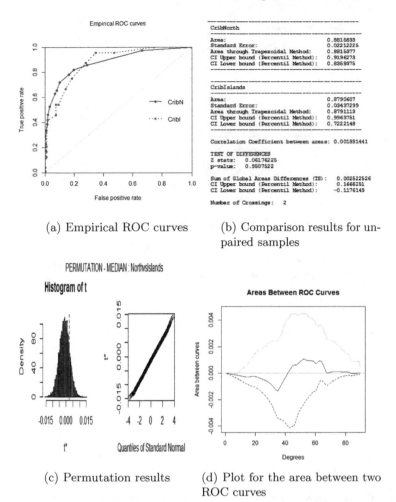

(a) Empirical ROC curves

(b) Comparison results for unpaired samples

(c) Permutation results

(d) Plot for the area between two
ROC curves

Fig. 2. Comp2ROC output of index CRIB. Comparison between North and Islands
NICUs.

The hypothesis to be tested is that there is no statistically significant difference between the AUC of the generated ROC curves. This hypothesis is represented as follows:

$$H_0 : area(1) = area(2) = area(3) = area(4) = area(5) = area(6)$$

The chi-square test ($\chi^2(5) = 1.40$, $Prob > \chi^2 = 0.9245$) reveals that there are
no significant differences, as we have concluded with ROCNPA and Comp2ROC.

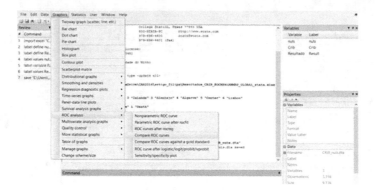

Fig. 3. Stata® menu for comparing ROC curves.

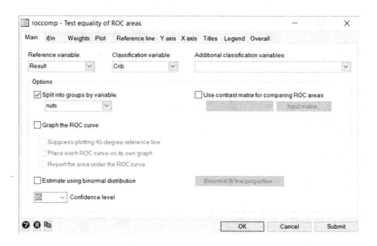

Fig. 4. Stata® window for obtaining comparative ROC curves.

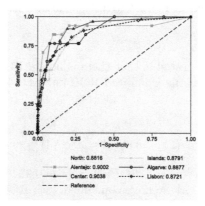

Fig. 5. Comparative empirical ROC curves (Stata® output).

Table 5. Comparative empirical ROC curves (Stata).

ROC				Asymptotic Normal
NUTS	Obs	Area	Std. Error	[95 % Conf. Interval]
North	782	0.8816	0.0221	[0.8382,0.9249]
Islands	142	0.8791	0.0344	[0.8116,0.9464]
Alentejo	215	0.9002	0.0664	[0.7700,1.0000]
Algarve	174	0.8877	0.0443	[0.8009,0.9745]
Center	514	0.9038	0.0230	[0.8587,0.9489]
Lisbon	1288	0.8721	0.0157	[0.8414,0.9028]

5   Discussion and Conclusion

This study presented the differences and similarities of three software programs on the comparison of ROC curves. The results show that all three provide good supporting software for comparing the ROC curves and calculating the associated quality measures. In terms of data entry, the three software programs are similar. All of them allow manual input or import of files with the .csv format. The production output is similar but uses different test statistics (particularly ROCNPA and .roccomp). ROCNPA and Comp2ROC have the advantage of being free software. While the ROCNPA and the Stata® allow comparison of more than two diagnostic systems, Comp2ROC only allows comparison of two systems. However, Comp2ROC has the great advantages of enabling the comparison between two ROC curves that cross each other.

Acknowledgments. The authors would like to thank the availability of the data by the RNMBP team.
This work has been supported by COMPETE: POCI-01-0145-FEDER-007043 and FCT - (Fundação para a Ciência e Tecnologia) within the Project Scope: UID/CEC/00319/2013.

References

1. Bamber, D.: The area above the ordinal dominance graph and the area below the receiver operating characteristic graph. J. Math. Psycology **4**, 387–415 (1975)
2. Braga, A.C., Costa, L., Oliveira, P.: An alternative method for global and partial comparison of two diagnostic system based on ROC curves. J. Stat. Comput. Simul. **83**, 307–325 (2013)
3. Frade, H., Braga, A.C.: Comp2ROC: R package to compare two ROC curves. In: Mohamad, M.S., et al. (eds.) 7th International Conference on PACBB. AISC, vol. 222, pp. 127–135. Springer, Heidelberg (2013)
4. Braga, A.C.: Compare two ROC curves that intersect. Repository CRAN (2015)

5. Coelho, S., Braga, A.C.: Performance evaluation of two software for analysis through ROC curves: Comp2ROC vs SPSS. In: Gervasi, O., Murgante, B., Misra, S., Gavrilova, M.L., Rocha, A.M.A.C., Torre, C., Taniar, D., Apduhan, B.O. (eds.) ICCSA 2015. LNCS, vol. 9156, pp. 144–156. Springer, Heidelberg (2015)
6. DeLong, E.R., DeLong, D.M., Clarke-Pearson, D.L.: Comparing the areas under two or more correlated receiver operating characteristic curves: a nonparametric approach. Biometrics **44**, 837–845 (1988)
7. Dorfman Jr., D.D., Alf Jr., E.: Maximum-likelihood estimation of parameters of signal-detection of confidence intervals-Rating-method data. J. Math. Psychol. **6**, 487–496 (1969)
8. Egan, J.P.: Signal Detection Theory and ROC Analysis. Academic Press, New York (1975)
9. Fletcher, R.H., Fletcher, S.W., Wagner, E.H.: Epidemiologia Clínica: Aspetos Fundamentales. Elsevier Espanha, Barcelona (2002)
10. Green, D.M., Swets, J.A.: Signal Detection Theory and Psychophysics (rev.ed). Robert E. Krieger Publishing Company, Huntington (1974)
11. Hanley, J.A., McNeill, B.J.: The meaning and use of the area under receiver operating characteristic (ROC) curve. Radiology **143**, 29–36 (1982)
12. Hanley, J.A., McNeill, B.J.: A method of comparing the areas under receiver operating characteristic curves derived from the same cases. Radiology **3**, 839–843 (1983)
13. Lusted, L.B.: Logical analysis in roentgen diagnosis. Radiology **74**, 178–193 (1960)
14. Martinez, E.Z., Louzada-Neto, F., Pereira, B.B.: A curva ROC para testes diagnósticos. Cadernos Saúde Coletiva **11**, 7–31 (2003)
15. Metz, C.E.: Basic principles of ROC analysis. Semin. Nucl. Med. **8**, 283–298 (1978)
16. Metz, C.E.: Statistical Analysis of ROC Data in Evaluating Diagnostic Performance. Multiple Regression Analysis: Applications in the Health Sciences. American Institute of Physics (1986)
17. Mourão, M.F., Braga, A.C.: Evaluation of the CRIB as an indicator of the performance of neonatal intensive care units using the software ROCNPA. In: Computational Science and its Applications, pp. 151–154 (2012)
18. Mourão, M.F., Braga, A.C., Almeida, A., Mimoso, G., Oliveira, P.N.: Adjusting covariates in CRIB score index using ROC regression analysis. In: Gervasi, O., Murgante, B., Misra, S., Gavrilova, M.L., Rocha, A.M.A.C., Torre, C., Taniar, D., Apduhan, B.O. (eds.) ICCSA 2015. LNCS, vol. 9156, pp. 157–171. Springer, Heidelberg (2015)
19. Swets, J.A., Pickett, R.M.: Signal Detection Theory and ROC Analysis in Psychology and Diagnostic: Collected Papers. Lawrence Erlbaum Associates, Inc., Mahwah (1996)

An Approach to Solve the Set Covering Problem with the Soccer League Competition Algorithm

Adrián Jaramillo[1(✉)], Broderick Crawford[1,2,3(✉)], Ricardo Soto[1,5,6(✉)],
Sanjay Misra[4(✉)], Eduardo Olguín[4], Álvaro Gómez Rubio[1],
Juan Salas[1], and Sebastián Mansilla Villablanca[1]

[1] Pontificia Universidad Católica de Valparaíso, Valparaíso, Chile
{adrian.jaramillo.s,alvaro.gomez.r,
juan.salas.f,sebastian.mansilla.v}@mail.pucv.cl,
{broderick.crawford,ricardo.soto}@pucv.cl
[2] Universidad Central de Chile, Santiago, Chile
[3] Universidad San Sebastian, Santiago, Chile
[4] Covenant University, Ota, Nigeria
sanjay.misra@covenantuniversity.edu.ng, eduardo.olguin@uss.cl
[5] Universidad Científica del Sur, Lima, Peru
[6] Universidad Autónoma de Chile, Santiago, Chile

Abstract. The Soccer League Competition algorithm (SLC) is a new nature-based metaheuristic approach to solve optimization problems. It gets its basis model from the interaction between soccer teams and their players in a soccer league competition, where each player (feasible solution) compete for victory and be the best player.

This paper presents a review of the underlaying SLC model and a practical approach to solve the Set Covering Problem using SLC and Python as programming language and tested over a widely OR-Library SCP benchmarks to obtain convergence capability and effectiveness of the implementation.

Keywords: Soccer League Competition · Combinatorial · Optimization · Set Covering Problem · Constraint satisfaction

1 Introduction

The Set Covering Problem (SCP) is a widely studied optimization problem present in many real-life scenarios like operations research, machine learning, planning, data mining, data quality and information retrieval as is discussed in [4]. Its main goal is to find the smallest sub-collection of items from a given universe so that its union is that universe and a set of constraints are met. As an example described in [3] considering a broadcasting context of a wireless or an ad-hoc network and the problem to find which of the nodes should forward messages.

Several techniques have been proposed to find the best solutions for complex scenarios of SCP as is discussed in [5–14, 19, 20]. In the last decades several metaheuristic approaches have arisen where exact methods do not offer effectiveness.

© Springer International Publishing Switzerland 2016
O. Gervasi et al. (Eds.): ICCSA 2016, Part I, LNCS 9786, pp. 373–385, 2016.
DOI: 10.1007/978-3-319-42085-1_29

Most of this methods are nature-based as collaborative work of ant colonies, swarm of bees, schooling of fishes, genetic mutations, and gravitational systems, among others.

Soccer League Competition (SCL) [16–18] is a newly metaheuristic approach based on the soccer competitions where the teams having the best players (players with a high performance) increases its victory chances in each match, and consequently, getting the winner title at the end of the season.

The existence of players with an exceptional performance encourage other players to improve their own performance to be, in turn, acquired by teams in better positions, or to act in a more relevant position inside the same team. In other scenarios, some situations will promote the teams with lower performance to acquired young talents, same as the substitution of players with poor performance. This dynamic generates a continuous improvement for all players who are part of each team, allowing the identification of the best player in each team anytime, together with the opportunity to choose the best player of whole season. This will be related with the best local solutions and the best global known solution, respectively.

We propose a derived model from SLC to solve optimization problems inside a binary search space, specially applied to solve the set covering problem. This derived model is built in Python programming language and it can lead to a wide variety of benchmarks.

The purpose of the solution is to bring a functional tool to solve general problems under the set covering problem approach and to assess their effectiveness by comparing the results against known OR-Library [1] SCP benchmarks.

2   The Set Covering Problem

As indicated in [21], given n sets, being S be the union of all the sets. An element is covered by a set if the element is in the set. A *cover* of S is a group of the n sets such that every element of S is covered by at least one set in the group. The Set Covering Problem is to find a *cover* of S of the minimum size.

Set Covering Problem defined as NP-hard problem establishes a set of m constraints over n decision variables in $\{0,1\}$ domain. Its formulation is as follow:

$$\min \quad C = \sum_{j=1}^{n} c_j x_j \tag{1}$$

s.a.:

$$\sum_{j=1}^{n} a_{ij} x_j \geq 1 \quad \forall i \in I = \{1, 2, ..., m\} \tag{2}$$

$$x_j \in \{0, 1\} \quad \forall j \in J = \{1, 2, ..., n\} \tag{3}$$

The main idea is to find the solution vector $\mathbf{X} = (x^1, x^2, ..., x^n) \in \{0,1\}^n$ such every constraint is verified and $C : R^n \to R$ is minimum.

3   The Soccer League Competition Algorithm

Identifying and defining some model elements are relevant to understand the relation between reality and the underlying mathematical model of SLC. The concepts *player, player power, team, team power, match, season* will be discussed in the follow sections.

Mapped as a soccer player, allow to define $\mathbf{X} = (x^1, x^2, ..., x^d)$ as a d-dimensional solution vector. Each dimension corresponds to a decision variable defined by the problem. Like a soccer league team consisting of several players, let's define a set of $N_{players}$ solution vectors in a solution space $\mathcal{S}$, then:

$$\mathbf{X}_i = (x_i^1, x_i^2, ..., x_i^d) \mid \mathbf{X}_i \in \mathcal{S} \\ i \in \{1, 2, ..., N_{players}\} \tag{4}$$

We identify two types of players: *fixed* and *substitute* players, designed as $\mathbf{F}_i$ and $\mathbf{S}_i$, respectively. A solution vector $\mathbf{X}_i$ will be considered as fixed or a substitute player according to certain ranking definition to be explained in the next sections. Each team $\mathcal{T}_k$ consists of a set of fixed and a set of substitutes players. In SLC all teams will have the same number of fixed players and substitutes players, N_f and N_s, respectively. Then,

$$|\mathcal{T}_k| = N_f + N_s \qquad k \in \{1, 2, ..., N_{teams}\} \tag{5}$$

For each player $\mathbf{X}_i$ we define a scalar value performance indicator pp_i called *player power*. Player power allows to compare performance among players. We define $PP : \mathbb{R}^n \to \mathbb{R}$ as the player power function:

$$pp_i = PP(\mathbf{X}_i) \tag{6}$$

where if two vectors $\mathbf{X}_j$ and $\mathbf{X}_k$ verifies $PP(\mathbf{X}_j) > PP(\mathbf{X}_k)$ then we say $\mathbf{X}_j$ has a better performance than $\mathbf{X}_k$ and it is a better known solution in relation to $\mathbf{X}_k$ for the problem we are solving. For a cost function $C(\mathbf{X})$ in an optimization problem looking for minimization fitness, we can define $PP(\mathbf{X})$ as follow:

$$PP(\mathbf{X}) = \frac{1}{C(\mathbf{X})} \tag{7}$$

On the other hand, for a maximization fitness problem, we will define:

$$PP(\mathbf{X}) = C(\mathbf{X}) \tag{8}$$

We identify the *star player* $\mathbf{SP}_k \in \mathcal{T}_k$ as the $\mathbf{X}_i$ with the best player power inside the team $\mathcal{T}_k$. In a similar way, we identify the *super star player* $\mathbf{SSP}$ as the player with the best power player regarding all players in the league.

Similar to player power calculation, we define the *team power TP* for the team $\mathcal{T}$ as the average value of power player regarding all its players (fixed and substitutes), i.e.:

$$TP = \sum_{\mathbf{X}_i \in \mathcal{T}} \frac{PP(\mathbf{X}_i)}{|\mathcal{T}|} \tag{9}$$

In SLC a match between two teams $\mathcal{T}_j$ and $\mathcal{T}_k$ defines always a single winner. The winner team is unknown until the match ends. The victory chance of $\mathcal{T}_k$ is directly proportional to its team performance and inversely proportional to opposing team performance.

Given TP_j and TP_k as the team power for $\mathcal{T}_j$ and $\mathcal{T}_k$ respectively, the probability of victory for $\mathcal{T}_j$ facing $\mathcal{T}_k$ is given as:

$$PV_j = \frac{TP_j}{TP_j + TP_k} \tag{10}$$

In the same way, the probability of victory for $\mathcal{T}_k$ is given as:

$$PV_k = \frac{TP_k}{TP_j + TP_k} \tag{11}$$

It is clear that:

$$PV_j + PV_k = 1 \tag{12}$$

All teams must confront one with another without repetition. Given a set of N_{teams} teams in competition the number of possible matches without repetition is given by the combinatorial statement:

$$\binom{N_{teams}}{2} = \frac{N_{teams}!}{2!(N_{teams} - 2)!} = \frac{N_{teams} * (N_{teams} - 1)}{2} \tag{13}$$

In a scenario for $\binom{N_{teams}}{2}$ matches, where all teams confront each other, and define winners and losers, the matches end with a single winner team, which include the best player of the league. This best player will correspond to the **SSP** associated with the best solution vector for the problem.

3.1 Movement Operators

SLC defines [16] four movement operator types: *imitation, provocation, mutation* and *substitution*. When a match is completed, each fixed player in the winner team attemps to improve its performance in order to become like team's super player or the league super star player. It is achieved moving $\mathbf{F}_i$ in direction to **SSP** or $\mathbf{SP}_k$, depending which destination improve the player power of $\mathbf{F}_i$.

As indicated in [16], we define the following **imitation operator** to get two news solution vectors $\mathbf{F}_{new_1}$ and $\mathbf{F}_{new_2}$:

$$\mathbf{F}_{new_1} = \mu_1 \mathbf{F}_k + \tau_1 (\mathbf{SSP} - \mathbf{F}_k) + \tau_2 (\mathbf{SP_k} - \mathbf{F}_k) \tag{14}$$

$$\mathbf{F}_{new_2} = \mu_2 \mathbf{F}_k + \tau_1 (\mathbf{SSP} - \mathbf{F}_k) + \tau_2 (\mathbf{SP_k} - \mathbf{F}_k) \tag{15}$$

where $\mu_1 \sim U(\theta, \beta)$, $\mu_2 \sim U(0, \theta)$, $\theta \in [0, 1]$, $\beta \in [1, 2]$ and $\tau_1, \tau_2 \sim (0, 2)$ are random numbers with uniform distribution. If $\mathbf{F}_{new_1}$ improve player power of $\mathbf{F}_k$ then it is replaced by $\mathbf{F}_{new_1}$. In the other case, if $\mathbf{F}_{new_2}$ improve the player power of $\mathbf{F}_k$ then it is replaced by $\mathbf{F}_{new_2}$. If none of both improve power player, then $\mathbf{F}_k$ remains without change.

In the winner team each substitute player tries to become a fixed player. This is achieved by improving its performance to have a similar level value to the average performance of the team's fixed players. In SLC algorithm this means moving $\mathbf{S}_k$ substitute player to the team's fixed player centroid $\mathbf{G}_k$. We calculate two new candidates for substitute player $\mathbf{S}_k$:

$$\mathbf{S}_{new_1} = \mathbf{G}_k + \chi_1(\mathbf{G}_k - \mathbf{S}_k) \tag{16}$$

$$\mathbf{S}_{new_1} = \mathbf{G}_k + \chi_2(\mathbf{S}_k - \mathbf{G}_k) \tag{17}$$

where $\chi_1 \sim U(0.9, 1)$, $\chi_2 \sim U(0.4, 0.6)$ are random numbers with uniform distribution. We define the d-dimension of the centroid vector G for the team $\mathcal{T}$ as follow:

$$\mathbf{G}_k^d = \frac{\sum\limits_{\mathbf{F}_i \in \mathcal{T}} \mathbf{F}_i^d}{N_f} \tag{18}$$

According (16) and (17), if $\mathbf{S}_{new_1}$ is a better player power value than $\mathbf{S}_k$ then it is replaced by $\mathbf{S}_{new_1}$. In the other case, if $\mathbf{S}_{new_2}$ is a better player power value than $\mathbf{S}_k$ then it is replaced by $\mathbf{S}_{new_2}$. If none of them improve their player power, then $\mathbf{S}_k$ is replaced by a new random generated solution vector.

Fixed players in the losing teams try to apply small changes to avoid repeating the previous failures. In this scenario, we will apply some mutation operator like Genetic Algorithm (GA). Some substitute players in the losing team are replaced by young talents. In SLC it is achieved by generating new substitutes vector as follow. We will choose two substitute vectors $\mathbf{S}_k$ and $\mathbf{S}_l$ in the losing team and calculate two new solutions vector for each one:

$$\mathbf{S}_{new_k} = \alpha \times \mathbf{S}_k + (\mathbf{1} - \alpha) \times \mathbf{S}_l \tag{19}$$

$$\mathbf{S}_{new_l} = \alpha \times \mathbf{S}_l + (\mathbf{1} - \alpha) \times \mathbf{S}_k \tag{20}$$

where $\alpha \in R^n$ is a vector with values in $\{0, 1\}$ defined randomly with uniform distribution, and $\mathbf{1}$ is the unitary vector in R^n.

If $\mathbf{S}_{new_k}$ is a better player power value than $\mathbf{S}_k$ then this is replaced for the new vector $\mathbf{S}_{new_k}$. In the same way, if $\mathbf{S}_{new_l}$ is a better power player value than $\mathbf{S}_l$ then it is replaced by the new vector $\mathbf{S}_{new_l}$.

3.2   Iteration Process and Convergence

The iteration starts defining $N_{teams} * (N_f + N_s)$ random solutions vectors (or players) and ranking them by player power from the highest to lowest value. First $N_f + N_s$ players at the top ranking corresponds to the first team, next $N_f + N_s$ players corresponds to team two and so on. In each team, first N_f players in the ranking corresponds to the team's fixed players, and the next N_f players corresponds to substitutes. First player at the top ranking corresponds to the SSP. First player at each team ranking corresponds to its SP.

Two teams faced define a winner and a loser. Each winner team fixed player try to imitate SSP and/or SP to improve his performance applying *imitation operator*. Each winner team substitute player try to improve their performance to be closer to the average power of the fixed players by *provocation operator*. The losing team apply changes to the player configuration by *mutation* and *substitution*. Next two teams are faced and so on. As a result of this dynamic, player power of each player could be changed and the raking in the same way. The best players (solution vectors with highest player) will be ranked at the top of the raking. SSP at the end of the seasons will correspond to the best solution to the problem.

4   Solving SCP by Using SLC

The algorithm implementation of SLC, regardless its technological support (i.e. runtime environment, programming language, among others), considers populations of feasible solutions (*teams*) which move through a search space by certain movement operators to explore and find the best regions. These movement operators essentially apply linear algebra over continuous space elements in order to determine new positions from already known solutions.

In a discrete scenario, especially in a binary space, concepts like centroid calculation, vector addition or scalar multiplication are no longer applicable as they are in a continuous space and some redefinitions are required as this paper attempts to show. We can define the player power function $PP(X)$ from the SCP cost function in (1) when the problem is looking for a maximum, or its inverse in case of minimum. The SCP constraint statements (2) and (3) can be build as a test-feasibility function with the capability to repair an unfeasible solution if necessary under an ADD/DROP covered columns approach.

The *imitation operator* in (14) and (15) in a binary space could be achieved by reducing Hamming distance instead vector arithmetic. We consider the follow statement to get two new positions for $\mathbf{F}$ considering a Hamming approach:

$$\mathbf{F}^d_{new_1} = \begin{cases} \mathbf{SSP}^d & \text{if } rand() \leq p_{imitation} \\ \mathbf{F}^d & \text{other case} \end{cases} \tag{21}$$

$$\mathbf{F}^d_{new_2} = \begin{cases} \mathbf{SP}^d & \text{if } rand() \leq p_{imitation} \\ \mathbf{F}^d & \text{other case} \end{cases} \tag{22}$$

where $rand() \sim U(0,1)$ is a random generated value with uniform distribution and $p_{imitation}$ is a probability of imitation defined as parameter as part of model.

In the *provocation operator* on a binary solution space a new approach is presented to obtain a centroid vector based on (18) and the probability for a dimension d to get 1 or 0 value.

$$\mathbf{BG}_k^d = \begin{cases} 1 & \text{if } \mathbf{G}_k^d \geq 0.5 \\ 0 & \text{other case} \end{cases} \tag{23}$$

In the *provocation operator* to calculate new positions as defined in (16) and (17) will generate a not feasible solution vectors because each x^d dimension value will be in the domain R instead $\{0, 1\}$. It requires a binarization operator to transform not binary solutions vectors into binary solutions. We could define a *binarization function* for a x^d dimension as follows:

$$B(x^d) = \begin{cases} 1 & \text{if } rand() \leq T(x(t)^d) \\ 0 & \text{other case} \end{cases} \tag{24}$$

being $T(x^d)$ the transfer function as defined in [15] as follows:

$$T(x^d) = \left| \frac{2}{\pi} arctan(\frac{\pi}{2} x^d) \right| \tag{25}$$

A *mutation operator* for fixed players could be considered as follows:

$$\mathbf{F}_{new}^d = \begin{cases} 0 & \text{if } rand() \leq p_{mutation} \\ 1 & \text{other case} \end{cases} \tag{26}$$

where $rand() \sim U(0,1)$ is a random generated value with uniform distribution and $p_{mutation}$ is a probability of mutation defined as parameter part of the model.

5   Python Implementation and Execution Results

Python 3.5.1 was used for coding internal data representation, imitation, provocation, mutation, substitution operators, test-feasibility and repair functions. A snippet including a general workflow in a single session match is shown as follow. $nTeam$ corresponds to number of teams of the execution. nFP corresponds to the number of fixed players. sFP corresponds to the number of substitute players. Each team is referenced by a 0-based index integer value. The *imitationFactor* and *mutationfactor* are decimal parameter values.

```
1   #Initial tasks
2   playerMatrix = BuildPlayerMatrix(nTeams, nFP, nSP)
3   playerMatrix.SetInitialPopulation()
4   playerMatrix.SortByPowerPlayer()
5
6   #Each element in the matches collection corresponds to
7   #pairs (indexTeam1, indexTeam2)
8   matches = playerMatrix.BuildTeamConfrontation()
9
10   for match in matches:
```

```
11      #We get team power for team 1 and team 2
12      teamPower1 = playerMatrix.GetTeamPower(match[0])
13      teamPower2 = playerMatrix.GetTeamPower(match[1])
14      pv_1 = teamPower1 / (teamPower1 + teamPower2)
15      pv_2 = teamPower2 / (teamPower1 + teamPower2)
16
17      #The match starts and will be a unique winner
18      if (random.uniform(0, 1) <= pv_1):
19          winnerTeamInx = match[0]
20          losserTeamInx = match[1]
21      else:
22          winnerTeamInx = match[1]
23          losserTeamInx = match[0]
24
25      #Imitation operator in the winner team
26      SSP = playerMatrix.GetSuperStarPlayer()
27      SP = playerMatrix.GetStarPlayer(winnerTeamIndex)
28
29      for FP in playerMatrix.GetPlayers(winnerTeamIndex, Type.Fixed):
30          FP1 = FP.ReduceHamming(SSP, imitationFactor)
31          FP2 = FP.ReduceHamming(SP, imitationFactor)
32
33          if (FP1.playerPower > FP.playerPower):
34              FP.ReplaceBy(FP1)
35          elif (FP1.playerPower > FP.playerPower):
36              FP.ReplaceBy(FP2)
37
38      #Provocation operator in the winner team
39      BG = playerMatrix.GetCentroide(winnerTeamInx, Type.Fixed)
40      for SP in playerMatrix.GetPlayers(winnerTeamInx, Type.Substitute):
41          SP1 = FP.ReduceHamming(BG, imitationFactor)
42
43          if (SP1.playerPower > SP.playerPower):
44              SP.ReplaceBy(FP1)
45          else:
46              SP.ReplaceBy( playerMatrix.GetNewPlayer() )
47
48      #Mutation operator in the loosing team
49      for FP in playerMatrix.GetPlayers(losingTeamInx, Type.Fixed):
50          FP.ReplaceBy( FP.ApplyMutation(mutationFactor) )
51
52      #Substitution operator in the loosing team
53      for SP in playerMatrix.GetPlayers(losingTeamInx, Type.Substitute):
54          FP.ReplaceBy( playerMatrix.GetNewPlayer() )
55  #End of matches
56
57  playerMatrix.SortByPowerPlayer()
58  #The best solution vector corresponds to SSP
59  SSP = playerMatrix.GetSuperStarPlayer()
60  ...
```

For data generated by the process, a relational database MySQL was used to store, summarize and do analysis. pypyodbc lib [2] served as middle-ware between Python and MySQL.

5.1   Execution Results

Benchmark test was performed in several servers and PC's with different process computing capability. From OR-Library [1], first 30 SCP benchmarks where tested on the Python implementation. The rest of SCP benchmarks test data was discarded in this stage of work by a low and poor convergence velocity of the Python implementation. The results obtained on each benchmark is summarized in the Table 1.

Table 1 include benchmark instance name, number of constraints and number of decision variables of the instance, best known optimun (Z_{BKS}) from the benchmark instance, min fitness value (Z_{min}) and max fitness value (Z_{max}) obtained by the implementation execution, average fitness value obtained by the implementation execution over 31 runs (Z_{AVG}) and the relative percentage deviation (RPD) defined as in (16).

$$RPD = \frac{Z_{min} - Z_{BKS}}{Z_{BKS}} * 100 \qquad (27)$$

For each benchmark instance 4.1 to 4.10, involving 200 constraints and 1000 decision-variables, the Python implementation was executed 31 times, using 7 teams, 11 fixed players and 5 substitute players, matching in 5 seasons, showing good RPD values regarding Z_{BKS} known optimum.

For each benchmark instance 5.1 to 5.10, involving 200 constraints and 2000 decision-variables, the Python implementation was executed 31 times, using 7 teams, 11 fixed players and 5 substitute players, matching in 5 seasons, showing good RPD values regarding Z_{BKS} known optimum.

For each benchmark instance 6.1 to 6.5, involving 200 constraints and 1000 decision-variables, the Python implementation was executed 31 times, using 7 teams, 11 fixed players and 5 substitute players matching in 5 seasons showing good RPD values regarding Z_{BKS} known optimum.

For benchmark instances A1 to A5, involving 300 constraints and 3000 decision-variables, it was not possible to obtain final evidence about convergence to the optimum Z_{BKS}, because first iterations showed a poor and extremely slow convergence to the Z_{BKS} known optimum.

5.2   Imitation Factor

A challenge to face is to define which value or values range is more suitable to use as imitation factor in the imitation and provocation operators. Testing over 31 executions of the implementation, one each of them using different imitation factor values in range [0.5 %, 95 %] by step 0.5, shows that this value has an important incidence in the imitation process.

Table 1. Experimental results of SLC resolving SCP benchmarks (4, 5, 6, A sets) from OR-Library.

Instance	Constraints	Var decisions	Z_{BKS}	Z_{min}	Z_{MAX}	Z_{AVG}	RPD
4.1	200	1,000	429	431	461	444.5	0.47
4.2	200	1,000	512	519	570	544.1	1.37
4.3	200	1,000	516	520	549	535.0	0.78
4.4	200	1,000	494	503	549	525.8	1.82
4.5	200	1,000	512	518	550	531.4	1.17
4.6	200	1,000	560	566	640	585.3	1.07
4.7	200	1,000	430	435	464	447.4	1.16
4.8	200	1,000	492	499	541	518.2	1.42
4.9	200	1,000	641	678	709	689.5	5.77
4.10	200	1,000	514	524	575	548.4	1.95
5.1	200	2,000	253	254	13,893	722.1	0.40
5.2	200	2,000	302	311	17,408	1,404.2	2.98
5.3	200	2,000	226	229	43,344	1,958.6	1.33
5.4	200	2,000	242	242	250	245.8	0.00
5.5	200	2,000	211	212	227	219.0	0.47
5.6	200	2,000	213	217	230	221.8	1.88
5.7	200	2,000	293	301	315	309.7	2.73
5.8	200	2,000	288	294	315	302.0	2.08
5.9	200	2,000	279	292	294	292.7	4.66
5.10	200	2,000	265	269	13,601	942.5	0.02
6.1	200	1,000	138	144	153	147.0	4.35
6.2	200	1,000	146	149	162	154.0	2.05
6.3	200	1,000	145	150	157	152.3	3.45
6.4	200	1,000	131	131	135	132.7	0.00
6.5	200	1,000	161	171	176	174.0	6.21
A.1	300	3,000	253	29,235	44,037	34,892.7	11,455.34
A.2	300	3,000	252	32,524	41,013	35,855.0	12,806.35
A.3	300	3,000	232	30,864	44,422	35,501.3	13,203.45
A.4	300	3,000	234	29,311	42,167	35,287.7	12,426.07
A.5	300	3,000	236	29,814	41,348	34,462.7	12,533.05

Figure 1 shows that imitation percent values closer to 0.5 % involves a significant quantity of solution vectors improving their fitness value by imitation operator. However values closer to 100 percent imply a significant reduction of the effectiveness of the imitation operator.

In the other hand, Fig. 2 shows that the repair operator, responsible to make feasible an unfeasible solution vector, has a work overload peak in the neighbourhood of 15 % imitation factor. For this stage we used a value of 20 % percent as factor imitation.

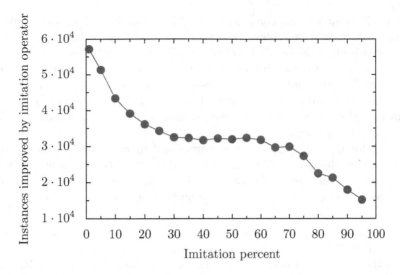

Fig. 1. Relation between instance's fitness improvement versus the imitation factor percent.

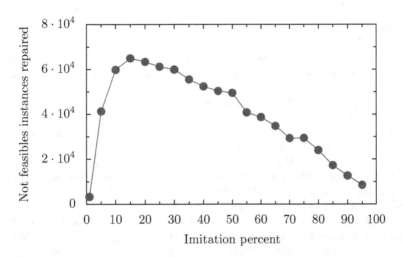

Fig. 2. Relation between repaired solution instances versus the imitation factor percent.

6   Conclusions

In this paper a variation of SLC to solve the SCP has been presented and implemented using Python to obtain computational results. It has also been applied in 30 OR-Library SCP benchmark sets to test the convergence capability of the implementation obtaining acceptable convergence up to 200 constraints and 2000 decision-variables. Benchmark test was performed in several servers and

PC's with different process computing capability, but in I7 Intel processor with 16GB in RAM it was possible to run simultaneously up to 10 instances of the algorithm without overload CPU (closer to 87 % CPU workload).

Time consumption is an important issue to face for benchmarks with a big decision-variables size and parallel-processing scenarios is a good alternative, but it is necessary to modify the SLC model to take the most to work in this kind of scenario and get capability convergence using complex benchmark like A1 to A5 OR-Library data sets.

As additional feature in the algorithm implementation, it is necessary to include some mechanisms to be able save and recover the status of the process of every instance in case of failure or power down, especially in scenarios of complex benchmark with long-time consumption processing.

Acknowledgements. Broderick Crawford is supported by Grant CONICYT/ FONDECYT/REGULAR/1140897. Ricardo Soto is supported by Grant CONI-CYT/FONDECYT/REGULAR/1160455. Sebastiásn Mansilla, Álvaro Gómez and Juan Salas are supported by Postgraduate Grant Pontificia Universidad Católica de Valparaiso 2015 (INF-PUCV 2015).

References

1. OR-Library a collection of test data sets for a variety of operations research (or) problems. http://people.brunel.ac.uk/~mastjjb/jeb/orlib/scpinfo.html. Accessed 30 Mar 2015
2. PyPyODBC a pure python odbc module by ctypes. https://pypi.python.org/pypi/ pypyodbc. Accessed 20 Jan 2015
3. Agathos, S.N., Papapetrou, E.: On the set cover problem for broadcasting in wire-less ad hoc networks. IEEE Commun. Lett. **17**(11), 2192–2195 (2013)
4. Cormode, G., Karloff, H., Wirth, A.: Set cover algorithms for very large datasets. In: Proceedings of the 19th ACM International Conference on Information and Knowledge Management, CIKM 2010, pp. 479–488. ACM, New York (2010). http://doi.acm.org/10.1145/1871437.1871501
5. Crawford, B., Soto, R., Cuesta, R., Olivares-Suárez, M., Johnson, F., Olguín, E.: Two swarm intelligence algorithms for the set covering problem. In: ICSOFT-EA 2014 - Proceedings of the 9th International Conference on Software Engineering and Applications, Vienna, Austria, 29–31 August, pp. 60–69 (2014)
6. Crawford, B., Soto, R., Cuesta, R., Paredes, F.: Using the bee colony optimization method to solve the weighted set covering problem. In: Stephanidis, C. (ed.) HCI 2014, Part I. CCIS, vol. 434, pp. 493–497. Springer, Heidelberg (2014)
7. Crawford, B., Soto, R., Monfroy, E.: Cultural algorithms for the set covering prob-lem. In: Tan, Y., Shi, Y., Mo, H. (eds.) ICSI 2013, Part II. LNCS, vol. 7929, pp. 27–34. Springer, Heidelberg (2013)
8. Crawford, B., Soto, R., Olivares-Suárez, M., Paredes, F.: Using the firefly opti-mization method to solve the weighted set covering problem. In: Stephanidis, C. (ed.) HCI 2014, Part I. CCIS, vol. 434, pp. 509–514. Springer, Heidelberg (2014)

9. Crawford, B., Soto, R., Palma, W., Paredes, F., Johnson, F., Norero, E.: The impact of a new formulation when solving the set covering problem using the ACO metaheuristic. Model. Comput. & Optim. in Inf. Syst. & Manage. Sci. AISC, vol. 360, pp. 209–218. Springer, Heidelberg (2015)
10. Crawford, B., Soto, R., Peña, C., Palma, W., Johnson, F., Paredes, F.: Solving the set covering problem with a shuffled frog leaping algorithm. In: Nguyen, N.T., Trawiński, B., Kosala, R. (eds.) ACIIDS 2015, Part II. LNCS, vol. 9012, pp. 41–50. Springer, Heidelberg (2015)
11. Crawford, B., Soto, R., Peña, C., Riquelme-Leiva, M., Torres-Rojas, C., Johnson, F., Paredes, F.: Binarization methods for shuffled frog leaping algorithms that solve set covering problems. In: Silhavy, R., Senkerik, R., Oplatkova, K.Z., Prokopova, Z., Silhavy, P. (eds.) Software Engineering in Intelligent Systems. AISC, vol. 349, pp. 317–326. Springer, Heidelberg (2015)
12. Crawford, B., Soto, R., Peña, C., Riquelme-Leiva, M., Torres-Rojas, C., Misra, S., Johnson, F., Paredes, F.: A comparison of three recent nature-inspired metaheuristics for the set covering problem. In: Gervasi, O., Murgante, B., Misra, S., Gavrilova, M.L., Rocha, A.M.A.C., Torre, C., Taniar, D., Apduhan, B.O. (eds.) ICCSA 2015, Part IV. LNCS, vol. 9158, pp. 431–443. Springer, Heidelberg (2015)
13. Crawford, B., Soto, R., Riquelme-Leiva, M., Peña, C., Torres-Rojas, C., Johnson, F., Paredes, F.: Modified binary firefly algorithms with different transfer functions for solving set covering problems. Software Engineering in Intelligent Systems. AISC, vol. 349, pp. 307–315. Springer, Heidelberg (2015)
14. Crawford, B., Soto, R., Torres-Rojas, C., Peña, C., Riquelme-Leiva, M., Misra, S., Johnson, F., Paredes, F.: A binary fruit fly optimization algorithm to solve the set covering problem. In: Gervasi, O., Murgante, B., Misra, S., Gavrilova, M.L., Rocha, A.M.A.C., Torre, C., Taniar, D., Apduhan, B.O. (eds.) ICCSA 2015. LNCS, vol. 9158, pp. 411–420. Springer, Heidelberg (2015)
15. Mirjalili, S., Lewis, A.: S-shaped versus v-shaped transfer functions for binary particle swarm optimization. Swarm Evol. Comput. **9**, 1–14 (2013)
16. Moosavian, N.: Soccer league competition algorithm, a new method for solving systems of nonlinear equations. Sci. Res. **4**, 7–16 (2014)
17. Moosavian, N.: Soccer league competition algorithm for solving knapsack problems. Swarm Evol. Comput. **20**, 14–22 (2015)
18. Moosavian, N., Roodsari, B.K.: Soccer league competition algorithm: a novel metaheuristic algorithm for optimal design of water distribution networks. Swarm Evol. Comput. **17**, 14–24 (2014)
19. Soto, R., Crawford, B., Olivares, R., Barraza, J., Johnson, F., Paredes, F.: A binary cuckoo search algorithm for solving the set covering problem. In: Vicente, J.M.F., Álvarez-Sánchez, J.R., López, F.P., Toledo-Moreo, F.J., Adeli, H. (eds.) IWINAC 2015, Part II. LNCS, vol. 9108, pp. 88–97. Springer, Heidelberg (2015)
20. Soto, R., Crawford, B., Vilches, J., Johnson, F., Paredes, F.: Heuristic feasibility and preprocessing for a set covering solver based on firefly optimization. In: Silhavy, R., Senkerik, R., Oplatkova, Z.K., Prokopova, Z., Silhavy, P. (eds.) Artificial Intelligence Perspectives and Applications. AISC, vol. 347, pp. 99–108. Springer, Heidelberg (2015)
21. Yang, Q., McPeek, J., Nofsinger, A.: Efficient and effective practical algorithms for the set-covering problem. In: Proceedings of the 2008 International Conference on Scientific Computing, CSC, 14–17 July 2008, Las Vegas, Nevada, USA. pp. 156–159 (2008)

Direct Sequential Based Firefly Algorithm for the α-Pinene Isomerization Problem

Ana Maria A.C. Rocha[1]([✉]), Marisa C. Martins[1],
M. Fernanda P. Costa[2], and Edite M.G.P. Fernandes[1]

[1] Algoritmi Research Centre, University of Minho, 4710-057 Braga, Portugal
{arocha,emgpf}@dps.uminho.pt, marisacunhamartins@gmail.com
[2] Centre of Mathematics, University of Minho, 4800-058 Guimarães, Portugal
mfc@math.uminho.pt

Abstract. The problem herein addressed is a parameter estimation problem of the α-pinene process. The state variables of this bioengineering process satisfy a set of differential equations and depend on a set of unknown parameters. A dynamic system based parameter estimation problem aiming to estimate the model parameter values in a way that the predicted state variables best fit the experimentally observed state values is used. A numerical direct method, known as direct sequential procedure, is implemented giving rise to a finite bound constrained nonlinear optimization problem, which is solved by the metaheuristic firefly algorithm (FA). A Matlab^{TM} programming environment is developed with the mathematical model and the computational application of the method. The results produced by FA, when compared to those of the `fmincon` function and other metaheuristics, are competitive.

Keywords: α-pinene isomerization · Parameter estimation · Direct sequential procedure · Firefly algorithm

1 Introduction

Bioinformatics expertise is usually required to address some issues related with slow and expensive processes from the biotechnology area. This type of skills may be used to understand the behavior of certain organisms or biochemicals, such as the case of the parameter estimation problem on the α-pinene isomerization dynamic model. A parameter estimation problem aims to find the parameter values of a mathematical model that gives the best possible fit with existing experimental data [1]. Thus, in a dynamic model based parameter estimation process, an objective functional that gives the mean squared error between the model predicted state values and the observed values (within a fixed time interval) is minimized, subject to a set of differential equations. In a dynamic system, the parameter estimation concept may lead to confusion since the parameters are not usually time-dependent in the mathematical model. Nevertheless, in a problem like this, the parameters are the decision variables of the optimization process, although they are constant throughout the entire simulation process.

© Springer International Publishing Switzerland 2016
O. Gervasi et al. (Eds.): ICCSA 2016, Part I, LNCS 9786, pp. 386–401, 2016.
DOI: 10.1007/978-3-319-42085-1_30

The general form of a first order ordinary differential equation (ODE) is given by

$$\frac{dy}{dt} = g(t, y) \tag{1}$$

where t is the independent variable (here, it is assumed that t represents time) and $y \equiv y(t)$ is the dependent variable. The equation is called ODE since the unknown function depends on a single independent variable and the order of the ODE is defined by the highest derivative order that appears in the equation. The equation is called linear if the function $g(t, y)$ is linear on y. Further, a solution of the ODE is a function $y(t)$ that satisfies the Eq. (1) for all values of t in the domain of y. Most of the time, we are not interested in all solutions to a differential equation, but only in a particular solution satisfying an extra condition. If the extra condition provides the value of y at the initial instant of time, the problem is called 'initial value problem'. Hereinafter the more compact notation $y'(t) = dy/dt$ is used.

Numerical methods for solving ordinary differential equations are discretization methods that compute approximations to the solution $y(t)$ at a finite set of points $t_0, t_1, t_2, \ldots, t_n$ of the independent variable interval $t_i \leq t \leq t_f$, where t_i and t_f are the initial and final values of the interval, respectively. A variety of numerical methods for solving a system of ODEs is available in the literature, being the most known the Euler's method, the Runge-Kutta method and the backward differentiation formula [2].

A dynamic mathematical model emerges when the optimization and control of industrial bioprocesses [3] are carried out. The construction of the model involves several stages. First, the objectives are defined based on theoretical and/or empirical knowledge of the process under study. In general, the mathematical model depends on a set of unknown parameters that require to be investigated. Second, the values for the parameters are estimated based on experimental data, assuming that the mathematical model simulates the process the best possible way. The goal of the parameter estimation problem is to calibrate the model so that it can reproduce the experimental results as close as possible. This is performed by minimizing an objective function that measures the goodness of the fit. Finally, the model with the estimated parameter values may be validated [4,5].

In general, the mathematical modeling of bioprocess engineering problems involves nonlinear dynamic equations as constraints and a nonlinear objective function to be optimized, giving the so called dynamic optimization (DO) problem. Nonconvex and multimodal functions frequently arise in DO. Moreover, some kind of noise and/or discontinuities may be present making the problem even more complex. Therefore, robustness and efficiency are crucial properties of an optimization solver so that it is capable of computing a good approximation to a global solution to the DO problem without an excessive computational effort.

Gradient-based local methods are generally very efficient to solve finite-dimensional constrained nonlinear programming (NLP) problems, although they can only deal with problems where the objective function and the constraint functions are continuously differentiable [6]. In general, they may lead to suboptimal solutions if multiple local optima are present.

The alternative to compute a global optimal solution to a NLP problem is to use a global optimization method [1, 7–10]. There has been a growing interest in developing algorithms that converge to the global optimal solution. Global search algorithms often emerge from the modifications introduced in local search algorithms. A common practice is to introduce some stochastic components to force the algorithm to diversify the search for other solutions and explore the search space. Then, they become stochastic algorithms. The stochastic component can be introduced in various form, such as for example, a simple random generation of solutions in the search space, and the use of random numbers to define the path of solutions. Most stochastic algorithms are also known as metaheuristics since they combine random components with historical knowledge of previous search in regions that have been explored and are in the neighborhood of local solutions. This is an important issue since premature convergence to local optimal solutions can be avoided. Metaheuristics are approximate methods or heuristics that are designed to search for good solutions, known as near-optimal solutions, with less computational effort and time than the more classical algorithms. Algorithms that simulate the behavior of animals in nature have been widely used to solve complex mathematical problems. Classical examples of this type of algorithms are the particle swarm optimization [11] and more recently the firefly algorithm (FA) [12].

Deterministic global optimization techniques have also been used to find a global solution that best reconciles the model parameters and measurements [13–15]. Although the convergence to the global optimal solution is guaranteed, a deterministic method is only feasible for a small number of decision variables. Since stochastic methods are able to converge rapidly to the vicinity of the global solution, although they are very expensive to locate a global solution with high quality, hybrid methods that combine global and local strategies have recently become very popular [16].

Since the problem of estimating the parameters of dynamic models of complex biological systems is becoming increasingly important, the contribution of the present study is focused on the implementation of the metaheuristic FA to solve the finite NLP problem that arises from using a numerical direct approach to locate a global optimal solution of a specific DO problem - the parameter estimation problem on the α-pinene isomerization. A powerful direct method that transcribes the dynamic model parameter estimation problem in a small finite-dimensional optimization problem through the discretization of the decision variables only is used.

The remaining part of this paper is organized as follows. In Sect. 2, we describe the α-pinene isomerization problem and in Sect. 3 we present the selected methodology for solving the α-pinene problem. Section 4 presents the results of the numerical experiments and comparisons and Sect. 5 contains the conclusions of the present study.

2 Isomerization of α-Pinene

The α-pinene isomerization was studied in 1973 by Box, Hunter, MacGregor and Erjavec in [17] and goals the estimation of a set of five velocity parameters $p_1, \ldots, p_5$ of a homogeneous chemical reaction. The reaction scheme for this chemical reaction that describes the thermal isomerization of α-pinene, y_1, to dipentene, y_2, and allo-ocimen, y_3, which in turn produces α- and β-pyronene, y_4, and a dimer, y_5, is shown in Fig. 1.

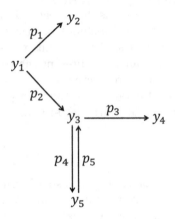

Fig. 1. α-pinene isomerization mechanism (adapted from [18]).

Fuguitt and Hawkins, in [19], observed the concentrations of the reactant and the four products in eight time instants of the interval $[0, 36420]$. After the chemical reaction orders having been defined, the formulation of the mathematical model describing the concentration of each compound over time may be set. In [20], a first order kinetics is assumed and the following system of first order ODEs to describe the dynamics of the concentration of the compounds over time is proposed:

$$\begin{cases} y_1' = -(p_1 + p_2)y_1 \\ y_2' = p_1 y_1 \\ y_3' = p_2 y_1 - (p_3 + p_4)y_3 + p_5 y_5 \\ y_4' = p_3 y_3 \\ y_5' = p_4 y_3 - p_5 y_5 \end{cases} \qquad (2)$$

for $t \in [0, 36420]$, with the initial conditions

$$y_1(0) = 100, \ y_2(0) = 0, \ y_3(0) = 0, \ y_4(0) = 0, \ y_5(0) = 0. \qquad (3)$$

The velocity parameters $p_1, \ldots, p_5$ are unknown but can be estimated through the minimization of an objective functional which measures the distance between the experimental observed values and the model predicted values over the entire time interval. Thus, the formulation of α-pinene isomerization problem is the following:

$$
\begin{aligned}
\min_{p} J(p) &\equiv \int_{0}^{t_f} (y(t) - y^{\text{obs}}(t))^T (y(t) - y^{\text{obs}}(t)) dt \\
\text{subject to } y'(t) &= g(t, y(t), p) \quad \text{ODEs listed in(2)} \\
p^{L} &\leq p \leq p^{U} \\
y(0) &= (100, 0, 0, 0, 0)
\end{aligned}
\tag{4}
$$

where $y(t)$ denotes the vector $(y_1(t), y_2(t), y_3(t), y_4(t), y_5(t))$ of the state variable functions (predicted by the model), J is the objective functional to be minimized, p is the vector that contains the parameters to be estimated, also referred to as the vector of the decision variables, y^{obs} contains the experimentally observed values of the state variables, p^{L} and p^{U} represent the vectors with the lower and upper bounds on the parameters, respectively, and $y(0)$ is the vector with the initial conditions. This type of parameter estimation problem belongs to a class of DO problems where the decision variables are not time-dependent.

3   Methodology

The α-pinene isomerization problem (4) consists of estimating a set of five parameters $(p_1, \ldots, , p_5)$ in a way that the state variables y_1, y_2, y_3, y_4, y_5 that satisfy the system of ODEs shown in (2), with initial conditions (3), best fit the experimentally observed state values. This dynamic system based parameter estimation problem is solved by a numerical direct method and the resulting finite bound constrained nonlinear programming (BCNLP) problem is solved to global optimality by a metaheuristic, known as FA. In the remaining part of this section, the numerical direct approach is briefly described and the main ideas behind the FA are presented.

3.1   A Direct Sequential Approach

To solve the dynamic system based parameter estimation problem, a numerical direct method is used. This type of method discretizes the problem (4) and applies nonlinear programming techniques to the resulting finite-dimensional optimization problem. Methods for solving DO problems like (4) can be classified into indirect methods and direct methods [21]. Indirect methods use the first order necessary conditions from the Pontryagin's Minimum Principle to reformulate the problem as a two-point boundary value problem. Although this is a widely used technique, the resulting boundary value problems may become difficult to solve specially if the problem contains state variable constraints [22,23]. In direct methods, the optimization (4) is performed directly. Depending on whether the system of ODEs are integrated explicitly or implicitly, two different approaches emerge, the direct sequential approach and the direct simultaneous approach, respectively. In the direct sequential approach, also denoted by single-shooting or control vector parametrization (specially if the decision variables are time-dependent), the optimization is carried out in the space of the input variables only. Given a set of values for the decision variables, the system of ODEs are accurately integrated (over the entire time interval) using specific numerical integration formulae so that the objective functional can be evaluated. Thus, the differential equations are satisfied at each iteration of the optimization procedure [22–24]. This is the main characteristic of the sequential approach, also coined as 'feasible path' approach. The strategy may lead to a slow convergence process since the differential equations are solved again and again. On the other hand, the direct simultaneous approach (orthogonal collocation, complete parametrization or full discretization) is computationally less expensive since an approximation to the solution of the ODEs is used instead. The optimization itself is carried out in the full space of discretized inputs and state variables leading to a large-dimensional finite NLP problem, in particular if a fine grid of points is required to obtain high level of integration accuracy. Furthermore, the number of time stages and collocation points as well as the position of the collocation points have to be chosen before using the NLP solver. Thus, efficient NLP solvers are crucial when solving the finite optimization problem. The ODEs are satisfied only at a finite number of time instants in the interval (at the solution of the optimization problem), which is why it is called 'infeasible path' approach [21, 22].

To summarize, the advantages of the direct sequential approach over the direct simultaneous are highlighted as follows:

- the NLP problem is relatively small-dimensional;
- an error control mechanism within the numerical integration formula is able to enforce the accuracy of the state variable values;
- the ODEs are satisfied at each iteration of the NLP algorithm, although this may lead to a computationally demanding process.

It is recognized that a direct sequential framework is easily constructed and may link reliable and efficient ODEs numerical integration and NLP solvers. However, its implementation requires repeated numerical integration of the ODEs. In the context of solving the herein studied optimization problem, the ODEs (model (2)) with the initial conditions (3) are solved by the Matlab[TM1] function ode15s. This function implements a variable order method for solving stiff differential equations and is appropriate since the equations include some terms that might lead to dramatic changes and oscillations in the solutions, in a small time interval, when compared to the interval of integration. The direct sequential optimization process can be described schematically as shown in Fig. 2. Therefore, the direct sequential approach transcribes the problem (4) into a small finite-dimensional BCNLP problem through the discretization of the decision variables $p_1, \ldots, p_5$ only, while the system of ordinary differential equations is embedded in the NLP problem.

The experimental data provided by [19] in the interval $t \in [0, t_f]$ consider the set of eight time instants $\{1230, 3060, 4920, 7800, 10680, 15030, 22620, 36420\}$ and the corresponding observed concentrations of the reactant and the

---

[1] Matlab is a registered trademark of the MathWorks, Inc.

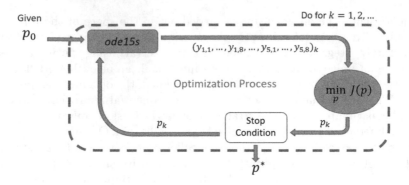

Fig. 2. Direct sequential process.

four products $(y_{1,i}^{\mathrm{obs}}, y_{2,i}^{\mathrm{obs}}, y_{3,i}^{\mathrm{obs}}, y_{4,i}^{\mathrm{obs}}, y_{5,i}^{\mathrm{obs}}, i = 1, \ldots, 8)$ are shown in Table 1. Thus, based on the information herein presented, the α-pinene isomerization parameter estimation problem is formulated as

$$
\begin{aligned}
\min_{p} \ J(p) &\equiv \sum_{j=1}^{5} \sum_{i=1}^{8} \left(y_j(t_i) - y_{j,i}^{\mathrm{obs}} \right)^2 \\
\text{subject to } y'(t) &= g(t, y(t), p) \qquad \text{ODEs listed in (2)} \\
p^{\mathrm{L}} &\leq p \leq p^{\mathrm{U}} \\
y(0) &= (100, 0, 0, 0, 0) \,.
\end{aligned}
\tag{5}
$$

Table 1. Experimental values of the concentration of the reactant and the four products, for eight time instants.

	1230.0	3060.0	4920.0	7800.0	10680.0	15030.0	22620.0	36420.0
y_1^{obs}	88.35	76.4	65.1	50.4	37.5	25.9	14.0	4.5
y_2^{obs}	7.3	15.6	23.1	32.9	42.7	49.1	57.4	63.1
y_3^{obs}	2.3	4.5	5.3	6.0	6.0	5.9	5.1	3.8
y_4^{obs}	0.4	0.7	1.1	1.5	1.9	2.2	2.6	2.9
y_5^{obs}	1.75	2.8	5.8	9.3	12.0	17.0	21.0	25.7

3.2   The Firefly Algorithm

In order to avoid the convergence to a local solution of a BCNLP problem without providing a good initial approximation, a global optimization (GO) method is usually selected [8,9]. There are deterministic and stochastic methods for GO [25,26]. Stochastic methods have been showing to be robust alternatives to the deterministic and exact methods. The stochastic alternatives to solve GO problems are usually general-purpose, population-based and are normally referred to as evolutionary algorithms since they are motivated by biological evolution [27]. The genetic algorithm [28, 29] is the most known evolutionary strategy, although other swarm intelligence based algorithms [30–33] are common in the literature. A recent strategy successfully used in biological and engineering applications is the FA [34].

The FA [12] is a stochastic approach for GO that is inspired by the collective behavior of fireflies, specifically in how they attract each other. Several variants of FA have been designed so far, but the main simple rules that define the firefly movement are the following [35]:

- all fireflies are unisex, meaning that any firefly will be attracted to other fireflies regardless of their sex;
- the brightness of a firefly is determined by the value of the objective function;
- attractiveness of a firefly is proportional to their brightness; thus for any two fireflies, the less brighter will move towards the brighter one; the attractiveness is directly proportional to the brightness but decreases with distance.

Hereafter, to represent the position of the firefly i the following vector notation is used $p^i = (p_1^i, \ldots, p_5^i)$. Initially, the positions of the population of m fireflies are randomly generated in the search space defined by the bound constraints on the decision variables, as follows

$$p_s^i = p_s^L + rand^i (p_s^U - p_s^L), \quad \text{for } s = 1, \ldots, 5 \text{ and } i = 1, \ldots, m \qquad (6)$$

where i stands for the firefly i in the population of m fireflies, s represents the component of the vector, p^L and p^U are the vectors of lower and upper bounds on the decision variables, and $rand^i$ is a vector of random numbers from a uniform distribution on the interval $[0, 1]$.

The movement of firefly i, in direction to a brighter firefly j is defined by

$$p^i = p^i + \beta(p^j - p^i) + \alpha \, rand^{i,j} \qquad (7)$$

where $0 \leq \alpha \leq 1$ is the randomness parameter and $rand^{i,j}$ is a vector of uniformly distributed numbers in the range $[-1, 1]$, and β is given by:

$$\beta = \beta_0 e^{-\gamma ||p^i - p^j||^v} \quad \text{for } v \geq 1, \qquad (8)$$

being β_0 the attraction when the distance is zero. The parameter $\gamma \in [0, \infty)$ characterizes the variation of the attractiveness and it is crucial to determine the algorithm convergence speed. When $\gamma \to 0$, attractiveness is constant which means that the brightness of the firefly can be seen from any position of the space. On the other hand, when $\gamma \to \infty$ the attractiveness is almost zero and each firefly moves randomly. Note that, in the movement described by (7), the second term is due to the attraction and the third term is due to randomization.

4   Numerical Experiments

In this section, the practical performance of the metaheuristic FA for solving the BCNLP problem that emerges from the implementation of the direct sequential procedure to the α-pinene isomerization problem, is analyzed. The mathematical model was coded in the Matlab programming language and the computational application of the direct sequential procedure with the FA was developed in Matlab programming environment. The computational tests were performed on a PC with a 2.7 GHz Core i7-4600U and 8 Gb of memory.

4.1   FA Parameter Tuning

First, an experimental study was conducted to tune some of the FA parameters, in the context of the α-pinene problem. The parameter in the randomization term is made to depend on the allowed maximum number of iterations and decreases from an initial value, α, to a near zero value. Four values are tested, ranging from the largest possible to a rather small one, $\alpha = \{1, 0.5, 0.25, 0.05\}$, being 0.25 the most common value used in the literature. The parameter γ is kept fixed over the iterative process. To analyze the effect of the attractiveness in the algorithm, four moderate values have been selected: $\gamma = \{2, 1, 0.5, 0.25\}$. Only eight combinations of α and γ have been tested in these experiments. A population of 20 points is used and the problem is solved 10 times with independent randomly generated initial populations. The iterative FA terminates when the absolute difference between the obtained solution, at the current iteration, and the best known optimal solution is below a small tolerance, for instance 1.0e-4, or when a maximum of 500 iterations is reached. The lower and upper bounds for the decision variables are $p_i^L = 0$ and $p_i^U = 5.0e\text{-}4$ for $i = 1, \ldots, 5$. The best known solution for this problem is $p_1^* = 5.9256e\text{-}5$, $p_2^* = 2.9632e\text{-}5$, $p_3^* = 2.0450e\text{-}5$, $p_4^* = 2.7473e\text{-}4$, $p_5^* = 4.0073e\text{-}5$, with the optimum objective function value $J(p^*) = 19.872$, as reported in [36].

Table 2 contains the results produced by the proposed algorithm and correspond to the best function value, J_{best}, the average of the function values, over the 10 runs, J_{avg}, the median of the function values, J_{med}, and the average number of iterations, $N.It_{\text{avg}}$. From the table, it can be concluded that combining $\alpha = 0.25$ with $\gamma = 1$ gives the best results in terms of best, average and median J values, and a lower average number of iterations. The run that produced the best value $J_{\text{best}} = 19.8772$ required 294 iterations and 5860 function evaluations. From the table, it is possible to conclude that the two best sets of results correspond to a small value of α, meaning that the randomness characteristic of the FA is somehow controlled, and to a moderate value of γ (0.5 and 1) which promotes the local search ability of the FA. We note that a very small value of α (for example, 0.05) is harmful since the algorithm is not capable of exploring the search space for the region where the global solution lies.

Figure 3 shows the evolution of the objective function along the iterations, relative to the best and worst runs for the case where $\alpha = 0.25$ and $\gamma = 1$. For comparative purposes, the evolution of the objective function of the worst run,

Table 2. Results obtained by FA, for different combinations of α and γ.

α	γ	J_{best}	J_{avg}	J_{med}	$N.It_{\text{avg}}$
1	1	20.0334	30.2361	28.2366	500
0.5	1	20.4490	25.7785	25.5231	500
0.5	0.5	19.9366	67.8867	29.4598	500
0.25	2	19.8819	44.2389	30.5656	490
0.25	1	19.8772	25.6777	20.7662	459
0.25	0.5	19.8970	27.0022	24.2394	500
0.25	0.25	19.9029	356.7275	180.7635	500
0.05	1	260.6788	979.0378	657.9491	500

for the case where $\alpha = 0.5$ and $\gamma = 0.5$ is also shown. As it can be seen the value $\alpha = 0.25$ produces a significant decrease in J during the early iterations.

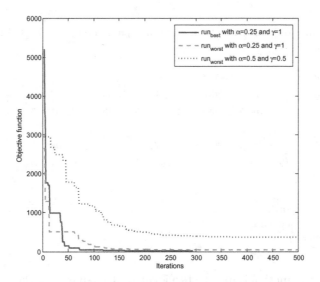

Fig. 3. Evolution of the objective function from the best and worst runs, when $\alpha = 0.25$ and $\gamma = 1$, and from the worst run, when $\alpha = 0.5$ and $\gamma = 0.5$.

4.2 Comparative Results

The second experiment aims to compare the results produced by the FA with those obtained by the `fmincon` function from the Matlab Optimization Toolbox, as well as with those achieved in [14,18,36–38]. Therefore, Table 3 contains a summary of the results – the best objective function value, the best parameter values, and the number of iterations, $N.It$, and the number of function evaluations, $N.f.ev$, of the reported solution – obtained by:

- FA (in the context of the direct sequential approach by invoking the function `ode15s` for the numerical integration of the ODEs);
- `fmincon` (in the context of the direct sequential approach with `ode15s` for the numerical integration of the ODEs);
- the hybrid scatter search methodologies proposed in [18,36] – therein denoted by SSm – (in the context of the direct sequential approach);
- the metaheuristic EM, implemented with a random local search procedure, in [37] (in the context of the direct sequential approach, with `ode45` for the numerical integration of the ODEs);
- the deterministic global optimization algorithm based on outer approximation proposed in [14] (in the context of orthogonal collocation on finite elements in $[0, 36420]$);
- the deterministic gradient-based solver 'filterSQP' (a sequential quadratic programming based on the filter set methodology) tested in [38] (in the context of 3-stage collocation method with uniform partition of 100 subintervals in $[0, 36420]$).

Table 3. Optimal objective function value and parameters: FA *vs* `fmincon` and the results in [14,18,36–38].

	Solution	Parameter vector	N.It	N.f.ev
FA	19.877	(5.928e−5, 2.962e−5, 2.041e−5, 2.764e−4, 4.046e−5)	294	5860
fmincon	19.929	(5.911e−5, 2.968e−5, 2.070e−5, 2.745e−4, 4.001e−5)	24	217
in [36]	19.872	n.a	n.a	1163[a]
in [18]	19.87	(5.926e−5, 2.963e−5, 2.047e−5, 2.745e−4, 3.998e−5)	9518	(*122*)[b]
in [37]	19.874	(5.926e−5, 2.963e−5, 2.047e−5, 2.743e−4, 3.991e−5)	147	3824
in [14]	19.87	n.a	2	(*8916*)[b]
in [38]	19.8721	n.a	n.a	(*28*)[b]

n.a. means not available.
[a] Average value over ten independent runs.
[b] CPU time in seconds.

The `fmincon` function implements a local gradient-based method and requires an initial approximation provided by the user. The sequential quadratic programming based on a quasi-Newton update formula to approximate the second derivatives of the objective function is used, and `fmincon` was invoked with its default settings. It is common knowledge that traditional local solvers may fail to reach a global optimal solution unless a good initial approximation is provided by the user. Thus, the initial approximation $p_i = 0, i = 1, \ldots, 5$ is used. The `fmincon` function achieved the minimum $J = 19.929$ after 24 iterations and 217 function evaluations. The best solution reported by FA $J_{best} = 19.8772$ (see Table 2) is close to the best known optimal solution, with a relative error of 0.026 %, while the solution obtained with `fmincon` has a relative error of 0.29 %. This is a confirmation that FA makes a comprehensive exploration of the search space, without

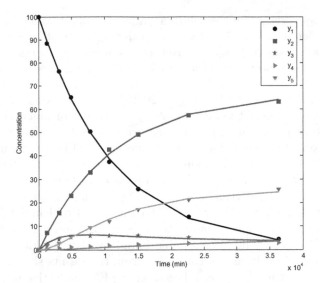

Fig. 4. Concentrations estimated by the model (full lines) and experimentally observed values (points alone) in the isomerization of α-pinene, based on the parameters produced by FA.

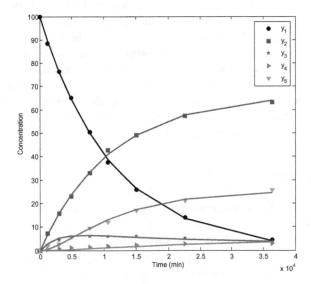

Fig. 5. Concentrations estimated by the model (full lines) and experimental values (points alone) in the α-pinene isomerization, based on the parameters obtained by `fmincon`.

requiring any good initial approximation. The scatter search algorithms proposed in [18,36] activate a gradient-based local search procedure (the `fmincon` function from Matlab) that is carried out from different vectors as initial points to accelerate convergence to the solution. In [36], the best value $J_{best} = 19.872$, the average value $J_{avg} = 24.747$, the worst value $J_{wor} = 68.617$ and the average number of evaluations $N.f.ev_{avg} = 1163$ (over ten executed runs) are reported. In [18], the initial approximation $p_i = 0.5, i = 1, \ldots, 5$ is used and the optimal value $J = 19.87$ is reached after 9518 iterations and 122 s. In [37], the best value $J_{best} = 19.874$, the median value $J_{med} = 20.325$, the worst value $J_{wor} = 21.806$, and the average number of iteration $N.It_{avg} = 147$ and of function evaluations $N.f.ev_{avg} = 4070$ (over ten executed runs) are reported. The outer approximation based algorithm presented in [14] converges to $J = 19.87$. Although only two iterations of the outer approximation strategy are required to reach the solution, the therein used approach may lead to computational burdens and the process is a very costly one (see details in Table 3). The problem at hand is also part of the well-known Large-Scale Constrained Optimization Problem Set (COPS) and was solved by a collocation method and several NLP solvers, for comparative purposes, in [38]. The minimum J value found by the well-known 'filterSQP' was 19.8721 after 28 s. It can be concluded from the results shown in the table that the direct sequential approach based on the FA is effective in reaching the optimal solution. Convergence to the solution may be accelerated and solution quality may be improved by incorporating a local search technique into FA.

Figure 4 shows the evolution over time of the concentrations of reactant and the four products for the α-pinene isomerization problem, considering the optimal values of the parameters obtained by the FA (shown in Table 3) – represented by full lines. The figure also shows the experimental data represented by points alone, as reported in [19]. As it can be seen the obtained optimal values for the parameters allow the model to reproduce almost exactly the experimental data. On the other hand, Fig. 5 shows the experimental data (represented by the points alone) and the values predicted by the model (represented by full lines) for the same time instants, considering the optimum values for the parameters produced by fmincon (as reported in Table 3).

5   Conclusions

We have shown that the parameter estimation problem on the α-pinene isomerization model can be easily solved by a direct sequential approach which transcribes the DO problem in a small finite-dimensional BCNLP problem. As a consequence, a stochastic global optimization technique can be applied to locate a global minimum of the goodness of fit objective function subject to bound constraints on the values of the parameters. The simple to understand and easy to implement metaheuristic FA has shown to be a reliable method to locate a global optimal solution of this challenging DO problem. Some preliminary experiments were carried out in order to tune the parameters α and γ of the FA, and it is shown that this issue is crucial to the convergence features of the algorithm. The implementation of FA was effective in a way that the produced results are comparable to those of previously used metaheuristics and outperform the results obtained by fmincon function from Matlab. The obtained solution is very close to the best known solution in the literature and allows the model to reproduce almost exactly the experimental data.

Future developments will be directed to the FA enhancing strategy by using an efficient but simple random local search procedure so that convergence could be accelerated.

Acknowledgments. The authors wish to thank two anonymous referees for their comments and suggestions. This work has been supported by COMPETE: POCI-01-0145-FEDER-007043 and FCT - Fundação para a Ciência e Tecnologia, within the projects UID/CEC/00319/2013 and UID/MAT/00013/2013.

References

1. Banga, J.R., Balsa-Canto, E., Moles, C.G., Alonso, A.A.: Improving food processing using modern optimization methods. Trends Food Sci. Tech. **14**(4), 131–144 (2003)
2. Grossmann, I.E. (ed.): Global Optimization in Engineering Design, Nonconvex Optimization and Its Applications, vol. 9. Springer Science & Business Media (1996)
3. Vanrolleghem, P.A., Dochain, D.: Bioprocess model identification. In: Van Impe, J.F.M., Vanrolleghem, P.A., Iserentant, D.M. (eds.) Advanced Instrumentation, Data Interpretation, and Control of Biotechnological Processes, pp. 251–318. Kluwer Academic Publ. (1998)
4. Rodríguez-Fernández, M.: Modelado e Identificación de Bioprocesos. Ph.D. thesis, University of Vigo, Spain (2006)
5. Rodriguez-Fernandez, M., Mendes, P., Banga, J.R.: A hybrid approach for efficient and robust parameter estimation in biochemical pathways. Biosystems **83**(2–3), 248–265 (2006)
6. Gill, P.E., Murray, W., Saunders, M.A., Wright, M.H.: Constrained nonlinear programming. In: Handbooks in Operations Research and Management Science, vol. 1, Optimization, pp. 171–210 (1989)
7. Banga, J.R., Balsa-Canto, E., Moles, C.G., Alonso, A.A.: Dynamic optimization of bioreactors: a review. Proc.-Indian National Sci. Acad. Part A **69**(3/4), 257–266 (2003)
8. Esposito, W.R., Floudas, C.A.: Global optimization for the parameter estimation of differential-algebraic systems. Ind. Eng. Chem. Res. **39**(5), 1291–1310 (2000)
9. Moles, C.G., Mendes, P., Banga, J.R.: Parameter estimation in biochemical pathways: a comparison of global optimization methods. Genome Res. **13**(11), 2467–2474 (2003)
10. Moles, C.G., Gutierrez, G., Alonso, A.A., Banga, J.R.: Integrated process design and control via global optimization: a wastewater treatment plant case study. Chem. Eng. Res. Des. **81**(5), 507–517 (2003)
11. Eberhart, R.C., Kennedy, J.: A new optimizer using particle swarm theory. In: Proceedings of the Sixth International Symposium on Micro Machine and Human Science, vol. 1, pp. 39–43, New York, NY (1995)
12. Yang, X.-S.: Firefly algorithms for multimodal optimization. In: Watanabe, O., Zeugmann, T. (eds.) SAGA 2009. LNCS, vol. 5792, pp. 169–178. Springer, Heidelberg (2009)

13. Lin, Y., Stadtherr, M.A.: Deterministic global optimization for parameter estimation of dynamic systems. Ind. Eng. Chem. Res. **45**(25), 8438–8448 (2006)
14. Miro, A., Pozo, C., Guillén-Gosálbez, G., Egea, J.A., Jiménez, L.: Deterministic global optimization algorithm based on outer approximation for the parameter estimation of nonlinear dynamic biological systems. BMC Bioinf. **13**, 90 (2012)
15. Polisetty, P.K., Voit, E.O., Gatzke, E.P.: Identification of metabolic system parameters using global optimization methods. Theor. Biol. Med. Model. **3**, 4 (2006)
16. Balsa-Canto, E., Peifer, M., Banga, J.R., Timmer, J., Fleck, C.: Hybrid optimization method with general switching strategy for parameter estimation. BMC Syst. Biol. **2**, 26 (2008)
17. Box, G., Hunter, W., MacGregor, J., Erjavec, J.: Some problems associated with the analysis of multiresponse data. Technometrics **15**(1), 33–51 (1973)
18. Rodriguez-Fernandez, M., Egea, J.A., Banga, J.R.: Novel metaheuristic for parameter estimation in nonlinear dynamic biological systems. BMC Bioinf. **7**, 483 (2006)
19. Fuguitt, R.E., Hawkins, J.E.: Rate of the thermal isomerization of α-pinene in the liquid phase. J. Am. Chem. Soc. **69**(2), 319–322 (1947)
20. Hunter, W., McGregor, J.: The estimation of common parameters from several responses: Some actual examples. Unpublished Report, Department of Statistics. University of Winsconsin (1967)
21. Srinivasan, B., Palanki, S., Bonvin, D.: Dynamic optimization of batch processes: I. Characterization of the nominal solution. Comput. Chem. Eng. **27**(1), 1–26 (2003)
22. Banga, J.R., Balsa-Canto, E., Moles, C.G., Alonso, A.A.: Dynamic optimization of bioprocesses: efficient and robust numerical strategies. J. Biotechnol. **117**, 407–419 (2005)
23. Schlegel, M., Marquardt, W.: Direct sequential dynamic optimization with automatic switching structure detection. In: Shah, S.L., MacGregor, J. (eds.) Dynamics and Control of Process Systems 2004 (DYCOPS-7), vol. 1, pp. 419–424. Elsevier IFAC Publ. (2005)
24. Biegler, L.T.: Nonlinear Programming: Concepts, Algorithms, and Applications to Chemical Processes. MOS-SIAM Series on Optimization (2010)
25. Zhigljavsky, A., Žilinskas, A.: Stochastic Global Optimization. Springer Optimization and its Applications. Springer, New York (2008)
26. Floudas, C.A., Akrotirianakis, I.G., Caratzoulas, S., Meyer, C.A., Kallrath, J.: Global optimization in the 21st century: advances and challenges. Comput. Chem. Eng. **29**, 1185–1202 (2005)
27. Tsai, K.-Y., Wang, F.-S.: Evolutionary optimization with data collocation for reverse engineering of biological networks. Bioinformatics **21**(7), 1180–1188 (2005)
28. Goldberg, D.E., Korb, B., Deb, K.: Messy genetic algorithms: motivation, analysis, and first results. Complex Syst. **3**(5), 493–530 (1989)
29. Holland, J.H.: Genetic algorithms. Sci. Am. **267**(1), 66–72 (1992)
30. Bonabeau, E., Dorigo, M., Theraulaz, G.: Swarm Intelligence: From Natural to Artificial Systems. Santa Fe Institute Studies on the Sciences of Complexity, Oxford University Press (1999)
31. Kennedy, J., Eberhart, R.C.: Swarm Intelligence. Morgan Kaufmann Publishers Inc., San Francisco (2001)
32. Millonas, M.M.: Swarms, phase transitions, and collective intelligence. In: Langton, C.G. (ed.) Artificial Life III, vol. XVII, pp. 417–445. Addison-Wesley, Reading (1994)
33. Yang, X.-S.: Nature-Inspired Metaheuristic Algorithms, 2nd edn. Luniver Press, Bristol (2010)

34. Yang, X.-S.: Biology-derived algorithms in engineering optimization. In: Olariu, S., Zomaya, A.Y. (eds.) Handbook of Bioinspired Algorithms and Applications, pp. 589–600. Chapman & Hall, Boca Raton (2005)
35. Yang, X.-S.: Firefly algorithm, stochastic test functions and design optimisation. Int. J. Bio Inspired Comput. **2**(2), 78 84 (2010)
36. Egea, J.A., Rodríguez-Fernández, M., Banga, J.R., Marti, R.: Scatter search for chemical and bioprocess optimization. J. Glob. Optim. **37**(3), 481–503 (2007)
37. Rocha, A.M.A.C., Silva, A., Rocha, J.G.: A new competitive implementation of the electromagnetism-like algorithm for global optimization. In: Gervasi, O., Murgante, B., Misra, S., Gavrilova, M.L., Rocha, A.M.A.C., Torre, C., Taniar, D., Apduhan, B.O. (eds.) ICCSA 2015. LNCS, vol. 9156, pp. 506–521. Springer, Heidelberg (2015)
38. Dolan, E.D., Moré, J.J., Munson, T.S.: Benchmarking Optimization Software with COPS 3.0. Argonne National Laboratory Technical report ANL/MCS-TM-273 (2004)

Extensions of Firefly Algorithm for Nonsmooth Nonconvex Constrained Optimization Problems

Rogério B. Francisco[1], M. Fernanda P. Costa[1],
and Ana Maria A.C. Rocha[2(✉)]

[1] Centre of Mathematics, University of Minho, Braga, Portugal
`rbf@estgf.ipp.pt, mfc@math.uminho.pt`
[2] Algoritmi Research Centre, University of Minho, Braga, Portugal
`arocha@dps.uminho.pt`

Abstract. Firefly Algorithm (FA) is a stochastic population-based algorithm based on the flashing patterns and behavior of fireflies. Original FA was created and successfully applied to solve bound constrained optimization problems. In this paper we present extensions of FA for solving nonsmooth nonconvex constrained global optimization problems. To handle the constraints of the problem, feasibility and dominance rules and a fitness function based on the global competitive ranking, are proposed. To enhance the speed of convergence, the proposed extensions of FA invoke a stochastic local search procedure. Numerical experiments to validate the proposed approaches using a set of well know test problems are presented. The results show that the proposed extensions of FA compares favorably with other stochastic population-based methods.

Keywords: Firefly algorithm · Constrained global optimization · Stochastic ranking

1 Introduction

In the last decades, different methods have been developed in order to solve a wide range of different kind of optimization problems. Metaheuristics are an important class of contemporary global optimization algorithms, computational intelligence and soft computing. The observation and study of nature and behavior of some living species have been served as inspiration for the development of new methods. A subset of metaheuristics, often referred to as swarm intelligence based algorithms, have been developed by mimicking the so-called swarm intelligence characteristics of biological agents such as birds, fish, humans among others. Swarm Intelligence belongs to an artificial intelligence subject that became increasingly popular over the last decade [1]. The three main purposes of metaheuristics are: to solve problems with low computational time, to solve large dimensional problems, and to obtain robust algorithms. In fact, metaheuristics are the most used stochastic optimization algorithms. In recent years, metaheuristic algorithms have emerged as global search approaches used for solving complex optimization problems. The most popular metaheuristic methods are

O. Gervasi et al. (Eds.): ICCSA 2016, Part I, LNCS 9786, pp. 402–417, 2016.
DOI: 10.1007/978-3-319-42085-1_31

Genetic Algorithm (GA) [2], Ant Colony Optimization [3], Particle Swarm Optimization (PSO) [4], Harmony Search [5] and Firefly Algorithm (FA) [6]. All of them are metaheuristic population-based methods. The FA, initially proposed by Yang, is one of the new metaheuristic techniques inspired by the flashing behavior of fireflies and was designed for solving bound constrained optimization (BCO) problems. This algorithm is inspired by the nocturnal luminous of the fireflies, mating and social behavior. The FA algorithm takes into account what each firefly notes in its line of sight in an attempt to move to a new location, which is brighter than its prior. Simulation results indicate that FA is superior over GA and PSO [7, 8]. Although the original version of FA was designed to solve BCO problems, many variants of this algorithm has been developed and applied to solve constrained problems from different areas. FA has become popular and widely used in many applications like economic dispatch problems [9, 10], mixed variable optimization problems [11–13] and multi-objective continuous optimization problems [14, 15]. A recent review and advances of the firefly algorithms are available in [16, 17].

In this paper, we aim to extend the FA for solving nonsmooth nonconvex constrained global optimization (CGO) problems. The mathematical formulation of the problem to be addressed has the form:

$$
\begin{aligned}
&\underset{x \in \Omega}{\text{minimize}} && f(x) \\
&\text{subject to} && g_k(x) \leq 0, \quad k = 1, \ldots, p \\
& && h_j(x) = 0, \quad j = 1, \ldots, m
\end{aligned}
\tag{1}
$$

where $f : \mathbb{R}^n \to \mathbb{R}$, $g : \mathbb{R}^n \to \mathbb{R}^p$ and $h : \mathbb{R}^n \to \mathbb{R}^m$ are nonlinear continuous functions, possibly non differentiable, and $\Omega = \{x \in \mathbb{R}^n : -\infty < l_b \leq x \leq u_b < \infty\}$, with l_b and u_b the vectors of lower and upper bounds on the variables, respectively. In (1), f, g and h may be nonconvex functions and many local minima may exist in the feasible region $\Omega_F = \{x \in \Omega : g(x) \leq 0, h(x) = 0\}$. In order to solve (1) two constraint-handling techniques based on feasibility and dominance rules and a global competitive ranking, are proposed.

The paper is organized as follows. Section 2 briefly presents some common constraint-handling techniques and the main ideas that motivated this work. Section 3 describes the original FA and in Sect. 4 we propose three extensions of FA for solving nonsmooth nonconvex CGO problems. The preliminary numerical experiments are reported in Sect. 5 and the paper is concluded in Sect. 6.

2 Constraint-Handling Techniques

In population-based methods, the widely used approach to deal with constrained optimization problems is based on exterior penalty methods [18–21]. In this type of approach, the constrained problem is replaced by a sequence of unconstrained sub-problems, defined by penalty functions. A penalty function consists of the objective function of the constrained problem combined with one additional term for each constraint (which is positive when the point is infeasible for that constraint and zero otherwise) multiplied by some positive penalty parameter. Making the penalty parameter larger along the iterative process, the constraints violation is more severely penalized, forcing in this way the minimizer of the penalty function to be closer to the feasible region of the original problem.

A well-known penalty function is the ℓ_1 exact penalty function, in which the terms that measure the constraints violation of a point x_i, are given by

$$\zeta(x_i) = \sum_{k=1}^{p} max\{0, g_k(x_i)\} + \sum_{j=1}^{m} |h_j(x_i)|.$$

Assuming that the bound constraints on the variables are guaranteed by the population stochastic method, at each iteration, the problem (1) is transformed into a BCO problem as follows:

$$\underset{x \in \Omega}{minimize} f(x) + \lambda \left(\sum_{k=1}^{p} max\{0, g_k(x_i)\} + \sum_{j=1}^{m} |h_j(x_i)| \right) \tag{2}$$

where $\lambda > 0$ is the penalty parameter. For a sufficiently large, positive value of $\lambda > 0$, one minimization of exact penalty function (2) will produce the solution of problem (1). However, in practice it is hard to determine a priori the λ values, being necessary to use rules for adjusting this parameter along the iterative process.

Despite the popularity of penalty methods regarding its simplicity and easy implementation, they have several drawbacks. The most difficult issue lies in finding the appropriate penalty parameter values λ, since they require a suitable fine tuning to estimate the degree of penalization to be applied. New penalty approaches in this field are constantly under research.

2.1   Global Competitive Ranking

Runarsson and Yao [22] proposed a constraint-handling technique called global competitive ranking, where an individual point x_i is ranked by comparing it against all other members in the population, for $i = 1, \ldots, N$ being N the population size. In this technique, first the objective function value, $f(x_i)$, and the constraints violation value $\zeta(x_i)$, are calculated, for all points of the population. Then, considering a minimization problem, these values are ranked separately in ascending order. In case of tied individuals, the same higher rank will be given. After giving ranks to all points, based on f and ζ, separately, the fitness function of each individual point x_i is computed by:

$$\Phi(x_i) = P_f \frac{I_{i,f} - 1}{N - 1} + (1 - P_f) \frac{I_{i,\zeta} - 1}{N - 1} \tag{3}$$

where $I_{i,f}$ and $I_{i,\zeta}$ are the ranks of point x_i based on the objective function f and the constraints violation ζ, respectively. P_f is the probability that the fitness is calculated based on the rank of the objective function. According to the authors of [22], the probability should take a value on $0 < P_f < 0.5$ in order to guarantee that a feasible solution may be found. The main goal of this technique is to strike the right balance between the objective function and the constraints violation. From (3), the best point of the population is the point that has the lowest fitness value. One drawback detected by the authors associated to this constraint handling technique was the need to use different values of P_f to solve different optimization problems. To prevent this drawback, using the same ranking process, we propose a new fitness function that does not depend on the probability value P_f.

2.2 Feasibility and Dominance Rules

Deb [23] proposed another constraint-handling technique that is based on biasing feasible over infeasible points. The constraints violation and the objective function values are used separately and optimized by some sort of order, where feasible points are always preferable to infeasible ones. This technique is based on three simple feasibility and dominance rules proposed for binary tournaments:

(i)   Any feasible point is preferred to any infeasible one.
(ii) Between two feasible points, the one having better objective function is preferred.
(iii) Between two infeasible points, the one having smaller constraint violation is preferred.

In this work, we propose a ranking scheme based on rules (i)–(iii) with the additional following new rule, that takes into account the number of violated constraints (nc):

(iv) Between two infeasible points, the one having smaller number of violated constraints is preferred.

Hence, when two points of the population are compared to see which one improves over the other, the rules (i)–(iv) are used. These rules can be mathematically stated in the following definition.

Definition 1. (Point y *improves over* point x)
Let x and y be two points in Ω. The point y improves over point x if the following condition holds:

$$(\zeta(x) > \zeta(y) \text{ or } nc(x) > nc(y)) \text{ or } (\zeta(x) = \zeta(y) = 0 \quad \text{and} \quad f(x) > f(y))$$

3 Firefly Algorithm

3.1 Standard Firefly Algorithm

FA is a stochastic population-based algorithm for solving BCO problems. In order to develop FA, some of the flashing characteristics of fireflies were idealized. Yang formulated FA by assuming three simple rules [6].

- All fireflies are unisex, meaning that any firefly will be attracted to other fireflies regardless of their sex.
- The brightness of a firefly is determined by the objective function value.
- Attractiveness between fireflies is proportional to their brightness but decreases with distance. For any two fireflies, the firefly with less bright will move towards the brighter.

In the description of the algorithm, the position of the firefly j will be represented by $x_j \in \mathbb{R}^n$ and firefly j is brighter than firefly i if $f(x_j) < f(x_i)$. Most of metaheuristics optimization methods are based on the generation of random initial population of feasible points. All points of the population are placed in the search space to guide the search to the best location. Thus, the FA applies a similar strategy and the random initial population of Ω is generated as follows:

$$x_{i_s} = l_{b_s} + rand_s(u_{b_s} - l_{b_s}), \quad s = 1, \ldots, n.$$

where $rand_s \sim U(0,1)$ is a uniformly distributed random number in $[0,1]$. After generating the initial population, the objective function values $f(x_i)$ for all points x_i, $i = 1, \ldots, N$; are calculated and ranked from lowest to largest value of f, and the iteration counter k is set to 1. In each iteration k, for each point x_i, the FA examines every point $x_j, j = 1, 2, \ldots, N$. If point x_i has higher objective function value than x_j (firefly j is brighter than firefly i), the firefly i moves towards the firefly j according to following movement equation:

$$x_i = x_i + \beta(x_j - x_i) + \alpha(rand_i - 0.5)S \tag{4}$$

where $rand_i$ is a vector of random numbers generated from a uniform distribution in $[0,1]$, α is a randomization parameter defined by the user, usually a number in the range $[0,1]$ and S (scale of the problem) is a problem dependent vector scaling parameter defined componentwise by $S = |l_b - u_b|$. The parameter β of (4) is the attractiveness between fireflies i and j, and is defined in terms of the monotonically decreasing negative exponential function as follows:

$$\beta(r) = \beta_0 e^{-\gamma \|x_i - x_j\|} \tag{5}$$

where $\|\cdot\|$ is the Cartesian distance between the fireflies i and j, and β_0 is the attraction parameter when the distance between themselves is zero. The variation of the attractiveness is defined by the control parameter γ. The value of parameter γ is crucial to determine the speed of the convergence and how the FA behaves. In theory, γ could take any value in the set $[0, \infty[$. When $\to 0$, the value of $\beta \approx \beta_0$, meaning that a flashing firefly can be seen anywhere in the search space and, when $\gamma \to \infty$, the attractiveness is almost zero in the sight of other fireflies and each firefly moves in a random way.

Finally, whenever a position of a point x_i is updated, the FA controls the bound constraints, i.e., the point x_i is projected onto the search space as follows:

$$x_{is} = \begin{cases} l_{is} & if & x_{is} < l_{is} \\ u_{is} & if & x_{is} > u_{is} \end{cases}$$

The pseudo-code of the standard FA is presented in the Algorithm 1.

Algorithm 1. Standard Firefly Algorithm

Data: k_{max}, α, β_0, γ

Set $k = 1$

Randomly generate a population of N fireflies, $x_i^k \in \Omega$, $i = 1, \ldots, N$

Based on $\{x_1^k, \ldots, x_N^k\}$ evaluate $f(x_i^k)$, $= 1, \ldots, N$

Rank the fireflies using the objective function values (from lowest to largest of f)

Set $x_{best}^k = x_1^k$ and $f_{best}^k = f(x_1^k)$

Compute the scaler parameter S as $|l_b - u_b|$

while $k \leq k_{max}$ **do**

  for $i = 1$ *to* N

    for $j = 1$ *to* N

      if $f(x_i^k) > f(x_j^k)$ **then**

        Compute the attractiveness β using (5)

        Move firefly i towards firefly j using (4)

      end if

    end for j

  end for i

Project x_i^k onto Ω, for all $i = 1, \ldots, N$

Evaluate $f(x_i^k)$, $i = 1, \ldots, N$

Rank the fireflies using the objective function values (from lowest to largest of f)

  Set $k = k + 1$

  Set $x_{best}^k = x_1^k$ and $f_{best}^k = f(x_1^k)$

end while

3.2   Dynamic Updates of the Parameters α, γ and S

The parameters α and γ affects the performance of FA. In the version of FA proposed in [11] to solve mixed variable structural optimization problems, the authors improved the solution quality by reducing the value of the parameter α with a geometric progression reduction scheme defined by $\alpha = \alpha_0 \theta^k$, where α_0 is the initial randomness scaling factor, $0 < \theta < 1$ is the reduction factor of randomization and k is the current iteration. In [12] the authors improved the quality of the solutions by reducing the randomness of the parameters α and γ. The computational experiments shown that they must take large values at the beginning of the iterative process and decrease gradually as the optimum solution is approached, to enforce the algorithm to increase the diversity and the convergence of the algorithm. In order to improve convergence speed and solution accuracy, dynamic updates of these parameters, which depend on the iteration counter of the algorithm, were defined. The parameter α is defined at each iteration k as follows:

$$\alpha^{(k)} = \alpha_{max} - k \frac{\alpha_{max} - \alpha_{min}}{k_{max}} \tag{6}$$

where α_{max} and α_{min} are the limits to an upper and lower level for α, k is the number of current iteration and k_{max} is the maximum number of iterations allowed. The parameter γ, used for increasing the attractiveness with k, is defined at each iteration k by the following dynamic update formula:

$$\gamma^{(k)} = \gamma_{max} e^{\frac{k \log(\frac{\gamma_{min}}{\gamma_{max}})}{k_{max}}} \tag{7}$$

where γ_{min} and γ_{max} are the minimum variation and maximum variation of the attractiveness, respectively.

In this paper we propose a dynamic update formula to compute the vector of scaling parameters with k, in order enhance the convergence of the proposed FA extensions. Thus, the vector S is dynamically updated in order to decrease with k as follows:

$$S^{(k)} = \frac{\left|(l_b - u_b) - (x_N^k - x_1^k)\right|}{k} \tag{8}$$

where $x_N^k - x_1^k$ is the vector of the ranges given by the positions between the best and the worst fireflies.

4   Constrained Firefly Algorithm

In this section, we present extensions of FA for solving nonsmooth nonconvex CGO problems. We propose two constraint-handling techniques based on feasibility and dominance rules and the global competitive ranking that are able to explore both feasible and infeasible regions.

4.1   Ranking Scheme Proposals

In the global competitive ranking (GR) proposed algorithm, after calculating $f(x_i)$, $\zeta(x_i)$ and $nc(x_i)$, for all points x_i of the population, the points are ranked considering separately the ascending order of $f(x_i)$ and $\zeta(x_i)$, $i = 1, \ldots, N$. Then, taking into account the ranking of all points, the fitness function of each point x_i, $i = 1, \ldots, N$; is computed by:

$$v(x_i) = \frac{I_{i,f} - 1}{N(N-1)} + nc(x_i)\frac{I_{i,\zeta} - 1}{N(N-1)} \qquad (9)$$

where $I_{i,f}$ and $I_{i,\zeta}$ are the ranks of point x_i based on the objective function f and the constraints violation ζ respectively. Finally, using the fitness function values $v(x_i)$, $i = 1, \ldots, N$, the N points of the population are ranked by comparing all pairs of points in at least N sweeps. The description of the proposed GR scheme based on fitness function (9) is presented in Algorithm 2.

Algorithm 2. GR

Compute I_f and I_ζ
for $i = 1$ *to* $N - 1$
  for $j = i + 1$ *to* N
    if $v(x_i) > v(x_j)$
        switch rank of firefly x_i with firefly x_j
      end if
    end for j
end for i

In the ranking scheme based on feasibility and dominance (FD) rules, first the objective function value, $f(x_i)$, the constraint violation value, $\zeta(x_i)$, and the number of constraints violated, $nc(x_i)$, are calculated for all points x_i of the population; $i = 1, \ldots, N$. Then, using the rules (i)–(iv) the N points of the population are ranked by comparing all pairs of points in at least N sweeps.

A formal description of the proposed ranking scheme based on the FD rules (i)–(iv) (Definition 1) is presented in Algorithm 3.

Algorithm 3. FD rules

---

for $i = 1$ *to* $N - 1$
  for $j = i + 1$ *to* N
    if x_j *improves over* x_i
        switch rank of firefly x_i with firefly x_j
      end if
    end for j
end for i

Both ranking schemes, the GR and FD rules, ensure that good feasible solutions as well as promising infeasible ones are ranked in the top of the population.

4.2   Local Search

In order to reach high quality solutions the proposed extensions of FA are designed to invoke, at the end of each iteration, a stochastic local intensification search procedure aiming to exploit the search region around the best firefly, x_{best}. This local search, presented in [24], is a random line search algorithm that is applied coordinate by coordinate to the best point of the population. The procedure can be described as follows. First, for a fixed parameter δ the procedure computes the maximum feasible step length

$$\Delta = \delta(\max_{1 \leq s \leq n}(u_{bs} - l_{bs})).$$

Then, for each coordinate $s(s = 1, 2, \ldots, n)$, a random number $\mu \sim U[0, 1]$ (uniformly distributed between 0 and 1) is selected as a step length and a trial point y is componentwise moved along that direction and a new position is obtained as follows

$$y_s = x_{best_s} + \mu\Delta.$$

When $y \notin \Omega$, the trial point is rejected and the search along that coordinate ends. If y improves over the best point x_{best} according to Definition 1, within $LSit_{max}$ iterations, the best point x_{best} is replaced by the trial point y and the search along that coordinate s ends. A description of the local search procedure is presented in Algorithm 4.

Algorithm 4. Local Search

Data: x_{best} (the best point of the population at iteration k), $LSit_{max}$, δ
$\Delta = \delta \max\limits_{1 \leq s \leq n}(u_{b_s} - l_{b_s})$
for $s = 1$ *to* n do
   Set $it = 1$
   while $it < LSit_{max}$ do
      Set $y = x_{best}$
      $y_s = x_{best_s} + \mu\Delta$, $\mu \sim U[0,1]$
      if y *improves over* x_{best} *and* $y \in \Omega$ **then**
         Set $x_{best} = y$
      else
         Set $it = LSit_{max} - 1$
      end if
      Set $it = it + 1$
   end while
end for

Algorithm 5. exts- FA

Data: $k_{max}, \alpha_{max}, \alpha_{min}, \gamma_{max}, \gamma_{min}$

Set $k{=}1$

Randomly generate a population of N fireflies, $x_i^k \in \Omega, \ i = 1, ..., N$

Based on $\{x_1^k, ..., x_N^k\}$ evaluate $f(x_i^k), \zeta(x_i^k), nc(x_i^k), \ i = 1, ..., N$

Rank the fireflies using GR (Algorithm 2) or FD rules (Algorithm 3)

Set $x_{best}^k = x_1^k$ and $f_{best}^k = f(x_1^k)$

While (stopping criteria is not met)

  Compute the randomization parameter $\alpha^{(k)}$ using (6)

  Compute the scale parameter $S^{(k)}$ using (8)

  for $i = 2$ to $N - 1$

    for $j = 1$ to $i - 1$

      Compute the attractiveness β using (5) and (7)

      Move firefly i towards firefly j using (4), and project onto Ω the trial position t_i

      Evaluate $f(t_i), \zeta(t_i), nc(t_i)$

      if t_i *improves over* x_i^k **then**

        Set $x_i^k = t_i$

      end if

    end for j

  end for i

  Set $t_N = x_1^k + \varepsilon$, where $\varepsilon \sim U(0,1)$ (vector of random numbers) and project onto Ω

  Evaluate $f(t_N), \zeta(t_N), nc(t_N)$

  if t_N *improves over* x_N^k **then**

    Set $x_N^k = t_N$

  end if

  Set $k = k + 1$

  Evaluate $f(x_i^k), \zeta(x_i^k), nc(x_i^k), \ i = 1, ..., N$

  Rank the fireflies using GR (Algorithm 2) or FD rules (Algorithm 3)

  Set $x_{best}^k = x_1^k$ and $f_{best}^k = f(x_1^k)$

  Invoke the Local Search (Algorithm 4)

end while

4.3 Extensions of FA

The proposed extensions of FA (herein denoted by exts-FA) use a population of points/fireflies to compute, at each iteration k, an approximate solution, x_{best}^k, to the problem (1). Along the iterative process, the exts-FA generate approximate solutions, x_{best}^k, that satisfy the bound constraints, with increasingly better accuracy.

In the standard FA, each firefly i moves towards the brighter fireflies. However, when a firefly i, located at x_i, moves as in standard FA, its brightness may decrease. To prevent this, after moving each firefly i in the direction of a brighter firefly j, the selection rule given by Definition 1 is applied. We remark that if the trial position lies outside the search space Ω, the point is projected onto Ω. Denoting by t_i the trial position, if t_i *improves over* x_i, t_i will be the current position of the firefly i for the next movement; otherwise, the position x_i will be maintained as current for the next movement. To further improve exts-FA, all fireflies will move according to (4) except the less bright firefly, x_N. The position of firefly N is replaced by a random movement position of the brightest firefly. The pseudo-code of exts-FA is given in Algorithm 5.

We will denote by FA1 the exts-FA with FD rules, and by FA2 the exts-FA with GR in the ranking of the points. In the context of the implementation of FA1, we also propose a new movement equation (instead of (4)) in which all fireflies will move towards the best one. We will denoted this implementation by FA1#.

4.4   Stopping Criteria

The algorithm stops when the following condition is reached:

$$\left(\left|f_{opt} - f_{best}\right| \leq 10^{-6} \text{ and } \zeta(f_{best}) \leq 10^{-6}\right) \text{ or } k > k_{\max} \tag{10}$$

where f_{opt} represents the known global optimal solution, kf_{best} is the objective function value of the best point of the population, k denotes the iteration counter and $k_{\max}$ is the maximum number of iterations allowed.

5   Experimental Results

In this section, we aim to investigate the performance of FA1, FA1# and FA2 when solving a set of nonlinear optimization problems. Thirteen benchmark global optimization test problems, with dimensions ranging from 2 to 20, chosen from [25] containing characteristics that are representative of what can be considered difficult when solving global optimization problems. Their characteristics are outlined in Table 1.

The two first columns display the name of the problem (Prob.) and the best known solution (f_{opt}), followed by the number of variables (n), the type of objective function (function), the number of inequality constraints (LI and NI, for linear and nonlinear inequality constraints, respectively), the number of equality constraints (LE and NE, for linear and nonlinear equality constraints, respectively), as reported in [25]. The feasibility ratio ρ, in the last column, is an estimate of the size of the feasible search space Ω_F to the size of the whole search space. In practice, ρ represents the degree of difficulty of each problem.

Table 1. Summary of main properties of the benchmark problems.

Prob.	f_{opt}	n	function	LI	NI	LE	NE	ρ (%)
G01	−15.000000	13	Quadratic	9	0	0	0	0.011
G02	−0.803619	20	Nonlinear	1	1	0	0	99.99
G03	−1.000500	10	Nonlinear	0	0	0	1	0.002
G04	−30665.538672	5	Quadratic	0	6	0	0	52.123
G05	5126.496714	4	Nonlinear	2	0	0	3	0.000
G06	−6961.813876	2	Nonlinear	0	2	0	0	0.006
G07	24.306209	10	Quadratic	3	5	0	0	0.000
G08	−0.095825	2	Nonlinear	0	2	0	0	0.856
G09	680.630057	7	Nonlinear	0	4	0	0	0.521
G10	7049.248021	8	Linear	3	3	0	0	0.001
G11	0.749900	2	Quadratic	0	0	0	1	0.000
G12	−1.000000	3	Quadratic	0	9	0	0	4.779
G13	0.053942	5	Nonlinear	0	0	1	3	0.000

The numerical experiments were carried out on a MacBook Pro (13-inch, Mid 2012) with processor 2.5 GHz and 4 Gb of memory. The algorithms were coded in Matlab® programming language, version 8.01 (R2013a).

Since the FA is a stochastic method, each problem was solved 20 times. The size of the population used was $N = 40$ fireflies and in the stopping criteria, defined in (10), the maximum number of iterations allowed was $k_{max} = 5000$ iterations. The initial parameters used to dynamically compute α and γ are: $\alpha_{max} = 0.9, \alpha_{min} = 0.01,$ $\gamma_{max} = 100$ and $\gamma_{min} = 0.001$. All equality constraints $h_j(x)$ have been converted into inequality constraints using $\left| h_j(x) \right| - \varepsilon \leq 0$, where $\varepsilon > 0$ is a very small violation tolerance. In our numerical experiments $\varepsilon = 10^{-4}$ is used for the problems G05, G11 and G13 and $\varepsilon = 10^{-6}$ for the remaining problems. The step length in the local search procedure is set to $\delta = 10^{-5}$, except those marked with (*) that are set to $\delta = 10^{-2}$. The maximum number of local search iterations allowed is $LSit_{max} = 10$.

Table 2 summarizes the numerical results produced by the exts-FA, namely the proposed FA1#, FA1 and FA2. The first column shows the name of the problem, followed by the acronym of the exts-FA implementation. The remaining columns present: the best (f_{best}), the mean (f_{mean}), the median (f_{med}), the standard deviation (SD) and the worst (f_{worst}) solution values obtained over the 20 runs.

We remark that our proposed algorithms were able to find feasible solutions for all the runs of all of thirteen benchmark tested problems. This is due to the fact that the proposed algorithms prioritize the search of feasible solutions before proceeding to the search of the global optimum value.

The proposed exts-FA were also able to achieve very good results in almost of the problems. In the G01, G03, G08, G11 and G12 problems, the FA1#, FA1 and FA2 implementations reached the known global optimal solution in all runs. Consequently the measures of mean, median and the worst of the objective function values are equal to the global optimum and the standard deviation is zero. For G05 and G13 problems, the optimal solutions obtained by FA1 are lower than the known optimum values. This

Table 2. Results produced by the FA1#, FA1 and FA2.

Prob.	exts-FA	f_{best}	f_{mean}	f_{med}	SD.	f_{worst}
G01	FA1#	−15.0000	−15.0000	−15.0000	0.0000	−15.0000
	FA1	−15.0000	−15.0000	−15.0000	0.0000	−15.0000
	FA2	−15.0000	−15.0000	−15.0000	0.0000	−15.0000
G02	FA1#	−0.4373	−0.2968	−0.2841	0.0557	−0.2405
	FA1	−0.4048	−0.2941	−0.2911	0.0398	−0.2414
	FA2	−0.4799	−0.3458	−0.3330	0.0631	−0.2488
G03	FA1#	−1.0005	−1.0005	−1.0005	0.0000	−1.0005
	FA1	−1.0005	−1.0005	−1.0005	0.0000	−1.0005
	FA2	−1.0005	−1.0005	−1.0005	0.0000	−1.0004
G04	FA1#	−30665.5385	−30538.1527	−30546.2560	96.7170	−30372.8244
	FA1	−30665.5386	−30663.8601	−30665.5382	7.5011	−30631.9925
	FA2	−30665.5385	−30660.8451	−30665.5382	14.4310	−30611.1510
G05	FA1#[a]	5126.7259	5402.6462	5274.4115	279.0658	5986.442138
	FA1	5126.3617	5157.5428	5128.489	66.0467	5401.545758
	FA2	5126.5175	5128.7370	5128.4245	2.0577	5133.3343
G06	FA1#*	−6961.8101	−6961.7905	−6961.7905	0.0130	−6961.7553
	FA1	−6961.8016	−6961.7747	−6961.7747	0.0172	−6961.7347
	FA2[a]	−6961.5405	−6959.8675	−6960.2069	1.2835	−6956.3551
G07	FA1#[a]	24.6203	26.0814	26.1657	0.7935	27.1457
	FA1	24.3571	24.5827	24.5456	0.1967	25.2044
	FA2	24.3273	24.3772	24.3663	0.0311	24.4368
G08	FA1#	−0.095825	−0.095825	−0.095825	0.0000	−0.095825
	FA1	−0.095825	−0.095825	−0.095825	0.0000	−0.095824
	FA2	−0.095825	−0.095825	−0.095825	0.0000	−0.095825
G09	FA1#	680.6445	680.7663	680.7867	0.0676	680.8671
	FA1	680.6348	680.6895	680.6659	0.0468	680.7948
	FA2	680.6328	680.6869	680.6821	0.0414	680.7852
G10	FA1#	7069.4263	8183.2192	7785.5967	1053.5456	10276.0255
	FA1[a]	7125.2110	7552.1114	7364.5715	486.5234	8872.8741
	FA2[a]	7073.7810	7171.9798	7145.6712	107.9335	7542.8818
G11	FA1#	0.7499	0.7499	0.7499	0.0000	0.749910
	FA1	0.7499	0.7499	0.7499	0.0000	0.749900
	FA2[a]	0.7499	0.7499	0.7499	0.0000	0.749900
G12	FA1#	−1.0000	−1.0000	−1.0000	0.0000	−1.0000
	FA1	−1.0000	−1.0000	−1.0000	0.0000	−1.0000
	FA2	−1.0000	−1.0000	−1.0000	0.0000	−1.0000
G13	FA1#[a]	0.054106	0.325382	0.435195	0.2509	0.817849
	FA1	0.053556	0.283400	0.434445	0.1924	0.444500
	FA2	0.053944	0.053960	0.053951	0.0000	0.054017

[a]Means the step length of $\delta = 10^{-2}$ in the local search procedure.

Table 3. Comparison of our study with others stochastic population-based methods

Prob.	FA1	FA2	[26]	[27]	[28]	[29]	[30]
G01	−15.0000	−15.0000	−14.7082	−15.0000	−15.0000	−15	−15.000
G02	−0.4048	−0.4799	−0.79671	−0.803515	−0.802970	−0.8036	−0.803202
G03	−1.0005	−1.0005	−0.9989	−1.0000	−1.0000	−1.0	−1.000
G04	−30665.539	−30665.539	−30655.3	−30665.539	−30665.500	−30665.5	−30665.401
G05	5126.3617	5126,5175	n.a.	5126.497	5126.989	5126.4981	5126.907
G06	−6961.8016	−6961,5405	−6342.6	−6961.814	−6961.800	−6961.8	−6961.046
G07	24.3571	24,3273	24.826	24.307	24.480	24.306	24.838
G08	−0.095825	−0.095825	−0.089157	−0.095825	−0.095825	−0.09582	−0.095825
G09	680.6348	680.6328	681.16	680.630	680.640	680.63	680.773
G10	7125.2110	7073.7810	8163.6	7054.316	7061.340	7049.25	7069.981
G11	0.749900	0.749900	0.75	0.750	0.75	0.75	0.749
G12	−1.000000	−1.000000	−0.999135	−1.000000	n.a.	n.a.	−1.000000
G13	0.053556	0.053944	0.557	0.053957	n.a.	n.a.	0.053941

n.a. means not available.

is related to the fact that the equality constraints of these problems were relaxed by a threshold value of $\varepsilon = 10^{-4}$. For the G04, G06, G07 and G09 problems the proposed extensions of FA produced very competitive results since they were able to obtain optimum values very close to the known optimum ones. On the other hand, for the problems G02 and G10 they were not able to reach the known optimum solution.

In general, the best performance was obtained with FA1 and FA2 implementations. Then, we analyze the performance of these two extensions when compared with five stochastic population-based global methods. In [26] the method incorporates a homomorphous mapping between an n-dimensional cube and the feasible search space. Runarsson and Yao in [27] present results of the original stochastic ranking method for constrained evolutionary optimization. In [28] a self-adaptive fitness formulation is used. The results reported in [29] were obtained with an adaptive penalty method with dynamic use of DE variants, while in [30] a self-adaptive penalty based genetic algorithm is used. Table 3 reports the best results found by these methods and by our proposed best implementations FA1 and FA2.

From Table 3 the competitiveness of FA1 and FA2 with the reported approaches is shown. The stochastic ranking in [27] produced very good results. However this algorithm was able to obtain feasible solutions only in 6 out of the 30 runs performed for the test problem G10. In [28] only 17 out of 20 runs produced feasible solutions for the problem G10 and 9 out of 20 runs for G05 problem, while the exts-FA obtained feasible solutions for all problems in all runs. In general, the proposed exts-FA is competitive as the reported algorithms in the related field.

6 Conclusions

The FA is a stochastic global optimization algorithm, inspired by the social behavior of fireflies and based on their flashing and attraction, which was originally designed to solve bound constrained optimization problems. In this paper we extend the FA to solve nonsmooth nonconvex constrained global optimization problems. The extensions of FA denoted by FA1 and FA1# incorporate the constraint-handling technique based on the feasibility and dominance rules, while FA2 uses a global competitive ranking combined with a new fitness function. Moreover, FA1# uses a movement equation where all fireflies move towards the best one.

Thirteen well known benchmark problems were used in order to test the performance of the implementations of the exts-FA. The numerical experiments show that the proposed exts-FA are competitive when compared with other stochastic methods. Further research will be directed to improve the results through testing other fitness functions combined with the GR scheme. Future developments may include solving problems with large dimensions.

Acknowledgements. This work has been supported by COMPETE: POCI-01-0145-FEDER-007043 and FCT – Fundação para a Ciência e Tecnologia within the projects UID/CEC/00319/2013 and UID/MAT/00013/2013.

References

1. Blum, C., Li, X.: Swarm intelligence in optimization. In: Blum, C., Merkle, D. (eds.) Swarm Intelligence: Introduction and Applications, pp. 43–86. Springer Verlag, Berlin (2008)
2. Tuba, M.: Swarm intelligence algorithms parameter tuning. In: Proceedings of the American Conference on Applied Mathematics (AMERICAN-MATH 2012), pp. 389–394, Harvard, Cambridge, USA (2012)
3. Holland, J.H.: Adaptation in Natural and Artificial Systems. University of Michigan Press, Ann Arbor (1975)
4. Kennedy, J., Eberhart, R.C.: Particle swarm optimization. In: IEEE International Conference on Neural Networks (Perth, Australia), pp. 1942–1948. IEEE Service Center, Piscataway (1995)
5. Geem, Z.W., Kim, J.H., Loganathan, G.V.: A new heuristic optimization algorithm: harmony search. Simulations **76**, 60–68 (2001)
6. Yang, X. S.: Nature-Inspired Metaheuristic Algorithms. Luniver Press (2008)
7. Dorigo, M.: Optimization, learning and natural algorithms, Ph.D. Thesis, Dipartimento di Elettronica, Politecnico di Milano, Italy (1992)
8. Eberhart, R.C., Kennedy, J.: A new optimizer using particle swarm theory. In: Proceedings of the Sixth International Symposium on Micro Machine and Human Science, pp. 39–43. IEEE Press, Nagoya (1995)
9. Horng, M.H., Liou, R.J.: Multilevel minimum cross entropy threshold selection based on the firefly algorithm. Expert Syst. Appl. **38**(12), 14805–14811 (2011)
10. Yang, X.S., Hosseini, S.S., Gandomi, A.H.: Firefly algorithm for solving non-convex economic dispatch problems with valve loading effect. Appl. Soft Comput. **12**(3), 1180–1186 (2012)
11. Gandomi, A.H., Yang, X.S., Alavi, A.H.: Mixed variable structural optimization using Firefly Algorithm. Comput. Struct. **89**(23–24), 2325–2336 (2011)
12. Costa, M.F.P., Rocha, A.M.A.C., Francisco, R.B., Fernandes, E.M.G.P.: Heuristic-based firefly algorithm for bound constrained nonlinear binary optimization. Adv. Oper. Res. **2014**, Article ID 215182, 12 (2014)

13. Costa, M.F.P., Rocha, A.M.A.C., Francisco, R.B., Fernandes, E.M.G.P.: Firefly penalty-based algorithm for bound constrained mixed-integer nonlinear programming. Optimization **65**(5), 1085-1104 (2016). doi:10.1080/02331934.2015.1135920

14. Yang, X.-S.: Firefly algorithms for multimodal optimization. In: Watanabe, O., Zeugmann, T. (eds.) SAGA 2009. LNCS, vol. 5792, pp. 169–178. Springer, Heidelberg (2009)

15. Yang, X.S.: Multiobjective firefly algorithm for continuous optimization. Eng. Comput. **29** (2), 175–184 (2013)

16. Fister, I., Fister Jr., I., Yang, X.-S., Brest, J.: A comprehensive review of firefly algorithms. Swarm Evol. Comput. **13**, 34–46 (2013)

17. Yang, X.-S., He, X.: Firefly algorithm: recent advances and applications. Int. J. Swarm Intell. **1**(1), 36–50 (2013)

18. Ali, M., Zhu, W.X.: A penalty function-based differential evolution algorithm for constrained global optimization. Comput. Optim. Appl. **54**(3), 707–739 (2013)

19. Barbosa, H.J.C., Lemonge, A.C.C.: An adaptive penalty method for genetic algorithms in constrained optimization problems. In: Iba, H. (ed.) Frontiers in Evolutionary Robotics, pp. 9–34. I-Tech Education Publications, Vienna (2008)

20. Mezura-Montes, E., Coello Coello, C.A.C.: Constraint-handling in nature-inspired numerical optimization: past, present and future. Swarm Evol. Comput. **1**(4), 173–194 (2011)

21. Lemonge, A.C.C., Barbosa, H.J.C., Bernardino, H.S.: Variants of an adaptive penalty scheme for steady-state genetic algorithms in engineering optimization. Eng. Comput. Int. J. Comput.-Aided Eng. Softw. **32**(8), 2182–2215 (2015)

22. Runarsson, T.P., Yao, X.: Constrained evolutionary optimization – the penalty function approach. In: Sarker, R., et al. (eds.) Evolutionary Optimization: International Series in Operations Research and Management Science, vol. 48, pp. 87–113. Springer, New York (2003)

23. Deb, K.: An efficient constraint-handling method for genetic algorithms. Comput. Methods Appl. Mech. Eng. **186**(0045–7825), 311–338 (2000)

24. Birbil, S.I., Fang, S.-C.: An electromagnetism-like mechanism for global optimization. J. Global Optim. **25**, 263–282 (2003)

25. Liang, J.J., Runarsson, T.P., Mezura-Montes, E., Clerc, M., Suganthan, P.N., Coello, C.A. C., Deb, K.: Problem definition and evolution criteria for the CEC 2006 special session on constrained real-parameter optimization. In: IEEE Congress on Evolutionary Computation, Vancouver, Canada, 17–21 July (2006)

26. Koziel, S., Michalewicz, Z.: Evolutionary algorithms, homomorphous mappings, and constrained parameter optimization. Evol. Comput. **7**(1), 19–44 (1999)

27. Runarsson, T.P., Yao, X.: Stochastic ranking for constrained evolutionary optimization. IEEE Trans. Evol. Comput. **4**(3), 284–294 (2000)

28. Farmani, R., Wright, J.: Self-adaptive fitness formulation for constrained optimization. IEEE Trans. Evol. Comput. **7**(5), 445–455 (2003)

29. Silva, E.K., Barbosa, H.J.C., Lemonge, A.C.C.: An adaptive constraint handling technique for differential evolution with dynamic use of variants in engineering optimization. Optim. Eng. **12**(1–2), 31–54 (2011)

30. Tessema, R, Yen, G.G.: A self adaptive penalty function based algorithm for constrained optimization. In: IEEE Congress on Evolutionary Computation (CEC 2006), pp. 246–253, Vancouver, Canada (2006)

Lie Algebraic Methods as Mathematical Models for High Performance Computing Using the Multi-agent Approach

Serge N. Andrianov and Nataliia Kulabukhova$^{(\boxtimes)}$

Saint-Petersburg State University, Saint Petersburg, Russia
sandrianov@yandex.ru, n.kulabukhova@spbu.ru

Abstract. In this paper we discuss some problems of the construction of mathematical models of dynamic processes to effectively carry out computational experiments using high performance computing systems (both parallel and distributed). The suggested approach is based on the Lie algebraic approach and the multi-agents paradigm. The Lie algebraic tools demonstrated high effectiveness in dynamical systems modeling. A matrix presentation for dynamical systems propagation allows to implement a modular representation of the objects of study as well as the corresponding operations. Moreover, the matrix formalism is based upon the multi-agent paradigm for modeling and control of complex systems for physical facilities. The corresponding codes are realized both in symbolic and numerical forms.

1 Introduction

In today's world the role of experimental physics is growing on one hand as a source of new knowledge in the various fields of science and technologies (not only in high-energy physics), and on the other – as a field of knowledge, in which new methods and tools for modeling complex processes of different nature are constantly generated. In this case under the complexity of the modeling process and the complexity of the governance structure we understand not only a huge number of control actions, but also the complexity of behavior of management subjects. Moreover, in physics of particle beams, plasma physics, and a number of other areas there is a three-tier control structure. The object of control is beam (ensemble) of particles and the subjects of the first level are presented by control mechanisms consisting of control units that provide the necessary (optimum) conditions for the evolution of the particle beam. Finally, second level control subjects are presented using information elements of the system which controls (based on certain principles) enormous number of control devices (magnets, multipole lenses, etc., diagnostic systems for plasma, beam etc.), see, for example, [1]. In particular, a common problem in the cyclic accelerators is the necessity for careful monitoring of certain characteristics of the particle beam for a long period of time with minimal loss of intensity of the beam, while ensuring the necessary degree of stability of the particle beam parameters and so

O. Gervasi et al. (Eds.): ICCSA 2016, Part I, LNCS 9786, pp. 418–430, 2016.
DOI: 10.1007/978-3-319-42085-1_32

on [2]. Unlike mechanical systems, in this class of problems we are dealing with complex behavior of the ensemble of particles (both regular and chaotic) that is in need of special mathematical tools, taking into account the collective nature of the beam (control subject). Regarding the controls (first level entities), they are usually realized with the aid of electromagnetic fields, which are distributed in space). The difficulty of formalization and management at this level is related to the spatial distribution of the control field, and is caused by the limited knowledge not only about the state of the ensemble of particles, but also about "field configurations". This knowledge restriction is determined mainly by some limitations in hardware installation of control elements and other factors which affect the quality control.

Taking into account the problems mentioned above, in this paper we consider a three-tier computing scheme: a physical object – a mathematical model – an information model. In other words, we can say that the distributed nature of a control object (particle beam) and control subject (in our case the electromagnetic field) leads to the necessity of building distributed information environment adequate for such physical features. Using technologies of multi-agent systems maximizes adequacy of the process of mapping of the physical model onto the relevant information model. Implementation of necessary computational experiments is carried out using parallel systems (for example, GPU-Accelerated Research Clusters) and distributed computing. It should be noted, that after construction of relevant information units we should incorporate this triune structure into the concept of the Virtual Accelerator, see, e.g. [3].

2   A Multi-agent Concept

The above mentioned reasons lead to the necessity of forming a management information system (the phase of control subject of the second level). This system has to provide the necessary tools for adaptation of electrophysical devices of standard control systems using a multiplicity of control parameters, as well as multiplicity devices that provide necessary information about the state of the control object – the particle beam. We should note that this hierarchy can also be used for modeling systems of control in biological systems, economics and other distributed (in time and space) systems. In other words, our goal is to create an information control system that is based on the rather limited amount of experimental data and can offer a variety of scenarios for implementation of "optimal" regimes of systems functioning.

In some papers published previously by the author the concept of Lego-objects was proposed (see, e.g., [5,6]), which operate as independent and specialized control objects, implemented in the form of special computer programs (including methods and technologies of artificial intelligence). The realization of similar concept is largely determined not only by the structure of subject and object management, but also by the methods used for mathematical models construction. The corresponding mathematical models should be adequate both to a description of physical problems, and to the corresponding operations in computational environment.

Indeed, the introduction of Lego-objects allows us to allocate relevant knowledge and resources among sufficiently "independent" intelligent objects (according to modern terminology – agents) for distributing rights in control procedures in whole, and also for process of problem-solving and for ensuring the implementation of some specified modes (an optimal in some sense).

At the same time, we reserve a narrow functional orientation of our agent (a Lego-object) on a separate part for the overall solution of the problem. However, the totality of similar agents, considered as a single system, gradually began to give way to universal integrity (autonomy of the control system as a whole). In other words, according to the concept of multi-agent technology (see [7]), for solution of this controversy we suggest a new method for control problems in complex physical systems.

In other words, we offer a decentralized system (in contrast to classical centralized control system), in which the control procedures are realized at the expense of local interactions between agents. At the same time, we should preserve functional orientation of our agent (a Lego-object) on a separate part of the overall solution of the problem narrow enough. However, the universality of such agents gradually begin to give way to universal integrity of an informational system as a whole. It should be noted, however, that incomplete universality of individual agents are caused by large variability of practical oriented tasks that need to be addressed in accelerator physics problems.

Usually, in the classical approach we use a set of well-defined algorithms which can help to find the best solution for the control systems of ensembles of particles (in accelerators, thermonuclear facilities – tokamaks, stellarator, etc.). But the complexity of the control systems for similar systems incites us to change of paradigm of control both in theoretical and practical aspects. The multi-agent paradigm can also decrease both the computational burden and the corresponding computational time. This also motivates us to the need for a concept of complex physical processes management using the intellectualization of the user interface.

Common understanding of the properties of the agent requires formulation of some criteria of optimality. The agent has the ability to interact with the environment. Thus, the agent is an entity that is installed in an external environment and can interact with it. This agent can and should make a decision for autonomous rational action to achieve defined goals. Under the intelligent agent we understand an entity featuring the following properties:

- *reactivity* – the agent perceives the environment and reacts to changes in it, performing actions to achieve the goals specified;
- *proactivity* – showing an agent behavior goals, showing initiative to commit actions aimed at achieving the given objectives;
- *sociality* – the agent interacts with other entities of the environment (first of all with other agents) to achieve formalized objectives.

Implementation of the first two properties separately can be achieved quite easily. However, trying to combine these two properties in a system (in specified proportions) researchers typically face a number of difficulties. Indeed, the operation of the agent is not effective enough both in the case of a rigid script execution for achieving the goal (in other words, ignoring the changes in the external environment) and in the absence of the ability to notice the need to adjust the plan. Here we can note that it is the result of inefficient work in response only to environmental stimulus neglecting the planning of targeted actions.

Implementing the third property can be a difficult problem, because here we can usually surmise not only communication with each other, but also certain elements of the "cooperation" with other entities of control process. In other words, it is necessary to provide separation between the objectives of specific agents and joint planning and coordination of actions aimed at achieving common goals. So, a "social behavior" requires from the agent the corresponding representations not only about goals of other agents but also about how they design scripts to achieve these goals. The analysis of control systems for complex physical processes leads us to the conclusion about the lack of a priori precision for all conditions of operation. This leads us to necessity of description of adaptive formalization of investigated problems, having such an attribute of the agent as adaptability, i. e. the ability to automatically adapt to uncertain and changing conditions in a dynamic environment. The agent operates in a complex, changing environment, interacting with other agents. This leads to the fact that the multiagent system is much more effective in comparison with a "standard" adaptive system, since similar system can be trained faster for operating more efficiently due to redistribution of functions and/or tasks between agents. The notion of complex systems is linked with the following fundamental ideas that directly affect the functioning multi-agent systems. We list these ideas ([7], for example) below:

- in complex systems there exist autonomous objects that interact with each other in carrying out their defined tasks;
- agents must be able to respond to changing conditions in the environment, in which they operate, and possibly change its behavior on the basis of the information received;
- complex systems, which are characterized in terms of emerging (temporary and/or permanent) structures, and are logically represented by schemes which are formed by the interaction between the agents;
- complex systems with emerging structures often oscillate between order and chaos;
- biological analogies (such as parasitism, symbiosis, reproduction, genetics, mitosis and natural selection) are often used when creating complex agent-based systems.

The concept of agents developed in the framework of multi-agent technology and multi-agent systems, automatically implies an active behavior of the agent, in other words the ability of the computer program itself to respond to external events and "choose" the appropriate actions. In view of the above, in the future we will stick to the following definition of an agent.

Definition 1. *An agent is a certain software unit capable of acting for the purpose of achieving the goals set by the user.*

Thus, returning to the concept of Lego-objects in view of the corresponding definitions, it can be argued that the Lego-object has dual nature. First – from mathematical point of view it describes the subject area (in our case, the ensembles of particles, electromagnetic fields, etc.). It's second "essence" (informational) presents the interaction of these objects in the process of computational experiments and in the process of creating a scenario for control of the behavior of the beam by using control devices (or rather, using electromagnetic fields generated by these devices). We should note that in modern literature devoted to accelerator physics the concept of the Virtual Accelerator is being actively discussed (see [3]).

In almost all publications quite traditional modeling and control systems are considered despite the ever-increasing requirements for physical experiments support.

However, the approaches developed in these articles are quite different. We can conclude that a plurality of characteristics of control devices of the particle beam on one hand, and complexity of strict optimality formalization, on the other hand, leads us to the need of using methods of multi-agent approach. In other words, we need to introduce a family of independent "executive agents" with different functionalities (with different specializations), that interact with each other in accordance with prescribed rules: synchronizing actions, exchange of information, automated adaptation to uncertain and variable conditions of the dynamic medium etc.

3   Classes and Subclasses of the Lego-Objects

Following the previously proposed concept of Lego-objects, we describe the hierarchical structure of mathematical objects that are used to build mathematical models of physical processes occurring in the accelerator facilities. We should note that many objects can also be used in other problems of computational physics, which are described in terms of nonlinear dynamical systems. We should note that the idea of multi-agent systems is actively implemented in many systems of mathematical modeling (see, for example, the package Mathematica [8]), indicating the relevance and effectiveness of this approach for a wide class of problems. So, following the concept described above, we have a three-tier structure of control: the control object (an ensemble of particles) – a system of control (the subjects of the first level) – the subject of the second level – an informational system, which realizes the control strategy. We should note that one of the most important principles to be followed is the universality of information units "gluing" at all levels, thus ensuring the universality and interoperability in terms of preserving the information structure. In this case, we mean the fact that the properties of a computational model for each information unit at all stages of the simulation should lead to the correct result in terms of performing computations of corresponding mathematical operations for solution of our practical problems.

In case of the evolution of particle beams in accelerators (or plasma particles in the corresponding systems), dynamics of the particles is described by nonlinear systems of motion equations. Furthermore, it is often necessary to take into account additional factors (e.g. effect of beam polarization, interparticle interaction and etc.). Note that each of said blocks can have different shapes which correspond to physical description of an object and can be processed according to the chosen model of computational process. If we can provide researchers with a sufficiently extensive set of facilities and operations at any stage of modeling, then researchers can replace corresponding modules (after comparing respective results with empirical ones) and thereby formulate certain additional conditions for an optimal dynamic regime or for the target result. In accordance with the concept of Lego-objects, we can modify necessary information items for execution of corresponding computational experiments. In other words, these information items describe the nature of intelligent agents, which are supplied with special attributes. Moreover they can be considered as meta-objects that can affect other objects, as well as possess a necessary set of instruments developed for interaction with the environment and other meta-objects.

3.1   Construction of Information Units

As shown in [4], the main type of agents from a mathematical point of view is a matrix with certain matching rules, which leads to the correct description of the beam evolution in the accelerator facility. At the same time, all the characteristics of the particle beam (control object) and external field (control sub-object) can also be expressed using the matrix representation, and the corresponding mathematical operations. Here we should mention such matrix operations as Kronecker sums and products. There is also a set of permitted associations of subagents, which is determined by the physical nature of the respective objects. Here is the view of the rules that define the necessary properties of the original physical object, and on the other – the rules by which the appropriate agent or subagent interacts with other agents (subagents) for optimal results. In this section, as an example, we consider the structure of the description of a controlling agent responsible for some physical element. In our example, we consider a magnetic or electric quadrupole as a control element. As mentioned above, the corresponding agent (possessing an integrity property from physical standpoint) from a mathematical point of view can be represented as a compound information essence, which is considered as a set of subagents that are not self-sufficient by themselves, but determine the integrity of the agent in whole (see Fig. 1).

The physical parameters that define configuration of the field on each interval and are included in the matrix $\mathbf{M}^{1k}$ represent information units, which are responsible for the appropriate subagents. Thus, we formulated the physical nature of the problem being solved in terms of information objects. In addition, we have to impose certain rules on the choice of necessary parameters and put

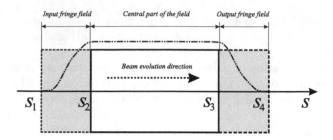

Fig. 1. The scheme of the modeling process interpretation.

them in our set of determining rules for "assembly" of subagents in a single agent, and thereby provide a physical meaning for all "aggregate" agents. The complete set of similar components should be classified by their physical interpretation, as well as their significance. This allows you to build a hierarchy of proper relations for corresponding interactions of all agents for solution of the investigated problem.

Thus, subagents which are responsible for the fringing fields can be used for describing possible configurations of fringe fields (satisfying to the Laplace equation). The corresponding admissible set of fringe fields should be determined using both experimental data and theoretical investigations, see, for example [5,9]. Accordingly, all components of the control field (the central part of the field, contribution of fringe fields – input and output parts) in a general field up to the N-order should be taken into account. This leads to the fact that the matrix for the total field is the product of the respective matrices. In other words, the matrix of the mapping for the total field of our quadrupole lens can be written up to the N-th order in the following form

$$\mathbb{M}^N(s_4|s_1) = \mathbb{M}^N(s_4|s_3)\mathbb{M}^N(s_3|s_2)\mathbb{M}^N(s_2|s_1) =$$

$$= \left\{\sum_{k=1}^{N}\mathbb{M}^{1k}(s_4|s_3)\right\}\left\{\sum_{k=1}^{N}\mathbb{M}^{1k}(s_3|s_2)\right\}\left\{\sum_{k=1}^{N}\mathbb{M}^{1k}(s_2|s_1)\right\}. \quad (1)$$

This multiplicative form allows us to calculate the total matrix as a product of partial matrices and can be constructed on all subintervals in according the rules described in [6] both in symbolic and numerical form. Necessary control over the quality of the computational procedures can be carried out using the methods set out in the articles [10, 11].

In addition, we have to not only define special rules for the choice of these parameters, but also preserve them in the form of corresponding rules for forming of subagents as a single agent. On the next step we should provide a physically meaningful "cross-linking" of all agents for description of rules. The corresponding set of similar rules must have the necessary physical interpretation. This allows us not only to build a hierarchy of relations for corresponding interactions of all agents for solution of our problem, but also to automate the process of forming corresponding matrices.

Here is the view on the rules that on one hand define the necessary properties of an original physical object and on the other – the rules by which the appropriate agent or subagent interacts with other agents (subagents) for optimal results. In addition we have to impose certain rules for the choice of these parameters as well as to transform them in a complex of corresponding rules for concatenating of subagents into an integral whole. We should note that the corresponding family of rules consists of two subfamilies. The first contains prohibitive rules of cross-linking, while the second defines flexible adjustment of rules for cross-linking of sub-agents into a single agent. In the next step we need to formalize the physical cross linking rules for all agents. This is necessary to "physical users" for formal descriptions of informational units using the corresponding physical interpretation.

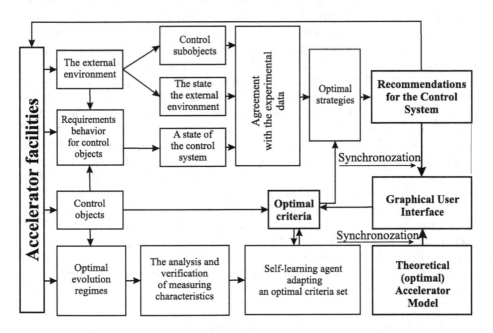

Fig. 2. The scheme of the modeling process interpretation.

So, the physical parameters that define the control field configuration on each interval and included in the matrix $\mathbb{M}^{1k}$ are incorporated as information parameters for appropriate ordered subagents. So the physical core of the problem of fringe fields can be interpreted in terms of some set of information objects. However, at first we have to impose certain rules for the choice of control parameters and then formulate the rules for corresponding "inclusion" of subagents into a single agent in according to chosen optimal strategy. So, we have reasonable physical restrictions for our agents and we have enough freedom to implement the search for optimal solutions. Naturally, the optimality criteria depend on the problem under consideration, but they can be clearly formalized in terms of the corresponding matrices M^{1k}. It should be noted that this allows you to search the optimal regimes not only using various methods of parametric optimization, but also using the corresponding physically intuitive and experimental knowledge. It should be noted that in the first case specified parameters lose their physical content, which greatly simplifies the optimization process using multi-agent technology. Typically, similar problems can be solved within the different concepts of global optimization (e.g., [12]), but a large number of parameters and the complexity of computational procedures (and thus significant computational cost) lead to the need to use alternative methods for finding optimal solutions.

Here we must make an observation that in this class of problems (management of complex distributed systems) an optimization problem itself is formulated as the task of finding not only the optimal solution (in some sense), but the best solution possible (taking into account the formalized tasks in the experiment). This approach is adequate to meet the concept of the Virtual Accelerator. The point is that in large control systems for large physical facilities (accelerators, tokamaks, and etc.) the control is not carried out in an online mode (the speed of behavior of the physical processes is significantly higher than response speed of corresponding control systems). It should also be noted that the training session for a control system, based on the multi-agent technologies can be realized much easier if the requirements for stability of the system under small variations of control parameters are met. The fact is that this property of stability may be incorporated into the properties of the agents (subagents) naturally.

Moreover, since there exists a memory effect of intelligent agents, then if we have a set of "successful" combinations it allows to find appropriate solutions using the accumulated information (memory effect or adjustment of intelligent agents) for certain class of problems. We should also point out the need for a mechanism to monitor the correctness of all implementation procedures. The general scheme of interaction between logical blocks in the search for optimal solutions based on the matrix formalism for dynamical systems and multi-agent approach is shown on Fig. 2.

3.2   The Architecture of Classes and Subclasses of Agents

It is well known that a distributed multi-agent system may be divided into two classes: one class is composed of systems of "highly intelligent agents", while the second class is based on the so-called group mind. It should be noted that this classification corresponds to two classes of opposite options for multi-agent systems. The above concepts of agents and sub-agents allow us to consider the proposed multiagent system as a mixed class with the dynamic nature of the formation process for the implementation of such a system. Thus, the proposed variants are used for formation of classes of agents and sub-agents may be implemented in a dynamically generated computing environment. Indeed, the first class is suitable for implementation on distributed systems with a fairly small number of agents, while the second class uses a relatively small number of servers. In this case the agents themselves are not complicated objects and therefore can be implemented using simple algorithms for their interaction. Indeed as such agents act they introduce subagents. Moreover, it's principally possible to introduction a hierarchy of agents according to complexity and scope of the operations. This approach "fits in" well with the above ideology based on one hand on the matrix formalism, and on the other – on the concept of Lego-objects (see Fig. 3).

To implement the approach described above, we considered various techniques, in particular, for multi-agent systems construction a set of ready-to-use components Magenta Toolkit technology[1] has been considered. The proposed set of components Magenta Toolkit is a software tool (written in Java) that greatly simplifies and accelerates multiagent systems development for a variety of complex tasks. Knowledge which are inherent in our area of problems (beam physics and plasma) are notable for not only the large number of interconnections, but also for high computational complexity (which is usually absent in the "standard" problems). One should also consider the large number of many individual factors specific to various controls, as well as the nonlinearity of the corresponding evolutionary processes (nonlinear dynamics of particle beams).

In constructing the necessary software, for knowledge required for its functioning can be divided into the domain knowledge and the knowledge needed for the task and/or decision-making. In our case, the subject area comprises a mathematical model of corresponding physical reality and certain management rules (including the optimal), based on the matrix formalism for description of particles dynamics in the accelerators. On the second step we should include methods and tools for decision-making and description of relationships between agents. To describe the subject domain we can use the Domain Ontology, while for the implementation of methods for decision-making – the ontology of "virtual market" (Virtual Market Ontology), see [13]. Based on the above, appropriate ontologies can be represented in the form of the corresponding circuit of Fig. 3. It should be noted that one of the features of ontologies application is to use them for solving important problems as well as to use the knowledge for reapplication. Indeed, the ontology as a general scheme for presentation and usage of knowledge can not only clearly be defined for agents working with it (as with the shared resource), but also shared between agents, as well as replicated. We should also note that the corresponding knowledges are generated according to the approach described above (see Fig. 2) and corresponding information should be stored in the special data and knowledge bases.

Thus, the use of ontology involves direct work with ontological user model. With expansion and evolution of ideas on the subject area during work, ontology should be updated as well. Accordingly, an important requirement for system maintenance of ontologies is to provide their support editing operations and expansion of the existing ontologies. In addition, the applicative use of ontologies involves interaction with ontological system of software agents, providing users requests processing and implementing the interaction of components of distributed applications. In other words, the above mathematical models under-

---

[1] http://www.magenta-technology.ru/.

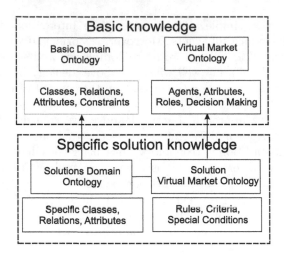

Fig. 3. The ontologies types.

lying Lego-objects technologies give users the ability to work with the ontological model, thus providing support for the necessary operations for creating, editing and ontology extension the subject area.

To support the operation and modernization of distributed intelligent systems, performing the operations described above (i. e., thereby functioning ontologies) must meet the following requirements:

- reliable semantic basis in determining the content of information;
- general logical theory, which consists of a "dictionary" and set of statements, implemented using some logical language;
- reference tool to provide the necessary communication both between users and computer agents, as well as between agents.

To implement the necessary operations we should use a special tool for editing ontologies (in particular, on the site of W3C several dozen tools for editing ontologies are provided).

3.3   The Problems of Data Processing

The approach described above also requires efficient data processing techniques. Specifics of the problems under consideration, and the proposed approach demand effective methods for solving the corresponding problems. Required data collection process becomes a very large and complex problem, and it gets difficult to process using relatively traditional database management tools or traditional data-processing applications.

In particular, similar problems include such special tools as data capture, monitoring, storage, retrieval, exchange, transfer, analysis, and visualization. The above-discussed concept of multi-agent systems is connected with the necessity of processing additional information obtained from the analysis of a large intermediate set of related data. Researchers are constantly faced with limitations in the computational procedures that are primarily associated with large volumes of data.

In the process of developing and implementing particle beams management systems, researchers are faced with the need to deal with the enormous number of data coming from the accelerator as a control system, and also from the beam itself. Thus, we obtain two sets of data that must be processed and implemented using the corresponding control processes. The first data set contains the data of numerical simulation, the second – the corresponding experimental data.

Effectiveness of these manipulations is largely determined by both modeling techniques used (in this case, the use of multi-agent systems within the concept of Lego-objects) and software methods implemented in the necessary procedures to process arrays of data obtained both in the working functioning and in numerical modeling.

The second set of data is associated with the measurement data, such as the monitoring control system and state of the beam. The harmonization of these data sets, their consolidation, storage and processing can be implemented effectively within the Big Data ideology. Moreover, since the physics accelerators become relatively common international mega projects, then the required data volume repeatedly increases.

In this case we have in mind the final data set (the end result of simulation) as well as an intermediate (current) one, which volume is much higher. It should be noted that the volume of such data is increased significantly in the study of particle beams dynamics with the influence of the self-field of the beam (for intense beams). Processing such data may be conducted within the concept of Virtual Accelerator.

4   Conclusion

In this article, we described a technique matching the physical essence of the modeled area of knowledge (in our case the beam physics in accelerators) with modern technologies, high performance computing through the use of multi-agent technologies. In the cited works, the solution of similar problems with usage of the matrix formalism for Lie algebraic methods has led to the possibility of introducing the concept of global optimization methods [12] in the structure of modern computing. Indeed, multiparametric character of the modern control systems for accelerator complexes leads to the need for use of multi-agent technologies as a tool for finding optimal solutions in multidimensional space of control parameters. The main advantages of multi-agent systems consist not only in processing of distributed information, but also in using the principles of intelligent processing of the corresponding information. Efficiency of implementation of multi-agent-based approach is largely determined by the choice of the mathematical tools. The matrix formalism for Lie methods allows to optimally adapt methods used for describing of the dynamics of particles in accelerators to the ideology of multi-agent systems. It should also be noted that the proposed mathematical and informational models necessitate the use of advanced technology in distributed data processing both in the processing and storage of data.

Acknowledgements. The work is supported by Saint Petersburg State University (project 0.37.155.2014).

References

1. Kosovtsov, M., Andrianov, S.N., Ivanov, A.N.: A matrix presentation for a beam propagator including particles spin. In: Proceedings of IPAC2011, San Sebastian, Spain, pp. 2283–2285. http://accelconf.web.cern.ch/AccelConf/IPAC2011/papers/wepc116.pdf
2. http://www.aps.anl.gov/epics/
3. Korkhov, V., Ivanov, A., Kulabukhova, N., Bogdanov, A., Andrianov, S.: Virtual accelerator: distributed environment for modeling beam accelerator control system. In: Proceedings of the 13-th International Conference on Computational Science and Its Applications, pp. 166–169
4. Kulabukhova, N., Andrianov, S., Bogdanov, A., Degtyarev, A.: Simulation of space charge dynamics in high intensive beams on hybrid systems. In: These Proceedings
5. Andrianov, S.N.: LEGO-technology approach for beam line design. In: Proceedings of EPAC 2002, Paris, France, pp. 1607–1609
6. Andrianov, S.N.: Dynamical Modeling of Particle Beam Control Systems. SPbSU, St. Petersburg (2004). (in Russian)
7. Kevin, Y.S.: Leyton-Brown Multiagent Systems. Algorithmic, Game-Theoretic, and Logical Foundations. http://www.masfoundations.org
8. http://demonstrations.wolfram.com/topic.html?topic=multi-agent+modeling&limit=20
9. Tereshonkov, Y., Andrianov, S.N.: Load curves distortion induced by fringe field effects in the ion nanoprobe. In: Proceedings of EPAC08, Genoa, Italy, pp. 1514–1516. http://accelconf.web.cern.ch/AccelConf/e08/papers/tupd039.pdf
10. Andrianov, S.N.: Symbolic computation of approximate symmetries for ordinary differential equations. Math. Comput. Simul. **57**(3–5), 139–145 (2001)
11. Andrianov, S.N.: Symplectification of truncated maps for hamiltonian systems. Math. Comput. Simul. **57**(3–5), 146–154 (2001)
12. Andrianov, S.N., Podzyvalov, E.A., Lvanov, A.N.: Methods and instruments for beam lines global optimization. In: Proceedings of the 5th International Conferenceon Physics and Control (PhysCon 2011), Leon, Spain, September 2011. http://lib.physcon.ru/file?id=d693c7ab0f17
13. Smirnov, S.V.: Ontology analysis of subject domains of modelling. In: Proceedings of the Samara Scientific Center RAS, vol. 3 (2002)

Spin-Coupling Diagrams and Incidence Geometry: A Note on Combinatorial and Quantum-Computational Aspects

Manuela S. Arruda[1,4](✉), Robenilson F. Santos[2,3], Dimitri Marinelli[4], and Vincenzo Aquilanti[4]

[1] Centro de Ciências Exatas e Tecnológicas,
Universidade Federal do Recôncavo da Bahia, Cruz das Almas, Brazil
manuelaarruda@gmail.com
[2] Instituto de Física, Universidade Federal da Bahia, Salvador, Brazil
[3] Instituto Federal de Alagoas, Campus Piranhas, Piranhas, Brazil
roferreirafs@gmail.com
[4] Dipartmento di Chimica, Biologia and Biotecnologie,
Università di Perugia, Perugia, Italy
dimitri.marinelli@gmail.com, vincenzoaquilanti@yahoo.it

Abstract. This paper continues previous work on quantum mechanical angular momentum theory and its applications. Relationships with projective geometry provide insight on various areas of physics and computational science. The seven-spin network previously introduced and the associate diagrams are contrasted to those of the Fano plane and its intriguing missing triad is discussed graphically. The two graphs are suggested as combinatorial and finite-geometrical "abacus" for quantum information applications, specifically for either (i)- a fermion-boson protocol, the hardware being typically a magnetic moiety distinguishing odd and even spins, or (ii)- a quantum-classical protocol, the hardware being materials (arguably molecular radicals) with both large and small angular momentum states.

Keywords: Angular-momentum · Spin-coupling · Projective-geometry

1 Introduction

This note belongs to a series in this journal on advances and applications of quantum mechanical angular momentum theory that has been evolved in various mathematical and computational subjects [1–16]. In a previous recent work [11], we consider how the four angular momenta a; b; c; d, and those arising in their intermediate couplings, x; y; z, form what we have defined a 7-spin network, emphasizing the projective geometry interpretation introduced by U. Fano and Racah [17] and by Robinson [18], see also Reference [19,20]. The interestingly symmetric nature of the network suggests mapping on the Fano plane, introduced by G. Fano in 1892 as the smallest nontrivial finite projective geometry.

© Springer International Publishing Switzerland 2016
O. Gervasi et al. (Eds.): ICCSA 2016, Part I, LNCS 9786, pp. 431–442, 2016.
DOI: 10.1007/978-3-319-42085-1_33

Two papers in preparation (Aquilanti and Marzuoli, to be published [21] and Robenilson *et al.* to be published [22]) consider issues from a similar perspective, such as that which will be developed in the present note. Reference [21] and [22] are devoted mainly to the Biedenharn-Elliot identity and the $9j$ symbol of angular momentum theory, the ten-spin network and Desarguesian geometry. Papers by Rau [23, 24], dealing in particular with applications to quantum computation provide a useful account of the involved geometrical and combinatorial features.

In the next section, the relationship with projective geometry is developed in detail. The last section concludes the paper with remarks on open problems, and two protocols for quantum information applications are proposed.

2   Spin Coupling Diagrams and Geometry

In presenting the connection between angular momentum theory and geometry, for the latter we choose to refer to one of best known and authoritative textbook by H. S. M. Coxeter [25]. Specific points will be directly referred to as propositions and pages of this book.

2.1   The Triangle of Couplings and the Quadrangle of Recouplings

Graphical representations of quantum angular-momentum coupling and recoupling coefficients have been an intriguing subject of studies for many years, long before they became of interest beyond the fields of spectroscopy and dynamics in nuclear, atomic and molecular physics, and of quantum chemistry. Modern investigations especially related to quantum gravity (after Ponzano and Regge [26], and Penrose [27]) and to quantum information are comprehended within the expanding mathematical subjects of spin networks, their q-extensions, the knot coloring and associated areas of research [Aquilanti and Marzuoli, to be published [21] and Robenilson *et al.* to be published [22]].

The graphical approaches most naturally arose from the observation arguably occurring to anyone dealing with the coupling of two angular momenta L and S to a resultant J that the associated triad of quantum numbers (SLJ) be represented either by

(i) the junction of three branches;
   or, alternatively, by
(ii) a triangle whose three sides are segments (possibly endowed with arrow heads) as graphical representations of vectors;
   or else by
(iii) three points on a line, as initiated by U. Fano and G. Racah in 1958, who introduced a projective approach, whereby connection with incidence geometry was first established [17].

The basic building block of angular momentum theory is the recoupling coefficient of Racah or equivalently the $6j$ symbol of Wigner: as amply exploited for example in [1–16], diagrammatic techniques permit visualizations of otherwise often obscure formulas. We will list features of the several alternative presentations, to be put into correspondence with those above:

(i) In his 1957 textbook, Edmonds [28] showed how the recouplings associated with the $6j$, the $9j$, and the $12j$ symbols could be represented, using the first of the above mentioned labeling schemes, by diagrams with no open ends. This approach has been elaborated by Yutsis and his collaborators [29] to such an extent that actual algebraic analyses can be carried out graphically: they have been exposed in the comprehensive handbook by Varshalovich and collaborators [30]. In this approach $6j$ symbols are represented by six branches whose ends terminate at four junctions of three branches: ostensibly, when branches are drawn as segments in turn being viewed as parts of infinite lines in the projective plane, they are perceived as complete quadrangles, i.e. four points and the six lines joining them. The $6j$ (Wigner's) symbols and the closely related Racah coefficients are at the foundations of the whole building. As we will resume in Sect. 2.2, their main properties – a pentagonal and a hexagonal relationship to be encountered e.g. in Reference [14] – permitted us to construct the useful icosahedral "abacus" to deal with $9j$ symbols, and a morphogenetic scheme [4] to generate explicit expressions for the higher $3nj$ symbols, $n > 3$ (and therefore the manipulation of the often complicated angular momentum coupling schemes). We remark that the content of such diagrams is combinatorial and of a topological significance.

(ii) The second approach that had been exploited in important semiclassical investigations was initiated by Racah in 1942 [31], Wigner in 1959 [32], and Ponzano and Regge in 1968 [26] (see also related references given in the bibliography providing a modern view) – that revealed the possibility of substantiating graphs with a metrical flavor, interpreting the triads as triangular faces of polyhedra in a three-dimensional space, or even of higher dimensionality polytopes (or simplices): specifically the $6j$ symbol is represented as a generically irregular tetrahedron. Indeed, if one looks at the two diagrams in Fig. 1, he cannot avoid noticing convenience to recognize the similarity and also a subtle significant difference: if the quadrangle in Fig. 1 is regarded as a planar projection of a tetrahedron, then its known Euler duality is recognised with respect to the exchange of roles of the four faces and of the four points, in this case representing respectively the triads and the incidences of three of the six lines, namely the six js of the symbol.

(iii) The projective approach was suggested by U. Fano and Racah [17] permitting them to demonstrate the Desarguesian nature of angular-momentum theory. It has been invariably followed by all those who studied the relationship between angular momentum theory and projective geometry (Robinson [18], Biedenharn–Louck [19], Judd [33] and Labarthe [20]). This was illustrated by drawing correspondences between relations satisfied by the nj symbols and various collinearity properties of the appropriate diagrams; in this third labeling scheme a triad is represented by three points on a line, and the $6j$ symbol appears as a complete quadrilateral, related to the complete quadrangle by the point-line duality in the projective plane (Fig. 1).

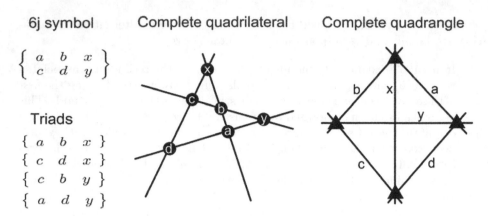

6j symbol

$$\left\{ \begin{array}{ccc} a & b & x \\ c & d & y \end{array} \right\}$$

Complete quadrilateral   **Complete quadrangle**

Triads

$$\{ \; a \quad b \quad x \; \}$$
$$\{ \; c \quad d \quad x \; \}$$
$$\{ \; c \quad b \quad y \; \}$$
$$\{ \; a \quad d \quad y \; \}$$

Fig. 1. The 6-j and its four triads as the complete quadrilateral (left) and the complete quadrangle (right), the 4_6 and 6_4 configurations of incidence geometry respectively, where M_n is a graph with M lines and n points. The association of spins to points and lines to triads originate from the projective geometry construction of Fano and Racah. The dual one (spins as line and their incidence points as triads) is emphasized in our work in order to establish a connection between the commonly used Yutsis graph and the physicist's perception of spins as (pseudo)-vectors. The latter view given to the $6j$ symbol is that of a generically non-planar tetrahedron, and is exploited in semiclassical asymptotics and in quantum gravity models.

A fourth representation, as an incidence graph combining the previous ones, is a novelty of recent papers and needs the geometrical detour sketched in [Aquilanti and Marzuoli, to be published [21] and Robenilson *et al.* to be published [22]]. For popular introductions to concepts of projective and finite geometries, see [34] and [35].

2.2   Incidence Structure and Projective Features

Discussions of aspects of angular momentum algebra and spin networks using any of the above alternatives should benefit of advantages of all of them: beneficial is also their ample interchangeability, due to appropriately individuated and employed duality properties. However, from the viewpoint of the remarkable simplicity of the relationships with the foundational aspects of geometry, preference to (i) is supported by strong evidence and its consistent use is followed in this paper. The purpose of these sections is to establish how extensive is the conceptual overlap of the two apparently separate subjects belonging to physical and mathematical areas.

Reference can be specifically made to Coxeter [25], Projective Geometry, Second Edition (the initial discussion is on p.15). In the axiomatic construction of projective geometry, the first two axioms are written using the unifying concept of incidence when the only primitive objects of the theory, "points" and "lines", meet:

Axiom I - *Any two distinct points are incident with just one line (Coxeter 2.13).*

Axiom II - *Any two lines are incident with at least one point (Coxeter 3.11).*

Any asymmetry between points and lines is effectively eliminated by their duality valid in projective geometry, [Coxeter, p. 25], but not in affine or Euclidean geometries. Denoting configurations consisting of points and lines by the number of points and the numbers of lines as subscripts, a triangle – three points and three lines – is 3_3 and therefore self-dual: however, it is too simple for a geometry with no angles nor lengths, and to proceed one assumes the existence of the "complete quadrangle".

Axiom III - *There exist four points of which no three are collinear (Coxeter 3.12).*

This is the simplest configuration allowing assignment of coordinates in the plane as emphasized by M. Hall [36], in 1943, the four points being denoted *e.g.* as (00), (01), (10), (11), using only the simplest of fields, F_2. The configuration of the complete quadrangle of Yutsis diagrams is 4_6, and that of its dual, the complete quadrilateral, is 6_4. Figure 1, establishes the connection between the $6j$ symbol and its two associated graphs.

An obvious labelling of the complete quadrangle and of the complete quadrilateral in terms of the corresponding labelling of the $6j$ symbols, is from now on used here, employing small case letters in italics and faithful for exhibiting the 24 permutational symmetries of the $6j$ symbols.

In the notation that we use an apparently slight lexicographically asymmetric notation still persists, originating from the fact that a basic facet of $6j$ symbols is their being elements of orthogonal matrices where indices x and y mark either rows or columns. Their ranges, given that entries must obey triangular relationships $|a - b| \leq c \leq a + b$ and cyclically, and therefore, by a theorem by Robert G. Littlejohn (private communication), are dictated by Regge symmetries, see [8,9]. Orthogonality and normalization have thus to be endorsed as a property of $6j$ symbols (this is the first of the three defining properties of the $6j$ symbols, according to [17]):

Property A of $6j$ symbols

Orthonormality:

$$\sum_x (2x + 1) \begin{Bmatrix} a & b & x \\ c & d & y \end{Bmatrix} \begin{Bmatrix} c & d & x \\ a & b & y' \end{Bmatrix} = \frac{\delta_{yy'}}{2y' + 1} \{a\ d\ y\}\{b\ c\ y\} \tag{1}$$

where $\{\cdots\}$ means that the three entries obey the triangular relationship.

2.3   The Quadrangle and Its Diagonal Triangle and the Seven-Spin Network

The other defining property of $6j$ symbols, according to Fano and Racah [17], Robinson [18], Biedenharn and Louck [19], is known as the Racah sum rule (1942) [31], the equation below being labeled according to our conventions:

Property B of $6j$ symbols

Additivity: (Racah sum rule)

$$\sum_{x}(-1)^{z+y+x}(2x+1)\begin{Bmatrix} a & b & x \\ c & d & z \end{Bmatrix}\begin{Bmatrix} c & d & x \\ b & a & y \end{Bmatrix} = \begin{Bmatrix} c & a & y \\ d & b & z \end{Bmatrix} \tag{2}$$

which, using Eqs. (1) and (2) becomes ([30], Eq. (23) p. 467)

$$\sum_{xy}(-1)^{x+y+z}(2x+1)(2y+1)\begin{Bmatrix} a & b & x \\ c & d & z \end{Bmatrix}\begin{Bmatrix} a & c & y \\ d & b & x \end{Bmatrix}\begin{Bmatrix} a & d & z' \\ b & c & y \end{Bmatrix}$$

$$= \frac{\delta_{zz'}}{2z+1}\{a\ d\ z\}\{b\ c\ z\} \tag{3}$$

The hexagonal illustration of the formulas, which involve seven angular momenta, is for example in Fig. 1 in [11]. (see also Sect. 3.1 of [14]).

Equation (3) shows the symmetry of the seven angular momenta and six triads which are seen to be involved. Their representation on the Fano plane is given in Fig. 2. Appropriate then appears to establish the connection to the projective geometric construction, which introduces the following axiom, referred to the so-called "diagonal triangle of a quadrangle":

Axiom IV - *The three diagonal points of a quadrangle are never collinear (Coxeter 2.17), also known as Gino Fano's axiom.*

Figure 2 illustrates in two equivalent ways the important association of a unique diagonal triangle to any quadrangle. The new three diagonal points are added at the incidence of the three couples of opposite lines, but Fano axiom forbids their collinearity: in the angular momentum interpretation, we admit the new triads acz and bdz. The z segment joins them, but doesn't correspond to any xyz triad! The resulting configuration contains the graphical properties of Racah's sum rule, since the addition of z results in the appearance of the two quadrangles corresponding to the two $6j$'s in the right-hand side of Property B. Such a configuration is often referred to as *anti*-Fano (at times also *quasi*-Fano): this apparently contradictory terminology is due to the fact that, admitting the extension of the z segment as a full line including also a new unphysical triad xyz, would generate a special configuration 7_7 violating G.Fano's axiom: but that one is indeed the configuration appearing in finite geometries initiated by him, who studied the effect of such violation. These finite geometries

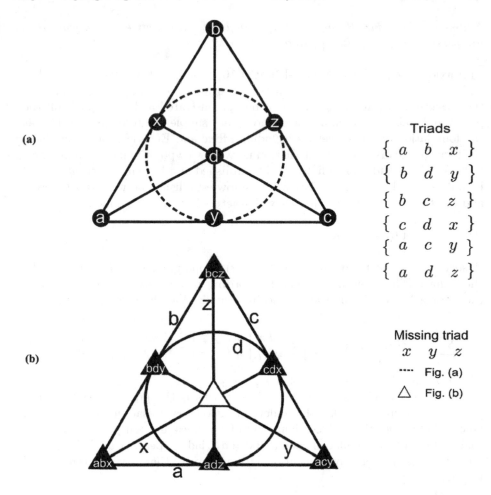

Triads

$$\{\, a \quad b \quad x \,\}$$
$$\{\, b \quad d \quad y \,\}$$
$$\{\, b \quad c \quad z \,\}$$
$$\{\, c \quad d \quad x \,\}$$
$$\{\, a \quad c \quad y \,\}$$
$$\{\, a \quad d \quad z \,\}$$

Missing triad

$$x \quad y \quad z$$

···· Fig. (a)

△ Fig. (b)

Fig. 2. The seven spin network. The Racah sum rule involving seven spins and six triads represented in the Fano plane in the dual views of the Fig. 1. In the upper one, the missing triad is a broken circle (a projective line). In the lower one, the missing triad is an empty triangle (a projective point).

had a wealth of developments and important applications in expanding fields. Not long ago Roger Penrose noted, in his influential book "The Road to Reality" [27]: "Although there is a considerable elegance to these geometric and algebraic structures, there seems to be little obvious contact with the workings of the physical world". However, this work conveys the message that this may not be so, and, as far as for computational science and block design are concerned, see [23, 24] and below.

Axiom V - *Desargues theorem: if two triangles are perspective from a point they are perspective from a line (Coxeter 2.32, p.19).*

This axiom is equivalent to the celebrated 10_{10} configuration of general validity and basic to planar projective geometry: it regards the impossibility of proving the Desargues theorem, that needs to be demonstrated at least a point off the plane, and so is a theorem in space, and a very simple one to be proven. It is at the foundation of all geometries including affine and Euclidean ones, and is at the focus of the discussion from the spin network viewpoint in papers in preparation [Aquilanti and Marzuoli [21], to be published and Robenilson *et. al* [22], to be published]. The striking relationship is now established between this theorem and the third and last basic property characterizing the 6 *j* symbol.

Property C of 6j symbols

Associativity: The third (and last defining) property of the 6j symbol is the Biedenharn-Elliot identity, relabeled according to our convention. It provides the associative relationship schematically visualized as the pentagons in [14].

$$\sum_x (-1)^{\phi}(2x+1) \begin{Bmatrix} a & b & x \\ c & d & p \end{Bmatrix} \begin{Bmatrix} c & d & x \\ e & f & q \end{Bmatrix} \begin{Bmatrix} e & f & x \\ b & a & r \end{Bmatrix} = \begin{Bmatrix} p & q & r \\ f & b & c \end{Bmatrix} \begin{Bmatrix} p & q & r \\ e & a & d \end{Bmatrix} \quad (4)$$

where $\phi = a + b + c + d + e + f + p + q + r + x$.

As we will further see elsewhere [21, 22], this is the property that, from a projective viewpoint, makes the underlying geometry Desarguesian [17,18,33]. One can see there the 10_{10} configuration of ten lines and ten points, observing that in Eq. (4) ten angular momenta and ten triads appear. A unified vision of *e.g.* a ten-spin network involving also relationship with the 9j symbol is discussed in [21,22].

3   Remarks and Two Protocols

A further important question demanding investigation concerns symmetry, The seven objects displayed in the complete Fano plane enjoy 168 permutational symmetries. The 6j symbols have 144 (permutational plus the intriguing Regge ones). Relationships are pointed out in [18–20]. However, detailed connections need to be established. Effectively in the construction of Fig. 1 one adds either to the six lines of a quadrangle an extra one or to the four points another two. Dually for the quadrilateral representation in the first case one has a missing line. This physically corresponds to the missing triad (x,y,z). correlating angular momentum theory to an *anti*-Fano geometry. For similar matters regarding more specifically the two 9j symbols and the corresponding ten-spin networks, see [21, 22].

Another remark will need to be considered regarding the next (and last) axiom on projectivity. Axiom VI (2.18 in Coxeter, pp. 31 and 93) appears as of no relevance here, and is actually superfluous in many geometries.

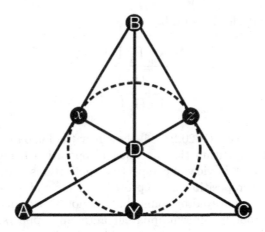

Fig. 3. The quantum-classical protocol. Points and lines are representative in the Fano graph for angular momenta and triads respectively, as in Fig. 2 and as in the extensively developed semiclassical angular momentum theory, where a distinction is made between large and small "spins", denoted as large or small case letters, respectively (scales for large and small are *e.g.* in units of Planck's constant h in atomic, molecular, nuclear physics or dependent on Immirzi parameter and cosmological constant in quantum gravity loop-type theories). The figure shows that a consistent picture can be constructed where six lines of the Fano graph are representative of admissible triads, yet the circle corresponding to the coupling is not, being a triad having two small spins, x and y, and a larger one, Z, violating the triangular rule for addition of angular momentum (two "small" spins cannot add to a "large" one).

Another remark is in order and concerns the final observation in Judd's 1984 paper that no example is known in angular momentum theory about role of Pappus theorem, which is arguably related to commutativity [33]. Judd's conjecture is that its non-occurrence is due to noncommutativity in quantum mechanics and as far as we know it has not been proven nor confuted yet.

The Fano configuration 7_7 and the Fano plane where the simplest of geometries can be constructed coordinating it by the finite field F_2 are nowadays very much investigated, for example as abacus of octonion calculus (See *e.g.* [23, 24]). We recall the definition of F_2 as an arithmetic endowed of two numbers, say 0 and 1, with addition table

+	0	1
0	0	1
1	1	0

and multiplication Pythagorean table

$$
\begin{array}{c||c|c}
\times & 0 & 1 \\
\hline\hline
0 & 0 & 0 \\
\hline
1 & 0 & 1
\end{array}
$$

The quoted references [17–20, 23–25] show how the Fano diagram is an "abacus" for this mod (2) arithmetic, the unusual operation being essentially the addition rule $1 + 1 = 0$. In view of perspective applications to quantum information we envisage the proposition of two protocols, Fano and *anti*-Fano, according to whether the q-bits conveyed by spins abide or not his axiom. Note once again that from this viewpoint a known apparent antinomy appears, the Fano finite geometry is *anti*-Fano, because contradicts his axiom!

Remarkably, we recognize two interesting ways of attribute to the seven-spin networks discussed in this paper a role of providing protocols for quantum computation. They are based on two simple theorems, that we propose in this work, the proofs being obvious and to be better detailed elsewhere:

(a) a strikingly simple finite geometrical "proof" on the boson or fermion nature of particles.
(b) the "proof" that angular momentum algebra is *anti*-Fano. The field F_2, consists of two elements only whose addition and multiplication tables are those of the numbers 0 and 1, but the exception is that, being a number system mod 2, one has $1+1=0$.

Theorem (a) follows from the fact that the property of numbers of being either even or odd generates an F_2 arithmetics, as immediately seen by direct verification, and corresponds to the boson or fermion nature of particles, carrying half-even or half-odd spins, respectively. It establishes what we can call *the Fermion-Boson protocol* for quantum computer simulations. Hardware would indeed be perspectively based on materials distinguishing even or odd spins.

Theorem (b) is based on a similar argument, but with subtly different implications. Now we exploit the properties of the correspondence between concurrence with triangularity, whereby encounter as a triad of two small spins and a large one is forbidden. This is also of interest for the semiclassical limits (Fig. 3). It establishes what we can call the *quantum-classical protocol* for quantum computer simulations. Hardware would be indeed perspectively be based on moieties distinguishing large or small angular momenta. Inspiration for the latter experimental configuration may be based on molecular environments, specifically even extremely common magnetic ones, such as O_2 and NO: these moleculas have small spins, possibly also electronic angular momentum, yet at normal temperatures high rotational angular momentum states can be easily populated.

Acknowledgments. Manuela Arruda is grateful to Brazilian CNPq for a post doctoral fellowship to the Perugia University. Vincenzo Aquilanti thanks Brazilian Capes for a Special Visiting Professorship at the Bahia Federal University, and Roger Anderson (Santa Cruz, California), Ana Carla Bitencourt and Mirco Ragni (Feira de Santana, Bahia, Brazil), Robert Littlejohn (Berkeley, California), Cecilia Coletti (Chieti, Italy), Annalisa Marzuoli (Pavia, Italy) and Frederico Prudente (Salvador, Bahia, Brazil) for inspiring and productive collaborations over the years. Vincenzo Aquilanti is grateful to the support of the Italian MIUR through the SIR 2014 Grant RBSI14U3VF.

References

1. Aquilanti, V., Bitencourt, A.P., da Ferreira, S.C., Marzuoli, A., Ragni, M.: Quantum and Semiclassical spin networks: from atomic and molecular physics to quantum computing and gravity. Phys. Scr. **78**, 58103 (2008)
2. Anderson, R., Aquilanti, V., Ferreira, C.S.: Exact computation and large angular momentum asymptotics of 3nj symbols: semiclassical disentangling of spinnetworks. J. Chem. Phys. **129**, 161101 (2008)
3. Aquilanti, V., Bitencourt, A.P., da Ferreira, S.C., Marzuoli, A., Ragni, M.: Combinatorics of angular momentum recoupling theory: spin networks, their asymptotics and applications. Theo. Chem. Acc. **123**, 237–247 (2009)
4. Anderson, R.W., Aquilanti, V., Marzuoli, A.: 3nj morphogenesis and semiclassical disentangling. J. Phys. Chem. **A 113**, 15106–15117 (2009)
5. Ragni, M., Bitencourt, A., da Ferreira, S.C., Aquilanti, V., Anderson, R.W., Littlejohn, R.: Exact computation and asymptotic approximation of 6j symbols. Illustration of their semiclassical limits. Int. J. Quantum Chem. **110**, 731–742 (2010)
6. Bitencourt, A.C.P., Marzuoli, A., Ragni, M., Anderson, R.W., Aquilanti, V.: Exact and asymptotic computations of elementary spin networks: classification of the quantum–classical boundaries. In: Murgante, B., Gervasi, O., Misra, S., Nedjah, N., Rocha, A.M.A.C., Taniar, D., Apduhan, B.O. (eds.) ICCSA 2012, Part I. LNCS, vol. 7333, pp. 723–737. Springer, Heidelberg (2012)
7. Aquilanti, V., Marinelli, D., Marzuoli, A.: Hamiltonian dynamics of a quantum of space: hidden symmetries and spectrum of the volume operator, and discrete orthogonal polynomials. J. Phys. A: Math. Theor. **46**, 175303 (2013)
8. Anderson, R.W., Aquilanti, V., Bitencourt, A.C.P., Marinelli, D., Ragni, M.: The screen representation of spin networks: 2D recurrence, eigenvalue equation for 6j symbols, geometric interpretation and hamiltonian dynamics. In: Murgante, B., Misra, S., Carlini, M., Torre, C.M., Nguyen, H.-Q., Taniar, D., Apduhan, B.O., Gervasi, O. (eds.) ICCSA 2013, Part II. LNCS, vol. 7972, pp. 46–59. Springer, Heidelberg (2013)
9. Ragni, M., Littlejohn, R.G., Bitencourt, A.C.P., Aquilanti, V., Anderson, R.W.: The screen representation of spin networks: images of 6j symbols and semiclassical features. In: Murgante, B., Misra, S., Carlini, M., Torre, C.M., Nguyen, H.-Q., Taniar, D., Apduhan, B.O., Gervasi, O. (eds.) ICCSA 2013, Part II. LNCS, vol. 7972, pp. 60–72. Springer, Heidelberg (2013)
10. Aquilanti, V., Marinelli, D., Marzuoli, A.: Symmetric coupling of angular momenta, quadratic algebras and discrete polynomials. J. Phys. Conf. Ser. **482**, 012001 (2014)
11. Marinelli, D., Marzuoli, A., Aquilanti, V., Anderson, R.W., Bitencourt, A.C.P., Ragni, M.: Symmetric angular momentum coupling, the quantum volume operator and the 7-spin network: a computational perspective. In: Murgante, B., et al. (eds.) ICCSA 2014, Part I. LNCS, vol. 8579, pp. 508–521. Springer, Heidelberg (2014)

12. Bitencourt, A.C.P., Ragni, M., Littlejohn, R.G., Anderson, R., Aquilanti, V.: The screen representation of vector coupling coefficients or Wigner 3j symbols: exact computation and illustration of the asymptotic behavior. In: Murgante, B., et al. (eds.) ICCSA 2014, Part I. LNCS, vol. 8579, pp. 468–481. Springer, Heidelberg (2014)
13. Aquilanti, V., Haggard, H., Hedeman, A., Jeevanjee, N., Littlejohn, R.: Semiclassical mechanics of the wigner 6j-symbol. J. Phys. A. **45**, 065209 (2012)
14. Aquilanti, V., Coletti, C.: 3nj-symbols and harmonic superposition coefficients: an icosahedral abacus. Chem. Phys. Lett. **344**, 601–611 (2001)
15. Biedenharn, L.C., Louck, J.D.: The Racah-Wigner Algebra in Quantum Theory. Encyclopedia of Mathematics and its Applications, 1st edn, pp. 353–369. Cambridge University Press, Cambridge (1981). Chapter 5.8
16. Anderson, R.W., Aquilanti, V.: The discrete representation correspondence between quantum and classical spatial distributions of angular momentum vectors. J. Chem. Phys. **124**, 214104 (2006)
17. Fano, U., Racah, G.: Irreducible Tensorial Sets. Academic Press, New York (1959)
18. de Robinson, B.G.: Group representations and geometry. J. Math. Phys. **11**(12), 3428–3432 (1970)
19. Biedenharn, L.C., Louck, J.D.: The racah-wigner algebra in quantum theory. In: Rota, G.C. (ed.) Encyclopedia of Mathematics and its Applications. Wesley, Reading, MA (1981)
20. Labarthe, J.-J.: The hidden angular momenta for the coupling-recoupling coefficients of SU(2). J. Phys. A. **33**, 763 (2000)
21. Aquilanti, V., Marzuoli, A.: Desargues Spin Networks and Their Regge-Regularized Geometric Realization. To be published
22. Santos, R.F., Bitencourt, A.C.P., Ragni, M., Prudente, F.V., Coletti, C., Marzuoli, A., Aquilanti, V.: Addition of Four Angular Momenta and Alternative 9j Symbols: Coupling Diagrams and the Ten-Spin Networks. To be published
23. Rau, A.R.P.: Mapping two-qubit operators onto projective geometries. Phys. Rev. A. **79**, 42323 (2009)
24. Rau, A.R.P.: Algebraic characterization of X-states in quantum information. J. Phys. A. **42**, 412002 (2009)
25. Coxeter, H.S.M.: Projective Geometry. Springer-Verlag, Heidelberg (1974)
26. Ponzano, G., Regge, T.: Semiclassical Limit of Racah Coefficients in Spectroscopy and Group Theoretical Methods in Physics, F. Block (1968)
27. Penrose, R.: The Road to Reality: a Complete Guide to the Laws of the Universe. Randorn House Group, London (2004)
28. Edmonds, A.: Angular Momentum in Quantum Mechanics. Princeton University Press, Princeton, New Jersey (1960)
29. Yutsis, A., Levinson, I., Vanagas, V.: Mathematical Apparatus of the Theory of Angular Momentum, Israel Program for Scientific Translation (1962)
30. Varshalovich, D., Moskalev, A., Khersonskii, V.: Quantum Theory of Angular Momentum. World Scientific, Singapore (1988)
31. Racah, G.: Theory of Complex Spectra. II. Phys. Review. **63**, 438–462 (1942)
32. Wigner, E.: Group Theory: And its Application to the Quantum Mechanics of Atomic Spectra. Academic Press, New York (1959)
33. Judd, B.: Angular-momentum theory and projective geometry. Found. Phys. **13**, 51–59 (1983)
34. Hilbert, D., Cohn-Vossen, S.: Anschauliche Geometrie. Published in English under the title Geometry and the Imagination. AMS Chelsea, New York (1952)
35. Sawyer, W.W.: Prelude to Mathematics. Penguin, London (1955)
36. Hall, M.: Projective planes. Trans. Amer. Math. Soc. **54**, 229–277 (1943)

Mobile Device Access to Collaborative Distributed Repositories of Chemistry Learning Objects

Sergio Tasso[1(✉)], Simonetta Pallottelli[1], and Antonio Laganà[2]

[1] Department of Mathematics and Computer Science, University of Perugia,
via Vanvitelli, 1, 06123 Perugia, Italy
{simonetta.pallottelli,sergio.tasso}@unipg.it
[2] Department of Chemistry, University of Perugia,
via Elce di Sotto, 8, 06123 Perugia, Italy
lagana05@gmail.com

Abstract. The educational open contents are having a very strong impact in research and training activities. In this paper we discuss the use of Internet to supply educational contents in open and free formats from distributed repositories containing educational and scientific resources. Starting from an existing federation of distributed repositories (G-Lorep) [1] and working on parameters defined by apps, we extended their functionalities to allow mobile devices access. The structure of the access, search, selection, download, use/reuse functions needed to manage learning objects on distributed repositories have also been revisited for adaptation to mobile use.

Keywords: Learning object · Repository · Mobile · App · E-content

1 Introduction

The paradigm represented by Learning Objects (LO)s is a new established e-learning pillar. LOs are, in fact, small, modular and indexable e-learning elements, having their own meaning and internal consistency in the respect of appropriate standards. The growing interest in LOs lies in the fact that they are semantically consistent objects and offer an adaptive way of creating ad hoc courseware. Moreover, the possibility of preserving and indexing the LOs within collaborative and distributed repositories, as is G-Lorep, makes them shareable and reusable [2–5] and gives them invaluable advantages over more popular e-learning tools.

G-Lorep can, in fact, establish relations, recalls and references between the different LOs stored in the repositories, and can build, therefore, a network of concepts. As a matter of fact, G-Lorep LOs are open educational resources, in the sense that they include not only contents and collections of contents, but also the tools needed for their collaborative management and fruition. In this respect, our most recent work has focused on coupling the versatility of G-Lorep with the ubiquitous connectivity of current mobile devices, like smartphones and tablets, to significantly enhance the potentialities of LOs [6].

O. Gervasi et al. (Eds.): ICCSA 2016, Part I, LNCS 9786, pp. 443–454, 2016.
DOI: 10.1007/978-3-319-42085-1_34

In recent years, in fact, the use of mobile devices has increased in many areas, and the devices themselves have technologically evolved into real computers becoming suitable for playing as innovative educational and training tools. This has attracted significant work from our side [7–12] most of which has been channeled through the activities of the European Chemistry Thematic Network (ECTN) [13, 14] that in the past 20 years has built through several European projects the main frame for educational activities of the related Association of Chemistry Higher education institutions. The two main educational tools developed for the activities of ECTN and its Association are EOL [15] and G-Lorep. The first, EOL, supports EChemTest® [16], a set of libraries of questions and answers on chemistry knowledge of different levels of competence going from Schools (General chemistry 1), to access to University (General chemistry 2), Chemistry Bachelor in Analytical, Biological, Inorganic, Organic and Physical (level 3), Master in Computational chemistry, Organic synthesis and Cultural heritage (level 4) leveraging on a network of National Test Centers (NTC) s, Accredited Test Sites (ATS)s and Agencies taking care of the administration of Self Evaluations Sessions (SES)s e-tests and offering Higher Education Institutions the possibility of a more objective evaluation of the students. The second, G-Lorep, supports a set of LOs concerned with Molecular Sciences and technologies composed and improved in a collaboratively fashion by the teachers of the ECTN partner institutions according to specific correspondence with educational credits an leveraging on a network of European distributed repositories offering an innovative ground for curricular education and life-long learning programs.

The objective of this paper is to present the evolution undergone by G-Lorep towards the creation of a Mobile LO Catalogue (MoLOC) system aimed at facilitating and enhancing interactive continuous education.

Accordingly:

in Sect. 2 the currently adopted environment G-Lorep is illustrated;

in Sect. 3 the main adaptations introduced by MoLOC in G-Lorep are singled out;

in Sect. 4 the LOs structuring for mobile is analyzed;

in Sect. 5 the LOs passage is described;

in Sect. 6 a use case for Chemistry LOs searching via app is discussed;

in Sect. 7 some conclusions are drawn and some guidelines for future work are given.

2 The Currently Adopted Environment: G-Lorep

G-Lorep (Grid Learning Object Repository) [7–12] is a project created as a collaborative endeavor by the Mathematics and Computer Science Department and the Department of Chemistry, Biology and Biotechnology of the University of Perugia. The goal was the implementation of a federation of repositories, in which the educational content is embodied in LOs to support e-learning activities like those of ECTN.

The federation of G-Lorep was designed (as illustrated in Fig. 1) with the purpose of:

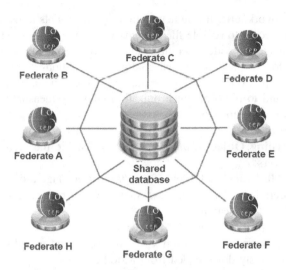

Fig. 1. *The G-Lorep network architecture*

- making available educational and scientific contents suitable for teaching and learning of a given community at large;
- allowing dynamical improvement of the available materials by properly storing and cataloguing revised versions.

The adopted platform consists of a hybrid architecture in which no server has privileges or specific abilities. Its basic component is a shared database accessible by all servers that stores a map of the federation and an index of its contents (though it is not mandatory that it is up for carrying out the most elementary operations). In this platform the servers are peer entities with a primary data source for external data.

The database contains the identifiers and addresses of all servers as well as the information about their status. It also contains the metadata of all the objects belonging to the federation that can be directly accessed and updated by all servers. When all servers and the shared database contain the same knowledge basis of the federation, they are considered as synchronized. G-Lorep adopts a distributed search that extends over the whole federation repositories and operates through the use of keywords controlling the management. This makes use of a taxonomy that results from a specialized Thesaurus whose categories are built out from the Dewey Decimal Classification (DDC) pattern [17, 18]. DDC is an updated international standard, a multi-discipline classification that covers all the relevant fields, allows to associate a description label to the subject numeric code and defines classes and subclasses on different specialization levels.

The G-Lorep communication paradigm is a Client/server model.

Each federated server hosting a CMS (Drupal [19]) provides the repository management with activities at back-end-level (such as: backup and access control) and, through a web portal, with different services for clients (like up/download LOs, file management) at front-end level.

The internet network, after the authentication, allows clients to use all the available facilities up to the access to remote file systems where the LOs are stored.

In order to manage the wide federation along with its contents, some new modules have been added to the standard Drupal configuration; among them a crucial role is played by **CollabRep** (Collaborative Repositories) that can be used to create, join and quit a federation, and to perform synchronization recovery measures in case of communication issues during updates.

Other modules are:

Linkable Learning Object (*LinkableObject*) to enable the creation of LOs and their upload to servers;

Search Linkable Learning Object (*SearchLO*) to manage a distributed search of LOs on wide federations. Users can use most of the metadata recommended by IEEE LOM to refine their search.

Taxonomy Assistant to manage LO classification.

Moodledata to download files from a Moodle LMS and to upload on a G-Lorep server the files previously downloaded from Moodle.

3 The MoLOC Adaptations of G-Lorep

In order to adapt G-Lorep to mobile access the access interface was modified so as to allow requests from related devices. This was achieved by adopting IT technologies allowing the system to adapt to the different size of the screens and to the different operating systems as sketched in Fig. 2.

smartphone
client

G-Lorep unipg
federated server

Personal
Computer
client

Fig. 2. *Access to G-Lorep services*

The related mobile application of MoLOC, called AppG-Lorep (see its schema in Fig. 3), allows access to the platform through the G-Lorep credentials consisting of username and password.

These credentials are already defined in the G-Lorep system because users are registered[1]. Accordingly, in the present version of AppG-Lorep, the login operation

---

[1] The user must already be registered at least in one of the G-Lorep repositories. The registration procedure is only available on a PC client at the moment.

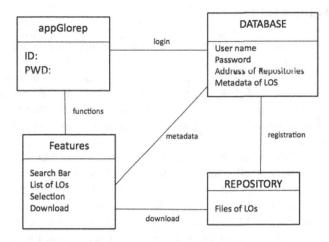

Fig. 3. *The AppG-Lorep schema*

consists only of a check on the compliance of the credentials with the content of the User table and access is permitted only to previously registered users.

In MoLOC the connection to the site database is established using an Async Task, that allows the interfacing to the database each time it is invoked.

The control is via a PHP script that returns two types of values for the correct credentials. When the credentials are incorrect, the user will have to repeat the login operation whereas when they are correct the user has access to the G-Lorep home, appropriately generated for the mobile device.

The home of G-Lorep shows the list of the latest LOs included in the repository. In this case the ListView function, which generates the list, displays the title, the description, and the author of each LO.

The metadata that is exchanged between the AppG-Lorep application and the database are in JSON format (sometimes JavaScript Object Notation) [20]. Each time the client requests access to the repository server the database responds with the data to be sent.

It is always possible to make from the G-Lorep home a search that will be carried through the SearchBar function. SearchBar allows the use of filters for the search.

Once identified and selected the LO, the application allows the download of the LO itself.

The download is only possible with certain LOs as specified in an ad hoc field of Requirements. After the download of the LO from a repository of the G-Lorep federation, the app can proceed to subsequent requests.

AppG-Lorep is available on mobile Android systems.

4 The LOs for MoLOC

The extensive functionality of the mobile devices, the optimum conditions of their networks, their widespread access to the web services have significantly improved the traceability of LOs. Also the quality of the audio and video players of mobiles (thanks also to the increasing size of their screens) have made popular their use for enjoying multimedia contents. For this reason, the step of adding the features needed to access G-Lorep from mobile environments represents a significant leap forward in the use of LOs in education and training.

Accordingly, we stress here the fact that when designing a LO one has from the very beginning keep in mind whether it will be accessed from a mobile or not. The IEEE LOM schema adopted by G-Lorep, in fact, does not directly support the description of resources with respect to their mobile nature. Therefore, if a mobile access is going to be used, LO content structures (e.g. audio, video, text, 3D virtual environments, augmented reality) and content types (e.g. guides, tutorials, test data, etc.) must contain specifications and delivery features of the various mobile devices

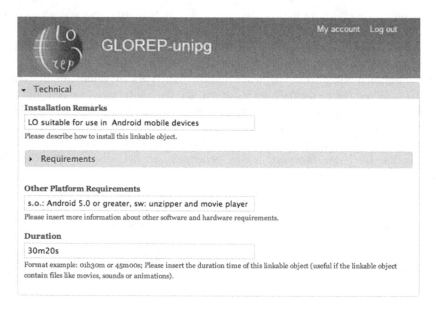

Fig. 4a. *LO typifying for mobile environments*

(e.g. smartphone, tablet, PDA, ect). This means that it is necessary to prepare in advance the corresponding specific settings for the LO access and visualization.

This can be performed separately on the G-Lorep *Create Linkable Object* form in which one can specify some metadata to characterize those objects that are meant to be used on mobile devices. In particular the LO owner can insert (see Fig. 4a):

- in the *Installation Remarks* field of the *Technical* Metadata Session the desired functions of the operating systems and what is needed to install this linkable object
- in the *Other Platform Requirements* field of the *Requirements* Sub-session more information about other software and hardware requirements
- and in the *Duration* field the duration time of this LO. This is particularly useful if the LO contains files like movies, sounds or animations.

In addition, the holder can also provide details about the *Educational* session (see Fig. 4b), and in particular he/she can specify some comments on how the linkable object is to be used:

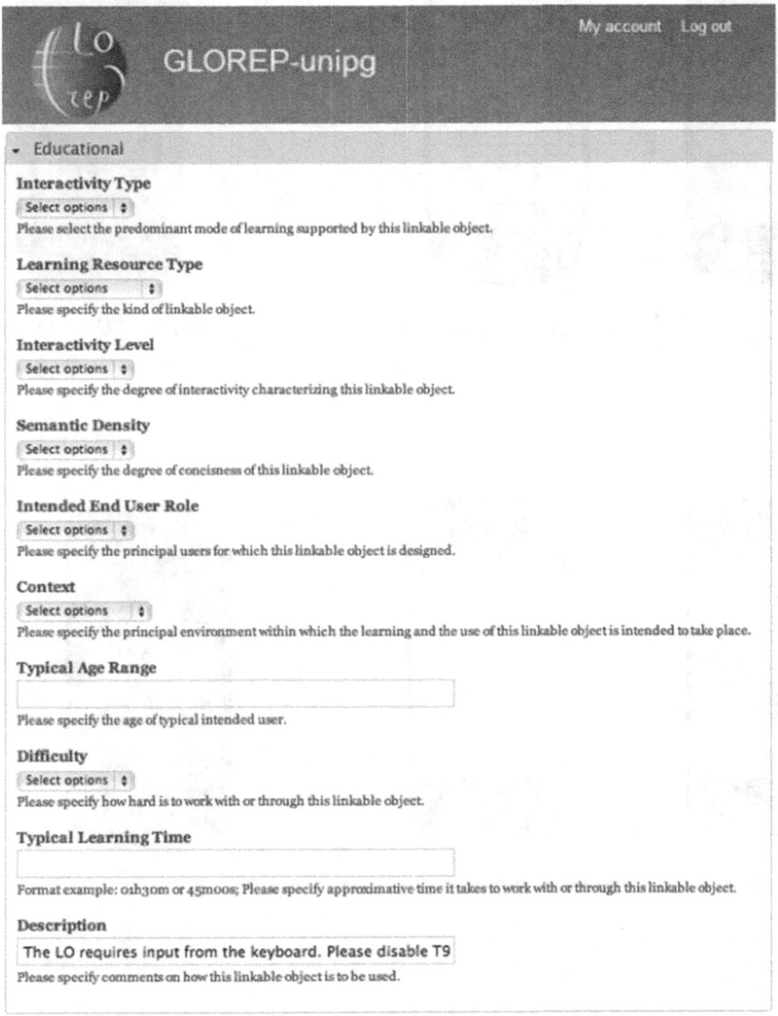

Fig. 4b. *LO typifying for mobile environments*

- in the *Interactivity Type* field (*Active, Expositive, Mixed*)
- in the *Learning Resource Type* field (*Exercise, Simulation, Questionnaire, Diagramm, Figure, Graph, Index, Slide, Table, Narrative text, Exam, Experiment, Problem Statement, Self Assessment, Lecture*)
- in the *Interactivity Level* field (*Very Low, Low, Medium, High, Very high*)
- and in the *Description* field (*free comment*).

Figure 4b: LO typifying for mobile environments

Fig. 5a. *Login screen*

Fig. 5b. *LOs list screen.*

Fig. 5c. *Download LO*

Fig. 6a. *Search screen*

Fig. 6b. *Search results screen.*

Fig. 6c. *Download LO.*

5   Description of the LO Passing

The application provides an interface suitable for various types of mobile devices. After the user starts the app, the interface connects the smartphone to the G-Lorep server and the login screen shows up (see Fig. 5a). The user must insert a valid login and the related password. When the user is inside the G-Lorep server, a list of recently entered LOs is displayed (Fig. 5b).

The user can select the desired LO from the list shown and require a visualization of its main metadata. Then he/she can choose a fast download (Fig. 5c).

The user can also fill a search field by combining one or more terms (Fig. 6a). The query is formalized and sent to the G-Lorep server. The result of the search is displayed in the app interface (Fig. 6b) sorted by relevance. Within the app, users can choose the LO by looking at the details of its metadata and finally download the chosen ones into the mobile device (Fig. 6c) and open them.

6   A Use Case: Searching Chemistry LOs in G-Lorep via App

As a typical use case of the above presented mobile version of G-Lorep, we consider here the *"Nanotube experiment"* a typical LO of the "Theory" module of the "Mechanisms and Dynamics of Chemical Reactions" course of the Chemistry Euromaster of the University of Perugia. In particular, the *"Nanotube experiment"* LO is associated with Sect. 5 "Large systems studies using classical mechanics" of the book "Chemical reactions: basic theory and computing" in preparation for the Springer series "Theoretical Chemistry and Computational Modeling". In the mentioned section the assemblage of Force Fields for systems made of a large number of atoms is illustrated and a typical application to the utilization of a popular single wall Carbon nanotube computational experiment is given. The related three steps (search result screen, LO download and LO Multimedia) of AppG-Lorep involved in the associated flow of a fluid (a ion in a water solution) inside the mentioned nanotube computational experiment are shown in Fig. 7.

This G-Lorep Multimedia LO previously created and uploaded (as detailed in Sect. 4[2]) by a registered user, can be viewed on any mobile media equipped with the Android operating system. In fact, even if this LO is stored in the G-Lorep server of the University of Perugia, a mobile user looking for educational scientific material related to nanotubes might from an Android device using the AppG-Lorep enter in the search form the keyword *nanotube*. The App searches in the database all the LOs having in the title and/or in the description the typed keyword. The search result offers the LO of the author Lagana' (Fig. 7a). At this point the user can select the *"Nanotube experiment"* LO and automatically start its download into his/her mobile device (Fig. 7b). Then, thanks to the video and media player invoked by the LO metadata, the user can visualize the video of the experiment as shown in Fig. 7c.

---

[2] For example this LO, which is overly interactive and detailed for mobile devices, could be changed to a more user-friendly movie.

Fig. 7a. *Search results screen.*

Fig. 7b. *Download LO.*

Fig. 7c. *multimedia LO.*

7 Conclusion and Goals for Future Work

In the present paper we discuss how we tackled the problem of adapting G-Lorep LOs to become MoLOC by exploiting the rapidly evolving mobile technologies. To overtake the limits of the original LOs for mobile e-learning platforms we have built a specific AppG-Lorep for this new user requirement. The AppG-Lorep allows access from Android mobile devices to the G-Lorep federation of distributed repositories of LOs. As a use case we have considered the chemical system formed by a single walled carbon nanotube and a ionic solution flowing through it. Although the method proposed for solving the problem is not yet general enough to make a straightforward and ubiquitous highly collaborative use of G-Lorep (due to the difficulty and the complexity of fully aligning highly interactive distributed systems) the proposed solution is a true leap forward in educational activities. This finds a clear confirmation on the present commitment of the largest European Chemistry educational community, ECTN, to adopt such evolution as a standard for its activities and as the reference instrument for supporting the internationalization of the educational system in Kazan. Moreover, along the same line two new proposals for the Moldavia educational system (CPAS: Cooperation and partnership in applied sciences for competences required by the labour market [21]) and the training and dissemination in the H2020 proposal for a European Research infrastructure for Molecular Sciences and Technologies (SUMO-CHEM: Supporting Research in Computational and Experimental Chemistry via Research Infrastructure [22]). As future work we are also planning to extend the app to other platforms (e.g. IOS) and to accompany LOs of a QR-code for a quick identification and tightly packing of the LOs through smart mobile devices.

Acknowledgements. The authors acknowledge ECTN (VEC standing committee) and the EC2E2 N 2 LLP project for stimulating debates and providing partial financial support. Thanks are due also to EGI and IGI and the related COMPCHEM VO for the use of Grid resources.

References

1. Official site G-Lorep (Online). http://glorep.unipg.it. January, 2016
2. Wiley, D.A.: The Learning Objects Literature, in Handbook of Research on Educational Communications and Technology, third edition (2008)
3. Ochoa, X., Duval, E.: Uantitative analysis of learning object repositories. IEEE Trans. Learn. Technol. 2(3), 226–238 (2009)
4. Never, F., Duval, E.: Reusable learning objects: a survey of LOM-based repositories. In: Proceedings of the Tenth ACM International Conference On Multimedia, pp. 291–294. ACM, New York, NY, USA ©(2002)
5. Brooks, C., McCalla, G.: Towards flexible learning object metadata. Int. J. Continuing Eng. Educ. Life Long Learn. 16(1/2), 50–63 (2006). Inderscience Publishers, ISSN 1560-4624 (Print), 1741-5055 (Online)
6. Tabuenca, B., Drachsler, H., Ternier, S., Specht, M.: OER in the Mobile Era: Content Repositories' Features for Mobile Devices and Future Trends. Europe eleaning papers, [Special issue] Mobile learning (2012)
7. Pallottelli, S., Tasso, S., Pannacci, N., Costantini, A., Lago, N.F.: Distributed and collaborative learning objects repositories on grid networks. In: Taniar, D., Gervasi, O., Murgante, B., Pardede, E., Apduhan, B.O. (eds.) ICCSA 2010, Part IV. LNCS, vol. 6019, pp. 29–40. Springer, Heidelberg (2010)
8. Tasso, S., Pallottelli, S., Bastianini, R., Lagana, A.: Federation of distributed and collaborative repositories and its application on science learning objects. In: Murgante, B., Gervasi, O., Iglesias, A., Taniar, D., Apduhan, B.O. (eds.) ICCSA 2011, Part III. LNCS, vol. 6784, pp. 466–478. Springer, Heidelberg (2011)
9. Tasso, S., Pallottelli, S., Ferroni, M., Bastianini, R., Laganà, A.: Taxonomy management in a federation of distributed repositories: a chemistry use case. In: Murgante, B., Gervasi, O., Misra, S., Nedjah, N., Rocha, A.M.A., Taniar, D., Apduhan, B.O. (eds.) ICCSA 2012, Part I. LNCS, vol. 7333, pp. 358–370. Springer, Heidelberg (2012)
10. Tasso, S., Pallottelli, S., Ciavi, G., Bastianini, R., Laganà, A.: An efficient taxonomy assistant for a federation of science distributed repositories: a chemistry use case. In: Murgante, B., Misra, S., Carlini, M., Torre, C.M., Nguyen, H.-Q., Taniar, D., Apduhan, B. O., Gervasi, O. (eds.) ICCSA 2013, Part I. LNCS, vol. 7971, pp. 96–109. Springer, Heidelberg (2013)
11. Tasso, S., Pallottelli, S., Rui, M., Laganá, A.: Learning objects efficient handling in a federation of science distributed repositories. In: Misra, S., Rocha, A.M.A., Torre, C., Rocha, J.G., Falcão, M.I., Taniar, D., Apduhan, B.O., Gervasi, O., Murgante, B. (eds.) ICCSA 2014, Part I. LNCS, vol. 8579, pp. 615–626. Springer, Heidelberg (2014)
12. Pallottelli, S., Tasso, S., Rui, M., Laganà, A., Kozaris, I.: Exchange of learning objects between a learning management system and a federation of science distributed repositories. In: Gervasi, O., Murgante, B., Misra, S., Gavrilova, M.L., Rocha, A.M.A.C., Torre, C., Taniar, D., Apduhan, B.O. (eds.) ICCSA 2015. LNCS, vol. 9156, pp. 371–383. Springer, Heidelberg (2015)
13. ECTN Network: General Chemistry and the Organic Chemistry courses. http://ectn-assoc. cpe.fr/. January 2016
14. Faginas Lago, N., Gervasi, O., Laganà, A., Tasso, S., Varella, E.: Special issue: ECHEMTEST + First Accreditation and Training Event. http://www.hpc.unipg.it/ojs/ index.php/virtlcomm/issue/view/14. January 2016

15. Faginas Lago, N., Gervasi, O., Laganà, A., Tasso, S., Varella, E.: Tools for e-Learning and e-Assessment: Glorep and EOL. http://www.hpc.unipg.it/ojs/index.php/virtlcomm/article/view/101. January 2016
16. EChemTest. Available: http://ectn-assoc.cpe.fr/echemtest/default.htm. January 2016
17. Dewey, (Online). http://www.oclc.org/dewey/. January 2016
18. Mitchell, J.S. et al.: Dewey Decimal Classification and relative index / devised by Melvil Dewey,v. 4, 23, Dublin, Ohio: OCLC (2011)
19. Drupal (online). http://www.drupal.org. January 2016
20. JSON (online). http://www.json.org. January 2016
21. CPAS proposal to EAC/A04/2015, Selection 2016, KA2. Cooperation for innovation and the exchange of good practices - Capacity building in the field of Higher Education
22. SUMO-CHEM proposal to INFRAIA-02-2017: Integrating Activities for Starting Communities – European Research Infrastructures including e-infrastructures

New Approach to Calculate Adiabatic Curves of Bound States and Reactive Scattering in Quantum Chemistry Problems

Fernanda Castelo Branco de Santana[1], Angelo Amâncio Duarte[2],
Mirco Ragni[2(✉)], Ana Carla Peixoto Bitencourt[2],
and Herman Augusto Lepikson[1]

[1] Federal University of Bahia, Salvador, Brazil
fernandacastelobs@gmail.com, herman@ufba.br
[2] State University of Feira de Santana, Feira de Santana, Brazil
{angeloduarte,mirco,ana.bitencourt}@uefs.br

Abstract. This paper presents a new approach to calculate adiabatic curves in quantum chemistry problem. Based on the Shifted Inverse, the proposed approach differs from the others especially because the initial guesses (eigenvectors and offsets) are not arbitrary. Instead, they are chosen in coherence with the Potencial Energy Surface and quantum number. Initial guesses are determined from the knowledge of experts about the problem under study. From the numerical experiments, it was noticed that the new approach has proved to be more efficient than the Divider and Conquer Method in cases where the problem requires that only some eigenvalues of interest need to be calculated, as often occurs in problems of quantum chemistry.

Keywords: Adiabatic curves · Shifted inverse power method · Eigenvalues · Eigenvectors · Inverse iteration

1 Introduction

In the literature, many works are devoted to the study of quantum dynamics of three and four-body problems. See [1] and references therein. Despite this, a robust, fast and general method that permits an accurate evaluation of kinetic and rate constants for reactions that present quantum effects does not exist. To understand the reason for this, two factors must be considered. The first aspect is that quantum effects in nuclear dynamics are generally difficult to be described with a reduced basis set. This implies that a large number of basis functions is necessary to obtain accurate results and large matrix representations are obtained. A direct consequence of this fact is that the solution of the Schrodinger equations is possible only employing supercomputers. The other aspect to be considered is that the calculation of the matrix elements involves integrals of oscillating functions that, in general are difficult to be calculated with accuracy.

An important but demanding advance (by the computational point of view) is the use of DVR-like representations. These type of representations have the important

© Springer International Publishing Switzerland 2016
O. Gervasi et al. (Eds.): ICCSA 2016, Part I, LNCS 9786, pp. 455–469, 2016.
DOI: 10.1007/978-3-319-42085-1_35

characteristic that no integrals involving the potential must be calculated. Moreover, the eigenvectors directly represent the value of the eigenfunction on the grid points used to build the matrix representation.

Some other advances are represented by the strategies for the calculation of eigenvalues and eigenvectors introduced in recent papers [2–9]. With the present work, we improve those advances introducing a general algorithm for the fast computation of adiabatic curves used in bond state calculations and elastic, inelastic and reactive scattering. This algorithm is based on the Shifted Inverse Power Method and on the use, as a starting point for the solution of the eigenproblem for a fixed value of ρ, of the eigenvalues and eigenvectors calculated for $\rho+\Delta\rho$. Also, we apply our method to the Stereodirect Representation obtained with the Hyperquantization algorithm that leads to a tridiagonal matrix representation of the monodimensional quantum hamiltonian. See references [10–12] for more details.

In the next section we present the theoretical background while in Sect. 3 our proposal is detailed. The article continues with Sect. 4 (Simulation Environment) where details about computational language and used hardware are discussed. Our results and conclusions are presented in Sects. 5 and 6, respectively. The bibliography ends the article.

2 Numerical Determination of Eigenvalues and Eigenvectors

Given an $n \times n$ matrix $\mathbf{A}$ and let $n > 1$ be a integer, an n-dimensional nonzero column vector $\mathbf{v}$ is called an *eigenvector* of $\mathbf{A}$ if there is a real number λ such that [13]

$$\mathbf{A}\mathbf{v} = \lambda\mathbf{v} \tag{1}$$

The number λ is called an *eigenvalue* of $\mathbf{A}$. The matrix $\mathbf{A} - \lambda\mathbf{I}$ in Eq. (2) is called characteristic matrix of $\mathbf{A}$

$$A - \lambda I = \begin{pmatrix} a_{11} - \lambda & a_{12} & \cdots & a_{1n} \\ a_{12} & a_{22} - \lambda & \ldots & a_{2n} \\ \vdots & \ldots & \ddots & \vdots \\ a_{n1} & a_{n2} & \cdots & a_{nn} - \lambda \end{pmatrix}, \tag{2}$$

where $\mathbf{I}$ is the $n \times n$ identity matrix and λ is a parameter. The characteristic polynomial of the matrix $\mathbf{A}$ is $p(\lambda) = |\mathbf{A} - \lambda\mathbf{I}|$, a polynomial of λ of degree n and its roots are the eigenvalues.

Numerical methods that perform direct expansion of the determinant for determining the characteristic polynomial of the matrix are inefficient with the exception of the cases where $\mathbf{A}$ is a low-order matrix or $\mathbf{A}$ has many zero elements. For this reason, usually eigenvalues and eigenvectors are found without using the calculation of the determinant. Numerical methods that solve eigensystems without calculating determinants are classified into three groups [14]:

- Methods that determine the characteristic polynomial: After determination of the characteristic polynomial, the eigenvalues are found from numerical methods for determination of polynomials zeros. The Leverrier Method and the Leverrier-Faddeev Method are examples of this group.
- Methods that determine some eigenvalues: They are also known as iterative methods, used when not all eigenvalues are of interest. Examples of this group are the Power Method and the Inverse Power Method.
- Methods that determine all eigenvalues: They are also known as Transformations Methods, methods of this group perform similarity transformation to calculate all eigenvalues. Rutishauser Method (LR) and Francis Method (QR) are examples.

The solution to problems that require calculating eigenvalues and eigenvectors usually takes place by three main steps [15]:

1. Reduction of the matrix to tri-diagonal form, typically using the Householder Rduction.
2. Solution of the real symmetric tri-diagonal eigenproblem.
3. Back transformation to find the eigenvectors for the full problem from the eigenvectors of the tridiagonal problem.

The strategy used by most algorithms most recently created to determine eigenvalues and eigenvectors of a matrix is to push the matrix towards its diagonally through sequences of similarity transformations, that is diagonalize the matrix [16, 17]. Done this, the diagonal elements are the eigenvalues and the columns of the accumulated transformation are the eigenvectors of the given matrix. If the problem to be solved does not require the determination of the eigenvectors, the algorithm ends before reaching the diagonal form of the matrix. In this case it is sufficient to transform the matrix to a triangular shape. A square matrix is called lower (upper) triangular if all the elements above (below) the main diagonal are zero. In this case the eigenvalues are directly the values of the main diagonal.

The chemical problems that are expected to extract the eigenvalues and eigenvectors presented in this paper are represented computationally by real symmetric matrices. Routines for the calculation of eigenvalues and eigenvectors of a real symmetric tridiagonal matrix are available (frequently used as a Black Box) in many package distributed under different type of license. In Lapack, for example, the solution is found via one of the following methods [15]:

- Bisection for the eigenvalues and Inverse Iteration for the eigenvectors
- QR algorithm
- Divide & Conquer method
- Multiple Relatively Robust Representations (MR^3 algorithm)

Table 1 lists these methods and their corresponding algorithmic complexities.

The approach to the determination of eigenvalues and eigenvectors proposed in this paper is an iterative method based on the Inverse Iteration (also known as Inverse Power Method). Subsections 2.1 and 2.2 provide a theoretical basis for the new proposal.

Table 1. Methods for determination eigenvalues and eigenvectors and theiralgorithmic complexities

Method	Algorithmic complexity
Bisection and Inverse Iteration	$O(n^3)$ in the worst case
QR	$O(n^3)$
Divide & Conquer	$O(n^3)$ in the worst case
MR^3	$O(n^2)$

2.1 The Power Method

Let A be an n x n real symmetric matrix. Assume that the eigenvectors of the matrix A span the vector space R^n, and all its eigenvalues are real; in this case, the components of the eigenvectors are also real. Considere that the matrix A has a single eigenvalue of largest absolute value. That is, assume that A has a real eigenvalue λ of multiplicity 1 such that for all other eigenvalues p it occurs that $|p| < |\lambda|$.

The Power Method enables the determination of the eigenvalue of largest absolute value of a matrix A and the eigenvector associated with the eigenvalue found without determining the characteristic polynomial. To describe the method assume that $\mathbf{x}^{(0)}$ be an arbitrary n-dimensional column vector and let the vectors v_i be linearly independent; arbitrarily, let it be [13]

$$\mathbf{x}^{(0)} = (1, 1, \ldots, 1)^T. \tag{3}$$

Let the vectors $\mathbf{x}^{(k)}$ be like

$$\mathbf{x}^{(k)} = \sum_{i=1}^{n} \alpha_i^{(k)} v_i \tag{4}$$

and $\mathbf{u}^{(k)}$ as follows

$$\mathbf{u}^{(k+1)} \underline{\underline{def}} A\mathbf{x}^k = \sum_{i=1}^{n} \lambda_i \alpha_i^{(k)} v_i. \tag{5}$$

Denoting by μ_{k+1} the component of largest absolute value of the vector $\mathbf{u}^{(k+1)}$, then

$$\mathbf{x}^{(k+1)} = \sum_{i=1}^{n} \alpha_i^{(k+1)} v_i = \frac{1}{\mu_{k+1}} \mathbf{u}^{(k+1)} = \sum_{i=1}^{n} \frac{\lambda_i \alpha_i^{(k)}}{\mu_{k+1}} v_i. \tag{6}$$

The step in Eq. (6) normalizes the vector $\mathbf{x}^{(k+1)}$. It ensures that the component of largest absolute value of $\mathbf{x}^{(k+1)}$ is 1. As the vectors v_i were assumed to be linearly independent, the Eq. (6) means that

$$\alpha_i^{(k+1)} = \frac{\lambda_i \alpha_i^{(k)}}{\mu_{k+1}} \tag{7}$$

for each i with $1 \leq i \leq n$. Thus,

$$\frac{\alpha_i^{(k+1)}}{\alpha_1^{(k+1)}} = \frac{\lambda_i \, \alpha_i^{(k)}}{\lambda_1 \, \alpha_1^{(k)}}. \tag{8}$$

Using the Eq. (8) repeatedly, one obtains

$$\frac{\alpha_i^{(k)}}{\alpha_1^{(k)}} = \left(\frac{\lambda_i}{\lambda_1}\right)^k \frac{\alpha_i^{(0)}}{\alpha_1^{(0)}}. \tag{9}$$

It was assumed that $|\lambda_i|$ is larger than any other eigenvalue. Thus,

$$\lim_{k\to\infty} \frac{\alpha_i^{(k)}}{\alpha_1^{(k)}} = \frac{\alpha_i^{(0)}}{\alpha_1^{(0)}} \lim_{k\to\infty} \left(\frac{\lambda_i}{\lambda_1}\right)^k = 0 \tag{10}$$

for each i with $2 \le i \le n$. According with Eq. (4), we have

$$\lim_{k\to\infty} \frac{1}{\alpha_1^{(k)}} x^k = \sum_{i=1}^n \frac{\alpha_i^{(k)}}{\alpha_1^{(k)}} v_i = v_1. \tag{11}$$

Since the component of the largest absolute value of $x^{(k)}$ is 1, it follows that the limit

$$\alpha_1 \underset{=\!=\!=}{\text{def}} \lim_{k\to\infty} \alpha_1^{(k)} \tag{12}$$

exists, and $\alpha_1 \ne 0$. Thus, for large n, $x^{(k)} \approx \alpha_1 v_1$. Since any nonzero scalar multiple of the eigenvector v_1 is an eigenvector, this means that $x^{(k)}$ is close to an eigenvector of A for the eigenvalue λ_i. This allows us to (approximately) determine λ_i and a corresponding eigenvector.

The Power Method should be applied if the goal is to find the eigenvalue of largest absolute value of a matrix A. The disadvantage of this method is that it calculates only one eigenvalue at a time [14].

2.2   The Inverse Power Method

The Inverse Power Method, also known as Inverse Iteration, allows the determination of the eigenvalue of lower absolute value of a matrix and the eigenvector associated with the eigenvalue found without determining the characteristic polynomial. For this, the Power Method must be applied to the matrix $(A - sI)^{-1}$ for some number s. If λ is an eigenvalue of A corresponding to the eigenvector v, then this vector is also an eigenvector of $(A - sI)^{-1}$ with eigenvalue $(\lambda - s)^{-1}$ [13].

If the eigenvalues of A are $\lambda_1, \ldots, \lambda_n$, then the eigenvalues of $(A - sI)^{-1}$ will be $(\lambda_1 - s)^{-1}, \ldots, (\lambda_n - s)^{-1}$. Thus, if λ_1 is the closest eigenvalue of A to s, $(\lambda_1 - s)^{-1}$

will be the eigenvalue of $(A - sI)^{-1}$ with the largest absolute value. Therefore, the eigenvalue with smaller absolute value will be $1/(\lambda_1 - s)^{-1}$.

It is not necessary to determine the inverse of $A - sI$ when the Inverse Power Method is applied. Instead, the LU-factorization of this matrix should be calculated, only once, to be used repeatedly in all iterations [13, 14].

The number s represents an offset to be applied to the matrix $(A - sI)^{-1}$. When $s \neq 0$, the method is known as Shifted Inverse Power Method. The eigenvalues $\lambda_i - s$ are shifted s units in real number line. The eigenvectors of $A - sI$ are the same of matrix **A**.

The Inverse Power Method should be applied if the goal is to find the smaller absolute eigenvalue of a matrix A. The disadvantage of this method is that it calculates only one eigenvalue at a time [14].

3 Proposal of New Approach

The new approach proposed in this paper for the determination of eigenvalues is based on the Shifted Inverse Power Method. The main contributions of the new approach are:

- The decrease of the convergence time in the calculation of eigenvalues - The initial shot of the vector $x^{(0)}$ is not chosen arbitrarily as the Eq. (3). It is determined from the knowledge of the expert in the problem under study.
- The estimation of offsets applied in matrix to the calculation of the i first smaller eigenvalues of interest – In the Shifted Inverse Power Method, for matrices of higher order 3 it is not easy to obtain appropriate values for the offsets to be applied in the matrix. In the new approach, these offsets are determined from the knowledge of the expert in the problem under study.

3.1 Quantum System Under Study

For obtaining information of a quantum system it is necessary to solve the Schrödinger equation. A solution of this equation is a wave function Ψ that is one eigenfunction of the Hamiltonian operator H and that corresponds to the total energy of the system [18]. Solving the Schrödinger equation involves finding the eigenvalues and eigenfunctions (eigenvectors) of the Hamiltonian operator of the system.

The Hamiltonian operator is a differential operator whose discretization by finite difference generates a matrix and the diagonalization of this matrix allows the determination of eigenvalues and eigenvectors. Thus, information of the physical system under study may be extracted.

The quantum system studied is one-dimensional, it is given as a function of ρ and only its kinetic energy is considered.The matrix representation of the Schrödinger equation that models the system is tridiagonal and its main diagonal and subdiagonal elements are presented in (13).

$$A_{ij} = \begin{cases} \frac{II}{2} + (I+\rho)(1+\rho) - 2t^2, i = j \\ \sqrt{(\frac{I}{2} - t + 1) * (\frac{I}{2} - t + 1 + \rho) * (\frac{I}{2} + t) * (\frac{I}{2} + t + \rho)}, |i-j| = 1 \end{cases} \quad (13)$$

The value of I is $I = l - 1$, where l represents the number of rows of the matrix and $t = -I/2$.

The equations in (13) must be solved for different values of ρ. Then, for each value of ρ, there is a different matrix for which the eigenvalues associated set should be calculated.

3.2   Determination of the Initial Guesses

The Schrodinger equation is solved for different values of ρ_t, where $-1 \leq t \leq max$ and max is the index of highest value that ρ may be. For each value of ρ_t there is an associated set of eigenvalues and eigenvector. Consider $\lambda'_i = 1/(\lambda_1 - s)^{-1}$, where $1 \leq i \leq m$ and m represents the total number of eigenvalues that a system has. Thus, λ'_i is the eigenvalue with smaller absolute value by the application of the Shifted Inverse Power Method. Eigenvectors associated to λ'_i are v'_i.

The new approach proposes that each set of eigenvalues are calculated considering the eigenvectors found in the previous iteration as the initial shot to the Shifted Inverse Power Method as illustrated in Fig. 1.

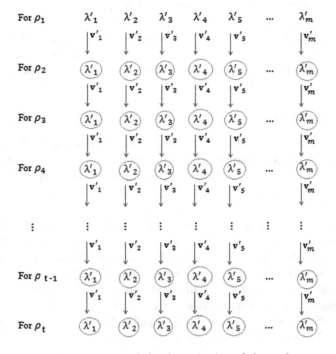

Fig. 1. New approach for determination of eigenvalues

Only in the first iteration, initial guesses for calculating of λ_i' are chosen arbitrarily. From the first set of eigenvectors calculated in the first iteration, the second set of eigenvalues and eigenvectors is estimated. From the second set of eigenvectors, the third set of eigenvalues and eigenvectors is estimated and so on, until the set of eigenpairs of the last iteration t is calculated.

After solving the Schrodinger equation for values of ρ_t by using the new approach all calculated eigenvalues may be grouped into a matrix $\mathbf{M}$ as in Eq. (14).

$$
M_{ij} = \begin{bmatrix}
\lambda_{11}' & \lambda_{12}' & \lambda_{13}' & \lambda_{14}' & \cdots & \lambda_{1m}' \\
\lambda_{21}' & \lambda_{22}' & \lambda_{23}' & \lambda_{24}' & \cdots & \lambda_{2m}' \\
\lambda_{31}' & \lambda_{32}' & \lambda_{33}' & \lambda_{34}' & \cdots & \lambda_{3m}' \\
\vdots & \vdots & \vdots & \vdots & \cdots & \vdots \\
\lambda_{(t-1)1}' & \lambda_{(t-1)2}' & \lambda_{(t-1)3}' & \lambda_{(t-1)4}' & \cdots & \lambda_{(t-1)m}' \\
\lambda_{t1}' & \lambda_{t2}' & \lambda_{t3}' & \lambda_{t4}' & \cdots & a_{tm}
\end{bmatrix} \tag{14}
$$

It is important to highlight that each set of eigenvaluesis increasingly ordered, i.e. $\lambda_{11}' < \lambda_{12}' < \ldots < \lambda_{1m}'; \lambda_{21}' < \lambda_{22}' < \ldots < \lambda_{2m}'; \ldots; \lambda_{t1}' < \lambda_{t2}' < \ldots < \lambda_{tm}'$. The elements of the columns of the resulting matrix $\mathbf{M}$ are the points of the adiabatic curves of the quantum system - each column of the $\mathbf{M}$ represents a curve.

It is assumed that the proposed approach may be more computationally efficient than the classical approaches available in the literature since that the classical algorithms calculate the eigenvalues from the initial guesses chosen arbitrarily. In the new approach, eigenvalues of the current iteration should be calculated from the eigenvectors associated with the eigenvalues of the previous iteration so that the processing time should decrease considerably due to faster convergence in the determination of the sought eigenvalues. The strategy of using vectors of the previous iteration was suggested by quantum chemistry experts who know the system and its behavior.

3.3 Estimation of Offsets

As previously mentioned, the Inverse Power Method calculates only one eigenvalue – the eigenvalue of smaller absolute value of a matrix. For calculating the remaining eigenvalues it is necessary to perform displacements in the matrix. Displacements allow determining the eigenvalues that were shifted by s units on the real line. The method becomes the Shifted Inverse Power Method.

Each time an offset$s$ is applied to the matrix, the closest eigenvalue to s is found. Thus, for determining the i first smaller eigenvalues of interest it is necessary to shift the matrix i times and applying the Shifted Inverse Power Method i times.

The new approach suggests that, for each iteration, the offsets that it will be applied in the input matrix are estimated from the eigenvalue previously found as illustrated in Fig. 2, that highlights the second iteration in order to exemplification.

The first offset in an iteration should be equal to the first eigenvalue of the set of eigenvalues calculated in the previous iteration. Then, the values previously calculated

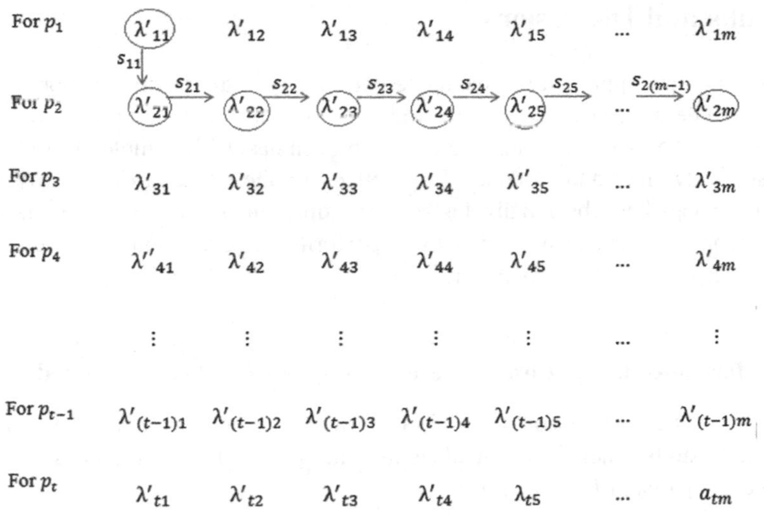

Fig. 2. Estimation of offsets based on the eigenvalues previously calculated

in the same iteration are used as a starting point to estimate the next offsets added to a Δs as presented in Eq. (15). The addition of Δs to the offset ensures that the eigenvalue obtained from applying the Shifted Inverse Power Method is not the same value of the eigenvalue used as starting point.

$$s_{ij} = \begin{cases} \lambda_{(i-1)j}, j = 1 \\ \lambda_{i(j-1)} + \Delta s, j \neq 1 \end{cases} \tag{15}$$

4   Simulation Environment

All programs necessary for the development of this study were developed in C language. The Linear Algebra Package Lapack has provided routines for applying the Divide & Conquer Method. Since Lapack was written in fortran, it was necessary to use standard C language API for integrating routines in Lapack with C language programs. API is available in Lapack Project website [19].

The program executions have been implemented in a machine with the following configuration: Rocks 6.1.1 operating system, dual six core CPU, with 16 Gigabytes of RAM and 2 Terabyte Hard Drive.

Execution times of the programs were obtained by the average of the times of five simulations, disregarding the highest and lowest value time in the calculating of the average.

5　Results and Discussions

Obviously the new approach has the same complexity than the Inverse Iteration since the difference between them is not the number of operations, but the initial shot used as a parameter of the methods. Thus, the new approach has $O(n^3)$ complexity in the worst case, just like Divider and Conquer. In practice, despite of equal complexity Divider and Conquer can be substantially faster, depending on the amount of deflation [9]. A comparison on their execution times was performed to evaluate the efficiency of the new approach proposed in this paper.

5.1　Performance in the Cases Where All Eigenvalues Are Calculated

Table 2 presents the execution times obtained applying both methods to the quantum system under study when all eigenvalues and all eigenvectors of matrices of different orders were calculated for $\rho = 100$.

Table 2. Execution times for calculating all eigenvalues

Matrix Order	Execution time of new approach in seconds	Execution time of divider and conquer in seconds
11	0.000113	0.000083
111	0.006215	0.010158
211	0.026953	0.047022
311	0.070398	0.140850
411	0.141545	0.293467
511	0.247824	0.530575
611	0.213742	0.874125
711	0.265028	1.341573
811	0.378031	1.908075
911	0.518563	2.669163
1011	0.692954	3.587960
1111	1.246532	4.747693
1211	1.177810	6.054190
1311	1.439677	7.598130
1411	1.768147	9.339617
1511	2.147429	11.435981

From Table 2, it is noted that, in cases where all eigenvalues are of interest, computationally the fastest method is Divider and Conquer. The graph of Fig. 3 presents the time curves obtained from the calculation of all eigenvalues of the matrices indicated in Table 2 according to the methods.

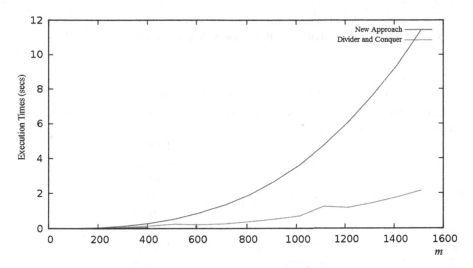

Fig. 3. Time curves for calculating all eigenvalues for matrices of different orders (Color figure online)

5.2 Performance in the Cases Where Some Eigenvalues of Interested Are Calculated

The new approach has been more efficient than the Divider and Conquer in cases where only some eigenvalues must be calculated. This is the reason for applying the method to problems of quantum chemiştry. Most applications in this field require that only a few eigenvalues are found, instead of all, and usually the first ones.

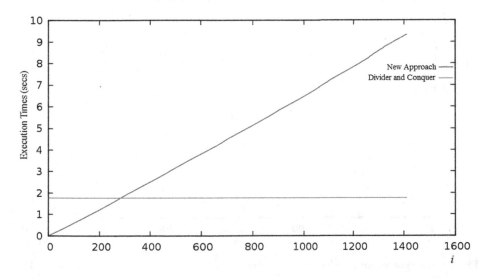

Fig. 4. Time curves for calculating the first i eigenvalues of interest for a matrix of order 1411

As an example, the obtained time curve from simulations of both methods to a matrix of order 1411 for calculating the *i* first values is illustrated in Fig. 4.

Table 3. Execution times for calculating some eigenvalues of interest

Amount of i first eigenvalues of interest	Execution time of new approach in seconds	Execution time of divider and conquer in seconds
10	0.056662	1.768156
30	0.114636	1.768166
50	0.286547	1.768195
70	0.408044	1.768215
90	0.533419	1.768235
110	0.662049	1.768255
130	0.782474	1.768275
150	0.904958	1.768295
170	1.027789	1.768315
190	1.152779	1.768335
210	1.279584	1.768355
230	1.407728	1.768375
250	1.536563	1.768395
270	1.667588	1.768415
283	1.758248	1.768428
284	1.765125	1.768429
285	1.771260	1.768430

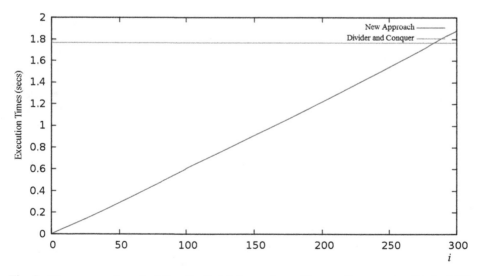

Fig. 5. Time curves for calculating the first *i* eigenvalues of interest for a matrix of order 1411

The Divider and Conquer method performs the calculation of all the eigenvalues, although the problem to be solved requires only that the eigenvalues of interest are found. The method uses a time of about 1,77 s to calculate the sample matrix eigenvalues, regardless of the number of values that actually need to be estimated. On the other hand, the obtained execution time from applying the new approach can vary, depending on the number of eigenvalues of interest. As noted in Table 3, the new approach is most suitable for being used to the sample matrix in cases where it is desired to calculate up to the first 285 values.

Figure 5 presents the comparison among the performances of methods in the case where the i first eigenvalues are calculated, with i ranging up to 300.

6 Conclusion

Studies conducted in the quantum chemistry field are directly beneficial to the industrial sector since they enable the development of new drugs, a better understanding of diseases, the development of new materials, among other activities. Especially with regard to processing time, simulations based on quantum theory are computationally very expensive due to the large amount of subatomic particles to be modeled computationally. In many cases, the simulations become even impossible [10]. Thus, researches have been developed to make them viable.

In this sense, this paper has proposed a new approach for reducing the computation time required for the determination of eigenvalues and eigenvectors in problems of quantum chemistry. The new approach is based on the Shifted Inverse Power Method, but uses the knowledge about the quantum problem under study to define the best initial guesses and to estimate the values of offsets to be applied to the input matrix.

From the numerical experiments, it was concluded that the new approach has proved to be efficient in cases where the problem requires only some eigenvalues must be calculated, as it is usual in problems of quantum chemistry.

It is important to point out that the execution times presented were obtained by the application of the proposed method to a monodimensional quantum problem. In further research, it is intended to model complex problems, such as the representation of the molecular geometry of three-dimensional triatomic sistems. Another task to be accomplished is to compare the performace of the new approach to the MR^3, currently considered as the faster algorithm for calculation of eigenvalues.

References

1. Albernaz, A.F., Aquilanti, V., Barreto, P.R.P., Caglioti, C., Cruz, A.C.P.S., Grossi, G., Lombardi, A., Palazzetti, F.: Interactions of hydrogen molecules with halogen containing diatomics from ab initio calculations. spherical-harmonics representation and characterization of the intermolecular potentials. J. Phys. Chem. A **134**, 130–137 (2016). doi:10.1021/acs.jpca.6b01718

2. Anderson, R.: Discrete orthogonal transformations corresponding to the discrete polynomials of the askey scheme. In: Murgante, B., et al. (eds.) ICCSA 2014, Part I. LNCS, vol. 8579, pp. 490–507. Springer, Heidelberg (2014)

3. Ragni, M., Bitencourt, A.C.P., Silva, A.E., Prudente, F.V.: Umbrella inversion energy levels of AB_3 like molecules for $J \geq 0$. In: Murgante, B., et al. (eds.) ICCSA 2014, Part I. LNCS, vol. 8579, pp. 538–553. Springer, Heidelberg (2014)

4. Anderson, R.W., Aquilanti, V., Bitencourt, A.C.P., Marinelli, D., Ragni, M.: The screen representation of spin networks: 2d recurrence, eigenvalue equation for $6j$ symbols, geometric interpretation and hamiltonian dynamics. In: Murgante, B., Misra, S., Carlini, M., Torre, C.M., Nguyen, H.-Q., Taniar, D., Apduhan, B.O., Gervasi, O. (eds.) ICCSA 2013, Part II. LNCS, vol. 7972, pp. 46–59. Springer, Heidelberg (2013)

5. Guimarães, M.N., Ragni, M., Bitencourt, A.C.P., Prudente, F.V., Prudente, F.V.: Alternative hyperspherical adiabatic decoupling scheme for tetratomic molecules: quantum two-dimensional study of the ammonia umbrella motion. Eur. Phys. J. D, At. Mol. Opt. Phys. (Print) **67**, 253 (2013)

6. Ragni, M., Prudente, F.V., Bitencourt, A.C.P., Barreto, P.R.P.: Analysis of vibrational modes of the P_4 molecule through hyperspherical variants of the local orthogonal coordinates: The limit of dissociation in dimers. Int. J. Quantum Chem. **111**, 1719–1733 (2011)

7. Bitencourt, A.C.P., Prudente, F.V., Ragni, M.: Roto-torsional levels for symmetric and asymmetric systems: application to HOOH and HOOD systems. In: Murgante, B., Misra, S., Carlini, M., Torre, C.M., Nguyen, H.-Q., Taniar, D., Apduhan, B.O., Gervasi, O. (eds.) ICCSA 2013, Part II. LNCS, vol. 7972, pp. 1–16. Springer, Heidelberg (2013)

8. Ragni, M., Littlejohn, R.G., Bitencourt, A.C.P., Aquilanti, V., Anderson, R.W.: The screen representation of spin networks: images of $6j$ symbols and semiclassical features. In: Murgante, B., Misra, S., Carlini, M., Torre, C.M., Nguyen, H.-Q., Taniar, D., Apduhan, B. O., Gervasi, O. (eds.) ICCSA 2013, Part II. LNCS, vol. 7972, pp. 60–72. Springer, Heidelberg (2013)

9. Lombardi, A., Pirani, F., Laganà, A., Bartolomei, M.: Energy transfer dynamics and kinetics of elementary processes (promoted) by gas-phase CO2-N2 collisions: Selectivity control by the anisotropy of the interaction. DOI:10.1002/jcc.24359 (2016)

10. Ragni, M., Lombardi, A., Barreto, P.R.P., Bitencourt, A.C.P.: Orthogonal coordinates and hyperquantization algorithm. The NH_3 and H_3O^+ umbrella inversion levels. J. Phys. Chem. A **113**, 15355–15365 (2009)

11. Ragni, M., Bitencourt, A.C.P., Prudente, F.V., Barreto, P.R.P., Posati, T.: Umbrella motion of the methyl cation, radical, and anion molecules. Eur. Phys. J. D, At. Mol. Opt. Phys. (Print) **70**, 60 (2016)

12. Ragni, M., Bitencourt, A.C.P., Prudente, F.V., Barreto, P.R.P., Posati, T.: Umbrella motion of the methyl cation, radical, and anion molecules. Eur. Phys. J. D, At. Mol. Opt. Phys. (Print) **70**, 61 (2016)

13. Máté, A.: Introduction to Numerical Analysis with C Programs. Brooklyn College of the City University of New York, New York (2004)

14. Franco, N.M.B.: Cálculo Numérico. Prentice Hall, São Paulo (2007)

15. Sunderland, A.: Performance of a New Parallel Eigensolver PDSYEVR on HPCx. Computational Science and Engineering Department, CCLRC Daresbury Laboratory. Technical report, HPCx Consortium. Warrington, UK (2006)

16. Press, W.H., Teukolsky, S.A., Vetterling, W.T., Flannery, B.P.: Numerical Recipes in C: the art of scientific computing. Cambridge University Press, Cambridge (2002)
17. Golub, G., Loan, C.: Matrix computations. The Johns Hopkins University Press, London (1996)
18. Heisberg, W.: Über quantentheoretische Umdeutungkinematischer und mechanischer Beziehungen. Z. Phys. **33**, 879–893 (1925). Paris
19. Lapack - Linear Algebra PACKage. http://www.netlib.org/lapack/

Continuous Time Dynamical System and Statistical Independence

Madalin Frunzete[1,2]([⊠]), Lucian Perisoara[1], and Jean-Pierre Barbot[2,3]

[1] Faculty of Electronics, Telecommunications and Information Technology,
Politehnica University of Bucharest, 1-3, Iuliu Maniu Bvd.,
Bucharest 6, Romania
madalin.frunzete@upb.ro
[2] Electronique et Commande des Systmes Laboratoire, EA 3649
(ECS-Lab/ENSEA) ENSEA, Cergy-Pontoise, France
[3] EPI Non-A INRIA, Lyon, France

Abstract. Dynamical systems can give information and can be used in applications in various domains. It is important to know the type of information which will be extracted. In an era when everybody is speaking and is producing information that can be referred as big data, here, the way to extract relevant information by sampling a signal is investigated. Each state variable of a dynamical system is sampled with a specific frequency in order to obtain data sets which are statistical independent. The system can provide numbers for random generators and the sequence obtained can be easily reproduced. These type of generators can be used in cryptography.

1 Introduction

Dynamical systems are systems with very complex behaviour and they are used in various applications, such as communications [4,10], control theory [13,16], music [3], meteorology and so on. Here, the dynamical system is analyzed by using statistical methods with the purpose of describing the statistical independence for two data sets. The case study is Rössler system, the continuous case [18].

The theory of chaotic systems has a relatively short history. Henri Poincaré (1854–1912) brings up the idea of chaos for the first time in 1903 while participating in a contest where one of the challenges was rigorous demonstration that the solar system as it is modeled by Newton's equations, is dynamically stable. This question is a generalization of the problem of three types [17]. Poincaré described the brand of chaos, depending on initial conditions as follows: "...even if it were the case that the natural laws had no longer any secret for us, we could still only know the initial situation approximately. If that enabled us to predict the succeeding situation with the same approximation, that is all we require, and we should say that the phenomenon had been predicted, that it is governed by laws. But it is not always so - it may happen that small differences in the initial conditions produce very great ones in the final phenomena. The

O. Gervasi et al. (Eds.): ICCSA 2016, Part I, LNCS 9786, pp. 470–479, 2016.
DOI: 10.1007/978-3-319-42085-1_36

meteorologists see very well that the equilibrium is unstable, that a cyclone will be formed somewhere, but exactly where they are not in a position to say; a tenth of a degree more or less at any given point, and the cyclone will burst here and not there, and extend its ravages over districts it would otherwise have spared..." [17].

Investigation of statistical independence in the context of chaotic systems appears to be in contradiction with their deterministic nature, while the stationarity of the chaotic signals is one thing generally accepted, [2]. Before analyzing the statistical independence, some preliminary investigations are necessary on the transient time - the time elapsed from the system initialization to its entrance in the stationarity region.

Analysis were performed on chaotic systems needed overall, the development of a theoretical and experimental instrument, [1,22] which is based on inter-disciplinary of the concepts. Thus, elements of information theory (information channel and associated sizes) are combined with a series of statistical methods (Smirnov tests, Kolmogorov-Smirnov tests and an original method [1] to verify the statistical independence in the case of continuous random variables which obey Gaussian law or not - the last method can be applied on all types of random variables, even if they obey an unknown type of probability law).

In Sect. 2 the investigated system will be presented; Rössler system will be discussed and the random process associated to this system is described. Here an experiment with different initial conditions for the system will give information from a statistical point of view.

In Sect. 3 the behaviour of the Rössler system will be described from the statistical point of view. The statistical aspects concern: the transient time and the probability law (related by using Smirnov test) and the statistical dependence/independence issue (treated by using on original test [11]).

2   System Investigated

The Rössler system (1), [18], a system of three non-linear ordinary differential equations originally studied by Otto Rössler. These differential equations define a continuous-time dynamical system that exhibits chaotic dynamics associated with the fractal properties of the attractor. This attractor has some similarities to the Lorenz attractor [14], but is simpler and has only one manifold.

$$\begin{aligned}
\dot{x} &= -y - z \\
\dot{y} &= x + ay \\
\dot{z} &= b + z(x - c)
\end{aligned} \tag{1}$$

For Rössler system the attractor, for parameters $a = 0.398$, $b = 2$ and $c = 4$ [20], is given in the Fig. 1(a). The figure was computed for the parameters $a = 0.398$, $b = 2$ and $c = 4$. The original Rössler paper says the Rössler attractor was intended to behave similarly to the Lorenz attractor, but also be easier to analyze qualitatively. "An orbit within the attractor follows an outward spiral close to the (x, y) plane around an unstable fixed point. Once the graph spirals out enough, a second fixed point influences the graph, causing a rise and twist in the z-dimension. In the time domain, it becomes apparent that although each variable is oscillating within a fixed range of values, the oscillations are chaotic. This attractor has some similarities to the Lorenz attractor, but is simpler and has only one manifold. Otto Rössler designed the Rössler attractor in 1976, but the originally theoretical equations were later found to be useful in modeling equilibrium in chemical reactions" [15].

The Rössler system was computed in Simulink and the corresponding scheme is presented in Fig. 1(b). In order to implement the system the following blocks were used: 3 integrators, 3 Function blocks and a multiplexor with 3 inputs and 3 outputs. For each integrator an external constant block was used in order to define the initial condition $(x0, y0, z0)$.

On this system, a random process was defined by considering $4 \cdot 10^4$ initial conditions with a uniform probability law in $(0, 1)$ interval. These initial conditions were given as external source for each integrator. The system was simulated for 10^5 seconds with 10 intermediary point for each second. Finally, for each trajectory, corresponding to the $4 \cdot 10^4$ initial conditions, were computed 10^6 samples.

All the data obtained from this system will be used in order to obtain some statistical information. This type of information correlated with other types of information, such as observability [8,21], can be used when using the system in applications such as random numbers or cryptography. Another important property is the singularity manifold for dynamical system which can create serious damage for data reconstruction [6].

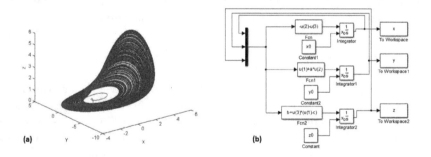

Fig. 1. (a) Rössler Attractor; (b) Rössler system in Simulink

3   Statistical Analysis for Dynamical System

The investigations for the statistical properties of Rössler system (1) were focused on the following:

- the computational measurements of the transient time - the time elapsed from the initial condition (initial state vector) of the system up to its entrance in stationarity.
- the evaluation of the minimum sampling distance which enables the statistical independence between two random variable extracted from the same state variable (x or y or z).

Note that, three random processes assigned to the three state variables are obtained for a fixed set of parameters $a = 0.398, b = 2, c = 4$, but different initial conditions generated in the $(0, 1)$ interval. The $(0, 1)$ interval is only an example, chosen in this paper, in order to obtain some statistical information for the investigated system.

Transient Time for the Rössler System

In order to use the system in cryptographic applications it is necessary to see if it fits the stationarity behaviour. The system in simulated by starting from a data set with uniform distributed law and it is tested if, after a time (transient time), the three state variable will fit a specific probability law - this probability law is unknown and this is the reason why the Smirnov test was chosen to test the transient time.

So, the Smirnov test [19] is applied for statistical analysis by measuring the transient time for each state variable of system (1) and by verifying if the probability law for each state variable x, y, z is the same at any moment.

The size of the experimental data sets was $N = 2 \cdot 10^4$ and the significance level $\alpha = 0.05$. For generating these data sets are needed $2N = 2 \cdot 10^4$ different initial conditions of the type $(x0, y0, z0)$ which were generated according to the uniform law in the domain $(0, 1) \times (0, 1) \times (0, 1)$.

The two data sets required by Smirnov test are obtained as follows:

- for a fixed state variable, the data set $(x_1, x_2, \ldots, x_N)$ is obtained from the system (1) by sampling the state variable at the iteration k_1. The values k_1 were chosen to be more than 10^3 seconds.
- the data set $(y_1, y_2, \ldots, y_N)$ is obtained from the system (1) by sampling the same state variable at the iteration $k_2 = 10^5$ (this iteration is considered in the stationarity region, regardless of the initial condition).

In order to have a solid decision the Smirnov test was resumed 500 times for each pair $(k_1; k_2)$ d different moments for sampling the system. The results are presented in Table 1. To understand what is given in this table by looking on the first line, at the iteration $k_1 = 10^4$ seconds in 17.3 % of times from the ensemble of 500, the Smirnov test indicates that the random variables X has the same probability law as the random variable Y.

Figure 2 presents the experimental cumulative distribution functions of the random variables extracted from the random processes assigned to each state variable of the systems. The random processes were sampled at $k_1 = \{10^3, 10^4, 5 \cdot 10^4\}$ seconds after computing the system (1) for N times.

Table 1. The proportion of accepting H_0 for Smirnov test (verifying probability law for each state variable of systems (1))

k_1 (seconds)	10^3	10^4	$5 \cdot 10^4$
x	16.6 %	24.8 %	35.4 %
y	15.8 %	17.4 %	36.2 %
z	15.2 %	16.4 %	25.8 %

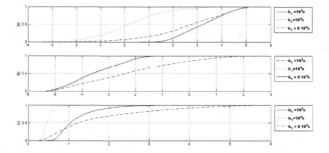

Fig. 2. Experimental Cumulative Distribution Function for Rössler system: (a) x state variable, (b) y state variable, (c) z state variable

Remark. The experimental results presented indicate that the statistical behaviour of the systems (1) does not fit the transient time. This type of system cannot be used in cryptographic applications because of a specific initial condition (which can be part of the key) the behaviour of the system gives information about the key. By interpreting the cumulative distribution function of data sets extracted from specific time series the eavesdropper can obtain information about the initial condition, so he can have a part of the key.

Statistical Independence

Dynamical systems and random number generators can be an approach useful in many applications. By considering the transient time analyzed above it seams that cryptographic applications are not in the area of interest for the mentioned approach. But the statistical independence in the context of chaotic behaviour is a subject of interest for applications of data coding or compressive sensing [23].

Here, a sampling distance which enables the statistical independence will be evaluated. The two jointly distributed random variables X and Y sampled from Rössler system which are investigated have an unknown probability law so it is necessary to do some operation in order to obtain the result for statistical independence. Sampling each state variable at the iteration k_1 a random variable X is obtained and a random variable Y results by sampling the same state variable at $k_2 = k_1 + d$, where d is the sampling distance under investigation for statistical independence.

The test procedure given in [1] was applied in order to evaluate the minimum sampling distance d which enables statistical independence between X and Y. The independence test was applied in a similar way as in [5,22]. The statistical independence in the context of Rössler map (discrete case) was discussed in [5].

To obtain the two sets $(x_1, x_2, \ldots, x_N)$ and $(y_1, y_2, \ldots, y_N)$ necessary for the statistical investigation it is necessary to have $2N$ different initial conditions. By sampling the random process assigned to a state variable at two specific moments, the two random variables X and Y corresponding to the mentioned case. Each of the two data sets has to comply with the *i.i.d.* statistical model. The N initial conditions of the type $x0, y0, z0$ were randomly chosen in the domain $(0, 1) \times (0, 1) \times (0, 1)$.

It is important to notice that when a single random process is concerned, each pair of values (x_i, y_i), $i = \overline{1, N}$ is obtained on the same trajectory at k_1 and $k_2 = k_1 + d$ moments.

The test has 2 hypotheses: the null hypothesis, H_0, meaning that the random variables X and Y are statistically independent, and the alternative hypothesis, H_1, meaning that the random variables X and Y are dependent. There are necessary two successive transforms because the probability law is unknown. For X the transform is $X \longrightarrow X' \longrightarrow U$ and for Y corresponds $Y \longrightarrow Y' \longrightarrow V$.

So, there will be 2 random variables uniformly distributed in the $(0, 1)$ interval (X' and Y'); the random variables U and V are standard normal random variables. So if U and V are independent also X and Y are independent. From Kolmogorov-Smirnov concordance test where the investigated variables have a law well known results a huge amount of data to verify the theoretical facts. In Smirnov test the amount of data is double.
The independence test algorithm applied in [7] is recalled here:

(a) For X and Y is obtained cumulative distribution functions $Fe_X(x)$ and $Fe_Y(y)$ from the experimental data sets.
(b) By the notation $x'_i = Fe_X(x_i)$ and $y'_i = Fe_Y(y_i)$ the data sets are uniformly distributed in $(0, 1)$, $i = \overline{1, N}$
(c) By using the inverse of the distribution function of the standard normal law on data $(x'_1, x'_2, \ldots, x'_i, \ldots, x'_N)$ and $(y'_1, y'_2, \ldots, y'_i, \ldots, y'_N)$ (of mean 0 and variance 1), the new sets obtained the transformation $(u_1, u_2, \ldots, u_i, \ldots, u_N)$ and $(v_1, v_2, \ldots, v_i, \ldots, v_N)$. By the theory, both u_i and v_i values are standard normally distributed. So, Pearson's independence test on Gaussian populations can be applied for U and V normal random variables.
(d) The correlation coefficient r between U and V and the t test value are computed:

$$r = \frac{\sum_{i=1}^{N}(u_i - \bar{u})(v_i - \bar{v})}{\sqrt{\sum_{i=1}^{N}(u_i - \bar{u})^2 \sum_{i=1}^{N}(v_i - \bar{v})^2}} \tag{2}$$

$$t = r\sqrt{\frac{N - 2}{1 - r^2}} \tag{3}$$

If the test value should obey the Student law with $N-2$ degrees of freedom then the null hypothesis H_0 is correct, so X and Y are independent. The $t_{\alpha/2}$ point value of the Student law of $N-2$ degrees of freedom is calculated by using the α significance level. If $|t| \leq t_{\alpha/2}$, then the two random variables X and Y could be independent. For a final decision it is necessary to follow the last step of the test. Otherwise, if $|t| \geq t_{\alpha/2}$, the test stops because it is clear that random variables X and Y are dependent.

(e) In this step U and V has to be proven jointly normally distributed and finally the decision is that X and Y are statistically independent. If U and V are not jointly normally distributed, X and Y are dependent. In order to decide upon the normal bivariate law, the test uses the visual inspection of the (u, v) scatter diagram. For the decision to accept H_0, the (u, v) scatter diagram has to fit to reference on (see Fig. 3).

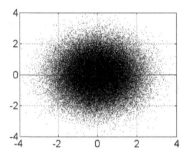

Fig. 3. The reference scatter diagram for two independent random variables

The experiments were done for N initial conditions following a uniform probability law in $(0, 1) \times (0, 1) \times (0, 1)$ intervals. The significance level was chosen $\alpha = 0.05$, thus the $\alpha/2$ point value of the Student law of $N - 2$ degrees of freedom is $t_{\alpha/2} = 1.96$. There were performed many tests and the results will be presented bellow. The results are finally organized in Table 2.

Some examples of numerical and visual results are recalled here:

– X and Y both sampled from the random process assigned to x state variable of Rössler system, with the sampling distance $d = 10^3$ seconds. The numerical results obtained by means of (2) and (3) are:

$$r = -0.0352 \Rightarrow t = -11.1380 \Rightarrow |t| \geq t_{\alpha/2}$$

Following the test procedure by the numerical results the conclusion is that the two random variables X and Y are statistically dependent for $d = 10^3$ seconds, so the test algorithm stops. The result is confirmed observing that the scater diagram in (u, v) coordinates, Fig. 4(a), differs from the respective reference scater diagram from Fig. 3.

– For the same state variable x of Rössler system, but the sampling distance $d = 10^4$ seconds, the numerical results:

$$r = -0.0051 \Rightarrow t = -1.6128 \Rightarrow |t| \leq t_{\alpha/2}$$

The test procedure decides that the two random variables are statistically independent. The correlation coefficient r has a small value so it can result a t value lower than the given threshold $t_{\alpha/2}$ and the scatter diagram in (u, v) coordinates, Fig. 4(b), resembles the respective reference scatter diagram in Fig. 3.

In order to have an accurate decision a Monte Carlo analysis is preformed, so a final conclusions on the minimum sampling distance which enables statistical independence is taken. The test procedure was resumed 500 times and the numerical results are summarized in Table 2 for system (1). The experiments were computed for $k_1 = 10^5$ seconds and $k_2 = k_1 - d$.

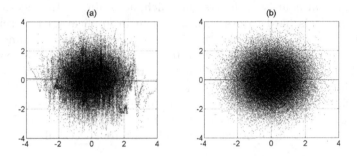

Fig. 4. Scater diagram for (u, v) coordinates and x state variable from system (1): (a) distance $d = 10^3$ s; (b) distance $d = 10^4$ s

Some of the numbers are explained bellow. On the last line of Table 2, the random variable X is obtained at iteration k_1 from the random process assigned to state variable z and the random variable Y is obtained at iteration k_2 from the random process assigned to state variable z, $X := z[k_1]$ and $Y := z_3[k]$. In this case the decision is that $d = 10^4$ seconds is the minimum sampling distance which enables statistical independence, because for $d \geq 10^4$ the proportion of acceptance of H_0 null hypothesis remains in the range $[93\,\%; 97\,\%]$. For $d = 10^4$ seconds, the proportion is $93.6\,\%$ (see last column of Table 2), so it is in the interval mentioned as the other bold values from the table.

According to Table 2, for each state variable a value $d = 10^4$ seconds is a minimum sampling distance which enables statistical independence. This result can be used in order to generate random numbers or to characterize the system behaviour. The results presented here can be used as preliminary results for other applications. Also, the visual investigation is necessary to be improved with some theoretical tests [12]. The results of these type of statistical instruments can be compared with other tests from the literature [9].

Table 2. The proportion of acceptance for H_0 null hypothesis for independence test applied on system (1)

d	(seconds)	10^2	10^3	$2 \cdot 10^3$	$5 \cdot 10^3$	10^4	$2 \cdot 10^4$	$5 \cdot 10^4$
X	Y							
$x(k_1)$	$x(k_2)$	0%	84.8%	89.8	90%	$\mathbf{93.2\%}$	$\mathbf{94.8\%}$	97%
$y(k_1)$	$y(k_2)$	15%	15.8%	89.8	87.6%	$\mathbf{97.2\%}$	$\mathbf{96.8\%}$	$\mathbf{95.8\%}$
$z(k_1)$	$z(k_2)$	60.4%	92.4%	87	$\mathbf{93.6\%}$	96%	95.6%	94%

4   Conclusions

The statistical independence in the context of continuous dynamical system is present and quite hard to be measured. The procedure applied gave information about the randomness in this type of system and recommends to extend the analysis by using different sets of parameters or initial conditions. An open problem is the stationarity of the system which, according to the measurements, is not present. It is hard to use such a system in application for secure data transmission but still can be used in order to generate random numbers. The applications with random number generators needs to rebuild the same sequence and the chaotic behaviour provide this opportunity. Starting from the same initial condition with the same set of parameters a chaotic system will have the same behaviour. The chaotic systems is useful if the user knows his statistical properties, if not the application can crush. Even if the implementation is quite simple the results obtained by using such systems are spectacular.

Acknowledgment. This work was supported by a grant of the Romanian Space Agency, Space Technology and Advanced Research (STAR) Programme, project number 75/29.11.2013.

References

1. Badea, B., Vlad, A.: Revealing statistical independence of two experimental data sets: an improvement on spearman's algorithm. In: Gavrilova, M.L., Gervasi, O., Kumar, V., Tan, C.J.K., Taniar, D., Laganá, A., Mun, Y., Choo, H. (eds.) ICCSA 2006. LNCS, vol. 3980, pp. 1166–1176. Springer, Heidelberg (2006)
2. Baptista, M.S.: Cryptography with chaos. Phys. Lett. A **240**(1–2), 50–54 (1998)
3. Chen, G., Han, B.: An audio scrambling degree measure based on information criteria. In: Proceedings of the 2nd International Signal Processing Systems (ICSPS) Conference, vol. 1 (2010)
4. Dogaru, I., Dogaru, R., Damian, C.: Fpga implementation of chaotic cellular automaton with binary synchronization property. In: Proceedings of the 8th International Communications (COMM) Conference, pp. 45–48 (2010)
5. Frunzete, M., Luca, A., Vlad, A.: On the statistical independence in the context of the rössler map. In: 3rd Chaotic Modeling and Simulation International Conference (CHAOS2010), Chania, Greece (2010). http://cmsim.net/sitebuildercontent/sitebuilderfiles/

6. Frunzete, M., Barbot, J.-P., Letellier, C.: Influence of the singular manifold of nonobservable states in reconstructing chaotic attractors. Phys. Rev. E **86**(2), 26205 (2012)
7. Frunzete, M., Luca, A., Vlad, A., Barbot, J.-P.: Statistical behaviour of discrete-time rössler system with time varying delay. In: Murgante, B., Gervasi, O., Iglesias, A., Taniar, D., Apduhan, B.O. (eds.) ICCSA 2011, Part I. LNCS, vol. 6782, pp. 706–720. Springer, Heidelberg (2011)
8. Frunzete, M., Luca, A., Vlad, A., Barbot, J.-P.: Observability and singularity in the context of roessler map. Univ. "Politehnica" Bucharest Sci. Bull. Ser. A Appl. Math. Phys. **74**(1), 83–92 (2012)
9. García, J.E., González-López, V.A.: Independence tests for continuous random variables based on the longest increasing subsequence. J. Multivar. Anal. **127**, 126–146 (2014)
10. Grigoras, V., Tataru, V., Grigoras, C.: Chaos modulation communication channel: A case study. In: Proceedings of International Symposium Signals, Circuits and Systems ISSCS 2009, pp. 1–4 (2009)
11. Hodea, O., Vlad, A.: Logistic map sensitivity to control parameter and its implications in the statistical behaviour. In: 2013 International Symposium on Signals, Circuits and Systems (ISSCS), pp. 1–4, July 2013
12. Hodea, O., Vlad, A., Datcu, O.: Evaluating the sampling distance to achieve independently and identically distributed data from generalized h énon map. In: 2011 10th International Symposium on Signals, Circuits and Systems (ISSCS), pp. 1–4, June 2011
13. Ivan, C., Serbanescu, A.: Applications of nonlinear time-series analysis in unstable periodic orbits identification - chaos control in buck converter. In: Proceedings of International Symposium Signals, Circuits and Systems ISSCS 2009, pp. 1–4 (2009)
14. Edward, N.: Lorenz.: Deterministic nonperiodic flow. J. Atmos. Sci. **20**(2), 130–141 (1963)
15. Peinke, J., Parisi, J., Rössler, O.E., Stoop, R.: Encounter with chaos: self-organized hierarchical complexity in semiconductor experiments. Springer Science & Business Media (2012)
16. Perruquetti, W., Barbot, J.-P.: Chaos in automatic control. CRC Press, Taylor & Francis Group (2006)
17. Poincaré, H., Goroff, D.: New Methods of Celestial Mechanics. AIP Press, Williston (1903)
18. Rössler, O.E.: An equation for hyperchaos. Phys. Lett. A **71**(2–3), 155–157 (1979)
19. Smirnov, N.: Table for estimating the goodness of fit of empirical distributions. Ann. Math. Stat. **19**(2), 279–281 (1948)
20. Soofi, A.S., Cao, L.: Modelling and forecasting financial data: techniques of nonlinear dynamics, vol. 2. Springer Science & Business Media (2012)
21. Letellier, C., Aguirre, L.A.: Interplay between synchronization, observability, and dynamics. Phys. Rev. E **82**(1), 016204 (2010)
22. Vlad, A., Luca, A., Frunzete, M.: Computational measurements of the transient time and of the sampling distance that enables statistical independence in the logistic map. In: Gervasi, O., Taniar, D., Murgante, B., Laganà, A., Mun, Y., Gavrilova, M.L. (eds.) ICCSA 2009, Part II. LNCS, vol. 5593, pp. 703–718. Springer, Heidelberg (2009)
23. Lei, Y., Barbot, J.-P., Zheng, G., Sun, H.: Compressive sensing with chaotic sequence. IEEE Signal Process. Lett. **17**(8), 731–734 (2010)

NCS-EC: Network Coding Simulator with Error Control

Aicha Guefrachi[✉], Sonia Zaibi, and Ammar Bouallègue

Communication System Laboratory Sys'Com, National Engineering School of Tunis,
Tunis El Manar University, Tunis, Tunisia
aicha.guefrachi@gmail.com

Abstract. This paper proposes NCS-EC, a new simulation framework entirely written in C. It is dedicated to evaluating the performances of inter-flow network coding based protocols and error correction codes in networks with errors and/or erasures. Our simulator's architecture is composed of three ordered layers, NC-network, NC-codec and NC-application. In the first one, the topology of network, the link models and the Network Coding procedure are defined. In the second, two network error correcting codes (KK and LRMC) are implemented. The application parameters and statistical analysis are given in the third layer. To do these operations, we developed a new C library (gf_lib) that operates in a finite field $\mathbb{F}_{q^m}$ with $q \geq 2$.

Keywords: Simulation · Network Coding (NC) · Error control · KK-code · LRMC

1 Introduction

Recently, Network Coding (NC) has emerged as an important information theoretic approach to improve the performance of both wired and wireless networks. In short, it is a technique which allows the intermediate nodes to combine several received packets into one coded packet for transmission instead of simply store and forward the individual packets. With network coding, the number of transmissions are reduced which improves network reliability [6] and increase throughput [8].

Various methods exist to combine packets. A linear combination is a sum of packets weighted by coefficients chosen in a fixed finite field. It was proved in [12], for a multicast case, that Linear Codes Multicast (LCM) are sufficient to reach the maximum capacity bounds. One construction method of such code is the algorithm proposed in [12]. The authors of [9] proposed to solve the network coding problem in an algebraic way. They found necessary and sufficient conditions for the feasibility of a given set of connections over a given network. These coding scenarios are centralized and require the knowledge of network architecture. The evolution of NC led to using random parameters and hence propose a distributed version of NC applicable to practical networks. In [5] Medard and

© Springer International Publishing Switzerland 2016
O. Gervasi et al. (Eds.): ICCSA 2016, Part I, LNCS 9786, pp. 480–490, 2016.
DOI: 10.1007/978-3-319-42085-1_37

Koetter show that a distributed design is sufficient in most cases. More precisely, if network nodes select linear mappings independently and randomly from inputs onto output links over some field, then the capacity is achieved with high probability [13, Theorem 2] if the field size is sufficiently large. Benefits of this approach are robustness to network changes or link failures. Thus, Random Linear Network Coding (RLNC) is arguably the most important class of network coding.

Unfortunately, the principle of network coding is not without its drawbacks. It is highly susceptible to errors caused by various factors. Corrupted packets may contaminate other packets when coded at the internal nodes, they may cause widespread error propagation. Thus, novel coding techniques are needed to solve this new error correction problem. We are interested in two network error correction (NEC) approaches: Koetter and Kschischang codes (KK-Codes) [10] and Lifted Rank Metric Codes (LRMC) [5] based on Gabidulin algorithm.

Up to now, the above approaches have been exhibited in theory in $GF(q^m)$. To the best of our knowledge many implementations of network coding have been reported (COPE [11], NECO [4], Lava [14]...) and only one introduces error control [7]. The authors give an efficient hardware implementations in Verilog of rank metric decoder architectures for RLNC with error control. These implementations have been designed for different parameters evaluation (area, power, and throughput). In all of these implementations, NC operations are performed in $GF(2^m)$.

Our development effort have been targeted toward developing a library that manipulates the operation on integers, matrix and polynomials over $GF(q^m)$. Based on this library, we introduce a new simulation framework that implement inter-flow NC with end to end error control. We will refer to such framework as NCS-EC (Network Coding Simulator-Error Correction). NCS-EC is entirely written in C and allows for the evaluation of network coding in networks with errors and/or erasures. Its main features include (1) definition of graphs representing the topology (source node, destination node, intermediate nodes, link parameters,...) (2) definition of network coding procedure (the coefficients of combinations are defined or generated randomly, the buffer management in each node...) (3) visualization of the network transmission operations, (4) implementation of KK-code and LRMC code and (5) statistical analysis (Bit Error Rate, Rank of received matrix, number of transmission...).

The remainder of this paper is organized as follows. In Sect. 2, we describe in more detail the NCS-EC framework. In Sect. 3, we explain the simulator components and representation details. In Sect. 4, we describe the used libraries in the development of our simulator. Some simulation results will be given in Sect. 5. Finally, Sect. 6 concludes the paper.

2   NCS-EC Framework

NCS-EC is a framework that implements NC in $GF(q^m)$ with an error control carried out by the implementation of previously mentioned error correction codes. NCS-EC is implemented in C and thus can be compiled in speed on any computer with a suitable compiler. The other important strengths of NCS-EC are the extensibility and the modularity of simulator composition.

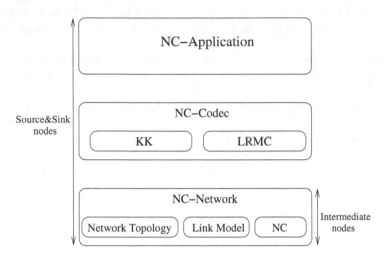

Fig. 1. General NCS-EC architecture.

The general architecture of NCS-EC is depicted in Fig. 1. It consists of three ordered layers, (1) NC-Network, (2) NC-Codec and (3) NC-Application which are analogous to upper layers of OSI model (Application, Transport and Network). There is an abstraction of the lower layers because all the packets will be transmitted directly from stacks. The operations performed in each layer are:

1. NC-Network layer:
    (a) Definition of network topology: number of nodes and type of each node (source, destination, intermediate), links between nodes,
    (b) definition of link models: each link can be a classical channel model (Gaussian channel, Gilbert-Elliot channel (GE), Rayleigh fading channel),
    (c) definition of NC protocols: Deterministic LNC (DLNC), RLNC, Conditional RLNC (CRLNC).
2. NC-Coded layer:
    (a) Applying an encoding at the source and a decoding at the destination using KK or LRMC code.
3. NC-Application
    (a) Generation of the transmitted message,
    (b) transmission (multicast) of packets,
    (c) saving and processing the statistics.

3   Simulator Components and Representation

In the following Section, we present an overview of the internal representation of the simulator's components, as well as of the statistics and simulation Output.

3.1   Network Topology

A network is defined by a graph with a set of nodes (source, destination and intermediate nodes) and links. The connections between nodes are described by a matrix whose entries are:

- 1 if there is a link between nodes
- 0 otherwise.

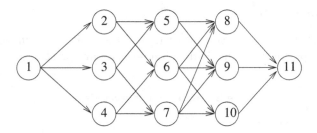

Fig. 2. Network topology with 11 nodes and 20 links.

Let's consider the network given in Fig. 2. The definition of such network is given in a header file as follows:

```
#define NB_LINKS 20
#define NB_NODES 11
#define NB_OUTPUT_LINK 3
int names[]={1,2,3,4,5,6,7,8,9,10,11}
```

The links between nodes are given by the following binary matrix

$$
\begin{pmatrix}
0 & 1 & 1 & 1 & 0 & 0 & 0 & 0 & 0 & 0 & 0 \\
0 & 0 & 0 & 0 & 1 & 1 & 0 & 0 & 0 & 0 & 0 \\
0 & 0 & 0 & 0 & 1 & 0 & 1 & 0 & 0 & 0 & 0 \\
0 & 0 & 0 & 0 & 0 & 1 & 1 & 0 & 0 & 0 & 0 \\
0 & 0 & 0 & 0 & 0 & 0 & 0 & 1 & 1 & 0 & 0 \\
0 & 0 & 0 & 0 & 0 & 0 & 0 & 1 & 1 & 1 & 0 \\
0 & 0 & 0 & 0 & 0 & 0 & 0 & 1 & 1 & 1 & 0 \\
0 & 0 & 0 & 0 & 0 & 0 & 0 & 0 & 0 & 0 & 1 \\
0 & 0 & 0 & 0 & 0 & 0 & 0 & 0 & 0 & 0 & 1 \\
0 & 0 & 0 & 0 & 0 & 0 & 0 & 0 & 0 & 0 & 1 \\
0 & 0 & 0 & 0 & 0 & 0 & 0 & 0 & 0 & 0 & 0
\end{pmatrix}
$$

Each link of the network is a channel model and has its own characteristics which are shown in Table 1 with their corresponding command line options.

Table 1. Channel command-line options

Channel model	Command options	Meaning
Gilbert-Elliot	–p_gb	transition probability from "good" to "bad" state
	–p_bg	transition probability from "bad" to "good" state
	–p_eg	error probability in "good" state
	–p_eb	error probability in "bad" state
Gaussian	- -g_snr	Signal to Noise Ratios(db) saved into a text file
Rayleigh	- -rl_FFd	maximum doppler frequency
	–rl_Ne	number of samples
	–rl_Te	sampling interval

3.2 Encoder Component

The first step in our simulation is to define the finite field in which all the operations are performed. The field characteristics are presented in Table 2.

Table 2. Finite field characteristics

Parameters	Meaning
q	field characteristic
m	field dimension
prim_poly	primitive polynomial used for construction of extended field

As mentioned above, the encoding component can be kk or LRMC. Each encoder has its mandatary parameters given by Table 3.

Table 3. Encoder parameters and their corresponding command-line options

Encoder	Parameters	Command options	Meaning
KK	k $(k \leq l)$	-k	message size
	l $(l \leq m)$	-l	number of independent elements $\mathbb{F}_{q^m}$
LRMC	k	-k	dimension of code
	n	-n	$k \leq n \leq m$: length of code

3.3 Transmission Protocol

If a source node has a message to send, a NEC coding is applied and resulting packets will be sent on every output link. Before starting the transmission, we must choose the NC protocol. If the DLNC is selected, the coefficients of linear combination are fixed in each intermediate node. These coefficients are determined by algebraic approach. In the case of the network 2, the coefficients in each intermediate node are:

$$[1], 3:[1], 4:[1], 5:\begin{bmatrix}1\\1\end{bmatrix}, 6:\begin{bmatrix}1\\0\end{bmatrix}, 7:\begin{bmatrix}1\\1\end{bmatrix}, 10:\begin{bmatrix}1\\1\end{bmatrix}, 8:\begin{bmatrix}0\\1\\0\end{bmatrix}, 9:\begin{bmatrix}1\\1\\0\end{bmatrix}.$$

Given that LNCC (Linear Network Coding Channel) in error-free case is $Y = HX$ where X, Y and H are transmitted, received and transfer matrix respectively [5]. Then, the transfer matrix H corresponding to DLNC in network 2 can be expressed as follows and is non-singular.

$$H = \begin{pmatrix}1\,0\,0\\0\,1\,0\\1\,1\,1\end{pmatrix}$$

If using RLNC, the coefficients are chosen randomly according to the selected distribution (Uniform, Poisson, Beta, Gamma...). Note that we cannot generate random variables by simulation but only pseudo-random ones. For the CRLNC case, if all the random coefficients generated at a node are null, then they have to be regenerated until at least one of them is not null. The command-line options corresponding to transmission protocol in our simulator are given in Table 4.

Table 4. Command-line options corresponding to Transmission Protocol

Command options	Meaning
-s	NC Protocol:
	0:RLNC
	1:DLNC
	2:CRLNC
-dist	the distribution used for random coefficients generation
	0:UNIFORM distribution
	1:POISSON distribution
	2:GEOMETRIC distribution
	3:BERNOULLI distribution
	4:BINOMIAL distribution
	5:EXPONENTIAL distribution
	6:GAMMA distribution
	7:BETA distribution
-E	if E=1 redundant packets will be sent
-L	number of redundant paquets

The transmission is by bit or symbol in a series of rounds as in [2]. During each round, a number of transmission steps is achieved. Each intermediate node of the network forwards packet that is a linear combination of packets inside their local stack. The transmission is stopped upon reaching the destination and then a decoding algorithm is applied to extract information within the limits of its correction capability.

3.4   Statistics and Simulation Output

The simulator outputs for given Monte-Carlo Value Iterations (MCVE) are Packet Error Rate (PER), Symbol Error Rate (SER), average number of received and independent received packets, average theoretical rank... These statistics are saved in a matlab file (.mat) and can be represented by curves which will be interpreted according to its structure. Some simulations results are given in Sect. 5.

4   Used Libraries

Based on the standard C libraries, we develop a library (gf_lib) that operates on finite fields (Galois Field (GF)) for any field characteristic (q) and dimension (m). It is declared in a number of header files. Each header file contains one or more function declarations, data type definitions, and macros. Functions definition is given in source files. The gf_lib structure is given in Fig. 3.

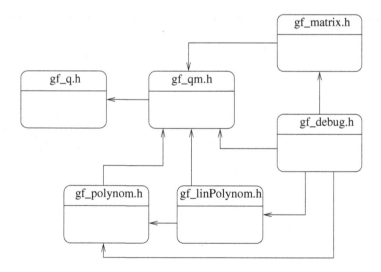

Fig. 3. gf_lib structure.

The "gf_q.h" file is the basic file containing $Gf(q)$ operations. The "gf_qm.h" includes "gf_q.h" file and contains the $GF(q^m)$ operations. As the titles indicate,

"gf_matrix.h", "gf_polynom.h" and "gf_linPolynom.h" handle matrix, polynomial and linearized polynomial over $GF(q^m)$, respectively. The "gf_debug.h"
defines the debug parameters of the simulator.

If an error occurs in the input parameters of simulator the help will be displayed directly indicating the bad argument. The help can be voluntarily shown
by adding the command line option (-h).

5   Simulation Results

We consider the network of Fig. 2 where all links are error-free. Then, only errors
introduced by network coding are considered (no errors from network links).
The error correction code applied is KK($l = 3, k = 2$). The table below gives the
results of simulation in terms of PER for two NC protocols (RLNC and CRLNC)
and different field characteristics.

Table 5. PER for random and conditional random network coding where $k = 2, l = 3$,
$m = 4$. The primitive polynomial for $\mathbb{F}_{q^m}$

q	prim_poly	PER RLNC	CRLNC
2	$1 + x^3 + x^4 (25)$	0,787402	0,104542
3	$2 + x + x^4 (86)$	0,442674	0,037985
5	$2 + 2x + x^2 + x^4 (662)$	0,169607	0,009050
7	$5 + 3x + x^2 + x^4 (2476)$	0,083465	0,003293

From Table 5, it can be seen that for both RLNC and CRLNC, an increase in
field characteristic decreases the PER and then increases the code performance.
We obtain an improvement in terms of packet error rate for $q = 7$, which varies
from 89.4 % in RLNC to 96.85 % in CRLNC. The larger the Galois field size q^m,
the better the performance of RLNC [3]. Indeed, the probability that transfer
matrix (formed by generated coefficients at intermediate nodes) is non-singular
increases as the field size increases.

In [13, Theorem 2], the authors gave a lower bound on the probability that a
random network code is valid (decoded). This probability is at least $(1 - \frac{d}{q^m})^\eta$,
where d is the number of receivers and η the number of links with associated
random coefficients (in our case we have $d = 1$ and $\eta = 20$). Table 6 shows a
comparison between this theoretical minimal probability and the experimental
probability of decoding for different values of q.

We consider now that every link of the network of Fig. 2 is a q−ary Gilbert
Elliot channel as described in [1].

Figure 4 shows the effect of adding our condition on the generated coefficients
(CRLNC) for two field sizes. For $q = 2$ we have, for a BER/link=0.08, an
improvement of about 53 %. Increasing q to 7 results in a small additional gain.

Table 6. Comparison between theoretical minimal probability and the experimental probability of decoding

	Theoretical minimal probability	Experimental probability	
q	RLNC	RLNC	CRLNC
2	0,27505	0,212598	0,895458
3	0,78001	0,557326	0,962015
5	0,96848	0,830393	0,99095
7	0,99170	0,916535	0,996707

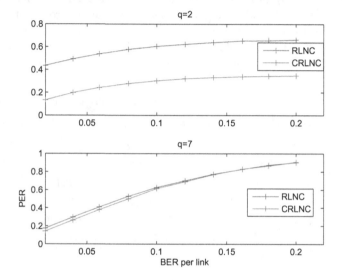

Fig. 4. Performance comparison between RLNC and DLNC in network 2 where l = 3, k = 2 and m = 4.(Color figure online)

In fact, the probability that all the coefficients are zero is higher for $q = 2$ compared to $q = 7$. Then applying the condition at intermediate node increases the number of independent packets and thus decreases the number of erasures. As given in [10], a code KK can correct t errors and ρ erasures provided that $\rho + t < l - k + 1$. Thus, decreasing the number of erasures increases the error correction capability which justifies the simulation results.

Let the network of Fig. 5 where each link is a gaussian channel. CRLNC is applied at intermediate nodes and the LRMC(8,4) is used for error correction. Figure 6 shows the performance in terms of SER(Symbol Error Rate) of lengthening LRMC(n, k). We vary its length and dimension by adding more data symbols while keeping redundancy $n - k$ fixed. Best results are observed for LRMC(8,4).

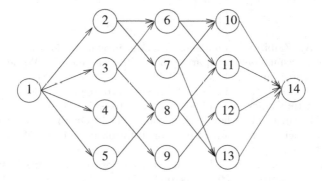

Fig. 5. Network topology with 14 nodes and 21 links.

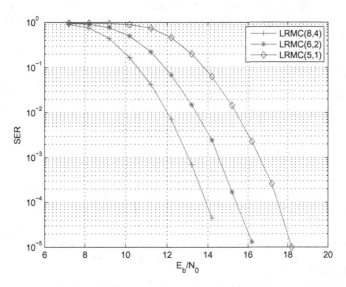

Fig. 6. Performance of lengthening code LRMC(n,k) for fixed $n - k = 4$.

6 Conclusion

We presented our simulator that implements inter-flow network coding with end to end error control using KK-code and LRMC based on Gabidulin code. The simulator design is based on a layered architecture which makes it extensible and easy to manipulate. We also gave some simulation results while varying network coding protocols, field characteristics and other parameters. The results verify the theoretical result that the larger the Galois field size q^m, the better the performance of RLNC.

References

1. Guefrachi, A., Zaibi, S., Bouallgue, A.: Conditional random network coding in the case of transmission over q-ary gilbert elliot channel. In: World Congress on Multimedia and Computer Science WCMCS, pp. 33–40 (2013)
2. Al Hamra, A., Barakat, C., Turletti, T.: Network coding for wireless mesh networks: a case study. In: Conference Publications, pp. 103–114 (2006)
3. Koller, C., Haenggi, M., Kliewer, J., Costello, D.J.: On the optimal block length for joint channel and network coding. In: Information Theory Workshop (ITW), pp. 528–532 (2011)
4. Ferreira, D., Serra, J., Lima, L., Prior, L., Barros, J.: NECO: NEtwork COding simulator. In: Simutools 2009 Proceedings of the 2nd International Conference on Secondary Ion Mass Tools and Technical, pp. 52–58 (2009)
5. Silva, D., Kschischang, F., Kötter, R.: A rank-metric approach to error control in random network coding. IEEE Trans. Info. Theor. 54(9), 3951–3967 (2008)
6. Ghaderi, M., Towsley, M., Kurose, J.: Reliability gain of network coding in lossy wireless networks. In: INFOCOM, 27th Conference on Computer Communication, pp. 816–820 (2008)
7. Chen, N., Yan, Z., Gadouleau, M., Wang, Y., Suter, B.W.: Rank metric decoder architectures for random linear network coding with error control. In: IEEE Transactions, pp. 296–309 (2012)
8. Ahlswede, R., Cai, N., Li, S.-Y.R., Yeung, R.W.: Network information flow. IEEE Inf. Theor. 46, 1016–1024 (2000)
9. Koetter, R., Médard, M.: An algebraic approach to network coding. IEEE/ACM Trans. Netw. 11, 782–795 (2003)
10. Koetter, R., Kschischang, F.: Coding for errors and erasures in random network coding. IEEE Trans. Inf. Theor. 54, 3579–3591 (2008)
11. Katti, S., Rahul, H., Hu, W., Katabi, D., Médard, M., Crowcroft, J.: Xors in the air: practical wireless network coding. In: ACM Conference of the Special Interest Group on Data Communication. (SIGCOMM 2006), pp. 243–254 (2006)
12. Li, S., Yeung, R., Cai, N.: Linear network coding. IEEE Trans. Inf. Theor. 46, 371–381 (2003)
13. Ho, T., Médard, M., Kötter, R., Karger, D., Effros, M., Shi, J., Leong, B.: A random linear network coding approach to multicast. IEEE Trans. Inf. Theor. 52, 4413–4430 (2006)
14. Wang, M., Li, B.: Lava: a reality check of network coding in peer-to-peer live streaming. In: INFOCOM 2007, 26th IEEE International Conference on Computer Communications, pp. 1082–1090 (2007)

Trends in Students Media Usage

Gerd Gidion[1], Luiz Fernando Capretz[2(✉)], Michael Grosch[1], and Ken N. Meadows[2]

[1] Karlsruhe Institute of Technology, Karlsruhe, Germany
{gidion,michael.grosch}@kit.edu
[2] Western University, London, ON, Canada
lcapretz@uwoa.ca, kmeadow2@uwo.ca

Abstract. Trends in media usage by students can affect the way they learn. Students demand the use of technology, thus institutions and instructors should meet students' requests. This paper describes the results of a survey where drivers in the use of media show continuously increasing or decreasing values from the first to the fourth year of study experience at the Western University, Canada, highlighting trends in the usage of new and traditional media in higher education by students. The survey was used to gather data on students' media usage habits and user satisfaction from first to fourth year of study and found that media usage increases over the years from first to fourth. The presentation of data using bar charts reveals a slight increase over the years in students owning notebooks or laptops off-campus and a significant increase from first to fourth year of students accessing online academic periodicals and journals. Another noteworthy finding relates to fourth year students being more conscious of the quality of information that they read on the Internet in comparison to students in first year, even though this is a slight year on year increase.

Keywords: Technological trends in education · Educational survey media usage habits · Educational survey · e-learning · Technology-enhance learning

1 Introduction

Digital media has changed students' learning environments and behaviors in higher education (Venkatesh et al. 2014). The NMC Horizon Report (Johnson et al. 2014) shows that a rapid and ubiquitous diffusion of digital media into higher education has led to changes in the students' learning environments and it has also influenced their learning behaviors. To keep pace with this changing learning environment, post-secondary institutions need to understand and analyze media usage behaviors of their students. This study, focusing on the media usage habits of students, was designed to provide an evidence base upon which reliable predictions can be made about future trends in media usage in higher education. The framework on which the research is based posits that the current teaching and learning methods are utilizing, and are influenced by, media that are a combination of traditional (e.g., printed books and journals) and new media (e.g., Wikipedia, Google). The current state of teaching and learning has been influenced by former media usage habits and these habits tend to change with the introduction of new media. In the future, teaching and

© Springer International Publishing Switzerland 2016
O. Gervasi et al. (Eds.): ICCSA 2016, Part I, LNCS 9786, pp. 491–502, 2016.
DOI: 10.1007/978-3-319-42085-1_38

learning will be influenced by the rising new media usage habits as well as by the current teaching and learning paradigms.

A second focus of this research is on media acceptance which provides an indicator of media quality from the participants' perspectives. Therefore, media quality is assessed by measuring the acceptance of the services used by students.

The Media Usage Survey was developed to provide researchers with a deeper and more detailed understanding of students' technology usage in learning and of possible environmental factors that may influence that usage.

2 Literature Review

Students tend to be early adopters of media and information technology, as they possess ample opportunities to access media, encouraged by their curiosity and self-learned skills. But they are not just passive users of technology, they are the designers and developers of the technology as well. For example, Google, the most commonly used search engine on the Internet, was created by Stanford students in the late 90s and Facebook was created by Harvard University students in 2004 and in less than ten years became one of the most successful Internet services worldwide.

Students in post-secondary education intensively use web services, such as Google, Wikipedia, and Facebook during their free time as well as for their studies (Smith et al. 2009). Current development in the so-called web 2.0 is often characterized by the increase in interactions between users (Capuruco and Capretz 2009). This is even more apparent with the more recent rise of collaborative media - in the form of social networking sites, file-sharing, wikis, blogs, and other forms of social network software.

Buckingham (2007) asserts that one cannot teach about the contemporary media without taking into account the role of the Internet, computer games, and the convergence between 'old' and 'new' media. Much of the popular discussion in this area tends to assume that students already know everything about the new media; they are celebrated as 'millenials' (Howe and Strauss 2000), or as "digital natives" (Prensky 2001) who are somehow spontaneously competent and empowered in their dealings with new media, although other researchers argue about lack of evidence (Bennett et al. 2008; Bennett and Maton 2010; Helsper and Enyon 2009). They learn to use these media largely through trial and error – through exploration, collaboration with others, experimentation and play.

Pritchett et al. (2013) examined "the degree of perceived importance of interactive technology applications among various groups of certified educators" (p. 34) and found that, in the involved schools, some groups seem to perceive Web 2.0 media as more important than others, e.g. participants of the survey with "an advanced degree and/or higher certification level" (p. 37).

Furthermore, mobile broadband Internet access and the use of corresponding devices, such as netbooks and smartphones, have fueled the boom of the social networks by students in higher education. Murphy et al. (2013) reported that in spite of the limitations in the formal university infrastructure, many students would like to use their mobile devices for formal learning as well as informal learning. Recent development in technology, with smartphones and tablets dominating the market in recent years have ensured that these devices have great functionality and enable interactivity, thus fulfilling the desire for both formal learning (Lockyer and Patterson 2008) and informal learning (Johnson and Johnson 2010).

There has been suggestion concerning the potential of this technological shift in students' learning and the real benefits of these technologies for learning (Johnson et al. 2014). There has been considerable research demonstrating the costs and benefits of using social, mobile, and digital technology to enhance teaching and learning; yet the research is not conclusive as to whether the use of these technologies leads to improved learning outcomes (Cusumano 2013). Klassen (2012) states: "If there is one thing I have learned the last ten years about the use of new technology in education, it is that the combination of old and new methods make for the best model."; and goes on to say: "Students will continue to seek out inspiring teachers. Technology alone is unlikely to ensure this, although it may make a lot of average teachers seem a lot better than they are!"

The usage of media at a university is a topic of interest for students, staff and faculty. There may be diverse interests and habits, but several interdependencies and interactions. The understanding of one of these scenarios has been the objective of a study by Kazley et al. (2013). They surveyed these groups and defined certain "factors that determine the level of educational technology use" (p. 68). They describe a model with increasing intensity/ quality of technology use, from beginners (using email and basic office software) to experts (using videoconference, virtual simulation tools etc.).

There is no doubt, however, that the integration of IT media and services in higher education appears to have led to substantial changes in the ways in which students study and learn (Dahlstrom 2012). Higher education institutions are cautious about investing in programs to provide students with mobile devices for learning, due to the rapidly changing nature of technologies (Alrasheedi and Capretz 2013a). The acceptance of technology-enhanced education by students has increased in recent years, but not all services are equally accepted (Alrasheedi and Capretz 2013b). It has become clear that simply using media and adopting e-learning does not necessarily make a difference in student learning. Rather, key factors for effective use of technology are pedagogy and the quality of the services (Alrasheedi and Capretz 2013c).

Moreover, we contend that, despite the highly contextual nature of e-learning studies, several characteristics are similar and the results could be developed into a framework for the assessment of the success of e-Learning. One such framework was presented by Ali et al. (2012), where learning contexts, learning experience, and design aspects were used to assess the success of mobile learning. Also, Capuruco and Capretz (2009) developed a social-aware framework for interactive learning.

In summary, the variety of media enriched informal learning processes is relevant. This perspective on the whole spectrum of media used for learning (printed, e-learning, digital, web 2.0, etc.) requires a certain theory-oriented empirical research approach to reach a deeper understanding about the media usage behavior of student in higher education.

3   Methods and Procedures

The integration of IT media and services in higher education has led to substantial changes in the ways in which students study and learn, and instructors teach (Johnson et al. 2014). This survey was designed to measure the extent to which media services are used in teaching and learning as well as to assess changes in media usage patterns. The survey is a landmark, as it is the first of its kind in Canada and represents an initial foray into the North American post-secondary sector.

The survey of students' media usage habits was conducted at Western University in London, Ontario, Canada, in 2013. The survey focuses primarily on the media usage habits of students. Based on an assessment of the way in which media use relates to teaching and learning, the identification of trends provides an evidence base upon which more reliable predictions can be made about future trends of media usage in higher education.

The survey is anonymous and consists of 150 items measuring frequency of media usage and user satisfaction with 53 media services, including:

- Media hardware such as Wi-Fi, notebooks, tablet computers, desktop computers, and smartphones;
- Information services, such as Google search, Google Books, library cata logues, printed books, e-books, printed journals, e-journals, Wikipe dia, open educational resources, and bibliographic software;
- Communication services, such as internal and external e-mail, Twitter, and Facebook;
- e-learning services and applications, such as learning platforms and wikis.

Additional variables were also evaluated such as some aspects of learning behavior, media usage in leisure time, educational biography, and socio-demographic factors, etc.

The survey tool was first developed in 2009 and used at Karlsruhe Institute of Technology (KIT) in Germany (Grosch and Gidion 2011). During the course of 15 follow-up surveys that were administered in a variety of countries, the original survey underwent customization, translation into several languages, and validation.

In this study, the survey was administered at The University of Western Ontario in London, Ontario, Canada to undergraduate students in the Winter of 2013 academic term. Having data from students from first to fourth years allows the researchers to compare their media usage to gain insight into possible discrepancies in their respective media cultures. These discrepancies could conceivably result in problems in the use of media for studying and teaching. The data for this survey was collected online using a well-established online survey tool: Unipark.

4   Results

Data collection occurred in two waves. In the first wave, 300 students in their first year in Engineering were surveyed in the Fall of 2012 (ESt1styear2012) but only 100 students completed the survey. Partial results involving instructors and students only in the

Faculty of Engineering were presented at the Canadian Engineering Education Association Conference (Gidion et al. 2013). A students satisfaction with media survey was also conducted and results are reported in (Gidion et al. 2014).

In the second wave of data collection, conducted between January 16th and February 15th, 2013, 19,978 undergraduate students (with 1 year, 2 years, 3 years, 4 and more years of study) from across the faculties were invited to respond to the survey and 1266 participated. Eight hundred and three students completed the survey (i.e., had a completion rate of more than 90 % of the survey items). While participants were from a broad spectrum of demographic characteristics and faculties, female students were disproportionately represented as in common in survey research (Sax et al. 2003). Otherwise, with some caveats, respondents are generally representative of the winter 2013 student population at Western.

4.1 Media Usage of Students by Academic Years

The results show that students most often attend class, followed by studying using a computer, and studying by themselves at home. Searching on the Internet for learning materials seems to be slightly more common than visiting libraries. Cooperative learning, compared to the other habits, seems relatively rare. The influence of cohort or study experience on the general habits to utilize media could have been interpreted by continuously developing usage frequencies. One of the few items in the complete questionnaire that shows (with the exception of the "beginners" from Engineering) a continuous increase is "study using a computer", but this happens anyway on a high level for all subgroups. This trend is presented in Fig. 1.

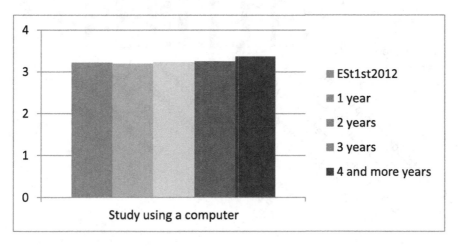

Fig. 1. Respondents mean use of a computer for studying, rated from "never" (0) to "very often" (4). (Color figure online)

4.2   Increasing Media Use

For a subset of media, there is an increase in their use across the four years of the degree. These media includes respondents' notebook/laptop off campus, printed books and printed handouts from instructors, Wikipedia, and Google books as well as online services from the universities library and bibliographic software. Interestingly, there are large increases over the course of the four years in the use of "e-versions of academic periodicals/journals", an increase the strength of which is not evident with printed versions of these journals. An Analysis of Variance (ANOVA) was performed to examine these results. Significant differences between the groups (program year) could be found concerning the usage frequency of own notebook/laptop off-campus, Computer labs on campus (e.g. Genlab), online services of the university library (central)/faculty library, e-versions of academic periodicals/journals and Wikipedia. Computer labs on campus are increasingly used in higher terms. This trend is depicted in Fig. 2.

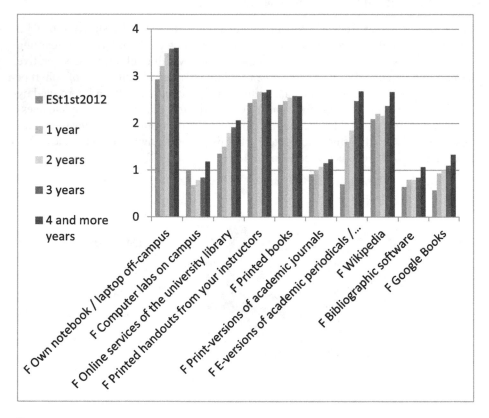

Fig. 2. Mean frequency of respondents use of different media for learning by year of program, rated from "never" (0) to "very often" (4). ['F' stands for Frequency] (Color figure online)

4.3 Decreasing Media Use

Looking at the decreasing media use, the two main trends concern online self- tests for studying and online exams for course grades. The ANOVA showed a significant result concerning the differences between the groups (year of program) for online exams (for grades in a course) and online self-tests with the frequency of use decreasing along the years. This decrease may be the result of the nature of learning assessments across the four years in some faculties, with assessments in the early years lending themselves more to self-testing (e.g., multiple choice examinations) than those often used in the later years (e.g., essay assignments and exams). There was not a significant effect for game-based learning applications. Facebook has a slightly (but significant) shrinking frequency, and on a lower level of frequency the items game-based learning applications, Google+ and virtual class in real-time as well as in non-real-time. This overall tendency is described in Fig. 3.

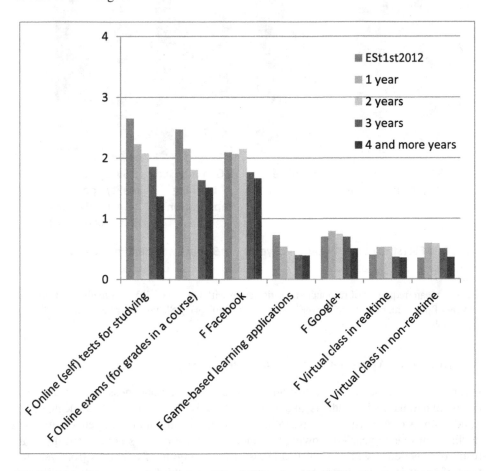

Fig. 3. Mean frequency of respondents use of different media for learning by year of program, rated from "never" (0) to "very often" (4). ['F' stands for Frequency] (Color figure online)

4.4   Increasing Satisfaction with Media Usage

There were no clear trends which suggest that students' satisfaction with their media use for learning increased across the four years. For the items "own notebook/laptop off campus" and "online dictionary", there are inconsistent increases across the time period. With "game-based learning applications" and "augmented reality applications" students from the first two years of study are less satisfied than students with three and more years at the university (see Fig. 4).

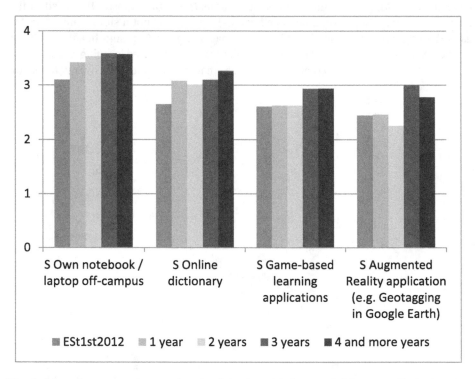

Fig. 4. Mean frequency of respondents' satisfaction with the use of different media for learning by year of program, rated from "never" (0) to "very often" (4). ['S' stands for Satisfaction] (Color figure online)

4.5   Decreasing Appreciation Regarding Media Usage

Figure 5 captures the answer to two questions reflecting the openness of instructors to the use of new media for students' and the instructors own teaching. There is a decreasing trend in terms of the instructors' performed use (significant) and their openness to new media from year to year – following the students perception. It is not clear if this is a result of experience with the finished terms or the changing of openness from year to year of study (e.g. that instructors are not as open for third year students media use compared with the first year students.

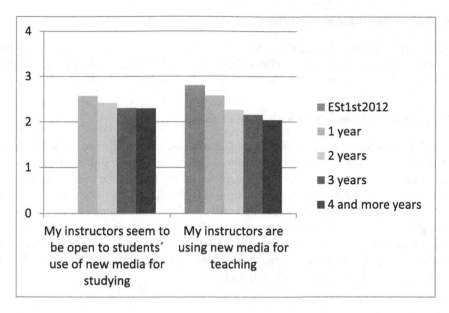

Fig. 5. Means of students' responses to specifically selected items to the question: to what extent do you agree/disagree with the following statements? [0 = strongly disagree, 4 = strongly agree] (Color figure online)

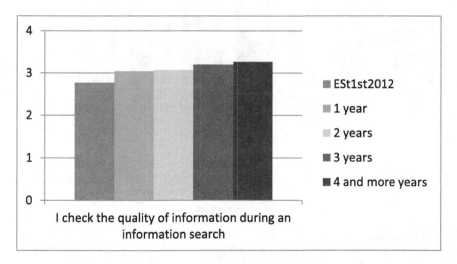

Fig. 6. Means of students' responses to a specifically selected item to the question: to what extent do you agree/disagree with the following statements? [0 = strongly disagree, 4 = strongly agree] (Color figure online)

4.6 Increasing Information Literacy with Media Usage

In the group of items with statements the results for students just came to one item with continuous increasing values: to check the quality of information during an internet search seems to be more relevant from year to year at the university (Fig. 6).

4.7 Decreasing Satisfaction with Media Usage

Several media showed a decreasing level of satisfaction from the first to the fourth year, specifically with mobile phones, own notebook/laptop on campus and wireless connection on campus (significant). With the exception of the first year students from Engineering, the respondents rated the universities website, the universities e-mail-account (significant), the Learning Management System (significant), online exams and video sharing websites (significant) as less satisfying from year to year (see Fig. 7). Again it is not possible to determine from these data if this decrease is the result of dissatisfaction because of the respondents' experience with the media or if the usage habits are different which is related to the respondents' satisfaction.

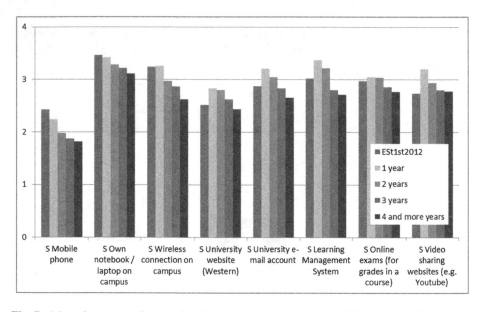

Fig. 7. Mean frequency of respondents' satisfaction with the use of different media for learning by year of program, rated from "never" (0) to "very often" (4). ['S' stands for Satisfaction] (Color figure online)

5 Conclusion

We examined media usage trends among students in higher education to provide an evidence base so that predictions can be made about future trends regarding students and their media usage and habits. Whilst current teaching and learning combines both old and new media, this study acknowledges the need for an analysis of the current changes in studying and learning based on the integration of IT and new media into higher education.

There is a great deal of value and potential in this research insofar as it presents current data on students' media usage and satisfaction in higher education. Overall the study demonstrates a trend of increased media usage by students. The use of technology increases from 1st to 4th year students, whereas satisfaction with technology decreases among 1st to 4 year student. However, first year student seem to be using social media more frequently than students in other years, whereas 4th year student appear to be more conscious about the quality of information they read on the Internet. A trend of special interest in the next few years will be mobile learning and the impact of BYOD (Bring Your Own Device) in the students' learning environment.

References

Ali, A., Ouda, A., Capretz, L.F.: A conceptual framework for measuring the quality aspects of mobile learning. Bull. IEEE Tech. Committee Learn. Technol. **14**(14), 31–34 (2012)

Alrasheedi, M., Capretz, L.F.: An m-learning maturity model for the educational sector. In: 6th Conference of MIT Learning International Networks Consortium (MIT LINC), Cambridge, MA, USA, pp. 1–10 (2013a)

Alrasheedi, M., Capretz, L.F.: Can mobile learning maturity be measured? a preliminary work. In: Proceeding of the Canadian Engineering Education Association Conference (CEEA 2013), Montreal, Canada, pp. 1–6 (2013b)

Alrasheedi, M., Capretz, L.F.: A meta-analysis of critical success factors affecting mobile learning. In: Proceedings of IEEE International Conference on Teaching, Assessment and Learning for Engineering, Bali, Indonesia, pp. 262–267 (2013c)

Bennett, S., Maton, K., Kervin, L.: The 'digital natives' debate: a critical review of the evidence. Br. J. Educ. Technol. **39**(5), 775–786 (2008). http://ro.uow.edu.au/cgi/viewcontent.cgi?article=2465&context=edupapers

Bennett, S., Maton, K.: Beyond the 'digital natives' debate: towards a more nuanced understanding of students' technology experiences. J. Comput. Assist. Learn. **26**(5), 321–331 (2010). doi:10.1111/j.1365-2729.2010.00360.x

Buckingham, D.: Media education goes digital: an introduction. Learn. Media Technol. **32**(2), 111–119 (2007)

Capuruco, R.A.C., Capretz, L.F.: Building social-aware software applications for the interactive learning age. Interact. Learn. Environ. **17**(3), 241–255 (2009)

Cusumano, M.: Are the costs of 'free' too high in online education? Commun. ACM **56**(4), 1–4 (2013)

Dahlstrom, E.: Study of undergraduate students and information technology. In: EDUCAUSE. Center for Applied Research, Louisville, Colorado, USA (2012). http://net.educause.edu/ir/library/pdf/ERS1208/ERS1208.pdf

Gidion, G., Capretz L.F., Grosch, G., Meadows, K.N.: Media usage survey: how engineering instructors and students use media. In: Proceedings of Canadian Engineering Education Association Conference (CEEA 2013), Montreal, p. 5 (2013)

Gidion, G., Capretz, L.F., Meadows, K., Grosch, M.: Are students satisfied with Media: a Canadian case study. Bull. IEEE Tech. Committee Learn. Technol. **16**(1), 6–9 (2014)

Grosch, M., Gidion, G.: Mediennutzungsgewohnheiten im Wandel - Ergebnisse einer Befragung zur studiumsbezogenen Mediennutzung (in German). Karlsruhe Institute of Technology, KIT Scientific Publishing, Karlsruhe, Germany (2011). http://digbib.ubka.uni-karlsruhe.de/volltexte/1000022524

Helsper, E., Eynon, R.: Digital natives: where is the evidence? Br. Educ. Res. J., 1–18 (2009). http://eprints.lse.ac.uk/27739/1/Digital_natives_%28LSERO%29.pdf

Howe, N., Strauss, W.: Millennials Rising: The Next Great Generation. Vintage, New York (2000)

Johnson, G.M., Johnson, J.A.: Dimensions of online behavior: implications for engineering e-learning. In: Iskander, M., Kaoila, V., Karim, M.A. (eds.) Technological Developments in Education and Automation, p. 6. Springer, Heidelberg (2010). doi:10.1007/978-90-481-3656-8_13

Johnson, L., Becker, S.A., Estrada, V., Freeman, A.: NMC Horizon Report 2014 Higher Education. New Media Consortium, Austin (2014)

Kazley, A.S., Annan, D.L., Carson, N.E., Freeland, M., Hodge, A.B., Seif, G.A., Zoller, J.S.: Understanding the use of educational technology among faculty, staff, and students at a medical university. TechTrends 57(2), 63–70 (2013)

Klassen, T.R.: Upgraded anxiety and the aging expert. Acad. Matters J. Higher Educ., Ontario Confederation of University Faculty Association, pp. 7–14 (2012)

Lockyer, L., Patterson, J.: Integrating social networking technologies in education: a case study of a formal learning environment. In: Proceedings of the 8th IEEE International Conference on Advanced Learning Technologies, pp. 529–533 (2008)

Murphy, A., Farley, H., Lane, M., Hafess-Baig, A., Carter, B.: Mobile learning anytime, anywhere: what are our students doing? In: 24th Australasian Conference on Information Systems (2013)

Pritchett, C.C., Wohleb, E.C., Pritchett, C.G.: Educators' perceived importance of web 2.0 technology applications. TechTrends 57(2), 33–38 (2013)

Prensky, M.: Digital natives, digital immigrants (2001). http://www.marcprensky.com/writing/Prensky%20%20Digital%20Natives%20Digital%20Immigrants%20-%20Part1.pdf

Sax, L.J., Gilmartin, S.K., Bryant, A.N.: Assessing response rates and nonresponse bias in web and paper surveys. Res. High. Educ. 44(4), 409–432 (2003)

Smith, S.D., Salaway, G., Caruso, J.G.: The ECAR study of undergraduate students and information technology. In: EDUCAUSE. Center for Applied Research, Louisville (2009)

Venkatesh, V., Croteau, A.M. Rabah, J.: Perceptions of effectiveness of instructional uses of technology in higher education in an era of Web 2.0. (2014). http://ieeexplore.ieee.org/stamp/stamp.jsp?tp=&arnumber=6758617

Short Papers

Computational Investigation of Heat Transfer of Nanofluids in Microchannel with Improper Insulation

Vai Kuong Sin[✉], Ka Kei Teng, and Wen Yue Deng

Department of Electromechanical Engineering,
University of Macau, Macau SAR, China
vksin@umac.mo

Abstract. Numerical simulation of heat transfer of nanofluids in rectangular microchannel has been performed. A micro-electro-mechanical system is built with two heating resistors embedded in a silicon substrate. Microchannel made from polydimethylsiloxane is planted in the substrate and nanofluids with different volume fractions are used as the coolant. Because of the not well-insulated micro-electro-mechanical system, heat generated by the heating resistors will be lost to the surrounding as well as carried away by the nanofluids through the microchannel. Heat transfer is analyzed at low Reynolds number for different nanofluid concentrations.

Keywords: Nanofluid · Convective heat transfer · Microchannel

1 Introduction

Nowadays heat exchange is an important problem in industry. Electronics devices, such as electronic cooling system, network server or integrated data center, have high requirement of cooling in tiny space. Convective heat transfer is a very effective method for this situation, heat flux of the order of 107 W/m^2 can be transferred in micro channels with a surface temperature below 71 °C [1]. The thermal efficiency can be improved in many ways. The most obvious methods are increase contact area or increase flow rate, but these methods may need to modify the device design or add extra pump. Another suggestion is to increase the conductivity of coolant, for example, mix high conductivity material into cooling liquid. Due to the technical limitation in the early days, the adding of micrometer scale particles in the coolant may lead to increase of friction and hence results in clogging of the microchannel. Lately, nanoscale particles are selected as constituents to add into the coolant. The most common nanoparticles that are used can be cataloged as ceramic particles, pure metallic particles, and carbon nanotubes (CNTs).

Nanofluid, a fluid containing nanometer-sized particles, has novel properties compared with its base fluid. Numerous literatures indicate that the nanofluids can increase the convective heat transfer coefficient and friction factor, but there are still some reports show that the heat transfer may not be improved. There are several experiments in the regard, but they did not show satisfactory evidence [2, 3].

© Springer International Publishing Switzerland 2016
O. Gervasi et al. (Eds.): ICCSA 2016, Part I, LNCS 9786, pp. 505–513, 2016.
DOI: 10.1007/978-3-319-42085-1_39

Enhancement of heat transfer is not obvious in [3]. Such a finding is different from many studies in the literature, although there are some studies which do show the increase of heat transfer by nanofluids [4, 5]. Further study on the energy budget of the system in [3] indicates that "heat loss" in the system is significant. Rate of heat carried away by the coolant is less than the heat generated by the electric power input, and the former is only a few percent to about fifty percent of the latter, depending on the Reynolds number of the flow. It indicates that the entire system in [3] is not well-insulated and portion of the heat is transferred to the surrounding air through the surfaces of the system. This is a common feature of many microfluidic cooling devices if they are not properly insulated. This research aims to investigate the convective heat transfer of nanofluids in microchannels which is embedded in a micro-electro-mechanical system (MEMS) that is not well-insulated and there is some heat loss to the surrounding through the outer boundaries of the MEMS. Since it is not easy to have good insulation and detailed temperature/heat flux measurement as the micro system built in [3], it would be much helpful for the research if a reliable numerical tool is developed in complementary to the experiments for a detailed understanding of the heat transfer phenomena in micro device.

2   Properties of Nanofluids

Commercial software COMSOL-Multiphysics is used to perform the heat transfer of nanofluids in microchannel. Navier-Stokes equations and energy equation are used to solve the conjugate heat transfer problem of nanofluids with different volume fraction in microchannel of MEMS. Physical properties of nanofluids such as density, thermal conductivity, specific heat capacity, and viscosity are required to specify for simulation. Calculation of all these properties is given in [6] and the information is reproduced in Table 1.

Table 1. Properties of nanofluids

Volume fraction	0 %	0.6 %	1.2 %	1.8 %
Thermal conductivity (W/m/K)	0.6070	0.61644	0.62599	0.63564
Specific heat (J/kg/K)	4181.3	4094.24	4010.43	3929.70
Density (kg/m^3)	998.57	1017.96	1037.34	1056.73
Dynamic viscosity (kg/m/s)	0.84e$-$3	0.84e$-$3	0.84e$-$3	0.84e$-$3

3   Simulation Setting

The experiment of [3] is modelled by COMSOL-Multiphysics and is given in Fig. 1. To complete this complex modelling, the following settings which are close to [3] as possible are applied in the modelling. A base silicon substrate with a size of 45 mm × 45 mm × 525 m is built first. Two heating resistors with size of 2 mm width, 30 mm long and 0.1 mm high are then embedded on the substrate. Two reservoirs with diameter of

2 mm are connected to the inlet and outlet of the microchannel separately. The micro channel has a length of 17.5 mm and a rectangular cross section areas of 100 m × 1000 m. The entire microchannel and reservoirs are sheltered with a poly-dimethylsiloxane (PDMS) channel cover. Property of silicon substrate and PDMS channel cover is given in Table 2. Bottom of the microchannel and the reservoirs which can be assumed to be the main surface for heat transfer by nanofluid is meshed as triangular element with maximum element size of $4e^{-5}$ m. There are thirty-five thousand triangular elements in the bottom of the microchannel and the reservoirs and it is extruded by fifteen layers to constitute the microchannel channel and reservoirs. The microchannel consists of over half million triangular prism elements. For the channel cover and silicon substrate, they are meshed with rectangular elements. The connection of triangular and rectangular elements can be merged by cutting the rectangle elements into two triangular prisms. Each of triangular can be converted into three tetrahedral elements. Total typical number of elements for the simulation is twenty-seven million. Details of the mesh arrangement can be found in [6].

In order to find out the amount of heat that is lost to the ambient air which surrounds the MEMS, an appropriate convective heat transfer coefficient of air has to be determined. It is obtained by using de-ionized (DI) water as the coolant and matching the numerical result with experimental data of [3]. At the assumption of room temperature of twenty-five degrees Celsius, it is found that heat transfer coefficient, $h = 12$ W/m²/K, which is in the range of natural convection coefficient between 2 and 25 W/m²/K [7].

The experiment of [3] has totally four sets of variable, they are channel cross section area (50 m × 50 m, 100 m × 100 m, and 100 m × 1000 m); Reynolds number, Re, which is based on mean velocity and hydraulic diameter of the microchannel (Re changes from 0.6 to 180); heating power (0.57 W to 1.376 W); nanofluid concentration or volume fraction (0 %, 0.6 %, 1.2 %, and 1.8 %).

4 Results and Discussion

For the purpose of duplicate the physical experiment of [3], especially the accurate estimation of the heat loss to the surrounding, a simulation model is run with conditions which is matching the experimental settings of [3] as possible. The model is a microchannel with cross section area of 100 m × 1000 m. DI water with no nanoparticles is the working fluid. Reynolds number is equal to one. Power generated by the two heating resistors is 0.57 W. Figure 2 shows the temperature variations along the microchannel. Solid lines represent simulation results in the top, center, and bottom of the channel. Also shown in the figure is the experiment data obtained at the bottom of the channel in [3]. Simulation results are compared well with those experimental founding except in the inlet of the channel. The big difference of the result in the inlet can be explained as that the location of the thermocouple in the inlet for temperature measurement in [3] is not exactly in the inlet of the channel. Instead, it is somewhere upstream in the reservoir and above the bottom as stated clearly in [3]. Therefore it has no meaning to compare the results at that particular location. A slightly increase of the fluid temperature along the flow direction of the channel is found as expected.

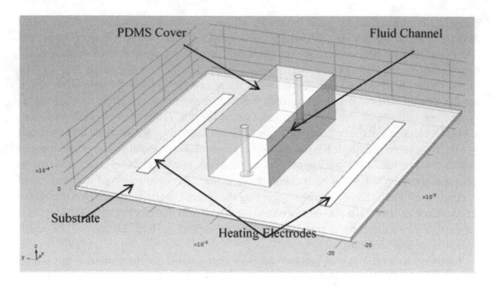

Fig. 1. Scale drawing for the simulation model

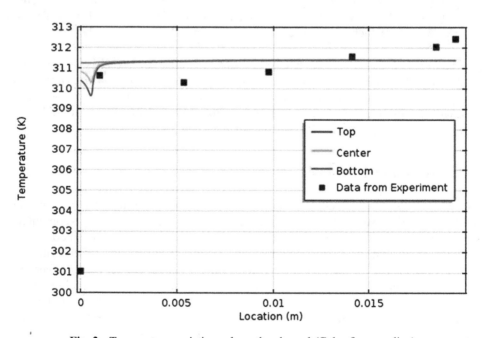

Fig. 2. Temperature variations along the channel (Color figure online)

Temperature distribution on the surface of the substrate and channel cover is given in Fig. 3. Temperature contour on the surface of the substrate is shown in Fig. 4. From these two figures it is noted that heat generated by the two heating resistors is transferred through the substrate to the channel cover by conduction. Heat conduction from

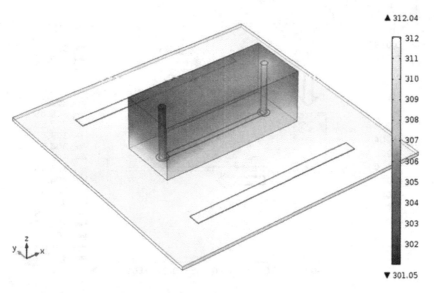

Fig. 3. Temperature distribution on the surface of the substrate and channel cover (Color figure online)

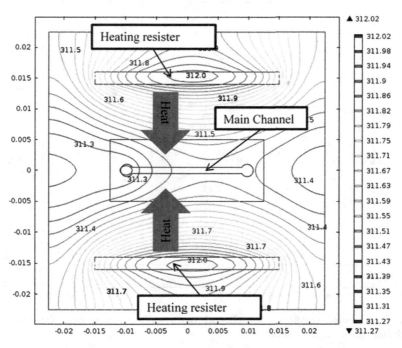

Fig. 4. Temperature contour on the surface of the substrate (Color figure online)

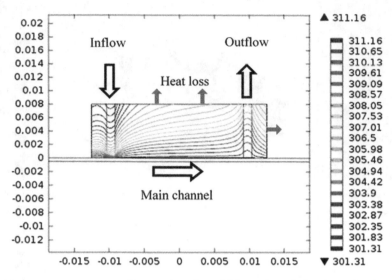

Fig. 5. Temperature contour on the central section which cut through the channel and reservoirs. (Color figure online)

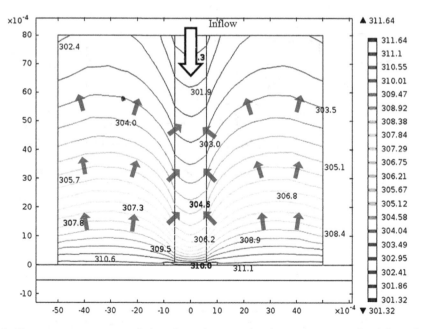

Fig. 6. Temperature contour of the channel cover in the region near the inlet of the microchannel (Color figure online)

the bottom of the channel cover to the side walls as well as to the top surface of the channel cover is also observed. Eventually, all heat at the surface of the substrate and

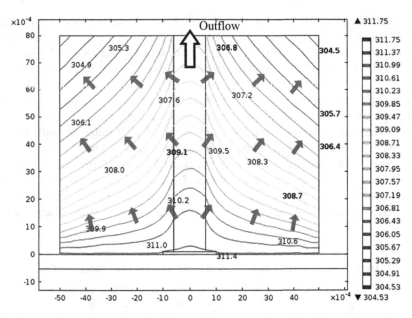

Fig. 7. Temperature contour of the channel cover in the region near the outlet of the microchannel (Color figure online)

Table 2. Property of silicon substrate and PDMS cover

Material	Silicon substrate	PDMS cover
Density (kg/m³)	2329	0.97
Thermal conductivity (W/m/K)	130	0.15
Specific heat (J/kg/K)	700	1460

channel cover will be transferred through convection to the ambient air and this is the heat loss. Figure 5 exhibits the temperature contour on the vertical section of the channel cover, which is cut through the two reservoirs and microchannel. It is clearly seen that heat is transferred by heat conduction from the bottom to the top of the channel cover. At the same time, heat is also transferred by conduction from the region which is close to the outlet of the microchannel to the region that is close to the inlet of the microchannel. This is due to the fact that heat from the bottom of the microchannel is carried away by the nanofluid through heat convection from the inlet to the outlet. And that makes the temperature in the channel cover near the outlet is higher than that near the inlet by around six degree Celsius on average. This can also be observed in Figs. 6 and 7 which gives temperature distributions in the region which is close to the inlet and outlet respectively. These two figures show clear that heat is being transferred from the channel cover to the nanofluid in the region near the inlet, while heat is transferred from the nanofluid back to the channel cover in the region near the outlet.

To study the effect of heat transfer of nanofluids in microchannel, simulation of nanofluids with Re = 1 for different volume fractions has been performed and the results are shown in Table 3. In Table 3, Q_f represents the heat carried away by the nanofluid. Q_{HR} is the heat generated by the two heating resistors. Efficiency of the heat transfer of the nanofluid is defined as the ratio of Q_f over Q_{HR}. Also shown in the table are the experimental data of [3] for comparison. Table 3 indicates that the efficiency of heat transfer of nanofluid is small. For certain volume fraction, i.e., 1.2 % and 1.8 %, the heat transfer of nanofluid is even worse than the pure DI water for both simulation results and experimental finding. This low efficiency means that most of the heat is lost to the ambient. This small effect of heat transfer of nanofluid may due to the small Reynolds number being used in the simulation. Because of the limit of time, Results in simulation with just one cross section area of 100 m × 1000 m, one Reynold number, Re = 1, heating power between 0.57 W to 0.64 W have presented and discussed in this paper.

Table 3. Energy carried away by nanofluid with different volume fractions

Volume fraction		0	0.6 %	1.2 %	1.8 %
Q_{HR} (W)		0.57	0.58	0.65	0.64
Simulation	Q_f [W]	0.021	0.022	0.020	0.020
	Efficiency	3.72 %	3.74 %	3.13 %	3.15 %
Experiment [3]	Q_f (W)	0.023	0.024	0.023	0.023
	Efficiency	4.10 %	4.14 %	3.57 %	3.59 %

5   Conclusions and Future Work

Computational study of heat transfer of nanofluid in microchannel embedded in MEMS which is not properly insulated has been performed. Numerical results for small Reynolds number, Re = 1, indicates that heat transfer efficiency is not obvious. This finding is in consistent with those experimental data available in the literature. It is suggested that further investigation should be made for higher Reynolds number as well as more proper insulation in order to better understanding the behavior of heat transfer of nanofluid in microchannel.

Acknowledgments. The research is supported by the Research Committee of University of Macau under Multi-Year Research Grant MYRG2016-00074-FST. We would also like to thank Professor U. Lei from National Taiwan University for providing experimental data for comparison and giving this research so many valuable suggestions and comments.

References

1. Jung, J.Y., Oh, H.S., Kwak, H.Y.: Forced convective heat transfer of nanofluids in microchannels. Int. J. Heat Mass Transf. **52**, 466–472 (2008)

2. Ho, C.J., Wei, L.C., Li, Z.W.: An experimental investigation of forced convective cooling performance. Appl. Therm. Eng. **30**, 96–103 (2010)
3. Chen, Y.M.: Experiment on forced convection of nanofluids in micro channel. Master Thesis of National Taiwan University (2013)
4. Wang, X.Q., Mujumdar, A.S.: Heat transfer characteristics of nanofluids – a review. Int. J. Therm. Sci. **46**, 1–19 (2007)
5. Murshed, S.M.S., Leong, K.C., Yang, C.: Thermophysical and electrokinetic properties of nanofluids – a critical review. Appl. Therm. Eng. **28**, 2109–2125 (2008)
6. Deng, W.Y.: Simulation of force convection of nanofluids in micro channel. Master Thesis of University of Macau (2015)
7. Cengel, Y.A., Ghajar, A.J.: Heat and Mass Transfer Fundamentals and Applications. McGraw Hill Education, New York (2011)

Modelling the MSSG
in Terms of Cellular Automata

Sara D. Cardell[1]([✉]) and Amparo Fúster-Sabater[2]

[1] Instituto de Matemática, Estatística e Computação Científica,
UNICAMP, Campinas, Brazil
`sdcardell@ime.unicamp.br`
[2] Instituto de Tecnologías Físicas y de la Información (CSIC),
144, Serrano, 28006 Madrid, Spain
`amparo@iec.csic.es`

Abstract. The modified self-shrinking generator is a non-linear cryptographic sequence generator designed to be used in hardware implementations. In this work, the output sequence of such a generator is obtained as one of the output sequences of a linear model based on Cellular Automata. Although irregularly decimated generators have been conceived and designed as non-linear sequence generators, in practice they can be easy modelled in terms of simple linear structures.

Keywords: Modified self-shrinking generator · Cellular automata · rule 102 · rule 60 · Stream cipher · Cryptography

1 Introduction

Nowadays stream ciphers are the fastest among the encryption procedures. They are designed to generate, from a short key, a long sequence (*keystream sequence*) of seemingly random bits. Typically, a stream cipher consists of a keystream generator whose output sequence is bit-wise XORed with the plaintext (in emission) to obtain the ciphertext or with the ciphertext (in reception) to recover the original plaintext. Some well known designs in stream ciphers can be found in [4,13].

Most keystream generators are based on maximal-length Linear Feedback Shift Registers (LFSRs) [6]. Such registers are linear structures characterized by their length L, their characteristic polynomial $p(x)$ and their initial state IS (currently the key of the cryptosystem). Their output sequences, the so-called PN-sequences, are usually combined in a non-linear way in order to break their inherent linearity and produce sequences of cryptographic application. Combinational generators, nonlinear filters, irregularly decimated generators or LFSRs with dynamic feedback are just some of the most popular keystream generators that can be found in the literature [11,12].

Irregularly decimated generators produce good cryptographic sequences characterized by long periods, adequate correlation, excellent run distribution, balancedness, simplicity of implementation, etc. The underlying idea of this kind

© Springer International Publishing Switzerland 2016
O. Gervasi et al. (Eds.): ICCSA 2016, Part I, LNCS 9786, pp. 514–520, 2016.
DOI: 10.1007/978-3-319-42085-1_40

of generators is the irregular decimation of a PN-sequence according to the bits of another one. The result of this decimation is a binary sequence that will be used as keystream sequence in the cryptographic procedure of the stream cipher. Inside the family of irregularly decimated generators, the self-shrinking generator (SSG) was first introduced by Meier and Staffelbach in [10]. This keystream generator is a simplified version of the shrinking generator (SG) introduced by Coppersmith et al. in [2] that used two different LFSRs. In contrast to the SG, the SSG is based on a unique maximal-length LFSR that generates via a self-decimation process a pseudorandom binary keystream.

In [5] the authors modelled the output sequence of the SSG, the so-called self-shrunken sequence, by means of cellular automata (CA) that used rules 90 and 150. Later, in [1] a new family of cellular automata modelled this sequence through simpler and shorter cellular automata that used rule 102.

The modified self-shrinking generator (MSSG) was recently discover by Kanso [8]. It involves a unique maximal-length LFSR and uses an extended decimation rule based on the XORed value of a pair of bits. In this work, we model the modified self-shrunken sequence in terms of CA by using rule 102 too.

The paper is organized as follows: in Sect. 2 we provide some fundamentals and basic concepts. Section 3 shows how to model the modified self-shrunken sequence via cellular automata. In Sect. 4, we see how to recover the self-shrunken sequence by using a fixed number of intercepted bits. Finally, conclusions in Sect. 5 end the paper.

2   Preliminaries

Notation and basic concepts that will be used throughout the work are now introduced.

2.1   Modified Self-shrinking Generator

The **modified self-shrinking generator** (MSSG) was introduced by Kanso in 2010 for hardware implementations [8]. It is a special case of the self-shrinking generator [10], where the PN-sequence generated by a maximum-length LFSR is self-decimated. Here the decimation rule is very simple and can be described as follows: Given three consecutive bits $\{u_{2i}, u_{2i+1}, u_{2i+2}\}$, $i = 0, 1, 2, \ldots$ of a PN-sequence $\{u_i\}$, the output sequence $\{s_j\}$ is computed as

$$\begin{cases} \text{If } u_{2i} + u_{2i+1} = 1 \text{ then } s_j = u_{2i+2} \\ \text{If } u_{2i} + u_{2i+1} = 0 \text{ then } u_{2i+2} \text{ is discarded.} \end{cases}$$

We call the $\{s_j\}$ sequence as the **modified self-shrunken sequence**. If L is the length of the maximum-length LFSR that generates $\{u_i\}$, then the linear complexity LC of the corresponding modified self-shrunken sequence satisfies:

$$2^{\lfloor \frac{L}{3} \rfloor - 1} \leq LC \leq 2^{L-1} - (L - 2),$$

and the period T of the sequence satisfies:

$$2^{\lfloor \frac{L}{3} \rfloor} \le T \le 2^{L-1}$$

as proved in [8]. As usual, the key of this generator is the initial state of the register that generates the PN-sequence $\{u_i\}$. Moreover, the characteristic polynomial of the register, $p(x)$, is also recommended to be part of the key.

Next a simple illustrative example is introduced.

Example 1. Consider the LFSR of length $L = 3$ with characteristic polynomial $p(x) = 1 + x^2 + x^3$ and initial state $IS = (1\ 0\ 0)$. The PN-sequence generated is $1001110\ldots$ with period $T = 2^3 - 1$.

Now the modified self-shrunken sequence can be computed as follows:

$$R: \underbrace{1\ 0\ \textbf{0}\ 1\ 1\ \cancel{1}}_{1}\ \underbrace{0\ 1\ \textbf{0}}_{0}\ \underbrace{0\ 1\ \textbf{1}}_{1}\ \underbrace{1\ 0\ \textbf{1}}_{1}\ \underbrace{0\ 0\ \cancel{1}\ 1\ 1\ \cancel{0}}_{0}\ldots$$

This sequence $0011\ldots$ has period $T = 4$ and its characteristic polynomial is $p_s(x) = (1 + x)^3$. Thus, the linear complexity of this modified self-shrunken sequence is $LC = 3$. ■

2.2 Cellular Automata

Cellular automata (CA) are discrete models where the contents of the cells (binary in our work) are updated following a function of k variables [14] called *rule*. The value of the cell in position i at time $t + 1$, notated x_i^{t+1}, depends on the value of the k neighbour cells at time t. If these rules used in the CA are composed exclusively by XOR operations, then the CA is said to be **linear**. In this work, the CA considered are **regular** (every cell follows the same rule) and **null** (null cells are considered adjacent to extreme cells). For $k = 3$, the rule 102 is given by:

$$\textbf{Rule 102: } x_i^{t+1} = x_i^t + x_{i+1}^t$$

111	110	101	100	011	010	001	000
0	1	1	0	0	1	1	0

According to Wolfram's terminology, the name rule 102 is due to the fact that 01100110 is the binary representation of the decimal number 102. In Table 1, we can find an example of a linear regular null 102-CA with initial state $(0\ 0\ 1)$.

CA have been used for many cryptographic applications. In fact, many authors have proposed stream ciphers based on CA [3, 7].

3 Modelling the Modified Self-shrunken Sequence in Terms of CA

The aim of this section is to construct a family of CA that generates the modified self-shrunken sequence as one of their output sequences.

Table 1. Example of linear regular null one-dimensional 102-CA of length 3

102	102	102
0	0	1
0	1	1
1	0	1
1	1	1
$\vdots$	$\vdots$	$\vdots$

The characteristic polynomial of the modified self-shrunken sequence has the form $p_s(x) = (1 + x)^{LC}$, where LC is the linear complexity of the sequence. Notice that the characteristic polynomial of the self-shrunken sequence has the same form [5].

Lemma 1. *Let $\{s_i\}$ be a binary sequence whose characteristic polynomial is $(1 + x)^t$. Then, the characteristic polynomial of the sequence $\{u_i\}$, where $u_i = s_i + s_{i+1}$, is $(1 + x)^{n-1}$.*

Now, we can introduce the next result.

Theorem 1. *Given a modified self-shrunken sequence, there exists a linear, regular, null 102-CA that generates such a sequence in its most left column. The length of such a CA is LC.*

Example 2. Consider the primitive polynomial $p(x) = 1 + x^3 + x^5$. Consider the LFSR with $p(x)$ as characteristic polynomial and initial state $IS = (1\ 0\ 0\ 0\ 0)$. The corresponding modified self-shrunken sequence is given by:

$$0\ 0\ 1\ 1\ 0\ 0\ 1\ 1\ 1\ 0\ 1\ 1\ 0\ 1\ 0\ 0$$

This sequence has period $T = 16$ and it is possible to check that its characteristic polynomial corresponds to $p_s(x) = (1 + x)^{13}$. Therefore, the linear complexity of this sequence is $LC = 13$, and there exists a CA of length 13 that generates it (see Table 2). ■

The other sequences in the CA have a well determined structure:

- The sequence in the most right column is the identically 1 sequence.
- Then, there are 2^{i-1} sequences of period 2^i, for $1 \le i \le L - 2$.
- Finally, there are $LC - 2^{L-2}$ sequences of period 2^{L-1} (including the modified self-shrunken sequence).

This is also due to the form of the characteristic polynomial, which is the same as that of the self-shrinking generator [1].

In Table 2, it is possible to check that the most right sequence is the identically 1 sequence. Next, there are one sequence of period 2, two sequences of period 4, four sequences of period 8 and five sequences of period 16.

On the other hand, it is worth noticing that rule 60, given by

Table 2. 102-CA of length 13 that produces the modified self-shrunken sequence generated in Example 2

102	102	102	102	102	102	102	102	102	102	102	102	102
0	0	1	0	0	0	0	0	1	1	1	1	1
0	1	1	0	0	0	0	1	0	0	0	0	1
1	0	1	0	0	0	1	1	0	0	0	1	1
1	1	1	0	0	1	0	1	0	0	1	0	1
0	0	1	0	1	1	1	1	0	1	1	1	1
0	1	1	1	0	0	0	1	1	0	0	0	1
1	0	0	1	0	0	1	0	1	0	0	1	1
1	0	1	1	0	1	1	1	1	0	1	0	1
1	1	0	1	1	0	0	0	1	1	1	1	1
0	1	1	0	1	0	0	1	0	0	0	0	1
1	0	1	1	1	0	1	1	0	0	0	1	1
1	1	0	0	1	1	0	1	0	0	1	0	1
0	1	0	1	0	1	1	1	0	1	1	1	1
1	1	1	1	1	0	0	1	1	0	0	0	1
0	0	0	0	1	0	1	0	1	0	0	1	1
0	0	0	1	1	1	1	1	1	0	1	0	1

Rule 60: $x_i^{t+1} = x_{i-1}^t + x_i^t$

111	110	101	100	011	010	001	000
0	0	1	1	1	1	0	0

generates exactly the same CA sequences but in reverse order. For example, in Table 3, we have a 60-CA that generates the same sequences as those of the 102-CA in Table 1. In brief, we have defined two different linear, regular, null 102-CA and 60-CA able to generate the modified self-shrunken sequence.

Table 3. Example of linear regular null one-dimensional 60-CA of length 3

60	60	60
1	0	0
1	1	0
1	0	1
1	1	1
⋮	⋮	⋮

4 Recovering the Modified Self-shrunken Sequence from Intercepted Bits

Assume LC is the linear complexity of the modified self-shrunken sequence. Given $2 \cdot LC$ intercepted bits, it is possible to determine the shortest LFSR that generates such a sequence by means of the Berlekamp-Massey algorithm [9].

We know that there exists a CA of length LC that generates the modified self-shunken sequence. Besides, we know that its most right sequence is always the identically 1 sequence. Therefore, it is enough to intercept $LC - 1$ bits of the modified self-shrunken sequence to recover the initial state of the CA and, consequently, to recover the complete sequence. Notice that this quantity is half the needed bits to apply the Berlekamp-Massey algorithm. Note that due to the recent design of this generator, no other approaches have been developed yet.

In Example 2, the modified self-shrunken sequence had period 16 and linear complexity 13. In Table 4, we can see that intercepting 12 bits of the self-shrunken sequence (bits in bold), we can recover the initial state of the CA (in grey) and, thus, the complete sequence.

Table 4. Bits needed to recover the initial state of the 102-CA given in Example 2

102	102	102	102	102	102	102	102	102	102	102	102	102
0	0	1	0	0	0	0	0	1	1	1	1	1
0	1	1	0	0	0	0	1	0	0	0		
1	0	1	0	0	0	1	1	0	0			
1	1	1	0	0	1	0	1	0				
0	0	1	0	1	1	1	1					
0	1	1	1	0	0	0						
1	0	0	1	0	0							
1	0	1	1	0								
1	1	0	1									
0	1	1										
1	0											
1												

5 Conclusions

Cryptographic generators based on irregular decimation were conceived as non-linear sequence generators. However, the sequences generated by these type of generators can be modelled as the output sequences of linear CA.

In this work, it is shown that the sequences generated by the modified self-shrinking generator are also output sequences of one-dimensional, linear, regular and null cellular automata based on rules 102 and 60. At the same time, the number of intercepted bits required by this 102/60 CA is half the number of bits needed by the Berlekamp-Massey algorithm to reconstruct the original sequence.

A natural extension of this work is the generalization of this procedure to many other cryptographic sequences, the so-called interleaved sequences, as they present similar structural properties to those of the sequences obtained from irregular decimation generators.

Acknowledgment. The work of the first author was supported by FAPESP with number of process 2015/07246-0. The work of the second author was supported by both Ministerio de Economía, Spain, under grant TIN2014-55325-C2-1-R (ProCriCiS), and Comunidad de Madrid, Spain, under grant S2013/ICE-3095-CM (CIBERDINE).

References

1. Cardell, S.D., Fúster-Sabater, A.: Linear models for the self-shrinking generator based on CA. J. Cell. Automata **11**(2–3), 195–211 (2016)
2. Coppersmith, D., Krawczyk, H., Mansour, Y.: The shrinking generator. In: Stinson, D.R. (ed.) CRYPTO 1993. LNCS, vol. 773, pp. 22–39. Springer, Heidelberg (1994)
3. Das, S., RoyChowdhury, D.: Car30: a new scalable stream cipher with rule 30. Crypt. Commun. **5**(2), 137–162 (2013)
4. eSTREAM: the ECRYPT Stream Cipher Project, ECRYPT II, eSTREAM portfolio. http://www.ecrypt.eu.org/stream/
5. Fúster-Sabater, A., Pazo-Robles, M.E., Caballero-Gil, P.: A simple linearization of the self-shrinking generator by means of cellular automata. Neural Netw. **23**(3), 461–464 (2010)
6. Golomb, S.W.: Shift Register-Sequences. Aegean Park Press, Laguna Hill (1982)
7. Jose, J., Das, S., Chowdhury, D.R.: Inapplicability of fault attacks against trivium on a cellular automata based stream cipher. In: Wąs, J., Sirakoulis, G.C., Bandini, S. (eds.) ACRI 2014. LNCS, vol. 8751, pp. 427–436. Springer, Heidelberg (2014)
8. Kanso, A.: Modified self-shrinking generator. Comput. Electr. Eng. **36**(1), 993–1001 (2010)
9. Massey, J.L.: Shift-register synthesis and BCH decoding. IEEE Trans. Inf. Theory **15**(1), 122–127 (1969)
10. Meier, W., Staffelbach, O.: The self-shrinking generator. In: De Santis, A. (ed.) EUROCRYPT 1994. LNCS, vol. 950, pp. 205–214. Springer, Heidelberg (1995)
11. Menezes, A.J., van Oorschot, P.C., Vanstone, S.A.: Handbook of Applied Cryptography. CRC Press, Boca Raton (1996)
12. Paar, C., Pelzl, J.: Understanding Cryptography. Springer, Heidelberg (2010)
13. Robshaw, M., Billet, O. (eds.): New Stream Cipher Designs. LNCS, vol. 4986. Springer, Heidelberg (2008)
14. Wolfram, S.: Cellular automata as models of complexity. Nature **311**(5985), 419–424 (1984)

A Novel Trust Update Mechanism
Based on Sliding Window
for Trust Management System

Juanjuan Zhang[1,2(✉)], Qibo Sun[1], Ao Zhou[1], and Jinglin Li[1]

[1] State Key Laboratory of Networking and Switching Technology,
Beijing University of Posts and Telecommunications, Beijing, China
zhangjuanjuan_815@hotmail.com,
{qbsun,aozhou,jlli}@bupt.edu.cn
[2] Science and Technology on Information Transmission
and Dissemination in Communication Networks Laboratory, Beijing, China

Abstract. In the dynamic distributed network environment, such as P2P, trust evaluation has become an important security mechanism. In order to support the network entities with different behavioral characteristics, the researchers have put a lot of effort into developing various trust management mechanisms. Due to the mobility of the network entities and the dynamic change of the context, there are many factors that can affect the final trust. Therefore, the node trust varies with time. However, most current researches lay more emphasis on building an effective and robust trust management system while ignoring the time-attenuation characteristics in trust updating. Therefore, the evaluation result is inaccuracy and makes the on-off attack possible. Considering the fast-fall and slow-rise time-attenuation characteristics, we propose a novel trust updating mechanism (NTUM) based on sliding window. Experimental results show that our trust update mechanism outperforms other compared mechanisms.

Keywords: Trust evaluation, dynamic distributed network, sliding window · Time factor · Trust update mechanism

1 Introduction

Trust [1, 2] is originally used in social science to describe the relationship between human beings, and is now an essential concept in security management for the dynamic distributed network. However, due to the mobility of the network entities and the dynamic change of the context in distributed network, it is too difficult to obtain the comprehensive trust description information for the network entities.

The existing trust mechanisms [3, 4] are usually updated through simple accumulation process or just considering the time-aware weight of the historical trust records, which make on-off attack possible. In on-off attack, the attacker may alternatively behave well and badly to stay undetected while disrupting the trust management system.

To address this problem, we propose an effective and extensible trust updating mechanism based on time sliding-window, which incorporates into two essential

© Springer International Publishing Switzerland 2016
O. Gervasi et al. (Eds.): ICCSA 2016, Part I, LNCS 9786, pp. 521–528, 2016.
DOI: 10.1007/978-3-319-42085-1_41

properties of trust relationship: time-attenuation characteristics and "fast-fall, slow-rise" updating pattern. In addition, a trigger mechanism is proposed to update the sliding window and integrated evaluation of trust records within the window.

The contribution of this paper includes: (1) We propose a novel trust update mechanism based on time sliding-window. (2) We propose two trust update triggering method based on time and event to manage the trust records within the sliding window. (3) A "fast-fall and slow-rise" updating pattern and a time-based forgetting factor is proposed to control the change tendency of trust incremental and integrate the filtered trust records within the sliding window. The experimental results show that our trust update mechanism outperforms other compared mechanisms.

The remainder of this paper is organized as follows. In Sect. 2, we present the technical detail of the update mechanism. In Sect. 3, we describe the experimental results. Finally, conclusions are given in Sect. 4.

2   Related Work

In order to resolve the security problem in network environments, the conception of trust management was first introduced by Blaze et al. [5]. The key idea of trust evaluation is to recognize the imperfection trust information in the network. In a trust management system, the node trust is affected by many factors. Therefore, the trust is not necessarily constant but may changes over time. Hence, how to update the trust has become a key problem in trust management. Several trust updating mechanisms have been put forward by researchers.

The concept of trust updating was first introduced by Liu and Issarny [6]. Shouxu and Jianzhong [7] introduced the belief factor to integrate local trust with global trust. Li et al. [8] proposed an objective trust management system (OTMF) based on Bayesian probability. Experiences in the OTMF are rated by two symbols: α for normal behaviors and β for misbehaviors. The sum of α and β represents the weight of the experience. The updating mechanism is a simple positive and negative evidence accumulation process. For each published trust message, a timer is set to trigger the updating process. Jiang et al. [9] introduced a distributed trust system named EDTM. The trust value is calculated with the consideration of communication trust, energy trust and data trust. Based on the integrated trust, this system uses a simple time sliding window to update the trust value and calculate the weights of the historical trust. Kanakakis et al. [10] proposed a trust update mechanism based on machine learning. This mechanism updates the trust of nodes according to the segment they belong to.

3   Updating Mechanism Based on Sliding Window

The framework of our novel trust updating mechanism (NTUM) is shown in Fig. 1. We will present the technical detail of each module in Fig. 1 in this section.

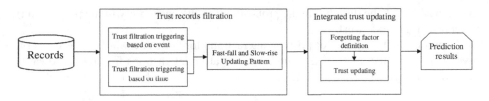

Fig. 1. Framework of NTUM.

3.1 Definitions

We now define some notations that will be used throughout the paper.

Definition 1: Trust record. In this paper, the two-tuples *record* = *{T$_i$, time$_i$}* is used to denote the *i-th* trust record between two nodes in the time period *time$_i$*, where *T$_i$* is the trust value from the subject node to the object node.

Definition 2: The size of the sliding window N. The size of the sliding window limits the amount of considered trust records. The value of N can be configured according to the specific requirements.

Definition 3: The valid time span of the trust record time_valid. The trust relationship between nodes is not necessarily static, but changes dynamically. Due to various incidents and gathered trust records in a certain time period, it is possible to specify that the weight of the latest trust records is larger than the older trust records. When the difference between the generating time of the latest trust record and an older trust record is larger than the valid time span *time_valid*, the older record can be taken as an expired record and will be removed outside the window.

Definition 4: Trust value level. For illustration purpose, in this paper we classify trust values into four levels: distrust, uncertain, good and complete. The relationship between trust values and trust levels is as shown in Table 1.

Table 1. Trust Level

No	Trust value	Level
1	[0, 0.5)	distrust
2	[0.5, 0.6]	uncertain
3	(0.6, 0.9)	good
4	[0.9, 1]	complete

3.2 Trust Value Calculation

In a distributed network, the trust that node A has on B is obtained through interaction records between them. Our research focus on how to update the trust based on the existed evaluation system, not on the trust evaluate system itself. Therefore, we employ the subjective logic trust management system proposed by Jøsang et al. [11] in this paper. The subjective logic trust management system proposed 3-tuple (*b*, the belief in the proposition; d, the disbelief in the proposition; u, the uncertainty in the proposition) to represent the node trust.

Suppose r is the number of positive interactions for node A towards node B over a given period of time, and s is the number of negative interactions for node A towards node B over a given period of time. The value of b, d, u can be calculated as follows.

$$\begin{cases} b = \dfrac{r}{r+s+C} \\[2mm] d = \dfrac{s}{r+s+C} \\[2mm] u = \dfrac{C}{r+s+C} \end{cases} \tag{1}$$

where the constant C denotes the event uncertainty in the network and we set C to 1 in this paper. Then the trust value that node A has on node B over a period of time $time_i$ can be calculated by the following:

$$T_i = b + \frac{1}{2}u. \tag{2}$$

3.3   Trust Records Filtration Based on the Sliding Window

Our update mechanism updates node trust value based on all the valid records within the sliding window. In consideration of both the maximum capacity of sliding window buffer and the attenuation temporal characteristics of trust, the filtration operations for the records within the window can be triggered automatically in two conditions: new trust record occurs between nodes in the latest time period (trust filtration triggering based on event) and current valid record expires (trust filtration triggering based on time).

(1) Trust filtration triggering based on event.
Take the subject node A and the object node B for example. As is shown in Fig. 2, suppose the size of sliding window N is 8 and R19 is the latest trust record between A and B. When an observation record for B is obtained by A, the time sliding window moves toward the right, the latest record R20 would add into the rightmost of window. The leftmost record R12, which is the oldest record, moves outside. However, the removed records are not exactly discarded, but would be taken into account in the final integrated trust update phase, which will be descripted in Sect. 3.

(2) Trust filtration triggering based on time.
As shown in Fig. 3, when there is no interaction between two nodes for a long period of time, some trust records within the sliding window would expire with the passage of time. The update mechanism will perform the following test to detect if each record is expired or not.

$$time_N - time_i \geq time\_valid \tag{3}$$

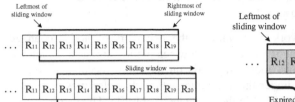

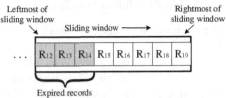

Fig. 2. Trust update triggering based on time.

Fig. 3. Trust update triggering based on time.

where $time_N$ denotes the update time of the latest record within the sliding window, and $time_i$ denotes the computing time of the i-th trust record within the sliding window. If the i-th record has passed the test, the i-th record is regarded as an expired trust record. Suppose the record R_{15} is the first detected expired record in the records sequence. As the records are obtained in chronological order from left to right, these records $\{R_{13}, R_{14}\}$ on its left are all regarded as the expired records, and would be removed.

To maintain the size of sliding window, we fill the free space on the left with uncertain trust records with trust value T, $time = time_{12}$. When there is no interaction between A and B in a long time, the old trust relationship might no longer be valid. This usually means that the trust value decreases over time, which is in accordance with time-attenuation characteristics of trust.

(3) Fast-fall and Slow-rise Updating Pattern.
As the trust is not fixed but changes over time, trust value would decline rapidly because of interaction failure. However, the increase of trust should be very slow. That is, the update of trust value should be characterized by slow rise and fast fall. We would take the record removed outside the window scope into account in the final integrated trust updating. The updated trust value can be calculated as:

$$T'_{new} = T_{old} + (T_{new} - T_{old}) \times \alpha^{(T_{new} - T_{old})}. \tag{4}$$

where the T_{new} is the trust value in the latest record, T_{old} is the trust value in the record to be removed outside the sliding window result from update triggering method based on time or event. T'_{new} is the updated trust value. Here we set the $\alpha^{(T_{new} - T_{old})}$ as the regulatory factor to adjust the change range of trust. For Eq. (4), we give more weight to trust incremental $(T_{new} - T_{old})$ when $T_{new} < T_{old}$ as opposed to the case which $T_{new} > T_{old}$. That is, the fast fall update operation with regulatory factor bigger than 1 can increase fall range of trust value, the slow rise update operation with regulatory factor smaller than 1 can slow down the increase range, which satisfies the fast-fall and slow-rise property.

3.4   Integrated Trust Updating Based on the Filtered Trust Records Within the Sliding Window

In the dynamic distributed network environment, the behaviors of nodes changed endlessly over time, namely that trust relation between nodes are also changing. Therefore, it is reasonable to place more emphasis on the more recent trust record within the sliding window. Thus under this time-attenuation characteristics of trust, a time-based forgetting factor ω_i is defined in the value interval [0, 1]. The factor is used to compute the weight of i-th trust record within the sliding window according to the update time of i-th record $time_i$ and the start time $time_1$, and the forgetting factor is given by:

$$\omega_i = e^{-(\Delta t)^k} = e^{-(time_i - time_1)^k}, k \geq 1. \tag{5}$$

The constant k determines the decay rate of the trust value over time interval Δt, and can be assigned by the subject node based on its perception about the change.

We now construct the integrated trust evaluation T_{total} at the current time based on all the N trust records at different time $time_i$. The integrated trust of the node can be calculated as follows:

$$T_{total} = \frac{\sum_{i=1}^{N} \omega_i \times T_i}{\sum_{i=1}^{n} \omega_i}. \tag{6}$$

where ω_i is the i-th time-based forgetting factor, which is used to weight the i-th direct trust value according to its time interval with the first record. T_i is the trust value stored in the i-th record within the current sliding window.

4   Evaluation

In our experiment, there is a network comprised of 100 nodes simulated by OMNeT ++. There are two types of nodes in the network: the normal nodes and the malicious nodes. The normal nodes would act honestly and provide honest information. The malicious nodes, which act as on-off attackers, may alternatively behave well and badly to stay undetected. The attacker will behave well in the start of interaction. When it has a high degree of trust, it begins to cheat the node who intends to have interaction with it and launch attack. To show the effectiveness of our trust update mechanism, we compare the mechanism proposed in this paper with the update mechanism of EDTM [9].

Figure 4 shows the trust update results in given period of time. The rise and fall of trust value denotes the effect result from the normal and malicious behaviors, respectively. From this figure, we can see that in the EDTM, the trust of the object node falls into the distrust level at 40 s and is regarded as a malicious node. However, after several update processes, the trust of the malicious node increases through intended positive actions. The node can be admitted as a reliable one to the network. For our

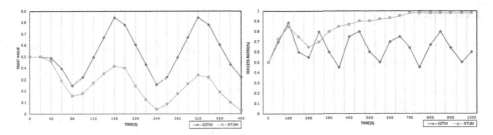

Fig. 4. Comparison of the trust values.   **Fig. 5.** Comparison of success ratios.

update mechanism, we give larger weight to malicious behaviors as opposed to normal ones. Therefore, the downswing of trust value is significantly greater than that of EDTM. For the fast-fall and slow-rise updating pattern, the trust value of a malicious one rises slower and cannot easily clean up because of the historical bad behaviors. Hence, update mechanism based on time sliding window with fast-fall and slow-rise updating pattern can recognize malicious node effectively while defending the on-off attack.

Figure 5 shows the comparison of the success ratios. The total number of nodes in the network is set as 100 and there are 50 % malicious ones. We select 10 nodes each time to interact with. If the selected node transfers the data successfully, the interaction is success. Otherwise, the interaction fails. The higher the success ratio of one mechanism is, the better its performance is. The result shows that the success ratio of EDTM fluctuates over time, while the success ratio of our mechanism is much higher and stable after the early period. The experimental result indicates that our update mechanism effectively reduces the influence of malicious nodes.

5   Conclusion

This paper presents a novel trust update mechanism based on time sliding-window for trust management system. The mechanism consists of two kinds of filtration method for managing trust records within the sliding window. In addition, a fast-fall and slow-rise updating pattern and a time-based forgetting factor are designed to control the trust decay rate and improve the evaluation accuracy. The efficiency of our proposed approach is validated by the experiments. The experimental results show that our proposed approach can improve the trust evaluation accuracy for network entities and outperform other approaches in terms of success rating. For future work, we will consider the various malicious behaviors in distributed network and design a set of trust update mechanisms to cope with these malicious attacks.

Acknowledgment. The work is partly supported by Science and Technology on Information Transmission and Dissemination in Communication Networks Laboratory.

References

1. Ganeriwal, S., Balzano, L.K., Srivastava, M.B.: Reputation-based framework for high integrity sensor networks. J ACM Trans. Sensor Netw. **4**(3), 1–37 (2008)
2. Gao, W., Zhang, G., Chen, W., Li, Y.: A trust model based on subjective logic. In: Fourth International Conference on Internet Computing for Science and Engineering (ICICSE), 2009, pp. 272–276. IEEE (2009)
3. Govindan, K., Mohapatra, P.: Trust computations and trust dynamics in mobile adhoc networks: a survey. J. Commun. Surv. Tutorials **14**(2), 279–298 (2012)
4. Jøsang, A.: A logic for uncertain probabilities. Int. J. Uncertainty Fuzziness Knowl. Based Syst. **9**(03), 279–311 (2001)
5. Blaze, M., Feigenbaum, J., Lacy, J.: Decentralized trust management. In: Proceedings of 1996 IEEE Symposium on Security and Privacy, pp. 164–173. IEEE (1996)
6. Liu, J., Issarny, V.: Enhanced reputation mechanism for mobile ad hoc networks. In: Jensen, C., Poslad, S., Dimitrakos, T. (eds.) iTrust 2004. LNCS, vol. 2995, pp. 48–62. Springer, Heidelberg (2004)
7. Shouxu, J., Jianzhong, L.: A reputation-based trust mechanism for P2P e-commerce systems. J. softw. **18**(10), 2551–2563 (2007)
8. Li, R., Li, J., Liu, P., Chen, H.H.: An objective trust management framework for mobile ad hoc networks. In: IEEE 65th Vehicular Technology Conference, VTC 2007, pp: 56–60. IEEE (2007)
9. Jiang, J., Han, G., Wang, F., Shu, L., Guizani, M.: An efficient distributed trust model for wireless sensor networks. J. Parallel Distrib. Syst. **26**(5), 1228–1237 (2015)
10. Kanakakis, M., van der Graaf, S., Kalogiros, C., Vanobberghen, W.: Computing trust levels based on user's personality and observed system trustworthiness. In: Conti, M., Schunter, M., Askoxylakis, I. (eds.) TRUST 2015. LNCS, vol. 9229, pp. 71–87. Springer, Heidelberg (2015)
11. Jøsang, A., Hayward, R., Pope, S.: Trust network analysis with subjective logic. In: Proceedings of the 29th Australasian Computer Science Conference, vol. 48, pp. 85–94. Australian Computer Society, Inc. (2006)

Some Problems of Fuzzy Networks Modeling

Kirill E. Garbuzov$^{(\boxtimes)}$

Department of Mathematics and Mechanics, Novosibirsk State University,
Novosibirsk, Russia
kirill.e.garbuzov@gmail.com

Abstract. The paper reviews some problems of using fuzzy logic in telecommunications networks modeling. Application of general methods from the theory of fuzziness are described. "Fuzzy prediction" problem is formulated and some ideas of solving it are stated.

Keywords: Telecommunications network · Flow network · Fuzzy logic · Fuzzy number

1 Introduction

Development of communications systems provides us with a big number of new optimization problems concerning modern infrastructure, transport and computer networks. Some problems are successfully solved with the help of classic methods but in other cases the new approach is necessary.

The problem of measuring network reliability is really important for modern modeling. The classic way is to use probabilistic approach where each unreliable component has probability of failure. But this method has its own cons since sometimes there can't be found sufficient statistical data for representing model parameters or even expert opinion has to be used. Futhermore, other problems connected to telecommunications networks are using exact values of initial parameters (capacities, costs, etc.) which can't always correspond to the real situation.

Another approach to represent such models is to introduce fuzzy numbers as model parameters which allows us to use the variety of instruments of fuzzy logic for solving these problems [1–3]. Different types of fuzziness create easy ways to initiate model parameters for every possible situation. Moreover, using fuzzy logic in networks modeling allows us to formulate the problems which can not be considered in standard probabilistic approach.

In this work the concept of fuzzy set is used. It can be represented as classic set with membership function having values between 0 and 1 (basically, an extension of the multi-valued logic). Examples of possible implementation of the particular type of fuzzy numbers (triangular numbers) are given. Moreover, the new "Fuzzy prediction" problem is formulated and the solution is suggested.

O. Gervasi et al. (Eds.): ICCSA 2016, Part I, LNCS 9786, pp. 529–535, 2016.
DOI: 10.1007/978-3-319-42085-1_42

2   Basic Definitions

2.1   Fuzzy Parameters and Operations

Definition 1. Let U be the classic set. Then fuzzy set A is defined as:

$$A = \{\langle x, \mu_A(x)\rangle : x \in U\} \tag{1}$$

where $\mu_A(x)$ membership function of U generally taking any value in $[0, 1]$.

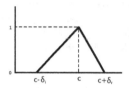

Fig. 1. Fuzzy triangular number

Definition 2. Fuzzy triangular number is three real numbers $\langle \delta_l, c, \delta_r \rangle$ which define the membership function:

$$\mu_A(x) = \begin{cases} 1 - \frac{c-x}{\delta_l}, x \in [c - \delta_l, c] \\ 1 - \frac{x-c}{\delta_r}, x \in (c, c + \delta_r] \\ 0, otherwise \end{cases} \tag{2}$$

Membership function of a triangular number is shown on the Fig. 1. We consider $\langle 0, c, 0 \rangle$ a standard real number.

Arithmetical operations between fuzzy triangular numbers can be defined in many ways. In this work we are using such way to perform addition and substraction:

$$\begin{aligned} \langle \delta_{1l}, c_1, \delta_{1r} \rangle + \langle \delta_{2l}, c_2, \delta_{2r} \rangle &= \langle \delta_{1l} + \delta_{2l}, c_1 + c_2, \delta_{1r} + \delta_{2r} \rangle \\ \langle \delta_{1l}, c_1, \delta_{1r} \rangle - \langle \delta_{2l}, c_2, \delta_{2r} \rangle &= \langle \delta_{1l} + \delta_{2r}, c_1 - c_2, \delta_{1r} + \delta_{2l} \rangle \end{aligned} \tag{3}$$

This is the popular way of performing soft computing on triangular numbers. Such operations have the properties of associativity and commutativity. Moreover, $\langle 0, 0, 0 \rangle$ is the identity element (but the operation of substraction doesn't define proper inverse elements, thus we are working with the monoid of fuzzy triangular numbers).

Furthermore, this way of representing operations allows us to consider $\langle 0, c, 0 \rangle$ as a standard real number. The big drawback of such approach is very fast growing of triangular number boundaries which leads to the growth of entropy.

Another popular way to perform operations on triangular numbers is to consider the field of triangular numbers with operations of addition and multiplication defined using isomorphism $R \rightarrow R^+$:

$$\langle \delta_{1l}, c_1, \delta_{1r} \rangle + \langle \delta_{2l}, c_2, \delta_{2r} \rangle =$$
$$= \langle \delta_{1l} \cdot \delta_{2l}, c_1 + c_2, \delta_{1r} \cdot \delta_{2r} \rangle$$
$$\langle \delta_{1l}, c_1, \delta_{1r} \rangle \cdot \langle \delta_{2l}, c_2, \delta_{2r} \rangle =$$
$$= \left\langle e^{ln(\delta_{1l}) \cdot ln(\delta_{2l})}, c_1 \cdot c_2, e^{ln(\delta_{1r}) \cdot ln(\delta_{2r})} \right\rangle \qquad (4)$$
$$\langle \delta_l, c, \delta_r \rangle^{-1} =$$
$$= \left\langle e^{\frac{1}{ln(\delta_l)}}, \frac{1}{c}, e^{\frac{1}{ln(\delta_r)}} \right\rangle$$

These operations are not always usable in describing network models since they don't seem "natural" enough.

Relational operators can be defined as follows:

$$\langle \delta_{1l}, c_1, \delta_{1r} \rangle < \langle \delta_{2l}, c_2, \delta_{2r} \rangle \leftrightarrow c_1 + \frac{\delta_{1r} - \delta_{1l}}{4} < c_2 + \frac{\delta_{2r} - \delta_{2l}}{4} \qquad (5)$$

The values in the right part are "centers of mass" of triangular numbers, thus both centers and boundaries are used in comparing fuzzy triangular numbers.

Fuzzy number membership function looks much like a non-normalized probability density function. The biggest difference is the way of performing arithmetical operations. Instead of using the ones listed above we could have tried to treat these fuzzy numbers as continuous random variables with special density functions. But in this case the set of triangular numbers would not be closed under the traditional operations on random variables. Consequently, using fuzzy numbers simplifies models but at the same time it provides a various set of tools for different purposes.

For each fuzzy number we can also introduce fuzziness - some function showing us how much information we can get from this parameter. One of the purposes of this function is calculating informativeness - a value which shows us how much influence has the change of fuzziness of this particular element to the fuzziness of the whole system. For example, logarithmic entropy can be used as such function:

$$d(A) = \int_U S(\mu_A(x)) \, dx \qquad (6)$$

where $S(y) = -y \cdot ln(y) - (1 - y) \cdot ln(1 - y)$ - Shannon function.

For triangular numbers we have:

$$d(A) = \frac{1}{2}(\delta_l + \delta_r) \qquad (7)$$

Using logarithmic entropy is an easy way to measure the uncertainty of most of the types of fuzzy numbers. But at the same time it ignores some information about the membership function, such as the ratio of the left and the right parts of the fuzzy triangular number as shown on the Fig. 2.

Informativeness comparing problem can be introduced on this simple example [6] (here we can consider these triangular numbers as probabilities of presence for the graph elements). On the Fig. 3 (left) we compare connection possibility (fuzzy

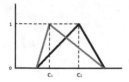

Fig. 2. Triangular numbers with the same entropy

analogue of connection probability) of two subgraphs connected by a bridge before and after changing the fuzziness of this bridge. At the beginning, the fuzzy number corresponding to objective function value equals $\langle 0.1, 0.245, 0.1 \rangle$. If we now change the fuzzy parameter of bridge to $\langle 0, 0.5, 0 \rangle$ connection possibility value becomes equal to $\langle 0.05, 0.245, 0.05 \rangle$.

On the Fig. 3 (right) we are calculating the possibility of connection of two subgraphs. Without changing parameters this value equals $\langle 0.1, 0.8, 0.1 \rangle$. Alternately modifying fuzzy numbers corresponding to both edges of the bridge, we get different results. It means that these elements are not equal in terms of informativeness.

Consequently, informativeness of elements can be affected not only by their own fuzzy parameters but also by the structure of the whole network and parameters of other edges. It is obvious that we only need to compare the values of informativeness of elements, so we can try to use different heuristic algorithms.

Fig. 3. Informativeness of network elements

Converting different types of fuzzy parameters is another important problem concerning the calculation of fuzziness. If we get network parameters from different sources, they can be represented in different types of fuzzy numbers. So we have to convert them to one common type and keep the most information they initially have.

2.2   Flow Network

Flow network (or transportation network) is a directed graph $G = \langle V, E \rangle$ where each edge $(u, v) \in E$ has non-negative capacity $c(u, v) \geq 0$ and flow $f(u, v)$, in this case represented by fuzzy triangular numbers. We consider vertices with numbers 0 and $n - 1$ as source s and sink t. Moreover, the network has non-negative cost $d(u, v)$ of sending the flow across the edge [4,5].

It is easy to show that using fuzzy triangular numbers doesn't break the important properties of flow networks, such as Max-flow min-cut theorem stating that in flow network the maximum amount of flow passing from the source to the sink is equal to the minimum capacity, or the following property of network cuts:

$$\sum_{v \in V} f(s,v) = \sum_{v \in A} \sum_{u \in B} f(u,v)$$
$$\forall A, B \subset E : A \cap B = \emptyset, A \cup B = E \tag{8}$$

The minimum-cost flow problem is to find the cheapest way to send a certain amount d of flow through a flow network:

$$\sum_{u,v \in V} w(u,v) \cdot f(u,v) \to min$$
$$f(u,v) \leq c(u,v)$$
$$f(u,v) = -f(v,u)$$
$$\sum_{v \in V} f(u,v) = 0 \forall u \in V \tag{9}$$
$$\sum_{v \in V} f(s,v) = d$$

3 "Fuzzy Prediction" Problem

Informal description of "Fuzzy prediction" problem can be stated as follows. New edges with fuzzy capacity and flow are added to the initial transportation network with certain amount of flow (not necessarily maximal). New distribution of flows in the network must be found using fuzzy operations (although initial parameters of the flow network were not fuzzy).

One of the examples of the "Fuzzy prediction" problem in practice is the real road network. Since we have statistics on the current state of the roads, an initial non-fuzzy flow network model can be constructed. Now if we want to predict its state after adding a new connection (bridge, for example), we can not use the exact values to set its parameters since there is no statistical information on this imaginary edge. One way to do it is to use fuzzy parameters based on expert opinion about this new connection. And this is basically a description of the "Fuzzy prediction" problem mentioned above.

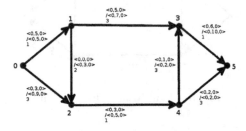

Fig. 4. Initial flow distribution

Figure 4 shows the example of the initial flow distribution in the flow network. Each edge has three parameters: current flow, capacity and cost of sending the flow across the edge. Initial capacities and flows are triangular numbers with no dispersion which is equal to real numbers. The amount of flow in this network is equal to $\langle 0, 8, 0 \rangle$.

Suggested algorithm is based on the famous Busacker-Gowen algorithm for solving the minimum-cost flow problem. Modifications are adapting the new algorithm for working with fuzzy numbers and keeping predicted flow on the new edges if it is possible. The brief description is as follows:

0. Preparation of the initial network: current flows through each edge are equal to $\langle 0, 0, 0 \rangle$, capacities of the new edges are equal to their stated flows (and their current flows are $\langle 0, 0, 0 \rangle$). Current network flow $V' = 0$. New edges are marked.

1. Forming graph of modified costs G_f using the following rules:
   (a) Set of vertices in G_f is equal to the set of vertices in G;
   (b) If graph G has $f(u, v) > \langle 0, 0, 0 \rangle$, $f(u, v) < c(u, v)$, then G_f has straight edge with weight $d(u, v)$ and reverse edge with weight $-d(u, v)$ between the vertices u and v;
   (c) If graph G has $f(u, v) = \langle 0, 0, 0 \rangle$, then G_f has straight edge with weight $d(u, v)$ between the vertices u and v;
   (d) If graph G has $f(u, v) = c(u, v)$, then G_f has reverse edge with weight $-d(u, v)$ between the vertices u and v;

2. Calculating minimum path containing one of the marked edges not exceeding their capacity in graph G_f. Calculating value $\epsilon = min\{c(u_i, v_i) - f(u_i, v_i), f(u_j, v_j), V - V'\}$ on the corresponding path in G where indices i mark straight edges and indices j mark reverse edges in G_f.

3. $V' += \epsilon$. If $V' = V$ then next step else return to step 1.

4. Recovering capacities and flows of marked edges.

Figure 5 shows the distribution of flows after adding the new edge between vertices 2 and 3.

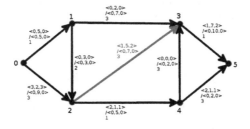

Fig. 5. Modified flow distribution

The cheapest path in the modified graph can be found using any appropriate algorithm (e.g. Bellman-Ford algorithm). It is important to note that triangular numbers with their borders on negative semiaxis should be treated as normal but we have to keep in mind that those negative values are unreachable during defuzzyfication.

This algorithm was implemented using C++ programming language with help of the Boost library. Debugging and testing were hold on the randomly generated networks with the maximum of 15 vertices. One of the main future objectives is to use this algorithm on some models of real communications networks.

We have to keep in mind that the resulting fuzzy flows are showing only the possible situation based on expert predictions. Usage of fuzzy numbers not only shows the "best" and the "worst" possible situations (borders of the resulting triangular numbers) but also indicates the "distribution" and shows the most possible values (centers of the resulting triangular numbers).

4   Conclusion

The usage of fuzzy models allows to present new problems which can not always be formulated using only classic approach. The other example of such problem is the problem of finding the most informative network element [6].

The next big goal is to extend this theory on hypernetworks and to introduce the problems of structural fuzziness.

Acknowledgment. The author would like to thank his scientific advisor Dr. Alexey S. Rodionov.

References

1. Zadeh, L.A.: Fuzzy logic and the calculi of fuzzy rules and fuzzy graphs. Multiple-Valued Logic **1**, 1–38 (1996)
2. Zadeh, L.A.: A note on Z-numbers. Inf. Sci. **181**, 2923–2932 (2011)
3. Mordeson, J.N., Nair, P.S.: Fuzzy Graphs and Fuzzy Hypergraphs. Physica-Verlag, Heidelberg (2000)
4. Heineman, G.T., Pollice, G., Selkow, S.: Network flow algorithms. In: Algorithms in a Nutshell, pp. 226–250. Oreilly Media (2008). Chap. 8
5. Ahuja, R.K., Magnanti, T.L., Orlin, J.B.: Network Flows: Theory Algorithms and Applications. Prentice Hall, Englewood Cliffs (1993)
6. Garbuzov, K.E., Rodionov, A.S.: Some problems of fuzzy modeling of telecommunications networks. In: Proceedings of the International Conference on Mathematical Methods, Mathematical Models and Simulation in Science and Engineering (2015)

A Transparent Accelerating Software Architecture for Network Storage Based on Multi-core Heterogeneous Systems

Qiuli Shang[1,2(✉)], Jinlin Wang[1], and Xiao Chen[1]

[1] National Network New Media Engineering Research Center,
Institute of Acoustics, Chinese Academy of Sciences, Beijing, China
{shagnql,wangjl,xxchen}@dsp.ac.cn
[2] University of Chinese Academy of Sciences, Beijing, China

Abstract. With the development of software-defined storage (SDS), virtualization, and converged system, network storage systems are evolving towards service convergence and function diversity. For the demands on both protocol acceleration and transparent development, this paper proposes a transparent accelerating software architecture for network storage based on multi-core heterogeneous systems. The architecture accelerates the network storage performance transparently, namely enhances the versatility and compatibility of network storage systems based on multi-core heterogeneous systems. Therefore, it has two aspects of advantage including (1) acceleration for protocol processing and (2) agility and maintainability for network service deployment. The experimental test based on prototype platform shows that the performance density of the architecture exceeds the traditional Linux scheme.

Keywords: Network storage · Multi-core processor · Heterogeneous system · Software architecture

1 Introduction

Nowadays, the rapid growth of network bandwidth and Internet traffic puts forward higher requirement on the I/O performance of network storage systems. Thumb's law points out that, it takes 1 Hz CPU capacity to process 1bit network data by traditional software manner of network protocol implementation [1]. For example, processing 10 Gbps bidirectional network storage I/O traffic will consume 20 GHz computing capacity, which is the ability that traditional CPUs on storage server rarely have.

By the features of parallel processing and hardware acceleration, multi-core network processor (NP) can support data processing for network storage protocols and TCP/IP in high-speed network environment. Thus, multi-core NPs are becoming the mainstream platforms of next generation network storage systems, high-performance network servers, and network edge devices [2, 3].

With the development of software-defined storage (SDS), virtualization, and converged system, network storage devices are evolving towards system generalization [4, 5]. Currently, besides I/O performance, a high-end network storage system should have

© Springer International Publishing Switzerland 2016
O. Gervasi et al. (Eds.): ICCSA 2016, Part I, LNCS 9786, pp. 536–546, 2016.
DOI: 10.1007/978-3-319-42085-1_43

the characteristics as follow. First, service convergence. Network storage servers are often acting as streaming media servers, TCP/IP acceleration agents, or intelligent security gateways in adjunct [6]. Second, function diversity. It should support functions including multi-service QoS (Quality of Service) classification, failover and recovery, auto-tiering, virtualization, thin provisioning, full disk encryption, data de-duplication and compression, service continuity, object storage, etc. Meanwhile, with the emerging of new services and applications in the Internet, network storage systems are required to be more agile and flexible to deploy new services [7].

However, due to the dedicated hardware design of multi-core network processors, developers have to handle and manage the underlying hardware (HW) resources directly, such as processing cores, physical memory, inter-core message, as well as co-processors, which leads to long development cycle, lack of portability and maintainability.

Therefore, it is a crucial problem to design a software architecture based on multi-core network processor which not only can accelerate network storage processing in high-speed environment, but also can provide transparent development and management interfaces to ensure deploying new services agilely and flexibly.

For the demands on both protocol acceleration and transparent development, this paper proposes a transparent accelerating software architecture for network storage based on multi-core heterogeneous systems. The transparent accelerating architecture provides general development and configuration interfaces for upper-level applications by general-purpose operation system, and provides data plane network protocol stacks accelerating by network processing operating system. By this way, the architecture accelerates the network storage performance transparently, namely enhances the versatility and compatibility of network storage system based on multi-core NP.

2 The Transparent Accelerating Software Architecture Based on Heterogeneous Systems

2.1 Heterogeneous System

Without loss of generality, the architecture in this paper uses multi-core network processor Cavium OCTEON as prototype platform. The heterogeneous system architecture divides the processing cores into two sets: one carries a general-purpose operating system which is Linux operating system (OS) in this paper; the other carries the network processing operating system which is the simple execution environment (SE) provided by Cavium OCTEON. Since the heterogeneous systems running on common physical memory, the inter-system communication between Linux and SE includes two types: the signaling interaction via inter-core message mechanism, and the data interaction via shared memory mechanism. Figure 1 demonstrates the proposed heterogeneous system architecture based on multi-core processors.

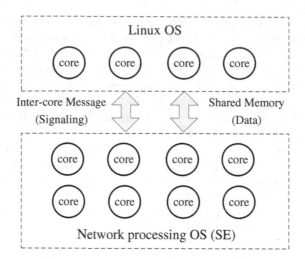

Fig. 1. The heterogeneous system architecture based on multi-core processors.

2.2 System Architecture

Based on the heterogeneous system architecture mentioned, a transparent accelerating software architecture for network storage based on multi-core heterogeneous systems is proposed shown by Fig. 2. The architecture achieves separation and decoupling of data plane and control/management plane for network storage system.

The data plane running on SE system can handle the underlying HW resources e.g. co-processors, physical shared memory, and further avoid many mechanisms in Linux kernel which hindering high-speed network protocol processing, such as interrupts, process scheduling, memory copy, etc. Therefore, there can be higher processing efficiency for network protocol stacks.

The control/management plane running on Linux system reduces the difficulty of service development and deployment, and also simplifies system management and configuration, which ensures the mentioned service convergence and function diversity of network storage systems.

By this way, the architecture can accelerate the network storage I/O performance transparently, namely enhances the versatility and compatibility of network storage systems based on multi-core NP.

2.3 System Module

The transparent accelerating network storage architecture in Fig. 2 includes several main parts:

Communication Adaption Layer. As the key module of the architecture, the communication adaption layer is responsible for the communication between heterogeneous operating systems including signaling interaction by inter-core message mechanism and data interaction by shared memory mechanism. The Linux communication adaption layer

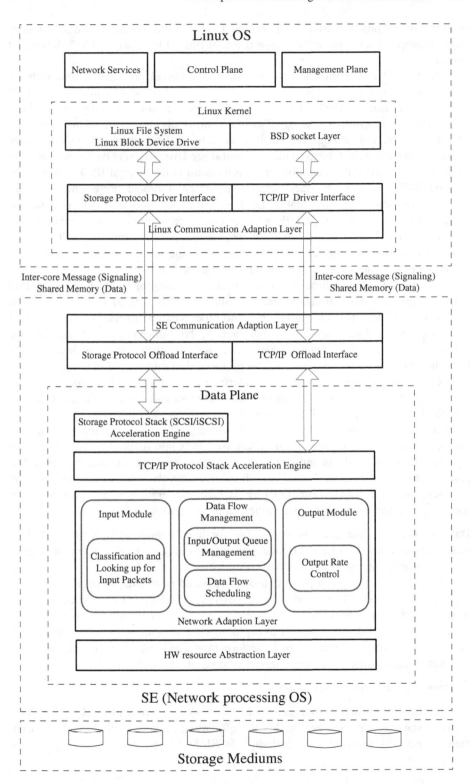

Fig. 2. The transparent accelerating software architecture based on multi-core heterogeneous systems.

includes storage protocol driver interface and TCP/IP driver interface. The SE communication adaption layer includes storage protocol offload interface and TCP/IP offload interface.

The TCP/IP driver interface running in Linux kernel provides general-propose TCP/IP socket interfaces through standard BSD layer. Instead of calling the TCP/IP stack of Linux kernel directly, the TCP/IP driver interface communicates with the data plane in SE by encapsulating the signaling of function calls into inter-core messages, and copying the protocol data into shared physical memory. Conversely, the TCP/IP driver interface also parses the inter-core message and shared memory sent by SE into function callback and protocol data, and further submits them to the general BSD socket layer.

The TCP/IP offload interface running in SE environment parses the inter-core messages and shared memory sent by Linux into function calls and protocol data, and further calls the underlying TCP/IP protocol stack acceleration engine in data plane. Conversely, the TCP/IP offload interface also encapsulates the function callbacks and protocol data submitted by TCP/IP acceleration engine into inter-core messages and shared memory, and further sends them to the Linux side.

Similarly, the storage protocol driver interface running in Linux Kernel provides general-propose storage protocol interfaces such as SCSI, iSCSI to the upper-level modules including Linux file system (FS) and block device drivers. Instead of calling the storage protocol stack of Linux kernel directly, the storage protocol driver interface communicates with the data plane in SE by encapsulating the signaling of function calls into inter-core messages, and copying the protocol data into shared physical memory. Conversely, the storage protocol driver interface also parses the inter-core messages and shared memory sent by SE into function callbacks and protocol data, and further submits them to the general Linux FS or device drivers.

The storage protocol offload interface running in SE environment parses the inter-core messages and shared memory sent by Linux into function calls and protocol data, and further calls the underlying storage protocol stack acceleration engine in data plane. Conversely, the storage protocol offload interface also encapsulates the function callbacks and protocol data submitted by TCP/IP acceleration engine into inter-core messages and shared memory, and further sends them to the Linux side.

Data Plane. The data plane is responsible for accelerating network protocol stack processing and scheduling packet flows. It includes protocol stack acceleration engine, network adaptation layer, and hardware resource abstraction layer.

The performance of data plane is usually required to be high throughput and low latency, which is vital to network data processing in high-speed environment. Therefore, the data plane runs in SE environment to facilitate handling the underlying hardware resources directly in order to exploiting the advantage of multi-core network processor. To keep balance of both performance and scalability, P-SPL model is selected as the multi-core topology model of data plane [8, 9].

- *Protocol stack acceleration engine.*

The protocol stack acceleration engine includes storage protocol stack and TCP/IP protocol stack acceleration engine. It's the offloading implementation in SE environment of network protocol stack including SCSI/iSCSI protocols, TCP/IP protocols, which accelerates the processing performance based on the characteristics of multi-core network processor.

- *Network adaptation layer.*

The network adaptation layer is responsible for network interface initialization, packet input/output, and flow management. It includes input module, output module, and data flow management module. The input module receives packet from network interface card (NIC) and operates packet filtering, classifying and looking-up. To be mentioned, the filtering can be accelerated by the HW CRC co-processor of OCTEON, and the classifying and looking-up can be accelerated by the HW hash co-processor of OCTEON. The output module submits the packets sent by network stack to NICs by precise timing and rate control. The data flow management module maintains input/output queues and schedules data flows to corresponding queues according to specific upper-level requirements.

- *Hardware resource abstraction layer.*

The hardware resource abstraction layer abstracts and provides friendly HW access interfaces of multi-core platform to upper-level modules and developers, including co-processor interfaces, memory management interfaces, inter-core message interfaces, etc.

Control/Management Plane and Network Service Module. Control plane provides configuration of data plane such as configuring protocol stack acceleration engine parameters, NIC arguments, flow scheduling algorithms, etc. Management plane provides interactive interfaces, and also monitors, collects and uploads operation information of the network storage system. Network services including SDS, cloud service and streaming media server, etc., are evolving towards convergence and diversification.

Due to the service convergence and function diversity of current network storage systems, it requires control/management plane to be general-propose and portable, and requires the development and deployment of network services to be agile and maintainable.

Therefore, the control/management plane and network service module are implemented in Linux. Furthermore, the transparent accelerating architecture defines and provides enough general system interfaces, development tools, and dedicated customizable platform interfaces, which facilitates transparent control, management and exploitation of the data plane in SE environment.

2.4 Technical Advantage

The advantages of the proposed transparent accelerating architecture in this paper can be summarized as two key words:

Transparent. Based on multi-core heterogeneous systems, it makes the underlying data plane in SE transparent to the control/management plane and network service module in Linux, which meets the requirements of current network storage systems on service convergence and function diversity.

Accelerating. For high-speed storage I/O performance, the acceleration techniques implemented on data plane includes the following.

- *Zero copy.*

Unlike Linux kernel-based network protocol stack implementation, the packet operations in this architecture are all based on the shared memory pools between Linux, SE and NICs. Instead of copying packet between different modules, it merely transfers the unique physical memory address and offset pointer to avoid the overhead of memory copy.

- *Interrupt to poll.*

In high-speed network environment, the large number of interrupts is an important overhead that restricting the performance of network processing in Linux kernel [10]. The architecture deals with the packets and messages between SE data plane and NICs by polling to ease the overhead of interrupts.

- *Protocol operation accelerating.*

With the aid of the co-processors including Hash, CRC, regular expression, encryption and decryption, the architecture offloads some key protocol computing in order to reducing CPU usage.

- *Sending rate control.*

The timing mechanism of data plane in SE is called timer work which is more accurate than the clock interrupt-based timing of Linux kernel. Therefore, the system can send packets with accurate rate control via timer work, which helps to smooth sending traffic bursts, and further ensures the friendliness to downstream network devices.

3   Experiment and Analysis

3.1   Experiment Setup

This paper uses "Hili" server based on Cavium OCTEON multi-core NP as the experimental environment. It tests storage I/O performance and performance density of the transparent accelerating software architecture based on multi-core heterogeneous systems ("heterogeneous scheme" for short) by comparing with the traditional Linux kernel-based architecture ("Linux scheme" for short). The test uses typical SCSI/iSCSI as the storage protocol. The iSCSI initiator and target are "Hili" server and disk array BW2000 respectively. The specific hardware parameters are shown in Table 1 below.

Table 1. Hardware parameters

	Hili	BW2000
Processor	Cavium OCTEON 5860	Intel
Processor architecture	MIPS64	X86
Processor specification	800 MHz*16	–
Memory	8 GB	–
Disks	–	12*1TB
NICs	1GE*10	1GE*6

Both heterogeneous scheme and Linux scheme are implemented on Cavium OCTEON multi-core NP. In heterogeneous scheme, the control/management plane and network service module in Linux are running on 2 processing cores, and the data plane in SE are running on other cores. In Linux scheme, the whole network storage system in Linux is running on the same set of processing cores. The specific operating parameters are shown in Table 2 below.

Table 2. Operating parameters

	Heterogeneous scheme	Linux scheme
Processing core distribution	Linux: 2 cores SE: 2, 4, 6, 8, 10, 12, 14 core	4, 6, 8, 10, 12, 14, 16 cores
Memory	Linux: 2 GB SE: 6 GB	8 GB
Linux kernel version	2.6.21	2.6.21
Open-iSCSI version	2.0-873	2.0-873
SE version	Cavium SDK 1.7.3	–

3.2 Experiment Result

Figure 3 illustrates the experimental results for the two schemes on network storage I/O throughput, wherein the horizontal axis is the number of running cores. On the same number of total processing cores, the I/O performance of heterogeneous scheme is 8.3 % higher than Linux scheme on average. Due to allocating two for Linux, the number of data plane cores in heterogeneous scheme is less than Linux scheme, which implies that the former has even higher performance density.

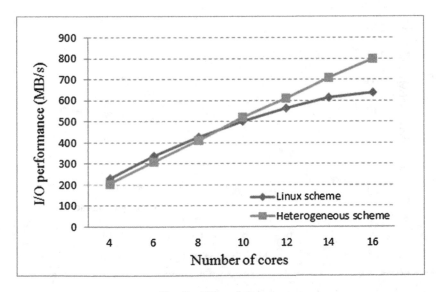

Fig. 3. I/O performance.

Table 3. Resource utilization

	Heterogeneous scheme	Linux scheme
CPU utilization	Linux: 4 %–11 % = 7.5 % SE: 90 %	80 %
CPU overhead	N is the number of processing cores, 2*800 Hz*7.5 % + (N-2)*800*100 %	N*800 Hz*80 %
Memory utilization	Linux: 17 %–33 % = 25 % SE: 100 %	100 %
Memory overhead	2G*25 % + 6G*100 % = 6.5G (Bytes)	8G*100 % = 8G (Bytes)

As the number of processing cores increasing, the advantage of heterogeneous scheme becomes obvious. When the core number exceeds 10, the performance of heterogeneous scheme is 16.9 % over Linux scheme.

The experimental result also presents that the performance growth of Linux scheme tends to slow down along with processing core number increasing, namely the scalability is dropping. However the performance of heterogeneous scheme remains near-linear growth, which indicates much better multi-core scalability.

The resource utilization of the schemes is shown in Table 3. On CPU utilization, the Linux scheme keeps 78 %–85 %, while the Linux side of heterogeneous scheme is 4 %–11 % because Linux is not involved in network protocol processing. Due to the difficulty to supervise CPU utilization of SE environment, we take 90 % as the upper bound. On memory utilization, in order to getting the performance maximum value, the Linux scheme keeps 100 %, while the SE side of heterogeneous scheme is also 100 %, and the Linux side is 17 %–33 % because it is out of data processing.

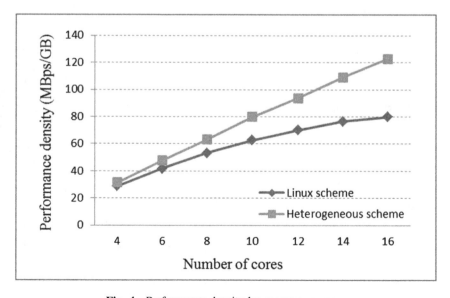

Fig. 4. Performance density by memory resource

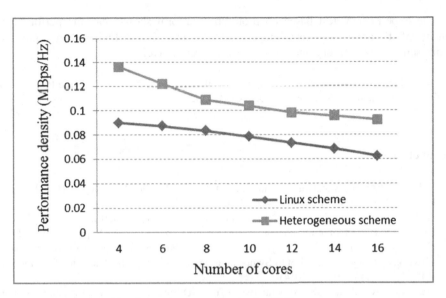

Fig. 5. Performance density by CPU resource

Figures 4 and 5 present the performance density by memory resource and CPU resource respectively. Compared with Linux scheme, the heterogeneous scheme has obvious superiority on performance density by both CPU computation resource and memory resource. For performance density by memory resource, the heterogeneous scheme exceeds Linux scheme by 32.5 % on average, and further the advantage is more significant with core number increasing. For performance density by CPU resource, the heterogeneous scheme exceeds Linux scheme by 53.2 % on average, and the mean value is 1.3bit/Hz that breaks the limit of Thumb's law.

4   Conclusion

For the demands of network storage systems on both protocol acceleration and transparent development, this paper proposes a transparent accelerating software architecture for network storage based on multi-core heterogeneous systems. The transparent accelerating architecture provides data plane network protocol stacks accelerating by network processing operating system, and general development and configuration interfaces for upper-level applications by general-purpose operation system.

The architecture accelerates the network storage I/O performance transparently, namely enhances the versatility and compatibility of network storage systems based on multi-core NP. Therefore, the architecture has two aspects of advantage including (1) acceleration for protocol processing and (2) agility and maintainability for service deployment.

The experimental test based on prototype platform shows that, the performance density of the architecture is averagely 32.5 % and 53.2 % higher than traditional Linux scheme by memory and CPU utilization respectively.

Acknowledgments. This research is supported by the "Strategic Priority Research Program" of the Chinese Academy of Sciences (No. XDA06010302).

References

1. Yeh, E., Chao, H., Mannem, V.: Introduction to TCP/IP offload engine (TOE). 10 Gigabit Ethernet Alliance, 10GEA, pp. 18–20 (2002)
2. Halfhil, T.R.: Netlogic broadens XLP family. Microprocessor Report **24**(7), 1–11 (2010)
3. Hussain, M.R.: Octeon multi-core processor. Keynote Speech of ANCS, pp. 611–614 (2006)
4. Gonzalez-Ferez, P., Bilas, A.: Tyche: an efficient Ethernet-based protocol for converged networked storage. In: 2014 30th Symposium on Mass Storage Systems and Technologies (MSST), pp. 1–11. IEEE (2014)
5. Darabseh, A., Alayyoub, M., Jararweh, Y.: SDStorage: a software defined storage experimental framework. In: IEEE International Conference on Cloud Engineering (IC2E), pp. 341–346 (2015)
6. Zheng, W.M., Shu, J.W.: Next generation distributed & intelligent network storage. World Telecommun. **17**(8), 16–19 (2004)
7. Chen, Y.L., Yang, C.T., Chen, S.T.: Environment virtualized distributed storage system deployment and effectiveness analysis. In: 2015 Second International Conference on Trustworthy Systems and Their Applications (TSA), pp. 94–99. IEEE (2015)
8. He, P.C., Wang, J.L., Deng, H.J.: On architecture of multi-core packet processing system. Microcomput. Appl. **9**, 12–20 (2010)
9. Guo, X.Y., Zhang, W., Wang, J.L.: Multieore frame of sending real-time streaming data for video-on-demand system. J. Chin. Mini-Micro Comput. Syst. **32**(7), 1301–1316 (2010)
10. Wu, W., Crawford, M.: Potential performance bottleneck in Linux TCP. Int. J. Commun. Syst. **20**(11), 1263–1283 (2007)

A Feature Selection Method of Power Consumption Data

Changguo Li[✉], Yunxiao Zu, and Bin Hou

School of Electronic Engineering, Beijing University of Posts and Telecommunications,
Beijing, China
lichangguo@bupt.edu.cn, zuyx@mail.tsinghua.edu.cn,
robinhou@163.com

Abstract. It is of great significance for the power supply enterprise to analyze the electrical power consumption data. However, there is no effective way to separate abnormal data. This paper presents a classification method for high dimensional power consumption data classification. Based on the information theory, the proposed method consists of two parts, feature selecting and classification of selected features by logistic regression. The experimental results below show that the method has a lower computational complexity than that of data classification without pretreatment, and higher efficiency and reliability than random feature selection.

Keywords: Classification · Feature selection · Logistic regression · Power consumption data

1 Introduction

Data displayed in electric meters represents the electrical power consumption of users over a period of time. For some users, the data may change greatly in a short time because of their growing demand for electricity, the failure of meters or otherwise. Extracting such abnormal data from a massive data is very vital to power supply enterprises. Based on the analysis of abnormal data, the power supply strategy can be changed to meet power demands in the future, and the corresponding maintenance for lines and equipment can be carried out. Usually, the raw data are high-dimensional, thus how to classify those high dimensional power consumption data is the key to solve the problem.

With the growing of the amount of data, the traditional processing mode is no longer suitable for massive data analysis. Hadoop [1] provides MapReduce architecture for parallel computing and Hdfs, a file system for distributed storage. To analyze the data will be more efficient using Hadoop. What is more, the open source Mahout library contains a large number of data mining algorithms based on Hadoop. All of which facilitate our work. Common classification algorithms include logistic regression [2], naive Bias classification, support vector machine (SVM) [3], etc. The logistic regression can obtain the coefficients of features after searching for the optimal solution of the likelihood function [4], and the coefficients can be used for subsequent classification. Naive Bayes algorithm bases on Bayesian theorem. Get the classification results throw the formula below:

© Springer International Publishing Switzerland 2016
O. Gervasi et al. (Eds.): ICCSA 2016, Part I, LNCS 9786, pp. 547–554, 2016.
DOI: 10.1007/978-3-319-42085-1_44

$$\max\{p(y_1|x), p(y_2|x), p(y_3|x) \dots p(y_n|x)\} \tag{1}$$

Support vector machine is suitable for classification of high dimensional data, but the calculation process is relatively complex. Besides, the algorithm has not been implemented in Mahout library. In this paper, the data to be processed is a series of multidimensional vector $x_1, x_2, x_3 \dots x_n$. Each x_i is power consumption data of an ID of a period. The amount of days may be 200, 300 or 400 so the dimension of the vector is correspondingly to 200, 300 or 400. We use c_m to present the column of a vector, a complete column is defined as a feature, which is t_m. Logistic regression is easy to be understood, well-suited for processing digital files. Naive Bayes algorithm needs to evaluate $p(x|y_n)$ and $p(y_n)$ for calculating $p(y_n|x)$. The statistical method is word frequency statistics, which is not sensitive to the number, not suitable for the classification mentioned in this paper.

Unlike the traditional logistic regression classification, firstly, we pre-process the data to select more representative columns as input features for logistic regression classification, during which we compare the information of each feature through the information theory [5] to find out t'_m with most information. This method, not only reduces the computational complexity but also improves the efficiency of classification. Excessive features may lead to model instability, while insufficient features cannot present enough information for modeling. Details of this method will be described in Sect. 2. The experimental results show that compared with the traditional logistic regression classification, the computation time is greatly reduced, and the classification accuracy is improved. The structure of this paper is organized as follows: in the Sect. 2, the process of feature selection is introduced in detail. The Sect. 3 describes the experiments and the results, and the Sect. 4 draws conclusions.

2   Feature Selection Method

2.1   Logistic Regression

Logistic regression is commonly used in binary classification problem. The logistic regression model can be expressed as follows.

$$p(Y = 1|x) = \pi(x) = \frac{1}{1 + e^{-g(x)}} \tag{2}$$

$$g(x) = \beta_0 + \beta_1 x_1 + \beta_2 x_2 + \dots + \beta_\rho x_\rho \tag{3}$$

Apparently, the key to solve the model is to obtain the value of $\beta_0, \beta_1, \beta_2 \dots \beta_\rho$ which maximize the value of the likelihood function. The function can be expressed as:

$$L(\beta) = \ln[l(\beta)] = \sum_{i=1}^{n} \{y_i \ln[\pi(x_i)] + (1 - y_i) \ln[1 - \pi(x_i)]\} \tag{4}$$

There are many methods of searching the optimal solution of the likelihood function, such as Gradient Descent Method [6], Conjugate Gradient Method, Quasi-Newton Method [7]. In operation, some features should be specified for logistic regression. That is c_m, the column of x_i. Although you can specify that all the features are predictors, but this greatly increased the complexity of the algorithm, and lead to a less stable model, which will cause the success rate of classification declining.

2.2   Feature Selection Method Based on Information Theory

Characteristics of Power Consumption Data. By analysis of power consumption data, it is easy to find that the data of most users is relatively stable. Only small part of them has a large absolute value and floating scope. The characteristic above means the data to be classified is imbalanced. For this question, we are more concerned about improving the classification accuracy of minority classes, because separating the abnormal data record from a large number of power consumption data is much meaningful. For this reason, we try to select the features that can reflect the characteristic of the abnormal data rather than containing less such information. Figure 1 shows the number of the abnormal data records and the normal data records in the data set. The amount of abnormal data records is only less than 300 but the normal data records is over 2000. Figure 2 shows the typical normal records and abnormal records. Compared to the normal records data the abnormal records are smaller and less volatile.

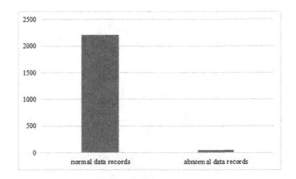

Fig. 1. Number of each type of power consumption data

Calculation Information and Select Feature. In order to effectively select the features, we define the concept of information for every column of the abnormal records. Firstly, in the abnormal records, the larger the value of a feature is, the smaller the probability occurrence will be. and the amount of information of this record will be greater. Secondly, according to information theory, in one abnormal record, the sum of the probability of all columns should be 1. Based on the above, we define information of features of the abnormal data records as:

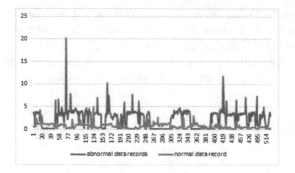

Fig. 2. Typical data of the normal and abnormal records (Color figure online)

$$h(m) = \log_2\left[1/p(m)\right] \tag{5}$$

where

$$p(m) = \left(\frac{1}{v_m}\right) \Big/ sum\left(\frac{1}{v_m}\right) \tag{6}$$

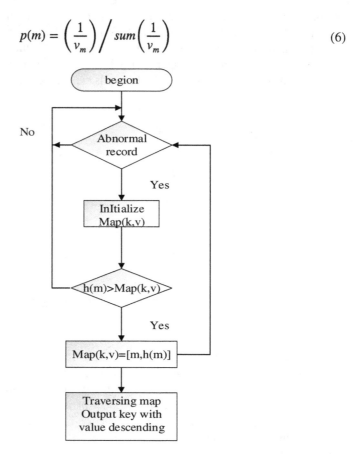

Fig. 3. Progress of feature selection

The detail of the algorithm are as follows:

- check whether the record is abnormal, if not, check the next,
- for the abnormal data record, the initialize value of information through making $map_m(k, v) = (m, 0)$,
- calculate information of feature $h(m)$,
- compare $h(m)$ and $map_m(k, v)$. If $h(m) > map_m(k, v)$, $map_m(k, v) = [m, h(m)]$,
- traverse the map and find out keys with top n value.

The data of normal columns are very close and they contain little information. What is more as the number of records is very large, it is important to simplify the calculation process. In the final step, n features with most information will be output. So the performance has a great relationship with parameter n. Figure 3 is the flow chart of feature selection.

3   Experiments Design and Results

3.1   Experiment Data Set

The experiment data is big clients' power consumption data of "Changge" City, Henan Province from January 1th, 2013 to December 31th, 2014. Data is true and reliable. Depending on the experimental requirement, the dimension of the vector can be changed. For example, if we need to analysis data of one month, a 30-dimensional vector will be selected; if we need to analysis that of one year, a 365-dimensional vector will be selected. The data are derived from oracle database. And data format is text, which is common in classification and easy for program to process. The format of data is shown in Table 1.

Table 1. The form of power consumption data

Lab	V1	V2	V3	V5	...
0	0	2.6	1.29	0.53	...
0	0.86	1.05	0.67	1.75	...
1	1.94	7.87	12.73	4.49	...

3.2   Experiment Settings

As the amount of data grows, processing data in Big Data mode is common. In order to complete the experiment better, we build a five-node Hadoop cluster, with a master node and four slaves nodes. Mahout was deployed on Hadoop platform which implements many classification algorithms, such as Logistic regression and Naive Bayes classification. The structure of the experimental environment is shown in Fig. 4.

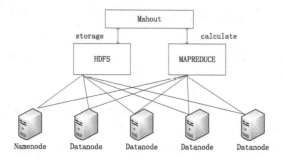

Fig. 4. Hadoop and Mahout

To examine the effect of n to classification performance, we define feature selection coefficient.

$$r = n/\beta \tag{7}$$

where n is the number of features selected and β is the number of all the features. The classification performance is tested as r and β change. In experiment 1, the dimension of the vector is set to 200, 300 and 400. Under different dimension c, the influence of r is different. We compared the experiment results and find out the best r to different dimension. We also try to analysis the relationship between r and the vector's dimension.

In experiment 2, set the total dimension of the vectors as 300, and the number of features vectors as 20 (from experiment 1, we find the number of features which results in the best performance). Compare training time, classification time and AUC (Area Under Curve) [8] under three different situations. The situations are:

- (A) select 20 features in the way this paper proposed by.
- (B) randomly selected 20 features.
- (C) use all the columns without any selection.

3.3 Experiment Results

The results of experiment 1 are shown in Fig. 5. The X-axis is feature selection coefficient r; the Y-axis is AUC. AUC is a parameter to represent the comprehensive performance of a classifier. Generally, the greater the value of AUC, the better the classifier will be. It is obvious to see that if the dimension is 200, the optimal r is 0.07; if the dimension is 300, the optimal r is 0.06; if the dimension is 400, the optimal r is 0.05. The optimum feature selection coefficient changes with the total dimension. For one certain r, more features will be selected as the number of dimension grows. For one certain classification issues, excessive features easily lead to instability of the model, and worse classifier performance. Thus, with the increase of dimension, r tends to be smaller.

Results of experiment 2 are shown in Table 2. If classification is carried out with no feature selection, the time it cost is much more than classification with feature selection, which only consume about a half of the time of the first. Moreover, taking 300-dimensional data as input as logistic regression lead to the classifier less stable and performance

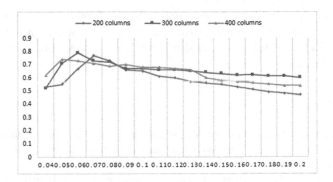

Fig. 5. Result of experiment 1

degradation. When comparing *classification time* of classifying with selecting features method in this paper and that of random feature selecting method, the difference will be little. However, disparity of *AUC* is greatly. Features selected as this paper described contains more information, which contributes to far better performance than classifying with random selection features.

Table 2. Result of experiment 2

No.	Training time	Classification time	Auc
A	24119	1046	0.78
B	22790	1423	0.47
C	58858	3588	0.53

4 Conclusions

For multidimensional power consumption data classification, we propose a prepro-cessing method which will select features as inputs for logistic regression. Our work reduces the time required for classification. As data dimension grows, feature selecting become even more critical for classification. In feature selection, we use the concept related with information theory described in Eq. (5). $p(m)$ represent the probability of t_m and $p(m)$ is inversely proportional to the value of c_m. Such a definition is in line with normal cognition. The probability of a column with a bigger value appears should be smaller, and the amount of information contained in such columns shall be greater. With the information of all columns is calculated and updated, and we list the columns in descending order by the amount of information. Output columns with top n maximum $p(m)$, we get t'_m.

Furthermore, we test the influence of parameter n on the performance of classification algorithm. Find out the appropriate feature selection coefficient for data of different dimension. At the same time, we test the performance of the proposed algorithm in 3 different cases. Experiments showed that selecting features for high-dimensional data classification greatly reduces calculation time and improves the classification perform-ance. The algorithm proposed in this paper is good performance. However, this paper is just a try for improving high-dimensional data classification. In the pre-processing stage, method of Principal Component Analysis [9] is a potential choice. In the classification stage, SVM and Adaboost can also be applied to further improve the performance of classification.

References

1. Huang, Y., Lan, X., Chen, X., Guo, W.: Towards model based approach to hadoop deployment and configuration. In: 2015 12th Web Information System and Application Conference (WISA), Jinan, pp. 79–84 (2015)
2. Dong, Y., Guo, H., Zhi, W., Fan, M.: Class imbalance oriented logistic regression. In: 2014 International Conference on Cyber-Enabled Distributed Computing and Knowledge Discovery (CyberC), Shanghai, pp. 187–192 (2014)
3. Karatsiolis, S., Schizas, C.N.: Region based Support Vector Machine algorithm for medical diagnosis on Pima Indian Diabetes dataset. In: 2012 IEEE 12th International Conference on Bioinformatics & Bioengineering (BIBE), Larnaca, pp. 139–144 (2012)
4. Byram, B., Trahey, G.E., Palmeri, M.: Bayesian speckle tracking. Part I: an implementable perturbation to the likelihood function for ultrasound displacement estimation. IEEE Trans. Ultrason. Ferroelectr. Freq. Control $60(1)$, 132–143 (2013)
5. Xu, J., Ma, X., Shen, Y., Tang, J., Xu, B., Qiao, Y.: Objective information theory: a Sextuple model and 9 kinds of metrics. In: Science and Information Conference (SAI), London, pp. 793–802 (2014)
6. Liu, J., Wang, H., Yi, D., Sun, L.: Seismic reflectivity inversion by a sparsity and lateral continuity constrained gradient descent method. In: 2012 IEEE International Conference on Service Operations and Logistics, and Informatics (SOLI), Suzhou, pp. 236–240 (2012)
7. Honda, T., Kohira, Y.: An acceleration for any-angle routing using quasi-newton method on GPGPU. embedded multicore/manycore SoCs (MCSoc). In: 2014 IEEE 8th International Symposium on Aizu-Wakamatsu, pp. 281–288 (2014)
8. Srinivasulu, A., SubbaRao, C.D.V., Jeevan, K.Y.: High dimensional datasets using hadoop mahout machine learning algorithms. In: 2014 International Conference on Computer and Communications Technologies (ICCCT), Hyderabad, p. 1 (2014)
9. Nie, B., et al.: Crowds' classification using hierarchical cluster, rough sets, principal component analysis and its combination. In: International Forum on Computer Science-Technology and Applications, IFCSTA 2009, Chongqing, pp. 287–290 (2009)

In-band Busy Tone Protocol for QoS Support in Distributed Wireless Networks

Xin Zhou[1,2(✉)] and Changwen Zheng[1]

[1] Institute of Software, Chinese Academy of Sciences, Beijing, People's Republic of China
{zhouxin,changwen}@iscas.ac.cn
[2] University of Chinese Academy of Sciences, Beijing, People's Republic of China

Abstract. Although the physical transmission rate increases rapidly, the quality of service (QoS) support for real-time traffic is still a challenge. 802.11 EDCA can only provide service differentiation, but not ensure the QoS as expected because of the uncertain short-term prioritized access and restrained high-priority traffic capacity problems. This paper proposes a distributed MAC protocol called in-band busy tone (IBT) protocol to address both the problems simultaneously. It uses the in-band busy tone reservation and preemptive scheduling schemes to provide the deterministic prioritized channel access, and improves the real-time traffic capacity through the optimized CW settings. Simulation results demonstrate that the IBT performs much better than EDCA on supporting the QoS of the real-time traffic.

Keywords: 802.11 · EDCA · Busy tone · QoS · Priority · Capacity

1 Introduction

Multimedia traffic (VoIP, video etc.) has occupied more than 64 % of all Internet traffic since 2014, and grows about 20 % per year [1]. Unlike non-real-time traffic, multimedia traffic imposes strict Quality of Service (QoS) requirements, such as sufficient throughput, low and stable end-to-end delay, and low packet loss rate.

The 802.11 EDCA is the most famous protocol to support QoS for distributed wireless networks. It classifies the upper layer frames into four Access Categories (ACs) according to the traffic types, and uses a set of AC specific parameters, such as Arbitrary Inter-Frame Space (AIFS), Transmission Opportunity (TXOP) and Contention Window (CW), to provide the service differentiation [2]. However, the EDCA can only provide prioritized access in the long-term scale, but for a single transmission, lower-priority (LP) traffic can still transmit packets prior to the higher-priority (HP) traffic [3], and thus affect the QoS of HP traffic especially when the load of the LP traffic is heavy. This is called the uncertain short-term prioritized access problem.

Besides, the EDCA can only support a limited number of concurrent HP traffic, no matter how much the physical transmission rate is. As we known, the CW size of the HP traffic is generally small for priority differentiation, but the packets are more likely to collide in this situation, especially when the station number is big. We call it the restrained high-priority traffic capacity problem.

© Springer International Publishing Switzerland 2016
O. Gervasi et al. (Eds.): ICCSA 2016, Part I, LNCS 9786, pp. 555–562, 2016.
DOI: 10.1007/978-3-319-42085-1_45

Many studies extending the EDCA can be found in the literatures. Authors of [4–6] improve the QoS ability by adjusting the AC specific parameters, authors of [7] reduces the MAC headers of aggregated frames to improve the efficiency of small packet traffic (i.e. VOIP), authors of [8, 9] present queue-adaptive scheduling mechanisms to employ differentiation among the same AC. However, all of them don't solve the uncertain short-term prioritized access problem.

Some works have been done to ensure the prioritized access. The scheme in [10] assigns different AIFS values to several flows, which can send immediately after AIFS stage without any other defer. It indeed maintains the QoS well for the very limited number of flows, but completely neglects the others' service. Prioritized Idle Sense (PIS) algorithm [11] provides a fairer scheme, which tunes the CW sizes of the HP traffic to a smaller target dynamically, but it can only differentiate a few ACs without decreasing the throughput.

Besides employing different priority parameters, several researchers also utilize the busy tone scheme to provide absolute prioritized access [12–15]. As best as we know, [12] is the first literature that uses busy tone for QoS support. It makes use of two out-of-band busy-tone signals to provide prioritized access, but the supported priority levels are limited due to the lack of busy tone channels. In [14], the authors propose an in-band busy tone scheme, which will send busy tone during the whole backoff stage and transmit packets if the channel is idle after that. But the packet delay is large since it prefers the station with the largest backoff counter, instead of the smallest, to access the channel. The DPCA protocol [15] alleviates this disadvantage by transmitting two slot-length busy tones after AIFS stage, but it does not mention the restrained high-priority traffic capacity problem.

In this paper, we propose a novel distributed MAC protocol called In-band Busy Tone (IBT) protocol, which uses the busy tone (BT) reservation and preemptive scheduling schemes, to provide the deterministic prioritized channel access and eliminate the restrained high-priority traffic capacity problem. The rest of the paper is organized as follows. Section 2 presents the IBT protocol in detail. Section 3 investigates the performance through simulations. Conclusion is drawn in Sect. 4.

2   In-band Busy Tone Protocol

As shown in Fig. 1, the access process of the proposed scheme is very similar to the EDCA, except the busy tone transmission at the end of the AIFS stage. In the 802.11 EDCA, a station defers for a random backoff interval after sensing that the channel is idle for AIFS period. However, in the IBT, a station sends a slot-length in-band BT at the last slot of the AIFS stage. If a station detects a BT transmission before its own, it will stop the forthcoming BT transmission and the backoff procedure until a packet transmission occurs. Here, the BT works like a deterministic priority announcement. As long as at least one higher AC exists, all the lower ACs will sense the busy tone and defer. This scheme ensures that the highest AC can always access the channel firstly without being affected by the lower ACs. If two stations with the same priority traffic send BTs simultaneously, they will contend for channel in the normal random backoff stage.

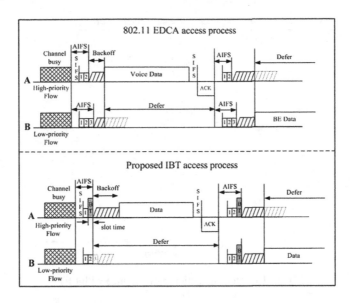

Fig. 1. Medium access process comparison between the IBT and EDCA

For the right operation of the proposed scheme, we should assume that the higher AC always has a smaller AIFS number (AIFSN), i.e.

$$AIFSN[i] = AIFSN[i + 1] - 1, 0 \leq i \leq AC_{max} - 1 \tag{1}$$

where smaller i denotes the higher priority, and AC_{max} is the largest value of i.

Although the access process of the IBT protocol is only described for the basic access mode (DATA/ACK) in this paper, it should be noted that it is also suitable for the RTS/CTS access mode, as it only changes the process before the random backoff stage. The features and advantages of the IBT protocol are described in the following.

2.1 Single Tone's Transmission and Detection

We suggest the sine wave as the busy tone, because of its simplicity and good performance on detection sensitivity and time cost. In addition, collisions between sine signals cannot change their spectrum feature, and thus don't affect the detection. Based on the discussion in [13], 5us is sufficient for the single tone detection. To verify this conclusion, we implement the FFT transaction for the single tone detection through MATLAB simulator. In the evaluation, the receiver bandwidth is 20 MHz, FFT sample frequency is 40 MHz, and the detection time is 4us. As shown in Fig. 2, the spectrum feature of the single tone is very clear even when SNR = −5 dB, and thus can be easily detected.

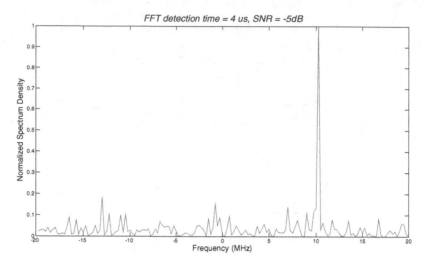

Fig. 2. Busy tone detection performance with FFT

2.2 Preemptive Internal Scheduling

The EDCA uses the internal collision resolution function to coordinate the internal AC flows [2]. Each AC within a station contends for the channel as an external contention between different stations. So, LP ACs may be scheduled prior to the HP ACs as same as the external uncertain short-term prioritized access problem.

For the IBT protocol, a preemptive scheduling scheme is proposed for the internal ACs coordination. The higher-priority ACs can break the transmission of the lower as long as the lower is not sending right now. Whenever the channel becomes idle from busy, the MAC module always schedules the transmission of the first packet with the highest priority, even if the LP frame is deferring to send or retransmit. This scheme provides the prioritized access as firmly as possible.

2.3 Optimized CW's Setting

For the EDCA, the CW size is used to differentiate the traffic priorities, so, the CW*max* of the HP AC is generally less than or equal to the CW*min* of the LP [2]. However, it is not necessary for the IBT, because the prioritized access does not rely on the CW settings, but the busy tone reservation and preemptive scheduling schemes. As a result, the CW size of each AC can be set independently according to its own condition. This is the key note to resolve the restrained high-priority traffic capacity problem.

In our previous work, it has been proved that the optimal CW size is proportional to the number of active stations [16], i.e.

$$CW_{opt} \approx \left\lceil \frac{1 + \sqrt{2q+1}}{2} M \right\rceil, q = \frac{T_{\text{Data}} + SIFS + T_{\text{Ack}} + AIFS}{slot} \tag{2}$$

where M is the active stations number. In this paper, to show the improvement by busy tone scheme more fairly and convincingly, we still utilize the same backoff algorithm as the EDCA. The only change is the CW settings according to the formula.

3 Simulation Analysis

In this section, we use NS-2 simulator to evaluate the performance of the proposed IBT protocol. As a comparison, the EDCA and DPCA are also implemented. In the simulations, stations are randomly scattered in a square area of 100 m by 100 m. The PHY and MAC layer parameters are picked up from the 802.11 g listed in Table 1. Three types of traffic (AC_VO, AC_VI, and AC_BE) are produced with parameters listed in Table 2.

Table 1. PHY and MAC parameters used in the simulations

bandwidth	20 MHz, OFDM
bit rate	54 Mbps
slot	9 us
SIFS	16 us
AIFS	SIFS + $AIFSN$*slot
PLCP overhead	Preamble + Header = 20 us
MAC overhead	Header + FCS = 40 Octets
Data packet size	Depend on the IP frame length of traffic
ACK	14 Octets
Retry limit	7

Table 2. Traffic related parameters

Parameters		AC_VO	AC_VI	AC_BE
Encoder		G.729a	H.264	none
IP frame length		80	1350	1500
Inter arrival time (ms)		40	10	1
Queue buffer (packets)		64	64	64
MAC data rate (Mbps)		0.024	1.112	12.32
$AIFSN$	for EDCA	2	3	6
	for IBT	2	3	4
	for DPCA	2	3	4
$CWmin$	for EDCA	3	7	15
	for IBT	31	31	31
	for DPCA	3	7	15
$CWmax$	for EDCA	7	15	1023
	for IBT	255	255	1023
	for DPCA	7	15	1023

G.729a is assumed as the voice encoder, which encodes 10 ms voice information into a single frame of 10 octets. Through putting four voice frames into one UDP packet and adding the overhead of IP/UDP/RTP headers, one voice IP frame is 80 octets. H. 264 is assumed as the encoder for 720P video with the data rate about 1Mbps. BE traffic is saturated that every station always has a packet of AC_BE to transmit.

The main performance metrics of interest are throughput, average delay, and packet loss rate (PLR). Throughput is the transmission rate in the view of MAC layer, including the IP and MAC overhead. Delay is the period between the packet arrival time at the MAC layer and the receive time of the ACK frame. PLR is the ratio of the dropped packets number to the generated number. In the simulations, packets are dropped due to three reasons: (1) collisions or bit errors; (2) unacceptable delay according to the QoS requirements; (3) queue buffer overflow. So, PLR is the most important metric for the real-time traffic.

As recommended in the ITU G.114, a maximum of 100 ms one-way latency and 1 % packet loss are required for VOIP [8]. Any voice calls that don't fulfill these requirements should be dropped. Thus, we can get the VOIP capacity that is defined as the maximum number of simultaneous voice calls supported by the network. The QoS requirements for the video traffic are similar except that the one-way latency is set to 200 ms [10].

3.1 Single AC Scenario

In the single AC scenarios, each station has only one single traffic flow with the same type. Figure 3(a)–(c) shows the performance when the VOIP is the only traffic in the network. The VOIP capacity is only 10 flows for the EDCA and DPCA protocol, because the PLRs of them raise rapidly due to the packet collision. Since the IBT has already used the busy tone and preemptive scheduling to provide the deterministic prioritized access, it can increase the CW bound of the real-time traffic without any priority disruption. Extended study shows that, when the CW is {31,1023}, the VOIP capacity of IBT protocol is more than 150, which is about 15 times as many as EDCA and DPCA. At the same time, the throughputs of EDCA and DPCA are also smaller than the IBT due to the increasing collision probability.

Figure 3(d)–(f) shows the performance for video traffic. The performance pattern is similar as the VOIP. Because of the higher data rate of video traffic, the maximum concurrent video flows number is much smaller than the VOIP. The simulation results show that the video flow capacity for EDCA is 15, whereas the capacity for IBT is 22, which is increased by about 45 %.

3.2 Mixed AC Scenario

In the mixed AC scenarios, each station has three flows, one for each traffic type (AC_VO, AC_VI and AC_BE). Figure 4(a) also shows that, for the IBT, the LP traffic can scarcely access the channel before the QoS requirements of HP traffic have been fulfilled. However, without the preemptive internal scheduling, the DPCA cannot provide deterministic prioritized access even if it uses a similar busy tone scheme. As shown in Fig. 4(c), the PLRs of EDCA and DPCA exceed the QoS upper bounds when

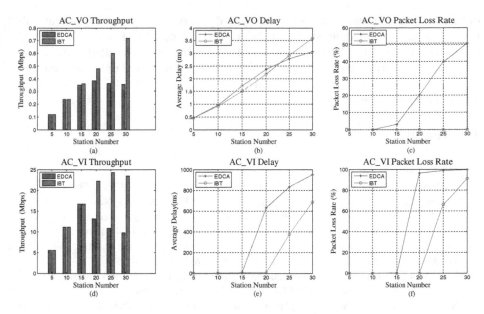

Fig. 3. Performance in the single AC scenarios, according to the station number. (Color figure online)

the stations number is more than 5. But for the IBT, the QoS requirements of the mixed flows are always fulfilled until the stations number is more than 22.

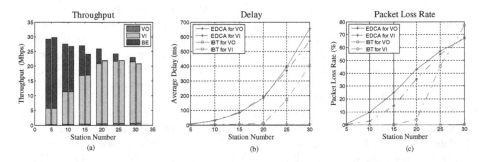

Fig. 4. Performance in the mixed AC scenarios, according to the station number. (Color figure online)

4 Conclusion

In this paper, we have proposed a novel distributed MAC protocol named IBT, which uses the busy tone and preemptive scheduling schemes to solve the uncertain short-term prioritized access and restrained high-priority traffic capacity problems. It only uses one slot-length in-band busy-tone to announce the priority without any additional bandwidth or hardware cost, consumes much less energy, and supports much more concurrent real-time flows. Simulation results show that, for the IBT protocol, the LP traffic can scarcely access the channel before the requirements of HP traffic have been fulfilled, and the real-time traffic capacity is improved by more than 4 times in the mixed traffic scenarios. In the future, more extensive study will be done about the backoff rules and the capture effect of the busy tone.

References

1. Meeker, M.: 2015 Internet Trends. KPCB report (2015)
2. IEEE Standard 802.11e: Wireless LAN Medium Access Control (MAC) and Physical Layer Specifications, Amendment 8: MAC Quality of Service Enhancements, November 2005
3. Charfi, E., Chaari, L., Kamoun, L.: PHY MAC enhancements and QoS mechanisms for very high throughput WLANs: a survey. IEEE Commun. Surv. Tutorials 15(4), 1714–1735 (2013)
4. Park, E., Kim, D., Choi, C., So, J.: Improving quality of service and assuring fairness in WLAN access networks. IEEE Trans. Mob. Comput. 6(4), 337–350 (2007)
5. Xiao, Y., Li, H., Choi, S.: Protection and guarantee for voice and video traffic in IEEE 802.11e wireless LANs. In: IEEE/ACM INFOCOM, pp. 2153–2163 (2004)
6. Xiao, Y., Li, H., Choi, S.: Two-level protection and guarantee for multimedia traffic in IEEE 802.11e distributed WLANs. ACM Wireless Netw. 15(2), 141–161 (2009)
7. Saif, A., Othman, M., et al.: Impact of aggregation headers on aggregating small MSDUs in 802.11n WLANs. In: IEEE ICCAIE, pp. 630–635 (2010)
8. Hammouri, M., Daigle, J.: A distributed scheduling mechanism to improve quality of service in IEEE 802.11 ad hoc networks. In: IEEE ISCC, pp. 1–6 (2011)
9. Gupta, A., Lin, X., Srikant, R.: Low-complexity distributed scheduling algorithms for wireless networks. IEEE/ACM Trans. Network. 17(6), 1846–1859 (2009)
10. Hoffmann, O., Schaefer, F., et al.: Prioritized medium access in ad-hoc networks with a SystemClick model of the IEEE 802.11n MAC. In: IEEE PIMRC, pp. 2805–2810 (2010)
11. Nassiri, M., Heusse, M., Duda, A.: A novel access method for supporting absolute and proportional priorities in 802.11 WLANs. In: IEEE/ACM INFOCOM, pp. 1382–1390 (2008)
12. Yang, X., Vaidya, N.: Priority scheduling in wireless ad hoc networks. In: IEEE/ACM MobiHoc, pp. 71–79, June 2002
13. Banerjee, A., Tantra, J., et al.: A service/device differentiation scheme for contention-tone-based wireless LAN protocol. IEEE Trans. Veh. Technol. 59(8), 3872–3885 (2010)
14. Jiang, H., Wang, P., Zhuang, W.: A distributed channel access scheme with guaranteed priority and enhanced fairness. IEEE Trans. Wireless Commun. 6(6), 2114–2125 (2007)
15. Kim, S., Huang, R., Fang, Y.: Deterministic priority channel access scheme for QoS support in IEEE 802.11e wireless LANs. IEEE Trans. Veh. Technol. 58(2), 855–864 (2009)
16. Zhou, X., Zheng, C., He, X.: Adaptive Contention Window Tuning for IEEE 802.11. IEEE ICT, pp. 1–6, April 2015

An IoT Application: Health Care System with Android Devices

Guanqun Cao and Jiangbo Liu[✉]

Computer Science and Information Systems Department, Bradley University,
Peoria, IL 61625, USA
jiangbo@bradley.edu

Abstract. With more and more widespread use of The Internet of Things (IoT), many industries, including health care industry, are being revolutionized in gathering and analyzing the data using IoT. In this article, we develop an IoT system that is capable of measuring saturation peripheral oxygen and pulse rate, and sending them to cloud side server. The development includes Amazon cloud server set-up and Android app development. The server is used for data storage, manipulation, analysis and access. The Android app is for device data collection, working as a bridge between the measuring devices and cloud side server. The advantage of this IoT system is specified and the future promising of this technology is analyzed.

Keywords: The Internet of Things · IoT · Bluetooth · Android · Cloud computing

1 Introduction

The Internet of Things (IoT) is the network featuring connections among physical objects or things embedded with electronics, software, sensors or connectivity. It enables those objects to exchange data, and hence, reach superior value and service. Like each computer in the Internet with a unique identifier, namely IP address, every physical node in the IoT can be uniquely identified through its embedded computing system that is capable to work with each other within the existing Internet infrastructure. Nowadays, IoT technology is widely spreading in media, environmental monitoring, manufacturing, energy management, medical and health care systems, and many other industries. For example, earthquake and tsunami detectors, emergency centers and individual cell phones can be connected by IoT to provide immediate emergency service [1].

Before IoT, computers and the Internet almost completely relied on humans to input data and information, and most information was acquired and created through typing, recording, photographing and bar code scanning, which are mainly human activities. Here comes a problem: humans are restricted in gathering data from real world in respect of time, endurance and precision. We live in material world and we cannot feed on virtual information, because despite the importance of it, material is more substantial. Information technology is so dependent on processed information of humans that it records our thoughts rather than the studied objects themselves. If somehow, computers can be cognitive and gather information from real world without human beings, we can reduce

© Springer International Publishing Switzerland 2016
O. Gervasi et al. (Eds.): ICCSA 2016, Part I, LNCS 9786, pp. 563–571, 2016.
DOI: 10.1007/978-3-319-42085-1_46

loss and consumption yielded by tracking and quantifying them [2]. We will know when our cars need repairing, out houses need renovating and our bodies need examining. In a word, IoT has the potential to change the world at least as much as the Internet did.

There are many advantages of incorporating IoT into our lives, which can help individuals, businesses, and society on a daily basis [3]. For individuals this new concept can come in many forms including health, safety, financially, and every day planning. The integration of IoT into the health care system could prove to be incredibly beneficial for both an individual and a society. A chip could be implemented into each individual, allowing for hospitals to monitor the vital signs of the patient. By tracking their vital signs, it could help indicate whether or not serious assessment is necessary. With all of the information that is available on the Internet, it can also scare people into believing they need more care than what is really needed. Hospitals already struggle to assess and take care of the patients that they have. By monitoring individual's health, it will allow them to judge who needs primary attention. Health care systems are organizations, institutions, and resources that deliver health care services to meet the health needs of target populations [4]. In terms of the aspect of technology, it usually needs implementation of public health measurement, data gathering, data analysis, and other technological methods. With widespread use of IoT technology, health care industry is undergoing revolution. In an IoT-based health care system, devices gather and share information directly with each other and the cloud, making it possible to collect record and analyze new data streams faster and more accurately [5]. For instance, a health service center can send immediate rescue if they identify a patient with abnormally high heartbeat rate detected by an assigned device that is capable of sending data within the IoT infrastructure.

In this article, we develop a preliminary IoT-based health care system consisting of three nodes: a Nonin 3230 Bluetooth® Smart Pulse Oximeter for detecting peripheral oxygen saturation and pulse rate, a Google Nexus 9 Tablet as a data transferring freight station, and an Amazon Web Services (AWS) cloud server for remote data storage and manipulation. In this system, the oximeter and the tablet are assigned to a patient, while doctors and nurses can access the data through a LAMP (Linux+Apache+MySQL+PHP) query system on an AWS server. This system can be used in the real time patient care.

2   Architecture of the IoT System

The architecture of the IoT system is shown below in Fig. 1 that indicates and specified the major components of the system.

2.1   Nonin 3230 Bluetooth® Smart Pulse Oximeter

The Model 3230 is one of the first medical devices to incorporate Bluetooth Smart wireless technology. Bluetooth Smart helps to facilitate simple and secure connections to Bluetooth Smart Ready devices for vital information exchange over a secure wireless connection [6]. The device is worn on the fingertip with a display built in that shows the

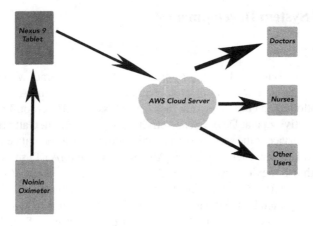

Fig. 1. The IoT system architecture

user's blood oxygenation [7]. The Model 3230 allows simple and secure connections to Bluetooth Smart-ready devices such as smart phones, tablets and telehealth solutions using iOS, Android and Windows 8. It is designed for patients who live with diseases such as chronic obstructive pulmonary disease (COPD), congestive heart failure (CHF) and asthma. These patients require frequent and accurate measurement of their SpO2 levels—the measurement of oxygen saturation in the blood [8].

2.2 Google Nexus 9 Tablet

The Nexus 9 is a tablet running Android 5.0 Lollipop, developed in collaboration between Google and HTC. [9] The reason for selecting this device for development is that beyond Android 4.3 (API Level 18), API support for Bluetooth Low Energy was introduced [10] and hence can work with Nonin Oximeter. Plus, Google Nexus is capable of Android OS update, while many Android tablets of other brands are not.

2.3 AWS Cloud Server

Amazon Web Services (AWS) is a collection of remote computing services, also called web services that make up a cloud computing platform offered by Amazon.com. These services are based out of 11 geographical regions across the world. The most central and well-known of these services are Amazon EC2 and Amazon S3. These products are marketed as a service to provide large computing capacity more quickly and cheaper than a client company building an actual physical server farm [11].

An AWS EC2 instance is used in the development for hosting a LAMP stack query system, which is able to receive data from the Android app and display gathered data on web application.

3   The IoT System Development

3.1   Nonin App

Nonin Provides the source code of the Android app that can interact with the oximeter. We use Android Studio to import this project and run the app on the Android tablet. As shown in the following picture, there are four Java class files: BluetoothLeService.java, DeviceControlActivity.java, DeviceScanActivity.java and SampleGattAttributes.java.

BluetoothLeService is the service for managing connection and data communication with a Bluetooth LE (low energy) Generic Attribute Profile (GATT) server hosted on a given Bluetooth LE device. And for a given BLE device, DeviceControlActivity provides the user interface to connect, display data, and display GATT services and characteristics supported by the device. The Activity communicates with BluetoothLe-Service, which in turn interacts with the Bluetooth LE API. DeviceScanActivity is the activity for scanning and displaying available Bluetooth LE devices. Finally, Sample-GattAttributes is the class including a small subset of standard GATT attributes for demonstration purposes.

Figure 2 shows the Nonin App operation when the oximeter is identified and the data sent from the oximeter, including saturation peripheral oxygen and pulse rate after the connection is listed.

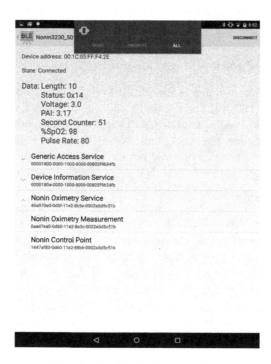

Fig. 2.   The Nonin App operation

Although we can acquire the data from the Android app, it is not capable of sending the data from the tablet to cloud server database with Nonin App. So we have to modify the source code to accomplish that.

3.2   AWS Set-up and LAMP Stack Construction

To set up an Amazon Cloud server, go to EC2 dashboard to create a server instance. Also install packages to build a LAMP stack web server. After that, create a database and a table to receive and store the data sent from the PHP file.

3.3   Develop Server Connection App

Since Nonin App cannot send data to the server, we need to modify the code to fulfill this function.

First, we need to add a method that returns the array of data we want from the oximeter in BluetoothLeService. Figure 3 shows the code of the method developed.

Fig. 3. BluetoothLeService Java method.

Then we created a new activity, DisplayMessageActivity, to test whether the method mentioned above can successfully extract the data. If so, it will display the extracted data sent from BluetoothLeService.

After the activity is created (see Fig. 4), code the method to display the data received from BluetoothLeService. Modify the on click method in DeviceControlActivity so that it will send the pulse rate to DisplayMessageActivity.

Also after the successful test, we need to create an activity, ServiceHandler, to receive data from DeviceControlActivity and send it to get_method.php on the server through get method. Then edit the on click method in DeviceControlActivity to make it send data to ServiceHandler instead of DisplayMessageActivity.

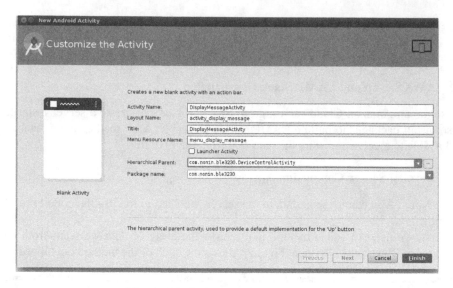

Fig. 4. DisplayMessageActivity created.

4　The IoT System Testing Results

Figure 5 illustrated the testing of the DisplayMessageActivity. It shows that the modified app can receive and display the data. In addition, there is an input field and a send button that can send pulse rate to DisplayMessageActivity once clicked.

The data collected is processed by the ServiceHandler and forwarded to the cloud server. Figure 6 shows the data on the oximeter after the successfully run of the IoT system.

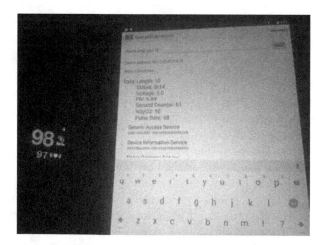

Fig. 5. The IoT system run.

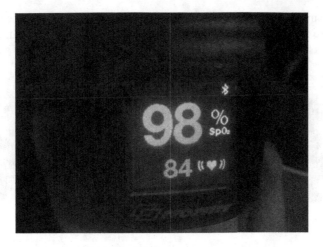

Fig. 6. Oximeter Reading.

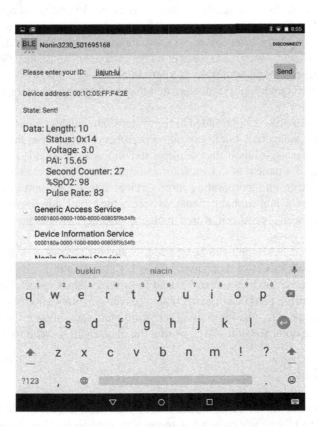

Fig. 7. The IoT system interface.

The user name entered in the user interface of the app along with other data will be sent to database for the healthcare workers for identify the user record in the database (see Figs. 7 and 8).

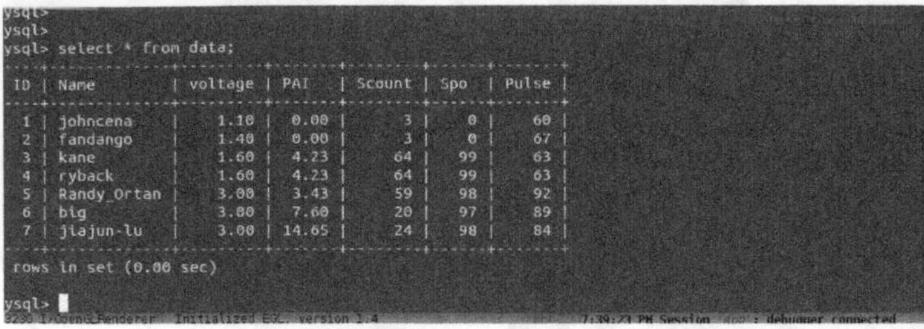
ysql>
ysql>
ysql> select * from data;

ID	Name	voltage	PAI	Scount	Spo	Pulse
1	johncena	1.10	0.00	3	0	60
2	fandango	1.40	0.00	3	0	67
3	kane	1.60	4.23	64	99	63
4	ryback	1.60	4.23	64	99	63
5	Randy_Ortan	3.00	3.43	59	98	92
6	big	3.00	7.60	20	97	89
7	jiajun-lu	3.00	14.65	24	98	84

rows in set (0.00 sec)

ysql>

Fig. 8. The IoT system database record.

5 Future Work

This system is successfully built and tested, but it can be extended further to support more functions. A website for database query and charts for statistical application can be very useful. Once we can successfully send the data from the oximeter to the database on the cloud server. Other users can access the database directly. We can develop a PHP site that allows users to query data by user name. Also, we can deploy Google Charts to display the records for the same user in a certain period and thus, we can see the trend of the user's health condition. For doctors and nurses, they can interfere quickly when necessary.

More devices are also can be added to the system. The health condition of a patient cannot be simply indicated by only saturation peripheral oxygen and pulse rate conditions. If other devices, like blood glucose meter and scale, can be added to the IoT system, health condition of a patient of diabetes, for example, can be reflected at more precise level. Thus, working on incorporating more devices into the system is needed. Plus, trying to improve the app so that patient can access the cloud side database from their tablet is vital as well, because it would increase patients' awareness of their health condition.

The IoT system developed can be applied to any measuring and data generating device. The abstract model of data passing flow is data generating device → tablet or phone → cloud server, and hence can be used for various applications. For example, a Bluetooth connectible piano can record how a player performs and then send data to cloud server via the data passing flow. Then computer can compare the data and the score, or generate other statistical data, so that the player will know how to improve.

6 Conclusion

Human activities are never a reliable data source generator. In case of measuring pulse rate, it will always take people quite a few seconds to measure it. Also, mismeasurement is never rare. But with measuring device, it would be a different story. Take the oximeter for example, it can let you know your pulse rate in three seconds and miscalculation seldom occurs. Plus, device is usually more honest because it is not easy to make up data with it. We developed this system based on that belief. It is hence expected that our system or the systems similar to ours will be more widely used in various industries in future.

References

1. Internet of Things. http://en.wikipedia.org/wiki/Internet_of_Things
2. Internet of Things (Chinese). http://zh.wikipedia.org/wiki/%E7%89%A9%E8%81%94%E7%BD%91
3. Advantages - The Internet of Things. http://sites.google.com/a/cortland.edu/the-internet-of-things/advantages
4. Health system. http://en.wikipedia.org/wiki/Health_system
5. How the Internet of Things Is Revolutionizing Healthcare. http://cache.freescale.com/files/corporate/doc/white_paper/IOTREVHEALCARWP.pdf
6. Nonin Medical's New Bluetooth® Smart Model 3230 Finger Pulse Oximeter Wins 2014 Bluetooth Breakthrough Award. http://www.prnewswire.com/news-releases/nonin-medicals-new-bluetooth-smart-model-3230-finger-pulse-oximeter-wins-2014-bluetooth-breakthrough-award-247159491.html
7. Nonin debuts Bluetooth 4 pulse oximeter. http://mobihealthnews.com/22213/nonin-debuts-bluetooth-4-pulse-oximeter/
8. Nonin News and Media. http://www.nonin.com/News.aspx?NewsID=159
9. Google Nexus. http://en.wikipedia.org/wiki/Google_Nexus#Nexus_9
10. Bluetooth | Android Developers. http://developer.android.com/guide/topics/connectivity/bluetooth.html
11. Amazon Web Services. http://en.wikipedia.org/wiki/Amazon_Web_Services

A TCP Traffic Smoothing Algorithm Based on Rate Pacing

Qiuli Shang[1,2(⊠)], Jinlin Wang[1], and Xiao Chen[1]

[1] National Network New Media Engineering Research Center,
Institute of Acoustics, Chinese Academy of Sciences, Beijing, China
{shagnql,wangjl,xxchen}@dsp.ac.cn
[2] University of Chinese Academy of Sciences, Beijing, China

Abstract. The rapid growth of streaming media puts great demand on network traffic smoothness. The traffic sawtooth waving caused by typical TCP congestion control method based on sending/congestion window and AIMD (Additive Increase and Multiplicative Decrease) constrains the transmission performance and the QoS (quality of service) of streaming media. To solve these problems, this paper proposes a unilateral TCP optimization method – TCP traffic smoothing algorithm based on rate pacing (TCP-SRP). Instead of using the traditional method based on window + AIMD, TCP-SRP achieves congestion control by pacing the packet transmission rate in order to avoiding the sawtooth waving and improving the smoothness. Further, TCP-SRP deals with packet loss and time out events via packet-loss event probability and time-out backoff strategy. Moreover, TCP-SRP enhances the bandwidth competitiveness in heavy congestion situation by improving the packet-loss event sensitivity. The experiment proves that, compared with TCP NewReno, the traffic smoothness of TCP-SRP increases by 26.5 %–38.6 % which gets even better in heavy congestion situation. The TCP friendliness of TCP-SRP is 0.36–0.58 which indicates good inter-protocol fairness.

Keywords: Streaming media · TCP · Traffic smoothing · Rate based pacing

1 Introduction

Nowadays, the rapid development of streaming media service in the Internet is proposing new challenge to the transport layer of network protocol stack. As the dominant transport layer protocol in the Internet, TCP was originally designed for traditional datagram service, which is already difficult to meet the requirements of streaming media communication.

The typical TCP algorithms often use congestion control method based on sending/congestion window and AIMD (Additive Increase and Multiplicative Decrease), namely "window + AIMD". It leads to the TCP transmission rate in steady state appearing sawtooth waving [1] which results in the decrease of traffic smoothness. On one hand, in micro, the severe fluctuation of transmission rate greatly affects the QoS (Quality of Service) of streaming media service. Some study suggested that [2], compared with delay, jitter is more likely to affect the play quality and user experience

© Springer International Publishing Switzerland 2016
O. Gervasi et al. (Eds.): ICCSA 2016, Part I, LNCS 9786, pp. 572–582, 2016.
DOI: 10.1007/978-3-319-42085-1_47

of streaming media, therefore streaming services tend to have less tolerance to jitter and higher requirement on smoothness. On the other hand, in macro, the sawtooth waving of traditional TCP traffic is the deep cause of the internal traffic burst and fluctuation in storage area networks [3]. Some study further noted that, the traffic waving of traditional TCP consumes too much buffer space of the switches and routers, and further causes the Incast problem [4].

To solve these problems, this paper proposes a unilateral TCP optimization method – TCP traffic smoothing algorithm based on rate pacing (TCP-SRP). Instead of using the traditional method based on window + AIMD, TCP-SRP achieves congestion control by pacing the packet transmission rate in order to avoiding the sawtooth waving and improving the smoothness. Further, TCP-SRP deals with the packet loss and time out events via packet-loss event probability and time-out backoff strategy. Moreover, TCP-SRP enhances the bandwidth competitiveness in heavy congestion situation by improving the packet-loss event sensitivity.

2   Related Work

2.1   Rate-Based Pacing Algorithm

Since streaming media services are sensitive to traffic jitter, the control congestion algorithms based on rate pacing, namely RBP (Rate-based Pacing) algorithms come into being, including TFRC [5], TEAR [6], RAAR [7], etc.

The classical RBP algorithm TFRC (RFC 5348: TCP Friendly Rate Control) [5], as well as its derivative protocol DCCP [8], achieves congestion control by means of TCP-friendly rate pacing. Instead of traditional method using sending/congestion window and AIMD mechanism, TFRC algorithm copes with packet loss via pacing the transmitting rate according to TCP traffic model (throughput formula), which ensures TCP friendliness and improves smoothness by avoiding traffic waving.

However, the limitations of TFRC include the following. (1) The traffic model of TFRC is based on TCP Reno, which is no longer suitable for the current Internet environment where TCP NewReno is dominant [9]. (2) Similar to TCP, TFRC uses a slow-start mechanism which also leads to fluctuation of transmitting rate and shortage of bandwidth utilization [1]. (3) Since the rate information of TFRC relies on the feedback of receive side (receiver), TFRC is a bilateral protocol that is difficult in large-scale deployment in the Internet. (4) Essentially, TFRC is still a UDP protocol without flow control and reliability assurance, which can hardly deal with the most common TCP scenarios.

2.2   TCP Traffic Model

RBP algorithm must ensure the TCP friendliness firstly. Common methods use TCP traffic model (throughput formula) to adjust the transmission rate, which ensures substantially the same throughput with the competitive TCP flows, namely TCP friendliness. TFRC algorithm [5, 10] is based on the traffic model of TCP Reno [11].

$$R(p) = \frac{1}{RTT\sqrt{\frac{2bp}{3}} + RTO\min\left(1, 3\sqrt{\frac{3bp}{8}}\right)p(1 + 32p^2)} \tag{1}$$

Among that, $R(p)$ is the throughput of TCP, that is, the sending rate of TFRC; p is the packet-loss event probability; b is the cumulative response factor, namely the number of packets confirmed by each ACK (Common TCP implementations set b to 1 or 2); RTT is the round trip delay; RTO is the time-out retransmission timer.

However, in today's Internet, TCP NewReno [12] is recognized as the most widely used TCP implementation [9]. Therefore, the proposed algorithm TCP-SRP refers to the traffic model of TCP NewReno [13] as the rate pacing formula.

$$R(p) = \frac{\frac{1}{p} + \frac{1}{E[n]}\left(\frac{1}{1-p} + 2^{1 + \log_2\frac{W}{4}} - 1\right)}{\left(\frac{W}{2} + 2 + (W-1)p\right)RTT + \frac{1}{E[n]}\left(\frac{f(p)}{1-p}RTO + \left(1 + \log_2\frac{W}{4}\right)RTT\right)} \tag{2}$$

Equation (2) is the traffic model of TCP NewReno. Among that,

$$f(p) = 1 + p + 2p^2 + 4p^3 + 8p^4 + 16p^5 + 32p^6 \tag{3}$$

$\frac{1}{E[n]}$ is the probability of ACK time-out caused by one single packet loss [13], therefore

$$\frac{1}{E[n]} = \min\left(1, \frac{\left(1 - (1-p)^3\right)\left(1 + (1-p)^3\left(1 - (1-p)^{W-3}\right)\right)}{1 - (1-p)^W}\right) \tag{4}$$

W is the congestion window size when packet-loss happens. Assume that the network environment has the characteristic of self-similarity, the packet-loss event probability p is approximately equal to the network packet loss rate q [13], then

$$W = \frac{10p^2 - 7p + \sqrt{105p^2 + 24p}}{4p^2 + 3p} \tag{5}$$

3 Methodology

As a unilateral TCP algorithm, TCP-SRP achieves congestion control by pacing the packet transmission rate in order to avoiding the sawtooth waving of TCP traffic and improving the smoothness of streaming media service. Figure 1 shows the design of TCP-SRP algorithm that includes three technical aspects.

- TCP-SRP ensures TCP friendliness by using the traffic model of TCP NewReno to pace the transmission rate.

- TCP-SRP deals with the packet loss and time out in poor network environment via packet-loss event probability and time-out backoff strategy.
- TCP-SRP enhances the bandwidth competitiveness in heavy congestion situation by improving the packet-loss event sensitivity.

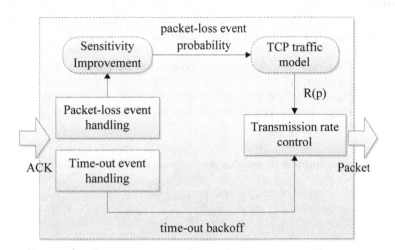

Fig. 1. The design of TCP-SRP algorithm.

3.1 Transmission Rate

For TCP friendliness, TCP-SRP uses the traffic model in (2) at sending side (sender). Since $R(p)$ is function of packet-loss event probability p, TCP-SRP paces the transmission rate according to p.

In implementation, TCP-SRP sets a timer to control the sending rate. Assume that the transmission rate of network interface is *Rate*, the TCP packet size is *Size*, then it launches a timer with *Period* = *Rate/Size*. When the timing ends, it starts another new timing of the next packet, which keeps the packet sending rate at *Rate*.

In traditional TCP algorithms, there is a slow-start process that is essentially the detection to unpredictable events, namely the limitation of network capacity. Instead of using slow-start detection, TCP-SRP can perceive the initial state via TCP traffic model. Since common Internet traffic has the characteristic of self-similarity [14], the packet-loss event probability p is approximately equal to the network packet loss rate q, then $R(q)$ should be the initial sending rate of TCP-SRP. When the algorithm starts running, the transmission rate $R(p)$ and the packet-loss event probability p can be calculated by the sender.

3.2 Packet-Loss Event Probability

The calculation of packet-loss event probability p of TCP-SRP algorithm is derived from TFRC [5]. The difference is that, TFRC calculates the p at receiver and feedbacks to the sender by ACK (Acknowledgement), but TCP-SRP is implemented at sender.

The packet-loss event means the event that the sender receives 3 duplicative ACK, namely 3-dup ACK. The difference between the first packet sequence numbers of the two successive packet-loss events is referred to be the packet-loss event interval. Packet-loss event probability is the reciprocal of the weighted moving average of the recent n packet-loss event intervals. The period n is usually set as $n = 8$ [15].

Specifically, firstly it records the 8 recent packet-loss event intervals $interval(i)$, where $i = 1, 2, 3, \ldots 8$, then it takes weighted moving average of the intervals with weighted value $weight(i)$, and finally it gets reciprocal of the weighted moving average.

$$\begin{cases} weight(1) = weight(2) = weight(3) = weight(4) = 1 \\ weight(5) = 0.8 \\ weight(6) = 0.6 \\ weight(7) = 0.4 \\ weight(8) = 0.2 \end{cases} \qquad (6)$$

Among that, the formula of weights is:

$$\begin{cases} w_i = 1, \ 1 \leq i \leq n/2 \\ w_i = 1 - \frac{i-n/2}{n/2+1}, \ n/2 < i \leq n \end{cases} \qquad (7)$$

The weighted moving average of the recent n packet-loss event intervals is:

$$\hat{s} = \frac{\sum_{i=1}^{n} w_i s_i}{\sum_{i=1}^{n} w_i} \qquad (8)$$

Then the packet-loss event probability p is $p(n) = \frac{1}{\hat{s}}$.

3.3   Packet-Loss and Time-Out Event

The main task of TCP congestion control is to handle packet-loss and time-out events by window + AIMD, which is the root cause of the traffic sawtooth waving in steady state. TCP-SRP handles packet-loss and time-out events as following:

Packet-Loss Event. When the sender receives 3-dup ACK, namely a packet-loss event happens, the traditional TCP will halve the congestion window size and enter fast recovery/fast retransmission states. The TCP-SRP algorithm replaces the fast recovery/fast retransmit process by pacing the transmission rate based on the packet-loss event probability p and the traffic model $R(p)$, which handles the packet-loss events implicitly.

Time-Out Event. When the send timer goes off, namely a time-out event happens, the traditional TCP will reduce the congestion window size to $1MSS$ and enter slow-start state. Time-out event often implies serious network congestion [16]. When time-out event happens, TCP-SRP enters a time-out backoff state. In the backoff state, it reduces the transmission rate to $\alpha \cdot R(p)$, and launches a backoff timer $\beta \cdot RTT$. If there is a new

ACK time out within $\beta \cdot RTT$, it will restart the backoff timer; if not, it will exit time-out backoff state and set the transmission rate to $R(p)$ again. By this way, TCP-SRP further smoothes out the traffic fluctuation in the traditional slow-start growth.

3.4   Sensitivity Improvement

Some studies suggested that, the RBP algorithms such as TFRC go to a sharp performance decline in heavy-congestion environments, namely shortage of bandwidth competitiveness [16–18]. The cause is that, these algorithms are based on the packet loss probability p which is insensitive to network saturation or congestion. The insensitivity results in two aspects: (1) Instead of taking backoff action and reducing the transmission rate immediately, the RBP algorithms continues generating large packet-loss backlog, which further aggravates the network burden. (2) When the network congestion is perceived and the transmission rate is reduced according to time-out and packet-loss events, it remains at low transmission rate for a period of time because of remaining packet-loss backlog, even if the network condition gets better.

TCP-SRP algorithm uses a sensitivity improvement method based on traffic prediction to solve the problem of performance decline in heavy-congestion environment. RTT (Round Trip Time) consists of three parts: transmission delay of network link, processing delay of end node, and queuing/processing delay of network node. For a certain data flow, the first two parts are usually fixed. RTT is mainly depends on the queuing/processing delay of network node which can indicate the network traffic in current environment. In other words, RTT might be the only indicator for network condition that the sender can get in TCP protocol. Therefore, several studies demonstrated the feasibility and effectiveness of using RTT to predict the self-similar long range-dependent network traffic, and further to predict the occurrence of network congestion [7, 19–21].

Based on the traffic prediction by RTT, TCP-SRP adjusts the sensitivity of packet-loss event probability p, which makes the sender reduce the transmission rate and take backoff action in advance when network link approaching congestion.

Specifically, assume that the period of p is $n = 8$, TCP-SRP divides the network traffic into $n/2 = 4$ levels, namely the network congestion level L, which is calculated by:

$$RTT_A = \frac{RTT_c - RTT_{\min}}{RTT_{\max} - RTT_{\min}} \tag{9}$$

$RTT_{\max}$ and $RTT_{\min}$ are the maximum and minimum RTT measured by sender respectively, RTT_c is the current RTT. The normalized value RTT_A indicates the network traffic. As is shown in Table 1, the network congestion level L is calculated by RTT_A. L goes higher means the traffic is getting heavy, namely the network is more congested.

The algorithm uses the sensitivity factor γ to correct the number of statistics intervals of packet-loss event probability p. The correction number is $n' = \gamma \cdot n$.

Table 1. Congestion level

Congestion level	RTT_A	Sensitivity factor	Number of statistics intervals (n = 8)
1	[0, 0.60)	1	8
2	[0.60, 0.80)	0.75	6
3	[0.80,0.90)	0.5	4
4	[0.90, 1)	0.25	2

The sensitivity correction factors according to the congestion levels are shown in Table 1. When the congestion level goes higher, the calculation of packet-loss event probability is more sensitive by reducing the number of statistics intervals.

By this method, the sender can reduce the transmission rate and take backoff action by Eq. (2) more sensitively, which avoids the large packet-loss backlog and further eases the sharp decline of throughput caused by heavy congestion.

4   Evaluation

4.1   Experiment Setup

The experiment tests and evaluates the bandwidth friendliness and traffic smoothness of TCP-SRP. Since TCP NewReno is recognized as the most widely used TCP algorithm [9], this paper takes TCP NewReno as the comparison and background flows.

4.1.1   Friendliness
The TCP friendliness is represented by the inter-protocol fairness factor F_p [22].

$$F_p = \frac{R_p}{R_p + R_c} \tag{10}$$

R_p is the average throughput of protocol flow P, R_c is the average throughput of the competitive flows.

TCP-friendliness means that the average bandwidth occupation of the protocol flow does not exceed the competitive TCP flows in the same link [23]. Hence the flow with $F_p < 0.5$ is TCP-friendly to the other TCP flows in the network. If the F_p increases, the bandwidth competitiveness of protocol flow P will increase, and the TCP friendliness will decrease. However, when F_p is far less than the critical value 0.5, the flow P will be starved by the other competitors. It is generally considered that the flow with F_p between 0.35–0.65 can share the link bandwidth with the competitive flows fairly [24], which means good inter-protocol fairness. In particular, when $F_p = 0.5$, it has ideal fairness.

4.1.2   Smoothness
The traffic smoothness is represented by the *COV* (Coefficient of Variation) of the protocol flow throughput [22]. *COV* is defined in (11). *COV* indicates the traffic

fluctuation, the lower the *COV*, the lower the traffic burstiness, and the better the smoothness.

$$COV_p = \frac{\sqrt{\frac{1}{(t_t-t_s)/\delta} \sum_{i=1}^{(t_t-t_s)/\delta} \left(R_p(t_s + \delta \cdot i) - \bar{R}_p\right)^2}}{\bar{R}_p} \qquad (11)$$

t_s and t_t are the start time and end time of observation period respectively, δ is the step size, $\bar{R}_p$ is the average throughput in observation period, then $R_p(t_s + \delta \cdot i)$ is the throughput in the i-th observation interval.

This paper uses "Hili" server based on multi-core network processor as the experimental platform. The specific hardware parameters of Hili are shown in Table 2. In the experimental test, the link capacity is 10 Gbps, the *RTT* is generated randomly in 0–50 ms. The network saturation is adjusted by changing the concurrent TCP flows in the bottleneck link. Without loss of generality, it sets the proportion of test and contrast flow numbers to 1:1, and the remainder are background flows. The mentioned test flows are implemented by TCP-SRP, the contrast and background ones are by TCP NewReno.

Table 2. Hardware parameters of experimental platform

Parameter	Value
Processor	Cavium OCTEON 5860
Number of cores	16
Architecture	MIPS64
Frequency	800 MHz
Memory	8 GB
Network interface	10GE

4.2   Experiment Result

Figure 2 illustrates the experimental result on TCP friendliness, namely the inter-protocol fairness. Compared with the competitive flow TCP NewReno, the F_p of TCP-SRP is between 0.36–0.58, mean value is 0.43, which means good friendliness and fairness.

When the network congestion is light, namely concurrent flow number <25, the F_p of TCP-SRP is more than 0.5, which indicates that the bandwidth competitiveness exceeds TCP NewReno because of avoiding the slow-start process. As the congestion going heavy, when concurrent flow number >30, the F_p of TCP-SRP is less than 0.5, which means that the bandwidth competitiveness becomes weaker than NewReno.

As the network link going saturated, the packet-loss event probability based traffic model is insensitive to the congestion, which is the common problem of RBP algorithms. This paper has proposed a solution to the sensitivity problem of TCP-SRP. Figure 2 shows that, compared with the algorithm without sensitivity improvement

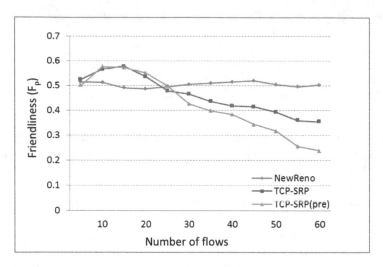

Fig. 2. TCP friendliness.

(TCP-SRP(pre)), the bandwidth competitiveness of TCP-SRP under heavy-congestion environment (concurrent flow number >50) increases by 36.3 %.

The traffic smoothness that indicates the fluctuation of transmission rate is an important index to streaming media transport protocol. Figure 3 illustrates the smoothness of TCP-SRP. As is mentioned, n is the statistical period of packet-loss event intervals which is the sensitivity of TCP-SRP to network situation changing. The *COV* of TCP-SRP is less than TCP NewReno by 26.5 % (n = 4) and 38.6 % (n = 8), which indicates the smoothness of TCP-SRP is much better.

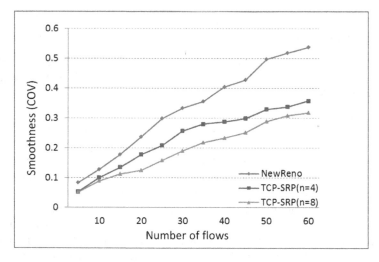

Fig. 3. Traffic smoothness.

As the concurrent flow number increasing, the network link goes saturated and the congestion goes heavy, then the traditional TCP implementations switches between slow-start and fast recovery/fast retransmission state repeatedly, which leads to the sawtooth waving of TCP traffic. When traffic fluctuation happens, the COV of TCP NewReno rises sharply and the advantage of TCP-SRP becomes even obvious. Figure 3 shows that, when concurrent flow number >50, the COV of TCP-SRP ($n = 8$) is less than NewReno by 45.3 %.

5 Conclusion

The sawtooth waving of TCP transmission rate caused by traditional congestion control method of sending/congestion window and AIMD leads to the decease of traffic smoothness, which constrains the transmission performance and the QoS of streaming media.

To solve these problems, this paper proposes a unilateral TCP optimization method – TCP traffic smoothing algorithm based on rate pacing (TCP-SRP). Instead of using the traditional TCP congestion control method by window + AIMD, TCP-SRP achieves congestion control by pacing the packet transmission rate in order to avoiding the sawtooth waving and improving the smoothness. Further, TCP-SRP deals with the packet loss and time out in poor network environment via packet-loss event probability and time-out backoff strategy. Moreover, TCP-SRP enhances the bandwidth competitiveness in heavy congestion situation by improving the packet-loss event sensitivity.

The experimental test based on prototype platform proves that, compared with TCP NewReno, the traffic smoothness of TCP-SRP increases by 26.5 %–38.6 % which is even better in heavy congestion situation. The TCP friendliness of TCP-SRP is 0.36–0.58 which indicates good inter-protocol fairness.

As a unilateral TCP algorithm, TCP-SRP provides a novel alternative scheme and optimization framework for streaming media transport layer. The deployment of TCP-SRP on Linux kernel and specific hardware platform might be the direction to be explored in the further.

Acknowledgments. This research is supported by the "Strategic Priority Research Program" of the Chinese Academy of Sciences (No. XDA06010302).

References

1. Sun, W.: Research on some problems of TCP friendly rate control protocol for streaming media. Ph.D. dissertation, Department of Information Science and Engineering, Northeastern University, Shenyang, China (2010)
2. Tulu, B., Chatterjee, S.: Internet-based telemedicine: an empirical investigation of objective and subjective video quality. Decis. Support Syst. **45**(4), 681–696 (2008)
3. Ghobadi, M., Ganjali, Y.: TCP pacing in data center networks. In: 2013 IEEE 21st Annual Symposium on High-Performance Interconnects, IEEE Computer Society, pp. 25–32 (2013)

4. Alizadeh, M., Greenberg, A., Maltz, D.A.: DCTCP: efficient packet transport for the commoditized data center. ACM Sigcomm Comput. Commun. Rev. **40**(4), 63–74 (2010)
5. Handley, M., Floyd, S., Padhye, J.: TCP friendly rate control (TFRC): protocol specification. Sonstiges **1**, 146–159 (2003)
6. Volkan, O., Injong, R., Yung, Y.: TEAR: TCP emulation at receivers – flow control for multimedia streaming. Department of Computer Science, pp. 1–24 (2000)
7. Liu, Y.H., Hu, Y., Zhang, G.Z.: Study on TCP-friendly protocol in self-similar traffic network. Chin. J. Comput. **27**(1), 42–51 (2004)
8. Kohler, E., Handley, M., Floyd, S.: Datagram congestion control protocol (DCCP). Internet Draft Internet Eng. Task Force, RFC **4340**(4), 206–207 (2006)
9. Zhang, J., Ren, F., Tang, L.: Modeling and solving TCP Incast problem in data center networks. IEEE Trans. Parallel Distrib. Syst. **26**(2), 478–491 (2015)
10. Floyd, S., Handley, M., Padhye, J.: Equation-based congestion control for unicast applications. ACM Sigcomm Comput. Commun. Rev. **30**(4), 43–54 (2000)
11. Padhye, J., Firoiu, V., Towsley, D.F.: Modeling TCP Reno performance: a simple model and its empirical validation. IEEE/ACM Trans. Network. **8**(2), 133–145 (2000)
12. Floyd, S., Henderson, T.: The NewReno modification to TCP's fast recovery algorithm. Expires **345**(2), 414–418 (1999)
13. Sun, W., Wen, T., Feng, Z.: Steady state throughput modeling of TCP NewReno. J. Comput. Res. Develop. **47**(3), 398–406 (2010)
14. Zhang, B., Yang, J.H., Wu, J.P.: Survey and analysis on the internet traffic model. J. Softw. **22**(1), 115–131 (2011)
15. Ren, F.Y., Lin, C., Liu, W.D.: Congestion control in IP network. Chin. J. Comput. **26**(9), 1025–1034 (2003)
16. Jiang, X., Wu, C.M., Jiang, M.: A congestion aware slow-start algorithm for TFRC protocol. Acta Electronica Sin. **37**(5), 1025–1029 (2009)
17. Jiang, M., Wu, C.M., Zhang, W.: Research of the algorithm to improve fairness and smoothness of TFRC protocol. Acta Electronica Sin. **37**(8), 1723–1727 (2009)
18. Aggarwal, A., Savage, S., Anderson, T.: Understanding the performance of TCP pacing. IEEE Comm. Lett. **3**, 1157–1165 (2006)
19. Li, Q.M., Xu, M.W., Yan, H.: A novel network congestion forecast method. J. Syst. Simul. **18**(8), 2101–2104 (2006)
20. Li, S.Y., Xu, D., Liu, Q.: Study on self-similarity traffic prediction and network congestion control. J. Syst. Simul. **21**(21), 6935–6939 (2009)
21. Wang, Y., Zhao, Q.C., Zheng, D.Z.: TCP congestion control algorithm on self-similar traffic network. J. Commun. **22**(5), 31–38 (2001)
22. Widmer, J., Handley, M.: Extending equation-based congestion control to multicast applications. In: ACM Sigcomm, pp. 275–285 (2001)
23. Widmer, J., Denda, R., Mauve, M.: A survey on TCP-friendly congestion control. IEEE Netw. **15**(3), 28–37 (2001)
24. Hassan, S., Kara, M.: Simulation-based performance comparison of TCP-friendly congestion control protocols. In: Proceedings of the 16th Annual UK Performance Engineering Workshop, pp. 1–11 (2000)

A Hybrid PSO and SVM Algorithm
for Content Based Image Retrieval

Xinjian Wang[✉], Guangchun Luo, Ke Qin, and Aiguo Chen

School of Computer Science and Engineering, University of Electronic Science
and Technology of China, Chengdu, China
wangxinjian_lw@163.com,
{gcluo,qinke,agchen}@uestc.edu.cn

Abstract. In order to improve the speed and accuracy of image retrieval, This paper presents a hybrid optimization algorithm which originates from Particle Swarm Optimization (PSO) and SVM (Support Vector Machine). Firstly, it use PSO algorithm, The image in the database image as a particle in PSO algorithm, After operation, return to the optimum position of the image. Secondly, use SVM to feedback the related images, Use the classification distance and nearest neighbor density to measure the most valuable image, After update classifier, choose the furthest point from the classification hyperplane as target image. Finally, the proposed method is verified by experiment, the experimental results show that this algorithm can effectively improve the image retrieval speed and accuracy.

Keywords: Image retrieval · PSO · SVM · Classification distance · Neighbor density

1 Introduction

With the rapid development of multimedia, Internet and computer vision technology, more and more digital images are generated everyday. Traditional annotation heavily relies on manual labor to label images with keywords, which unfortunately can hardly describe the diversity and ambiguity of image contents. Hence, content based image retrieval (CBIR) [1] has drawn substantial research attention in the last decade.

The main idea of CBIR is to retrieve within large collections images matching a given query due to their visual content analysis. In order to quickly find the desired image and improve the retrieval performance of the image retrieval system, a lot of optimization algorithms in the field of bionic intelligent computing are introduced into the image retrieval. Kennedy proposed the PSO algorithm [2], It has the characteristics of simple structure and fast operation, which has received more and more attention.

In CBIR, there is a mapping between low-level image features and high-level semantic features, as well as the diversity of user perception and image content, that is, "semantic gap" problem. In order to solve these problems, Relevance Feedback mechanism is introduced. According to relevant feedback, allowing users to evaluate the results and annotation retrieval, users and retrieval system to realize the interaction, can improve image retrieval accuracy. In recent years, with the relevance feedback of

© Springer International Publishing Switzerland 2016
O. Gervasi et al. (Eds.): ICCSA 2016, Part I, LNCS 9786, pp. 583–591, 2016.
DOI: 10.1007/978-3-319-42085-1_48

in-depth study, machine learning theory, such as neural networks, SVM [3–8], the related feedback for supervised learning or classification problem handling, has achieved good results. Based on statistical learning theory, SVM has good generalization ability and has many advantages in solving nonlinear and high dimensional pattern recognition problems.

In this paper, we proposed an image retrieval method based on PSO and SVM, which can improve the retrieval efficiency and improve the retrieval accuracy, namely, PSO-SVM.

2   PSO-SVM

2.1   PSO Algorithm

PSO algorithm is a popular random search algorithm. The basic solution is as follows: A group of particles with position and velocity properties are initialized at first, then, the particles move in the solution space and find the optimal solution by updating the position. In the iteration process, each particle is adaptively updated by its own testing and global cooperation.

If the initial size of the particles is n, x_i is its velocity, p_i is the personal best position, p_g is the global best position for PSO, The position and velocity of each particle are updated based on the following equations:

$$\begin{cases} v_i(t + 1) = v_i(t) + c_1 r_1(t)(p_i(t) - x_i(t)) + c_2 r_2(t)(p_g(t) - x_i(t)) \\ x_i(t + 1) = x_i(t) + v_i(t + 1) \end{cases} \tag{1}$$

Where, c_1 and c_2 called learning factors representing the cognition and social component, respectively, called acceleration coefficients. r_1 and r_2 are two random numbers, their range is [0, 1].

If $f(x)$ is the minimize objective function, the individual extremum of each particle is determined by the following equations:

$$p_i(t+1) = \begin{cases} p_i(t), \text{ if } f(x_i(t + 1)) \geq f(p_i(t)) \\ x_i(t+1), \text{ if } f(x_i(t + 1)) < f(p_i(t)) \end{cases} \tag{2}$$

The global extremum of all particles is determined by the following equations:

$$p_g(t) = \min(f(p_i(t), (i = 1, 2 \ldots n)) \tag{3}$$

2.2   SVM

The basic theory of SVM can be depicted by a typical two dimensional case shown in Fig. 1. In Fig. 1, H is the separating hyperplane, H1, H2 are parallel to H (they have the same normal) and no training points fall between them; the margin of a separating hyperplane is defined as H1 + H2. The optimal separating hyperplane what you call

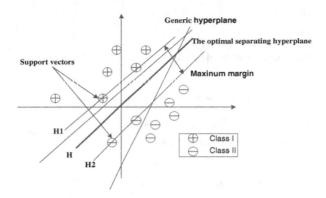

Fig. 1. Schematic diagram of SVM

not only can separate the two categories of samples exactly (the ratio of training errors is 0), but also has the maximal margin. Thus the problem of optimal separating hyperplane can betransformed a constraints problem.

For the training sets:

$\{(x_1,y_1),(x_2,y_2),\ldots,(x_n,y_n)\}$, where $x_i \in R^d$ and $y_i \in \{\pm1\}$, to get the relation between the input x_i and output y_i, it can seek an optimal function $f(x)$ by SVM training, For the linear situation, the form of function is $f(x) = wx + b$, $w \in x$, $b \in R$. In order to get an optimal function, it needs a minimum w, then the above problem can be described as an optimization problem:

$$\begin{cases} \min & \frac{1}{2}||w||^2 \\ s.t & y_i[w \cdot x_i + b] - 1 \geq 0 \end{cases} \tag{4}$$

To get the estimations of w and b, (4) can be transformed to the primal objective function (5)

$$\begin{cases} \min\left\{ -\sum_{i=1}^{n} \alpha_i + \frac{1}{2}\sum_{i,j=1}^{n} \alpha_i\alpha_jy_iy_jk(x_ix_j) \right\} \\ s.t. \quad \sum_{i=1}^{n} \alpha_iy_i = 0 \\ \forall i\, 0 \leq \alpha_i \leq c \end{cases} \tag{5}$$

Where α_i are Lagrangian multipliers, c is a user chosen constant, and $k$is a kernel function. Among various kernel functions, the Gaussian kernel is the most widely used one:

$$k(x_ix_j) = e^{-\gamma||x_i-x_j||} \tag{6}$$

Select the appropriate constant c, An optimal solution is obtained

$$\alpha^* = (\alpha_1^* \alpha_1^*, \alpha_2^*, \ldots, \alpha_n^*)^T \tag{7}$$

Choose a positive component of α^*, $0 < \alpha_l^* < c$

$$b^* = y_j(1 - \alpha_j^*/c) - \sum_{i=1}^{n} \alpha_l^* y_i k(x_i \cdot x_j) \tag{8}$$

Classification function can be defined as:

$$f(x) = sign(\sum_{i=1}^{n} \alpha_i^* \cdot y_i k(x_i \cdot x + b^*)) \tag{9}$$

According to the labeled images, SVM is divided into relevant and irrelevant samples by learning a hyper plane with the largest interval. One side of the classification is related to the sample, the other side is not related to the sample. Although the number of sample points near optimal classification plane less, but it contains very important information to establish the classification, These data are recorded as Critical data, expressed as $\Gamma_k(D)$.

For a more accurate description, we use classify distance and Neighbor density to measure $\Gamma_k(D)$, classify distance and density of neighboring can be defined as follows:

Definition 1 classify distance: it is the distance of Sample to hyperplane. If the is the classification hyperplane, $U = \{u_i\}_{i=1}^{n}$ is the unlabeled samples, the classify distance of sample can be expressed as:

$$d(u_i) = \left| \frac{f(u_i)}{\|w\|} \right| \tag{10}$$

Definition 2 Neighbor density: it is the Sample density information. The larger neighbor density, the greater the amount of information.

In Fig. 2, A circle with " + " represents the positive samples, with "−" represents the negative samples, hollow circular represents the unlabeled samples, f is the classification boundary. In all unlabeled samples, close to the border classify samples A and B have the greatest uncertainty. For the A point, although the distance far from the classification boundary B points, but its neighbor density is larger, so the comprehensive consideration, the first choice of A points as a key data sample. C point is the most remote from the classification of the positive samples, C point is the most satisfied with the user to find the target image.

If $d(u_i, u_j)$ expressed a Euclidean distance of any of the two samples within U, $D(U)$ is the distance of the samples which can be expressed as following equations:

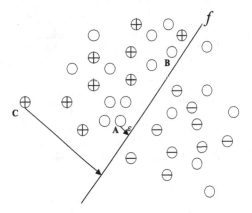

Fig. 2. Schematic diagram of optimal sample selection

$$D(U) = \frac{2}{n(n-1)} \sum_{i=1}^{n-1} \sum_{j=i+1}^{n} d(u_i, u_j) \tag{11}$$

Sample u_i is the neighborhood of the sample set can be expressed as following equations:

$$Neighbor(u_i) = \left\{ u_j \middle| u_j \in U \backslash u_i, d(u_i, u_j) < D(U) \right\} \tag{12}$$

The data set M is composed of a sample of $d(u_i) < \varepsilon$ (ε is the minimum threshold that can be set), N is the number of samples in the neighborhood of the sample u_i, the neighbor density can be expressed as:

$$p(u_i) = \frac{1}{N} \sum_{\varphi_i \in Neighbor(u)} k(u_i - \varphi_i) \tag{13}$$

Critical samples can be expressed as:

$$\Gamma_k(D) \leftarrow \arg \max_{u_i \in M} p(u_i) \tag{14}$$

2.3   Proposed PSO-SVM-RF Approach

The flow chart of PSO-SVM-RF development is illustrated in Fig. 3.

As shown in the Fig. 3, is the total number of images in the database, V_i is the position transformation velocity of the current image, p_i is the personal image best position, and p_g is the global image best position. To extract the feature set, SMFD

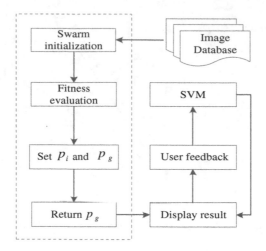

Fig. 3. Flow chart of the proposed system.

descriptors are used [9]. it extracts the corner information of the object by dividing the object contours, counts the corner points in raster coordination system to obtain the shape matrix, then analyses the periodic variation and its characteristics, expands it by column as a vector and performs Fourier transform, constructs the shape of a matrix Fourier descriptors by taking the part where the Fourier transform modulus is greater than the average value, and finally measures the similarity among the images with Euclidean distance. SMFD is determined by the following equations:

$$SMFD = u(|a_0|, |a_1|, \ldots, |a_r|), |a_i| \geq \sum_{j=1}^{N} |a_j| \bigg/ N, i \in (1, r) \qquad (15)$$

Where N is the number of the shape matrix curve transform coefficients, usually it's the several times the power of 2, r is the radial radius.

If the image q is the query image, image i is any image of database image. The similarity measurement of image q and image i can use the Euclidean distance of *SFMD* to describe.

$$D(SFMD_q, SFMD_i) = \sqrt{(a_0^q - a_0^i)^2 + (a_1^q - a_1^i)^2 + \ldots + (a_r^q - a_r^i)^2} \qquad (16)$$

The value is the adaptation value of each image particle, the smaller the value, themore similar the two images.

The description of the PSO-SVM algorithm process is shown in Table 1:

Table 1. PSO-SVM algorithm

Input: q % Query image; D % Image feature database

Initialize: $L \leftarrow \varnothing$; %labeled data; $U \leftarrow \varnothing$; %unlabeled data; k ; % an input kernel, e.g. an RBF kernel

$\qquad T$; % number of iterations

Produce:

1.the particle swarm initialization, The random position and velocity of the first generation image are initialized;

2.The adaptive value of the first generation of each particle is calculated according to equation (16);

3.Calculation of the minimum fitness value and its corresponding position of the the he first generation particle;

$\qquad$ 3.1 The minimum value of the first generation is the maximum of the global population that is P_g ;

$\qquad$ 3.2The optimal position of each particle in the first generation is p_i

4.Beginning from the first generation, update the position according to (1), update the velocity according to (2);

5.If reach the end condition, go to step 6, otherwise go to step 4;

6.The T generation optimal value of the i particle is located, The optimal value of the corresponding particles are sorted, and the corresponding image is stored to the set U ;

7.The R images in the U with the highest similarity values are chosen and return to the user.

$$S = sign(R) ; L \leftarrow S ; U = U \setminus S$$

8. If the user is satisfied with the current retrieval , then stop the algorithm, else go to the step 9;

9. $f \leftarrow SVM_{train}(L)$ %Build SVM classifier initial f on L;

10. for $t = 1$ to T do

$\qquad$ Get the $\Gamma_k(D)$, using (14); % Extract the most valuable uncertain samples on U ;

11. :Label the samples using experts on $\Gamma_k(D)$

12. Retraining SVM on new label set and obtain the new f ;

13. Update the data set

$\qquad$ 13.1 $L = L \cup \Gamma_k(D)$; 13.2 $U = U \setminus \Gamma_k(D)$;

14. end for

15. select the top N positive samples that have the largest distances to the final f ;

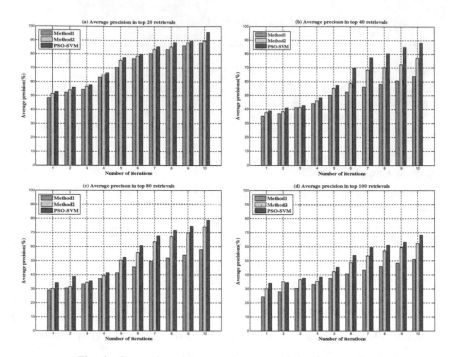

Fig. 4. Comparison for the performance (Color figure online)

3   Experimental Results and Analysis

In order to prove the effectiveness of the proposed algorithm, In this paper, we compared our algorithm with reference [10] (denoted as Method1) and reference [11] (denoted as Method2) in the same images set. This test used the MPEG7_CE-Shape-1_Part_B. It has 1400 images divided into 70 classes where each class contains 20 images.

Performance of the PSO-SVM is measured using precision. Precision are calculated using following expressions:

$$p = r/n \tag{17}$$

Where r is the number of relevant images, n is the total number of retrieved images.

Figure 4 shows the performance comparison between the proposed algorithm and other algorithms under the same condition, Selected top20, Top40, top80 and top100 retrievals. From the results of experimental data in the figure, the proposed method on the retrieval performance was significantly better than the other two methods. From the beginning of the second round of feedback, PSO-SVM algorithm has a larger upgrade, and with the increase of the number of feedback, the advantage is more obvious. Because this paper uses the particle swarm algorithm and SVM combination method, particle swarm optimization algorithm to improve the image retrieval speed, SVM related feedback to ensure the accuracy of the retrieval. When the database is large, the image features are more, the advantage is more obvious. Compared with the Method1 for simple PSO data image operations, the average performance increased by 10 %. Compared with Method2, the image feature extraction is simple, and can find the optimal image to the maximum, the performance is better than the Method2 average increase of 5 %.

4   Conclusion

In this paper, a method of image retrieval based on PSO and SVM is proposed. In this method, the PSO algorithm is proposed to select the representative image data in the image data, which can effectively reduce the range of image search and improve the speed of image retrieval. Then, the SVM method is used to improve the accuracy of image retrieval, which is related to the feedback of the results of image retrieval. By comparing the performance of the methods mentioned in the other literature, the results show that the method can significantly improve the speed and accuracy of image retrieval.

References

1. Datta, R., Joshi, D., Li, J., Wang, J.Z.: Image retrieval: ideas, influences, and, trends of the new age. ACM Comput. Surv. **40**(2), 1–60 (2008)
2. Kennedy, J., Eberhart, R.C.: Particle swarm optimization. In: Proceedings of IEEE International Conference on Neural Networks, vol. IV, pp. 1942–1948. IEEE, Piscataway (1995)
3. Fan, R.E., Chen, P.H., Lin, C.J.: Working set selection using second order information for training SVM. J. Mach. Learn. Res. **6**, 1889–1918 (2005)
4. Keerthi, S.S., Shevade, S., Bhattacharyy, C.: Improvements to Platt's SMO algorithm for SVM classifier design. Neural Comput. **3**, 637–649 (2002)
5. Keerthi, S.S., Giibert, E.G.: Convergence of a generalized SMO algorithm for SVM classifier design. Mach. Learn. **46**, 351–360 (2002)
6. Platt, J.: Fast training of support vector machines using sequential minimal optimization. In: Advances in Kernel Methods: Support Vector Learning. The MIT Press, Cambridge (1998)
7. Osuna, E., Frenud, R., Girosi, F.: An improved training algorithm for support vector machines. In: Proceedings of IEEE Workshop on Neural Networks for Signal Processing, pp. 276–285. IEEE, New York (1997)
8. Zhang, L., Lin, F., Zhang, B.: Support vector machine learning for image retrieval. In: Proceedings of IEEE International Conference on Image Processing, pp. 721–724 (2001)
9. Wang, X., Luo, G., Qin, K.: A composite descriptor for shape image retrieval. In: International Conference on Automation, Mechanical Control and Computational Engineering, pp. 759–764 (2015)
10. Broilo, M., De Natale, F.G.B.: A stochastic approach to image retrieval using relevance feedback and particle swarm optimization. IEEE Trans. Multimedia **12**(4), 267–277 (2010)
11. Imran, M., Hashim, R., Noor Elaiza, A.K., et al.: Stochastic optimized relevance feedback particle swarm optimization for content based image retrieval. Sci. World J. **2014**(2014), 752090–752091 (2014)

A Separate-Predict-Superimpose Predicting Model for Stock

Xiaolu Li[(✉)], Shuaishuai Sun, Kaiqiang Zheng, and Hanghang Zhao

Beijing University of Posts and Telecommunications,
Beijing, People's Republic of China
xlli113932@163.com,
{sun_2017,zhaohanghang}@bupt.edu.cn,
1695130462@qq.com

Abstract. The purpose of this research is to propose a more precise predicting model, the Separate-Predict-Superimpose Model, for time series, especially for the stock price and the stock risk than the established predicting method. In this model, time series are separated into three parts, including trend ingredient, periodic ingredient and random ingredient. Then the different suitable predicting methods are applying to predict different ingredients to receive accurate outcome. Ultimately, the final predicting result is superimposed by the three ingredient predicting outcome. The wavelet analysis, combination predict method, exponent smoothness method, Fourier Transform, fitting analysis and Autoregressive Moving Average (ARMA) are adopted in this model.

By applying the model to predict the Shanghai Composite Index, China National Petroleum Corporation stock price and risk and comparing with other predicting method, a conclusion can be made that this model can fit various characteristic time series and achieve a more precise result.

Keywords: The Separate-Predict-Superimpose model · Combination predict model · Wavelet analysis · Fourier transform · Fitting analysis

1 Introduction

While for stock investors, the more accurate the future stock price and stock risk forecast, the lower loss is obtained from the stock investment. Therefore, the study of predicting model for stock time series based on its intrinsic properties is of great theoretical significance and application prospect.

Because of the complexity of the stock information data, computer-aided data processing is expected to apply to analyze and process the large amount of stock information data, instead only relying on experience and intuition of relevant experts or practitioners for judgment. In this paper, on the basis of the existing predicting model, a Separate-Predict-Superimpose Model is proposed via computer aided processing and data mining.

O. Gervasi et al. (Eds.): ICCSA 2016, Part I, LNCS 9786, pp. 592–601, 2016.
DOI: 10.1007/978-3-319-42085-1_49

2 Separate-Predict-Superimpose Model

In terms of the characteristics of stock price series is a time series, therefore the methods of time series analysis are able to apply. By analysis of stock price time series, the unitary character of the time sequence, overall trend, certain period of time and small random fluctuation, is discovered. Virtually, the internal factors reflect the character of three parts of stock time series.

To raise the forecast accuracy, a Separate-Predict-Superimpose Model is proposed by this paper, based on the analysis of the characteristics of the stock price time series, combined with the established single stock price prediction model, for instance, time series model.

Separate-Predict-Superimpose Model separates a stock price time series into three ingredients, including the ingredient of time tend, quasi period and randomization.

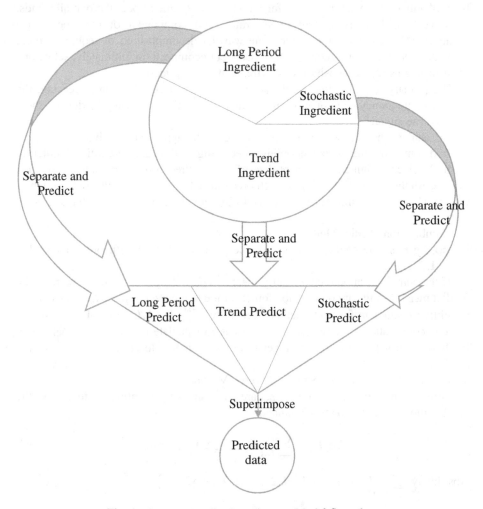

Fig. 1. Separate-Predict-Superimpose Model flow chart

For each ingredient, a suitable predicting model is adopted to predict, afterwards, the superposition of three ingredients is to acquire predicted sequence.

2.1 Separate-Predict-Superimpose Model Flow Chart

(Figure 1)

2.2 Separate-Predict-Superimpose Model Algorithm

As is shown in the Separate-Predict-Superimpose Model Flow Chart, this model is introduced by three steps as follows:

Step 1: Separate and predict trend ingredient
The evolvement of social economic for a considerable time impact the overall trends.

Wavelet analysis is applied to separate trend component of the stock price sequence. Above all, carry out five scale wavelet decomposition of sequence, then, reconstruct low frequency coefficient of wavelet decomposition, ultimately, select one scale in five as the composition of the trend of stock price series.

The primary reason is that, as the scale of wavelet decomposition increasing, the more high frequency components will be filter out, so that acquiring the development trend of the signal.

To achieve higher prediction precision overall, apply the combination predict method containing the index smoothing forecasting method and overfitting predicting method, which weighted coefficient determined by Prediction Error Sum of Squares Inverse Method. Single predicting methods can achieve the limited prediction precision, what's more, different methods suit different stages of different characteristics of trend.

- Combination Predict Method

Supposing m single predicting methods, the predicted result by ith single method is $f_i, i = 1, 2, \ldots, m$.

If the combination predicted result f meet $f = l_1 f_1 + l_2 f_2 + \cdots + l_m f_m$, naming the predict method as linear combination forecast model, among $l_1, l_2, \ldots, l_m$ represent the weighting coefficient for each method, generally $\sum_{i=1}^{m} l_i = 1, l_i \geq 0, i = 1, 2, \ldots, m$.

The core matter of combination forecast is to calculate the weighing coefficient, Prediction Error Sum of Squares Inverse Method adopted to effectively improve the prediction accuracy.

- Prediction Error Sum of Squares Inverse Method

The greater prediction error sum of squares, the smaller weighting coefficient in the combination forecast model. Command:

$$l_i = E_{ii}^{-1} / \sum_{i=1}^{m} E_{ii}^{-1}, \quad i = 1, 2, \ldots, m \tag{1}$$

Absolutely, $\sum_{i=1}^{m} l_i = 1, l_i \geq 0, i = 1, 2, \ldots, m$. $E_{ii} = \sum_{t=1}^{N} e_{it}^2 = \sum_{t=1}^{N} (x_t - x_{it})^2$.

x_{it} represents the predicted value of ith single method in tth time, x_t represents the actual value sequence $\{x_t, t = 1, 2, \ldots, N\}$ of the same forecast object, N represents the length of time, $e_{it} = (x_t - x_{it})$ represents the predicting error of ith single method in tth time.

Step 2: Separate and predict periodic ingredient
Certain periodic reflects stock price changing with seasons.

Apply Fourier Transform in rest ingredient to acquire the frequency practically energy concentrated, represented the roughly period of time. Fit the rest ingredient via cycle fitting to gain the quasi periodic fitting equation used to separate and predict periodic ingredient.

Step 3: predict virtual random ingredient
Small fluctuation may be due to short-term economic fluctuations and the psychological factors of investment.

The almost random sequence primary containing high frequency components after removing the periodic component from the rest ingredient. The autoregressive moving average (ARMA) method which is one of the time series prediction method, is applied to predict the high frequency primary random sequence.

Step 4: superimpose three ingredient
The superimposing process is to sum the three ingredient predicted value to acquire the eventual predicted result of stock series.

2.3   Evaluation of Separate-Predict-Superimpose Model

The Mean Square Error (MSE) and Mean Square Coefficient of Variation (MCV) are adopted as evaluation criterion. The smaller MSE and MCV reflect the higher precision of predicting model.

$$\mathbf{MSE} = \sqrt{\frac{\sum_{i=1}^{n} (\tilde{p}_i - E(\tilde{p}_i))^2}{n}} = \sqrt{\frac{\sum_{i=1}^{n} (\tilde{p}_i - p_i)^2}{n}} \tag{2}$$

To eliminate the dimension influence on the results of evaluation, adopting dimensionless mean square from the rate the MCV:

$$\mathbf{MCV} = \frac{MSE}{\bar{p}} \tag{3}$$

Absolutely, $\bar{P} > 0, \bar{P}$ represents the expectation price of one stock, $\tilde{p}_i$ and p_i respectively represent the predicted value and the actual value.

3   Experiment and Comparison of Separate-Predict-Superimpose Model

For evaluation the predicting performance of proposed model, by applying the model to predict the Shanghai Composite Index, China National Petroleum Corporation stock price and risk and comparing with other predicting method, a conclusion can be made that this model can fit various characteristic time series and achieve a more precise result.

3.1 Predicting Stock Price

3.1.1 Shanghai Composite Index

The 160 monthly stock prices of Shanghai Composite Index from November, 2000 to February, 2016 are used as experimental dataset. The Separate-Predict-Superimpose Model and overfitting predicting method are respectively applied to predict the stock index.

- Separate-Predict-Superimpose Model (Tables 1 and 2)
- Overfitting Predict Method (Fig. 2)

Table 1. Separate-Predict-Superimpose Model

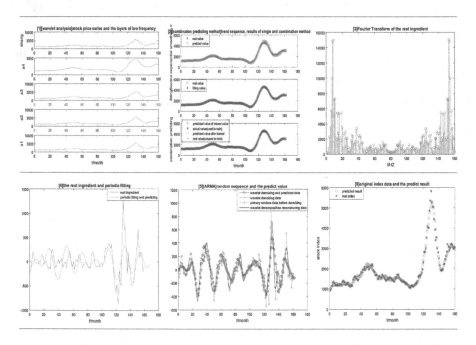

Table 2. Comparison of the two model

Evaluation criterion	Separate-Predict-Superimpose	Overfitting predicting
MSE	98.8246	280.3750
MCV	0.0756	0.6875

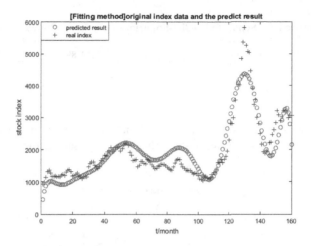

Fig. 2. Overfitting predict method

Table 3. Separate-Predict-Superimpose Model

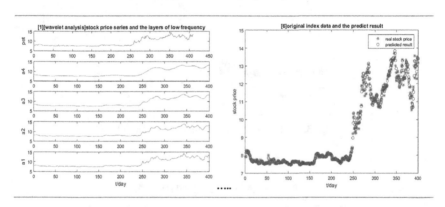

- Comparison of the two model

By contrast the evaluation criterion of these two methods, Separate-Predict-Superimpose Model achieved a higher precision satisfactory result.

3.1.2   China National Petroleum Corporation Stock Price

The 400 daily stock prices of China National Petroleum Corporation from November 18th, 2013 to July 6th, 2015 are used as experimental dataset. The Separate-Predict-Superimpose Model and the index smoothing forecasting method are respectively applied to predict the stock price.

- Separate-Predict-Superimpose Model (Table 3)
- Index smoothing forecasting method
- Comparison of the two model

Evaluation Criterion	Separate-Predict-Superimpose	Index Smoothing
MSE	0.7304	1.896
MCV	0.0787	0.5302

By contrast the evaluation criterion of these two methods, Separate-Predict-Superimpose Model achieved a higher precision satisfactory result.

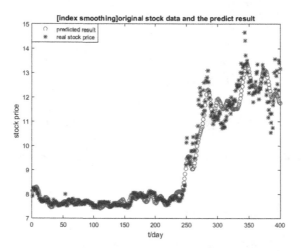

Fig. 3. Index smoothing forecasting method

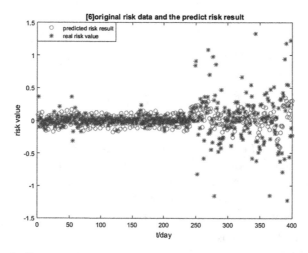

Fig. 4. Separate-Predict-Superimpose Model for stock risk predicting

3.2 Predicting Stock Risk

The dramatic appreciation or depreciation indicate the stock risk. Use the difference value of adjacent two values, for instance, the China National Petroleum Corporation stock price to predict the next arriving time of next stock risk by applying the Separate-Predict-Superimpose Model (Fig. 4).

As is shown in the Fig. 3, the Separate-Predict-Superimpose Model can predict the risk of the stock price.

4 Conclusion

The advantage of the Separate-Predict-Superimpose Model can be achieved from the experiment above.

It is true that this model can be applied to predict various growing characteristic timing sequence, not only the specific company stock price or risk, but also the overall stock index, for based on analysis of the characteristics of stock time sequence, combined with the existing predicting method, establishing a fresh stock time sequence predicting model.

5 Program Code

Here is primary code of the Separate-Predict-Superimpose Model by Matlab R2015a.

```
% the Separate-Predict-Superimpose Model
%using db5 to separete the signal into n-1 layers
[c,l]=wavedec(shang0,n-1,'db5');
for i=1:n-1
    a=wrcoef('a',c,l,'db5',n-i);
    if i==1
        aaa4=a;
    end
    if i==2
        aaa3=a;
    end
    if i==3
        aaa2=a;
    end
    if i==4
        aaa1=a;
    end
end
[c,l]=wavedec(shang,n-1,'db5');
 aa4=wrcoef('a',c,l,'db5',4);
%                 -------------------------Combination        Predicting
Model--------------------------
%-------------------Index Smoothing Forecasting Method----------------
load fa;
fadian=aaa4';
yyy=fa';
```

```
yt=fadian; n=length(yt);
alpha=0.3; st1(1)=yt(1); st2(1)=yt(1);
for i=2:n
st1(i)=alpha*yt(i)+(1-alpha)*st1(i-1);
st2(i)=alpha*st1(i)+(1-alpha)*st2(i-1);
end
a=2*st1-st2
b=alpha/(1-alpha)*(st1-st2)
yhat=a+b;
%                   ----------------------Overall            Fitting
Prediction--------------------------------------
    x=1:1:162;
f = a0 + a1*cos(x*w) + b1*sin(x*w) +  a2*cos(2*x*w) + b2*sin(2*x*w) +
a3*cos(3*x*w)  + b3*sin(3*x*w)  +   a4*cos(4*x*w) + b4*sin(4*x*w)   +
a5*cos(5*x*w)  +  b5*sin(5*x*w)  +  a6*cos(6*x*w)  +  b6*sin(6*x*w)  +
a7*cos(7*x*w) + b7*sin(7*x*w) +a8*cos(8*x*w) + b8*sin(8*x*w);
% ----------- The prediction error sum of squares reciprocal method
(Combination method)------------------------
l1=1/E1/sum([1/E1 1/E3])
l3=1/E3/sum([1/E1 1/E3])
sum([l1 l3])
%                     ------------------------------Combination
Predicting--------------------------------
ff=l1*yyy(1:160)+l3*f(1:160) ;
fff=l1*yyy(160:161)+l3*f(161:162);
% --------------------------Fourier  Transform  and  Periodic  Fitting
Predict------------
y=shang0-aaa4;
Xk=fft(y);
x=0:1:167;
f = a0 + a1*cos(x*w) + b1*sin(x*w) + a2*cos(2*x*w) + b2*sin(2*x*w) +
a3*cos(3*x*w) + b3*sin(3*x*w) +a4*cos(4*x*w) + b4*sin(4*x*w);
%--------------------fitting yc--------------------------
yc=y'-f(1:160);
x=yc;
n=5;
[c,l] = wavedec(x,n,'sym4');
% 2nd level wavelet coefficient reconstruction
xdet = wrcoef('d',c,l,'sym4',3);
% 2nd level approximation coefficient reconstruction
xapp = wrcoef('a',c,l,'sym4',3);
% ---------------------arma:-----------------------------------


xapp=xapp';
appdata = iddata(xapp,[],1);
% fit ARMA(2,3)
model = armax(xapp,[2 3]);
% predict 50 steps ahead
yf = forecast(model,appdata,2);
yf = get(yf);
y = [xapp; cell2mat(yf.OutputData)];
% -------------------Superimpose--------------------------
gupyuc0=ff+f(1:160)+xapp(1:160)';
eee=(gupyuc0-shang0').^2;
MSE=sqrt(sum(eee)/160)
MCV=MSE/mean(shang0)
```

Acknowledgements. This work is supported by the Beijing University of Posts and Telecommunications college student innovation fund (No. 151).

References

1. Box, G.E.P., Jenkins, G.M.: Time Series Analysis: Forecasting and Control. Holden-Day, San Francisco (1976)
2. Cao, L., Tay, F.E.H.: Support vector machine with adaptive parameters in financial time series forecasting. IEEE Trans. Neural Netw. **14**(6), 1506–1518 (2003)
3. Valeriy, G., Supriya, B.: Support vector machine as an efficient framework for stock market volatility forecasting. CMS **3**(2), 147–160 (2006)
4. Yeh, C.Y., Huang, C.W., Lee, S.J.: A multiple-kernel support vector regression approach for stock market price forecasting. Expert Syst. Appl. **38**(3), 2177–2186 (2011)
5. Sharifia, K.E., Hussainb, F.K., Saberic, M., Hussaind, O.K.: Support vector regression with chaos-based firefly algorithm for stock market price forecasting. Appl. Soft Comput. **13**, 947–958 (2013)
6. Tsai, F., Lin, Y.C., Yen, D.C., Chen, Y.M.: Predicting stock returns by classifier ensembles. Appl. Soft Comput. **11**(2), 2452–2459 (2011)
7. Ediger, Akar, S.: ARIMA forecasting of primary energy demand by fuel in Turkey. Energy Policy **35**, 1701–1708 (2007)
8. Friedman, J.H.: Multivariate adaptive regression splines (with discussion). Ann. Stat. **19**, 1–141 (1991)
9. Huang, A.Y.: Asymmetric dynamics of stock price continuation. J. Bank. Financ. **36**(6), 1839–1855 (2012)
10. Johnstone, D.J.: Economic interpretation of probabilities estimated by maximum likelihood or score. Manag. Sci. **57**(2), 308–314 (2011)

Performance and Resource Analysis on the JavaScript Runtime for IoT Devices

Dongig Sin and Dongkun Shin[✉]

Department of Electrical and Computer Engineering, Sungkyunkwan University, Suwon, Korea
{dongig,dongkun}@skku.edu

Abstract. The light-weight JavaScript frameworks such as IoT.js, DukServer, and Smart.js provide the asynchronous event-driven JavaScript runtime for low-end IoT device. These frameworks are designed for memory-constrained systems such as IoT devices. To evaluate the performance of these frameworks, existing JavaScript benchmarks are not suitable considering that the use cases of IoT device are mainly to execute a simple task generating sensor and network I/O requests. In this paper, we propose several IoT workloads to evaluate the performance and memory overhead of IoT systems, and evaluate several light-weight JavaScript frameworks. In addition, we evaluated the effectiveness of multi-core system for JavaScript framework.

Keywords: JavaScript engine · Internet of Things · Low-memory · Server-side JavaScript framework · Iot platform

1 Introduction

The JavaScript (JS) programming language is widely used for web programming as well as general purpose computing. JavaScript has the advantages of easy programing and high portability due to its interpreter-based execution model. Recently, various event-driven JavaScript frameworks have been introduced such as Node.js, which is based on Google V8 JavaScript engine and the libuv event I/O library [1]. To distinguish from the JS engine of web browser, these event-driven JS frameworks are called server-side JS framework since they can be used for implementing server computers.

The event-driven JavaScript environment is useful for implementing IoT systems which should handle many sensor and network I/O events. However, Node.js is designed for server computer system, and thus it is not optimized for resource usage. Therefore, Node.js is not suitable for IoT devices which have low-end processors and limited memory space in order to reduce product cost and power consumption. Accordingly, various light-weight JavaScript engines requiring small footprint and runtime memory have been proposed, such Duktape [2], JerryScript [3], and V7 [4].

In this paper, we propose several IoT workloads and evaluate the performance and resource usage of various event-driven JavaScript runtimes.

© Springer International Publishing Switzerland 2016
O. Gervasi et al. (Eds.): ICCSA 2016, Part I, LNCS 9786, pp. 602–609, 2016.
DOI: 10.1007/978-3-319-42085-1_50

2   Related Work

Recently, there have been several efforts trying to analyze performance of the server-side JavaScript framework for server computer system. Ogasawara [5] conducted context analysis of server-side JavaScript applications on the Node.js, and found that little time is spent on dynamically compiled code. Zhu et al. [6] present an analysis of the microarchitectural bottlenecks of scripting-language-based server-side event-driven applications. In addition, much work has been done to analyze and to improve the performance of JavaScript engine used in web browsers [7–10].

The prior works did not focus on the server-side JavaScript framework for memory-constrained devices and the lightweight JavaScript engine. Our work studies the lightweight server-side JavaScript framework and handles the workloads for the IoT environment, not for the web browser.

3   Light-Weight JavaScript Engine

The JavaScript engines for web browsers generally use the just-in-time (JIT) compiler technique for high performance. For instance, V8 engine [11] monitors the execution frequencies of JavaScript functions compiled by the base compiler at runtime. If a function is frequently executed, it is compiled by the optimizing compiler, called Crankshaft, through the techniques of hidden class and inline caching. However, these optimization techniques are not suitable for low-end IoT devices since they require a large amount of memory.

Therefore, several light-weight JavaScript engines are introduced, which support only the minimal functions of JavaScript language following ECMA standard. Considering memory-hungry IoT devices, the size of heap memory should be limited and aggressive garbage collection is necessary. As shown in Table 1, the binary file sizes of light-weight JavaScript engines are up to 125 times less than that of V8 engine.

Table 1. Binary file size of JavaScript engine

	V8	Duktape	JerryScript	V7
Size (KB)	21504	192	172	1228

Figure 1 shows the performances and memory consumptions of several JavaScript runtimes while executing SunSpider benchmark. V8 provides up to 300 times better performance compared with the light-weight JS engines. However, the memory consumption of V8 is significantly higher than those of light-weight JS engines. This is because V8 allocates a large memory pool for high performance. On the contrary, the performances of light-weight JS engines are significantly low since they do not adopt the JIT optimizing compiler. However, they require only several mega-bytes of run-time memory since the heap memory is limited and the aggressive garbage collection reclaims unused object memory within a short time interval. The memory optimization techniques further degrade the performance.

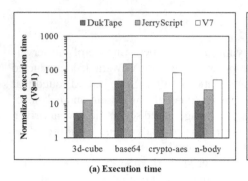
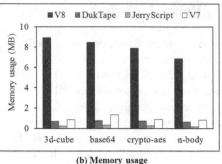

(a) Execution time   (b) Memory usage

Fig. 1. Execution time and memory usage of SunSpider benchmark

However, the SunSpider benchmark is designed to measure the performance of JS engine for high-end devices, and thus it includes computing and memory-intensive applications such as 3D cube rotation or encoding/decoding [12]. These applications are not suitable for measuring the performance of IoT device which handles only sensor data and network I/O requests. Consequently, in order to evaluate the performance and memory overhead of light-weight JS engine, we need IoT-specific workloads.

Based on JS engines, several event-driven JS frameworks are announced. Node.js is a platform built on V8 JS engine for easily building network applications. Node.js uses an event-driven, non-blocking I/O model that is useful for data-intensive applications. It is possible to extend the functionality of Node.js via package module called NPM. Many open source modules are available for NPM.

Although Node.js is versatile and it is based on high-performance V8 JS engine, it is inadequate for resource-constraint IoT devices. In this paper, we focus on light-weight JS frameworks for IoT devices. IoT.js has a similar architecture with Node.js [13]. However, it uses a light-weight JS engine called JerryScript. IoT.js also supports the package module. However, only a small number of packages are available currently. DukServer uses the Duktape JS engine. It supports C socket-based communication, with which DukServer can communicate other native functions. Therefore, the external native functions should be integrated with DukServer at build time. Currently, only the HTTP-server application is included at DukServer. Smart.js, relying on V7 engine, supports the binding to the network and hardware native API. In addition, the device firmware can be called from JS applications for bare metal execution.

4   Experiments

4.1   Experiment Environments

The HTTP servers are implemented with Node.js, IoT.js, DukServer and Smart.js. The hardware is ODROID-U3 which has 1.7 GHz Quad CPU and 2048 MB memory. We evaluate the performance and memory overhead of JS framework while running several IoT workloads. Four IoT service workloads are used as shown in Table 2. The workloads are designed considering the common scenarios of IoT systems.

Table 2. Workload specification

Name	Business logic	Feature
Query	Send a specific sensor data	Usual case
Collection	Transmit collected sensor data after sorting	Data-intensive
Compression	Transmit collected sensor data after compressing	Computing-intensive
Logging	Save sensor data to file	I/O-intensive

4.2 Performance and Memory Usage of IoT Workload

For performance comparison, the request handling times of the HTTP server implemented by JS framework are measured while client-side applications sends several requests to the server. In every workload, the first request handling time of Node.js is long compared with the following request handling times since the JS operations are not optimized at the first execution.

4.2.1 Query Workload

In IoT system, the most general scenario is to transfer specific sensor data requested by client. The Query workload reads sensor data and transfers them to client. Since the workload has no computing-intensive job, Node.js cannot benefit from the optimizing compiler of V8 JS engine as shown in Fig. 2(a). It shows rather performance degradation due to additional code execution for optimization. DukServer also shows longer request handling times than other frameworks. While V7 and JerryScript engines use a memory pool to assign a memory space for object, Duktape allocates the memory on-demand. Therefore, there is a memory allocation overhead whenever an object is created, even though the overall memory usage is lower than other schemes.

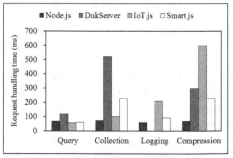

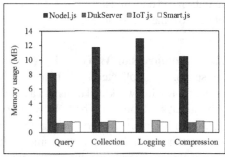

(a) Request handling time of IoT Workloads   (b) Memory usage of IoT workloads

Fig. 2. Performance and memory usage of IoT workload

4.2.2   Collection Workload

If an IoT device collects data from multiple sensors and processes them, a large size of memory space is required for the collected sensor data. The Collection workload collects, sorts, and transfers two hundreds of sensor values. For the sorting operation, many memory pages should be allocated. The performance of Collection workload is greatly affected by memory management technique of JS framework. Duktape has a high memory allocation overhead since it does not maintain memory pool, as shown in Fig. 2(a). JS framework needs a memory reclamation technique. There are two kinds of techniques used by current JS engines. The reference counting technique maintains the reference count of each object, and de-allocates the memory space of unused objects if the reference counts of them are zero. The second technique is garbage collection (GC) which is triggered when there is memory pressure. Once the GC is triggered, the mark-and-sweep operation is performed, which marks only referenced objects and then frees unmarked objects. Duktape uses both of the techniques, and saves the frequently accessed objects at hash table in order to reduce the search cost of garbage collection. On the other hand, V7 uses only the garbage collection, and manages the objects in a linear list, which causes a high search cost of garbage collection. Therefore, Smart.js shows a long request handling time due to the garbage collection cost.

4.2.3   Logging Workload

The IoT device can store the collected sensor data at storage device via file systems. The Logging workload stores two hundreds of sensor data at file system. Figure 2(a) shows that the request handling time of IoT.js is longer than other frameworks. The native file I/O operations are called by the JS applications via a native API binding technique. Since the native API binding techniques provided by each framework is different, the file I/O operations show different performances. The write API provided by IoT.js allocates a buffer object in order to transform the target data to a common type. Therefore, IoT.js shows poor performance in the file I/O intensive workload due to the overhead of buffer allocation and data transformation. In this workload, DukServer is excluded since it does not support the file system module.

4.2.4   Compression Workload

The sensor data collected by IoT device can be compressed for fast network transfer. The Compression workload transfers a hundred of sensor values after compressing them with the LZW algorithm. The light-weight JS frameworks show poor performances for the computing-intensive workload compared with Node.js.

Although the light-weight JS frameworks show worse performances than Node.js in most of the IoT workloads, the performance gap is less than several hundreds of milliseconds. However, the memory consumptions of light-weight JS frameworks are significantly lower than that of Node.js as shown in Fig. 2(b). Therefore, these frameworks are suitable for low-end IoT devices.

4.3 Performance on Multi-core Architecture

Recent embedded systems adopt multi-core processors for high performance. However, the single thread-based JavaScript framework cannot benefit from multi-core systems [6]. Moreover, multiple processor cores will cause high power consumption. Since the power consumption is also important metric for IoT devices, the power efficiencies of JS frameworks are observed.

For experiments, ODROID-XU3 is used, which has ARM Cortex A15 2.0 GHz Quad CPU and A7 1.5 GHz Quad CPU, called big.LITTLE architecture. The high-performance big cores support the out-of-order execution and use a high CPU clock. The low-performance LITTLE cores perform the in-order execution with a low CPU clock. We measure the request handling time while running the Compression and Logging workloads as shown in Fig. 3.

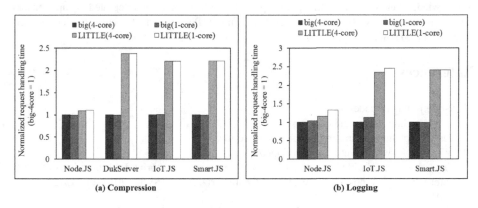

(a) Compression   (b) Logging

Fig. 3. Performance on multi-core system

For the Compression workload, the request handling times of light-weight JS frameworks are significantly reduced on the high-performance cores. However, there are no performance changes by the number of enabled cores since the JS frameworks use the single-thread execution model. The performance improvement by high-performance cores is insignificant at Node.js due to its optimizing compiler.

For the I/O-intensive Logging workload, the multi-core architecture improves the performance at Node.js and IoT.js. These frameworks support the asynchronous write API which uses I/O thread pool to prevent the main-thread to be blocked during the I/O operation [14]. Therefore, multiple cores can execute the main thread and I/O threads simultaneously. The multi-core processor improves the I/O performance of Node.js and IoT.js up to 12 % compared with the single-core system. On the contrary, Smart.js handles the I/O operations in the single main thread, and thus its performance is not improved by multi-core processor.

Although the multi-core processor can improve the performance of JS frameworks which use separated I/O threads, the performance improvement is not significant. Moreover, the multi-core processor can waste power without any performance improvement on computing-intensive workloads. Therefore, the single-core system with high performance will be more efficient than the multi-core system for IoT systems considering both performance and power.

5 Conclusion

This paper proposed several representative IoT workloads to evaluate the performance and memory overhead of IoT devices. We analyzed the features of different light-weight JavaScript frameworks with the IoT workloads. In addition, we evaluated the effect of multi-core system for JavaScript framework. Our analysis results will be useful for selecting or designing light-weight JavaScript framework and hardware systems for IoT applications.

Acknowledgements. This work was supported by the Center for Integrated Smart Sensors funded by the Ministry of Science, ICT & Future Planning as Global Frontier Project. (CISS-2011-0031863).

References

1. Node.js. https://nodejs.org
2. Duktape. http://www.duktape.org
3. JerryScript. https://samsung.github.io/jerryscript
4. V7. https://www.cesanta.com/developer/v7
5. Ogasawara, T.: Workload characterization of server-side javascript. In: Proceedings of IEEE International Symposium on Workload Characterization (IISWC) 2014, pp. 13–21. IEEE (2014)
6. Zhu, Y., Richins, D., Halpern, M., Reddi, V.J.: Microarchitectural implications of event-driven server-side web applications. In: Proceedings of the 48th International Symposium on Microarchitecture, pp. 762–774. ACM (2015)
7. Chadha, G., Mahlke, S., Narayanasamy, S.: Efetch: optimizing instruction fetch for event-driven web applications. In: Proceedings of the 23rd International Conference on Parallel Architectures and Compilation, pp. 75–86. ACM (2014)
8. Zhu, Y., Reddi, V.J.: WebCore: architectural support for mobile web browsing. In: Proceedings of International Symposium on Computer Architecture, p. 552 (2014)
9. Anderson, O., Fortuna, E., Ceze, L., Eggers, S.: Checked load: architectural support for JavaScript type-checking on mobile processors. In: Proceedings of 2011 IEEE 17th International Symposium on High Performance Computer Architecture (HPCA), pp. 419–430. IEEE (2011)
10. Halpern, M., Zhu, Y., Reddi, V.J.: Mobile CPU's rise to power: quantifying the impact of generational mobile CPU design trends on performance, energy, and user satisfaction. In: Proceedings of 2016 IEEE International Symposium on High Performance Computer Architecture (HPCA), pp. 64–76. IEEE (2016)
11. V8. https://developers.google.com/v8/
12. Tiwari, D., Solihin, Y.: Architectural characterization and similarity analysis of sunspider and Google's V8 Javascript benchmarks. In: Proceedings of 2012 IEEE International Symposium on Performance Analysis of Systems and Software (ISPASS), pp. 221–232. IEEE (2012)

13. Gavrin, E., Lee, S.J., Ayrapetyan, R., Shitov, A.: Ultra lightweight JavaScript engine for internet of things. In: Companion Proceedings of the 2015 ACM SIGPLAN International Conference on Systems, Programming, Languages and Applications: Software for Humanity, pp. 19–20. ACM (2015)
14. Tilkov, S., Vinoski, S.: Node.js: using javascript to build high performance network programs. IEEE Internet Comput. **14**(6), 80 (2010)

Enhancing the Reliability of WSN Through Wireless Energy Transfer

Felicia Engmann[1]([⊠]), Jamal-Deen Abdulai[2],
and Julius Quarshie Azasoo[1]

[1] School of Technology (SOT),
Ghana Institute of Management and Public Administration (GIMPA),
Accra, Ghana
{fapboadu, jazasoo}@gimpa.edu.gh
[2] Department of Computer Science, University of Ghana, Legon, Accra, Ghana
jabdulai@eg.edu.gh

Abstract. Communication reliability is the ability of the network to last long enough without breaking communication between neighbouring nodes that relay information to the final destination. The transfer of energy from the base station to largely spatially distributed sensor nodes in a network with concurrent data transmission is studied. Concurrent data and energy transfer methods require separation of their frequencies and optimum distance between nodes and Energy Transmitters (ETs). Three techniques of energy transfer: store and forward, direct flow transmission and a hybrid will be explored in an unequal clustering environment for concurrent data and energy transmission. The research seeks to explore the use of energy transmitters augmented with energy transfer techniques that redistribute energy in the network to ensure communication reliability. The need for an optimum energy threshold that will determine whether a node is "fit" to relay energy to neighbouring nodes is investigated.

Keywords: Energy transfer · Sensor nodes · Reliability · Energy transmitters

1 Introduction

In a typical sensor network, spatially distributed nodes scattered randomly across a field, transfer aggregated data to a base station located some distance away. The energy hole problem (Wafta et al. 2011) explains how nodes closer to the base station are depleted of energy. This is due to the many-to-one pattern of data transfer of data from peripheral nodes to the base station that are hopped from the nodes closer to the base station. Nodes close die faster and network connectivity will be lost while about 90 % (Lian et al. 2006) of nodes in the peripheral are still alive (Wafta et al. 2011). Unequal clustering ensures clusters at the peripheral have larger cluster members than clusters closer to the base station (Li et al. 2005). This is because cluster closer perform more intra-cluster and inter-cluster communications than peripheral nodes, depleting them of more energy. In a network, even with unequal clustering where clusters closest to the base station are smaller in sizes compared to clusters further away (Engmann et al.

© Springer International Publishing Switzerland 2016
O. Gervasi et al. (Eds.): ICCSA 2016, Part I, LNCS 9786, pp. 610–618, 2016.
DOI: 10.1007/978-3-319-42085-1_51

2015; Li et al. 2005) these near cluster heads will eventually die leaving some nodes with energy still alive but unable to forward data to the base station.

Wireless transfer of energy is the transfer of energy from one place to another without a wire. The ability of the nodes to transfer energy wirelessly ensures nodes remain operational for a longer time without the need to replace batteries (Naderi et al. 2015). Techniques in wireless energy transfer provide means of transferring energy from neighbouring nodes to nodes in the network that may request for it but they do not explore the means of harvesting energy (Bogue 2009).

In this energy transfer model understudy, the research considers an unequal clustering network with data being transferred from the nodes and aggregated to the base station by the cluster heads with concurrent transfer of energy from the neighbouring nodes in the cluster. It assumes an efficient energy distribution model amongst nodes in a cluster operating user fair constraints (Wang et al. 2010) and fair distribution without compromising on the reliability of the network in terms of data transmission. The reliability of the network is defined as the duration for a cluster to function properly (Wang et al. 2010). This is measured by the lifetime of the network which is the time elapsed until a first cluster member dies. The research assumes the cluster member that senses critical data cannot be selected a priori.

The concurrent transfer of energy and data from energy transmitters (ETs) to nodes have been studied in (Naderi et al. 2014, 2015; Roh and Lee 2014). This concurrent transfer generates interference leading to packet losses. The optimum distance between the nodes and the ET's that enables efficiently charging and also data communication was studied in (Naderi et al. 2014, 2015; Roh and Lee 2014) and encourages the introduction of multiple ET's to overcome the challenge of some nodes harvesting negligible energy from the ET's. Based on the quantified distance estimates for charging and data communication presented in (Naderi et al. 2014), we intend to analyse the three energy techniques (Wafta et al. 2011) that will be appropriate in ensuring efficient redistribution of energy in the network. The importance of separating the frequencies of energy and data for concurrent transfer, and the distance between energy transmitters and the nodes mitigate interferences, but (Naderi et al. 2014, 2015; Roh and Lee 2014) did not quantify their effect on a large network with fewer ETs. In Wafta et al. (2011), the three techniques of wireless energy transfer discussed did not include the effects of concurrent energy and data transfers on the reliability of the network.

The research considers a large network of randomly deployed nodes that require recharging of batteries by harvesting energy using energy transmitters. To reduce the number of energy transmitters used in the network, the research investigates three energy transfer techniques in clustering where nodes closer to the ETs may be charged directly and distance nodes are recharged by the transfer techniques.

1.1   Research Objectives

I. To obtain an optimum energy threshold of nodes that will determine whether a node has the ability to retransmit energy to neighbouring nodes.

II. To develop a method of redistributing harvested energy amongst nodes in the network that increase the network lifetime.

III. To propose an energy redistribution-clustering algorithm for nodes that has their batteries recharged wirelessly

2 Literature Review

Wireless transfer of energy is an emerging trend in wireless networks that presents several challenges (Naderi et al. 2014). Recent research in energy transfer investigates the harvest of ambient energy in the environment and the transfer of the energy to nodes in the network (Naderi et al. 2014, 2015; Roh and Lee 2014; Bogue 2009; Basagni et al. 2013; Anastasi et al. 2009). Commonly used is the electromagnetic radiation. The primary aim of nodes in WSN is transfer collected data from the environment and wireless send them to a base station located some distance away. Energy transfer in WSN proposes the use of these nodes to transfer energy wirelessly to nodes in the network to ensure maximum lifetime of the network.

The ability to transfer energy wirelessly without plugs or physical contact was explored by Kurs et al. (2007) using magnetic resonant coupling. Experimental results showed that the efficient energy transfer distance of nodes must be 2 m and does not require line of sight. They proposed that the power transfer distance of these nodes could be improved by silver-plating the coils.

The concurrent transfer of data and energy in wireless sensor networks comes with such challenges as the interference of multiple energy transmitters and the interference of frequencies of energy and data transmission (Naderi et al. 2014, 2015; Son et al. 2006). The limited charging range of RF energy imposes the concurrent and coordinated use of multiple ETs to power an entire network Naderi et al. (2015). A cross layer approach to handling the problem of interference and low throughput is proposed in Roh and Lee (2014). The method provides a joint solution considering routing, scheduling and power control in WSN. This was based on the assumption that with WSN with RF energy transfer, energy consumption is not a problem, but the transfer and harvesting of the energy which depends on location is.

The reliability of networks in RF powered sensor networks comes at a cost of interference and data throughput. An empirical study in Naderi et al. (2014) showed the importance of separating the energy transmission and data communication frequencies for different locations of ET's in a sensor network that supports higher throughput and minimizes interference. For a maximum distance of 5 m of separation between the receiver nodes and the ET, the ET's must transmit at an energy frequency greater than 917.6 MHz and lower than 912.2 MHz. Sensor nodes transmitting at 915 MHz could have greater throughput of packets sent and also be effectively charged.

The essence of separating the data and energy transfer frequencies and also the separation of the ETs to mitigate interferences is explored in same paper (Naderi et al. 2014). It was proposed that if the ET's frequency were separated by 2 MHz from the data communication frequencies, then the packet losses caused by the simultaneous data and energy transfers can be mitigated. For a single ET, using three locations ranges of nodes (C1, C2 and C3) away from a stationary ET, nodes within C1 closest to the transmitter have maximum charging rates with minimum interference with data communication whiles nodes further away at some distance (C3 and beyond) threshold receive so much interference data communication is negligible (Naderi et al. 2015). The maximum charging range for an ET is estimated at 5 m at a maximum power of 3 Watts. The research conducted was to show that the distance of nodes from the ET's affect the interference of signals and hence the throughput. It also provided an allowable separation between energy and data transmission frequencies of $\Delta f(e,d)$ where e and d are the frequencies of energy and data respectively. The value of $\Delta f(e,d)$ is 2 MHz. While multiple ET's ensure high energy transfer rates, they introduce interference among RF waves (Basagni et al. 2013).

Three methods of transferring energy that overcomes the power constraints in WSN by wirelessly charging nodes by single hop wireless energy transfer was studied (Wafta et al. 2011). These are the store and forward, direct flow and the hybrid. The store and forward technique required the charging of intermediate nodes and multi hoping this energy to neighbouring nodes. The direct flow technique receives transferred energy from the ET's and without charging its batteries, directly hops it to nodes in the network. For a small network of nodes, the direct flows outperforms the store and forward, but in a large network, receive and transmit losses makes it not feasible and efficient hence the use of the hybrid technique, which combines store and forward and the virtual circuit technique is most appropriate.

Unequal clustering algorithms form clusters where nodes closest to the base station have smaller number of nodes per cluster as compared to nodes further away (Basagni et al. 2013; Engmann et al. 2015). The use of distance and the residual energy parameters are used in forming cluster sizes for a large network but do not employ energy harvesting technique (Engmann et al. 2015). For data communication, I assume the use of unequal clustering and propose the implementation of the hybrid method of energy transfer to mitigate the dying of nodes, and hence increase the network lifetime.

The use of multiple ETs for harvesting energy in the network is carefully studied in Roh and Lee (2014). The 2D and 3D placement of multiple RF ETs for harvesting and recharging nodes based on distance in examined. A closed matrix form of power harvested at every distance in the network due to the concurrent energy transfer of multiple ETs and their performance of energy transfers through power outage probability, interference and harvested voltage, as a function of the wireless power received is analyzed. The results revealed that the network wide received power and interference power from concurrent energy transfers exhibit Log-Normal distributions and harvested voltage over the network follows a Rayleigh distribution. Much of the works reviewed (Naderi et al. 2014) considered single ET's and the constraints on distance and frequency. This opens the room for further research to be done considering the maximum preferred distance for charging nodes which is 5 m and the separation of frequencies of data and energy which is 2 MHz. Assuming the use of a single ET to constantly recharge nodes in its close region, I propose a redistribution method of recharging nodes that may be in the communication or interference regions. Considering "receive" and "charge" losses, the research will investigate the use of energy transfer techniques to mitigate the use of several ETs. To do this, an optimum threshold of energy on intermediate nodes must be considered which determines whether a node could retransmit energy without dying in the process.

3 Proposed Model

3.1 System Model

A randomly deployed static wireless sensor network with 100 nodes was considered with a section shown in Fig. 1 below. Assuming a controlled environment with no intermediate reflective objects, it is also assumed that the system is time slotted with a fixed time slot and a fixed amount of data is transmitted at a fixed data rate.

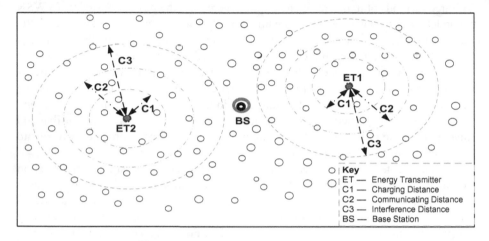

Fig. 1. Network setup of proposed solution

The Energy transmitters and sensor nodes are assumed to be equipped with Omni-directional isotropic antennas.

3.2 Proposed Solution

To obtain the optimal energy threshold of nodes to recharge other nodes outside the charging regions of the ETs, the Mica2 datasheet parameters that make use a frequency of 915 MHz with a default RF (data) transmission power of 0dBm and receiver sensitivity of −98 dBm was used. Motes then use a 38.4 Kbps data rate with Manchester encoding and non-coherent FSK modulation scheme (Naderi et al. 2014).

Considering the energy consumed during data transmission, taking equations from (Wang et al. 2010), for cluster members transmitting to their cluster heads.

$$T_i(P_i + P_{CT}) \leq \forall i \in \{1, 2, \ldots, N_d\} \tag{1}$$

where T_i is the operation time of the node duration, P_i is the transmit power of cluster member to the cluster head and P_{CT} is the circuit power consumption to transmit data.

For cluster heads transmitting during inter-cluster communication, the energy constraint is

$$P_{CR} \sum_{i=1}^{N_d} T_i \leq E_0 \qquad (2)$$

where P_{CR} is the circuit power consumption to receive data.

For energy transfer from ETs, our model assumes a direct flow technique for nodes within C1 region from the ETs (Wafta et al. 2011). The power distribution of energy from the transmitter to nodes will be

$$
\begin{aligned}
P_{total} &= P_{work} + P_{d_0} + P_{d_1} \\
P_{work} &= 2\Gamma_{charge}|A_{d_1}|^2 \\
P_{d_1} &= 2\left(\Gamma_{d_1} + \Gamma_{charge-loss}\right)|A_{d_1}|^2
\end{aligned}
\qquad (3)
$$

where P_{total} is the total power transmitted from the energy transmitter. Since the ETs receive constant supply of energy from the ambient environment, we assume there will be no constraint on the Power transmitted from them.

For nodes in the C2 and C3 regions, we assume the Store and Forward Technique on the assumption that nodes are equipped with rechargeable batteries (Wafta et al. 2011). Batteries will be charged to full capacity or the source energy reaches a threshold that will be determined later on in the paper. For a single source d_o, and intermediate node d_1 and a final node d_2, the total power distributed will be

$$
\begin{aligned}
P_{total} &= P_{work} + P_{d_0} + P_{d_1} + P_{d_2} \\
P_{work} &= 2\Gamma_{charge}|A_{d_1}|^2, P_{d_0} = 2\Gamma_{d_0}|A_{d_1}|^2 \\
P_{d_1} &= 4\Gamma_{d_1}|A_{d_1}|^2, P_{d_2} = 2\Gamma_{d_2}|A_{d_1}|^2
\end{aligned}
\qquad (4)
$$

P_{work} is the useful power delivered to the final node.

From the model in Fig. 1, to mitigate the interference of in the region of C3 and beyond, we assume a distance close to the base station referred to as "close" from the fuzzy logic algorithm from the paper in (Engmann et al. 2015). This distance will most likely be our region C3 and beyond (closer to the base station). All nodes in the C3 regions of neighbouring ETs will be made to receive energy but cannot transmit. Their energy lost due to absorption will be same as in P_{d2}.

From Eq. (4), the threshold of energy in nodes in C2 region that will enable concurrent data communication and energy transfer will be

For Cluster Members

$$T_i(P_i + P_{CT}) + P_0 - P_{charge} \qquad (5)$$

where P_0 is the total power in the node and P_{charge} is the Power needed to charge the batteries in the node.

For the cluster head

$$E_0 - P_{CR} \sum_{d=1}^{N_d} T_d + P_0 - P_{charge} \qquad (6)$$

Nodes in C1 are in 5 m away from the ETs and can be recharged directly (Naderi et al. 2014, 2015). Neighbouring nodes in C2 can be recharged with nodes in C1 within a hoping of 2 m (Wang et al. 2010) and this energy will be multi hoped until all nodes in the network are recharged.

For these intermediary nodes, the Power threshold for nodes for concurrent data communication and energy transfer will be as follows. Considering a node d_1

$$P_{d_1} = P_{total} - \left(\left(P_{charg} + P_{d_0} + P_{d_2} \right) \right) - P_{CR} \sum_{d=1}^{N_d} T_d \qquad (7)$$

where N_d is the is the total number of nodes in the cluster and T_d is the rate of the time duration for the operation of a node $_d$.

For nodes in C3, since they are required to receive energy but not transmit, we propose a direct flow technique for energy transfer. The energy constraints will be as same for nodes in C1.

Table 1. Fuzzy if-then mapping for competition radius

Distance	Node energy	Capacity to transfer
C1	Low	Request energy
C1	Medium	Receive energy
C1	High	Transmit energy
C2	Low	Request for energy
C2	Medium	Receive energy
C2	High	Transmit energy
C3	Low	Request for energy
C3	Medium	Receive energy
C3	High	Do not receive energy

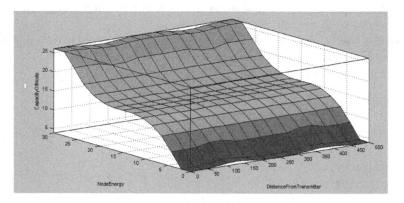

Fig. 2. Surface view of the proposed solution using fuzzy logic

Using the fuzzy logic, we simulate the energy threshold for the proposed algorithm using two parameters, the distance C1, C2 and C3, and the energy thresholds of values on Low, Medium and High (Table 1 and Fig. 2).

4 Conclusion

Recent studies in WSN try to overcome the energy problem where batteries had to be replaced occasionally. This is not feasible in human unreachable remote locations hence the need to harvest energy from the environment to power the nodes or to recharge their batteries. ETs charge nodes to a shorter distance away and hence the use of transfer techniques is conceptually explored for the recharging of distant nodes. Using fairness constraints and Shannon's channel capacity theorem (Wang et al. 2010) and energy transfer techniques (the direct flow and the store and forward techniques), we propose threshold values of energy for concurrent energy transmission based on their distance from energy transmission. We propose that given the different energy and distance thresholds, the number of ETs used in a given location would be reduced. The success of this research will provide a model for effective redistribution of energy between nodes and ensure that nodes do not die leaving the network unreliable and therefore resulting in a reliable Wireless Sensor Network (WSN).

Further work will be to simulate the findings using NS-2 simulator and compare the effectiveness of the proposed solution with other concurrent data and energy transfer techniques. We hope to also find out from the research the number of ETs that will be appropriate in a given area of WSN.

References

Anastasi, G., Conti, M., Di Francesco, M., Passarella, A.: Energy conservation in wireless sensor networks: a survey. Ad Hoc Netw. **7**(3), 537–568 (2009)

Ayatollahi, H., Tapparello, C., Heinzelman, W.: Transmitter-receiver energy efficiency: a trade-off in MIMO wireless sensor networks. In: 2015 IEEE Wireless Communications and Networking Conference (WCNC), pp. 1476–1481. IEEE, March 2015

Basagni, S., Naderi, M.Y., Petrioli, C., Spenza, D.: Wireless sensor networks with energy harvesting. In: Basagni, S., Conti, M., Giordano, S., Stojmenovic, I. (eds.) Mobile Ad Hoc Networking: Cutting Edge Directions, pp. 703–736. Wiley, Hoboken (2013)

Bogue, R.: Energy harvesting and wireless sensors: a review of recent developments. Sens. Rev. **29**(3), 194–199 (2009)

Engmann, F., Abdulai, J.D., Azasoo, J.Q.: Improving on the reliability of wireless sensor networks. In: 2015 15th International Conference on Computational Science and Its Applications (ICCSA), pp. 87–91. IEEE, June 2015

Kurs, A., Karalis, A., Moffatt, R., Joannopoulos, J.D., Fisher, P., Soljačić, M.: Wireless power transfer via strongly coupled magnetic resonances. Science **317**(5834), 83–86 (2007)

Lian, J., Naik, K., Agnew, G.B.: Data capacity improvement of wireless sensor networks using non-uniform sensor distribution. Int. J. Distrib. Sens. Netw. **2**(2), 121–145 (2006)

Li, C., Ye, M., Chen, G., Wu, J.: An energy-efficient unequal clustering mechanism for wireless sensor networks. In: 2005 IEEE International Conference on Mobile Adhoc and Sensor Systems Conference, p. 8. IEEE, November 2015

Naderi, M.Y., Chowdhury, K.R., Basagni, S.: Wireless sensor networks with RF energy harvesting: energy models and analysis. In: 2015 IEEE Wireless Communications and Networking Conference (WCNC), pp. 1494–1499. IEEE, March 2015

Naderi, M.Y., Chowdhury, K.R., Basagni, S., Heinzelman, W., De, S., Jana, S.: Surviving wireless energy interference in RF-harvesting sensor networks: an empirical study. In: 2014 11th Annual IEEE International Conference on Sensing, Communication, and Networking Workshops (SECON Workshops), pp. 39–44. IEEE, June 2014

Roh, H.T., Lee, J.W.: Cross-layer optimization for wireless sensor networks with RF energy transfer. In: 2014 International Conference on Information and Communication Technology Convergence (ICTC), pp. 919–923. IEEE, October 2014

Son, D., Krishnamachari, B., Heidemann, J.: Experimental study of concurrent transmission in wireless sensor networks. In: Proceedings of 4th International Conference on Embedded Networked Sensor Systems. pp. 237–250. ACM, October 2006

Wang, T., Heinzelman, W., Seyedi, A.: Maximization of data gathering in clustered wireless sensor networks. In: 2010 IEEE Global Telecommunications Conference (GLOBECOM 2010), pp. 1–5. IEEE, December 2010

Watfa, M.K., AlHassanieh, H., Selman, S.: Multi-hop wireless energy transfer in WSNs. Commun. Lett. IEEE 15(12), 1275–1277 (2011)

Senior Potential Analysis: A Challenge that Contributes to Social Sustainability

Teresa Guarda[1,2(✉)], Filipe Mota Pinto[3], Juan Pablo Cordova[4],
Maria Fernanda Augusto[5], Fernando Mato[4], and Geovanni Ninahualpa Quiña[4]

[1] Universidad Estatal Península de Santa Elena-UPSE, Santa Elena, Ecuador
tguarda@gmail.com
[2] Instituto Politécnico de Leiria- ESTG, Leiria, Portugal
[3] Universidad de las Fuerzas Armadas-ESPE, Sangolqui, Quito, Ecuador
filipe.mota.pinto.pt@gmail.com
[4] F&T Innovation, Leiria, Portugal
{jpcordova,fj.mato,gninahualp}@espe.edu.ec
[5] F&T Innovation, Leiria, Portugal
mfg.augusto@gmail.com

Abstract. Population aging is one of the greatest triumphs of humanity, also is both a major challenge. The number of older workers in the European Union (EU) will increase over the next decades, the active population of the EU points to a growth of approximately 16.2 % of the age group of 55–64 years between 2010 and 2030, being contrary to tendency of other age groups, with a reduction of 4 % to 5 %. This demographic change is due to the decline in fertility rates and increased life expectancy. The labor force in Europe will reach age levels never before registered in that older workers represent more than 30 % of the active population. In this paper, based on the questionnaire results, brought out by "harnessing the potential senior", which was applied in enterprises in Portugal. It is intended to address some of the concerns related to the aging population and the use of senior potential, facing this problem as a challenge that contributes to social sustainability. Thus, the senior potential should be seen as a very important contribution. In this paper, we intend extract the survey data from data bases, for knowledge discovery and analysis using the data mining tool Weka.

Keywords: Active aging · Senior potential · Data mining · Weka

1 Introduction

Given the structural changes in the age structure of Europe's population and increased life expectancy, longer working life imposes itself as an unavoidable necessity. The aging of the European population in general, and in particular Portugal, raises some issues, including the large number of older people will cause the collapse of health systems and social security?; how to help people remain independent and active as they age?; as the quality of life of older people can be improved?; how can we recognize and support the important role that older people play in your working life?

© Springer International Publishing Switzerland 2016
O. Gervasi et al. (Eds.): ICCSA 2016, Part I, LNCS 9786, pp. 619–626, 2016.
DOI: 10.1007/978-3-319-42085-1_52

This paper is organized as follows. In this introductory section is dedicated to the presentation and discussion of the data of this investigation a qualitative case study, which was developed based on online questionnaires sent to small and medium enterprises and located in in Portugal. This work aims to address some concerns related to the aging population and the use of senior potential. We will start by presenting a retrospective of the rapid global growth of the population over 60 years, especially in developing countries, addressing then the concept of active aging and the use of senior potential. The Sect. 2 discusses the concepts active aging, senior potential. The Sect. 3 is dedicated to data mining technique, and a specific tool, Weka. We pretend extract the survey data from databases, for knowledge discovery and analysis using the data mining tool Weka. In the following section (Sect. 4), we present the case study. The last section (Sect. 5) presents the conclusions and future work.

2   Background

Population aging is one of the greatest triumphs of humanity, also is both a major challenge. The World Health Organization (WHO) argues that countries can afford aging if governments, international organizations and civil society to implement policies and programs "active aging" to improve the health, participation and security of older citizens. These policies and programs should be based on the rights skills, needs, and preferences of older people [1].

The world population has grown from 5.7 to 7.2 thousand millions since 1994. Despite the slowdown in population growth, the projections of the UN suggest that worldwide, is the age group of people over 60 years, which shows an increase faster than any other age group. Between 1975 and 2025, we expect a growth of 223 %. It is planned for 2025, a total of approximately 1.2 billion people over 60 years. By 2050 there will be 2 billion, 80 % in developing countries [2]. The aging of a population is related to a decline in the number of children and young people and an increase in the proportion of people aged 60 or more years. The reduction in birth rates and increasing longevity will ensure the aging of the world population. It is estimated that by 2025, 120 countries will achieve full birth rates below replacement level, which is a substantial increase compared with 1975, when only 22 countries had a lower total fertility rate or equal to the replacement level, and the current number is 70 countries [2]. The aging population has been associated with the more developed regions of the world. Of the 10 countries with over 10 million inhabitants and where the number of elderly is higher than the number of people in other age groups, nine are in Europe. Expected little change in the ranking until 2025, when individuals aged 60 or over will constitute approximately 1/3 of the population of countries like Japan, Germany and Italy, followed closely by other European countries [2].

The aging population has been a gradual process, accompanied by a steady socio-economic growth for many decades and generations. While developed countries became rich before they grow old, developing countries are aging before obtaining a substantial increase in their wealth [3].

2.1 Active Aging

The term "active aging" was adopted by the WHO in the late '90 s, trying to convey a more inclusive message than "healthy aging" and to recognize, in addition to health care, other factors that affect how individuals and aging population [4]. The approach of active aging is based on the recognition of human rights of older people and the principles of independence, participation, dignity, care and self-realization established by the UN.

According to World Health Organization (WHO), "active aging applies to both individuals and the population groups. It allows people to realize their potential for physical, social and mental well-being throughout the life course, and that these people participate in society according to their needs, desires and capabilities; at the same time provides protection, security and care when needed" [1].

People who remain healthy as they age face fewer problems to continue working. With the aging population, there will be more pressure for public policy change, especially if individuals reach old age in good health and still able to work. This would help to offset the rising costs of pensions and retirement, as well as the cost of medical and social assistance.

In Portugal, given the low birth and death rates, the increase in aging population has been pronounced [5]. At this juncture, it is vital to plan and take action to help older people to remain healthy and active and is not a luxury but an urgent necessity. It's time to promote active aging.

2.2 Senior potential use

Based on initiatives that support and promote the senior entrepreneurship in some European countries it is intended to provide reflection and call to discuss this phenomenon as a challenge and opportunity for Portugal. The fact that the study focused primarily be on the European continent and present senior entrepreneurship as an opportunity for Portugal does not prevent from the Portuguese case, one may think about this possibility in other countries that will soon be feeling the effects of an aging its workforce.

The degree of importance that people attach to the work ethic is variable. Age, for example, has strong influence in the way they viewed the work ethic. Presently younger workers (17–26 years) seem less understanding about the work ethic, than older workers. Senior (40 to 65) show greater adherence to these ethics. The generation gap may be due in part to changes in the amount of work required [6].

The entrepreneurial business creation is highly connected with the growth of OECD (Organization for Economic Cooper) economies [7]. Thus, the senior entrepreneurship can emerge as a solution that allows greater flexibility in the labor force participation of older people, taking into account their needs and desires, such as allowing follow contributing to economic and social development of society, relieving, thus, the increase in social security costs [8].

3   Data Mining and Weka

We pretend extract the survey data from databases, for knowledge discovery and analysis using the Weka data mining tool.

Data Mining (DM) concerns the non-trivial extraction of identifying patterns in the data, valid, new, potentially useful and understandable from data in databases [9].

The main difference between the DM and other tools for data analysis is in how they explore the relationships between the data. While the various analysis tools available the user builds hypotheses about specific relationships and can corroborate them or refute them through the output produced by the tool used, the process of DM is responsible for creating hypotheses, ensuring greater speed, autonomy, and reliability of the results [10]. Data mining presents six steps in the knowledge discovery: Data cleaning – removal of inconsistent data and noise; Data integration – if necessary, a combination of different sources of data; Data selection – all relevant data for analysis is retrieved from the databases; Data transformation – The consolidation and transformation of data into forms appropriate for mining; Data mining – The use of intelligent methods to extract patterns from data; Pattern evaluation – Identification of patterns that are interesting. The DM aims to build data models. There are many algorithms available, each with specific characteristics. The principal activities of the DM are [11, 12]: Predictive modeling: these models are built starting from the set of input data (independent variables) for the output values (dependent variables) which may be developed in two different ways depending on the type of output; Classification: a learning function that allows associating each data object one of a finite set of pre-defined classes and user; Regression: learning a function that maps each data object in a continuous value; Descriptive modeling: discover groups or categories of data objects that share similarities and help in the description of data sets in given space; Dependencies modeling: is a model that describes relevant associations or dependencies between certain data objects; Deviations modeling (analysis and detection): tries to detect the most significant deviations from measurements and / or past behaviors considered as reference. The selection of data mining activities is directly dependent on the objectives set initially.

The tool chosen was the Weka, which provides the environment Weka Experiment Environment, and it is appropriate to make comparisons between the performance of various data mining algorithms, allowing select one or more algorithms available in the tool and then analyzing the results to determine if a classifier is statistically better than the other. The other environment available in the Weka tool is the Explorer, which allows the selection and implementation of a classifier algorithm at a time. Thus, the result of the experiment is the accuracy of a round of the algorithm. Comparison of the results of the classifiers is not performed automatically with the previous environment [13].

4   Study Case

In Portuguese society, increasing the statutory retirement age, and the number of problems arising from the approval of the Basic Law on Social Security, which provides for increased retirement for 66 years, is an important indication that the active aging and the development of effective strategies to support it need to be discussed seriously by the whole society. In this context arises the following research question: The use of senior potential in the current European situation can promote social sustainability?

To performing this work, was distributed an online questionnaire survey, which is an integral part of this study. The questionnaire was sent to small and medium-sized companies based in Portugal. According to the data of Census 2012, the active population in Portugal, is mostly an aging population (56 %), with the age group "65+" with 35 %, and the "55–64" at 21 % [14].

The data collection method used is based on the use of questionnaire surveys, applied via the web, which are often used to obtain data on a wide range of issues [15]. Data from surveys after submission were stored in a MySQL database, to posterior cleaning and preprocessing, and used to generate the DM models originate.

The gain information is used to assess how much an attribute influences algorithm classification criteria. The attributes of the base were ordered as follows: organization's relationship with its stakeholders; organizational commitment; knowledge management; and organizational effectiveness.

The following algorithms were selected: OneR; JRip; Decision Table; J48; Random Forest; SimpleLogistic; MultilayerPerception; NaiveBayes; and BayesNet. The methods of classification used by the selected algorithms: rule learning; decision table; decision tree; linear logistic regression models; probabilistic models; artificial neural network model; simple probabilistic classifier based on the application of Bayes' theorem.

In this study, three experiments were conducted in order to compare the performance of data mining algorithms applied to the problem domain. The analysis identifies which algorithms are more suitable for data mining. The OneR algorithm was chosen as baseline. The first experiment was performed in the Weka Explorer environment; eight classifiers algorithms presented above were selected. The database was divided into sets 10 using a cross validation method. The average was calculated of the accuracies obtained in each round of the classifiers. We found that none of the algorithms used in this experiment, was significantly better or worse than OneR. All proposed algorithms presented showed in Fig. 1 are very similar mean values between 77.16 to 78.98.

Classifiers	OneR	JRip	Decision Table	J48	Random Forest	Simple Logistic	Multilayer Perception	NaiveBaye	BayesNet
Accuracy	78.98	77.12	78.23	77.83	77.16	78.24	77.01	78.77	78.65

Fig. 1. First experiment algorithms

The second experiment was conducted in Weka Experiment Environment, we used the database and the selected algorithms, and the Train / Test Percentage split to divide the database in order to check whether the change in the form of training set and test selection affected the performance of the classifiers. We concluded that none of the algorithms used in this experiment was significantly different from baseline. The results of evaluations of the performance showed mean score of classifiers between 78 and

82 %. This is a strong indication that the attributes used are sufficient to execute the prediction that the use of senior potential in the current European situation can promote social sustainability. Comparing the first two experiments, we observed very similar results for most classifiers. The biggest difference in performance was 1.98 % (Fig. 2).

Classifiers	OneR	JRip	Decision Table	J48	Random Forest	Simple Logistic	Multilayer Perception	NaiveBaye	BayesNet
Accuracy	78.00	78.20	79.50	78.67	78.50	79.50	78.01	82.00	81.30

Fig. 2. Second experiment algorithms

The third experiment was carried out in the Weka Explorer environment, and the eight selected algorithms are individually loaded and executed. It supplied the test set option was chosen, such as the training set and test selection. In this experiment, the training set consists of 2/3 of the base and third base for testing. We note that the accuracy of classifiers for the test set shows a major difference compared with previous experiments. Comparing the results of three experiments, it is found that the accuracy varies on average between 76 % to 81 % (Fig. 3).

Classifiers	OneR	JRip	Decision Table	J48	Random Forest	Simple Logistic	Multilayer Perception	NaiveBaye	BayesNet
Accuracy	81.00	78.20	76.00	80.50	80.45	80.89	75.21	81.20	80.76

Fig. 3. Third experiment algorithms

5 Conclusion and Future Work

The study included 408 participants, from small and medium-sized companies based in Portugal, with 61.8 % of the participants of males and 38.2 % of females. The age of participants varies between 20 and 65 (Mo = 33, Med = 40.8) and was segmented by quintiles, represented by five age groups. Participants were grouped into the following age groups: 19.6 % with 30 or fewer years, 20.6 % at the age of between 31 and 34 years, 22.5 % at the age between 35 and 43 years, 18 6 % with aged between 44 and 54 years, and 18.6 % with 55 or more years. It is found that 50.8 % of male participants and 66.6 % of female have higher education.

In this paper, three experiments were carried out where the algorithms were applied rating on a database of respondents from small and medium-sized Portuguese companies. The experiments returned data with average accuracy ranging from 76 to 81 %. The performances obtained by data mining algorithms, from the simplest to the most complex, were similar.

The number of older workers in the European Union (EU) will increase in the next decades. This trend in the EU's active population points to a growth of about 16.2 % of the age group of 55–64 years between 2010 and 2030, and contrary to the trend in other age groups, with a reduction of 4 % to 5 %. This demographic shift is due to the decline in fertility rates and increased life expectancy. The labor force in Europe will reach age levels never before recorded in that older workers represent more than 30 % of the active population. In the EU, currently the employment rates of older workers (55–64 years old) are less than 50 %., But more than 50 % of older workers suspend their professional activity before the mandatory retirement age, by a variety of factors. It is urgent, better and longer working careers to finance and support the longer life of European citizens [16].

Age is not an obstacle to productivity and not the development of organizations. This well-managed can be an important asset. The age (longevity), said its new rules, both for employers and employees, taking these to face this challenge. In this context, human resource strategies should also be rethought. Without new strategies will not be overcome in time, the contingencies of the crisis.

In Portugal, the growth of senior businessmen as well as being associated with the strong dynamics of business creation in the last decades of the twentieth also reflects the fall in younger age groups, as well as the lack of alternatives. Between 1991 and 2011, the total volume of senior business increased 95 %, a value higher than the growth of entrepreneurs under 50 years of age, it is expected that the volume of senior entrepreneurs continue to increase [17].

Not being age an obstacle to the development of organizations. It is up to organizations proactively address this challenge. Thus, the use of senior potential in the current European situation can promote social sustainability. There are good references in the EU, such as Finland and Norway.

As future work, we intend to conduct this study in other countries of the European Union.

Acknowledgment. The authors thank Prometeo Project of SENESCYT (Ecuador) for financial support.

References

1. WHO. Envelhecimento ativo: uma política de saúde (2002)
2. UN. A Projeção da População Mundial: Revisão de 2012 (2012)
3. Kalache, A., Barreto, S.M., Keller, I.: Global ageing: The demographic revolution in all cultures and societies. The Cambridge handbook of age and ageing (2005)
4. Kalache, A., Kickbusch, I.: A global strategy for healthy ageing. World Health **50**, 4–5 (1997)
5. Correia, J.: Introdução à Gerontologia. Universidade Aberta, Lisboa (2003)
6. IEFP. DIRIGIR. Condições e Satisfação no Trabalho, 04/05 2012
7. Mendes, F.R.: Conspiração Grisalha. CELTA, Oeiras (2005)
8. Kautonen, T., Down, S., South, L.: Enterprise support for older entrepreneurs: the case of PRIME in the UK. Int. J. Entrepreneurial Behav. Res. **14**, 85–101 (2008)
9. Piatetsky-Shapiro, U., Fayyad, G., Smyth, P., Uthurusamy, R.: Advances in Knowledge Discovery & Data Mining. Cambridge: The AAAI Press/The MIT Press (1996)
10. Portela, F., Santos, M.F., Machado, J., Abelha, A., Silva, Á., Rua, F.: Pervasive and intelligent decision support in intensive medicine – the complete picture. In: Bursa, M., Khuri, S., Renda, M. (eds.) ITBAM 2014. LNCS, vol. 8649, pp. 87–102. Springer, Heidelberg (2014)

11. Povel, O., Giraud-Carrier, C.: Characterizing data mining software. Intell. Data Anal. **5**, 1–12 (2001)
12. Peixoto, R., Portela, F., Santos, M.F.: Towards a pervasive data mining engine - architecture overview. In: Advances in Intelligent Systems and Computing (WorldCist 2016 - Pervasive Information Systems Workshop), vol. 445, pp. 557–566 (2016)
13. Frank, R.R., Hall, E., Kirkby, M., Bouckaert, R., Reutemann, P., Seewald, A., Scuse, D.: WEKA manual for version 3–7–12 (2015)
14. PORDATA. PorData - Base de Dados Portugal Contemporâneo (2015). www.portdata/ Municipios
15. Lima, E.: Metodologia da pesquisa científica (2009)
16. Ilmarinen, J.: Aging and work: An international perspective. Johns Hopkins University Press, Baltimore (2009)
17. AEP. O Envelhecimento Ativo e os Empresários Seniore (2011)

Sustainable Planning:
A Methodological Toolkit

Giuseppe Las Casas and Francesco Scorza$^{(\boxtimes)}$

Laboratory of Urban and Regional Systems Engineering, School of Engineering,
University of Basilicata, 10, Viale dell'Ateneo Lucano, 85100 Potenza, Italy
{giuseppe.lascasas,francesco.scorza}@unibas.it

Abstract. This paper proposes a methodological appraisal developed in order to face the recurring conflict between environment and economic development assessed in a specific implementation context: the Val d'Agri in Basilicata (Italy), where protection and exploitation requests include the safeguard of particular traditions and culture of communities located in inland areas of the Apennines, as well as environmental resources (natural, agricultural, historical and artistic resources). In this view, we selected strategies in order to promote innovation required by the self-centred part of the development and we referred to a right of rationality of choices, based on: equity, efficiency and conservation of resources; accountability instruments and uncertainty sharing means to which connect the integrated cycle of assessment/governance. The Logical Framework Matrix (LFM) was proposed for this purpose and, through the derived call, applicants will have to work on the basis of a set of procedures oriented to the construction of ended and integrated supply chains.

Keywords: Planning · Tools and techniques · Territorial specialization · Regional development · Impact assessment

1 Introduction

This paper examines territorial organization choices defined during LISUT activities supporting the administration of the Project Val d'Agri, with reference to the formation of Inter-Municipal Structure Plan, still in progress.

Val d'Agri is placed in a territorial context where water landscapes take on an unique role: while on one side they ensure the supply for water systems in a wide region, even outside Val d'Agri area, on the other side they enrich landscapes and open realistic use perspectives for tourism purposes. These opportunities has to be compared with the presence of oil fields whose mining and processing generates significant impact risks.

We cannot ignore current political and judicial events affecting Val d'Agri territory, but we think that in spite of these highly current events, the technical and methodological instances at the base of this work do not change: accountability, systematic and independent frameworks of knowledge, sharing of uncertainty and transparency remain the basis of our research project.

© Springer International Publishing Switzerland 2016
O. Gervasi et al. (Eds.): ICCSA 2016, Part I, LNCS 9786, pp. 627–635, 2016.
DOI: 10.1007/978-3-319-42085-1_53

The strong demand of employment opportunities, the tragic demographic crisis, the vulnerability of the building stock make the refusal of opportunities offered by concession proceeds very difficult.

In particular – at local level – on one hand we find people that consider hydrocarbon revenues as a chance to reverse impoverishment and economic and socio-demographic trends, besides the possibility to increase political approval, and on the other hand, there are ranchers, farmers and part of the scientific community that invoke at least the precautionary principle.

The composition of these conflicts makes the case study an example of the complexity challenge through a rational approach.

In this framework, the developed methodology expects to implement an integrated strategy characterized by significant steps of shared and joint processing.

It corresponds with the search for the three safeguard principles that we consider as the logical basis of our proposal:

i.   efficient allocation of resources
ii. equity in the distribution of opportunities
iii. protection of non-renewable resources
     and it expects to test the research of an a-priori rational logic, in which the targets-products-activities-means-input connection becomes clear, starting by problems and through the Logical Framework Approach (LFA) implementation.

This methodology is based on an operative vision of the Falaudi's "proceduralist approach" [1] that – besides the desired scenario - proposes a process to define and to monitor objectives and strategies and promotes synergies in order to ensure the concentration of efforts on a few well-defined directions.

The strengths of this approach are the three above-mentioned principles and the explicit sharing of objectives: this represents Popper's demarcation principle that, by following Falaudi [1], identifies what can be called "a good plan."

Conclusions regard possible application and perspectives for improving and supporting regional development planning considering the exploitation of open data sources and spatial analysis.

2   Bounded Rationality, Incrementalism and Sustainable Strategic Planning

At the end of the seventies, the "classical" period of the so-called *systemic approach*[1], relevant elements of dissatisfaction about the transition from analysis to project persisted.

Such transition remained predominantly linked to the optimization attempt connected to Operative Research [6] and to the flourishing production of simulation models (cfr. [7]).

---

[1] Among the authors, besides the best known McLoughlin [2] and Ghadwick [3], we find a very interesting reconstruction of Wegener [4], proposing a survay 10 years later Lee's article [5].

According to our approach [8], the rationality of decisions about citizens' needs and aspirations and the use of common goods and non-renewable resources must be considered as a citizen's right and so a prerequisite in the development of plan proposals. An approach whose method focuses on:

- collective learning processes that feed themselves through the awareness of the interaction system complexity connected to social fabric, economy and environment[2];
- governance processes that could be applied after the definition of objectives, means and activities, logical links between the achievement of the desired scenario and available means, an adequate system of indicators measuring effectiveness and efficacy
- The references of this approach are:
- from the technical point of view, the Logical Framework Approach, included SODA, about whom we'll talk at the point 3;
- from the point of view of the legal feasibility, the GPRA (Government Performance and Result Act of 1993 of the United States) [12, 13];
- from a theoretical point of view, our main reference is Faludi, whose proposal considers the transition from a static concept of planning, that adopts technical knowledge to the development of a desired future scenario, to a dynamic vision, focused on the decision as a process [1].

In this research of rationality we propose the following presentation and comparison of three instruments, already applied in heterogeneous environments: cognitive maps, ontologies, the logical framework approach of the objectives (LFA).

3 Toolkit for Renewing Planning and Governance: 'Cognitive Maps' and SODA; Ontologies; Logical Framework Approach

'Cognitive maps' are useful elaborations to represent the perception of a specific object/domain of interest for an individual or, more precisely according to the specific feature of this work, for an "actor", or a group of actors. They are simplified and expressive abstractions based on the identification of concepts interconnected through links that express hierarchies and mutual dependencies.

In methodological terms, the use of mental or cognitive maps is part of Problem Structuring Methods [14]: a family of support methods at the first stage of the decision process that intends to achieve the involvement of mixed groups, in a complex context, with the aim to help participants. We are therefore at an early stage of the planning process.

---

[2] Cfr. the conspicuous production of Roy Benard's group with reference to what was produced in the field of understanding and modeling of decision processes. Among others, see Roy [9]; Ostanello and Tsukias [10]; Las Casas [11].

Among different techniques, the SODA (Strategic Option Decision Analysis) is a method to identify problems, based on the use of cognitive maps as a support to explain and record individual or collective opinions as a reference point for the discussion of a group, led by a facilitator. This is a useful technique for including different and often conflicting opinions on a specific issue[3]. It is based on Kelly's personal construct theory [15] according to which each individual, influenced by his own experience and culture, is a representation of reality through a system of concepts linked by different connections depending on the considered complexity degree.

When concepts are expressed and organized on a map, connected with each other to form propositions, learning is simplified by the presence of these relationships, which help to link new concepts to concepts already set in the personal wealth of knowledge [14].

The use of cognitive maps responds to a more general need of knowledge structuring with reference to a domain of detailed study. Therefore, the usefulness of this instrument belongs to the management of consultation/participation processes and of the interaction space assessment [16], where the vision, not necessarily formalized, should include multiple points of view by identifying key concepts ordered through relationships.

This is a not much formalized approach, in which facilitator's role contributes to establish general agreement on a comprehensive view of a particular domain.

Considering recent experiences [17–19] the application of ontologies in planning processes and territorial management, and more generally to governance processes of place-based development, represents a research field that connects ICT tools with operative procedures and deals with the problem of interoperability between databases for this purpose. Instead, what we call "ontological approach" [20] is an attempt to link processes of knowledge construction by integrating the program structure, the system of actors, resources and context.

Starting from the concept of ontology as a meta-model of reality, or rather, of the domain examined, where concepts and logical connections are used as part of the interpretive model and as generators of rules and constraints of relation system, we consider the following definition of "ontology": "formal and explicit description of a domain of interest" in which:

- "Description" is a form of knowledge representation;
- "Formal" means "symbolic" and "that can be mechanised";
- "Explicit" means all concepts used and constraints on their use are explicitly defined;
- Domain: "a certain subset of the system, faced from a certain point of view" [21].

In our methodological proposal, in a rational approach for plan process, in the synthesis phase, the Logical Framework (LF) helps to organize the logic of plan activities in order to simplify their assessment during the different phases of the Project Cycle Management.

---

[3] Lack of facilities or organizations, but also of efficiency, equity and respect of non-renewable resources.

LF represents the hierarchy of objectives in a grid composed by at least four rows and four columns [22–24] (Fig. 1):

Intervention logic	Objectively verifiable indicators			Sources of verification	Assumptions
1. Overall Objective	Context Analysis: 1. Obj.'s Pertinence 2. Obj.'s Relevance	Efficacy Indicators	Effectiveness Indicators		
2. Project Purposes					
3. Results/outcomes					
4. Activities	5. Inputs				
Preconditions					

Fig. 1. Log frame scheme [25]

Our proposal aims at emphasizing the principles of effectiveness and efficiency of public expenditure and at clarifying coherence and relevance of policy choices for the context of implementation, according to the research of a context or place-based policy [25], by introducing a specific box in the Logframe Matrix.

Generally, assessment is intended as the identification of policy effects, in connection with the given objectives and constraints (terms of reference).

Aune [26] warns us of the danger: "*Form over substance*". As for aids to enterprises, the "form" of LFA often replaces the "substance". In fact, in the widest applications of LFA, the "compiling of matrix" beyond the utility levels required by the project [27] can represent the victory of form over substance. Coleman [28] argues that LFA approach is an "aid to think" rather than a set of procedures.

4 Case Study

It seems to be evident that the dramatic point around which the decision-making process is tangled up is the connection among protection issues, exploitation and endogenous development of territory.

In the case of Val d'Agri, this aspect concerns research, extraction, transportation, processing of hydrocarbons and their link with natural and agricultural system.

We recorded the highest levels of uncertainty on which the debate, or better the dispute, became livelier about this key topic. These regard:

- objective, updated and forecasting data on the location and extent of activities;
- impacts on the air;
- effects on surface and underground hydrology system;

- effects on the quality of water transported by water schemes or reintroduced into surface water bodies;
- impacts on inhabitants' health;
- the allocation of revenues;
- future of the area after the depletion of deposits;
- direct and indirect impacts on employment, for activities related to hydrocarbon processing but also for agriculture and tourism;
- desires of a population that leaves the valley with a dramatic trend and achieves aging rates which, for some centres, do not foretell the permanence of inhabitants in the near future.

As for these and other problems, we cannot ask the question in a radical way: **yes oil/no oil.** Conversely, we have to attempt the research of appropriate context or place-based compromises, based on the sharing of information and uncertainty. Only through awareness and sharing of uncertainty sources, debate and negotiation can develop a process whose main product is information [29]. Financial resources from oil production could support such difficult research even if the total amount is not fully predictable for the future.

The proposed *vision* is based on the hypothesis of an area where the cultivation of oil fields and subsequent processes can coexist with the preservation of the most important share of natural features and the reinforcement of traditional activities and other innovations connected with local peculiarity.

We propose in fact a territory where the main mountain crests that enclose the valley and more internal ones, that include the greater and more significant natural areas, crown a valley where, conversely, the oil centre, pipelines and other innovative activities are dominant.

The hydrological system of the highest quotas will be protected and subject to monitoring procedures and aquifers will be effectively defended. The water collected in Pertusillo reservoir will be conveyed in a suitable pre-treatment plant before being introduced into the aqueduct.

Above all the idea to explore, produce and process oil in Val d'Agri must correspond to the exclusion of all other specified locations, avoiding the dissemination of plants without limits whose negative impact to land degradation has proved to be very serious.

The price paid in terms of environment is certainly high and it will be balanced by a development through which the activities of the valley may be partially sacrificed to the inevitable impacts (but not only those) derived from the exploitation of hydrocarbons.

Therefore, the measures of prevention, mitigation and precaution will be pre-condition for the reinforcement of an image, only dreamed nowadays, in which the weakest economic asset based on niche products, such as certified productions, or on the highly sensitive landscape natural elements and cultural heritage can be exploited through an integrated offer that links hiking, culture, food and wine to proposals such as the widespread museum of energy, widespread welcome and for the elderly (Table 1).

Table 1. Strategic functions and types of intervention.

Function	Typology				
	a	b	c	d	e
Nature tourism	Walking tours	Horse	Cycle tourism	Environmental education	Equipment
Cultural and religious tourism	Museum	Fruition of energy museum	Festivals	Religious celebrations	Libraries and homeland history
Tourist accommodation	Hotel	B&B	Farmhouse		
Health and old age	Senior housing	Health aid			
Recovery and re-use for residences and services	Recovery and re-use for leisure and socialization	Recovery and re-use for culture, entertainment and training	Education in the field of recovery and re-use and energy efficiency		
District/Energy museum	Oil refining and monitoring	Energy history and economy	Technological park and innovative R.E.S. systems	The water mills	
Industry	Energy Production	Electrochemistry	Metallurgy	Electronics	
Agro-Zootechnics	IGP bean	Ovine slaughter, preserves and dairy products	Podolica slaughter, preserves and dairy products	Restaurants	Logistics
Mobility and communications	Freight transport	Purifiers and water schemes	ICT for people transport	Organization LPT	Connection with tourist demand

5 Conclusions

A complex problem, for the harshness of conflicts, has been faced through instruments of plan rationality. This is an a-priori rationality based on the methodological assumption according to which a good plan can be managed only if an explicit, falsifiable system of goals and links between goals and strategies are available.

The next months will tell us if the proposed strategies are coherent with the requirements of a shared fulfilment.

For this purpose, some techniques were thought despite the risk to stiffen the process: in our opinion, this depends on the care and time that we can devote to their development through strong social interactions, free from political conditioning. These connections can generate a kind of collective intelligence in order to better develop the knowledge both of events and of dynamics, above all those from which all needs and aspirations, and awareness evolve [30].

References

1. Faludi, A.: A Decision-Centred View of Environmental Planning. Elsevier, Amsterdam (1987)
2. McLoughlin, J.B.: Urban and Regional Planning: A Systems Approach, p. 329. Faber, London (1969)
3. Chadwick, G.: A Systems View of Planning: Towards a Theory of Urban and Regional Planning Process. Pergamon Press, Oxford (1971)
4. Wegener, M.: Operational urban models: state of the art. J. Am. Plan. Assoc. **59**, 17–29 (1994)
5. Lee, D.B.: Requiem for large-scale models. J. Am. Inst. Planners **39**, 163–178 (1973)
6. Friend, J.K., Jessop, W.N.: Local Government and Strategic Choice: An Operational Research Approach to the Processes of Public Planning. Tavistock Publications, London (1969)
7. Wilson, A.: New roles for urban models: planning for the long term. Reg. Stud. Reg. Sci. **3** (1), 48–57 (2016). doi:10.1080/21681376.2015.1109474
8. Las Casas, G.B., Sansone, A.: Un approccio rinnovato alla razionalità nel piano. In: Depilano, G. (a cura di) Politiche e strumenti per il recupero urbano. EdicomEdizioni, Monfalcone (2004)
9. Roy, B.: Méthodologie Multicritère d'Aide à la Décision. Economica, Paris (1985)
10. Ostanello, A., Tsoukias, A.: An explicative model of' public interorganizational interactions. Eur. J. Oper. Res. **70**, 67–82 (1993). North-Holland
11. Las Casas, G.B.: criteri multipli e aiuto alle decisioni nel processo di piano (2010). https://www.researchgate.net/publication/268262221_CRITERI_MULTIPLI_E_AIUTO_ALLE_DECISIONI_NEL_PROCESSO_DI_PIANO. Accessed 29 Apr 2016
12. Goverment Performance and Result Act, United States (1993)
13. Archibugi, F.: la pianificazione sistemica: strumento della innovazione manageriale nella pa, negli usa e in europa (2002). http://win.progettosynergie.it/pianificazionestrategica/formarchibugi/L0.1/letture/pianificazione_sistemica.pdf. Accessed Apr 2016

14. Las Casas, G.B., Tilio, L.: Seismic risk reduction: a proposal for identify elements enhancing resilience of territorial systems. In: Popovich, S., Elisei, Z., (eds.) Proceedings REAL CORP 2012 Tagungsband 14–16 May 2012, Schwechat (2012). ISBN: 978-3-9503110-2-0
15. Kelly, G.: The Psychology of Personal Constructs. Norton, New York (1995)
16. Casas, G.B., Tilio, L., Tsoukiàs, A.: Public decision processes: the interaction space supporting planner's activity. In: Murgante, B., Gervasi, O., Misra, S., Nedjah, N., Rocha, A.M.A., Taniar, D., Apduhan, B.O. (eds.) ICCSA 2012, Part II. LNCS, vol. 7334, pp. 466–480. Springer, Heidelberg (2012). doi:10.1007/978-3-642-31075-1_35
17. Zoppi, C.: Ontologie ed analisi territoriale: un'introduzione al tema. Scienze Regionali 2011 (2), 121–123 (2011)
18. Rabino, G.: Le ontologie nella società dell'informazione. Scienze Regionali 2011(2), 125–131 (2011). doi:10.3280/SCRE2011-002006
19. Las Casas, G.B., Scorza, F.: Redo: applicazioni ontologiche per la valutazione nella programmazione regionale. Scienze Regionali 10(2), 133–140 (2011). doi:10.3280/SCRE2011-002007
20. Scorza, F., Casas, G.B., Murgante, B.: That's ReDO: ontologies and regional development planning. In: Murgante, B., Gervasi, O., Misra, S., Nedjah, N., Rocha, A.M.A., Taniar, D., Apduhan, B.O. (eds.) ICCSA 2012, Part II. LNCS, vol. 7334, pp. 640–652. Springer, Heidelberg (2012)
21. Gruber, T.: Toward principles for the design of ontologies used for knowledge sharing? Int. J. Hum. Comput. Stud. 43(5–6), 907–928 (1995). doi:10.1006/ijhc.1995.1081
22. UsAid Guidelines, Australian Agency for International Development (2009)
23. European Commission: Europe Aid Co-operation Office General Affairs Evaluation. Manual Project Cycle Management, Bruxelles (2001)
24. NORAD (Norwegian Agency for Development Cooperation): The Logical Framework Approach (LFA) (1999)
25. Las Casas, G.B., Scorza, F.: Un approccio "context-based" e "valutazione integrata" per il futuro della programmazione operativa regionale in Europa. In: Lo Sviluppo Territoriale Nell'economia Della Conoscenza: Teorie, Attori Strategie. Collana Scienze Regionali, vol. 41 (2009)
26. Aune, J.B.: Logical Framework Approach. Development Methods and Approaches, 214 (2000)
27. Bakewell, O., Garbutt, A.: The Use and Abuse of the Logical Framework Approach, p. 27. Swedish International Development Cooperation Agency (Sida), Stockholm (2005)
28. Coleman, G.: Logical framework approach to the monitoring and evaluation of agricultural and rural development projects. Proj. Appraisal 2, 251–259 (1987)
29. Las Casas, G.B.: L'etica della Razionalità. In: Urbanistica e Informazioni, vol 144, (1995). ISSN: 0392-5005
30. Las Casas, G., Murgante, B., Scorza, F.: Regional local development strategies benefiting from open data and open tools and an outlook on the renewable energy sources contribution. In: Papa, R., Fistola, R. (eds.) Smart Energy in the Smart City, pp. 275–290. Springer International Publishing, Switzerland (2016)

A Transnational Cooperation Perspective for "Low Carbon Economy"

Alessandro Attolico[1] and Francesco Scorza[1,2(✉)]

[1] Province of Potenza, Planning and Civil Protection Office,
P.zza M. Pagano, 85100 Potenza, Italy
alessandro.attolico@provinciapotenza.it,
francesco.scorza@unibas.it
[2] Laboratory of Urban and Regional Systems Engineering,
School of Engineering, University of Basilicata,
10, Viale dell'Ateneo Lucano, 85100 Potenza, Italy

Abstract. Many of Europe's local/regional actors struggle with developing targeted, implementation-oriented policies addressing low carbon challenges. This work describes the project LOCARBO, already in the early stage of implementation, financed by the INTERREG EUROPE Program. The project is mainly oriented at improving regional programming and main policy instruments in order to innovate low carbon strategies looking at the best European practices properly transferred in each implementation context. A peculiarity of the approach promoted by the project is the involvement of citizens in order to spread sustainable behaviours based on a formal 'green' agreement among users, local administrations and business/services operators. The challenge to involve and motivate stakeholders (especially energy consumers) is perceived broadly as a major problem for public authorities. Motivation and awareness of consumers are of high significance to influence their behaviour and support more conscious energy decisions.

Keywords: Energy efficiency · Covenant of majors · EU 2020 · New cohesion policy · Regional planning

1 Introduction

Many of Europe's local/regional actors struggle with developing targeted, implementation-oriented policies addressing low carbon challenges.

International cooperation projects represent an opportunities for local administrations to face common problems of the EU regional framework in a common way, exchanging experiences, case studies and discussing failures and inefficiency. This work describes the project LOCARBO, already in the early stage of implementation, financed by the INTERREG EUROPE Program. The province of Potenza leads the partnership reinforcing a former effective experience in transnational cooperation.

The project is mainly oriented at improving regional programming and main policy instruments in order to innovate low carbon strategies looking at the best European practices properly transferred in each implementation context.

© Springer International Publishing Switzerland 2016
O. Gervasi et al. (Eds.): ICCSA 2016, Part I, LNCS 9786, pp. 636–641, 2016.
DOI: 10.1007/978-3-319-42085-1_54

A peculiarity of the approach promoted by the project is the involvement of citizens in order to spread sustainable behaviours based on a formal 'green' agreement among users, local administrations and business/services operators.

The long term perspective looks at exploiting energy sustainable policy as a development drivers at regional and local level.

The challenge to involve and motivate stakeholders (especially energy consumers) is perceived broadly as a major problem for public authorities (see also [1]). Motivation and awareness of consumers are of high significance to influence their behavior and support more conscious energy decisions.

The overall objective of LOCARBO is improving policy instruments targeting demand-driven initiatives to increase energy efficiency related to the built environment. This is to be achieved by finding innovative ways for regional/local authorities to support energy consumers' behaviour change. LOCARBO is unique in focusing its activities on bottom-up initiatives and mainly because of the approach to handle 3 thematic pillars (services, organizational structures and technological solutions) in a fully integrated way.

The Province of Potenza looks at the implementation of the Territorial Master Plan as a way of concrete application of LOCARBO results especially in the weakest underdeveloped areas.

To look at inclusive energy policy making represents a form of innovation in governance models and practices and needs for new effective procedures and practices to be integrated in the traditional administrative structure concerning territorial planning.

Apart form the technological innovation promoted by the project [2], one of the main output is the Local Implementation Plan: a strategic document affirming the route for transferring Learned Best Practices in the local policy framework, promoting investments and ensuring sustainable performance for the local communities [3].

In this work the main component of the project are described including the main pillars the research and the operative contributions expected in the short term implementation.

Conclusions regard the perspectives of operative implementation as a contribution for the territorial planning [4–6] as a mainstream component of the Provincial Structural Master Plan (PSP).

2 Objectives and Priorities

In this section are described the general project objectives. Such strategic schema is oriented to the three main pillars of the project: (i) Supplementary services and products offered by authorities; (ii) Innovative cooperation models; (iii) Innovative smart technologies.

While the overall objective is to improve policy instruments targeting demand-driven initiatives to increase Energy Efficiency related to the built environment and users. This is to be achieved by finding innovative ways for regional/local authorities to support energy consumers' behaviour change. The project will focus on exploring the way local policy instruments can be improved so they can combine and roll out innovative practices linked to the following thematic pillars. Each thematic pillar includes sub-objectives:

(1) Supplementary services and products offered by authorities, i.e. energy consultancy services to end users and energy ambassadors (fast track training for individuals to carry out easy-to-use energy assessments), building on the track record and remarkable results of key LOCARBO partners.

(2) Innovative cooperation models: Local energy communities have demonstrated effectiveness in numerous EU regions to foster an active involvement of stakeholders, especially energy consumers. Local energy communities refer to groups of economic/civil actors cooperating on a common goal linked to EE/RES. Similarly, grassroots initiatives present innovative community-led solutions to efficiently respond to local bottlenecks in sustainability and represent the interests and values of communities involved.

(3) Innovative smart technologies: Europe is split concerning the penetration of smart/ICT technologies such as energy management systems or smart meters. The aim is thus to (1) support the spread of intelligent technologies in new Member States with typically lower penetration rates (2) systematically collect and analyse data and feed-back information to support fact-based regional policy making, as a joint issue for almost all EU regions.

The project aims at improving policy instruments targeting energy efficiency of the built environment through regional and local authorities supporting energy consumers' behavior change. The improvements targeted are structured into 3 interlinked Thematic Pillars (TP): (1) supplementary services and products offered by authorities, (2) innovative cooperation models and (3) innovative smart technologies. The three-leg structure with interactions both within and across pillars together with the two-way interactions between the regional and interregional levels (stakeholder involvement) strengthens the cross-thematic exchange and enriches the interregional knowledge portfolio.

The implementation procedure is based on the following steps:

- Elaboration of regional state-of-the-art analysis related to the targeted policy instrument and sets its key ambitions framework;
- Learning and knowledge transfer process partners collecting and sharing good practices and assessing the transferability and possible adaptation methods;
- Interregional Site Visits followed;
- Action Plans development organised in a way to ensure that interregional exchange findings.

A key element of information and knowledge exchange are divided on different levels as following: (1) individual learning through direct participation of partners' staff members in the project's activities; (2) organisational learning through Institutional Learning Platforms; (3) stakeholder learning through continuous involvement of stakeholders during the project's lifetime.

The continuous involvement of stakeholders, especially Managing Authorities, serves to transfer relevant knowledge and experience in capitalization and the consequent Action Plans can be guaranteed by incorporating the expertise and needs of the relevant stakeholders.

3 A Peculiar Context

In the case of Basilicata Region the energy production and consumption I characterized as follows:

The supply chain is divided in: Fossil sources and Renewable Energy Sources (RES). Each component presents criticalities and local conflicts with identity values and resources.

Concerning the Fossil Sources Energy supply, the territory of the Basilicata Region is characterized by a considerable fossil sources availability, which is partly exploited.

The Basilicata Region is top of the national ranking for its oil production thanks to the 4.3 million tons of oil extracted up to 2006, while as concerns the natural gas production, the Basilicata Region territory is the only one in Italy still steadily increasing (natural gas production from seabed minings: 10 % of national production; natural gas production from onshore extractions: 47 % of national production).

By analyzing the time series of the energy consumptions on the Basilicata Region territory, it can be said that most of the electric power produced in the Basilicata Region up to 2006 derived from the fossil fuel use, with a preponderance for the natural gas, which shows a use of 98 % compared to other fossil sources n late 2006; this reflects the national trends.

Ten thermal power plants fuelled by fossil fuels were in operation to 2005, for a total of 283 MWe and of 1.1185 GWht of gross production.

The Energy demand in Basilicata Region has not a very high amount for end-use. From the comparison with the national energy demand value, it is easy to see that the Lucanian consumptions represent less than 1 % of the national consumptions.

The historical data on the regional electricity consumptions show a steadily increase of the demand during the past twenty years. This growth is being driven mostly by the industry development, but a significant increase, in relative terms, has occurred also in the other reference sectors. This puts the Basilicata Region in contrast with the average trends of the national consumptions, which are almost regular.

The Basilicata Region had an electricity gross production of 2.238 GWh in 2010; 2.171 GWh were intended for the consumptions. At the end of 2010 the gross consumptions was estimated at 3.174 GWh. On this basis, the Basilicata Region is not meeting the consumptions for a 30.1 % share of the regional demand. This latter value was about 52.2 % just two years earlier. The energy deficit reduction has been achieved thanks to the increase of the electricity production from renewable sources.

As concerns the future scenarios on the energy demand, the forecast in the RES report describe a tendential increase of 8.5 % of the total consumptions (electricity and heat). This outcome confirms that the Basilicata Region is experiencing a period of growth and development in comparison to the other Italian regions.

4 Project Innovation

Among the numerous projects that try to increase energy efficiency in the built environment LOCARBO is unique in focusing its activities on bottom-up initiatives but mainly because of the approach to handle the 3 thematic pillars in a fully integrated way.

Partners are aware that only by bringing pieces of the puzzle together (services, organizational structures and technological solutions) regional policies on EE can be successful. The core idea of how local/regional authorities can become involved in these initiatives is a novel approach in this subject.

A strong added value is that LOCARBO starts development of Action Plans almost from the start of the project to leave ample time for their elaboration. Moreover, Action Plans will be developed using review pairs of partners that will reinforce realistic ambition setting and feasible action planning.

5 Conclusions

Project objectives and implementation procedure, described synthetically in this work, holds two different level of operational contribution: on one hand it represents an effective example of framework implementation for lessons learned in a best practices exchange project; on the other fits within a territorial planning instrument (the PSP of the Province of Potenza) orienting general policies towards specific implementation projects following an integrated approach [7].

Such contributions has an added values for coordinating the activities of local municipalities in a common strategy looking at local sustainable development.

The feasibility of the strategic lines implementation refers to financial sources identified among the intervention lines of the regional development programmes linked with the 2014–2020 New Cohesion Policy sources [8–10].

The perspective regards the integration of such strategic schema in the framework of Territorial Resilience strategy promoted by the Province of Potenza in cooperation with UNISDR [11, 12]. The Province developed a multi-stakeholders participatory process in order to include a number of stakeholders (crf. [13]). In fact energy efficiency and energy sustainability represents components and means of territorial security.

LOCARBO expected impacts regards a renewed approach in managing local development according with an inclusive approach centered on energy end-users.

References

1. Scorza, F.: Smart monitoring system for energy performance in public building. In: Gervasi, O., Murgante, B., Misra, S., Gavrilova, M.L., Rocha, A.M.A.C., Torre, C., Taniar, D., Apduhan, B.O. (eds.) ICCSA 2015. LNCS, vol. 9156, pp. 767–774. Springer, Heidelberg (2015)
2. Scorza, F., Attolico, A., Moretti, V., Smaldone, R., Donofrio, D., Laguardia, G.: Growing sustainable behaviors in local communities through smart monitoring systems for energy efficiency: RENERGY outcomes. In: Murgante, B., Misra, S., Rocha, A.M.A., Torre, C., Rocha, J.G., Falcão, M.I., Taniar, D., Apduhan, B.O., Gervasi, O. (eds.) ICCSA 2014, Part II. LNCS, vol. 8580, pp. 787–793. Springer, Heidelberg (2014). doi:10.1007/978-3-319-09129-7_57

3. Las Casas, G., Lombardo, S., Murgante, B., Pontrandolfi, P., Scorza, F.: Open data for territorial specialization assessment territorial specialization in attracting local development funds: an assessment. Procedure based on open data and open tools. TeMA J. Land Use, Mobility Environ. (2014). Print ISSN 1970-9889|online ISSN 1970-9870 [in Smart City. Planning for Energy, Transportation and Sustainability of the Urban System Special Issue, June 2014, Editor-in-chief: Papa, R.]
4. Frazier, T.G., Thompson, C.M., Dezzani, R.J., Butsick, D.: Spatial and temporal quantification of resilience at the community scale. Appl. Geogr. **42**, 95–107 (2013). doi:10.1016/j.apgeog.2013.05.004
5. Birkmann, J.: Risk and vulnerability indicators at different scales: applicability, usefulness and policy implications. Environ. Hazards **7**(1), 20–31 (2007). doi:10.1016/j.envhaz.2007. 04.002
6. Tilio, L., Murgante, B., Di Trani, F., Vona, M., Masi, A.: Resilient city and seismic risk: a spatial multicriteria approach. In: Murgante, B., Gervasi, O., Iglesias, A., Taniar, D., Apduhan, B.O. (eds.) ICCSA 2011, Part I. LNCS, vol. 6782, pp. 410–422. Springer, Heidelberg (2011). doi:10.1007/978-3-642-21928-3_29
7. Las Casas, G., Murgante, B., Scorza, F.: Regional local development strategies benefiting from open data and open tools and an outlook on the renewable energy sources contribution. In: Papa, R., Fistola, R. (eds.) Smart Energy in the Smart City, pp. 275–290. Springer, Switzerland (2016)
8. Scorza, F.: Improving EU cohesion policy: the spatial distribution analysis of regional development investments funded by EU structural funds 2007/2013 in Italy. In: Murgante, B., Misra, S., Carlini, M., Torre, C.M., Nguyen, H.-Q., Taniar, D., Apduhan, B.O., Gervasi, O. (eds.) ICCSA 2013, Part III. LNCS, vol. 7973, pp. 582–593. Springer, Heidelberg (2013)
9. Tajani, F., Morano, P.: An evaluation model of the financial feasibility of social housing in urban redevelopment. Property Manag. **33**(2), 133–151 (2015). ISSN 0263-7472
10. Morano, P., Tajani, F., Locurcio, M.: Land use, economic welfare and property values: an analysis of the interdependencies of the real estate market with zonal and macro-economic variables in the municipalities of Apulia Region (Italy). Int. J. Agric. Environ. Inf. Syst. **6**(4), 16–39 (2015). ISSN 1947-3192
11. Johnson, C., Blackburn, S.: Advocacy for urban resilience: UNISDR's making cities resilient campaign. Environ. Urbanization **26**(1), 29–52 (2014). doi:10.1177/0956247813518684
12. Djalante, R.: Review article: adaptive governance and resilience: the role of multi-stakeholder platforms in disaster risk reduction. Nat. Hazards Earth Syst. Sci. **12**(9), 2923–2942 (2012). doi:10.5194/nhess-12-2923-2012
13. Bäckstrand, K.: Multi-stakeholder partnerships for sustainable development: rethinking legitimacy, accountability and effectiveness. Eur. Environ. **16**(5), 290–306 (2006). doi:10. 1002/eet.425

Assessing Sustainability: Research Directions and Relevant Issues

Francesco Scorza[1](✉) and Valentin Grecu[2]

[1] School of Engineering, Laboratory of Urban and Regional Systems Engineering,
University of Basilicata, 10, Viale dell'Ateneo Lucano, 85100 Potenza, Italy
francesco.scorza@unibas.it
[2] Department of Industrial Engineering and Management, Lucian Blaga University of Sibiu,
Sibiu, Romania
valentin.grecu@ulbsibiu.ro

Abstract. The growing research debate concerning sustainability and its applications in interdisciplinary domain represents a conjunction point where basic and applied science (scientific computation and applications in all areas of sciences, engineering, technology, industry, economics, life sciences and social sciences), but also qualified practitioners, compare and discuss advances in order to substance what we consider a the future perspective: "applied sustainability". A relevant issue in order to compare and benchmark different position is the "sustainability performance assessment". It means to discuss in a general view critical aspects and general issues in order to propose research directions and common parameters (indicators) to exchange and disseminate results and milestones in "sustainability" applications.

Keywords: Sustainability · Policy and planning · Development

1 To Define a Position: A Tentative

After United Nations 2030 development agenda setting-up Sustainable Development Goals (SDGs), expected to be one of the main influencing paper for mainstream global development policies in the coming years, and on the basis of the previous objective framework MDGs (Millennium Development Goals) the focus on sustainability increased both in theoretical perspectives but mainly in practical applications in every field of human action.

It is already possible to register an increasing global effort on "renewed sustainable development" with influences and constraints at multiple scale (from global action to national, regional and local dimension).

We refer to "renewed sustainable development" as a former concept, widely argued in multiple scientific domains (form planning [1] and management to decision science, from environmental science to economics and econometrics, from social science to operative research), but applied with increasing awareness in everyday human activity. So the envisaged renovation belongs more to the consciousness in taking into account

© Springer International Publishing Switzerland 2016
O. Gervasi et al. (Eds.): ICCSA 2016, Part I, LNCS 9786, pp. 642–647, 2016.
DOI: 10.1007/978-3-319-42085-1_55

the need to verify sustainability as a balance between the use of resources and their reconstitution in a proper time-frame.

Most promising application domains belongs to the operative application in decision and policy making: knowledge management [2, 3]; land use and risk assessment [4 6]; organization management [7].

Such new starting point comes after excellent failure: Kyoto protocol is the main defendant on trial. It demonstrated how global agreement on challenging objectives could be undermined if human communities play the sole role of "les agìs" in such process.

2 Renovation Opportunities from the Challenge of Complexity

Renovation means – of course – innovation: through 'SDGs' UN launched a permanent call for innovation where sustainability becomes a transversal value to be measured in order to define effective intervention process in every resource-consuming sector.

Such assumption re-launches the "Challenge of Complexity". In other words it forces the research and the technical application towards interdisciplinary and, in particular, it asks for rigorous assessment methods in order to promote comparisons, sorting criteria, producing lessons learned from previous applications or outstanding projects.

If we focus the perspective in which 'assessing sustainability' means 'assessing long term impacts' on environmental or human resources, a requested innovation is to deliver products or supply chain models with required necessary features in a resource-scarce domain. If we consider social sustainability, the inclusiveness degree of social dynamics and policies represents an up to date indicator to be defined especially in the current EU development policy-making. If we point on sustainability assessment in anthropic practice: agriculture, industry, land use, urban development (including infrastructure), environmental risks, energy and/or – widely - climate change represents domains both for academic investigations and operative application regarding decision making, production, market and governance. But the list could be as longer as we enlarge the scope of subject areas or implementation domains. That's the case if will focus on the hot spots: energy, water, and housing, environment, food, soil consumption, urban sprawl, technological innovations, social inclusion, as well as on their combinations.

The scientific debate we intend to stimulate will pay particular attention to methodologies, research reports, case study assessment concerning the various combinations of these and other areas in a multi- and interdisciplinary way.

General research questions to be answered to:

– how to enhance effectiveness in policy making, planning, development programs etc.? looking at actions or procedures based on (or derived by) SDGs or widely applying sustainability principles.
– Assessing sustainability through place based approach: innovation in methods and practices.
– Does an assessment matrix help/exist? Comparing different quantitative and qualitative approaches in sustainability evaluation.

– To learn form failures and to discuss success examples: the critical appraisal of on going concrete practices

3   Why Be Sustainable?

The interrelated challenges of financial instability, resource constraints, systematic degradation of eco-systems and social inequity redefine the overall conditions for business in the twenty-first century. An organization faces new demands in order to address these challenges, but also great opportunity for innovation. Although is sometimes neglected, the design and re-design of business models is an important aspect of innovation.

According to França [8], redesigning business models has been identified as a greater source of lasting competitive advantage than new products and services per se. Most of the managers that have reported benefits from becoming sustainable, as their company's sustainable activities have added profits, also say that these activities have led to business model changes.

The economic reasons that can be understood by managers from a selfish profit-oriented perspective are doubled by the international context. The State of the World 2004 report focuses on the consumer society and it argues that sustainable development initiatives have not resulted in overall lower use of resources and foresee even a higher consumption rate, given the need for economic development in the poorest countries that are to achieve the Millennium Development Goals and eradicate extreme poverty and hunger [9]. These reports stress the idea that consumption without limits is a huge threat to the planet.

To avoid such outcomes, there is an urgent need to consider whether and how, notions of sustainable development can be incorporated into the management of business [10].

4   Indicators for Sustainability Performance Assessment

Indicators have been defined in a number of different ways: the Dictionary of Environment and Sustainable Development [11] defines an indicator as: "a substance or organism used as a measure of air or water quality, or biological or ecological well-being".

The ISO 14000 [12] series defines an environmental indicator as: "a specific expression that provides information about an organisation's environmental performance, efforts to influence that performance, or the condition of the environment."

The OECD [13] provides another useful definition of an indicator as: "a parameter or a value derived from parameters, which provides information about a phenomenon. The indicator has significance that extends beyond the properties directly associated with the parameter values. Indicators possess a synthetic meaning and are developed for a specific purpose."

Even if the term might appear to be vague, indicators have been widely used for monitoring and assessment of numerous environmental impacts of operations, and are increasingly used in social and economic arenas [14]. To date the emphasis of the vast majority of indicators has been placed on reporting, rather than management of impacts on mining on sustainable development. Consequently, to date, the most important criteria that define useful indicators are the capacity to simplify, quantify, analyse and communicate otherwise complex and complicated information, and the ability to make particular aspects of a complex situation stand out and thus reduce the level of uncertainty in the formulation of strategies, decisions or actions.

'Indicators arise from values (we measure what we care about), and they create values (we care about what we measure)' [15]. The main feature of indicators is their ability to summarise, focus and condense the enormous complexity of our dynamic environment to a manageable amount of meaningful information [16]. By visualizing phenomena and highlighting trends, indicators simplify, quantify, analyse and communicate otherwise complex and complicated information [14].

There is a widely recognised need for individuals, organisations and societies to find models, metrics and tools for articulating the extent to which, and the ways in which, current activities are unsustainable [17]. This need arises on multiple layers ranging from supra-national (e.g. the negotiation of protocols for environmental protection), national (e.g. via some version of "greening" GDP) and sub-national levels (e.g. in regional development forums) [18].

5   Conclusions

The focus on the "triple bottom line" that addresses issues related to the environmental impact, social responsibility and economic performance can determine the impact of industry on the environment and on the society. As presented above, many companies are addressing sustainable development and have different approaches in doing so. In order to achieve these objectives there is a need not only to re-think the practices in the industry, but also to re-define the instruments used to measure and monitor the achievements that have been made in the transition process towards sustainability.

The classical, standard financial indicators that have been used to assess the business effectiveness have been completed by sustainable performance assessment indicators, given the increased pressure and demand for sustainable practices, thus sustainability reports have become a trend in the corporate reporting [19].

As presented elsewhere [7], the sustainable organization needs to take into account several aspects that have been synthetized in performance indicators. These indicators not only measure the economic performance, as they used to do until recently, but also assess social responsibility and environmental performance. They are known as sustainability indicators and translate sustainability issues into quantifiable measures with the ultimate goal to address key sustainability concerns [20] and to provide information on how the company contributes to sustainable development [21].

Thus, it is clear that it is essential for any company to have integrated information on sustainable development for the decision-making process, as it is very complicated to rely on too many indicators.

References

1. Las Casas, G., Murgante, B., Scorza, F.: Regional local development strategies benefiting from open data and open tools and an outlook on the renewable energy sources contribution. In: Papa, R., Fistola, R. (eds.) Smart Energy in the Smart City, pp. 275–290. Springer International Publishing, Berlin (2016)
2. Scorza, F., Casas, G.L., Murgante, B.: Overcoming interoperability weaknesses in e-government processes: organizing and sharing knowledge in regional development programs using ontologies. In: Lytras, M.D., Ordonez de Pablos, P., Ziderman, A., Roulstone, A., Maurer, H., Imber, J.B. (eds.) WSKS 2010. CCIS, vol. 112, pp. 243–253. Springer, Heidelberg (2010). doi:10.1007/978-3-642-16324-1_26
3. Scorza, F., Las Casas, G.B., Murgante, B.: That's ReDO: ontologies and regional development planning. In: Murgante, B., Gervasi, O., Misra, S., Nedjah, N., Rocha, A.M.A.C., Taniar, D., Apduhan, B.O. (eds.) ICCSA 2012, Part II. LNCS, vol. 7334, pp. 640–652. Springer, Heidelberg (2012)
4. Amato, F., Pontrandolfi, P., Murgante, B.: Supporting planning activities with the assessment and the prediction of urban sprawl using spatiotemporal analysis. Ecol. Inf. **30**, 365–378 (2015)
5. Amato, F., Maimone, B.A., Martellozzo, F., Nolè, G., Murgante, B.: The effects of urban policies on the development of urban area. Sustainability **8**(4), 297 (2016)
6. Amato, F., Martellozzo, F., Nolè, G., Murgante, B.: Preserving cultural heritage by supporting landscape planning with quantitative predictions of soil consumption. J. Cult. Herit. (2016, in press)
7. Grecu, V.: The global sustainability index: an instrument for assessing the progress towards the sustainable organization. ACTA Univ. Cibiniensis **67**(1), 215–220 (2015)
8. França, C.L.: Introductory Approach to Business Model Design for Strategic Sustainable Development. Blekinge Institute of Technology, Karlskrona (2013)
9. The Intergovernmental Panel on Climate Change Report (www.ipcc.ch). The United Nations Millennium Forum Declaration reports (www.un.org/millennium/declaration). Millennium Ecosystem Assessment Reports (www.milleniumassessment.org) and UNEP's Fourth Global Environment Outlook: environment for development report (www.unep.org/gco/gco4/)
10. Birkin, F., Polesie, T., Lewis, L.: A new business model for sustainable development: an exploratory study using the theory of constraints in Nordic organizations. Bus. Strategy Environ. **18**(5), 277–290 (2009)
11. Gilpin, A.: Dictionary of Environment and Sustainable Development. Wiley, Hoboken (1996)
12. Corbett, C.J., Kirsch, D.A.: International diffusion of ISO 14000 certification. Prod. Oper. Manag. **10**(3), 327–342 (2001)
13. OECD – Organization for Economic Co-operation and Development, Natural Resource Accounts. Environmental Monographs no 84, OECD, Paris OSM. Interactive forum on bond release in arid and semi-arid areas. Office of Surface Mining, U.S. Department of the Interior. Denver, Colorado, September 1996
14. Warhurst, A.: Sustainability indicators and sustainability performance management. Mining, Minerals and Sustainable Development [MMSD] project report, 43 (2002)
15. Meadows, D.H.: Indicators and information systems for sustainable development (1998)
16. Godfrey, L., Todd, C.: Defining thresholds for freshwater sustainability indicators within the context of South African Water Resource Management. In: 2nd WARFA/Waternet Symposium: Integrated Water Resource Management: Theory, Practice, Cases, Cape Town, South Africa. Practice, Cases, Cape Town (2001)

17. Singh, R.K., Murty, H.R., Gupta, S.K., Dikshit, A.K.: An overview of sustainability assessment methodologies. Ecol. Ind. **9**(2), 189–212 (2009)
18. Ramachandran, N.: Monitoring Sustainability: Indices and Techniques of Analysis. Concept Publishing Company, New Delhi (2000)
19. GRI-Global Reporting Initiative. Sustainability Reporting Guidelines 2002 on Economic, Environmental and Social Performance. Global Reporting Initiative, Boston, USA (2002). http://www.globalreporting.org
20. Azapagic, A.: Developing a framework for sustainable development indicators for the mining and minerals industry. J. Cleaner Prod. **12**, 639–662 (2004)
21. Azapagic, A., Perdan, S.: Indicators of sustainable development for industry: a general framework. Trans. IChemE Part B Proc. Saf. Environ. Prot. **78**(4), 243–261 (2000)

Author Index

Printed in the United States
By Bookmasters